CW01500608

# CONTENTS

CONTENTS

# ILLUSTRATIONS

## Figures

## Table

# CONTRIBUTORS

**Julie Anderson** is Reader in History at the University of Kent. She has published a number of articles on the intersection of the history of disability and medicine and she is the author of *War, Disability and Rehabilitation: Soul of a Nation* (Manchester University Press, 2011). Her current work centres on histories of orthopaedics and ophthalmology, and her forthcoming book *The Science of Seeing: Vision and the Modern World in Britain 1900–1950* will be published by Manchester University Press.

**Roberta Bivins** is a historian of medicine at the University of Warwick. Her early research examined the cross-cultural transmission of medical expertise, particularly in relation to global and alternative medicine (*Acupuncture, Expertise and Cross-Cultural Medicine*, 2000; *Alternative Medicine? A History*, 2007). Since 2004, funded by the Wellcome Trust, she has studied the impacts of immigration and ethnicity on post-war British health, medical research and practice (*Contagious Communities: Medicine, Migration and the NHS in Post War Britain*, 2015). New research examines the cultural history and influence of the British National Health Service since 1948. Other research interests include the domestication of medical technologies, and the intersections of history and policy.

**Fay Bound Alberti**, DPhil, FRHS, is Senior Research Fellow in History at Queen Mary University of London, where she co-founded the Centre for the History of Emotions. She has lectured at universities throughout the UK including York, Manchester and UCL. Fay has also published widely on the themes of medical and cultural history between the ancient world and the present day and her special interests are gender, the body and the mind. Her work has appeared in journals including *Isis*, *The Lancet* and *History Workshop Journal* and her most recent books are *Matters of the Heart: History, Medicine and Emotion* (Oxford University Press, 2010) and *This Mortal Coil: The Human Body in History and Culture* (Oxford University Press, 2016). Fay is currently researching the cultural and medical history of transplantation. In addition to her historical research, Fay has considerable expertise in the fields of philanthropy and grant making, having worked as Head of Philanthropy for Lisbet Rausing and the Arcadia Fund as well as Acting Head of Medical Humanities for the Wellcome Trust.

**Elma Brenner** is Specialist, Medieval and Early-Modern Medicine, at the Wellcome Library, London. She previously undertook research at the University of Cambridge and the Pontifical Institute of Mediaeval Studies, Toronto. Her research examines the medical and religious culture of medieval France and England, particularly responses to leprosy and mental illness in the region of Normandy. She has co-edited *Memory and Commemoration in Medieval Culture* (with Meredith Cohen and Mary Franklin-Brown, Ashgate, 2013) and *Society and Culture in Medieval Rouen, 911–1300* (with Leonie V. Hicks, Brepols, 2013). Her book, *Leprosy and Charity in Medieval Rouen*, was published by Boydell & Brewer in December 2015.

**Helen Bynum,** PhD, read Human Sciences at UCL before becoming a historian of medicine and science. After a post-doctoral fellowship and Wellcome Award Lectureship in the history of medicine at the University of Liverpool she switched to freelance writing and speaking. She is an honorary research associate in the department of anthropology at UCL. Her book *Spitting Blood: The History of Tuberculosis* (Oxford University Press, 2012; pbk 2015) developed from a longstanding interest in the history of tropical medicine and the rise of drug resistance in malaria and tuberculosis.

**David Cantor** is a historian in the Office of History at the National Institutes of Health, Bethesda, Maryland. He is the editor of *Reinventing Hippocrates* (Ashgate, 2002) and *Cancer in the Twentieth Century* (Johns Hopkins University Press, 2008), co-editor (with Christian Bonah and Matthias Dörries) of *Meat, Medicine, and Human Health in the Twentieth Century* (Pickering and Chatto, 2010) and co-editor (with Edmund Ramsden) of *Stress, Shock, and Adaptation in the Twentieth Century* (University of Rochester Press, 2014). He is series editor (edited collections) of *Social Histories of Medicine*, published by Manchester University Press.

**Havi Carel** is Professor of Philosophy at the University of Bristol, where she also teaches medical students. Her research examines the experience of illness and of receiving healthcare. She was recently awarded a Senior Investigator Award by the Wellcome Trust, for a five-year project entitled 'Life of Breath' (with Jane Macnaughton, Durham University). She recently completed a monograph for Oxford University Press, provisionally entitled *Phenomenology of Illness*. She previously published on the embodied experience of illness, well-being within illness and patient–clinician communication in the *Lancet, BMJ, Journal of Medicine and Philosophy, Theoretical Medicine and Bioethics, Medicine, Healthcare and Philosophy*, and in edited collections. Havi is the author of *Illness* (2008, 2013), shortlisted for the Wellcome Trust Book Prize, and of *Life and Death in Freud and Heidegger* (2006). She is the co-editor of *Health, Illness and Disease* (2012) and of *What Philosophy Is* (2004). She also uses film in teaching and has co-edited a volume entitled *New Takes in Film-Philosophy* (2010). She also co-edited a special issue of *Philosophy* on 'Human Experience and Nature' (2013).

**Elena Carrera** received her DPhil from the University of Oxford. She is Senior Lecturer in Hispanic Studies and co-founder of the Centre for the History of Emotions at Queen Mary University of London. Her publications include *Teresa of Avila's Autobiography: Authority, Power and the Self in Mid-Sixteenth-Century Spain* (Legenda, 2005) and the edited collections *Madness and Melancholy in Sixteenth- and Seventeenth-Century Spain* (2010) and *Emotions and Health, 1250–1700* (Brill, 2013).

**Samuel Cohn, Jr.** is Professor of Medieval History at the University of Glasgow, an honorary fellow of the Institute for Advanced Studies in Humanities (University of Edinburgh), and a fellow of the Royal Society of Edinburgh. Over the past 15 years he has focused on the history of popular unrest in late medieval and early-modern Europe and on the history of disease and medicine. His latest books include: *The Black Death Transformed: Disease and Culture in Early Renaissance Europe* (Edward Arnold and Oxford University Press, 2002); *Popular Protest in Late Medieval Europe: Italy, France, and Flanders* (Manchester University Press, 2004); *Lust for Liberty: The Politics of Social Revolt in Medieval Europe, 1200–1425* (Harvard University Press, 2006); *Cultures of Plague: Medical Thinking at the End of the Renaissance* (Oxford University Press, 2010); and *Popular Protest in Late Medieval English Towns* (Cambridge University Press, 2012). He is presently on leave funded by a three-year 'Major Research Fellowship' from the Leverhulme Trust on the project, 'Epidemics: Waves of Disease, Waves of Hate from the Plague of Athens to AIDS'.

**Katrina Ford** is an honorary research associate in the History Department at the University of Auckland. She received a PhD from the University of Auckland in 2014 for her thesis on the history of bacteriological science and 'microbial mentalities' in New Zealand. Since completing her thesis, she has worked as a teaching fellow in the History Programme at the University of Waikato, teaching the history of science and empire. Her research interests include the history of public health, the history of agriculture and rural–urban connections, and food history. In 2015, she received a Dame Joan Metge Post-Doctoral Award to carry out research into the social and cultural history of pasteurization.

**Katherine Foxhall** is a lecturer in Modern History at the University of Leicester, UK. From 2011 to 2014 she held a Wellcome Trust Fellowship in Medical History and Humanities in the Department of History, King's College London. She is the author of *Health, Medicine and the Sea: Australian Voyages, c. 1815–1860* (Manchester, 2012) as well as research articles about colonial vaccination, medical experimentation, quarantines, and migraine. Her research interests span the social and cultural history of health and illness, colonial medicine and migration, imprisonment and institutions. She is currently completing a book on the history of migraine, and continues to develop work in the history of migration, punishment and quarantine.

**Arthur W. Frank** lives in Calgary, Alberta. He is Professor Emeritus at the University of Calgary and Professor at Betanien University College, Bergen, Norway. He also teaches at the Center for Narrative Practice in Boston. He specializes in illness experience, narrative, ethics, and body studies. His books include *At the Will of the Body* (1991/2003), a memoir of illness; *The Wounded Storyteller* (second edition, 2013); *The Renewal of Generosity* (2004), and *Letting Stories Breathe: A Socio-narratology* (2010). He has been Visiting Professor at universities in Japan, Britain, and Canada; he is an elected fellow of The Royal Society of Canada and in 2008 was recipient of their award for bioethics. He lectures internationally on clinical practice, health-care education, and narrative. Among many editorial positions, he is a contributing editor to *Literature and Medicine.*

**Jana Funke** is Advanced Research Fellow in Medical Humanities at the University of Exeter. Her research cuts across modernist literary studies and the history of sexuality and explores constructions of gender and sexuality in late nineteenth- and early twentieth-century literature, medicine and science. She has published various journal articles and book chapters and is the co-editor of *Sex, Gender and Time in Fiction and Culture* (Palgrave, 2011). Her critical edition *The World and Other Works by Radclyffe Hall* (Manchester University Press, 2015) presents a range of previously unpublished writings by Radclyffe Hall together with a substantial introduction. She is also writing a monograph entitled *Sexual Modernism: Femininity, Development, and Sexual Science*. In 2014, Jana was selected to participate in the AHRC- and Wellcome-Trust-funded New Generations in Medical Humanities Programme and currently holds a Wellcome Trust Joint Investigator Award to direct a five-year research project on 'The Cross-Disciplinary Invention of Sexuality: Sexual Science beyond the Medical' (2015–2020).

**Mónica García** is Associate Professor at the Universidad del Rosario (Colombia). She was trained as a physician and historian at the Universidad Nacional de Colombia and received her PhD in Science and Technology Studies at the University of Edinburgh. Her recent publications include: 'Debating Disease in Nineteenth-Century Colombia: Causes, Interests and the Pasteurian Therapeutics', *Bulletin of the History of Medicine*, Volume 89, Number 2, 2015, 293–321; 'Mortality Rates or Socio-medical Indicators? The Work of the League of Nations on Standardizing the Effects of the Great Depression on Health', *Health Policy and Planning*, 29, 2014, 1–11; and 'Producing Knowledge about Tropical Fevers in the Andes: Preventive Inoculations and Yellow Fever in Colombia, 1880–1890', *Social History of Medicine*, Volume 25, Number 4, 2012, pp. 830–47.

**Sam Goodman** is Lecturer in English and Communication at Bournemouth University. He is currently researching the intersection between medicine and colonial Anglo-Indian society and culture, with particular emphasis on literary representations of illness and treatment. He is a member of the New Generations in Medical Humanities programme run by Durham University and supported by the Arts and Humanities Research Council and the Wellcome Trust, and is a BBC/AHRC New Generation Thinker. He is the author of *British Spy Fiction and the End of Empire* (Routledge, 2015), and a co-editor of *Medicine, Health and the Arts: Approaches to the Medical Humanities* (Routledge, 2013) with Victoria Bates and Alan Bleakley.

**Christoph Gradmann** took his doctorate in 1992 in Hannover. After working in Heidelberg and Berlin, he became Professor in History of Medicine at the University of Oslo in 2006. In 2013–14 he was Leverhulme Professor in the History of Medicine at the University of Oxford. He has published extensively on the history of infectious disease and experimental medicine in the nineteenth and twentieth centuries. From nineteenth-century German medical bacteriology he moved on to investigate the standardization of biological medicines in the twentieth century. Recent work has been on the history of antibiotic resistance and global health. He is the head of the University of Oslo's Master Program in International Community Health. His publications include: 'Re-Inventing Infectious Disease: Antibiotic Resistance and Drug Development at the Bayer Company 1945–1980', *Medical History*, 60 (2016); 'Sensitive Matters: The World Health Organization and Antibiotic Resistance Testing, 1945–1975', *Social History of Medicine*, 26, 2013, pp. 555–74; 'Magic Bullets and Moving Targets: Antibiotic Resistance and Experimental Chemotherapy 1900–1940', *Dynamis* 31, 2011, pp. 29–45; with Jonathan Simon (eds), *Evaluating and Standardizing Therapeutic Agents 1890–1950* (Palgrave Macmillan, 2010); and *Laboratory Disease: Robert Koch's Medical Bacteriology* (Johns Hopkins University Press, 2009).

**R. J. Hankinson** is Professor of Philosophy and Classics at the University of Texas at Austin. He has published numerous articles on many aspects of ancient philosophy and science; his books include *The Sceptics* (1995), *Cause and Explanation in the Ancient Greek World* (1998) and *Galen on Antecedent Causes* (1998). He has also edited *The Cambridge Companion to Galen* (2008).

**Mark Harrison** is Professor of the History of Medicine and Director of the Wellcome Unit for the History of Medicine at the University of Oxford. He is also a visiting professor at Kyung-hee University in Seoul and Peking University in Beijing. He has written many articles and books on the history of disease and medicine, chiefly relating to war, imperialism and commerce. His work includes numerous studies of epidemic and pandemic diseases, particularly his most recent book: *Contagion: How Commerce Has Spread Disease* (Yale University Press, 2012). He is currently working on epidemic diseases and medicine in imperial and military contexts and on two interdisciplinary research projects. One is a study of malaria in Asia, funded by a Wellcome Trust Senior Investigator Award; the other, a research programme entitled 'The Human Factor: Infectious Disease and Collective Responsibility', is supported by the Oxford Martin School and examines the relationship between disease control and ideas of civic responsibility.

**Brian Hurwitz** trained as a family doctor in central London where he practised for 30 years. Since 2002 he has been Professor of Medicine and the Arts at King's College London where he directs the Centre for the Humanities and Health, which is funded by the Wellcome Trust. The Centre is a multidisciplinary unit offering research training at masters, PhD, MD, and post-doctoral levels. Based in the Department of English at King's, Brian's research interests include narrative studies in relation to medical practice, ethics, law, and the logic and literary shape of clinical case reports. Prior to his current position he was Professor of Primary Health Care and General Practice at Imperial College London.

**Mark Jackson** is Professor of the History of Medicine at the University of Exeter. He was Senior Academic Adviser (Medical Humanities) at the Wellcome Trust, 2013–16, and was a member of the History Sub-Panel for REF 2014. He has taught modules in the history of medicine and the history and philosophy of science for over 20 years at undergraduate and post-graduate levels, and has also been involved in teaching medical history to GCSE and A level students. His books include *New-born Child Murder* (1996), *The Borderland of Imbecility* (2000), *Infanticide: Historical Perspectives on Child Murder and Concealment 1550–2000* (ed., 2002), *Allergy: The History of a Modern*

*Malady* (2006), *Health and the Modern Home* (ed., 2007), *Asthma: The Biography* (2009), *The Oxford Handbook of the History of Medicine* (ed., 2011), *The Age of Stress: Science and the Search for Stability* (2013), *The History of Medicine: A Beginner's Guide* (2014, short-listed for the Dingle Prize), and *Stress in Post-War Britain, 1945–85* (ed., 2015). He is currently writing a book on health and well-being in middle age, entitled *The Midlife Crisis: A History* (under contract with Reaktion).

**Richard C. Keller** is Professor of Medical History and Bioethics at the University of Wisconsin-Madison, where he is also Associate Dean in the International Division. He is the author of *Fatal Isolation: The Devastating Paris Heat Wave of 2003* (Chicago, 2015), *Colonial Madness: French Psychiatry in Colonial North Africa* (Chicago, 2007), and *Enregistrer les morts, identifier les surmortalités: Une comparison Angleterre, États-Unis et France* (with Carine Vassy and Robert Dingwall; Rennes: Presses de l'EHESP). He is the editor, with Warwick Anderson and Deborah Jenson, of *Unconscious Dominions: Psychoanalysis, Colonial Trauma, and Global Sovereignties* (Duke, 2011), and editor with Sara Guyer of a forthcoming issue of *South Atlantic Quarterly* on 'Life after Biopolitics'. He was awarded the William Koren, Jr. Prize from the Society for French Historical Studies, and his work has been supported by the National Science Foundation, the Andrew W. Mellon Foundation, the French Ministry of Health, the City of Paris, and the Wisconsin Alumni Research Foundation, among other agencies.

**Richard A. McKay** is a Wellcome Trust research fellow in the Department of History and Philosophy of Science at the University of Cambridge. In 2011 he completed his doctorate in history at the University of Oxford; his thesis focused on the emergence, dissemination, and consequences of the idea of 'patient zero' during the early North American HIV/AIDS epidemic. An article based on this research appeared in the *Bulletin of the History of Medicine* in 2014 and his book, *Patient Zero: Public Health, the Media, and the Making of the North American AIDS Epidemic*, will be published by the University of Chicago Press. His current research explores the processes by which health workers, gay rights activists, and other groups and individuals became increasingly attentive to venereal disease transmission and sexual health among men who had sex with men; it focuses on the middle decades of the twentieth century in Canada, the United States, and England. In addition to his academic research, he is a certified career, academic and life coach at Rich Life Coaching (www.richlifecoach ing.co.uk), where he assists researchers, writers, and other thought leaders to live, work, and retire to their fullest potential.

**Martin D. Moore** is an associate research fellow in the Centre for Medical History at the University of Exeter, working as part of a team on the Wellcome Trust Senior Investigator Award, 'Lifestyle, health and disease: changing concepts of balance in modern medicine'. His current project investigates how, in twentieth-century Britain, patients and medical teams developed and coded strategies for maintaining bodily balance with differing sets of social priorities and cultural assumptions. Prior to his current post, he gained his PhD at the University of Warwick (2010–14). His thesis explored the historic connections between the development of new systems of disease management in diabetes care, and the emergence of medical management in twentieth-century British clinical and public health medicine.

**Robert Peckham** is Associate Professor of History and Co-Director of the Centre for the Humanities and Medicine at the University of Hong Kong. His most recent book is *Epidemics in Modern Asia* (Cambridge University Press, 2016).

**Catherine Rider** is a senior lecturer in Medieval History at the University of Exeter. She is a specialist on the history of magic, medicine, and reproduction in the medieval period with particular interests in the Church's attitude to magic; the relationship between magic and medicine; and the history of infertility and childlessness. Her publications include *Magic and Impotence in the Middle Ages* (2006) and *Magic and Religion in Medieval England* (2012).

**Akihito Suzuki** is a professor of History at the School of Economics of Keio University in Tokyo. He studied history of science at University of Tokyo and history of medicine at the Wellcome Institute for the History of Medicine in London. He has published extensively on the history of psychiatry in England and Japan, as well as on the history of infectious diseases in Japan. His publications include: *Madness at Home: The Psychiatrist, the Patient and the Family in England 1820–1860* (University of California Press, 2006); 'Measles and the Transformation of the Spatio-Temporal Structure of Modern Japan', *Economic History Review*, 62(2009), 828–56; and 'Smallpox and the Epidemiological Heritage of Modern Japan: Towards a Total History', *Medical History*, 55 (2011), 313–18. He is now preparing a book on mental illnesses in early twentieth-century Tokyo based on an extensive archive of a private psychiatric hospital, which will examine psychiatrists and patients in the contexts of the rise of modernity and the persistence of tradition.

**Alannah Tomkins** is Reader in History at Keele University and has published research on social history topics including parish poor relief, charity, and professionalization in medicine. She is the author, with Lisetta Lovett, of the textbook *Medical History Education for Health Practitioners* (Radcliffe: London, 2013). Her most recent research has concerned turbulent medical careers, and her book *Medical Misadventure in an Age of Professionalization, 1780–1890* will be published by Manchester University Press in 2017.

**David M. Turner** is Professor of History at Swansea University. He has published widely on the social and cultural history of early-modern England and is a specialist in the history of disability. He is the editor (with Kevin Stagg) of *Social Histories of Disability and Deformity* (Routledge, 2006) and the author of *Disability in Eighteenth-Century England: Imagining Physical Impairment* (Routledge, 2012), which won the Disability History Association Outstanding Publication Prize for 2012. He was also academic advisor to BBC Radio 4's *Disability: A New History* (2013). Turner currently leads the project, *Disability and Industrial Society: A Comparative Cultural History of British Coalfields 1780–1948*, funded by a Wellcome Trust Programme Award (2011–16), and is writing a book on disability in the nineteenth-century coal industry with Daniel Blackie.

**Robert Weston** is an honorary research fellow at the University of Western Australia. His principal field of research is European medical history in the early-modern period. His book, *Medical Consulting by Letter in France, 1665–1789*, was published by Ashgate in 2013; he has also published chapters and journal articles on medical consultations by letter, the history of disease, masculinity, the use of violence in early-modern medicine, and the role of emotions in medical practice. He is a fellow of the Royal Australian Institute of Chemistry and Member of the Royal Society of Chemistry.

**Abigail Woods** is Professor of the History of Human and Animal Health at King's College London. She trained in veterinary medicine at Cambridge University and in the history of science, technology and medicine at Manchester University. Before moving to KCL in 2013, she worked for eight years at Imperial College London. Her research interests encompass the modern British history of animal health and welfare, the evolution of veterinary medicine, their connections with human health and medicine, and the history of livestock agriculture. She is currently President of the World Association for the History of Veterinary Medicine. She has participated in several inter-disciplinary research projects, and engages actively with a wide range of audiences in order to bring historical lessons to bear on the present.

**Michael Worboys** is Emeritus Professor in the History of Science, Technology and Medicine in the Centre for the History of Science, Technology and Medicine at the University of Manchester. His doctoral and post-doctoral work focused on the history of science and the British Colonial Empire. His work on tropical medicine and parasitology led him to wider studies on the history of communicable diseases and microbiology, focusing on the period 1850 to 1940. He has published a monograph on the development and spread of germ theories of diseases in the late nineteenth century and on specific infections: smallpox in India with Sanjoy Bhattacharya and Mark Harrison, rabies with Neil Pemberton, and fungal disease with Aya Homei. He is currently working on two projects: (i) the place of the dog in the history of the biological and biomedical sciences, and how this work, especially in physiology, genetics and psychology, has shaped human–dog relationships; (ii) laboratory–clinic relations in the second half of the twentieth century, developing historical perspectives on translational medicine.

**Dominik Wujastyk** holds an appointment at University of Alberta as Professor and Singhmar Chair in Classical Indian Society and Polity. His early training was in Sanskrit and classical Indian studies, at Oxford University. He has held appointments at the Wellcome Institute for the History of Medicine, the Wellcome Centre for the History of Medicine at UCL, and at the Universities of Leiden and Vienna, with visiting positions at the Universities of Zurich and Helsinki and the University of Texas at Austin. His single-author books include *Metarules of Paninian Grammar* and *The Roots of Ayurveda*, and he has published and edited works on a wide range of topics in Indian cultural history, including the history of Indian science and medicine. He is currently co-authoring a book on the history of yoga postures (with Philipp Maas), and is continuing to research the history of medical thought in pre-modern India.

# ACKNOWLEDGEMENTS

This book was not my idea. Some years ago, Eve Setch, Senior Editor for modern history at Routledge, asked me whether producing a volume on the history of disease would be a worthwhile venture and whether I would be interested in editing it. Although I was then under some pressure to complete a manuscript on the history of stress, the prospect did appeal to me. I am deeply grateful to Eve both for inviting me to contribute to the Routledge history series in this way and for her advice and support throughout the process. My role as editor has also been greatly facilitated by the administrative skills of Amy Welmers, who has coordinated a range of complex issues relating both to the text and the illustrations. I am grateful to Rachel Singleton at Swales & Willis and Linda Smith for their meticulous proof-reading.

No book of this nature can be realized without the enthusiasm, expertise and congeniality of its contributors. From start to finish, collaboration with the scholars whose work appears in the following pages has been intellectually and socially rewarding. I have learned much that was previously unknown to me and developed friendships (some virtual) that will, I hope, continue for some years to come. Of course, other colleagues and friends at Exeter and elsewhere have informed my approaches to the history of medicine and I would particularly like to thank members of the research team with whom I currently work: Ali Haggett; Martin Moore; Ayesha Nathoo; Fred Cooper; Natasha Feiner; Nicos Kefalas; and Claire Keyte. For all of us in the Centre for Medical History at Exeter and for many of the contributors to this volume, the support of the Wellcome Trust remains invaluable.

Although focused on the past, this book is written with the future of historical scholarship in mind. My own days of intellectual endeavour are beginning to feel increasingly autumnal, but I hope that this collection offers the next generation of scholars a sense of vernal optimism. With the future explicitly in mind, this book is dedicated to Ciara, Riordan and Conall, whose intelligence and enthusiasm continue to shine.

# 1

# PERSPECTIVES ON THE HISTORY OF DISEASE

*Mark Jackson*

There is arguably no more important object of historical enquiry than the fluctuating manifestations and meanings of disease. Agents of pain, distress and death throughout the history of the world, diseases have not only reflected the customs, behaviours and lifestyles of populations, but have also in turn impacted on the lives and fortunes of individuals, families, communities, and nations. On the one hand, patterns and experiences of disease have been shaped by the demands of trade, by the import, export and consumption of commodities, by the transport of animals and the migration of peoples, and by the effects of warfare, work and living conditions.[1] Such links between disease and circumstance betray the impact of hunger, poverty, overcrowding, and sometimes leisure and luxury, on health and happiness. On the other hand, the symptoms and consequences of disease have in turn dictated personal and collective destinies, determining the outcome of conflicts, the inheritance of thrones and family estates, and the capacity to impose or resist values, norms and judgements. The designation of certain physical and emotional states as diseases has been used to marginalise, disenfranchise and subordinate sections of the population, to liberate or legitimate particular classes and professions, and to create and protect identities, agency and power.

In spite of our awareness of their historical significance and in spite of unrelenting efforts to prevent, control or eradicate diseases across periods and cultures, disease is difficult to define in either theoretical or practical terms. Diseases are not fixed entities: they change biologically in response to environmental conditions; and they are understood, experienced and responded to in strikingly different ways within different cultural and social contexts. What constitutes a specific disease in one place and time might be disputed in another, reflecting not only contrasting ecological circumstances and lifestyles, but also distinctive theoretical models, cultural values, social and political constraints, technological capabilities, and individual and collective expectations. As economic and political powers have shifted, so too has the locus of authority to name, define and manage disease. From this perspective, the history of disease (and the wider history of medicine) must necessarily be a social and cultural history, one that situates diseases in their immediate and historical contexts rather than privileging disease as a category external to, and unmediated by, social conventions and material conditions.[2]

Explanations, experiences and patterns of disease have directed the focus of historians of medicine as well as the efforts of biomedical scientists, governments, charities, and health-care services. The impact of disease on historical scholarship can be understood, in the broadest possible sense, as political. Both Allan Brandt and Virginia Berridge have emphasised the manner in which the challenges and fears created by the AIDS epidemic transformed the history of medicine as much as they changed the aspirations and paradigms of public health.[3] Similarly, the accelerating epidemiological shift from infectious to non-infectious degenerative disease across the twentieth century encouraged a turn towards histories of the chronic scourges of modern populations, such as cancer, heart disease, allergies, diabetes, and arthritis, which together accounted for an increasingly large proportion of illness and death particularly in industrialised regions of the world.[4] Like scientists and doctors, then, historians of medicine are clearly influenced by the 'questions, approaches, and concerns' of the present, as much as by those of the past.[5] The link between disease and historical research can also be personal. As the prefaces and acknowledgements of many scholarly monographs testify, an historian's choice of subject, period and place is often, perhaps necessarily, dictated by their own experiences and interests as well as by a spectrum of contemporary medical and political concerns.[6]

Disease has occupied a pivotal place in most histories of medicine. In some cases, a relatively well-recognised disease has been the explicit focus of analysis: fine studies exist of diseases as diverse as cholera, smallpox, salmonella, plague, tuberculosis, lung cancer, malaria, polio, silicosis, leprosy, diabetes, gout, syphilis, madness, haemophilia, breast cancer, influenza, sickle cell anaemia, mania, and AIDS.[7] Although there are methodological and conceptual challenges posed by writing histories of what are generally regarded as distinct and discrete diseases,[8] the best studies move beyond naïve notions of diseases as natural categories to explore the manner in which the naming and framing of diseases have been social and cultural enterprises that carry significant personal and political weight. In other histories of medicine, disease lingers either abstractly or concretely in the shadows of studies that have focused more directly on the processes of medical professionalisation, the development of state and charitable welfare services, the rise of national institutions and international health agencies, the politics of imperialism, and the impact of economic and environmental inequalities on health and well-being in a variety of geographical, cultural and chronological settings.[9] Written from an alternative perspective than more intentional histories of disease, these studies too reveal the contingent and contested nature of medical knowledge, health policies, clinical practices, and patient experiences.

Although much has been accomplished by historians tracing and situating patterns of disease in the past, there remain significant empirical, conceptual and historiographical challenges to the historical reconstruction of models, patterns and experiences of disease. In a provocative essay intended to set out a methodology for exploring the interlinkages between biology and culture in histories of disease, Charles Rosenberg pointed out some decades ago that we 'need to know more about the individual experience of disease in time and place, the influence of culture on definitions of disease and of disease in the creation of culture, and the role of the state in defining and responding to disease'.[10] As the chapters in this volume testify, in the intervening years historians have energetically addressed many of the substantive questions raised by Rosenberg. In particular, scholars have begun critically to utilise a richer array of sources, including case notes, personal correspondence, memoirs and visual media, in order to reveal the expectations and experiences of patients and their doctors and to integrate patient narratives into broader contextual studies.[11] Yet, in spite of successful engagement with much of Rosenberg's research agenda, historians have remained divided on an appropriate or consistent conceptual approach to writing histories of disease. While individual contributions in this volume give witness to the depth and diversity of recent scholarship, the aim of this introduction is to establish some of the key historiographical perspectives that have dominated the field in order to provide a framework for engaging with the substantive studies that follow.

### Framing histories of disease

Histories of medicine and disease have always been methodologically diverse. Older, more traditional histories of medicine tend to regard disease as a stable ontological category, determined by immutable biological characteristics and largely unaffected by shifting social and cultural contexts. In *Disease and History*, for example, first published in 1972 and appearing in a revised form in 2000, Frederick Cartwright and Michael Biddiss explore 'some of the many maladies which have afflicted the world' in order to 'illustrate their effect not only upon historically important individuals but also upon peoples'.[12] In a sweeping narrative that stretches from the ancient to the modern world, the authors trace in turn the roles of the Black Death, syphilis, smallpox, typhus, cholera, tuberculosis, malaria, haemophilia, 'mob hysteria' in Nazi Germany, and AIDS in shaping major military, political, colonial, and social events. Cartwright and Biddiss's approach is not without its merits: written evocatively for a general audience, it effectively situates health, medicine and disease within a rich tapestry of historical continuities and change, reminding historians of medicine to engage fully with the social, political and economic histories of their period.

There are, however, limitations to this approach. In the first place, many broad essentialist histories of medicine, like *Disease and History*, focus almost exclusively on the impact of human disease on Western societies, ignoring vast regions of the world in which animal and human diseases were manifest, experienced and understood in different ways. Secondly, such histories sometimes adopt judgemental tones in their evaluation of past practices and behaviour. Cartwright and Biddiss, for example, too readily dismiss Galen's 'huge collection of noisome and useless remedies' and blame the modern resurgence of syphilis in Britain and North America unproblematically on 'the sheer carelessness of young people' during the 1960s, without reflecting more closely on the contextual determinants of either clinical efficacy in ancient Greece and Rome or sexual behaviour in the modern world.[13]

More significantly, however, in the present context, such works display a tendency to retrospective diagnosis, that is, a propensity to project current scientific knowledge onto the past without considering contemporary diagnostic fashions and disputes. Used either to diagnose the diseases of great men, such as Socrates, Beethoven, Napoleon, George III, Charles Darwin and Frederic Chopin, or to establish continuities in the aetiology and pathology of what are regarded as singular diseases, retrospective diagnosis has been widely criticised by social historians of medicine. Both 'pathographies' of individuals and 'biographies' of disease have been dismissed as populist and presentist accounts of the past, as insufficiently attentive to change across time, or as forms of unsubstantiated anachronistic speculation. There is some truth in these accusations. Retrospective diagnosis certainly carries ontological and epistemological dangers, serving to reduce illnesses to biological categories alone, to reify and venerate current diagnostic nosologies, and to ignore the cultural contexts of health and disease.[14]

Yet, retrospective diagnosis need not be as antithetical to the disciplinary conventions of history as critics have suggested. History is retrospective: we write about the monarchy, the Crusades, the working classes, the Church, the Labour party, fascism, the media and colonialism in the past and present as if these labels and concepts too have some categorical continuity and stability across time and without forever needing to question or qualify subtle (or even manifest) semantic and structural shifts. Historians regularly impose present knowledge on past beliefs and experiences, and the impacts of subjective personal perspectives are often explicitly acknowledged by scholars. To argue that retrospective diagnosis prioritises biological knowledge of disease at the expense of objective 'historical understanding', as some scholars have insisted, is a disciplinary conceit: the history of medicine is no less 'historically fashioned' or less subjective than the modern science and medicine that historians often wish to denigrate.[15] Perhaps more importantly, however, critics of retrospective diagnosis restrict the potential for history to help answer pivotal questions about continuities as well as change: How do we account for the adoption of shared languages of disease in different periods and different cultures? How do we explain commonalities (as well as differences) in experience, of breathing difficulties, pain and mental distress for example, across time and space? And how do we understand bio-archaeological evidence that demonstrates the presence in skeletal remains of micro-organisms that possess the same genetic characteristics as those found in patients now? Although we clearly need to recognise the historical contingency of diagnostic and recording practices and the instability of disease categories, uncritical rejection of retrospection and the unrestrained pursuit of relativist accounts of disease limit the capacity of historians to contribute to debates about the emergence and spread of epidemics, for example, or their ability to interrogate and evaluate national and international strategies for disease control.[16]

Trenchant criticism of retrospective diagnosis has not completely eroded scholarly interest in writing 'biographies of disease'. Indeed, some expansive and well-received studies of singular diseases (or disease categories) have been published in high-profile series in recent years.[17] However, in an attempt to overcome the problems associated with linear narratives of medical progress and to align the history of medicine more comfortably with the analytical tools of social and cultural history, since the 1980s historians of medicine have elaborated new methodological frameworks that more clearly problematise the production of medical knowledge and emphasise the social, political, economic and cultural dimensions and determinants of health and illness. The most well-recognised approach, neatly elaborated by Ludmilla Jordanova in a seminal article published in *Social History of Medicine* in 1995, is social constructionism. Drawing on the analytical and political perspectives of the history and philosophy of science, the sociology of knowledge, cultural anthropology, Marxist histories, and feminist critiques of medical authority, social constructionism emphasises 'the ways in which scientific and medical ideas and practices are shaped in a given context'. According to Jordanova, the challenge for historians of medicine is to 'conceptualize, explain and interpret the processes through which this happens' by using the same 'ideas and frameworks that other historians employed, such as state, class, imperialism, patronage and so on.'[18]

Social constructionism has impacted significantly on histories of medical ideas, studies of professionalisation and institutional development, and accounts of the design, implementation and reception of health policies. It has made possible more sophisticated analyses of doctor–patient relationships, the division of medical labour, and the cognitive dimensions of medicine.[19] However, while it certainly allows us to understand diagnosis as a social process, one that is relational and culturally specific, social constructionism has contributed less obviously to the history of disease, since it tends to minimise the biological parameters of disease and diminish historical interest in everyday, lived experiences of illness, pain and death. An alternative historiographical framework that more specifically addresses the challenges raised by writing histories of disease is Charles Rosenberg's concept of 'framing'. In a series of publications in the 1980s and 1990s, Rosenberg rejected the belief that diseases can be readily reduced to stable organic entities, arguing for a more complex and fluid understanding of how diseases have been 'framed' historically:

> Disease is at once a biological event, a generation-specific repertoire of verbal constructs reflecting medicine's intellectual and institutional history, an aspect of and potential legitimation for public policy, a potentially defining element of social role, a sanction for cultural norms, and a structuring element in doctor/patient interactions.[20]

Deliberately distancing his approach from 'social constructivism', a term that he regarded as tautological and overly associated with particular forms of cultural criticism, Rosenberg encouraged historians to use the 'less programmatically charged metaphor' of framing to challenge the plausibility of medical models, to expose more convincingly the ways in which 'framing options are not equally available' to everyone, to recognise the role of diseases as 'actors' in social processes, and to re-evaluate claims that medicalisation is necessarily part of 'an oppressive ideological system'. Rosenberg's formulation is broadly seductive. His emphasis on the ways in which diseases are 'framed' in biological and cultural terms allows us to sharpen our focus on the 'nexus between biological event, its perception by patient and practitioner, and the collective effort to make cognitive and policy sense out of those perceptions'.[21] It also reminds us to acknowledge aspects of the history of medicine that can otherwise be overlooked or disregarded: the 'process of disease definition itself'; and the consequences of those definitions in terms of individuals, communities, and structures of care.[22]

Rosenberg's approach was not entirely original, but owed much to the historical contextualism of the Russian-born historian of medicine Owsei Temkin (1902–2002), particularly to his careful delineation between suffering, illness and disease and their formative impacts on medicine.[23] Nor is the notion of framing necessarily less politically charged than social constructionism, since both offer similar exhortations to reconcile biology and culture in some way, or at least to recognise more overtly the cultural dimensions of biological knowledge.[24] It is the case, however, that Rosenberg provided a more explicit framework for writing histories of disease, as opposed to histories of medicine. In particular, he set out many of the key questions and domains of enquiry that have preoccupied contemporary scholars. Although some of the lacunae that Rosenberg identified have been partially addressed in recent years, many remain to be fully explored. We still need to understand in greater detail: the cultural specificity of models of disease and their real-world consequences; the evolution of governmental and professional strategies for monitoring and addressing patterns of disease, mortality and morbidity; the organisational, bureaucratic, ethical and technological determinants and effects of scientific investigation and medical care; the commonality and diversity of personal experiences of disease; the construction and performative power of heterogeneous narrative forms that incorporate subjective, cognitive and emotional, as well as biological, dimensions of health and illness; and the means by which individuals, families and communities mobilise both history and biography to challenge dominant political paradigms and normative scientific models of disease.[25] It is these facets of disease in the past and present, as well as the rapprochement between biology and culture that they invite, that have framed this volume.

## Perspectives on disease

Just as there is no single way of defining, interpreting, experiencing or managing disease, there is similarly no ideal method for writing histories of disease. The chapters in this book clearly demonstrate that historians have employed a variety of sources, methods and approaches to reveal the changing manifestations and meanings of an eclectic range of diseases across time and space. While they acknowledge at once the cultural complexities, ambiguities and contingency of diagnosis and treatment, contributors to this volume also recognise the realities of suffering, illness, pain and death caused by disease. Indeed, integrating these two perspectives, one focusing more heavily on the political and epistemic aspects of categorising disease and the other on the personal and existential dimensions of illness, provides a fruitful opportunity to reconcile alternative historiographical ambitions and to understand relativist and essentialist approaches to illness and disease as two sides of a similar coin.[26]

At one level, historical studies demonstrate that 'disease' has always been an elusive category that 'does not exist as a social phenomenon unless we agree that it does'.[27] From this perspective, the processes of identifying, naming, explaining and treating disease, or of distinguishing between normal and pathological states, are culturally specific and contingent on a diversity of social, political, professional, economic and technological factors.[28] As several chapters in this volume indicate, one of the key tools of analysis for historians, sociologists, philosophers and anthropologists of medicine interested in how knowledge about disease has been constructed and legitimated is the notion of 'medicalisation', a term that has been used variably to describe the expansion of medical expertise and authority, the manner in which elements of everyday life (including birth, sex and death) become increasingly subject to scientific explanation and management, and the conversion of previously social and moral problems into diseases.[29] In many histories of medicine and disease, medicalisation has been construed as a repressive force, one that has served historically to discriminate, marginalise, stigmatise, control and punish certain sections of the population, to demonise and pathologise forms of behaviour deemed to be deviant or transgressive, to promote and consolidate conspiratorial alliances between doctors, the state and the pharmaceutical industry, and increasingly to reinforce the reactionary political aspirations of neo-liberalism. Medicine, at least in Western countries it has been argued, has gradually (and perhaps inappropriately) displaced religion and law as a repository of truth and as the principal architect of social values, cultural norms, and political prescriptions.[30]

It is relatively straightforward to identify case studies that not only reveal the cultural contingency of diagnosis, but also support political critiques of medical imperialism and the dominance of medical models of disease. During the course of the twentieth century, for example, experiences of anxiety and stress, symptoms of the menopause, diverse forms of sexuality, educational difficulties, parenting and hyperactivity all became subject to bio-scientific interpretation and medical management, at least in the industrialised Western world.[31] Propelled partly by concerns about preserving social stability and economic productivity in the face of sweeping cultural change and technological development, but also by demands from individuals, families and advocacy groups, the identification and legitimation of new diseases and medical intervention into facets of everyday life have on occasion effectively served to highlight needs and to mobilise the medical, educational and welfare resources required to address previously under-recognised inequalities in health-care. At the same time, however, medicalisation appears to have had its costs. Determined by presumptions about class, gender and race and by relatively unreserved beliefs in the ability of science and technology to improve lives, the process of medicalisation appears to have ceded inordinate levels of power to the medical profession to define and manage social problems, reinforced the control of diagnostic technologies and treatment options by multinational corporations, facilitated entrepreneurial exploitation of suffering, and silenced individual narratives of illness even while promoting individual responsibility for health.[32]

Recently, there have been attempts to move beyond preoccupations with medicalisation as a negative force, or at least to mobilise that notion in more neutral and constructive ways. Arguing that medicalisation 'has become a cliché of critical social analysis' that obscures as much as it reveals about medicine and disease, Nikolas Rose has suggested that the concept should be regarded as a point of historiographical departure, rather than arrival: 'The term medicalisation', he wrote in 2007, 'might be the starting point of analysis, a sign of the need for an analysis, but it should not be the conclusion of an analysis'.[33] With this in mind, humanities scholars have turned increasingly to alternative understandings and accounts of disease in the past and present, exploring it not merely in terms of the prevailing cultural norms that shape scientific and medical knowledge, but also in terms of what it is (or what we believe it is) that makes us feel ill, generates physical and emotional pain, and causes death. Regarding disease as a lived experience that constrains our ability to function and frustrates the pursuit of happiness requires fresh perspectives that draw on and interrogate historically contingent cultures of health and sickness, as well as examining and situating individual and collective narratives of pain and illness. As Mark Harrison has pointed out in a reflective reappraisal of global histories of health, medicine and disease, one of the benefits of focusing on health and illness, rather than exclusively on disease, is that it increases our awareness of, and attention to, the subjective and material elements of experience, a perspective that is also beginning to generate new paradigms of public health.[34] Of course, personal experiences of health and disease are no less culturally contingent than medical knowledge, but attention to the subjective elements and impacts of disease offers opportunities to escape from the relativism that infects some histories of medicine.

These two elements of diseases and their histories (the epistemological and the phenomenological) are not disconnected, but they are at times and in critical ways distinguishable. There are many historical examples of contradictions between scientific knowledge of disease and personal experiences of illness. It is evident from patient narratives, for example, that throughout the twentieth century some people suffered from chronic fatigue in a manner that impinged on every aspect of their lives, curtailing their capacity to work, to pursue leisure activities, or to sustain relationships and family responsibilities. Yet, there remained doubts about the validity of their symptoms, at least in terms of whether or not they were suffering from a legitimate 'disease', that is, a condition that, from the perspective of orthodox Western medicine, had a discernible aetiological and pathological identity. Labelled during the 1980s and 1990s as myalgic encephalomyelitis (ME), chronic fatigue syndrome (CFS), or more dismissively as 'yuppie flu', chronic fatigue has been variably understood as a post-viral syndrome, a neurological or neuromuscular disorder, or a manifestation of psychiatric illness or personality disorder.[35] Confusion about the cause, conflicts between patients and practitioners about the 'reality' of the condition, and scepticism about the biological determinants of chronic fatigue reveal much about late-twentieth-century accounts of disease: the ascendancy of expert over lay explanations of disease; the clinical, political and economic hegemony of somatic over psychological illnesses; and the especially potent matrix of biomedical reductionism that has dictated the legitimacy of disease categories and the distribution of resources in the modern Westernised world.

It should be clear that any history of disease needs to incorporate multiple perspectives, to consider the existential as well as the epistemological, the individual and epidemiological as well as the conceptual, ideological and technological dimensions of health, illness and disease in specific historical and cultural settings. The overall structure of this volume and its particular contributions reflect this challenge. The sections that follow aim to address some of the deficits in knowledge and methods identified by Rosenberg in 1989, by examining in turn the models, patterns, technologies and narratives of disease that have shaped, and been shaped by, the experiences, beliefs, emotions, norms, values, knowledge, lifestyles, bodies and environments of past and present populations. The themes explored throughout the book are inter-related and complex: models, patterns, technologies and narratives have not been mutually exclusive, but have co-constituted contemporary understandings, experiences and consequences of disease. Nor are these categories of analysis simple products of history; rather, they serve here as heuristic devices that enable historians and other scholars in the humanities to acknowledge the multiple determinants and elements of disease.

One of the key challenges for historians of medicine has been to explain both continuities and changes in models of disease. In the late nineteenth and early twentieth centuries, it was customary to regard diabetes, a disease that had been described clearly in ancient Egyptian, Indian and Greek manuscripts in terms of the production of excessive amounts of sweet urine, as a product of emotional stress. In various studies of emotions and disease, the American surgeon George Washington Crile (1864–1943), for example, argued that the emotional difficulty of adapting to the pace and stress of modern life led to the discharge of nervous energy, organic changes in the brain, liver and adrenal glands, and the appearance of somatic disease: 'Chronic emotional stimulation', he wrote in 1914, 'may fatigue or exhaust the brain and may cause cardiovascular disease, indigestion, Graves' disease, diabetes, and insanity even'.[36] Crile's formulation of emotions and disease owed much to the work of the Harvard physiologist Walter Bradford Cannon (1871–1945), whose laboratory studies of traumatic shock, digestion, the autonomic nervous system, the adrenal medulla and homoeostasis similarly suggested a connection between the emotional stress of job insecurity, bereavement, war and infidelity, on the one hand, and the appearance of organic diseases such as diabetes, thyroid disease and asthma, on the other.[37] Crile and Cannon were not alone in linking emotions to diabetes. Not only was diabetes thought to be more common among Jewish businessmen, who supposedly possessed greater emotional lability and were subject to extreme job stresses, but the presence of sugar in the urine was also regarded as indicative of emotional disturbance: 'It is interesting to be able to measure the power of emotion in terms as tangible as ounces of sugar', wrote the American physician R. T. Woodyatt in 1927.[38]

Models of diabetes that foregrounded the impact of emotions were gradually, but only partially, superseded in the twentieth century by organic accounts of aetiology and pathogenesis following the isolation and purification of insulin in 1922.[39] Increasingly, scientific and clinical attention focused on the accurate identification of raised blood sugar and early intervention with insulin injections to preclude the long-term cardiovascular and neurological consequences of the disease. Understood more clearly in terms of genetic and lifestyle factors, rather than chronic emotional stimulation, diabetes became one of the archetypal diseases of modern civilisation, a condition thought to be triggered by the rising consumption of sugar and fast foods, escalating levels of obesity, and the sedentary nature of Western ways of life. How do we accommodate these shifting approaches to diabetes in our histories of medicine? How do we balance stories of scientific advances, which have identified many of the genetic and metabolic features of diabetes with greater clarity, against awareness that understandings and definitions of, as well as attitudes to, diabetes have been framed by the values attached to specific lifestyle choices, certain body shapes, and levels of individual responsibility? How do we merge the biological and cultural, the epidemiological and personal, dimensions of diabetes in order to explain shifting models, patterns, experiences and treatments of the disease? Chapters in Part I of this volume address precisely these kinds of questions, exploring in turn ancient Greek and ayurvedic models of disease that emphasised the importance of humoral balance to health, adjacencies and conflicts between religious, magical and medical paradigms, the fluctuating fortunes of theories of contagion, shifting appreciation of the role of emotions in mental illness, and the medicalisation of deviance.

Diseases have rarely respected local, regional or national borders. Infectious diseases have commonly followed the passage of animals, people and consumables. Arguably more complex aetiologically, a number of chronic non-infectious diseases appear to have spread globally through the adoption and adaptation of Western lifestyles, including the consumption of tobacco, alcohol, refined sugars and saturated fats and increased levels of environmental pollution. Interpreting patterns of disease can be difficult. Prior to the mid-nineteenth century, hay fever was a rare condition that was thought initially to be triggered by sunshine, physical exertion and exposure to grasses, flowers and animals and was diagnosed predominantly among the educated, elite classes of Western industrialised countries. During the second half of the century, hay fever appeared to be becoming increasingly common among all classes, a trend attributed by the British general practitioner and homoeopath Charles Blackley (1820–1900) to greater medical and popular attention to the disease rather than to increased prevalence. Along with some American physicians, Blackley identified pollen as the principal immediate cause of hay fever and advocated avoidance by retreating to the coast or the mountains, or remaining indoors, during the summer months.[40]

It is difficult to explain the subsequent explosion of hay fever among modern populations, affecting men and women, poor and rich, and people in the East and West almost equally. In the late 1980s, the epidemiologist David Strachan proposed a theory to explain historical and contemporary patterns of hay fever that merged biological and cultural factors. 'Of the 16 perinatal, social and environmental factors studied', he wrote, 'the most striking associations with hay fever were those for family size and position in the household in childhood.'[41] The more children there were in a family and the greater number of older siblings, the lower the prevalence of hay fever. What became known as the 'hygiene hypothesis' provided a speculative explanation for social, geographical and historical trends:

> Over the past century, declining family size, improvements in household amenities, and higher standards of personal cleanliness have reduced the opportunity for cross infection in young families. This may have resulted in more widespread clinical expression of atopic disease, emerging earlier in wealthier people, as seems to have occurred for hay fever.[42]

Although Strachan later expressed doubts about the validity of the hygiene hypothesis,[43] the possibility that there was a downside to modern Westernised lifestyles (in terms of living standards, domestic environments, diet, and indeed the expanding powers of scientific medicine) attracted attention from clinicians, epidemiologists, journalists and the public keen to explain trends in both infectious and non-infectious diseases, a process that perhaps itself served to promote anxiety about certain diseases and increase the likelihood of diagnosis. Yet, questions still remain about the relative contributions of biological, environmental and socio-cultural factors to historical patterns of diseases. In Part II of this volume, contributors explore such issues from a variety of historical and historiographical perspectives. Taking into account the cultural determinants of diagnosis and lived experiences of disease, chapters in this section examine the appearance, spread and political significance of pandemics, animal diseases, plague, cholera, yellow fever and rickets, as well as the role of race, social geography, political ideology, built environments, and climatic conditions in shaping patterns and explanations of disease in various regions of the world.

In 2002, Charles Rosenberg examined the pivotal role of diagnosis in defining and managing disease during the closing decades of the twentieth century. Linked closely to prevalent beliefs that diseases constituted specific ontological entities (evident, for example, in successive issues of the *Diagnostic and Statistical Manual of Mental Disorders* and the *International Classification of Diseases*), modern diagnostic practices, he argued, had become 'increasingly technical, specialized, and bureaucratized'.[44] Diagnosis was not limited to a self-contained medical sphere of activity, but was also 'a bureaucratic and emotional necessity', Rosenberg suggested, 'not easily disaggregated from the technical capabilities' of contemporary societies.[45] Partly determined by efforts to address established and emerging patterns of sickness, new technologies also made possible the creation of new disease categories and new forms of treatment that have reshaped our beliefs and expectations in, and our commitment to, medical authority. In some ways, then, medical technologies have themselves expanded the range of diseases either by fragmenting older nosologies or by classifying previously inchoate collections of symptoms into more clearly defined, if perhaps arbitrary, disease categories. From this perspective, the history of disease is at least in part the history of diagnosis.

Examples of the indispensability (or what Rosenberg referred to as the 'tyranny') of diagnosis and the consequences of technological and bureaucratic encroachments into medical practice are legion. The elaboration of allergy (or hypersensitivity) as a distinct collection of diseases with a shared pathological basis in the early twentieth century, for example, was a direct product of clinical and laboratory investigations of serum sickness caused by the immunisation protocols for infectious diseases that had been developed in Europe in the wake of the rising professional and political appeal of germ theories of disease in the late nineteenth century.[46] The subsequent division of hypersensitivity reactions into four types according to demonstrable immunological mechanisms was not only driven by dissatisfaction with what were regarded as loose clinical definitions of allergy, but also determined by techniques developed in the laboratories of British pathologists.[47] In 1967, the identification of IgE as the antibody responsible for conditions such as hay fever and food allergies was made possible only by the availability of innovative methods of radioimmunoassay.[48] In all of these cases, technological invention led to new disease classifications, more targeted forms of treatment, fresh administrative approaches to measuring, preventing, and managing disease, and opportunities for commercial exploitation as pharmaceutical companies endeavoured to expand markets and boost profits. Contributors to Part III reflect on the social and cultural dimensions of the scientific and bureaucratic technologies that have served to redefine our understandings and experiences of disease from the eighteenth through to the early twenty-first century. Chapters focus in turn on disability and prosthetics, pain and rehabilitation, cosmetic surgery, cancer screening, medical bacteriology, tuberculosis, the management of diabetes, and venereal disease amongst homosexually active men prior to the advent of AIDS.

The French novelist Marcel Proust (1871–1922) suffered his first severe attack of asthma when he was nine years old, while he was out walking with his family in the Bois de Boulogne. Along with a number of other complaints, asthma dominated Proust's life, leading to prolonged periods in bed, shaping some of the characters in his novels, driving him to consult prominent European physicians, and prompting his attempts to control the symptoms of gasping, coughing and choking with anti-asthma cigarettes and fumigations of his cork-lined Parisian apartment with stramonium.[49] Proust recounted his experiences of asthma in letters to his friends and his mother, to whom he wrote on an almost daily basis until her death in 1905.[50] Recognising the emotional, as well as environmental, triggers of his condition, he compellingly described the 'violence and tenacity' of acute asthma attacks, the incessant 'gasping for breath' that sometimes prevented him from writing, his inability to sleep, and his desperate pursuit of more effective treatments. Proust's portrayals of asthma were clearly contingent, shaped by the nature and boundaries of contemporary medical knowledge as well as by his own temperament and circumstances. At the same time, however, his descriptions of asthma bear remarkable similarities both to ancient Greek and Roman accounts of asthma and to the results of early-twenty-first-century surveys of the symptoms, frequency and severity of asthma attacks amongst modern populations. From this perspective, it would appear that, in spite of changing medical models and treatments, the physical symptoms, existential impact, descriptive language and symbolic significance of asthma have remained relatively constant.[51]

Proust's literary and epistolary narration of his asthma reminds us that there are alternative, subjective narratives of illness that we need to accommodate in our histories, philosophies and sociologies of disease. Analysing letters, patient case books, newspapers, digital media, memoirs and novels as 'literary and epistemic genres' that develop 'in tandem with scientific practices' allows us to glimpse lived experiences of disease, to explore the ways in which burdens of disease have been borne, to recognise the manner in which dominant models, norms and values are adopted, adapted and contested by patients, families and health-care practitioners, and perhaps more expansively to merge biography with history.[52] Chapters in the final section of this volume pay attention to those dimensions of illness and disease that are too frequently ignored or evaded in many histories of medicine. Mobilising sources and methods from across the humanities, contributors examine in turn the construction of links between leprosy and identity in the Middle Ages, the practice of consulting by correspondence in early-modern France, the construction of clinical narratives of 'the shaking palsy' or what became known as Parkinson's disease, the richness of digital narratives of migraine, the role of doctors in framing the psychological illnesses suffered not only by their patients but also by themselves, shifting notions of contagion in twentieth-century literature, the impact of combining visual images and text in representations of disease in graphic novels, and ways of reflecting on the challenges of living well within the constraints of illness and disability. As Havi Carel points out in the final chapter, careful analysis of narratives of health and illness offers scholars opportunities to reconcile tensions between generalised knowledge claims, on the one hand, and individual experiences, on the other, and to understand more fully the elements of both change and continuity in the history of disease.

## Conclusion

This volume is neither an encyclopaedia nor a complete compendium of diseases.[53] Rather it presents a series of case studies, which together form a kaleidoscope of perspectives on the models, patterns, technologies and narratives of disease that can be identified in the past and that, to some extent, continue to structure our present. While most chapters focus on the manifestations and meanings of particular conditions or on the geographical and chronological circulation and subjective experiences of certain diseases, as an entity in itself disease often appears only at the margins of analysis. What emerges from these histories is a picture of disease as integral to societies, as shaping but not necessarily confining individual and communal lives in complex ways, as one of the key objects of personal anxiety, professional enquiry, and political action. Although it cannot be comprehensive in its coverage, the volume's scope is nevertheless broad, exploring human and animal diseases in different time periods, cultures and environments, and from different methodological perspectives. Its aim is not merely to provide historians, sociologists, philosophers and anthropologists with an overview of how we can approach the history of disease, but also to provide health-care professionals, students, and a wider public with insights into the social construction of disease categories, the cultural contexts of health and illness, the pluralities of healing practices and policies, and the variety of documentary, visual and literary sources that can be used to understand past and present accounts and experiences of disease.[54]

The book is also intended to serve as a provisional route map for future scholars. Although the history of medicine has matured and expanded in recent decades, there remain critical issues to address particularly in relation to the history of disease, which has sometimes been diminished by territorial disputes between disciplines and unproductive arguments about the perils of retrospective diagnosis.[55] The chapters in this volume suggest ways of reflecting more closely on the everyday experiences of health-care practitioners, patients and families, and on the role of language and ideas in structuring those experiences. They indicate the value of thinking more carefully about the changing and contrasting meanings attached to terms such as health, illness, disease, and well-being in the past and about the variety of sources and analytical tools that can be exploited to understand more fully the ways in which models, patterns, technologies and narratives of disease are reciprocally constituted. As the final chapters demonstrate, historians have much to learn from other disciplines, especially from sociological, philosophical and anthropological studies of disease, but also from the perspectives afforded by the biological sciences, psychology and clinical medicine. Overall, the book therefore comprises an argument for the assimilation and integration, rather than fragmentation, of sources and methods in order to discern more clearly the intriguing history of disease.

## Notes

1 For recent magisterial overviews of the links between commerce, migration, warfare and disease, see: M. Harrison, *Contagion: How Commerce Has Spread Disease*, London and New Haven: Yale University Press, 2013; M. Harrison, 'Disease and world history from 1750', in J. M. McNeill and K. Pomeranz (eds), *The Cambridge World History: Volume 7: Production, Destruction and Connection, 1750–Present*, Cambridge and New York: Cambridge University Press, 2015, pp. 237–58. For some older works on similar themes, see: A. W. Crosby, *The Columbian Exchange: Biological and Cultural Consequences of 1492*, Westport: Greenwood Press, 1972; F. F. Cartwright and M. D. Biddiss, *Disease and History*, Gloucestershire: Sutton Publishing Limited, [1972] 2000; W. H. McNeill, *Plagues and Peoples*, Garden City, NY: Anchor Press, 1976; and A. W. Crosby, *Ecological Imperialism: The Biological Expansion of Europe, 900–1900*, Cambridge: Cambridge University Press, 1986.

2 According to Allan Brandt, 'the distinction between medical history and social history' was already becoming 'increasingly obscure' by the 1990s – A. Brandt, 'Emerging themes in the history of medicine', *Milbank Quarterly*, 69, 1991, 199–214. Whether that trend has continued is arguable, as the history of medicine has increasingly emerged as a distinct genre of historical scholarship.

3 Brandt, 'Emerging themes in the history of medicine', 210–11; V. Berridge and P. Strong (ed.), *AIDS and Contemporary History*, Cambridge: Cambridge University Press, 1993; V. Berridge, 'Researching contemporary history: AIDS', *History Workshop Journal*, 38, 1994, 227–34; V. Berridge, *AIDS in the UK: The Making of Policy, 1981–1994*, Oxford: Oxford University Press, 1996; V. Berridge, 'AIDS', in R. Cooter and J. Pickstone (ed.), *Medicine in the 20th Century*, Amsterdam: Harwood Academic Publishers, 2000, pp. 687–707.

4 G. Mitman, *Breathing Space: How Allergies Shape Our Lives and Landscapes*, New Haven: Yale University Press, 2007; R. Tattersall, *Diabetes: The Biography*, Oxford: Oxford University Press, 2009; M. Jackson, *Asthma: The Biography*, Oxford: Oxford University Press, 2009; F. Bound Alberti, *Matters of the Heart: History, Medicine and Emotion*, Oxford: Oxford University Press, 2010; C. Timmermann, *A History of Lung Cancer: The Recalcitrant Disease*, London: Palgrave, 2013; G. Weisz, *Chronic Disease in the Twentieth Century: A History*, Baltimore: Johns Hopkins University Press, 2014.

5 Brandt, 'Emerging themes in the history of medicine', 211.

6 For some recent examples, see: Mitman, *Breathing Spaces*, pp. ix–xv; A. Tone, *The Age of Anxiety: A History of America's Turbulent Affair with Tranquilizers*, New York: Basic Books, 2009, p. x; M. Smith, *Hyperactive: The Controversial History of ADHD*, London: Reaktion, 2012, pp. 7–11.

7 M. McDonald, *Mystical Bedlam: Madness, Anxiety, and Healing in Seventeenth-Century England*, Cambridge: Cambridge University Press, 1981; C. E. Rosenberg *The Cholera Years: The United States in 1832, 1849, and 1866*, Chicago: University of Chicago Press, [1962] 1987; R. J. Evans, *Death in Hamburg: Society and Politics in the Cholera Years 1830–1910*, Oxford: Clarendon Press, 1987; L. Bryder, *Below the Magic Mountain: A Social History of Tuberculosis in Twentieth-Century Britain*, Oxford: Oxford University Press, 1988; B. Bates, *Bargaining for Life: A Social History of Tuberculosis, 1876–1938*, Philadelphia: University of Pennsylvania Press, 1992; F. M. Snowden, *Naples in the Time of Cholera, 1884–1911*, Cambridge: Cambridge University Press, 1995; K. Ott, *Fevered Lives: Tuberculosis in American Culture since 1870*, 1996; R. Porter and G. S. Rousseau, *Gout: The Patrician Malady*, New Haven: Yale University Press, 1998; P. Baldwin, *Contagion and the State in Europe 1830–1930*, Cambridge: Cambridge University Press, 1999; K. Wailoo, *Dying in the City of the Blues: Sickle Cell Anemia and the Politics of Race and Health*, Chapel Hill: University of North Carolina Press, 2001; C. Rawcliffe, *Leprosy in Medieval England*, London: Boydell Press, 2006; R. A. Aronowitz, *Unnatural History: Breast Cancer and American Society*, Cambridge: Cambridge University Press, 2007; S. Pemberton, *The Bleeding Disease: Hemophilia and the Unintended Consequences of Medical Progress*, Baltimore: Johns Hopkins University Press, 2011; S. K. Cohn, Jr., *Cultures of Plague: Medical Thinking at the End of the Renaissance*, Oxford: Oxford University Press, 2011; A. Hardy, *Salmonella Infections, Networks of Knowledge and Public Health in Britain 1880–1975*, Oxford: Oxford University Press, 2015; and M. H. Green (ed.), *Pandemic Disease in the Medieval World: Rethinking the Black Death*, Philadelphia: ARC Medieval Press, 2016.

8 See the discussion in the next section below.

9 A. S. Wohl, *Endangered Lives: Public Health in Victorian Britain*, London: Dent, 1983; R. Spree, *Health and Social Class in Imperial Germany: A Social History of Mortality, Morbidity and Inequality*, Oxford: Berg, 1988; C. Jones, *The Charitable Imperative: Hospitals and Nursing in Ancien Régime and Revolutionary France*, London: Routledge, 1989; M. Vaughan, *Curing Their Ills: Colonial Power and African Illness*, Stanford: Stanford University Press, 1991; D. Arnold, *Colonizing the Body: State Medicine and Epidemic Disease in Nineteenth-century India*, Berkeley: University of California Press, 1993; L. Manderson, *Sickness and the State: Health and Illness in Colonial Malaya, 1870–1940*, Cambridge: Cambridge University Press, 1996; B. Harrison, *Not Only the 'Dangerous Trades': Women's Work and Health in Britain, 1880–1940*, London: Taylor and Francis, 1996; C. C. Sellers, *Hazards of the Job: From Industrial Disease to Environmental Health Science*, Chapel Hill: University of North Carolina Press, 1997; D. Porter, *Health, Civilization and the State: A History of Public Health from Ancient to Modern Times*, London: Routledge, 1998; P. Weindling, *Epidemics and Genocide in Eastern Europe 1890–1945*, Oxford: Oxford University Press, 2000; R. Rotberg (ed.), *Health and Disease in Human History*, Cambridge, MA: MIT Press, 2000; P. W. J. Bartrip, *The Home Office and the Dangerous Trades: Regulating Occupational Disease in Victorian and Edwardian Britain*, Amsterdam: Rodopi, 2002; G. N. Grob, *The Deadly Truth: A History of Disease in America*, Cambridge, MA: Harvard University Press, 2002; A. Bashford, *Imperial Hygiene: A Critical History of Colonialism, Nationalism and Public Health*, London: Palgrave Macmillan, 2003; R. Stevens, *Medical Practice in Modern England: The Impact of Specialization and State Medicine*, Piscataway, NJ: Transaction Publishers, [1966] 2003; M. J. Dobson, *Contours of Death and Disease in Early Modern England*, Cambridge: Cambridge University Press, [1997] 2003; Roger French, *Medicine before Science: The Rational and Learned*

*Doctor from the Middle Ages to the Enlightenment*, Cambridge: Cambridge University Press, 2003; M. Harrison, *Disease and the Modern World: 1500 to the Present Day*, London: Polity Press, 2004; L. Nash, *Inescapable Ecologies: A History of Environment, Disease, and Knowledge*, Berkeley: University of California Press, 2006; J. Henderson, *The Renaissance Hospital: Healing the Body and Healing the Soul*, New Haven: Yale University Press, 2006; M. Gorsky and S. Sheard (eds), *Financing Medicine: The British Experience since 1750*, London and New York: Routledge, 2006; P. Razzell, *Population and Disease: Transforming English Society, 1550–1850*, London: Caliban Books, 2007; J. N. Hays, *The Burdens of Disease: Epidemics and Human Response in Western History*, New Brunswick: Rutgers University Press, 2009; M. Harrison, *Medicine in an Age of Commerce and Empire: Britain and its Tropical Colonies, 1660–1830*, Oxford: Oxford University Press, 2010; M. Jackson (ed.), *The Oxford Handbook of the History of Medicine*, Oxford: Oxford University Press, 2011.

10  C. E. Rosenberg, 'Disease in history: Frames and framers', *Milbank Quarterly*, 67, 1989, 1–15.

11  For example, see: J. Andrews and A. Scull, *Customers and Patrons of the Mad-Trade: The Management of Lunacy in Eighteenth-Century London*, Berkeley: University of California Press, 2003; G. Davis, 'The Cruel Madness of Love': Sex, Syphilis and Psychiatry in Scotland, 1880–1930, Amsterdam: Rodopi, 2008; A. E. Tomkins, 'Workhouse Medical Care from Working-Class Autobiographies, 1750–1834', in J. Reinarz and L. Schwarz (eds), *Medicine and the Workhouse*, Rochester: University of Rochester Press, 2013, pp. 86–102.

12  F. F. Cartwright and M. Biddiss, *Disease and History*, 2nd edition, Gloucestershire: Sutton Publishing Limited, [1972] 2000, p. 1.

13  Ibid., pp. 21, 62.

14  For discussion of some of the issues, see: A. Karenberg, 'Retrospective diagnosis: use and abuse in medical historiography', *Prague Medical Report*, 110, 2009, 140–45; R. Cooter, 'The life of a disease?', *Lancet*, 375, 2010, 111–12; O. Muramoto, 'Retrospective diagnosis of a famous historical figure: ontological, epistemic, and ethical considerations', *Philosophy, Ethics, and Humanities in Medicine*, 9, 2014, available at http://www.peh-med.com/content/9/1/10, accessed 6 January 2016.

15  Cooter, 'The life of a disease?'

16  For a balanced and incisive discussion of these issues, see: M. Harrison, 'A global perspective: Reframing the history of health, medicine and disease', *Bulletin of the History of Medicine*, 89, 2015, 639–89. Responses to this article by Alison Bashford, J. R. McNeill, and Kavita Sivaramakrishnan appear in the same volume.

17  Both Johns Hopkins University Press and Oxford University Press publish well-regarded series entitled 'Biographies of Disease' and edited by senior scholars in the field. See, for example: R. M. Packard, *The Making of a Tropical Disease: A Short History of Malaria*, Baltimore: Johns Hopkins University Press, 2007; D. Healy, *Mania: A Short History of Bipolar Disorder*, Baltimore: Johns Hopkins University Press, 2008; C. Hamlin, *Cholera: The Biography*, Oxford: Oxford University Press, 2009; S. L. Gilman, *Obesity: The Biography*, Oxford: Oxford University Press, 2010; H. Bynum, *Spitting Blood: The History of Tuberculosis*, Oxford: Oxford University Press, 2012.

18  L. Jordanova, 'The social construction of medical knowledge', *Social History of Medicine*, 7, 1995, 361–81.

19  Ibid., 375–7.

20  Rosenberg, 'Disease in history', 1.

21  Ibid., 4.

22  Ibid., 4.

23  See, for example, a number of Temkin's essays on these issues in O. Temkin, *The Double Face of Janus and Other Essays in the History of Medicine*, Baltimore: Johns Hopkins University Press, 1977.

24  C. E. Rosenberg, 'What is disease? In memory of Owsei Temkin', *Bulletin of the History of Medicine*, 77, 2003, 491–505.

25  Rosenberg, 'Disease in history', 14.

26  There have, of course, been many attempts to define disease in ontological and functional terms from a philosophical perspective. For a relatively recent, but perhaps unconvincing, discussion of some of the approaches, see: R. Cooper, 'Disease', *Studies in History and Philosophy of Biological and Biomedical Sciences*, 33, 2002, 263–82.

27  Ibid., 2.

28 For a seminal discussion of the contingency of ideas of the normal and pathological, see G. Canguilhem, *Le normal et le pathologique*, Paris: Presses Universitaires de France, 1966, translated into English in 1978.

29 R. A. Nye, 'The evolution of the concept of medicalization in the late twentieth century', *Journal of the History of the Behavioral Sciences*, 39, 2003, 115–29. Perhaps the earliest, but certainly the most influential, formulation of medicalisation is Ivan Illich, *Medical Nemesis*, London: Calder & Boyars, 1974.

30 For an early critique of this argument, see Irving Kenneth Zola, 'Medicine as an institution of social control', *Sociological Review*, 20, 1972, 487–504. For a series of provocative reflections on medicalisation, see the contributions by Faith McLellan, Jonathan Metzl and Rebecca Herzig, Nancy Tomes, Nikolas Rose, and Troy Duster in *Lancet*, 369, 2007, 627–8, 697–704.

31 A. V. Horwitz, *Anxiety: A Short History*, Baltimore: Johns Hopkins University Press, 2013; M. Jackson, *The Age of Stress: Science and the Search for Stability*, Oxford: Oxford University Press, 2013; M. Lock, *Encounters with Aging: Mythologies of Menopause in Japan and North America*, Berkeley: University of California Press, 1993; L. Foxcroft, *Hot Flushes, Cold Science: A History of the Modern Menopause*, London: Granta, 2009; M. M. Gullette, *Aged by Culture*, Chicago: University of Chicago Press, 2004; Smith, *Hyperactive*; M. Smith, *An Alternative History of Hyperactivity: Food Additives and the Feingold Diet*, New Brunswick: Rutgers University Press, 2011. See also the chapter in this volume by Jana Funke.

32 For a critique of commercial exploitation of concerns about stress, for example, see: S. Doublet, *The Stress Myth*, Sydney: Ipsilon Publishing, 1999; A. Patmore, *The Truth about Stress*, London: Atlantic Books, 2006.

33 Nikolas Rose, 'Beyond medicalisation', *Lancet*, 369, 2007, 700–2.

34 Harrison, 'A global perspective'. For a discussion of the fifth wave of public health, which emphasises the subjective and emotional dimensions of health and disease, see P. Hanlon, S. Carlisle, M. Hannah, D. Reilly, and A. Lyon, 'Making the case for a "fifth wave" in public health', *Public Health*, 125 (2011), 30–6.

35 For some late twentieth-century explorations of the manifestations and possible causes of fatigue syndromes, see: S. Wessely and R. Powell, 'Fatigue syndromes: A comparison of chronic "postviral" fatigue with neuromuscular and affective disorder', *Journal of Neurology, Neurosurgery and Psychiatry*, 52, 1989, 940–8; S. Wessely, 'The measurement of fatigue and chronic fatigue syndrome', *Journal of the Royal Society of Medicine*, 85, 1992, 189–90; G. MacLean and S. Wessely, 'Professional and popular views of chronic fatigue syndrome, *British Medical Journal*, 308, 1994, 776–7; S. Wessely, 'The epidemiology of chronic fatigue syndrome', *Epidemiological Reviews*, 17, 1995, 139–51; and contributions to a special issue of the *British Medical Bulletin*, 47, 1991, 793–1005. There have been few sustained historical discussions of fatigue, but see: E. Shorter, *From Paralysis to Fatigue: A History of Psychosomatic Illness in the Modern Era*, New York: Simon and Schuster, 1970; S. Wessely, 'History of postviral fatigue syndrome', *British Medical Bulletin*, 47, 1991, 919–41; A. Rabinbach, *The Human Motor: Energy, Fatigue, and the Origins of Modernity*, Berkeley: University of California Press, 1992; A. K. Schaffner, 'Exhaustion and the pathologization of modernity', *Journal of Medical Humanities*, 2014, available at http://link.springer.com/article/10.1007/s10912-014-9299-z, accessed 4 January 2016; A. Schaffner, *Exhaustion: A History*, New York: Columbia University Press, 2016. Changing experiences, biomedical accounts and occupational health implications of fatigue, particularly amongst pilots and doctors, are the focus of Natasha Feiner's current doctoral research at the University of Exeter.

36 G. W. Crile, *The Origin and Nature of the Emotions*, Claremont, SC: BiblioBazaar, [1915] 2006, preface, pp. 87–8. See also G. W. Crile, *The Thyroid Gland*, Philadelphia: W. B. Saunders Co., 1922.

37 Cannon wrote extensively on the role of emotions in disease, but in relation to diabetes see in particular: W. B. Cannon, 'The mechanism of emotional disturbance of bodily functions', *New England Journal of Medicine*, 198, 1928, 877–84; W. B. Cannon, *Bodily Changes in Pain, Hunger, Fear and Rage*, New York: D. Appleton-Century, 1939, pp. 66–8; W. B. Cannon, *Digestion and Health*, London: Martin Secker and Warburg, 1937, pp. 144–5.

38 Quoted in Cannon, 'The mechanism of emotional disturbance', 882.

39  M. Bliss, *The Discovery of Insulin*, Chicago: University of Chicago Press, 1984; Tattersall, *Diabetes*.

40  See the discussion in M. Jackson, *Allergy: The History of a Modern Malady*, London: Reaktion, 2006, pp. 56–65.

41  D. Strachan, 'Hay fever, hygiene and household size', *British Medical Journal*, 299, 1989, 1259–60.

42  Ibid.

43  D. P. Strachan, 'Family size, infection and atopy: The first decade of the "hygiene hypothesis"', *Thorax*, 55 Supplement, 2000, S2–10.

44  C. E. Rosenberg, 'The tyranny of diagnosis: Specific entities and individual experience', *Milbank Quarterly*, 80, 2002, 237–60, at 237.

45  Ibid., 256.

46  Jackson, *Allergy*, pp. 34–44.

47  Ibid., pp. 122–5.

48  Ibid., pp. 125–6.

49  Proust's asthma is discussed at greater length in Jackson, *Asthma*, pp. 1–9.

50  For examples of Proust's letters about asthma, see G. D. Painter (ed.), *Marcel Proust: Letters to His Mother*, London: Rider and Company, 1956.

51  For an extensive discussion of elements of constancy and change in the diagnosis and experience of asthma across cultures, time and place, see Jackson, *Asthma*, passim.

52  For a discussion of 'epistemic genres' in medicine, see Gianna Pomata, 'The medical case narrative: distant reading of an epistemic genre', *Literature and Medicine*, 32, 2014, 1–23. Reflections on the challenges and benefits of merging biography and history can be found in: C. Steedman, *Landscape for a Good Woman*, London: Virago, 1986; L. Abrams, 'Liberating the female self: epiphanies, conflict and coherence in the life stories of post-war British women', *Social History*, 39, 2014, 14–35.

53  The most comprehensive encyclopaedia of disease is K. F. Kiple (ed.), *The Cambridge World History of Human Disease*, Cambridge: Cambridge University Press, 1993.

54  The value of teaching history to medical students has been the subject of considerable discussion. For recent overview of debates and an argument for the importance of history, see D. S. Jones, J. A. Greene, J. Duffin, and J. H. Warner, 'Making the case for history in medical education', *Journal of the History of Medicine and Allied Sciences*, 70, 2014, 623–52. See also the contributions to F. Huisman and J. H. Warner (eds), *Locating Medical History: The Stories and Their Meanings*, Baltimore: Johns Hopkins University Press, 2004.

55  Over recent years, concerns have been expressed about a perceived 'crisis' in medical history and about the political and intellectual sterility of the discipline. Some of these concerns have been overly pessimistic and conceited, but they do raise questions about the boundaries, politics and purpose of history that historians of medicine and disease should not ignore. For the most compelling and provocative reflections, see: R. Hayward, '"Much exaggerated": The end of the history of medicine', *Journal of Contemporary History*, 40, 2005, 167–78; R. Cooter, 'After death/after-"life": The social history of medicine in post-postmodernity, *Social History of Medicine*, 20, 2007, 441–64; J. Toms, 'So what? A reply to Roger Cooter's "after death/after-'life': the social history of medicine in post-postmodernity"', *Social History of Medicine*, 22, 2009, 609–15; and the various contributions to a special issue of *Medical History*, 55, 2011, on the future of the field.

## Select bibliography

A. Brandt, 'Emerging themes in the history of medicine', *Milbank Quarterly*, 69, 1991, 199–214.

F. F. Cartwright and M. D. Biddiss, *Disease and History*, Gloucestershire: Sutton Publishing Limited, [1972] 2000.

M. Harrison, *Disease and the Modern World: 1500 to the Present Day*, London: Polity Press, 2004.

M. Harrison, *Contagion: How Commerce Has Spread Disease*, London and New Haven: Yale University Press, 2013.

M. Jackson (ed.), *The Oxford Handbook of the History of Medicine*, Oxford: Oxford University Press, 2011.

L. Jordanova, 'The social construction of medical knowledge', *Social History of Medicine*, 7, 1995, 361–81.

K. F. Kiple (ed.), *The Cambridge World History of Human Disease*, Cambridge: Cambridge University Press, 1993.

R. A. Nye, 'The evolution of the concept of medicalization in the late twentieth century', *Journal of the History of the Behavioral Sciences*, 39, 2003, 115–29.

C. E. Rosenberg, 'Disease in history: frames and framers', *Milbank Quarterly*, 67, 1989, 1–15.

C. E. Rosenberg, 'The tyranny of diagnosis: specific entities and individual experience', *Milbank Quarterly*, 80, 2002, 237–60.

C. E. Rosenberg, 'What is disease? In memory of Owsei Temkin', *Bulletin of the History of Medicine*, 77, 2003, 491–505.

# Part I

# MODELS

# 2

# HUMOURS AND HUMORAL THEORY

*R. J. Hankinson*

He knew the cause of everich maladye,
Were it hoot, or coold, or moyste, or drye,
And whet engendred and of what humour –
He was a verray, parfit praktisour
　　　　　(*Canterbury Tales*, Prologue 419–22,
　　　　　　　　the 'Doctour of Phisik')

I fear it is too choleric a meat.
　　　　　(*The Taming of the Shrew*, IV.iii)

The 'Theory of Humours' was of remarkable longevity. Its origins stretch back to the Hippocratics of the fifth century BC and perhaps earlier; certainly the language of 'humours' has a long pre-theoretical history. The theory was codified into the canonical four (blood, phlegm, yellow and black bile) largely as the result of the enormous influence of Galen. The humours found their way into Arabic medicine in what we think of as the Dark Ages, and then back into mediaeval and Renaissance Europe. Although not unchallenged by further developments from the seventeenth century onwards, nonetheless vestiges of the theory as a medico-physiological doctrine survived at least into the nineteenth century in Western medicine and arguably beyond. There are still Galenic physicians in Delhi who deploy its categories and its influence is detectable in ayurvedic medicine generally.[1] And it lives on, albeit with a slightly archaic patina, in the characterization of character types: sanguine, phlegmatic, bilious, melancholic. This chapter concentrates on the early history of the ideas as they developed in Classical Greece and eventually triumphed in the Galenic synthesis of the second century AD.

## The idea of a humour and the origins of element theory

The word we translate in this context as 'humour' is the Greek word *khumos*. Originally the term meant 'juice' or 'flavour' (and sometimes 'sap') and it continued to bear such non-technical senses throughout antiquity, in medical as well as in more general writing. But at some point, it began to take on a technical colouring, albeit one related to the original senses of 'juice' and 'flavour'. Unfortunately that colouring, as well as the specific substances that fall under its purview and their respective roles, is extremely variable. Part of the purpose of this chapter will be to exemplify that chromatic range, but also to account for how the theory came to acquire what were

21

to become its central, classical characteristics. Moreover, just as the term *khumos* need not refer to humours in any theoretical sense, such items may be designated in other ways and with other terminology.

Matters are further complicated by their association in many (but by no means all) texts with the fundamental qualities (Hot, Cold, Wet and Dry), which are sometimes viewed as being more or less equivalent to substances, and with the standard (although again by no means ubiquitous) four elements of Greek element theory, Earth, Air, Fire and Water. These associations are most clearly developed in the work of Galen, but there are hints of them a good deal earlier. From the very beginnings of Greek natural investigation, thinkers were concerned with trying to account for the basic structures of things, their changes and interactions, frequently in terms of fundamental opposites, although this came into conflict with the alternative view that everything was a modification of a single elemental stuff.

The earliest known 'Presocratic', Thales (fl. c. 585 BC), held that water was in some, still disputed, sense basic. A little later, Anaximenes asserted that everything was a modification of air, caused by relative condensation and rarefaction associated with degrees of cold and heat.[2] Later, impressed by its evident dynamism, Heraclitus made fire basic. Both of the latter allowed for the other 'elements', but as fundamentally derivative of the basic one. By (partial) contrast, before Anaximenes, Anaximander had made his basic stuff an undifferentiated 'indefinite', which took on physical properties as a result of the constant motility in things. In the fifth century, a fully developed four-element theory was introduced by Empedocles: earth, air, water and fire were basic (and unalterable, hence genuinely elemental), and the furniture of the visible world was caused by their mixing in various ratios under the influence of two opposing cosmic forces, which he called 'love' and 'strife', which were basically associating and disassociating forces. Finally, in this context, the Pythagoreans, in addition to investigating musical harmony and refusing to eat beans, liked to see the world as an inter-play of opposites.

None of this is directly relevant to medicine, much less specifically to humoral theory, although Empedocles had a reputation as a healer, and the early writers, in particular Anaximander and Empedocles, were interested in physiology and animal generation. But the general intellectual climate matters, and one pre-Hippocratic figure in particular stands out: Alcmaeon of Croton. An associate of the Pythagoreans, according to Diogenes Laertius (8.83) 'most of what he wrote concerns medicine, yet he sometimes also deals more generally in natural science, saying "most human things come in pairs"'. Aristotle (*Metaph.* 986a29–34) confirms this, but says that 'he speaks not like them [the Pythagoreans] of a determinate list of oppositions, but rather of any whatever, such as black and white, sweet and bitter, good and bad, big and small'. Recalling the original sense of *khumos* as 'flavour', that list is backed up by a later report in Plutarch:

> Alcmaeon says that health is preserved by equality among the powers: wet and dry, hot and cold, bitter and sweet, and so on; monarchy among them produces diseases, since dominance of either pair is destructive. Disease is caused *by* an excess of heat or cold, *from* an excess or deficiency of nutrition, and *in* the blood, marrow or the brain. Health is the proportional blending of the qualities.
>
> (Plutarch, *Physical Opinions* 911a;
> Barnes, 2001, 37, = Fr. 24 B 4 Diels/Kranz)

That fragment, in its language, its generality (and indeed its vagueness) foreshadows much of what we find in the developed Hippocratic tradition. Particularly striking is the careful causal analysis of immediate and remote cause, and location of the pathological effects.

Two other Presocratics deserve brief mention here, as prefiguring later humoral developments.[3] The influential fifth-century Pythagorean Philolaus of Croton[4] talks of blood, bile and phlegm as 'the origins of diseases' (*Anon.Lond.* 18.8,[5] = A27 Huffman, 1993): blood, when thickened or thinned; phlegm, attributed to what he calls 'rains',[6] and which is associated (uncharacteristically) with heat; and bile, which is a 'residue (*ikhor*) of flesh' and not associated with the liver (Huffman, 1993, 304–5). In Homer (*Iliad* 5.364) *ikhor* is what flows in the veins of the gods, but in the Hippocratics denotes a serious, often harmful, residual fluid. Whether we choose to call it humoral or not, this is a pathological theory. These items are not natural constituents of a normally functioning body. Even the blood is corrupted. Similarly for Thrasymachus of Sardis, bile, phlegm and pus, his chosen three, are pathological rather than constitutional; they are corruptions of blood engendered by heat and cold (*Anon Lond.* 9.42).[7] The distinction between *pathological* humoral theory, which these accounts at least foreshadow, and *constitutive* humoral theories, where the items are naturally present in a properly functioning body, and cause trouble only if they become disproportionate, is fundamental. In many cases it is unclear which type of theory (if either) we are dealing with.[8] But the canonical, developed theory to be found in Galen and fathered by him on what he takes to be the genuine Hippocrates, represented paradigmatically in the treatise *Nature of Man*, is fundamentally constitutive in nature. Our task is to see how he got there. Let us now, then, turn to the Hippocratics.

## The Hippocratic corpus: the case of 'ancient medicine'

More than 60 texts have come down to us from antiquity under the name 'Hippocrates'.[9] A mixed, indeed mutually inconsistent bunch, many were already recognized in antiquity to be spurious. That there was a historical Hippocrates, who lived roughly between 450 and 360 BC, is certain. More than that, we cannot say. The texts range from treatises on diagnosis and cure, prognosis, clinical history, the general association of prevalent epidemics with seasonal and climatic conditions, health and nutrition, embryology, physiology, and the setting of bones, to those concerned with medical etiquette, including the celebrated *Oath*. Some are largely descriptive (*Epidemics, Aphorisms*), some more concerned with causality and symptomology (*Diseases, Affections*), others much more theoretical (such as *Prognosis* and *Regimen*), some a mixture (*Airs, Waters, Places*), some more general and some (*Sacred Disease*) more particularly focused. Some are polemical and defensive (*On the Art*), arguing that medicine's failure to achieve 100 per cent success does not entail that it is not a science and no better than guesswork, evidently a common accusation (*Regimen in Acute Diseases* 1–9; *Nature of Man* 1). One of the most obscure and compressed texts in the corpus is actually called *Humours*, but it rarely uses the term *khumos*, and then only to cover pretty much any type of internal fluid.

The physiological, pathological and therapeutic approaches differ enormously as well. *Breaths*, for example, is resolutely monistic.[10] *Nature of Man* argues that monism is incoherent; but among the pluralist theories there is much disagreement as well. In

*Regimen*, fire and water are fundamental, the relations between them accounting not only for general physiological types and sexual differentiation, but also for quickness of wit (fire) and retentiveness (water: *Regimen* 1.35–6). Fire is paradigmatically hot and dry, while water is wet and cold; the connection between elements and qualities becomes specific. *Fleshes* privileges heat, earth (which is cold and dry) and air (which is thick and moist).[11]

However, the important *Ancient Medicine* argues that postulations involving Hot, Cold, Wet and Dry are irrelevant and indeed incoherent (Chs 13–16); one simply needs to infer inductively from nutritional experience. The author talks of *khumoi*, but he simply means empirically detectable kinds of flavour, such as sweet and bitter, which he associates with certain causal properties, observable externally in such processes as tanning, which involves astringency (*Ancient Medicine* 15). As for Alcmaeon, health consists in none of the humours becoming too powerful:

> the ancients . . . never imagined heat, coldness, wetness or dryness either harmed or were necessary for health; they attributed disease to some factor stronger and more powerful than the human body which it could not overcome. Every quality is most powerful when most concentrated – sweetness when sweetest, bitterness when bitterest, sharpness when sharpest, and so on. The existence of such qualities in human bodies was observed along with their harmful effects . . . when they are properly mixed and compounded, they are neither observable nor harmful, but when one is separated out it becomes apparent and harmful. Equally all foods which are unsuitable and harmful for us have some such property: they are either bitter or salt or sharp, or have some other strong, undiluted quality. . . . Strength, growth, and nourishment come from what is well-mixed and contains no strong, undiluted element.
>
> (*Ancient Medicine* 14)

The concern with diet is characteristically Hippocratic, but the rejection of theoretical Heat, Cold, and so on in favour of empirically observable, if generalizable, properties is striking.[12] This is not a humoral *theory* in the sense later to be investigated, but it does involve the language of humours. This is why such care is needed in analysis. Here, humours just are substances whose properties are empirically detectable, generally as types of flavour. Later, he writes that 'men do not become feverish because of heat alone, nor is this the only cause of the bad condition'; rather things are hot and bitter or hot and acid for example, and it is these 'humours' which do the work (17). Moreover, 'all other illnesses that are caused by acrid or undiluted humours come about in the same way, and decline as the humours are digested and mixed' (18). The general idea is that health is restored by destroying the power of the individual dominant pathogen:

> those humours of the throat that cause hoarseness and sore throats, those which cause erysipelas or pneumonia, are at first salt, moist and acrid, and during this period the diseases flourish. When the discharges become thicker and milder, and lose their acidity, the fevers cease as well as the other harmful effects of the disease . . . all human ailments are caused by 'powers' (*dunamies*); thus when there is an efflux of bitterness which we call 'yellow bile',[13] nausea, fever and weakness grip the patient; when he gets rid of it,

purged either spontaneously or by medication, manifestly both the pain and the fever disappear. But if the bitter particles are undissolved, undigested and uncompounded, pain and fever cannot be prevented in any way.

*(Ancient Medicine* 19)

Similarly, 'pungent and acrid humours' produce derangement, internal pain and confusion, until they can be 'purged or diluted with other humours'. This is generalizable: 'a man is in the best possible condition when there is complete coction[14] and repose, with no particular power in evidence' (19). The author defines a 'power' as 'an intensity and strength of humours' (22), having acknowledged that human constitutions evidently differ: some, for instance, cannot tolerate cheese because of a particular 'hostile humour', but this is not a universal human deficiency. Consequently, 'we must investigate each of these humours, their effect upon humans, and their relations with each other'. Thus humoral vocabulary began to appear in Hippocratic medicine, without remotely implying the existence of any developed 'humoral theory'.

## Bile and phlegm

Bile and phlegm appear frequently in Classical period Hippocratic texts, but again in a variety of roles.[15] In particular, there is no consensus as to whether they are natural constituents of the body, or whether rather they are intrinsically pathological. Take the case of *Affections* 1, which states baldly:

All human diseases occur as a result of bile and phlegm; bile and phlegm produce diseases when in the body one of them becomes too wet, too dry, too hot or too cold. Bile and phlegm suffer these things from food, drink, exertion, wounds, smells, sights, sounds, sex and heat and cold; this occurs when any of the things mentioned are applied to the body at the wrong time, against normal practice, in too great an amount or too strong, or not enough and too weak.[16]

This implies that bile and phlegm are natural bodily components, which only become pathological when exacerbated by the four qualities as a result of inopportune antecedent causes. The rest of the treatise bears this out. Certain types of headache, for example, are caused by excessive phlegm, which should be purged (*Affections* 2), and similarly with earache and swellings of the throat, tongue or gums (4). By contrast, acute feverish conditions, including phrenitis, a condition involving derangement, are due to excess bile (10–11). Some conditions are attributed indifferently to either, particularly in *Diseases* 1:[17] ardent fever attacks particularly the bilious, but also the phlegmatic (29).[18]

There is a good deal of indeterminacy, inconsistency even, behind the confident attitudes; and while heat, dryness and bile, cold and wetness and phlegm, tend to go together, vagueness persists. Consider *Diseases* 1.24:

Chills . . . occur most frequently and severely when either bile or phlegm or both are mixed in the same place with blood, more so in the case of phlegm, since it is the coldest thing in man, while blood is the hottest, and bile colder than blood . . . thus combined with blood, they make it congeal.

Bile and phlegm are both relatively cooler than blood, and thus both damage its efficacy, albeit to different degrees. So far, so good. But phrenitis is apparently caused by bile *heating* the cerebral blood, 'which contributes the greatest part . . . to intelligence' (*Diseases* 1.30; cf. 34). Melancholics can also suffer phrenitis-like symptoms:

> When their blood is disordered by bile and phlegm they are also out of their mind, and even raving; it is similar in phrenitis, but the madness and derangement are less severe to the extent to which this bile [i.e. presumably yellow bile] is weaker than the other one.
>
> (1.30)

Here we have an apparent reference to the mysterious black bile, the *melaina kholê* that survives in our own vocabulary of despondency and depression. Thus, in a sense, this treatise presents all four of the canonical humours of theory; but their roles are very different. Bile and phlegm are primarily pathogenic, causing harm by damaging the vital blood, while 'the other bile' is little more than an afterthought.

Similar views are expressed in *Sacred Disease* (Chs 8, 18), which is particularly concerned to demonstrate that seizure-disorders, often attributed to divine visitation, are in fact the result of phlegm accreting in the brain, interfering with the natural operation of cerebral blood, and causing disturbance and loss of consciousness and paralysis (10–13). Individuals are classified according to their prevailing humour; the naturally bilious are unaffected by these seizures, but are prone to more aggressive forms of derangement (18).

*Airs, Waters, Places*, which is concerned with the effects of climate and environment (and indeed political organization) on constitutions and character types, expresses comparable views. Thus those who live in cool, north-facing cities drink hard, cold water, tend to be sinewy and wiry and are prone to constipation; and 'they will tend to be bilious rather than phlegmatic' (*Airs* 4), while mild springs and hot summers predispose 'those of a phlegmatic constitution' to fever (10). This text also hints at the existence of black bile; if the autumn is dry with northerly winds:

> it is particularly advantageous to those of a phlegmatic and humid nature and to women, but very dangerous to the bilious, since they dry out too much and are prone to dry ophthalmia and extended acute fevers, and in some cases to melancholy conditions. For the bulk of the humid, watery part of the bile is dried up and consumed, leaving the thickest and most acrid part, and similarly with the blood.
>
> (*Airs* 10)

But if this really is black bile, it seems to be a noxious by-product of other natural bodily constituents, rather than an independent constituent component.

## Black bile

The term 'dark (or black) bile', *melaina kholê*, occurs in a number of places not associated with a fully developed humoral theory. Sometimes, as in *Diseases* 1.30, it is associated with melancholy; but it is usually far from clear in particular whether it represents a categorically separate substance from, rather than simply a modification of, regular yellow bile, which elsewhere is described as taking on a variety of colorations according to its particular type, although primarily associated with the yellow-green hue.[19] In *Places in Man* excess internal bile is indicated by such a skin colour (10, 16, 28; but cf. 41), a sign sometimes explicitly associated with jaundice.[20] At *Epidemics* 2.3.6,[21] the author writes: 'I think that the sanguine and the bilious tend to acid reflux and perhaps for those patients it ends in black bile', again suggesting that black bile is a pathological by-product of some harmful process; and this is consistent at least with other mentions: for example at 4.16, 'in the autumn, Eumenus' wife vomited black bile', as also happened to Hermophilus' son (5.40). At 5.22, Apellaeus of Larissa is said to have vomited both red and black bile, which again suggests that it is a mere variant. The author of *Regimen in Acute Diseases* (61) writes: 'the sharpness obtained from vinegar is more beneficial to those with bitter bile than those with black bile. . . . Black bile is lightened, expelled and diluted, for vinegar brings up black bile';[22] here too the blackness apparently signifies a particular contingent property of 'ordinary' bile rather than some genuinely distinct substance. The nearest we get to a suggestion that black bile is a proper, independent substance in *Epidemics* comes at 6.6.1: 'a small blood vessel which is heated when full of blood produces burning heat and immediately separates it off: in the fatty, yellow bile, in the bloody, black bile'.[23] Clearly we do not yet have anything like an established physiological constituent theory of which black bile is a fundamental component.

From the fourth century onwards, 'melancholy' came to have a more central and well-defined role in medical and psychological theory. A long section of the pseudo-Aristotelian *Problems* devoted to the issue begins:

> Why is it that all who have become eminent in philosophy or politics or poetry or the arts are clearly of an atrabilious temperament, and some of them to such an extent as to be affected by diseases caused by black bile, as is said to have happened to Heracles among the heroes?
>
> (*Problems* 30.1 953a10–14)

Here we discover perhaps the earliest expression of the association between creativity and depression. The imbalance cannot be crippling (954b21–27); but a modicum of the disposition is the recipe for a kind of genius. The author recognizes a variety of distinct melancholic types: melancholics whose black bile is naturally cold are 'dull and stupid', while those who have a great deal of hot black bile become 'frenzied, or clever, or erotic, or easily moved to anger and desire' (954a31–34).[24] Creative melancholia is its own sort of mean, but one of a naturally excessive condition.

## The emergence of constitutive theories

There is no real hint of genuine four-humour theory here. In connection with melancholy, 'black bile' has a broad range of meanings, and indeed is more often associated with burned ordinary bile. Although Diocles of Carystus, a rough contemporary of Aristotle, apparently subscribed to some version of a four humour theory, and attributed melancholy to black bile (Fr. 109 vdE),[25] the details are obscure and sometimes contradictory. Rufus of Ephesus, a Hippocratic doctor of the first century AD who wrote a treatise on melancholy,[26] was also a four-humour theorist,[27] but he regarded the causes of the condition as largely psychological: 'violent thoughts and worries may make one succumb to melancholy; no-one who devotes too much effort to thinking about some science can avoid ending up with it' (Fr. 35 Pormann). This echoes *Epidemics*, where melancholy is associated with excess bile and phlegm (6.6.14) and linked with epilepsy (6.8.31), and particularly *Aphorisms*: 'If there is fear and despondency which last for a long time, this is melancholy' (6.23).[28]

Even in genuinely constitutive theories, black bile does not always figure. In the influential embryological treatise, *Seed*, and *Nature of the Child*,[29] four humours are recognized as being fundamental to organogenesis and development, as well as to the maintenance of the animal:

> Seed is . . . secreted from the whole body, from the hard parts as well as the soft, and from the total bodily fluid.[30] This fluid has four forms: blood, bile, water and phlegm. All four are innate in man, and are the origin of disease. I have already discussed these forms, and how and why diseases arise from them.
>
> (*Seed* 3)

But by far the most influential Hippocratic constitutive humoral theory is that found in the first eight chapters of *Nature of Man*.[31] The opening is polemical: 'I shall not claim that man is wholly air or fire or water or earth, or indeed anything that is not evidently a human constituent' (1). This is usually interpreted as being a general attack on any element theory as applied to human physiology. Galen, however, takes it rather to target monistic theories, and monisms do come under immediate fire: 'they all say that what is, is one, but disagree as to its name, one saying everything is air, another fire, another water and another earth'. Here monism of the Anaximenean sort is under attack; part of the complaint is that even if they were right about fundamental elemental monism, their choice of element or name would be arbitrary, a criticism which goes back to Anaximander.

In the next chapter, the author turns from general cosmology to medicine:

> Some say that man is blood, some bile, a few phlegm[32] . . . They all agree that man is a unity, however they call it, which changes form and power under the compulsion of hot and cold, and becomes sweet, bitter, black and white and so on.
>
> (*Nature of Man* 2)

The fundamental anti-monistic argument is that pain, indeed sensation in general, requires the interaction between distinct and alterable things (Galen elsewhere employs this argument against atomism: *Natural Faculties* II 26–30 Kühn). Equally, if monism were true there should be only one sort of illness:

> But there are many things in the body which are such that, when they are unnaturally heated, cooled, dried or moistened by one another, engender diseases. Consequently, the forms of diseases are many, and their cure is multiple too.
>
> (*Nature of Man* 2)

Sophisticated monists would have no difficulty countering this; they are not committed to there not being contingent, phenomenal distinctions of property among the various states of the basic substances, and these may well require distinct forms of treatment. But according to the author of *Nature of Man*, for change of any kind to be explicable, something other than the changing substance, whatever its basic structure and status, needs to be in the picture: 'generation cannot take place from a single thing; for how could something, being single, generate anything, if it is not mixed with something else?' (*Nature of Man* 3). There must be some affinity of general type between what is changed and what does the changing, and the qualities responsible in excess for pathological change must be moderately proportioned as well. Furthermore, 'when a man's body dies, each thing must return to its own nature, the moist to the moist, the dry to the dry, the hot to the hot and the cold to the cold'. Galen takes this to refer to post-mortem dissolution into the traditional elements, even though here and elsewhere only the qualities are explicitly mentioned.[33]

At the beginning of chapter 4 of *Nature of Man*, our author lays his constitutive cards on the table: 'The human body has within itself blood and phlegm and yellow and black bile, and these are the nature of the body, and because of them it suffers and is healthy'. The humours differ in their essential natures:

> Phlegm in no way resembles blood, nor blood bile, nor bile phlegm. Indeed how could they resemble one another, when their colours are evidently dissimilar to sight, and nor do they appear similar when touched by the hand? For they are neither similarly hot or cold or dry or wet. Since they are so far apart from one another in form and power, they cannot be one, given that fire and water are not one.[34]

This is (supposedly) demonstrated by the action of various drugs. Specific purgatives first of all draw out their target humour at all times, even if the quantities vary, according to season, and the age and constitution of the patient (*Nature of Man* 5). But once this has been completed, further administration of the drugs will attract first the humour closest to them in constitution, then the others:

> Whenever someone drinks a drug which attracts bile, he first vomits bile, and then phlegm; and then in addition to these black bile, and finally pure blood.
> (*Nature of Man* 6)

This too suggests that these humours are real, persistent constituents of things. Moreover, these specifics attract first their targets in their purest forms, and then the ones mixed with impurities. This is significant too. Galen holds that pure 'elemental' blood differs from what we actually have in our blood vessels, which contains an admixture of the other humours.[35] Equally, none of the humours exhibits its associated qualities in pure form, but their basic qualitative natures are easily discernible (7). Hence, each tends to dominate in the season whose characteristics it resembles: phlegm increases in winter, yellow bile in summer, and so on, although phlegm is still strong in spring, even as the blood increases, and yellow bile in the autumn, as the natures of seasonal diseases confirms: dysenteries followed by nosebleeds in spring and summer, for instance.

All of this broadly follows the tradition exemplified by *Epidemics, Air, Water, Places* and *Sacred Disease* concerning the seasonal variability of endemic diseases; but our author holds that they are proximately caused by imbalance of the four constitutive humours. The body:

> is particularly healthy when these things maintain a balance of their powers and quantities in relation to one another, and in particular when they are mixed together. It suffers when one of them becomes either too small or too great, or is separated in the body and is not mixed with all the others. For necessarily, whenever one of them is separated and stands on its own, not only will the place from which it has been separated become diseased, but the one which it now occupies and flows into will cause pain and distress on account of its being overfilled.
>
> (*Nature of Man* 4)

Even so, each humour (and quality) is always present to some degree:

> Just as the whole year has a share in all of them, the hot, the cold, the dry and the wet, he [sc. man] would not persist for any length of time without all of the things contained in the cosmos, but if any one were to fail, all would disappear; for all are composed and nourished from one another by the same necessity. Equally, if one of these congenital components of a man were to fail, he would no longer be able to live.
>
> (*Nature of Man* 7)

A proof of this (allegedly) is that if you give the same individual the same general emetic at different times of the year, he will vomit 'the most phlegmatic matter in the winter, the moistest in the spring, the most bilious in the summer and the blackest in the autumn' (7).

## The Galenic synthesis

Galen believed that a comprehensive physical and physiological theory could be forged from four-quality theory, four-element theory and humoral theory, where the elements are related to pairs of qualities (water is cold and wet, air wet and hot, fire hot and dry, earth dry and cold),[36] and in turn produce the humours. According to Galen, this synthesis was entirely attributable to Hippocrates by way of *Nature of Man*. Early in his career, Galen wrote *Elements according to Hippocrates* in defence of this view and he never wavered from it.[37] The following passage sums the case up:

> In addition to this [the four elements] there are four qualities, pure cold, dryness, heat and moisture. These are not elements either of man or anything else, but rather principles: this was confused by the earlier thinkers, who failed to distinguish the concepts of principle and element. . . . But the two things are evidently distinct from one another, the latter being the least part of the whole, the former that into which this least part is conceptually

reducible. For fire cannot itself be divided into two bodies and show itself to be a mixture of them, and nor can earth, water or air. But one may distinguish conceptually between the substance of the thing which changes, and the change itself, since the changing body is not the same as change which takes place in it. The changing body is the substrate, while the change in it occurs because of the exchange of the qualities; when pure heat is generated in it, fire is created, and similarly air when it receives pure moisture.

(Galen *On Hippocrates' 'Nature of Man'*[38] XV 30–31 Kühn)

Intermediate medical writers, such as Diocles and Herophilus,[39] referred to the four humours, although their theories were distinct and more complex. Galen characteristically tries to ascribe it to a variety of figures he otherwise admires or generally approves of, such as Mnesitheus and Praxagoras; but the latter lists eleven humours (including a 'glassy humour', a term which Galen himself uses to refer to the matter of the eye), while the former apparently only uses the term *khumos* in its original non-theoretical sense.[40] Typical of Galen's general approach is the following:

Not only Plato[41] but also Aristotle, Theophrastus and other pupils . . . [of theirs] were great admirers of Hippocrates' account of the humours, just as were the most reputable ancient doctors, Diocles, Pleistonicus, Mnesitheus, Praxagoras Phylotimus and Herophilus.

(*Doctrines of Hippocrates and Plato* V 686 Kühn)[42]

Elsewhere, with rather more justification, Galen attributes quality theory to his great predecessors with similar lists (for example, *Natural Powers*[43] II 110–11 Kühn); indeed he takes quality theory, as applied to animal physiology, to entail constitutive humoral theory (116–17). Although this text does not explicitly mention black bile, Galen remarks that if there are humours associated with the combinations hot and wet (blood), hot and dry (yellow bile) and wet and cold (phlegm), why should there not be one assimilated to cold and dry? (130–1). But there is, according to (unnamed) 'intelligent doctors and philosophers'; it is named black bile, and associated with autumn and old age.

Humours are defined by their associated qualities: 'yellow bile is hot and dry in power, black bile dry and cold; blood is moist and hot, while phlegm is moist and cold' (*Causes of Diseases* VII 21–2 Kühn). As such they resemble the elements, although 'no animal is absolutely hot like fire or absolutely wet like water' (*Mixtures* I 510 Kühn).[44] In *Hipp.Elem.* I 481 Kühn, Galen writes: 'Hippocrates too spoke of them [the four qualities] in his *Nature of Man*, saying that the particular and proximate elements of our body are the four humours, while the common elements of things are wet, dry, hot and cold'. In Galen's own physics, not to mention his nosology and therapeutics, it is the qualities that predominate. The humours, at any rate as a constitutive physiological theory, are underplayed, which may reflect Galen's qualms about its general coherence given the apparently anomalous role of blood, but can also be accounted for by the fact that, for Galen, hot, cold, wet and dry are the fundamental causal properties, and the action of the humours, proximate though they may be to what they affect, are fundamentally to be accounted for in terms of the qualities they exhibit. Consider the following text:

[Hippocrates] says that the body has been generated from the four elements, referring to them by their active qualities, dry, wet, hot and cold. But he did not construct his account of disease in terms of them, since they are in the body only potentially, not in actuality;[45] but rather what is generated from them by way of nutriment, blood, phlegm and yellow and black bile [is actual]. Yellow bile is analogous to fire, black bile to earth, and phlegm to water, thus yellow bile is hot and dry in potentiality like fire, black bile cold and dry like earth, phlegm cold and wet like water. The airy element in animals' bodies is visibly close to its proper nature only in respiration. . . . The balanced mixture of the four elements is blood in the precise sense.

(*Doctrines* V 675–6 Kühn)

Here the ambivalence regarding blood is evident. Galen distinguishes between pure elemental blood, and what he slightly misleading calls 'blood in the precise sense', that is what flows in the vascular system and is in fact a mixture of all four elemental humours, although with pure blood predominating in the properly balanced mixture. This is linked with his embryological theory; the matter for the developing foetus is supplied by the menstrual blood, which contains all four humours:

Plants are generated from the four elements as do their fruits and seeds . . . these plants, and animals too are food for humans, and from them are generated blood, phlegm and the two biles. . . . Blood is the most abundant – that is why it alone is visible in the veins, and there is a small amount of each of the others, This is the actual blood, the matter of our generation, the blood that is composed from the four humours and gets its name from the one which predominates. We are generated from it when we are conceived in the uterus, and out of it receive our first formation, and the subsequent articulation, growth and maturation for the parts thus formed.

(*Doctrines* 671–2)

At any event, *Mixtures*, which outlines Galen's fundamental physiological physics, is primarily concerned with the qualities and their relations, and humours as such only make brief appearances, a feature that is true to a surprising extent elsewhere in his work:

There are four qualities with the capacity to act on and be acted upon by one another: heat, cold, dryness and moisture. . . . But of the six pairings that are theoretically possible between four things, two are physically impossible: a body cannot be at once wet and dry, or hot and cold.[46] There thus remain four pairings of the mixtures: two wet and two dry, distinguished in terms of heat and cold.[47]

(*Mixtures* I 518 Kühn; cf. 510–18)

There are eight generic ways of deviating from the healthy norm, characterized in terms of imbalance of the qualities; and Galen condemns 'the most distinguished of our predecessors, both doctors and philosophers' for 'ignoring the well-balanced mixture' (*Mixtures* 518–19). Qualities can be relative as well as absolute: to describe a condition as hot and wet is to say that it is hotter and moister than it ought to be (only fire is absolutely hot). Galen's opponents rightly suppose that spring is the most favorable season, but that is because it exhibits no excesses of any kind, rather than being, as they think, *unnaturally* warm and wet (524–7). This amounts to saying that we, in our natural condition, by comparison with the totality of the world's furniture, are relatively warm and moist. Excesses of heat and moisture, on the other hand, 'so far from being the characteristics of spring or good mixture in general, in fact constitute the worst possible state of the ambient air' (529), one associated with putrefaction and epidemic disease (529–33).[48]

Quite consistently, blood is the most natural humour, and is associated with warmth and moisture, while the others are more prone to imbalance:

> These men [Hippocrates, Aristotle, Praxagoras, Phylotimus and 'many others'] demonstrated that when nutriment is altered in the veins by the innate heat, blood is produced by its proper proportion, and the other humours by its disproportion.
>
> (*Natural Powers* II 117 Kühn)

Although small quantities of the other humours are necessary for health, they more readily tend towards pathological excess and their production is in a sense accidental, which is why many doctors have thought them to be only pathological, and not part of the animal's natural constitution:

> The most useful and particular of the humours is the blood. Black bile is a kind of sediment or dreg of this, and is thus colder and thicker than blood. Yellow bile is considerably hotter. The coldest and wettest of all animal humours is phlegm.
>
> (*Mixtures* I 693 Kühn)

The comparison of black bile with a sediment is found in *Black Bile* (V 112 Kühn), a text which emphasizes the visual differences between different varieties of the same humour: thicker and thinner, paler and ruddier blood (106–7); and various different consistencies and colours of phlegm and yellow bile, depending upon what they happen to be mixed with (108–10). But blood, no matter what its colour, always coagulates; and bile and phlegm have their essential properties as well. Which leaves black bile, properly so-called, to be distinguished from other dark-coloured residues observed in vomit and faeces by its own distinctive features. It is invariably thick, among other things, as well as being acid (110–12; cf. 135). Much of *Black Bile* is devoted to attacking Erasistratus, a frequent target, for daring to overlook black bile 'as a normal constituent of the blood vessels' and the diseases its excess causes, as well as failing to see that the spleen is the organ which naturally attracts it. But even Galen allows that the case of black bile is complex and problematic:

> The melancholic humour clearly exhibits different kinds of composition. One kind is like the sediment of blood and clearly shows itself as being quite thick, like wine lees. . . . [This one] I call 'melancholic humour' or 'melancholic blood', for I think one should not yet call it black bile. . . . This thick

melancholic humour, just like the phlegmatic one, sometimes causes cases of epilepsy when it is trapped where the cavities of the brain . . . have their outlets. But when it is present in excessive quantities in the very body of the brain, it causes melancholy, just as the other kind of black bile, the one produced by the burning of yellow bile, results in bestial delirious hallucinations, both with and without fever, when it clogs the body of the brain. This is also why one kind of phrenitis, namely that arising from pale bile, is more moderate, whereas the other one which arises from yellow bile is more severe. There is still another kind of delirium, both bestial and melancholic, that arises from the burning of yellow bile.

(Galen, *Affected Parts* VIII 176–8 Kühn)

In this context, one may well wonder just how well worked-out Galen's theory really was.

## Afterlife

Yet well worked-out or not, Galen's position gradually became canonical, and the endemic medical disputes in which he had played a crucial part dissipated over time. While alternatives did remain,[49] they gradually faded in importance. This was the medicine taken up and further investigated in the Islamic world, from Baghdad and Persia to Spain, towards the end of the first millennium and beyond; a key text in this process was Avicenna's systematizing *Canon of Medicine* of 1025. The doctrine gradually filtered back into the Christian West in the Middle Ages, initially in Latin translations of Arabic versions of Galen and Avicenna and finally by way of recovered Greek texts. As my epigraph attests, it was the basis for the practice of Chaucer's 'Doctour of Phisik'. Well after the rejection of Galen's fundamental anatomical physiology as a result of the work of Vesalius and Harvey, humoral theory remained influential throughout the Renaissance and, although it was to some extent supplanted in the eighteenth century by a revived, more empirical Hippocratism, it continued to exert an influence, albeit no longer a constitutive one, well into the nineteenth century, evident in the practice of bloodletting. Although not precisely humoral, the related and equally ancient miasmatic theory of epidemics such as the suggestively named cholera and malaria, survived into the middle of the nineteenth century before germ theories of invasive and contagious pathogens began to triumph.[50] Humoral theories and their associated attitudes and residual practices were remarkably long-lived, and the terminology of humours survives in our lingering vocabulary of character types, such as those present in Chaucer. Melancholy is still anatomized, albeit in a very different register from that of Burton; and if prone to it myself, I also like to think of myself as exhibiting a generally sanguine view of the world, all the while trying to remain phlegmatic in the face of adversity, and to temper my hot-tempered, bilious tendencies.

## Notes

1 See the discussion in Chapter 3 of this volume by Dominik Wujastyk.
2 The best English selection of the Presocratics is to be found in Barnes, 2001. For Thales' element theory, see Barnes, 2002, 11–12; for Anaximenes, 24–7; Anaximander, 18–19. 21–3;

for Heraclitus and fire, 54–5,60 (= Fr. 22 B 60 Diels/Kranz [DK]); for Empedocles four-element theory, 13–15 (Frs. 31 B 6, 16, 35, 71, 73, 75 DK). The primacy of air (which is identified with intelligence) is later championed by the late Presocratic, Diogenes of Apollonia (Frs 64 B 1–8 DK; Barnes, 2002, 254–7), and the Hippocratic *Breaths* (trans. Jones, 1923).

3  See the excellent survey of Nutton, 2004, 80.

4  His fragments are edited, translated and discussed in Huffman, 2006; see also Barnes, 2001, 176–81.

5  *Anonymus Londinensis*, a papyrus containing a précis of the history of medicine written by Aristotle's pupil Meno in the fourth century BC.

6  Interpreted by Huffman, 1993, 304, as metaphorically referring to an internal condensational process; however, see *Nature of Man* 7: 'when winter takes over, the bile, being cooled, becomes small in quantity, and the phlegm increases once again, both due to the abundance of the rains and because of the length of the nights'.

7  See Huffmann, 1993.

8  The theory of Menecrates is apparently constitutive, involving blood and bile as hot elements, breath and phlegm as cold, but he was physician to Philip of Macedon in the fourth century: *Anonymus Londinensis*. 19.15 ff.

9  The disagreement centres around what constitute separate treatises. For a good, if controversial, treatment see Jouanna, 1999.

10  See note 2 above.

11  Edited and translated in Potter, 1995.

12  'I do not think they [i.e. modern theoretical doctors] have ever found anything purely 'hot', 'cold', 'wet' or 'dry' without it sharing some other properties . . . there are many and various hot substances with many and various effects which may be contrary to one another. Everything has its own specific effect' (15).

13  This is one of the earliest references to what was to become one of the canonical 'humours' – but the early Hippocratic texts generally treat of phlegm and (yellow) bile as pathogens, or symptoms thereof.

14  Literally 'digestion'; i.e. a proper breaking down and processing of the constituent material.

15  'Phlegm' is not equivalent to, although it includes, mucus.

16  Edited and translated in Potter, 1988.

17  Edited and translated in Potter, 1988.

18  Cf. 1.17, 18 (erysipelas), 19 (lung tumours), 20, 21 (internal suppurations), 23 (fevers), 24 (chills), 25 (sweats), 26 and 32 (pleurisy), 27–8 (pneumonia).

19  See Galen *Commentary on Hippocrates' 'Nature of Man'* XV 35 Kühn: 'the passage refers to the fluids generated in severe diseases. For both greenish and grey bile (which they call woad-like), as well as a reddish and a grass-green one, along with some others that have no name, become apparent whenever the disease becomes putrefying'; and XV 75 Kühn, 'they refer to all the other biles with an adjective, calling them "rust-coloured" or "black", or "red", or "grass-green"'. At 66–7, Galen attributes these variations in perceptible qualities of the humours to the fact that they are not genuinely elements, but generated from them.

20  *Places in Man* is edited in E. M. Craik, *Hippocrates: Places in Man*, Oxford: Clarendon, 1998; translated in Potter, 1995. Craik (1998, 14) claims the text exhibits 'humoral theory at its most inchoate', but as she says herself, bile seems to be a product of diseases, rather than a natural constituent of the body.

21  The 7 books of *Epidemics* are a mixed bag, as has been known since antiquity. Books 1 and 3 are translated in Jones, 1923; 2 and 4–7 in Smith, 1994.

22  Translated in Jones, 1925.

23  Cf. 6.5.8, which claims that the colour of the tongue indicates the state of urine: greenish tongues are 'bilious; biliousness derives from fat; ruddy ones come from blood; black ones from black bile'; again, this is inconclusive.

24  On this, see Northwood, 1998.

25  Diocles' fragments are edited and translated with commentary in van der Eijk, 2000, 2001.

26  See Pormann, 2008.

27  Commended as such by Galen: *Black Bile* V 105 Kühn.

28  Edited and translated in Jones, 1931.

29 Although this has come down to us as two distinct texts, it probably forms a continuous treatise. It is translated, with the related *Diseases* 4 (which also adopts the same four-humour constitutive theory, with an associated account of disease as a result of imbalance among them: 1–7), in Lonie, 1981, and Potter, 2012.

30 I.e. male semen. In common with many other Hippocratic writers (see *Regimen* 1.26–31), and indeed Greek theorists in general (including Galen), the female too produces a kind of semen which is deposited into the womb.

31 *Nature of Man* is edited with French commentary and translation in Jouanna, 2001; English versions in Jones, 1929.

32 Note that the author clearly interprets the earlier theories as being constitutive, although it is not clear which ones (if any in particular) he is referring to.

33 In Galen's view, 'the purely hot' just *means* fire. At *Elements* I 457–70 Kühn, he recalls forcing his teacher Athenaeus to concede that there was a difference between 'heat' considered as a predicate, and what is purely hot, which you might as well call 'fire' (and equally with the other qualities and their associated elements).

34 The rare mention of fire and water suggests that our author does indeed subscribe to some sort of element theory, albeit perhaps one involving only two components, like that of *Regimen*.

35 According to Galen, this is why maternal blood functions as a universal nutrient to the growing foetus: *On Hippocrates' 'Nature of Man'* 73–4, 94 Kühn.

36 This theory is owed to Aristotle, *Generation and Corruption* 2.3, 330a30–331a6, esp. 331a1–6.

37 For details, see Hankinson, 2008; *Elements* (I I 413–508 Kühn) is edited with notes and English translation in De Lacy, 1996.

38 Galen's attribution of the text to Hippocrates is contentious: *Anonymus Londinensis* attributes *Nature of Man* to Hippocrates' successor Polybus.

39 See Von Staden, *Herophilus: The Art of Medicine in Early Alexandria*, Cambridge: Cambridge University Press, 1989, 242–7.

40 See Nutton, 2004, 125 ff.

41 Plato mentions bile and phlegm as primary causes of diseases at *Rep.* 564b–c; see also *Timaeus* 82e–5e; the elements are also implicated; where bile is said to come in a variety of colours, including black, which is more acidic – but this is caused by putrefying flesh (83b); while bile is 'stale blood' (85d). *Pace* Galen, this is not constitutive theory.

42 Edited and translated in De Lacy, 1978.

43 II 1–214 Kühn; edited and translated in Brock, 1916.

44 I 509–694 Kühn; English trans. in Singer, 1997.

45 The qualities exist in the body 'only potentially' in that they are exemplified not in themselves as such, but as properties of the humours; however in *Mixtures*, Galen makes much of the fact that substances are cold hot, cold etc. in virtue not of their immediate perceptible qualities but in terms of their causal powers; chilled wine is hot, heated poppy-juice refrigerant (I 658–9, 674 Kühn).

46 It can be hot and cold in respect of different parts; but this would not involve a *mixture* of the qualities; this would relate to what Galen calls an 'uneven distemper', on the subject of which he wrote a short text of the same name: VII 733–52 Kühn.

47 The theory of the eight possible types of imbalance ('distemper') is central to Galen's pathology and therapeutics: *Therapeutic Method* X 103–4, 463–5 Kühn; and *Differences of Diseases* VI 848–55 Kühn.

48 Galen is thus resistant to mere apriorism: what matters is not typological neatness but truth.

49 Caelius Aurelianus wrote his textbook of anti-humoral medical Methodism, *Chronic and Acute Diseases*, in the fifth century, although he followed very closely the model of the first century Soranus.

50 On theories of contagion, see Chapter 5 by Michael Worboys.

## Select bibliography

J. Barnes, *Early Greek Philosophy*, London: Penguin Classics, 2002.

Brock, A. J. (1916) *Galen: On the Natural Faculties* Loeb Classical Library, London/Cambridge, MA: Harvard University Press.

De Lacy, P. H. (1978) *Galen: On the Doctrines of Hippocrates and Plato* (*PHP*) (2 vols., ed., trans.): *CMG* V 4,1,2, Berlin.

De Lacy, P. H. (1996) *Galen: On the Elements according to Hippocrates* (*Hipp.Elem.*) (ed., trans., and comm.): *CMG* V 1,2, Berlin.

Diels, H. and Kranz, W. (1952) *Die Fragmente de Vorsokratiker*, 6th edn, Berlin: Weidmann.

Eijk, P. H. van der (2000f) *Diocles of Carystus* I Text and Translation, Leiden: E. J. Brill.

Eijk, P. H. van der (2001) *Diocles of Carystus* II Commentary, Leiden: E.J.Brill.

Hankinson, R. J. (2008) 'Philosophy of nature', in R. J. Hankinson, ed., *The Cambridge Companion to Galen*, Cambridge: Cambridge University Press.

Huffman, C. (1993) *Philolaus of Croton: Pythagorean and Presocratic*, Cambridge: Cambridge University Press.

Jones, W. H. S. (1922) *Hippocrates* I Loeb Classical Library, London/Cambridge MA: Harvard University Press.

Jones, W. H. S. (1923) *Hippocrates* II Loeb Classical Library, London/Cambridge MA: Harvard University Press.

Jones, W. H. S. (1931) *Hippocrates* IV Loeb Classical Library, London/Cambridge MA: Harvard University Press.

Jouanna, J. (1999) *Hippocrates*, trans. M. B. Devoise, Baltimore: Johns Hopkins University Press.

Lonie, I. M. (1981) *The Hippocratic Treatises 'On Generation', 'On the Nature of the Child', 'Diseases IV': A Commentary*, Berlin: De Gruyter.

Northwood, H. (1998) 'The melancholic mean: The Aristotelian *Problema* 30.1', in *Paideiai* World Congress of Philosophy, Boston.

Nutton, V. (2004) *Ancient Medicine*, London: Routledge.

Pormann, P. (2008) *Rufus of Ephesus On Melancholy*, Tubingen: Mohr Siebeck.

Potter, P. (1988) *Hippocrates* V Loeb Classical Library, London/Cambridge MA: Harvard University Press.

Potter, P. (1995) *Hippocrates* VIII Loeb Classical Library, London/Cambridge MA: Harvard University Press

Potter, P. (2010) *Hippocrates* IX Loeb Classical Library, London/Cambridge MA: Harvard University Press

Potter, P. (2012) *Hippocrates* X Loeb Classical Library, London/Cambridge MA: Harvard University Press

Temkin, O. (1973) *Galenism*, Baltimore: Johns Hopkins University Press.

Singer, P. N., *Galen: Selected Works*, Oxford: Oxford University Press, 1997.

Smith, W. C. (1979) *The Hippocratic Tradition*, Ithaca NY/London: Cornell University Press.

Smith, W. C. (1994) *Hippocrates* VII Loeb Classical Library, London/Cambridge MA: Harvard University Press.

# 3

# MODELS OF DISEASE IN AYURVEDIC MEDICINE

*Dominik Wujastyk*

The monk Moliya Sīvaka once approached the Buddha and asked him whether it was true, as many said, that all pleasant, painful, or neutral sensations were the results of past deeds, of karma.[1] The Buddha replied to Sīvaka that people who held this view were over-generalizing. In fact, he said, some pain arises from bile, some from phlegm, some from wind, some from humoral colligation, some from changing climate, some from being ambushed by difficulties, some from external attacks, and some, indeed, from the ripening of karma.[2] This is the first moment in documented Indian history that these medical categories and explanations are combined in a clearly systematic manner, and it is these very eight factors which later become the cornerstones of the nosology of classical ayurveda.

The use of the expression 'humoral colligation' (Pali *sannipātika*) in the Buddha's list is particularly telling. This is not just an ordinary item of vocabulary. It is a keyword, a technical term from ayurvedic humoral theory. In classical ayurvedic theory, as received by us from medical encyclopaedias composed several centuries after the Buddha, 'humoral colligation' is a category of disturbance in which all three humours are either increased or decreased simultaneously. Because therapy usually depends upon manipulating the humours in such a way that an increase in one is cancelled by a decrease in another, it is impossible to use normal therapies to counteract humoral colligation. That is why it is such an especially dangerous diagnosis. Epilepsy, for example, is described in the *Compendium of Caraka* (see below) as displaying various symptoms including frequent fitting, visions of bloody objects and drooling, according to the predominance of wind, bile or phlegm respectively. But if the patient shows all the symptoms at once, then the condition is called 'colligated'. 'Such a condition,' says the author, 'is reported to be untreatable.'[3]

The formality of the vocabulary in the Buddha's list of causes of pain suggests that he was consciously referring to a form of medicine that had a theoretical underpinning. This impression is increased by the presence at the end of the chapter of a verse summary of these eight causes of pain.

> Bile, phlegm and wind,
> colligation and seasons,
> irregularities, external factors,
> with the maturing of karma as the eighth.[4]

This verse looks very much like a citation from a formal medical work, or a medical mnemonic. But it is only several hundred years after the Buddha's time that we see this theoretical system worked out explicitly in the ancient Indian medical literature that has survived until today.[5]

The causes of disease listed by the Buddha constitute eight quite distinct models of disease aetiology. But as medical thought evolved in South Asia, the scholarly authors expanded these aetiological categories to include an even wider range of concepts. Different schools evolved, with varied emphases and disagreements. It is in this Buddhist canonical literature from the last centuries BCE that we see the first references to a developed theory of disease, and the terminology used is parallel to that found in the later, formalized ayurvedic literature.[6] Older religious literature, especially the *Rig Veda* and the *Atharva Veda*, contained prayers and invocations for health and against illness, but the models of disease behind this earlier material are theoretically unsophisticated, and are not the product of a professionalized class of healers.[7] Of those streams of medical practice in ancient India that did develop scholarly narratives, the most prominent was – and is – ayurveda, 'the knowledge of longevity'. Several encyclopaedias of ayurvedic medicine are known to have existed in early antiquity, although only three have survived to the present time, the *Compendium of Caraka*, the *Compendium of Bheḷa* and the *Compendium of Suśruta*.[8]

## Primary sources

The *Caraka* is one of the most important surviving works of classical Indian medicine. Composed probably in the first or second century CE, it is an encyclopaedic work in Sanskrit discussing many aspects of life, philosophy and medicine.[9] The *Compendium of Suśruta* dates from perhaps a century or two after the *Caraka*, and is a similarly encyclopaedic work, although its classifications, theories and therapies are not identical with those of the *Caraka*. This fact reveals something of the ferment of ideas that surrounded early Indian medicine. The *Bheḷa* survives in only a single damaged manuscript, and since its content is often similar to the *Caraka* it will not be cited further in this chapter. However, out of a vast later literature on medicine one further work bears special mention, the *Heart of Medicine* by Vāgbhaṭat (fl. ca. 600 CE, in Sindh). The compendia preceding Vāgbhaṭa are full of interesting materials on medical philosophy, and present the contradictory views of various doctors and have complex, multi-layered histories of manuscript transmission. Vāgbhaṭa absorbed these older materials, understanding them comprehensively and deeply, and produced a skilful and entirely plausible synthesis of the ancient works. The *Heart of Medicine* was widely and justifiably accepted as the clearest and best work on medicine in post-classical times, and was adopted as the school text for medical education. Students educated in the traditional manner were expected to learn Vāgbhaṭa's work by heart, and there are still physicians in India today who know the work word-perfect. Where the ancient works sometimes show ayurvedic doctrine still in formation, the *Heart* represents a settled medical orthodoxy. For that reason, it is often easier to cite the *Heart* when aiming for a clear view of standard ayurvedic doctrine. However, in doing so, one loses the plurality and complexity of the older works. For example, when defining the causes of disease, Vāgbhaṭa said:

The under-use, wrong use, or overuse of time, the objects of sense, and action, are known to be the one and only cause of illness. Their proper use is the one and only cause of health. Illness is an imbalance of the humours; freedom from illness is a balance of the humours. In that regard, illness is said to be of two kinds: it is divided into internally caused and invasive. And their location is of two types, according to the distinction between body and mind. Passion and dullness are said to be the two humours of the mind.[10]

This is a fine, concise and orderly statement of the causes of disease. But it is selective, omitting several older classification schemes from the *Caraka* and *Suśruta* compendia that cut across these categories in awkward ways. Vāgbhaṭa has made it all make sense; but we lose the historical view of a tradition forming its theories out of the messy processes of debate and evaluation.

In what follows, I will describe a series of the most prominent models of disease that were developed in ancient India and that formed part of the ayurvedic tradition. But Indian scholar-physicians were flexible and adaptive in their thinking, and did not hesitate to absorb models from many popular sources, for example from women's experience in the birthing house, from herb-collectors or from religious practice. The tradition was not mono-vocal, and many subsidiary models exist in the ancient literature.

## The equality of humours

In about 50 CE, the author Aśvaghoṣa wrote in his *Life of the Buddha* that when the young Gotama was still searching for liberation and met his first great teacher, Ārāḍa Kalāma, they greeted each other and asked after each other's health.[11] More precisely, what the Sanskrit says is: 'They politely asked each other about the equality of their humours.'[12] Aśvaghoṣa did not explain this expression; it was intended to be a simple account of the normal etiquette of greeting. In the first century CE, then, an author could take for granted that an audience would know perfectly well what it meant to have 'equal humours'.

As we saw earlier, the Buddha too referred to the three humours and their colligation as primary causes of disease. In the even older Vedic literature from the second millennium BCE onwards, we see oppositions between hot and cold as principles governing various aspects of life, and these can be connected with the later doctrines of bile and phlegm. It is possible that wind as a third category was a latter addition to the theory, though further research is needed into the early history of these ideas.[13] As formal medical doctrine evolved, the doctrine of three humors became the central explanatory model for disease for ayurveda, the scholarly medicine of India.[14]

In spite of the clarity and dominance of the three-humour theory, one sometimes senses that the model was being stretched, or applied as a veneer over older, folk traditions. For example, towards the end of the *Caraka* there occurs a description of treatments for disorders of the three most sensitive and important danger-points of the body, the heart, the bladder and the head. The text says:

When blood and wind are vitiated because of the unnatural suppression of urges, then indigestion and the like cause the brain-tissue to be vitiated and to coagulate. When the sun rises, liquid matter slowly flows out, because of the rapid heating. As a result, there is sharp pain in the head during the day, that increases as the day goes on. With the ending of the day, and the resulting thickening of the brain-tissue, it calms down completely. That is termed, 'The Turning of the Day'.[15]

The therapy for this affliction includes the application of fats to the head, with a poultice, including the meat of wild animals sprinkled with ghee and milk. The patient is to drink the ghee of milk boiled with peacock, partridge, quail, and be given a nasal infusion of milk that has been boiled eight times. These therapies are not explained or justified in terms of humoral theory; they are just stated. Other parts of the medical classics are like this too, notably the section of the *Suśruta* that deals with poisons, where therapies are normally recommended for symptoms without reference to any connecting theoretical model.[16]

I have used the word 'equality' in the context of this Indian humoral model. This translates the Sanskrit word *sāmya* which also means 'eveness' (as in the numbers 2, 4, 6, etc.), 'smoothness' (as in a road), and related meanings. It does not exactly mean 'balance'. For that, there is another lexical group of words, derived from the Sanskrit root √*tul.* As with English 'balance', the Sanskrit *tulā* means both a gadget for weighing things and the idea of resemblance or equal measure. This is not a term that is ever used in Sanskrit literature in connection with the humours. As readers we are accustomed to the idea of 'the balance of the humours' from Greek and later European narratives about humoral medicine. However, the characteristically Greek geometric harmony of the four humours and the hot–cold/dry–wet oppositions does not exist in the three-humour Indian model. Therefore, the metaphor of balance (*tulā*) is not present in the Indian model, but rather the metaphor of equal quantities of liquid in a cup, perhaps.

## Affinity

The concept of affinity, or wholesomeness to the individual, appears frequently in ayurvedic theory in the context of models of disease, but has thus far been little explored by medical historians.[17] 'Affinity' translates the Sanskrit word *sātmya*.[18] This is a compound of *sa-* 'with' and *-ātmya* 'relating to oneself'. Etymologically, then, *sāt-mya* implies 'connectedness with one's self'. In sentences, the word indeed means 'natural', 'inherent', 'wholesome', 'agreeable to one's constitution', or 'having an affinity to one's self'. In grammatical compounds, the word is most commonly joined to the word *oka-* (or *okas*) 'home, refuge'.[19] This is a puzzling collocation that I am not at present able to explain clearly. It seems to be a way of reinforcing the sense of *sātmya*, so *okasātmya* might be 'fundamental acquired affinity'. The *Caraka* described *okaḥsātmyam* as 'that which becomes suitable through habituation'.[20]

The word *sātmya* also commonly occurs in bigrams with season (*ṛtu*), place (*deśa*) and food (*anna*), giving a sense of the semantic spread of the term. It is the personal suitableness or appropriateness to a person of a particular season, place or food: one's affinity for a place, time, or diet. The ayurvedic model of 'affinity' is conceptually interesting for several reasons, amongst which is the fact that this suitability or wholesomeness-to-oneself is *acquired*, not inherent. The classical Sanskrit treatises develop this idea in some detail, describing graduated processes whereby a patient's bad habits may be attenuated and new, better habits inculcated.[21] The end of this process is that the patient gains an affinity (*sātmya*) for the new, better habit. It becomes, literally, 'second nature'. So the concept is similar to, but different from, the European idea of 'nature', as in one's personal disposition or temperament.[22] Nature, in this Western sense, is immutable, it is who we are. By contrast, ayurvedic affinity is malleable, the patient is trainable. It is nature, in a sense. Foods, places and seasons are natural to a patient, they agree with him or her. But this naturalness may be harmful and undesirable. In that case, it can be changed, and an unwanted bad habit can be transformed into a new, beneficial naturalness.

Starting from this conceptual apparatus, the Sanskrit medical tradition had quite a bit to say about, for example, alcoholism and inappropriate diets. The tradition here took a more sophisticated approach to habit-change than that implied in the ordinary-language English expression 'breaking habits', with its connotation of immediacy and rupture.[23] The *locus classicus* for the term *sātmya* is its characterisation in the *Caraka*. As an example of bad habits, and inappropriate dietary habit, the *Caraka* described how people from villages, cities, market towns, or districts may be habituated to the excessive use of alkali. Those who use it all the time may develop blindness, impotence, baldness, grey hair and injury to the heart. Examples of such people are the Easterners and the Chinese:[24] 'Therefore, it is better for those people to move by steps away from the affinity for that. For even an affinity, if it is gradually turned away from, becomes harmless or only slightly harmful.'

The *Caraka* here used the term affinity, for the *bad* habit. The full description of the term was as follows:

> Affinity (*sātmya*) means whatever is appropriate to the self. For the meaning of affinity is the same as the meaning of appropriateness (*upaśaya*). It is divided into three kinds: superior, inferior, and average. And it is of seven kinds, according to the savours taken one by one and all together. In that context, the use of all the savours is superior, and a single savour is inferior. And standing in the middle between superior and inferior is the average. One should encourage the inferior and the average types of affinity towards the superior type of affinity, but only in step-by-step manner. Even if one has successfully achieved the affinity characterised by all the savours, one should adhere to using good things only, after considering all the [eight previously described] foundations for the special rules of eating.[25]

This model linked the theory of affinity with that of the savours, or flavours (*rasa*), sweet, sour, salt, bitter, pungent and astringent. The discussion of these six savours in ayurvedic literature is detailed and pervasive.[26] It forms an intrinsic part of almost all therapeutic regimes. There is some ambiguity in the tradition as to whether savours are qualities that inhere in foods and medicines, or whether they are actual substances, like modern 'ingredients'. But in either case, savours were understood to increase or decrease the quantity of the humours, and therefore formed one of the primary tools for manipulating humoral balance. Also notable in the above passage is the introduction of a combinatorics of levels of potency. This tendency to introduce a simple form of mathematics into medical thinking occurs in other contexts too, and seems to be an approach to a quantitative grasp of theoretical categories.[27]

Later physicians continued to use affinity as a model of disease and therapy. The author Candraṭa (fl. ca. 1000 CE), commenting on the medical treatise of his father Tīsaṭa, brought together the discussions of several earlier authors on the topic, and added his own ordering and explanation.[28] Candraṭa seems to have had a penchant for creating clear lists and classifications, and developed a theory of nine types of affinity together with a scale of their relative strengths.

## Raw residues

In a chapter about the relationship between health and diet, the *Caraka* introduced the concept of the undigested residue of food that has been consumed in too great quantities.[29] In this model, the belly contains food, liquid, and humours (wind, bile, and phlegm). Consuming too much food and drink causes pressure to build up on the humours, and they become simultaneously irritated. These irritated humours then merge with the undigested mass of food and cause solidification, vomiting, or purging. The three humours each produce their own set of pathological symptoms including, amongst others, stabbing pain and constipation (wind), fever and flux (bile), and vomiting and loss of appetite (phlegm). The corruption of the undigested residues (Sanskrit *āma*) may also be caused by eating various kinds of bad food, and also by eating while experiencing heightened negative emotions or insomnia. A distinction is then articulated between two kinds of undigested residue: laxative and costive. Therapy for costive conditions can be hard because the treatments indicated may be mutually contradictory. (This is a problem that occurs periodically in ayurvedic therapy, and is also the danger inherent in the humoral colligation mentioned above.) Where corrupted residues are considered treatable, the therapies include the administration of hot saline water to induce vomiting, and then sweating and suppositories to purge the bowels. The model of disease used here involves the inflammation of undigested residues of food. The Sanskrit word is *āmas* (nominative), which means 'raw'. It is cognate with the classical Greek word ὠμός that has the same meaning, and it is striking that a similar doctrine about undigested residues also appears in the work *On Medicine* (Ἰατρικά) by the Greek author known as *Anonymus Londinensis* preserved in a papyrus datable to the first century CE.[30]

The idea of pathological residues continued to be used and to evolve alongside other etiological ideas throughout the history of medicine in South Asia. The ancient *Compendium of Suśruta* noted the opinion of some experts who asserted that raw residues were one of the forms of indigestion, and discussed its classification and interaction with diet.[31] The ayurvedic commentator Gadādhara, in the eighth or ninth century, regarded undigested residues to be humours, on the grounds that they themselves caused corruption or else because they became connected with corrupted humours.[32] The concept of pathological residues had a particular attraction for medieval yoga practitioners. The *Ayurveda Sutra* is a unique work probably composed in the seventeenth century. Although its title would seem to be that of an ayurvedic work, it is in fact a syncretic work that attempts for the first time to combine yoga and ayurveda into a single therapeutic regime. However, its account of ayurveda is idiosyncratic, and it presents the model of raw residues as being the source of all diseases: 'One should not retain raw residues, for raw residue is the beginning of all diseases, says the Creator. Curtailing it is health. Someone who is healthily free of residues, reverences the self.'[33] The idea of raw residues as poisons has played well into modern and global New Age fusions of ayurveda and yoga in the twentieth and twenty-first centuries, where the theory overlaps with ideas about non-specific blood toxins and therapies based on cleansing and purgation.[34]

## Errors of judgement

In the final analysis, according to the *Caraka*, all disease is caused by errors of judgement, or failures of wisdom. The *Caraka*'s term for 'judgement, wisdom' (*prajñā*) is well known from Indian philosophical writing and was especially taken up by later Buddhists as signifying the kind of wisdom that came from realizing that all existence is ultimately empty of permanent essence. But the *Caraka* has its own more specific definition of wisdom, as we shall see below. The *Caraka*'s word for 'error' (*aparādha*) is an ordinary-language word signifying all kinds of mistakes, offences, transgressions, crimes, sins or errors. When, in the third century BCE, King Aśoka commanded that his edicts be carved on rocks across India, he warned his readers that the stonemasons might make mistakes in carving the lettering, they might make *aparādhas*.[35] How did the *Caraka* unpack this concept of errors of judgement?

First of all, the *Caraka* defined wisdom as the combined powers of intelligence (*dhī*), will-power (*dhṛti*), and memory (*smṛti*).[36] These powers may become impaired in different ways. As an example of impaired intelligence, the classical authors cited errors such as mistaking something permanent as temporary, or something harmful as helpful, etc. Poor will-power would be exemplified by a lack of self-control in the face of sensual enjoyments which are unhealthy. Faulty memory was exemplified when a person's mind becomes so confused by passion or darkness, that they cease to be able to see things as they really are, and they cannot remember what should be remembered. Thus, erroneous mental processes lead a person to engage in several types of faulty activity that develop into a cascade of problems ending in illness.

In the *Caraka*'s core model of disease, an error of judgement – faulty intelligence, will-power or memory – leads to the over-use, under-use or abuse of the senses, of action or of time.[37] Wrong use of the *senses* would include listening to sounds that are too loud (over-use), looking at objects that are too small (under-use), or smelling a corpse (abuse). The sense of touch was treated as a special case, because the *Caraka* considered it to be the fundamental sense working in and through all the other senses, permeating the mind and the objects of cognition. Because of this, an unwholesome association of any sense and its object could be understood as an abuse of touch, and as a conduit by which the external world could adversely affect the inner being of a person. Wrong uses of *action* include similarly categorized inappropriate uses of body, mind and speech.[38] The overuse of *time* would include experiencing unseasonably intense weather – winters that are too cold or summers that are too hot. The under-use of time is the inverse: winters that are not cold enough, and so on. The abuse of time would be experiencing winters that are hot and sunny, or summers that are snowy and cold.

The *Compendium of Suśruta*, the other major ancient medical encyclopaedia, does not mention the concept of 'errors of judgement'. Rather, the *Suśruta* presents a quite different taxonomy of illness groups.

## Diseases of body, environment, and the supernatural

In its first book, the *Suśruta* set out a general classification of medical afflictions.[39] Pain, that defines illness when in a patient, is divided into three categories: pertaining to the body (*ādhyātmika*), pertaining to the physical world (*ādhibhautika*), and pertaining to non-physical causes (*ādhidaivika*). The first category, the bodily, was further broken down into ailments set in motion by the forces of conception, of birth, or of deranged humours. Diseases caused at the time of conception were caused by faulty sperm or female conceptual blood, and included diseases of pallid skin and haemorrhoids.[40] Diseases of birth were related to the mother's diet and behaviour during pregnancy. If she were undernourished or her pregnant cravings were denied, then the child might suffer from disabilities such as lameness, blindness, deafness, and dwarfism. Deranged humours could arise from anxiety or from faulty diet or behaviour, and could arise in the stomach or digestive tract. Deranged humours could affect the body or the mind. The second major category, traumas from the physical world, included physical assaults by animals or weapons. The third category, the non-physical, included ailments set in motion by time, such as exposure to seasonal extremes of temperature, or by supernatural causes such as curses and magic spells, or by processes of natural insult such as starvation and senility.

The *Suśruta* thus placed humoral medicine, such an important part of medical explanation in ayurveda in general, in a relatively minor location in its grand scheme of disease causation. In spite of this, the *Suśruta* went on to emphasize elsewhere that the three humours are the very root of all diseases, because their symptoms can be seen, their effects witnessed, and because of the authority of learned tradition. The *Suśruta* cited a verse from some older unidentified work that stated: 'inflamed humours flow around in the body and get stuck because of a constricted space, and that becomes the site at which a disease arises'.[41] Thus, the *Suśruta* seems to have expressed a certain tension between its classificatory scheme of disease causation and the widespread dominance of the humoral theory.

A good example of overlapping models in another author is Vāgbhaṭa's account of fever, the first and most serious of diseases discussed by all the ayurvedic treatises. Vāgbhaṭa started with references to the mythology of the god Śiva, but segued rapidly into a humoral narrative:[42]

> Fever is the Lord of diseases. It is evil, it is death, the devourer of energy, the terminator. It is the fury born from Śiva's third eye, which destroyed Dakṣa's sacrifice. It consists of the confusion that is present at birth and death.[43] It is essentially a high temperature and it arises from bad conduct. Called by many names, it is cruel and exists in creators of all species. It is of eight types, according to the way the humours come together singly, in combination, or as being of external origin. Thus, the impurities, corrupted each by its own particular irritant, enter the stomach. They then accompany the crude matter, and block the ducts. They then drive the fire out of the place of digestion and to the exterior. Then, together with it, they snake through the whole body, heating it, making the limbs hot, and bringing about a fever. Because the ducts are obstructed, there is usually no sweat.

Its first signs are lassitude, uneasiness, a heaviness of the limbs, dryness of the mouth, loss of appetite, yawning, and watery eyes.

There is friction of the limbs, indigestion, breathlessness, and excessive sleepiness. There are goose pimples, flexion, cramp in the calf muscles and fatigue. The patient is intolerant of good advice and has a liking for sour, pungent and salty things, and a dislike for sweet foods. He also dislikes children. He is extremely thirsty. For no reason, the patient likes or dislikes noises, fire, cold, wind, water, shade or heat.

Following these signs, the fever becomes manifest.

This passage interestingly combines humoral corruption, the displacement of digestive fire, the flow of humors in the body and the existence of ducts that become blocked. It is not unusual to find several etiological explanations side-by-side in the texts. It seems that in addition to setting out to describe the facts of medical theory, the texts may also have functioned as toolboxes of ideas that physicians could use in order to construct the medical narrative appropriate for a particular patient in a particular situation.

## Invasive diseases

In one of its several classificatory schemes, the *Caraka* asserted that there are three kinds of disease:

The three diseases are the internally caused, the invasive (*āgantuka*), and the mental. Thus, 'internal' is what arises out of the body's humours; 'invasive' is what arises from creatures, poison, wind, fire or wounding. And 'mental' is brought about by not getting what one wants, or getting what one does not want.[44]

Invasive diseases include those that arise out of demonic possession, poison, wind, fire, and assault. In all of these cases, according to *Caraka*, good judgement is violated. The *Suśruta* uses the category of 'invasive' disease to talk about foreign objects that have to be surgically removed, including shards of iron, bamboo, tree, grass, horn and bone, and especially arrows.[45]

When, in the sixteenth century, ayurvedic authors first began to grapple with the problem of syphilis, it was classified as an invasive disease.[46] The disease was first described in India by Bhāvamiśra in his sixteenth-century work *Bhāvaprakāśa*. Bhāvamiśra said that the disease was widespread in a country called 'Phiraṅga' – France or the Franks – and that therefore experts called it the 'Phiraṅga disease'.[47] Although Bhāvamiśra classified it as an invasive disease, caught by carnal contact with those who had the disease, he noted that the humours were also involved and that the expert physician would diagnose the disease by means of noting the signs displayed by the humours.

## Epidemic disease

Kāmpilya, a town on the Ganges close to 20° n 80° e, was the ancient capital of Pañcāla. The *Caraka* located a dialogue about epidemic disease at this place. The protagonists of the debate had to struggle with a major theoretical problem. Almost all models of disease in classical Indian medicine explain illness as some form of dysfunction of the patient's unique, personal constitution. Yet, in an epidemic one witnesses many people suffering the same disease, in spite of their varying personal constitutions. 'How can one single disease cause an epidemic all at once amongst people who do not have the same constitution, diet, body, strength, dietary disposition, mentality, or age?', asked the *Caraka*.[48] The discussion of this point came to the conclusion that when the air, waters, places and times are corrupted or discordant, then diseases would arise at the same time and with the same characteristics, and they would cause the epidemic destruction of a locality. Corrupted air, for example:

> fails to correspond with the appropriate season; it is stagnant; it is too mobile; it is too harsh; it is too hot, cold, dry, or humid; it is overwhelmed with frightful howling, with gusts clashing together too much; it has too many whirlwinds; it is contaminated with antagonistic (*asātmya*) smells, fumes, sand, dust, or smoke.

Waters become turbid and are abandoned by wildlife. Corrupted places are full of mosquitoes, rats, earthquakes, and bad water. Time goes wrong when the seasons display inappropriate features. Underlying these physical causes, however, is a moral causality. Errors of judgement are the ultimate cause of epidemics, according to the *Caraka*. These errors lead to unrighteousness (*adharma*) and bad karma.

> Thus, when the leaders in a district, city, guild, or community transgress virtue, they cause their people to live unrighteously. Their subjects and dependants from town and country, and those who make their living from commerce, start to make that unrighteousness grow. The next thing is that the unrighteousness suddenly overwhelms virtue. Then, those whose virtue is overwhelmed are abandoned even by the gods. Next, the seasons bring calamity on those whose virtue has thus been overwhelmed, on those who have unrighteous leaders, on those who have been abandoned by the gods.[49]

War is also implicated in the disruption of society that can lead to epidemic disease.[50] The whole discussion of epidemic disease is concluded, in the *Caraka*, by a narrative about social decay during the present degenerate age of man, caused by a primal accumulation of excess in the original Golden Age.

> Because they had received too much, their bodies became heavy. Because of this corpulence, they became tired. From tiredness came apathy, from apathy accumulation, from accumulation, ownership. And ownership led to the appearance of greed in that Golden Age.

A causal chain of further vices led to the diseases and social decay that the author saw at his time, including epidemic disease.[51]

## Contagion

Contagion plays almost no role in classical ayurvedic theory. A small number of references in the literature suggest that the idea was not completely absent, but it certainly played no major part in the general understanding of disease in ancient India.[52] The *Suśruta*, a century or two later than the *Caraka*, said:

> Skin disease, fever, consumption and conjunctivitis as well as secondary diseases are communicated from person to person by attachment, contact of the limbs, breath, eating together, lying or sitting down together, or from sharing clothes, garlands or makeup.[53]

This verse occurs at the end of a chapter, and the topic is not taken further. Similarly, a passage from the sixth-century *Heart of Medicine* noted the possibility of disease contagion through close proximity to others:

> Almost all diseases are contagious (*saṃcārin*) through the habit of touching, or of eating and sleeping together and the like. Especially ailments of the skin and eyes.[54]

This sounds like a strong observation of the model of contagion. However, it is an isolated verse in the middle of a chapter about other things. The very next statement in the text relates to worms, and the author did not return to the subject of contagion.

The philosopher Prajñākaragupta (fl. 750–810) casually used the following example to illustrate the behaviour of a crazy person: 'The first person to go into a place like that gets sick . . . If that's the case, I will not go in first. I will go in later.'[55] The model of disease implied in this eighth-century gnomic jest is that presence in a particular place may cause disease. The verbal noun 'go into' (Skt. *praveśa*) suggests actual entry into an enclosed place, rather than just arrival at a country, such as might be experiencing an epidemic. The notion of contagion is not explicit, but the connection of disease with place is certainly present. Small increments in developing the idea of contagion took place in the discussion of the *Caraka* and the *Suśruta* by the eleventh-century Bengali medical genius, Cakrapāṇidatta, who connected the ideas of contagion, unrighteousness, epidemics and skin diseases like leprosy.[56] But the idea never gained traction amongst traditional physicians until ayurveda began to be influenced by European medical ideas of disease in the nineteenth century.

## Conclusion

The traditional ayurvedic medicine of India is often represented as being an herbal medicine underpinned by a three-humour theory. This is true, but it is an inadequate account of ayurveda's complexity. The three-humour theory pervades much of the theory of disease, but it is displaced or overlapped by other equally important theories. The idea of affinity, for example, influenced diet and cookery, and had a profound influence on cuisine and dietary regime in South Asia. The notion of raw residues led to the widespread development of purging therapies.

Not all health practices in South Asia developed a self-aware scholarly tradition of reflection and theorization. Undocumented folk practices at the village or family level both informed scholarly practice and were influenced by it. Ayurvedic treatments of poisoning in the ancient treatises, for example, are almost devoid of humoral theory and mostly link symptoms with therapies, without an intervening layer of theory. A fine example of a completely isolated rationalization that may have originated as a folk belief occurs in the *Caraka*:[57]

> If the path of a patient's phlegm is corrupted by poison . . . and he breathes like a dead man and may die, then . . . one should apply the meat of goat, cow, water buffalo, or cock to an incision in the man's head like a crow's foot. Then, the poison moves across into the meat.

Other bizarre explanations occasionally intrude into the otherwise stately narratives of the ancient encyclopedias.

Across a large diverse society and through a period of more than three millennia, a multiplicity of diagnostic and therapeutic models developed in India. This plurality has persisted into the present time, with the Government of India providing official recognition, support and regulation for Ayurveda, Yoga, Unani, Siddha, Tibetan and Homoeopathic medicine, albeit at a much lower financial degree than for Modern Establishment Medicine.[58] The popularity and currency of applied ayurveda in the contemporary world has not always helped the appreciation of the heterogeneity of the disease models that existed historically, because most accounts of ayurveda have aimed at simplifying matters in order to reach a general audience, or have been confessional or promotional in purpose. The exploration of the disease models that were proposed in early Indian medicine remains a fertile area for research and clarification and will grow through the application of historical sensitivity and the close study of original sources.

## Notes

1 The word 'sensation' (Pali *vedanā*) is used in this passage to cover both experiences in general and painful experiences in particular.

2 This account occurs in the *Saṃyutta Nikāya, Vedanā-saṃyutta*, #21. Pali text edition by Leon Feer, ed. *Saṃyutta-Nikāya*. London: Henry Frowde for the Pali Text Society, 1884–1898; Internet Archive: pt4samyuttanikay00paliuoft: v. 4, 230–31, a translation by Bhikkhu Bodhi, *The Connected Discourses of the Buddha: A Translation from the Pāli*. Somerville, MA: Wisdom Publications, 2000: v. 2, 1278–9; Hartmut Scharfe, 'The Doctrine of the Three Humors in Traditional Indian Medicine and the Alleged Antiquity of Tamil Siddha Medicine', *Journal of the American Oriental Society* 119 (1999), 609–29, at 613. Translations given in this chapter are my own, but I also give references to published translations where available.

3 Ca.ni.8 (1–4); Priya Vrat Sharma, ed. and tr. *Agniveśakṛtā Carakapratisaṃskṛtā Carakasaṃhitā = Carakasaṃhitā. Agniveśa's Treatise Refined by Caraka and Redacted by Dṛḍhabala*, Varanasi, Delhi: Caukhambha Orientalia, 1981–1994, 1:294–5.

4 *Saṃyuttanikāya* 36.21 (Feer, *Saṃyutta-Nikāya*: v. 4, 231): *pittaṃ semhaṃ ca vāto ca/sannipātā utūni ca/ visamaṃ opakkamikam/ kammavipākena aṭṭhamīti//*.

5 It has been much discussed whether the Pali Canonical sermons represent the actual words of the Buddha or are a later re-creation by the monks of the early councils. My own judgement is that we can take the sermons as being more or less identical to the Buddha's words. For some convincing arguments, see Alexander Wynn, 'The Historical Authenticity of Early Buddhist Literature: A Critical Evaluation', *Wiener Zeitschrift für die Kunde Südasiens* XLIX (2005), 35–70, which suggests that the views he expressed are datable to the period before his death in about 400 BCE.

6 Scharfe, 'The Doctrine of the Three Humors', 612 ff. cites several other Pali passages that take the humoral model of disease for granted.

7 Kenneth G. Zysk, *Religious Healing in the Veda: With Translations and Annotations of Medical Hymns from the Rgveda and the Atharvaveda and Renderings from the Corresponding Ritual Texts*, Philadelphia: American Philosophical Society, 1985, surveys the medical concepts recoverable from the Veda.

8 A survey of over a dozen authorities whose works are lost is provided by Gerrit Jan Meulenbeld, *A History of Indian Medical Literature* (henceforth *HIML*), Groningen: E. Forsten, 1999–2002. 5v: 1A, pp. 689–99.

9 *HIML* 1A, pt. 1, for a comprehensive overview and discussion. Introduction and selected translations in Dominik Wujastyk, *The Roots of Āyurveda: Selections from Sanskrit Medical Writings*, 3rd edn. London and New York: Penguin Group, 2003, pp. 1–50.

10 Ah.sū.1.19–21 (in Wujastyk, *Roots*, p. 205).

11 Aśvaghoṣa's *The Life of the Buddha* was translated into Chinese in 420 CE and was widely popular across Asia in the first millennium. After being rediscovered in the nineteenth century, it is still read by Sanskrit students for its beautiful language and narrative construction. A recent edition and translation is that of Patrick Olivelle, *Life of the Buddha by Ashvaghosha*. Clay Sanskrit Library, New York: New York University Press & JJC Foundation, 2008. Arguments for dating Aśvaghoṣa to about 50 CE are given by Philipp A. Maas and Dominik Wujastyk, *The Original Āsanas of Yoga* (in preparation).

12 *Buddhacaritam* 12.3 (Olivelle, *Life of the Buddha*, p. 328). I take 'equality of their bodily elements (*dhātusāmyam*)' here to be equivalent to 'equality of their humours (*doṣasāmyam*)', since in medical theory 'equality (*sāmyam*)' is normally associated with the latter, not the former. Exceptions exist: inequality of the bodily elements/humours (*dhātuvaiṣyamyam*) is one definition of disease (*vikāra*) given by the *Caraka* at Ca.sū.9.4 (Sharma, *Caraka-Saṃhitā*: 1:62). Scharfe, 'The Doctrine of the Three Humors', 624, noted that: 'The older parts of the *Carakasaṃhitā* consider wind, bile, and phlegm in their natural state as elements (*dhātu*) and only in their riled condition as faults (*doṣa*)'. See further discussion by Philipp A. Maas. 'The Concepts of the Human Body and Disease in Classical Yoga and Āyurveda', *Wiener Zeitschrift für die Kunde Südasiens*, 51 (2008), 123–62, at 152 *et passim*.

13 Dominik Wujastyk, 'Agni and Soma: A Universal Classification', *Studia Asiatica: International Journal for Asian Studies* IV–V (2004), 347–70, explores the hot-cold, Agni-Soma, bile-phlegm parallelism that runs throughout ancient Indian culture.

14 G. J. Meulenbeld, 'The Characteristics of a Doṣa,' *Journal of the European Ayurvedic Society* 2 (1992): 1–5.

15 Ca.si.9.79–81 (Sharma, *Caraka-Saṃhitā*: 2:653). The term 'brain-tissue' translates Sanskrit *mastiṣka*, a word commonly translated as just 'brain', although the ayurvedic texts describe it as a fatty substance and do not connect it with cognition.

16 Su.ka, (Priya Vrat Sharma. *Suśruta-Saṃhitā, with English Translation of Text and Ḍalhaṇa's Commentary Alongwith* (sic) *Critical Notes*, Varanasi: Chaukhambha Visvabharati, 1999–2001, 3:1–102. See also Wujastyk, *Roots*, pp. 139–46.

17 A rare study of the concept in relation to the seasons is that of Francis Zimmermann. '*Ṛtu-sātmya*, le cycle des saisons et le principe d'appropriation', *Puruṣārtha: recherches de sciences sociales sur l'Asie du sud* 2 (1975), 87–105.

18 A useful dictionary entry is given by S. K. Ramachandra Rao and S. R. Sudarshan. *Encylopaedia of Indian Medicine*, Bombay: Popular Prakashan, 1985–1987: v. 2:184–5.

19 Lexical bigrams for *sātmya* were generated using Oliver Hellwig, *DCS: Digital Corpus of Sanskrit*. 1999-, http://kjc-fs-cluster.kjc.uni-heidelberg.de/dcs/index.php, accessed July 2015.

20 Ca.sū.6.49bc (Sharma, *Caraka-Saṃhitā*, 1:47).

21 See Ca.sū.7.36–38 (Wujastyk, *Roots*, pp. 7–8, 18; Sharma, *Caraka-Saṃhitā*, 1:51).

22 See the fine historical exploration of the concept by Clive Staples Lewis. 'Nature (with Phusis, Kind, Physical, etc.)', in *Studies in Words*. Cambridge: Cambridge University Press, 1960. Chap. 2. Ayurveda also has a concept of a person's immutable nature, the humoral disposition or temperament with which they are born (Sanskrit *prakṛti*).

23 However, as far as I can tell, there is no analogue in ayurvedic literature to the contemporary idea of approaching habit-change through removing reinforcers or triggers.

24 Ca.vi.1.17–20 (Sharma, *Caraka-Saṃhitā*, 1:304–5).

25 Ca.vi.1.20 (Sharma, *Caraka-Saṃhitā*, 305). Cf. Ca.vi.8.118 (Sharma, *Caraka-Saṃhitā*, 381).

26  Wujastyk, *Roots*, pp. 225 ff.
27  Cf. Dominik Wujastyk, 'The Combinatorics of Tastes and Humours in Classical Indian Medicine and Mathematics', *Journal of Indian Philosophy* 28 (2000), 479–95.
28  Tīsaṭācārya, *Tīsaṭācāryakṛtā Cikitsā-kalikā tadātmaja-śrīCandraṭapraṇītayā saṃskṛtavyākhyayā saṃvalitā, Aṅglabhāṣā-vyākhyātā tathā pariṣkartā ācārya Priyavrata Śarmā*, ed. by Priya Vrat Sharma. Vārāṇasī: Caukhambā Surabhāratī Prakāśana, 1987: 18–21, 224. On Candraṭa and his father Tīsaṭa, see Meulenbeld, *HIML*, IIA, pp. 122–5, 148–51.
29  The following discussion is based on Ca.vi.2 (Sharma, *Carakasaṃhitā*, 1:309–13).
30  *Anonymus Londinensis* VIII (W. H. S. Jones, *The Medical Writings of Anonymus Londinensis*, Cambridge: Cambridge University Press, 1947, p. 45 *et passim*). The word used in the papyrus is not in fact ὠμός but περίττωμα or περίσσωμα (Hermannus Diels, *Anonymi Londinensis ex Aristotelis Iatricis Menoniis et Aliis Medicis Eclogae*, Berolini: Georgii Reimeri, 1893). Internet Archive: p1anonymilondine03diel: index p. 103 for references. E. D. Phillips, *Greek Medicine*, Thames and Hudson, 1973, p. 127 pointed out that there are also parallels in ancient Egyptian and Cnidian beliefs about rotting residues in the body that require removal through purgation. See also: Vicki Pitman, *The Nature of the Whole: Holism in Ancient Greek and Indian Medicine*, Delhi: Motilal Banarsidass, 2006, pp. 113 ff.; and Vivian Nutton, *Ancient Medicine*, London and New York: Routledge, 2004, p. 42.
31  *Suśrutasaṃhitā*, sūtrasthāna 26.499–513 (Sharma, *Suśruta-Saṃhitā*, 1.556–9).
32  Reported by Vijayarakṣita (fl. 1100) on *Mādhavanidāna* 16.1–2 (Yādavaśarman Trivikrama Ācārya, ed. *MahāmatiŚrīMādhavakarapraṇītaṃ Mādhavanidānam ŚrīVijayarakṣita-Śrīkaṇṭhadattābhyāṃ Viracitayā Madhukośākhyavyākhyayā, Śrīvācaspativaidyaviracitayā Āṭaṅkadarpaṇavyākhyāyā viśiṣṭāṃśena ca samullasitam = Mādhavanidāna by Mādhavakara with the Commentary Madhukośa by Vijayarakṣita & Śrīkaṇṭhadatta and with Extracts from Āṭaṅkadarpaṇa by Vāchaspati Vaidya*, Vārāṇasī: Chaukhambha Orientalia, 1986: 133). Cf. *HIML* IA.379 f., & n.212. At least one manuscript of Vijayarakṣita's commentary attributes this remark to Gayadāsa, not Gadādhara (Yādavaśarman Trivikrama Ācārya, op. cit.: 133, variant 2).
33  Shama Sastry, (ed.), *Āyurvedasūtram Yogānandanāthabhāṣyasametam = The Āyurvedasūtram with the Commentary of Yogānandanātha*, Mysore: Government Press, 1922, 1.8–12. See *HIML* 2A.499–501.
34  See: Vaidya Bhagwan Dash and Manfred M. Junius. *A Handbook of Ayurveda*. New Delhi: Concept Publishing Co., [1983] 1987, pp. 34–6 *et passim* for an internalist view; and Jean M. Langford, *Fluent Bodies: Ayurvedic Remedies for Postcolonial Imbalance (Body, Commodity, Text)*, Durham, NC: Duke University Press, 2002, pp. 154–5 for an ethnological analysis.
35  K. R. Norman, 'The Languages of the Composition and Transmission of the Aśokan Inscriptions', in *Reimagining Asoka. Memory and History*, ed. Patrick Olivelle, Janice Leoshko, and Himanshu Prabha Ray, Delhi: Oxford University Press India, 2012, pp. 38–62: 59.
36  Ca.śā.1.98–109 (Sharma, *Caraka-Saṃhitā*, 1:406–77). The Sanskrit word translated as 'memory' is also used to mean 'mindfulness', a heightened state of awareness. This ambiguity runs through all uses of the term and requires context-sensitive reading.
37  Wujastyk, *Roots*, pp. 28–31.
38  Classical Indian medicine does not characterize the human as consisting of 'body, mind and spirit,' but rather 'body, mind and speech'.
39  This account is from Su.sū.24 (Sharma, *Suśruta-Saṃhitā*, 1:252–58).
40  The complexities of the ayurvedic theory of conception are explored by Rahul Peter Das, *The Origin of the Life of a Human Being: Conception and the Female According to Ancient Indian Medical and Sexological Literature*, Delhi: Motilal Banarsidass, 2003.
41  Su.sū.24.10 (Sharma, *Suśruta-Saṃhitā*, 1:257).
42  *Aṣṭāṅgahṛdayasaṃhitā* Ni.2.2–10 (Ananta Moreśvara Kuṃṭe and Kṛṣṇaśāstrī Rāmacandra Navare, eds. *Aṣṭāṅgahṛdayam, śrīmadvāgbhaṭaviracitam, sūtra-śārīra-nidāna-cikitsā-kalpa-uttarasthānavibhaktam śrīmadaruṇadattapraṇītayā sarvāṃgasuṃdaryākhyayā vyākhyayā samalaṃkṛtam. Kṛṣṇadāsa Āyurveda Sīrīja* 3. Muṃbayyām: Nirṇayasāgara Press, 1902: 243–4).
43  A commentator noted that this is why people cannot recall their actions in previous lives.
44  Ca.Sū.11.45 (Sharma, *Caraka-Saṃhitā*, 1:77; Wujastyk, *Roots*, 30).
45  Su.sū.26 (Sharma, *Suśruta-Saṃhitā*, 1:267 ff.).

46  Dagmar Wujastyk, 'Mercury as an Antisyphilitic in Ayurvedic Medicine', in *Asiatische Studien: Zeitschrift der Schweizerischen Asiengesellschaft = Études asiatiques: revue de la Société Suisse*, (in press), provides detailed study of this topic.

47  *Bhāvaprakāśa* Madhyakhaṇḍa 59 (Brahmaśaṅkara Miśra, ed. *Śrīmadbhiṣagbhūṣaṇa Bhāvamiśrapraṇītaḥ Bhāvaprakāśaḥ BhiṣagratnaśrīBrahmaśaṅkaramiśraśāstriṇā vinirmitayā 'Vidyotinī' nāmikayā Bhāṣāṭīkayā saṃvalitaḥ*. 3rd edn. Varanasi: Chowkhamba Sanskrit Series Office, 1961. 2v: II. 560–65).

48  Ca.vi.3.5 (Sharma, *Caraka-Saṃhitā*, 1:314–15).

49  Ca.vi.3.20.

50  Ca.vi.3.21.

51  Ca.vi.3.24–27.

52  See, for example: Rahul Peter Das, 'Notions of "Contagion" in Classical Indian Medical Texts', in *Contagion: Perspectives from Pre-Modern Societies*, ed. by Lawrence I. Conrad and Dominik Wujastyk. Aldershot, Burlington USA, Singapore, Sydney: Ashgate, 2000, pp. 55–78; and Kenneth G. Zysk. 'Does Ancient Indian Medicine Have a Theory of Contagion?', in *Contagion: Perspectives from Pre-Modern Societies*, ed. by Conrad and Wujastyk, pp. 79–95. For further discussion of contagion as a concept, see the chapter in this volume by Michael Worboys.

53  *Suśrutasaṃhitā* Nidānasthāna 5.33–34 (Sharma, *Suśruta-Saṃhitā*, 2:44).

54  *Aṣṭāṅgahṛdaya*, Nidānasthāna 14.41–4 (Kuṃṭe and Navare, *Aṣṭāṅgahṛdayam*, 297).

55  *Pramāṇavārttikālaṅkāra* 2.1.24 (Rāhula Sāṅkṛtyāyana, ed. *Pramāṇavārtikabhāshyam or Vārtikālaṅkāraḥ of Prajñākaragupta (Being a Commentary on Dharmakīrti's Pramāṇavārtikam)*. *Deciphered and edited*, Patna: Kashi Prasad Jayaswal Research Institute, 1943: 171). I am grateful to Eli Franco for drawing this passage to my attention (Eli Franco. 'Bhautopākhyāna or Dumb and Dumber: A Note on a Little-Known Literary Genre'. In press: f.n. 11).

56  Das, 'Notions of "Contagion"', pp. 63–4.

57  Ca.ci.23.66 (Sharma, *Caraka-Saṃhitā*, 2:371–2); Wujastyk, *Roots*, pp. 145, 158–9.

58  The Government department for indigenous medical systems is AYUSH (http://indianmedicine.nic.in/). In 2014, AYUSH was elevated from being a department of the Ministry of Health to being a full ministry in its own right. The politics of medicine in the post-independence period has attracted much scholarship, including: Dominik Wujastyk, 'Policy Formation and Debate Concerning the Government Regulation of Ayurveda in Great Britain in the 21st Century', *Asian Medicine: Tradition and Modernity*, 1.1 (2005), 162–84; Projit Bihari Mukharji, *Nationalizing the Body: The Medical Market, Print and Daktari Medicine*, London and New York: Anthem Press, 2011; and Rachel Berger, *Ayurveda Made Modern: Political Histories of Indigenous Medicine in North India, 1900–1955*, New York: Palgrave Macmillan, 2013.

## Select bibliography

Jolly, Julius, *Indian Medicine: Translated from German and Supplemented with Notes by C. G. Kashikar; with a Foreword by J. Filliozat*. 2nd edn. New Delhi: Munshiram Manoharlal Publishers, 1977.

Kutumbiah, P., *Ancient Indian Medicine*. Bombay, etc.: Orient Longman, [1962] 1999.

Majumdar, R. C., 'Medicine', in *A Concise History of Science in India*. Ed. D. M. Bose, S. N. Sen, and B. V. Subbarayappa. New Delhi: Indian National Science Academy, 1971. Ch. 4, pp. 213–73.

Mazars, G., *A Concise Introduction to Indian Medicine*. Vol. 8. Indian Medical Tradition. Delhi: Motilal Banarsidass, 2006.

Meulenbeld, G. J., *A History of Indian Medical Literature*. Groningen: E. Forsten, 1999–2002. 5v.

Sharma, P. V., *Āyurved kā Vaijñānik Itihās*. Vol. 1. Jayakṛṣṇadāsa Āyurveda Granthamālā. Vārāṇasī: Caukhambā Orientalia, 1975.

Wujastyk, D., *Well-Mannered Medicine: Medical Ethics and Etiquette in Classical Ayurveda*. New York: Oxford University Press, 2012.

Wujastyk, D., *The Roots of Āyurveda: Selections from Sanskrit Medical Writings*, 3rd edn. London, New York, etc.: Penguin Group, 2003.

Wujastyk, D., 'Indian Medicine', in *Oxford Bibliographies Online: Hinduism*, Oxford: Oxford University Press, 2011, http://dx.doi.org/10.1093/obo/9780195399318-0035 accessed on 22/09/2015.

Zimmermann, F., *The Jungle and the Aroma of Meats*. Berkeley: University of California Press, 1987.

Zysk, K. G., *Religious Healing in the Veda: With Translations and Annotations of Medical Hymns from the Rgveda and the Atharvaveda and Renderings from the Corresponding Ritual Texts*, Vol. 75, Transactions of the American Philosophical Society. Philadelphia: American Philosophical Society, 1985.

Zysk, K. G., *Asceticism and Healing in Ancient India: Medicine in the Buddhist Monastery*. New York and Bombay: Oxford University Press, [1991] 2000.

# 4

# RELIGION, MAGIC AND MEDICINE

*Catherine Rider*

Many societies have models of disease that are based on a physical understanding of how the human body works. Some of these are more familiar to modern readers than others. For example the humoral theory, which underpinned learned Western medicine from antiquity into the seventeenth century and beyond, looks very different from modern biomedicine but it is nonetheless based on an understanding of how certain physical factors within the body affect health: in this case, the four humours of blood, black bile, yellow bile, and phlegm.[1] However, physical models are rarely the only explanations for disease available or the basis for every kind of treatment. Instead they coexist with other models which can be described as religious and/or magical. An awareness of the difference between physical and non-physical models of disease goes back to the ancient world. For example, in around 400 BC the ancient Greek medical treatise *On the Sacred Disease*, which is part of the Hippocratic corpus, rejected religious explanations for epilepsy in favour of physical ones:

> I do not believe that the 'Sacred Disease' is any more divine or sacred than any other disease but, on the contrary, has specific characteristics and a definite cause. Nevertheless, because it is completely different from other diseases, it has been regarded as a divine visitation by those who, being only human, view it with ignorance and astonishment.[2]

Religious writers were sometimes equally negative about physical explanations and remedies, praising religious healing above medical cures. Thus a passage in the *Life* of the seventh-century Byzantine saint Theodore of Sykeon depicted the saint warning a sick person to 'Have done with doctors. Don't fall into their clutches: you will get no help from them. Be satisfied with this prayer and blessing and you will be completely restored to health.'[3]

In both of these passages religious and medical models of disease were presented as being in competition, and the authors sought to persuade the audience to reject the 'wrong' model in favour of what they believed was the correct one. However, for most people the situation was not so clear cut. Even the *Life* of Theodore of Sykeon was sometimes more positive about doctors or surgery.[4] For this author as for many others, physical and non-physical explanations and treatments could coexist or even complement each other, and sick people might try physical and non-physical cures simultaneously or move from one model of disease to another, with little sense that the two were incompatible. In many cases the priority is likely to have been to find a cure that worked rather than to analyse exactly how it worked.

This chapter will explore some of the religious and magical models of disease that have been offered in the past, and discuss the ways in which they interacted with one another and with 'medical' models of disease. It will start by discussing the difficult question of definition: what made an explanation or treatment 'religious,' 'magical' or 'medical'? Who drew those distinctions, and why did they do so? It will also discuss some of the sources available to study this topic. The chapter will then move on to explore two roles played by religious and magical models of disease. First, religion and magic could offer broad explanations for why diseases existed, and why they afflicted some people and not others. Second, religion and magic offered a range of cures for disease, which could be used alongside or instead of medical treatments.

## Defining religious, magical and medical understandings of disease

It is not always easy to categorize models and treatments of disease neatly as 'religious', 'magical' or 'medical'. Definitions of magic, religion, and medicine vary considerably across different societies and over time, and it is challenging to formulate distinctions that apply across more than one culture. In fact, the differences between religion, magic and science have been much debated in anthropology and sociology since the nineteenth century and scholars have offered many different views, which have been criticized in turn by later generations.[5] The issue has also attracted the attention of historians of medicine since at least 1915, when W. H. R. Rivers made 'Medicine, Magic, and Religion' the subject of the Fitzpatrick Lectures at the Royal College of Physicians in London.[6] I will not offer a single set of definitions here but will highlight some of the problems historians of disease face when discussing 'religious' or 'magical' healing, and describe some of the approaches taken by recent scholars.

Many readers of this book will have an intuitive sense of what constitutes a 'religious' or 'magical' approach to disease, in contrast to a 'medical' one, but these distinctions are culturally specific and in many societies they are difficult to draw firmly. In fact, several studies of healing in ancient societies have argued that modern terms such as 'magic', 'religion' and 'science' are not appropriate for historical analysis at all, because past cultures did not always draw clear distinctions between 'natural' and 'supernatural' processes or between magical, medical and religious cures. For example, a recent study of ancient Mesopotamian medicine has argued that modern distinctions between the natural and the supernatural or between magic and medicine do not help us to understand the surviving evidence.[7] In a similar vein, a study of ancient Greece has pointed out that the same Greek word, *pharmaka*, referred to a wide range of ways to affect the body, including medicinal drugs, poisons, charms, and spells which were recited over a patient. In addition to causing and curing illness, *pharmaka* could also be used for love magic.[8] Thus to categorize some of these *pharmaka* as magic and others as medicine could be criticized for imposing anachronistic concepts on the sources.

Even when distinctions between the natural and the supernatural or between magic and medicine are made in the sources, there are variations as to what is placed in these categories. In an important book on early-modern demonology, Stuart Clark has shown that educated physicians and theologians in this period often regarded demons as part of the natural world, and believed that they caused illness by manipulating physical processes that took place within the body. Demonic illnesses were therefore natural, and only God could perform actions that were truly supernatural.[9] As the seventeenth-century English physician John Cotta put it: 'Nature is nothing else but the ordinary power of God in al things created, among which the Divell being a creature, is contained, and therefore subject to that universall power.'[10] To take another example, many Western readers would now regard the use of astrology to diagnose and cure disease as not part of medicine, whereas educated physicians in the Middle Ages and the early-modern period, in both Europe and the Middle East, regarded astrology as an integral part of medicine and believed that it could provide crucial information about the future course of a disease and about when to administer remedies.[11] The definition of a 'medical' or 'natural' approach to disease therefore depends on changing understandings of the natural world and of what constituted science.

The boundary between magic and religion is similarly difficult to pinpoint. As with magic and medicine, certain writers in many periods have tried to draw a clear distinction between the two. For medieval and early-modern Christian theologians and other educated clergy, for example, the difference between a religious and a magical cure depended on the source of power which lay behind that cure. Religious cures appealed to God and the saints and drew on their power, whereas magical ones relied on the power of demons. It could be difficult, however, to apply these categories to individual remedies. A cure for toothache recommended by the early fourteenth-century English physician John of Gaddesden, who had studied medicine at Oxford, illustrates some of the problems:

> Write these words on the patient's jaw: '+ In the name of the Father, the Son, and the Holy Spirit, Amen. + rex + pax + nax + in Christ the Son.' And [the toothache] will instantly cease, as has often been observed.[12]

The + signs indicate where the reader should draw the sign of the cross. Were cures such as this one religious or should they be viewed as magic? On the one hand John of Gaddesden did not label this as magic, but rather included it in a medical treatise among many plant-based cures. The appeal to the Father, the Son and the Holy Spirit is a common Christian formula and other verbal cures from this period included a variety of other well-known prayers, including the Lord's Prayer, and appealed to one or more saints. On the other hand, the words 'rex pax nax' are more difficult to categorize as religious. 'Rex' (king) and 'pax' (peace) have meanings in Latin but 'nax' does not and the placing of the three words together seems to owe more to their sound than to the meanings of the words. Other charms went further and included words that had no clear meaning in any language. The use of unknown words or words that did not form comprehensible sentences was often cited as a characteristic of magic. Thus the thirteenth-century philosopher Thomas Aquinas, whose ideas about magic were influential in later centuries, deemed it magic to mix unknown words with well-known prayers, in case the unknown words were the names of demons.[13] However, John's inclusion of the cure shows that even well-educated physicians did not always agree with this strict position.

The boundary between religious and magical cures therefore depended to some extent on ideas about what was, and was not, an acceptable use of ritual: about, for example, whether it was appropriate to invoke the Father, the Son and the Holy Spirit to cure toothache. The answers varied between different writers and they also varied over time. For example, the practice of praying to a saint for a cure, or making a pilgrimage to a saint's shrine, was widespread in medieval Europe and has continued in Catholic tradition up to the present day. For some sixteenth-century Protestant reformers, however, the miracles which were said to have taken place at saints' shrines were either frauds or magic, drawing not on God's power but on the devil's.[14] To add to the complication, 'magic' has often been used as a term of insult, to discredit the activities of opponents – as sixteenth-century Protestants did with Catholic saints' shrines. Educated medical practitioners also attacked the cures offered by other practitioners as 'magic', as the anonymous author of *On the Sacred Disease* did when he denounced those who offered divine explanations for epilepsy as 'witch-doctors, faith-healers, quacks and charlatans'.[15]

Faced with these problems of definition and categorization, many historians prefer to respect the categories used in the societies they study and explore how particular diseases and cures would have been regarded within those societies, rather than relying on definitions which reflect modern ideas.[16] This avoids the danger of imposing anachronistic ideas about 'magic' or 'medicine' on past cultures, which may have seen things very differently. It also allows scholars to explore in detail where individuals within that society drew distinctions between religion, medicine and magic. There are, however, limitations to this approach. It tends to privilege the views of the educated over those of the uneducated, especially in societies where literacy is limited, because only the educated are likely to leave evidence of their definitions of religion, magic and medicine.[17] It can also make it difficult to draw comparisons between different societies because it emphasizes how culturally specific ideas and practices are, rather than focusing attention on the similarities between different cultures.

Other scholars have situated magic and religion, or magic and medicine, at opposite ends of a spectrum, rather than presenting them as distinct categories.[18] In particular several historians of early-modern Catholic Europe have chosen to view both religion and magic as part of a wider 'economy of the sacred' or 'system of the sacred', in which Christian rituals and words were used for a variety of purposes, including healing.[19] These rituals ranged from officially sanctioned prayers performed by clergy to informal actions performed by lay people. Educated clergy sometimes expressed anxiety about these unofficial uses of sacred power, labelling them as 'magical' or 'superstitious', but the demand for them was great and many priests and lay people continued to stretch the official boundaries of religion in the hope of curing illness or relieving the other difficulties of early-modern life.

As these different approaches show, there is no single 'right' way to define magic, medicine or religion when analysing historical sources; rather, different definitions can allow for different types of analysis. Historians working on magical and religious understandings of disease therefore need to be aware of the problems associated with the categories they choose so that they can make their own decisions about how to define the models of disease and treatments they discuss. Nevertheless, we should not see either magical or religious models of disease as irrational. Studies of ancient, medieval and early-modern magic, in particular, have argued that in these societies magic was a rational activity that made sense because a belief in magic fitted in with the assumptions that these cultures made about the world around them.[20] As one influential study of medieval magic explains:

> To conceive of magic as rational was to believe, first of all, that it could actually work (that its efficacy was shown by evidence recognized within the culture as authentic) and, secondly, that its workings were governed by principles (of theology or of physics) that could be coherently articulated.[21]

Similarly if you believe that God or the gods are able to influence the course of events on earth, it is rational to explain disease in religious terms and seek religious cures. To dismiss these responses as irrational is to misunderstand the people who found them a useful way to deal with difficult situations and believed that they were effective.

## Sources

Despite these challenges, there is much that can be done to explore religious and magical understandings of disease, and their relationship to medicine. A wide range of source material survives. Medical writings are an important source. Treatises like John of Gaddesden's compendium or the ancient Mesopotamian medical texts discussed by JoAnn Scurlock may include 'magical' or religious cures, and they may blame certain illnesses on the action of God or the gods, demons or magic. The significance of these 'magical' or 'religious' elements varies. In Gaddesden's treatise, and in most other learned medieval medical works, they play a comparatively minor role. In Scurlock's material, by contrast, they are much more prominent and harder to distinguish from physical cures. In addition to recommending magical or religious treatments, medical writers sometimes denounced the activities of other practitioners as magic, as the author of *On the Sacred Disease* did. These attempts to exclude 'magic' from medicine can tell us much about how medical practitioners regarded their own identity and how they tried to establish their superiority over practitioners who offered alternative views of disease.

The views of less well-educated medical practitioners and their clients are more difficult to recover but for the period after 1500, especially, it is possible to gain some understanding of individual cases. Across Europe secular and ecclesiastical courts, including in Catholic areas the Inquisition, prosecuted individuals for magical healing, although the number of people prosecuted probably represented a small proportion of the total number of healers.[22] These trial records contain much detail about views of disease and about the range of cures on offer. Since the nineteenth century, folklorists and anthropologists have also interviewed folk healers and recorded their practices and their understandings of disease, as well as interviewing the men and women who visit them.[23]

The writings of theologians and clergy, such as Thomas Aquinas's discussion of magic words in cures, are also useful in understanding how religious writers viewed the boundary between magic, religion and medicine, and they sometimes give us a very different perspective from the views found in medical works and trial records. Documents preserved by religious healing centres, such as shrines, go back to the ancient world and include details of who was cured and from what condition, which gives us useful information about attitudes and responses to disease. They can also describe the experiences of people from a range of different social groups, which allows us to look at the views of the uneducated as well as the educated, as long as we bear in mind that the testimonies were written up and shaped by educated scribes.

Sources also survive which articulate less mainstream ideas and practices. One example would be modern New Age healing guides, which often combine healing with religious themes and the search for personal growth.[24] There are even sources which give instructions for how to manipulate disease in ways that many people at the time would have deemed illicit: for example, magical texts from the ancient world onwards include instructions for how to cause, protect against or cure diseases and other forms of harm caused by magic.[25] In addition to the types of source discussed here, almost any source which records individuals' attitudes to and experiences of disease can help us to understand religious and magical views, including diaries, letters, and works of fiction. Often it is necessary to bring more than one type of source together to gain an understanding of how most people may have viewed the role of religion or magic in causing and curing disease.

## Religious and magical explanations for disease

One function performed by religious and magical models of disease – like medical ones – is to offer broad frameworks within which diseases and other health conditions can be understood. In contrast to most medical frameworks, however, religious and magical explanations can seek to answer different questions. One is the difficult question of why diseases exist at all. These questions are explored in most detail by religious writers. For example, early Christian thinkers debated why a good God allowed people to suffer from disease.[26] The answers that they offered existed alongside the humoral medicine which was derived from ancient Greece, but they nevertheless remained a distinct kind of explanation for disease and its place in the world.

Religious models of disease can also offer answers to another, related but perhaps more pressing, question: Why do diseases happen to certain people and not others? The answers to this have taken a variety of forms. In some cases disease could be seen as God's punishment for sin. For Christian and Jewish thinkers there were precedents for this in the Old Testament: for example God briefly struck Moses' sister Miriam with a skin disease after she criticized him (Numbers 12:9–15). The idea that diseases could be divine punishments did not disappear in later centuries. Stories setting out how sinners were struck with disease or sudden death as punishment for their misdemeanours circulated throughout the Middle Ages and continued to be told into the sixteenth and seventeenth centuries. In these stories the disease often fitted the sin, striking the fingers of thieves and the genitals of lechers.[27] These stories were lurid and unsubtle but they did convey a clear message that diseases, especially sudden and gruesome diseases, could occur as retribution for sinful behaviour. They rested on a widespread belief in God's providence, which assumed that God intervened regularly in the world, in order to correct sinners, bless the righteous, and dispense justice.[28]

There were also more positive ways of interpreting disease. Christian writers sometimes emphasized that God might use disease as a test for those He favoured, often taking their inspiration from the Old Testament book of Job. In this book God permitted Satan to afflict the pious Job with a series of misfortunes and illnesses in order to test his faith, but Job continued to trust in God, who eventually restored his fortunes. The story was retold numerous times in later centuries and the retellings conveyed the message that illness could be sent by God as an opportunity for spiritual growth.[29] For medieval Christians this was linked with the belief in Purgatory, which held that suffering was necessary in order to atone for sins committed during life. Disease and physical suffering could therefore offer the chance to atone for sins before death and lessen the time spent in Purgatory afterwards. Moralists might even praise people who asked for disease in order to expiate their sins: thus the twelfth-century Norman monk Orderic Vitalis tells of a monk, Ralph, who after a misspent youth became concerned about his soul and 'humbly implored God to afflict his body with incurable leprosy so that his soul might be cleansed of its foul sins'.[30] Most people probably did not go so far, but the idea persisted that disease was part of God's plan for individuals and could purify the soul if it was endured in the right spirit.[31]

A belief in magic can also offer answers as to why disease occurs, and why it happens to one person and not another. One of the classic anthropological studies of magic, E. E. Evans-Pritchard's *Witchcraft, Oracles and Magic among the Azande*, first published in 1937, argued that the witchcraft beliefs of the Azande, who lived in what is now southern Sudan, were important precisely because they offered explanations for misfortunes, including disease. Evans-Pritchard argued that the Zande linked many (but not all) deaths from disease to witchcraft. Witchcraft did not explain everything: the Zande recognized that diseases existed as part of the world around them and sought to treat them. According to Evans-Pritchard, what witchcraft offered was an extra layer of explanation. It could explain, for example, why one person caught a disease when another did not, and why one person died from a disease while another recovered.[32] Zande witchcraft beliefs also offered a way to manage disease: once witchcraft was identified as the cause, the kinsmen of the sick person could then take measures to identify the witch and ask him or her to cure the illness.[33] Influenced by anthropological studies like Evans-Pritchard's, historians of witchcraft in early-modern Europe, such as Robin Briggs, have also stressed that European witchcraft beliefs functioned, in part, as a way of explaining and managing disease, with many accusations designed to pressurize the 'witch' into offering a cure.[34]

Most of the explanations described above could be applied to any disease and often to other misfortunes as well, but certain conditions seem to have been particularly likely to attract religious and magical explanations. Carole Rawcliffe has noted that medieval religious writers tended to portray plague as a collective punishment for sins committed by the community, while they tended to present leprosy and mental illnesses as punishments for individual sins – although they often included a wider range of conditions under these headings than modern physicians would.[35] These conditions were sometimes singled out because they were mentioned in the Bible as having been caused by God or by demons. Thus Miriam's skin disease in Numbers 12 was often interpreted by medieval commentators as leprosy, while in the New Testament Jesus is shown healing a number of people who were possessed by demons whose symptoms matched medieval understandings of 'insanity' and epilepsy.[36] In some cases similar beliefs existed in non-Christian cultures. In particular, as we have seen, epilepsy was linked to gods and demons by some healers in ancient Greece, even though the author of *On the Sacred Disease* rejected the idea.

Sudden and horrific diseases also seem to have attracted religious explanations. When the Black Death spread across Europe in 1348, killing up to a third of the population, moralists sometimes described it as a punishment for sin in general, and on occasion singled out particular sins: a number of English chroniclers blamed it on the aristocracy's enthusiasm for participating in tournaments.[37] The idea that plague was a divine punishment persisted into later centuries and was not unique to Christian writers. In the early sixteenth century, the Spanish Jew Ilyās ibn Ibrāhīm, writing for the Ottoman court, noted that 'whenever God wants to punish sinners, He deliberately sends the sickness called the plague' and for that reason doctors were 'lost and helpless' to cure it.[38] Appearing suddenly in the late fifteenth century with devastating consequences, syphilis was likewise often presented as a divine punishment, and not surprisingly as a punishment for sexual immorality in particular.[39]

Even these diseases and conditions were never viewed exclusively in religious terms, however. Medieval and early-modern physicians offered many physical explanations for both plague and syphilis, including corrupt air, the malign influence of the planets, the condition of the individual's body which might render him or her more vulnerable to the disease, and increasingly they also discussed the possibility of contagion.[40] Often they presented these physical causes as compatible with the view that plague and syphilis were, ultimately, sent by God. Thus the fifteenth-century friar Thomas Moulton, whose plague treatise was printed several times in the sixteenth century, implied that God worked through natural (in this case astrological) causes to bring about the plague. Not every author was so willing to present physical and divine causes as inter-linked, however: some medical writers emphasized physical means for preventing infection from the plague and passed quickly over religious explanations, while moralists criticized medical pamphlets for neglecting the plague's spiritual causes and remedies.[41] Similarly not every case of 'insanity', epilepsy or leprosy was attributed to divine punishment or demonic possession. Ancient and medieval physicians offered a wide range of physical explanations for these conditions, often tracing them to imbalances in the body's humours, sometimes in turn provoked by poor diet or other lifestyle problems.[42] Religious writers likewise did not attribute every case of epilepsy or insanity to possession, and many Christian authors who recorded the miraculous cures performed by saints distinguished carefully between epileptics, the 'insane' and the possessed.[43]

Conversely under the right circumstances any illness or accident could be interpreted as a divine punishment. A medieval account of the miracles performed by St Foy, compiled at her shrine in Conques in the south of France, recorded a number of cases in which people who failed to show respect to the saint or the monks who tended her shrine were punished with disease. Benedict, a servant sent by his master to seize the monks' wine, was struck with paralysis and died a few days later. A noblewoman who went on pilgrimage to Conques but hid her gold ring so as to avoid having to make an offering to the saint was punished with fever, and only recovered when she offered Foy the ring.[44] Here the speed with which the illness or death followed the sin is emphasized, and it is sometimes accompanied by a vision of the saint which made the connection clear. In these cases the circumstances surrounding the disease seem to have been more important than the specific disease involved.

Historians of early-modern witchcraft have often drawn similar conclusions. Certain conditions seem to have been linked to magic more than others, and witness testimonies in witch trials frequently describe witchcraft illnesses as strange or unfamiliar.[45] The mysterious nature of these conditions therefore seems to have encouraged non-physical explanations. Nevertheless, as with punitive miracles, the events surrounding the onset of the disease were also important and any disease might, under the right circumstances, be presented as the direct result of magic. In particular, if the illness followed a quarrel with a suspected witch then witchcraft was more likely to be suspected.[46]

In most cases the choice of a religious or magical model for disease did not mean rejecting other models. It was quite possible to believe that a particular illness had been caused by God or witchcraft, and simultaneously to believe that many illnesses had physical causes which could respond to medicine. From ancient Greece onwards, medical writers offered physical explanations for all kinds of diseases based on humoral theory, and these models were taken up by educated physicians in Europe and the Middle East in later centuries. However, as we have seen, many of these physicians did not reject the idea that God or magic might ultimately cause illnesses by manipulating their physical causes. In other cases people might move between religious, magical and medical models of disease. For example, accounts of cunning folk, popular healers who specialized in the cure of witchcraft, suggest that sick people who approached them often did so after home remedies and conventional physicians had failed.[47]

The relative weight placed on religious, or magical, explanations for disease varies over time and between different cultures. Often these changes reflect wider changes in medical knowledge, such as the rise of biomedical models of disease based on contagion, bacteria and viruses. In the West these have, from the eighteenth century onwards, relegated religious and magical explanations of disease to a more marginal position. For example, the discovery of the plague bacillus, *Yersinia pestis*, in 1894 caused this to become the pre-eminent model for plague in the West, although there is still debate about whether this particular bacillus was responsible for all earlier outbreaks of what our sources call 'plague'.[48] However, it would be too simplistic to speak of a linear progression from 'religious' and 'magical' understandings of disease to 'medical' ones. Religious models of disease continue to be important to many people. Thus anthropological work on the Catholic shrine at Lourdes, in southern France, has noted that pilgrims are reminded of the spiritual benefits of physical suffering, which can bring sick pilgrims closer to Christ.[49] Other pilgrimage sites also continue to attract a variety of visitors seeking healing (among other things), including Iona in Scotland and Glastonbury in Somerset, which attract pilgrims with both Christian and New Age interests.[50] Under the right circumstances magical explanations can also increase in importance. A good example of this has been highlighted by recent studies of the occult in Africa, which have stressed that witchcraft beliefs are often intimately linked with modern political and economic changes, and are not simply survivals of a 'traditional' culture.[51] Religious and magical explanations of disease have therefore often coexisted with physical ones in the past and they continue to do so in a variety of contexts.

## Models of cure

As well as explaining disease, religion and magic also offer cures. Again these cures may operate alongside medical cures, or act as an alternative to them. They range from simple and often undocumented activities, such as private prayer, to more time-consuming and elaborate practices, some of which will be discussed below. This section will survey some of these religious and magical cures, and explore the ways in which they interact with one another and with 'medical' cures.

One widespread way of seeking a religious cure is to appeal to a holy man or woman. Miraculous healing has long been an important part of Christian culture: the New Testament shows Jesus performing healing miracles, and from antiquity onwards one of the key attributes which caused someone to be recognized as a saint was an ability to heal the sick. As early as the fourth century AD, healing miracles formed a crucial part of the biographies of many saints. The mid-fourth-century *Life* of St Antony, a hermit living in the Egyptian desert, told how Antony cured many sick and possessed people who came to him for help, although the *Life* is careful to stress that the cures were worked by God and not by Antony himself.[52] The *Life* of Antony was one of several highly influential early Christian saints' lives and in later centuries healing continued to be a crucial activity for saints. In 1200, medieval popes began to oversee saint-making by instituting a formal canonization process, which required a saint's supporters to compile written evidence of the miracles he or she had performed, complete with witness statements. The majority of the miracles recorded were healing miracles, and the witness testimonies tell us much about medieval attitudes to disease, medicine, and religious healing.[53] Today saints and holy men and women are still deemed to have an important role as healers and pilgrimages to sacred places remain a method for seeking healing. The shrine at Lourdes is one well-known example but this is not only a Christian phenomenon. An anthropological study of infertility in Egypt by Marcia Inhorn has discussed how Egyptians suffering from infertility and other health problems may visit the tombs of Sufi healers to ask them to intercede with God, although not all Egyptians are in favour of this practice.[54]

Seeking religious healing is not in most cases a rejection of medicine. Pilgrims to Lourdes may come from hospitals, for example, and the seriously ill may be accompanied by doctors and nurses to provide specialist medical care.[55] Similarly the infertile women interviewed by Inhorn had often tried many different forms of treatment, including biomedicine, folk medicine, and prayer at religious shrines. Some rejected one kind of treatment in favour of another but many combined different kinds of therapy.[56] Medieval accounts of miraculous cures sometimes stressed how much better miracles were than conventional medicine, as we have seen, but most Christian writers were not hostile to medicine. More often they emphasized that God, who had created the world, gave medicinal substances and medical knowledge to man.[57] From the thirteenth century onwards, medieval canonization committees also respected medical knowledge and called in doctors to give expert advice on whether or not a cure was truly miraculous, as canonization processes still do.[58]

Religion has also played a key role in encouraging the provision of care for the sick. The first European hospitals were religious institutions, founded as acts of charity, so that the rich could benefit from the prayers of sick people who, because of their suffering, were deemed to be closer to Christ. Many medieval European hospitals provided food, rest and prayer rather than specialist medical attention but some Byzantine hospitals were employing physicians as early as the sixth century.[59] In the medieval Muslim world hospitals were likewise more medicalized: the 'Aḍudī hospital in Baghdad, founded in 982, was said to have employed 25 medical practitioners, including oculists and surgeons as well as physicians.[60] By 1500, hospitals had also begun to emerge as medical centres in Western Europe and the hospital of Santa Maria Nuova in Florence was one of the first to be run as a therapeutic centre by medical professionals.[61] Nevertheless the impulse behind providing this medical care remained religious and Santa Maria Nuova's statutes reminded the rector of the hospital and its officers 'to live together in piety for the salvation of their souls and the good of the hospital' and to 'shelter and tend the sick poor who come to the hospital as they would Christ Himself'.[62] Smaller and less well-funded hospitals were probably far less medicalized but they existed all across medieval and early-modern Europe, showing how religious and charitable impulses could complement the physical care of disease and the patronage of secular medicine.

It is less common to find cures that claim to be 'magical' since the term often carried negative connotations. Charms such as the one quoted from John of Gaddesden's medical treatise earlier in this chapter are rarely labelled 'magic' by the writers who recommend them: instead, Gaddesden and other medieval medical writers categorized them as 'empirical remedies' which could not be explained by the humoral theories which formed the basis of medieval medicine, but which had been proved by experience to work.[63] However, there was scope for debate about what exactly constituted a magical or empirical remedy, and as we have seen charms like Gaddesden's were sometimes labelled as 'magic' or 'superstition' by observers who took a stricter view. Often these were clergy concerned about prayers and rituals being used in ways that had not been officially approved, an issue which attracted particular attention from Christian theologians in fifteenth-century Europe and again after the Reformation.[64] Concerns about 'magical' uses of religious rituals were not confined to the Christian world. A number of medieval Muslim authors of Prophetic Medicine treatises (a tradition of medical writing based on the Qu'ran and the sayings of the Prophet) likewise rejected the use of written talismans and amulets to cure illnesses, although this rejection was not universal and other treatises advocated them.[65] Criticism of unexplained remedies might also come from educated physicians although they tended to dismiss them because they were ineffective rather than because they were magical. For example, the fourteenth-century French physician and surgeon Guy de Chauliac criticized charms and empirical remedies as 'mere tales'.[66] Despite these criticisms, however, charms and amulets continued to be widely used in later centuries and were probably deemed acceptable by many people.

Healers who offer a mixture of 'magical', medical and religious cures are also well documented in many societies. In England until the early twentieth century they were known as 'cunning folk' and they offered a range of services, including cures for illnesses caused by witchcraft and other diseases.[67] Similar healers have been found in many parts of Europe.[68] These practitioners were a diverse group. In seventeenth-century southern Italy, *magare* or wise women accused of magical healing are particularly prominent in the records of the Inquisition, but in 1565 the local priest of the town of Guagnano gave a much more varied list of 'enchanters' operating in the town which included both men and women:

> Andrea Cappuccella; Antonio Agliano, alias Pipici, enchants [*incantat*] the pains of animals; Clementia Memma enchants the bewitched; Don Giuseppe Memmo enchants the diseases of animals; Pompilio Candido enchants the chill fevers of men; Sister Avenia enchants the pains of joints; Sister Rosata enchants headaches.[69]

Although it is hard to tell how numerous these local healers were, they probably made up the majority of medical practitioners before the twentieth century.

It is difficult to know how these practitioners were regarded by the majority of the population. On the one hand, people were clearly prepared to consult them and to pay for their services. On the other hand, there seems to have been a certain level of ambivalence about their activities. The groups who were most active in defining and condemning magic – strict clergy and educated medical practitioners – labelled cunning folk as 'witches'.[70] This was perhaps not surprising, but cunning folk were also vulnerable to accusations made by others. There seems to have been a widespread sense that those who knew how to heal might also know how to harm and, in particular, if cunning folk failed to cure a bewitched person or made their illness worse, then they might be suspected of having caused the problem in the first place. Cases of this sort are not common but they nonetheless suggest that folk healers were not always regarded as benign or harmless practitioners.[71]

'Magical' and 'religious' cures can therefore be seen interacting with 'medical' ones in a variety of ways. The boundary between the three categories is not always easy to draw, and depends to a large extent on whose view is preserved. Clergy or educated physicians, who had the most interest in distinguishing supposedly licit from illicit forms of medical practice, provide the majority of our sources but they probably took a stricter view of what constituted 'magic' than many of their contemporaries. Nevertheless, debate and disagreement about the legitimacy of certain cures could also occur among other groups of people, as with the differing views of appealing to Sufi healers found by Marcia Inhorn in Egypt. In many societies these different kinds of cure were available simultaneously, and patients had some freedom to choose between them or to combine medical, religious and magical cures from a variety of different healers. A number of historians working on the pre-modern period have described this situation as a 'medical marketplace', although the concept has been criticized for (among other things) assuming that the average patient had a greater choice of healers than was perhaps the case.[72] It is true that many sufferers of disease did not have an unlimited choice of cures: their options were restricted by what was available locally and what they could afford. Nevertheless, we can see people like the Egyptian women interviewed by Inhorn or the Victorians who consulted cunning folk moving from one method of healing to another and choosing what they deem to be the most appropriate response to their condition, based on their own analysis of what has caused the problem.

Alongside this image of diversity, we can also see long-term changes. In Europe and the West more generally, a slow growth in the prestige of university medicine tended to ignore or marginalize magical and religious cures in favour of physical ones. This process took many centuries but its beginnings can be seen in the Middle Ages. In a study of fourteenth-century Aragon, Michael McVaugh has argued that there was a rise in the numbers of medical practitioners claiming university learning and there were signs that it was viewed as prestigious. Although the overall numbers of university-trained physicians remained small, kings and aristocrats who could afford this expensive kind of medicine increasingly chose to spend money on it and supported university faculties of medicine.[73] This does not mean that religious or magical cures disappeared, however. Rather, as with explanations for disease, the balance between magical, medical and religious cures varies across time and between different societies and changes are the product of wider forces which shape medical knowledge and practice.

## Conclusion

This chapter has explored some of the ways in which 'religious', 'magical' and 'medical' models of disease interact, and highlighted some of the complexities associated with researching this topic. It is clear that, in many societies, these models do not operate on separate planes. Religious, magical and medical explanations for disease can coexist and sometimes compete, and sufferers may move from one model to another depending on what is most easily accessible, what suits their circumstances and, perhaps most importantly, what they believe is most effective for their particular condition. Medical writers and clergy have often sought to draw sharp boundaries between religion, magic and medicine, but they have also acknowledged the existence of grey areas such as 'empirical remedies'. The boundaries between these categories, and the ways in which religion, magic and medicine interact, are thus very different in different societies, and there is room for more work on these concepts, the overlaps between them, and the ways in which academic medical theory or theological ideas influenced people's choice of remedies in practice.

There is also still work to be done specifically on religious and magical views of disease, which have been less studied than medical theories. For example, we know comparatively little about how medical writers in many periods perceived the religious, demonic, or magical aspects of disease. How did they think magical or religious causes of disease interacted with physical ones? Did they deem religious or magical issues relevant to their own practice of medicine and, if so, in what ways? Treatises by physicians exploring these questions have received some attention, for example from Stuart Clark,[74] but many remain under-studied. Religious writers' perspectives on medicine have received more attention, but again the interactions between religion and medicine in different periods and among different social groups is still an area which would benefit from further research.

In addition to in-depth studies of magic, religion and medicine in individual past societies, it would also be desirable to explore these issues comparatively. A comparative approach can help scholars to answer different questions, such as: Do religious or magical models gravitate particularly around certain kinds of disease? What determines this? Comparative approaches would also allow us to assess the influence of ancient religious traditions on later cultures, such as the Old Testament's depiction of leprosy and other diseases. Comparison would also help us to assess the impact of more recent changes in medicine. For example, the rise of biomedical models of disease from the nineteenth century onwards has clearly had an important effect on how disease is understood, but how has this manifested itself in different cultures, and what other factors are important? There is also scope for comparison when looking at responses to disease. Certain responses seem to be widespread, such as pilgrimage or the use of amulets, but how widespread are they and how far do they vary between different cultures? Tracking change over time, and between different geographical areas, would help to answer these questions.

A wide variety of sources exists, including medical treatises and collections of remedies, the writings of theologians, shrine-keepers, and other religious figures, and the testimonies of individual people who claimed to have experienced a miracle or thought their neighbour had made them ill. One of the challenges when using this varied body of source material is to bring together the views of medical practitioners, of other interested parties such as priests, judges, shrine-keepers, and of the many ordinary people who suspected witchcraft, prayed for a cure, or sought healing at a shrine. The diversity of these views also needs to be kept in mind: ideas about what constitutes a legitimate cure are rarely monolithic in any culture. Another challenge is to understand how unorthodox views of disease found in sources such as magical texts relate to more mainstream views. Much interesting work has already been done, but scope still exists to set medical models of disease into their wider context of alternative explanations and treatments and to explore the ways in which patients chose between them.

## Notes

1 For further discussion of humoral theory, see the chapter by Jim Hankinson in this volume.
2 'On the Sacred Disease', in G. E. R. Lloyd (ed.), *Hippocratic Writings*, trans. J. Chadwick and W. N. Mann, London: Penguin, 1978, p. 237.
3 Peregrine Horden, 'Saints and Doctors in the Early Byzantine Empire: The Case of Theodore of Sykeon', in W. J. Shiels (ed.), *The Church and Healing*, Studies in Church History, vol. 19, Oxford: Basil Blackwell, 1982, pp. 1–13, at p. 1.
4 Horden, 'Saints and Doctors', p. 1.
5 For an overview, see Stanley Tambiah, *Magic, Science, Religion and the Scope of Rationality*, Cambridge: Cambridge University Press, 1990.
6 Published as W. H. R. Rivers, *Medicine, Magic, and Religion: The FitzPatrick Lectures Delivered before the Royal College of Physicians of London in 1915 and 1916*, London: Kegan Paul, 1924.
7 JoAnn Scurlock, *Magico-Medical Means of Treating Ghost-Induced Illnesses in Ancient Mesopotamia*, Leiden: Brill, 2006, p. 82.
8 G. E. R. Lloyd, *In the Grip of Disease: Studies in the Greek Imagination*, Oxford: Oxford University Press, 2003, p. 10.
9 Stuart Clark, *Thinking with Demons: The Idea of Witchcraft in Early Modern Europe*, Oxford: Clarendon Press, 1997, pp. 152–5, 161–6.
10 Quoted in Clark, *Thinking with Demons*, p. 165.
11 Peter E. Pormann and Emilie Savage-Smith, *Medieval Islamic Medicine*, Edinburgh: Edinburgh University Press, 2007, pp. 154–6.
12 John Shinners (trans.), *Medieval Popular Religion: A Reader*, New York: Broadview Press, 1997, p. 286.
13 Catherine Rider, 'Medical Magic and the Church in Thirteenth-Century England', *Social History of Medicine* 24 (2011), pp. 92–107, at p. 97.
14 Helen L. Parish, *Monks, Miracles and Magic: Reformation Representations of the Medieval Church*, London: Routledge, 2005, p. 60.
15 Lloyd (ed.), *Hippocratic Writings*, p. 237.
16 For example Richard Kieckhefer, 'The Specific Rationality of Medieval Magic', *American Historical Review* 99 (1994), pp. 813–36 at pp. 824, 832–3.
17 Kieckhefer, 'Specific Rationality', 832.
18 For example Peregrine Horden, 'Sickness and Healing', in Thomas F. X. Noble and Julia M. H. Smith (ed.), *The Cambridge History of Christianity*, vol. 3, Cambridge: Cambridge University Press, 2008, pp. 416–32, at p. 420.
19 R. W. Scribner, 'Cosmic Order and Daily Life: Sacred and Secular in Pre-Industrial German Society', in K. von Greyerz (ed.), *Religion and Society in Early Modern Europe 1500–1800*, London: Allen and Unwin, 1984, p. 17, repr. in R. W. Scribner, *Popular Culture and Popular Movements in Reformation Germany*, London: Hambledon Press, 1987, p. 1; David Gentilcore, *From Bishop to Witch: The System of the Sacred in Early Modern Terra d'Otranto*, Manchester: Manchester University Press, 1992, pp. 2–6.

20 Clark, *Thinking with Demons*, p. viii; Scurlock, *Magico-Medical Means*, p. 78.

21 Kieckhefer, 'Specific Rationality', 814.

22 See, for example, Gentilcore, *Bishop to Witch*, pp. 141–8; Owen Davies, *Popular Magic: Cunning Folk in English History*, London: Hambledon Continuum, 2003, pp. 9–18.

23 See, for example, E. E. Evans-Pritchard, *Witchcraft, Oracles and Magic among the Azande*, 2nd edn, Oxford: Clarendon Press, 1950; Marcia C. Inhorn, *Quest for Conception: Gender, Infertility and Egyptian Medical Traditions*, Philadelphia: University of Pennsylvania Press, 1994.

24 See Wouter J. Hanegraaff, *New Age Religion and Western Culture: Esotericism in the Mirror of Secular Thought*, New York: State University of New York Press, 1998, pp. 42–61.

25 See for example John G. Gager, *Curse Tablets and Binding Spells from the Ancient World*, Oxford: Oxford University Press, 1992.

26 See Darrel W. Amundsen, 'Medicine and Faith in Early Christianity,' *Bulletin of the History of Medicine* 56 (1982), pp. 326–50, at pp. 333–8, repr. in Darrel W. Amundsen, *Medicine, Society and Faith in the Ancient and Medieval Worlds*, Baltimore: Johns Hopkins University Press, 1996, pp. 136–40.

27 Alexandra Walsham, *Providence in Early Modern England*, Oxford: Oxford University Press, 1999, p. 77.

28 Ibid., pp. 1–2.

29 Carole Rawcliffe, *Leprosy in Medieval England*, Cambridge: Boydell, 2006, pp. 56–7.

30 Ibid., p. 59.

31 David Harley, 'Spiritual Physic, Providence and English Medicine, 1560–1640', in Ole Peter Grell and Andrew Cunningham (eds), *Medicine and the Reformation*, London: Routledge, 1993, pp. 101–3.

32 Evans-Pritchard, *Witchcraft, Oracles and Magic among the Azande*, pp. 70–7.

33 Ibid., p. 90.

34 Robin Briggs, *Witches and Neighbours: The Social and Cultural Context of European Witchcraft*, 2nd edn, Oxford: Blackwell Publishing, 2002, pp. 62–3.

35 Rawcliffe, *Leprosy*, p. 53.

36 Rawcliffe, *Leprosy*, p. 49. For Jesus's healings see Matthew 8:28–33; Matthew 9:32–33; Matthew 12:22–23; Mark 5:1–17; Mark 9:14–29; Luke 4:31–37; Luke 8:26–39; Luke 11:14–20.

37 Rosemary Horrox, *The Black Death*, Manchester: Manchester University Press, 1994, pp. 95–7.

38 Ron Barkai, 'Jewish Treatises on the Black Death (1350–1500): A Preliminary Survey', in Roger French, Jon Arrizabalaga, Andrew Cunningham and Luis García-Ballester (eds), *Medicine from the Black Death to the French Disease*, Aldershot: Ashgate, 1998, pp. 6–25, at p. 16.

39 Kevin Siena, '"The Venereal Disease", 1500–1800' in Sarah Toulalan and Kate Fisher (eds), *The Routledge History of Sex and the Body*, London: Routledge, 2011, pp. 463–78, at pp. 467–8.

40 Horrox, *Black Death*, pp. 100–8; Jon Arrizabalaga, 'Medical Responses to the "French Disease" in Europe at the Turn of the Sixteenth Century', in Kevin Siena (ed.), *Sins of the Flesh: Responding to Sexual Disease in Early Modern Europe*, Toronto: Centre for Reformation and Renaissance Studies, 2005, pp. 33–55, at pp. 37–42. For a discussion of contagion, see Chapter 5 by Michael Worboys.

41 George R. Keiser, 'Two Medieval Plague Treatises and Their Afterlife in Early Modern England', *Journal of the History of Medicine and Applied Sciences* 58 (2003), pp. 292–324, at pp. 311–18.

42 Owsei Temkin, *The Falling Sickness: A History of Epilepsy from the Greeks to the Beginnings of Modern Neurology*, 2nd edn, Baltimore: Johns Hopkins University Press, 1971, pp. 51–64, 121–33; Michael W. Dols, *Majnūn: The Madman in Medieval Islamic Society*, Oxford: Oxford University Press, 1993, pp. 23–32; Rawcliffe, *Leprosy*, pp. 73–8.

43 Bernard Bachrach and Jerome Kroll, 'Sin and Mental Illness in the Middle Ages', *Psychological Medicine* 14 (1984), pp. 507–14, at p. 511; Peregrine Horden, 'Responses to Possession and Insanity in the Earlier Byzantine World', *Social History of Medicine* 6 (1993), pp. 177–94, at p. 186.

44 *The Book of Sainte Foy*, trans. Pamela Sheingorn, Philadelphia: University of Pennsylvania Press, 1995, pp. 60–1, 83–4.

45  Briggs, *Witches and Neighbours*, p. 61.
46  Ibid., p. 62.
47  Davies, *Popular Magic*, pp. 104–5.
48  See Lester K. Little, 'Review Article: Plague Historians in Lab Coats', *Past and Present* 213 (2011), pp. 267–90.
49  Andrea Dahlberg, 'The Body as a Principle of Holism: Three Pilgrimages to Lourdes', in John Eade and Michael J. Sallnow (eds), *Contesting the Sacred: The Anthropology of Christian Pilgrimage*, 2nd edn, Urbana and Chicago: University of Illinois Press, 2000, p. 38.
50  Marion Bowman, 'More of the Same? Christianity, Vernacular Religion and Alternative Spirituality in Glastonbury', in Steven Sutcliffe and Marion Bowman (eds), *Beyond New Age: Exploring Alternative Spirituality*, Edinburgh: Edinburgh University Press, 2000, pp. 83–104; W. Graham Monteith, 'Iona and Healing: A Discourse Analysis', in Sutcliffe and Bowman, *Beyond New Age*, pp. 105–17;
51  For an overview, see Henrietta L. Moore and Todd Sanders, 'Magical Interpretations and Material Realities: An Introduction', in Henrietta L. Moore and Todd Sanders (eds), *Magical Interpretations, Material Realities: Modernity, Witchcraft and the Occult in Postcolonial Africa*, London and New York: Routledge, 2001, pp. 1–27.
52  Athanasius, *Life of Antony*, trans. Caroline White, *Early Christian Lives*, London: Penguin, 1998, p. 45.
53  Robert Bartlett, *Why Can the Dead Do Such Great Things? Saints and Worshippers from the Martyrs to the Reformation*, Princeton: Princeton University Press, 2013, pp. 349–50.
54  Inhorn, *Quest for Conception*, pp. 109–10.
55  Dahlberg, 'Principle of Holism', p. 36.
56  Inhorn, *Quest for Conception*, pp. 95–7.
57  Amundsen, 'Medicine and Faith', p. 333; Amundsen, *Medicine, Society and Faith*, p. 135.
58  See Joseph Ziegler, 'Practitioners and Saints: Medical Men in Canonization Processes in the Thirteenth to Fifteenth Centuries', *Social History of Medicine* 12 (1999), pp. 191–225.
59  Peregrine Horden, 'Sickness and Healing' in Thomas F. X. Noble and Julia M. H. Smith (eds), *The Cambridge History of Christianity*, vol. 3, Cambridge: Cambridge University Press, 2008, p. 431.
60  Pormann and Savage-Smith, *Medieval Islamic Medicine*, p. 98.
61  Katharine Park and John Henderson, '"The First Hospital among Christians": The Ospedale di Santa Maria Nuova in Early Sixteenth-Century Florence', *Medical History* 35 (1991), pp. 164–88, at p. 169.
62  Park and Henderson, 'First Hospital among Christians', 176.
63  Lea T. Olsan, 'Charms and Prayers in Medieval Medical Theory and Practice', *Social History of Medicine* 16 (2003), pp. 343–66, at pp. 349–52.
64  Michael D. Bailey, 'A Late Medieval Crisis of Superstition?' *Speculum* 84 (2009), pp. 633–61, at p. 651.
65  Pormann and Savage-Smith, *Medieval Islamic Medicine*, pp. 150–1.
66  Olsan, 'Charms and Prayers', 349.
67  See Davies, *Popular Magic*, ch. 4.
68  For an overview see Willem de Blécourt, 'Witch-Doctors, Soothsayers and Priests: On Cunning Folk in European Historiography and Tradition', *Social History* 19 (1994), pp. 285–303, at pp. 299–301.
69  Gentilcore, *From Bishop to Witch*, p. 130.
70  Davies, *Popular Magic*, pp. 29–30
71  Ibid., p. 13.
72  See Mark S. R. Jenner and Patrick Wallis, 'The Medical Marketplace', in Mark S. R. Jenner and Patrick Wallis (eds), *Medicine and the Market in England and Its Colonies, c. 1450–c. 1850*, Basingstoke: Palgrave Macmillan, 2007, pp. 1–23.
73  Michael McVaugh, *Medicine before the Plague: Practitioners and Their Patients in the Crown of Aragon 1285–1345*, Cambridge: Cambridge University Press, 1993, pp. 78–87.
74  Stuart Clark, 'Demons and Disease: The Disenchantment of the Sick (1500–1700)', in Marijke Gijswijt-Hofstra, Hilary Marland and Hans de Waardt (eds), *Illness and Healing Alternatives in Western Europe*, London: Routledge, 1997, pp. 38–58.

## Select bibliography

Amundsen, Darrel W., *Medicine, Society and Faith in the Ancient and Medieval Worlds*, Baltimore: Johns Hopkins University Press, 1996.

Clark, Stuart, 'Demons and Disease: The Disenchantment of the Sick (1500–1700)', in Marijke Gijswijt-Hofstra, Hilary Marland and Hans de Waardt (eds), *Illness and Healing Alternatives in Western Europe*, London: Routledge, 1997, pp. 38–58.

Clark, Stuart, *Thinking with Demons: The Idea of Witchcraft in Early Modern Europe*, Oxford: Clarendon Press, 1997.

Davies, Owen, *Popular Magic: Cunning Folk in English History*, London: Hambledon Continuum, 2003.

Evans-Pritchard, E. E., *Witchcraft, Oracles and Magic among the Azande*, 2nd edn, Oxford: Clarendon Press, 1950.

Horden, Peregrine, 'Sickness and Healing' in Thomas F. X. Noble and Julia M. H. Smith (eds), *The Cambridge History of Christianity*, vol. 3, Cambridge: Cambridge University Press, 2008, pp. 416–32.

Olsan, Lea T., 'Charms and Prayers in Medieval Medical Theory and Practice', *Social History of Medicine* 16 (2003), pp. 343–66.

Pormann, Peter E., and Emilie Savage-Smith, *Medieval Islamic Medicine*, Edinburgh: Edinburgh University Press, 2007.

Rawcliffe, Carole, *Leprosy in Medieval England*, Cambridge: Boydell, 2006.

Shiels, W. J. (ed.), *The Church and Healing*, Studies in Church History, vol. 19, Oxford: Basil Blackwell, 1982.

Sutcliffe, Steven, and Marion Bowman (eds), *Beyond New Age: Exploring Alternative Spirituality*, Edinburgh: Edinburgh University Press, 2000.

Tambiah, Stanley, *Magic, Science, Religion and the Scope of Rationality*, Cambridge: Cambridge University Press, 1990.

# 5

# CONTAGION

*Michael Worboys*

One notable feature of the Ebola epidemic in West Africa that began in 2014 was uncertainty about how the disease spread, despite nearly four decades of scientific and epidemiological research. In the three countries affected and then in the rest of the world, the message went out from health officials that it was only possible to contract Ebola when the body fluids of an infected person directly entered another person, either through broken skin (a wound or rash) or mixing with the fluids of the mouth, eyes, or nose. This characterisation made Ebola a disease spread by contagion in the restricted sense that it was only possible to catch by direct contact with the body fluids of an infected person. Diseases transmitted in this way are very few in number; more generally contagion refers to diseases spread by close proximity to an infected person.

The perceived dangers of Ebola widened when experts pointed to the range of body fluids it might be caught from – blood, vomit, faeces, saliva, breast milk, sweat, and semen – and to the possibility that the virus could be carried on everyday and medical objects. In the public mind at least, Ebola moved into the broader category of infectious diseases where transmission could also be indirect or at a distance, mediated by air, water, food, or a vector. This understanding was symbolised in a common image of the crisis: health workers in full body isolation suits, which echoed the garments worn by physicians to protect against plague in the seventeenth century. Indeed, as this chapter will show, uncertainties over how 'catching diseases' spread have been constant themes in the history of medicine and disease throughout the modern era.

Contagion is a word rarely used in medicine in the twenty-first century. Nowadays doctors talk mostly of communicable or infectious disease, terms which include both direct and indirect transmission. However, the term was important historically and had two usages: (i) '*a* contagion' – the noun linked to the adjective 'contagious' when applied to a directly catching disease; and (ii) '*the* contagion' – the matter transmitted by contact, assumed to be a chemical or vital poison.

The changing and complex meanings of these terms for catching diseases led Margaret Pelling to choose the unusual title 'Contagion/Germ Theory/Specificity' for her chapter in Bynum and Porter's influential *Encyclopaedia of the History of Medicine.*[1] Pelling warned that the historiography of contagion had been framed inappropriately by two late-nineteenth-century medical developments, which she wished to avoid: the germ theory of disease and the notion of disease specificity. The starting point of her analysis was the complexities inherent in answering simple questions: Is the disease catching? How is it caught? Why is the disease caught by some people and not others? Doctors and the public have long recognised that on exposure not every-one develops disease. While twenty-first-century medicine explains this in terms of exposure to germs and the immune system fighting infection, in earlier times people looked to Providence, prayers answered and good fortune, while doctors thought in terms of the resistance offered by a person's constitution.

Vivian Nutton has questioned whether medical practitioners working in the tradi-tion of Hippocrates and Galen had a word for contagion.[2] Pre-modern assumptions about the body and disease militated against acceptance of a direct role for external forces in the generation of illness. The humoral model of the body saw most disease as arising internally, from imbalances in humours, due to inherited propensities and lifestyle, including everything from diet to morality. Types of sickness were named after their humoral origins: for example, choleraic conditions, like jaundice and diar-rhoea, were due to an excess of yellow bile, and melancholia related to an excess of black bile.[3] External factors were secondary and ranged from the spiritual and astrological, through seasonal and meteorological, to conditions in the immediate bodily environment. The nearest concept to modern notions of contagion was the notion of environmental triggers, but these acted in combination and mostly in non-specific ways, most often at a distance by 'influence' rather than by contact. The most well-known environmental factors were miasmas – foul, poisonous airs that acted var-iably, weakening the body, disturbing its function, or acting as catalysts of internal disturbances.

Pre-modern medicine tended not to regard diseases as entities or specific; rather they were seen as states of the whole humoral body. Thus, there were no diseases as such to be communicated.[4] There were a few exceptions that were regarded as direct contagions, but transmission was quite unlike modern notions. Rabies illustrates the point. Until the late nineteenth century there were two variants: rabies in animals and hydrophobia in humans.[5] Both conditions are now seen to be solely due to the trans-mission of a specific virus from animal to animal, or animal to human. Historically, hydrophobia could be 'caught' materially and mentally: both from the inoculation of a poison and the mental trauma of the bite and fear of its consequences. With the latter, agitation and anxiety upset the humours and led to the spontaneous produc-tion of the poison. Another way that doctors understood that hydrophobia could be 'caught' from a rabid animal was by *sympathies* between the bodies of man and dog.[6] However acquired or produced, the poison led to an excess of phlegm, evident in profuse salivation and strange behaviours. The poison was mostly thought to be analogous to snake and insect venom, although since it came from a fellow mammal it might contain an animal spirit. Celestial and seasonal influences were thought to have disturbed the humours of dogs; rabies was prevalent in high summer heat and after the rise of the Dog Star 'Sirius' in July. Given that some symptoms of hydropho-bia mimicked the behaviour of the rabid dog, many people assumed that some vital or spirit bond might remain between the rabid dog and hydrophobia sufferer. Hence, one way of preventing the development of hydrophobia was to the kill the rabid dog and break the metaphysical connection.

While principally a medico-social history of contagion, this chapter cannot ignore its political and commercial dimensions, but my task has been made easier as these aspects have been well covered in two major books: Peter Baldwin's *Contagion and the State in Europe, 1830–1930* (1989) and Mark Harrison's *Contagion: How Commerce*

*Spread Disease* (2012).[7] Both were influenced by Erwin Ackerknecht's seminal essay on 'Anticontagionism between 1821 and 1867', which argued for a direct connection between changing medical ideas about contagion and state attitudes to quarantines and *cordons sanitaires*.[8] Ackerknecht contended that during the first half of the nineteenth century the adoption of free trade policies led to the erosion of quarantines, a shift that was associated with new medical views on epidemics that emphasised aerial rather than person-to-person transmission. Considering a longer timescale, Baldwin argued against any strong link between political economy and quarantine policies, claiming that 'the imperatives of geoepidemiology' were most important. Harrison's *longue durée* history of the relations between commerce and disease revealed a complex picture, with varying relations between politics, economics and medical ideas at different times and in different places.[9]

Though rarely used in contemporary medicine, the term 'contagion' has current cultural resonance. Public discourses on Ebola were said to show 'political contagion', with politicians, medical agencies, and experts vying to shape policy and actions, albeit after an initial period of neglect. The most common figurative use in the early twenty-first century is 'financial contagion', widely used for the spread of sentiment across global stock markets. This usage joined longer-standing worries about 'moral contagion' and 'psychological contagion', within and between social groups.[10] In popular culture, contagion is used as a shock term to emphasise the immediacy and potency of disease threats, for example, in films entitled 'Contagion' released in 2002 and 2011. The latter was promoted with posters headed 'NOTHING SPREADS LIKE FEAR'. This sentiment was echoed in August 2014 with Ebola, when a report in *Time* magazine observed that the fear of contagion was spreading more rapidly and insidiously than the disease itself.[11]

This chapter charts changing medical and public health understandings of, and practices for, the prevention, control and treatment of communicable diseases. My approach is to focus on exemplary communicable diseases in different periods that illustrate key shifts in ideas and practices. My case studies move from the sixteenth to the twenty-first century, with overlaps in time and with some diseases appearing several times. As the Black Death and plague are discussed elsewhere in the volume,[12] this chapter starts with the first major contagions of the early-modern era: the 'Great Pox' and smallpox. The chapter then explores in turn competing views and policies relating to the control of yellow fever and cholera, the development and spread of germ theories of disease in the nineteenth century, focusing on syphilis, sepsis and pulmonary tuberculosis, and the delineation of viral diseases such as polio, influenza and HIV-AIDS in the twentieth century. The chapter closes with brief reflections on emerging and re-emerging communicable diseases.

## Poxes great and small, 1500–1800

The poxes were amongst the first diseases linked to specific poisonous matter and to be understood as entities. They were defined principally by eruptions of pus through the skin, which often left pockmark scars in those who recovered. They were also experienced as amongst the most readily communicable. There were a variety of forms, defined by the size, pattern and frequency of eruptions, but a key moment was the arrival of the 'Great Pox' in Europe in the late fifteenth century.[13] First reported in the French army during the invasion of Naples in 1495, its symptoms were pustules in and on the genital organs, which later became general and caused severe illness. For many decades it was also known as the 'French Disease' and historians have debated whether it was modern syphilis.[14]

Girolamo Fracastoro coined the term syphilis in 1530 in his epic poem *Syphilis sive morbus gallicus* (*Syphilis or The French Disease*), in which a shepherd boy named Syphilus insulted the Sun God Apollo and was punished with a horrible affliction.[15] In 1546 Fracastoro published *On Contagion*, which is now widely cited because he used the phrase 'seeds of disease', which some historians have seen as an anticipation of nineteenth-century bacterial germ theories of disease.[16] The aim of *On Contagion* was to advance the claim that plague, and by implication syphilis, should be regarded as catching.[17]

Fracastoro maintained that the origin or first cause of syphilis was a virus, which then meant poison, venom or pus, entering the body from polluted air, where it had been spontaneously produced by adverse planetary alignments, corrupted environments and, in the fictional case of Syphilus, the acts of the Gods.[18] The poison entered the body through the lungs and pores of the skin, from where it spread, either to spark putrefaction or generate the imbalance that led to an excess of phlegm and lesions. From affected individuals, the poison could be transmitted, especially by sexual intercourse where bodily friction produced heat and opened pores. However, Fracastoro argued that disease would only tend to be engendered in susceptible individuals, with vulnerability determined by their physical, moral and spiritual state. Hence, intercourse with prostitutes was seen to be especially favourable to its transmission. Fracastoro's metaphor of 'seeds of disease' found little favour with his contemporaries, as majority medical opinion was that the syphilis poison was chemical, not living. This latter view informed treatments and encouraged the preference for antidotes, such as mercurial salts. Presenting principally as an external affliction of the skin, syphilis was mostly treated by surgeons, whose approach to diseases tended to be local and to see them as entities – the technical term for this is the ontological concept of disease.[19]

As well as exemplifying a disease that was an entity and separate from the body, syphilis changed the social position of endemic contagious diseases. Firstly, local government agencies attempted to halt its spread by breaking the cycle of transmission, most commonly by trying to regulate prostitution.[20] This made so-called fallen women also carriers of moral contagion and a corrupting influence on men, and a threat to family life. Pregnant women passed on the disease to unborn children, showing hereditary contagion. In adults, syphilis was a chronic rather than an acute disease, debilitating its sufferers over years, producing disfiguring, stigmatising lesions. It was said to be concentrated amongst paupers, beggars, the dissolute and fallen women, marking these groups as sources and symbols of contagion.[21]

Historical demographers argue that until the sixteenth century, smallpox was a childhood disease and one amongst the many fevers with high infant and child mortality.[22] In the early seventeenth century, however, it became endemic in adults, with epidemic peaks. Retrospective speculation as to why this occurred has focused on changes in susceptibility and patterns of exposure due to urbanisation. We are unlikely ever to have definitive answers to such questions, but it is clear that contemporaries saw the 'new' smallpox as highly contagious, from direct contact with, or proximity to, a pustule-covered sufferer. Individual and social responses of isolation and quarantine were based on those used with the plague, with measures such as the fumigation of dwellings, cleansing and disinfecting targeted towards individuals and their families.[23]

In the eighteenth century, a new smallpox preventative measure – inoculation, or variolation after the Latin name for smallpox, *Variola* – was brought to Europe.[24] This procedure, largely practised by lay people and often dismissed by medical practitioners as folk healing, was based on the experience that someone infected with smallpox was protected from future infection. Inoculation involved rubbing into skin, opened by scarifying or cutting, a small account of pus from a sufferer. The aim was to induce a mild infection and hence future protection. Two points were significant about the practice: first, the assumption of direct contagion; and second, a dose effect with the inoculated matter that suggested that the body could learn to tolerate the poison. The use of inoculation increased through the eighteenth century, spurred on by the growth in the number and severity of smallpox epidemics. The ontological status of smallpox as a distinct disease was due to its external origin, confirmed by inoculation, well-marked symptoms, and specific course.

At the end of the eighteenth century, a different method of protection against smallpox was introduced by Edward Jenner, a physician from Gloucestershire in England.[25] Jenner had observed that milkmaids who had suffered cowpox, known as *Vaccinia*, a mild disease in humans transmitted from cows, never subsequently developed smallpox. He inoculated children first with material from cowpox pustules, which he termed vaccination, and then smallpox pustules. He found that vaccinated children did not develop smallpox. The practice spread rapidly, becoming the medical sensation of the age: news of the procedure spread around Britain and soon the rest of the world, as did supplies of his vaccine. Vaccination was first practised in America in 1800, India in 1802 and New South Wales in 1804.[26]

The importance of smallpox and vaccination in the history of disease is that from the early nineteenth century, doctors increasingly defined contagion in the restricted sense of transmission by direct contact. Diseases caught by indirect, mediated and at-a-distance transmission were termed infectious and many old and new catching diseases were placed in this category. One result was that many doctors and public officials, who became known as anticontagionists, questioned the use of policies of quarantine and *cordons sanitaires* to control epidemics.[27]

## Yellow fever, cholera and anticontagionism, 1700–1900

The arrival of yellow fever in epidemic form in the Caribbean and North American British Colonies in the eighteenth century posed problems for officials and doctors. It was not a new disease as such, but had become more virulent and liable to sporadic epidemics. The story of yellow fever can be illustrated by the experience of one city – Philadelphia, the capital of the United States of America at independence.[28] The first important epidemic that quarantines failed to keep at bay occurred in 1762. One of the city's leading doctors and politicians, Benjamin Rush, argued that yellow fever flourished in certain meteorological conditions and in cramped ill-ventilated environments, and that quarantines and isolation actually promoted such conditions. Rush later argued that the epidemics were best combatted by social and sanitary improvements to the city. Philadelphia merchants opposed this view largely because they were anxious about the consequences for trade of admitting that their city was filthy and unhealthy.[29]

In the event, pragmatism prevailed and elements of both policies were implemented, with greater adaptation to the dangers posed by specific groups, places and circumstances. This 'middle way' was also followed when yellow fever struck East coast cities again in the 1820s, with assessments of the commercial impact of different measures. The political rhetoric and battles over when to institute quarantines and how rigidly to apply them continued, but medical and sanitary policies became more nuanced, reflecting a position on epidemic spread and control policies that historians have termed 'contingent contagionism'.[30] This view was not an articulated position as such; rather it followed the experience that the contagiousness of a disease varied and was influenced by many factors ranging from large-scale meteorological conditions to the susceptibility of individuals.

Asiatic cholera in the 1820s and 1830s was a new disease that had spread east from India, crossing the Middle East and Europe, reaching America in 1832.[31] The initial response in Europe and North America was to try and halt its 'invasion' by quarantines and, when these failed, by the isolation of sufferers and the disinfection of their homes. In the first wave in the early 1830s, quarantines and isolation were widely adopted, but to little effect. Anticontagionists promoted the view that Asiatic cholera spread in aerial miasmas that carried an 'epidemic influence'. This might be material or immaterial, but would always evade quarantines and *cordons sanitaires* and excite disease in local conditions of filth. A particular feature of the pattern of incidence of cholera was that it arose simultaneously in unconnected locations, which was further evidence against contagion.[32]

Asiatic cholera returned to the West in the late 1840s, along with political revolutions. If there was a connection, contemporaries saw cholera as consequence rather than cause of political and social unrest. Quarantines were then only favoured in southern European states; in Northern Europe and North America anticontagionism was dominant. Doctors and officials saw the epidemic as originating in, and spreading in the air from, local conditions of filth, overcrowding, pollution and poor ventilation, aided by social vice and intemperance. The anticontagionist's answer was cleanliness and Godliness. Sanitary measures focused on removing waste, cleaning streets, purifying the air (or at least deodorising it), supplying clean water, and controlling foul trades. The public complained that they were too poor and powerless to act on such advice and that their health would only improve when they had higher wages to afford better housing, good food and time for moral improvement.[33]

Epidemiological studies in the 1840s showed that cholera was, to use Charles Rosenberg's term, 'portable': as well as aerial spread, goods and people could carry the 'epidemic influence'.[34] It was in this context, using epidemiological methods, that John Snow suggested in 1849 that cholera was spread by faecally contaminated water.[35] His approach followed the adoption of scientific observations, experiments and speculation in medicine, which led to greater interest in the nature of 'epidemic influences'. Many candidates – animal, vegetable and mineral – were proposed and there was still speculation of immaterial influences, such as electricity.[36] There was, however, a favourite theory, that the poison was a 'zyme' – a large organic chemical that could catalyse fermentation-like changes, or 'zymosis', in susceptible bodies.

The notion of zymosis came from the work of the German chemist Justus von Liebig and was popularised by the British epidemiologist William Farr.[37] Farr was influential because he held the post of Registrar General from 1839 to 1879 and classified the new class of zymotic diseases as either 'epidemic, endemic or contagious'. He regarded smallpox, hydrophobia and syphilis as contagious in the restricted sense of person-to-person spread, while most zymotics were epidemic or endemic. The latter categories included: typhus, plague, influenza, cow pox, glanders, sepsis, erysipelas, puerperal fever, measles, scarlet fever, whooping cough, dysentery, diarrhoea, and cholera. Doctors had noted analogues between progress of fermentation and infection: heat – fever; bubbling – shaking and shivering; expansion – eruptions; reaction ending – recovery.[38] Typically zymotic diseases were acute infections, with symptoms of fever, fatigue, specific local lesions and standard course. Most were self-limiting, as it was believed zymes exhausted the nutrients in the body on which they depended. Zymes were also seen to be responsible for putrefaction: the breakdown of dead tissue, which surgeons met in wounds and could lead to local sepsis or blood poisoning. Septic matter was highly infectious and potent; a mere prick from a lancet or sharp bone fragment was a common cause of death amongst surgeons.

When cholera returned in the mid-1850s, John Snow again advanced his waterborne theory. Farr was in part persuaded, but only to the extent of adding water to airborne transmission. There were further cholera pandemics in every decade of the nineteenth century. By the 1865–66 pandemic, there was much more scientific interest in the nature of the zymes. Investigators identified many contenders for cholera's zyme or 'cholerine', which was variously referred to as a germ, miasm, animalcule, microphyte, seed, fungus, and virus. Epidemiological investigations continued and the Third International Sanitary Conference in 1866, attended by state sanitary officials and leading investigators, saw inconclusive discussions of the value of quarantines. However, in most ports quarantines had been largely replaced by inspections of ships, cargoes and people, with variable times and sites of isolation depending on calculations of risk and benefit.[39]

The cholera pandemic of the early 1880s spread from India only as far as the Mediterranean, halted by quarantines or unfavourable environmental conditions depending on whether you were a contagionist or anticontagionist. Significantly, it attracted a new breed of scientist, searching to discover not the nature of the cholerine zyme, but the causative bacterium or germ. The most famous germ hunter was Robert Koch, who travelled to Egypt and India, identifying a specific cholera bacillus in Calcutta in 1883. However, it was Koch's discoveries of the specific germs of septic infection in the late 1870s and pulmonary tuberculosis in 1882 that was to radically change medical and popular understandings of contagious and infectious diseases.

## Germs, 1860–1900

In 1864 the British Parliament passed legislation to control the spread of venereal diseases in five garrison towns. The statutes were entitled the Contagious Diseases Acts, a euphemism for legislation to control syphilis and gonorrhoea.[40] In Farr's classification, only syphilis, smallpox and rabies were 'contagious' diseases and it is interesting that each was subject to legislative controls: syphilis with the inspection of prostitutes; smallpox with compulsory vaccination; and rabies with the muzzling of dogs. All the measures were seen as coercive exercises of state authority, and actively resisted in the name of personal liberties. The Contagious Diseases Acts assumed that the carriers of contagion were female, stigmatising certain women as diseased and fallen.[41] The measures were analogous to quarantines, with individual prostitutes isolated in lock hospitals until cleared of the disease. Syphilis was seen to be the main target of the measures, but it was difficult to diagnose, being regarded as 'the great imitator' because its symptoms were similar to many other diseases. The poison or 'virus' of syphilis was seen as a potent and peculiar zymotic agent; for while it produced immediate genital sores, it could persist in the body to produce all manner of lesions and morbid changes, including insanity when the brain was affected. Transmission from mother to foetus was known as congenital syphilis, which had a classic triad of signs: notched incisors, inflamed eyes and deafness, but sufferers endured many other symptoms and a life of disability and stigmatisation.[42] Doctors worried particularly about hereditary contagion and familial degeneration.

In 1866 the British government passed further legislation to control contagious diseases, the Contagious Diseases (Animals) Act, which aimed to control and stamp out cattle plague (rinderpest).[43] Historians have not explored the link, if any, between the Acts, but the association certainly underlined for contemporaries the high degree of communicability of both diseases. The cattle plague legislation also covered foot-and-mouth disease, glanders, pleuropneumonia, sheep-pox and sheep scab, which were all framed as imported and highly contagious, to be controlled by quarantines, isolation and slaughter. The legislation was not about protecting individual animals, rather the national herd, with the implication that these diseases were damaging the livestock economy and exports. Critics of the measures, at home and abroad, complained that they were another example of using quarantines for economic and political ends; in this instance, protecting British farmers from foreign competition and maintaining the higher market value of disease-free herds.

Livestock diseases at the time were also subject to laboratory investigation to identify their zymotic agents. The assumption of communicability, by direct contact or proximity, led to speculation that the agents were *contagia viva* – living disease-causing organisms. Literally and metaphorically they were said to be the germs of disease: its cause and beginning.[44] This research drew upon Louis Pasteur's work, which associated microorganisms with diseases in silkworms, and his speculation that microorganisms might be the germs of contagious and infectious diseases, in what was styled the germ theory of infection. This notion led Joseph Lister to develop an analogous germ theory of putrefaction, the process in which decaying dead tissue produced local sepsis in wounds and systemic infections such as septicaemia.[45]

Lister assumed that *contagia viva* reached wounds via the air and by touch. Hence, he experimented to develop techniques to destroy the germs before they reached wound tissue and in wounds themselves. First, he altered his surgical methods to remove dead tissue, the seat of putrefaction. Next, his chosen weapon to destroy invading germs was carbolic acid, previously used to deodorise and disinfect drains. Lister's methods were elaborate and painstaking. He soaked all instruments and dressings in carbolic acid, then operated in a spray of weak acid, before dressing the wound with an air tight, acid-impregnated dressing to keep out any germs circulating in hospital air. Styled as 'antiseptic surgery', Lister's methods were seen as protecting the body from invasion, or put another way, quarantining or isolating wounds with a barrier of chemical disinfectant. Antiseptic surgery was controversial amongst surgeons for a generation. A minority believed that disease-germs came from degenerative changes within the body rather than from without. A larger group held that it was better to work in a germ-free, aseptic environment and that carbolic acid delayed healing by damaging healthy tissue. However, Lister's methods or variations on them were widely adopted in the 1870s and 1880s and were based upon what John Burdon Sanderson called 'the intimate pathology of contagion'.[46]

Although Lister was only concerned to stop germs causing putrefaction in dead or damaged tissues, his ideas and practices promulgated the idea that disease germs were everywhere, waiting to invade the body. Antisepsis spawned a large number of commercial anti-germ products and technologies, all of which emphasised the dangers to individuals, families and society of disease germs. Contagion was often represented by images of disease germs, mostly bacteria, but also vectors such as flies, dust, spitting, and poor hygiene. All this was represented to the public in what Nancy Tomes has called the 'Gospel of Germs', in official propaganda, commercial advertising, antiseptic products and new institutions such as isolation hospitals and disinfecting stations.[47]

The first line of defence against smallpox remained vaccination, which was recommended by governments across the Western world and made compulsory in many. New epidemics, public resistance to vaccination and germ theories led to governments to invest in other anti-germ measures.[48] Isolation hospitals were built for smallpox sufferers, typically in out-of-town locations to avoid or minimise contagion.[49] In the event, the incidence of smallpox waned and the hospitals were used for diseases such as scarlet fever, diphtheria, and typhoid fever, and later measles and whooping cough.[50] Although it was never explicit policy, public health agencies turned isolation hospitals into institutions for sick children, while also reframing childhood infections as epidemic, contagious and serious. Hospitals offered treatment and care, as well as isolation, being important sites for the development of antitoxin and antisera therapy of bacterial diseases, which produced 'miracle cures' in children with diphtheria.[51]

Community health also became important. Medico-sanitary agencies targeted childhood infections and other communicable diseases, seeking notification and contact tracing. The clothing and soft furnishings of sufferers were taken to disinfection stations for high temperature washing, while teams visited homes to sanitise with chemical disinfectants. In the 1900s a new type of germ-harbouring person was discovered – the healthy carrier. Robert Koch demonstrated that in some people disease germs did not produce disease but continued to be harboured in the body, allowing those individuals to remain sources of infection.[52] The most famous was so-called 'Typhoid Mary', Mary Mallon, who worked as a cook and was alleged to have caused outbreaks of typhoid fever for many years in cities on the East Coast of America.[53] She was deemed responsible for an outbreak of the disease in New York in 1907 and was held in an isolation clinic for three years. She broke the terms of her release and worked again as a cook in 1915, starting another outbreak. This time she was arrested by the police and quarantined in an isolation clinic until her death in 1938. Her story, retold many times, came to symbolise the hidden dangers of infection in the modern world and the difficult balance in sanitary policies between the health of communities and the freedom of individuals.

In the 1860s, pulmonary tuberculosis, also known as consumption and phthisis, was the largest single cause of death in Europe and North America, accounting for one in seven of all deaths and one in four of adult deaths. Doctors and the public understood the disease to 'run in' families. However, it was not communicable generation-to-generation like syphilis; rather individuals inherited a vulnerable constitution, often termed a tubercular diathesis: William Farr, for example, classified it with cancer as a constitutional disease, not as zymotic. While the hereditary view prevailed in North America and Northern Europe, in Mediterranean countries it was seen as contagious, albeit contingently so. Support for this view came in the 1860s from veterinary laboratories in France, when Jean Antoine Villemin produced the disease by inoculating susceptible animals with tuberculous tissue. However, critics claimed that he had only produced 'artificial tuberculosis' and that it was difficult to see how deep-seated tubercular matter could find its way from body to body. The idea that tuberculosis was contagious was opposed by doctors at consumption hospitals, who reported that their staff rarely if ever succumbed to the disease despite constant exposure.[54] They pointed out that tuberculosis was quite unlike zymotic diseases, which were specific and ran a predictable course; pulmonary tuberculosis was one of the most variable of all diseases, liable to wax and wane. Also, epidemiological data showed that its mortality was constant month-to-month, which suggested that internal rather than external forces were the exciting cause.[55]

Robert Koch's famous paper 'On the aetiology of tubercular disease', in which he announced the identification of the tubercle bacillus, was published in April 1882. Acknowledging Koch's 'revolutionary' work as evidence that tubercular diseases were communicable, some doctors believed that there was no precedent for 'so sudden and complete casting aside of tradition'. Most physicians seem to have accepted the 'reality' of Koch's bacillus, but synthesised its properties with existing ideas on pathogenesis and their clinical experience. The commonest way of harmonising the *bacillus* with 'the fact of heredity' was to suppose that 'physico-chemical changes must precede botanical aggression' – a new form of contingent contagionism expressed metaphorically as 'seed and soil': the human 'soil' having to be susceptible to the 'seed' of disease.[56] Koch was also the key figure in a second radical change, the development of bacteriological laboratories, and the use of microscopy, culturing and animal experimentation to identify the microorganisms that were the essential cause of specific diseases.

## Viruses, vaccines and antibiotics, 1900–60

Researchers in the new bacteriological laboratories, principally in Germany and France, identified the microorganisms of many contagious and infectious diseases between the mid-1870s and the 1900s. The great age of bacteriological discovery and the establishment of germ theories of disease had begun with the publication of Robert Koch's work on anthrax in 1876 and new discoveries came regularly. However, the germs of a number of diseases were not readily found, most notably Farr's 'contagious' trio of smallpox, syphilis and rabies. Syphilis joined the ranks of bacterial infections in 1905, but the germs of smallpox and rabies, while manipulable in the laboratory in animal experiments, could not be cultured or seen by microscopy. In the twentieth century, the germs of these diseases were found to be ultra-microscopical and termed viruses, giving an old word a new meaning.

By the early decades of the twentieth century, in Europe at least, major epidemics of contagious and infectious diseases were in abeyance. The diseases with the highest levels of morbidity and mortality were endemic: measles, chickenpox, diphtheria, scarlet fever and mumps in children, and pulmonary tuberculosis and pneumonia in adults. Mortality rates had fallen steeply since the mid-nineteenth century due to improved standards of living and hygiene, along with sanitary improvements and medical advances.[57] However, the sense that contagious and infectious diseases were being tamed was challenged in the early twentieth century by new viral epidemics: first, by the emergence of poliomyelitis epidemics in children and then in 1918–19 by the Great Influenza Pandemic.[58]

While not a new disease, the incidence and severity of poliomyelitis grew from the 1890s and peaked with a major epidemic on the East Coast of the United States in 1916.[59] Children were most susceptible and tragedy struck in two ways – untimely deaths and permanent crippling of limbs and spine. In 1908 Karl Landsteiner isolated a possible causal germ. The disease was catching, but exactly how it spread remained uncertain, which created the space for scientific speculation and heightened public fears. One of the achievements of bacteriologists in the late nineteenth century had been the specification of the bodily portals of entry of germs, but precisely how polio infected sufferers remained a mystery when it struck New York in 1916. Anxieties were heightened by the sporadic and seemingly random incidence of cases. People, air, water, food, flies and pets were all proposed as conduits of infection. Public and civic responses echoed earlier epidemics. Sufferers were isolated and minority communities blamed and shunned. Families fled the cities and public authorities undertook sanitary clean-ups and disinfection. Assaults on house flies and vermin were ordered. State and federal agencies organised educational campaigns on how to best to avoid infection, which, reflecting aetiological uncertainties, were general rather than specific, doing little for public confidence. Indeed, public health authorities were aware that polio and fear of polio were equal problems.

The 1918–19 influenza pandemic was of a different order epidemiologically. There were 50 million deaths worldwide.[60] There was seemingly no mystery about what it was and how it spread. Following a previous influenza epidemic in 1890–91, Richard Pfeiffer had isolated the *Bacillus influenza* and its symptoms of chills, fever, headache, cough and fatigue were well recognised, as was the tendency, unlike with the common cold with which it was often confused, to develop secondary pneumonia.[61] Influenza was understood to spread from person to person by infected droplets in coughs and sneezes, and contracted from surfaces contaminated with nasal mucus and phlegm. Medical advice was to stay away from overcrowded, ill-ventilated and dirty places, and avoid mental anxiety. Trains, trams and buses were seen to be dangerous places and cinemas were closed. Those who developed symptoms were instructed to stay indoors, maintain personal hygiene and go to bed. Doctors had few if any direct means to attack the influenza germ; hence, they advised a good diet to strengthen the body's powers of resistance and the avoidance of alcohol and other vices.[62]

In the 1920s and in response to the epidemics of polio and influenza, medical research laboratories across the Western world turned their attentions to these diseases, with both proving to be due to viruses, not bacteria. One goal was to try and develop vaccines, not least because one of the medical successes of the Great War was claimed to be the protection given to soldiers by anti-typhoid vaccine. With polio, a key goal was to determine how the infection spread to better direct preventive and control policies. Research in the United States favoured the virus spreading through the air by inhalation in the nasal cavity, placing it close to the brain for its neurological effects. Preventive advice was framed accordingly and only changed in the late 1940s when, after the virus became easier to handle in the laboratory, the overwhelming importance of the gut as a portal of entry was shown. Research also showed that until the later nineteenth century, infection had been near universal in babies and infants. This had produced only sub-clinical illness and, crucially, the development of immunity. Paradoxically, improved water and food hygiene brought less exposure to a virus that was ubiquitous, meaning that fewer babies and infants had acquired immunity from low level exposure, which made them susceptible to serious paralytic illness or death when older. The moral of this story for agencies seeking to manage communicable diseases was the importance of 'soil' as well as 'seed', and that the best hope of controlling polio was to replace natural acquired immunity with vaccine-induced, artificial immunity.

The first widely used polio vaccine was developed by Jonas Salk, using a killed virus as the active agent, but it was displaced within a few years by an attenuated, live virus vaccine, developed by a team led by Albert Sabin.[63] The story of the rivalry between these two scientists has been told many times, but its significance for the history of communicable diseases was the ultimate success of vaccines in controlling polio. This achievement convinced the public and health policy-makers that vaccines were the most effective way to combat communicable diseases. Indeed, vaccines became the basis for global eradication campaigns, achieved with smallpox in 1977, planned for polio in the 1990s, with near success in the 2010s, and rinderpest in 2011.

The status of the *Bacillus influenza* was challenged throughout the 1920s, and a specific viral cause was not confirmed until 1933.[64] It was at this time that culturing viruses in living tissue became routine, a factor that together with the development of electron microscopy, allowed viruses to be 'seen' for the first time. The influenza virus was not as tractable as others, as attempts to produce a vaccine proved disappointing. Parallel work on the virus of the common cold was equally unproductive, but researchers were able to demonstrate that both diseases were principally communicable by droplet infection. Such work was the basis for public health education, most famously in the Second World War with campaigns to spread the message that 'Coughs and Sneezes Spread Diseases'. The other great symbol of contagion in the 1930s and 1940s was the fungal infection athlete's foot, which was another disease of hygiene, readily caught in swimming baths and showers.[65]

There were also attempts in the 1930s to revive the research programme of the German bacteriologist Paul Ehrlich, who in the 1900s had searched for so-called 'Magic Bullets' – chemical drugs that would selectively kill disease-germs leaving the infected person's cells untouched. In 1907, he produced the first such drug, Salvarsan, which was effective against the bacillus causing syphilis. However, being an arsenical compound it was toxic and its use carried considerable risks. A second generation of 'Magic Bullets' was introduced in the mid-1930s with sulphonamide drugs, which killed the streptococci that caused septicaemia, childbed fever, and erysipelas, as well as the gonococcus of gonorrhoea. The third generation of 'Magic Bullets' were antibiotics, which brought about a revolution in the treatment of communicable diseases.[66] The wonder drug of the early 1940s was penicillin, which had a wide spectrum of activity against bacteria, and by the end of the decade it had been joined by many others: aureomycin, streptomycin, chloramphenicol, erythromycin and tetracycline.[67]

Antibiotics were the medical sensation of the middle decades of the twentieth century and radically changed how doctors and the public thought about communicable diseases. Apart from the limited use of sulphonamides, clinicians had previously only been able to offer symptomatic and palliative treatments. To the public at least, with antibiotics it seemed that doctors had a cure for every infection. This view was encouraged by doctors' liberal prescription of the drugs, even though they knew that they were ineffective against viral diseases and that antibiotic resistance was an issue.[68] Nonetheless, communicable diseases were not the threat that they had been. Diseases like syphilis lost much of their menace, if not their moral stigma, when they could be cured by a short course of penicillin. The decline of pulmonary tuberculosis was accelerated with a combination of three antibiotic drugs, leading to the closure of sanatoria, which had symbolised its contagiousness.[69] The insurance against infection offered by antibiotics emboldened surgeons to develop more radical and reconstructive procedures, most famously total hip replacement, which relied upon antibiotics and strict aseptic practices.[70] Attempts to develop antiviral drugs were less successful, although effective vaccines against many common bacterial and viral diseases were developed. These have had a major impact on childhood infections in first world countries; the run of childhood infections (measles, mumps, whooping cough, rubella, diphtheria) that had previously been a rite of passage to adolescence have been reduced to very low levels of incidence and severity. However, such diseases remain serious in third world countries.

## Emerging and re-emerging infections

In 1981, Richard Krause warned against medicine lowering its guard against communicable diseases. He argued that genetic variation and ecological change would continually alter the balance of advantage between humans and bacterial, viral and other pathogens.[71] The term 'emerging infection' for these diseases did not become widely used for another decade when it was popularised by Joshua Lederberg, but already in 1981 Krause warned of new infections in tropical Africa: Lassa fever (1969), Marburg haemorrhagic fever (1976), and Ebola (1977), and at home Legionnaires' disease (1976).[72] He wrote at the exact moment when HIV-AIDS was being recognised. It was not long after that the term 're-emerging infections' was coined, to refer to diseases that had been well controlled medically, but were again increasing in prevalence. The best known was pulmonary tuberculosis, which had a growing incidence in first world countries due to multi-drug-resistant strains of the bacillus, immigration from countries with high prevalence, new types of immuno-compromised patients, and poverty in marginalised groups.[73] In third world countries, in addition to drug resistance and poverty, urbanisation, industrialisation and greater exposure in insanitary conditions were the important factors.[74]

In the last quarter of the twentieth century it was HIV-AIDS that raised the spectre of contagion and changed the medical and public landscape with regard to communicable diseases.[75] The syndrome was defined as a disease entity – Acquired Immune Deficiency Syndrome (AIDS) in 1982 and a causative retrovirus was identified in 1983. There were rival claims about the identity and character of the virus, before scientists agreed upon an organism styled the human immunodeficiency virus (HIV) in 1986. Scientists demonstrated that the virus was only transmitted in body fluids and needed to enter the blood stream directly. However, there remained many unknowns: which body fluids; where and how easily could it enter; what objects could carry it; and when were people contagious? Public concern about the communicability of AIDS first surfaced in the summer of 1983 and quickly deepened. The first event, in what became a panic, was when the wife of a haemophiliac man was infected, prompting an editorial in the *Journal of the American Medical Association* to conjecture that 'routine household contact' might spread AIDS.[76] Then, the US conservative lobby group the Moral Majority published a report under the title 'Homosexual Disease Threatens American Families', with a photograph of a family wearing surgical facemasks, illustrating fears of disease and moral contagion.[77] As in earlier epidemics, the groups associated with the syndrome were stigmatised and ostracised, particularly homosexual men, with AIDS labelled the 'Gay Plague'.

Government public health responses in Western countries began in earnest in 1987. Major educational campaigns tried to normalise HIV-AIDS, as the disease was now called, to be seen as a communicable disease, albeit spread in quite unfamiliar ways.[78] Responses to the early HIV-AIDS epidemic set new standards in public health education in terms of trying to teach the complexities of disease transmission and in the frankness of advice, most clearly in the messages about 'Safe Sex' and 'Clean Needles'. HIV-AIDS mortality in Western countries fell after the mid-1990s, reflecting safer behaviours and new treatments, though the numbers of people who were HIV-positive continued to rise. Nonetheless, this was regarded as a medical and public health success story, not least against the dire predictions of mortality and morbidity of the mid-1980s.

Awareness of the scale of the incidence of HIV-AIDS in the rest of the world was slower to develop and responses much weaker.[79] The World Health Organization started its Global Program for AIDS in 1987, with epidemiological studies soon revealing that over 85 per cent of infections and deaths in the world were in sub-Saharan Africa. It was clear that the dynamics of the epidemic were different there. The principal means of transmission was heterosexual intercourse and most of the countries affected had neither the resources nor expertise to respond at anywhere near the level required. On every Continent and in every country, cultural, religious and political factors also shaped preventive and treatment strategies.[80] Africa was painted as the AIDS continent in another way. Epidemiological and genetic evidence pointed to its origins there, when the retrovirus jumped from chimpanzees to humans, perhaps as early as the 1920s.[81]

## Conclusion

This review of the history of contagion has revealed both continuities and discontinuities in medical ideas and practices from the Great Pox in the fifteenth century to Ebola in the twenty-first. In the fifteenth century, the poxes were unusual in being regarded as directly 'catching' person to person. This was in marked contrast to most diseases, which were seen to have internal origins, or those which were acquired or 'caught' at a distance from 'epidemic influences'. Over time, and particularly from the late nineteenth century, not only were a greater number of contagious and infectious diseases recognised, but their specific modes of transmission were also delineated and causative pathogens identified. Knowledge changed practice, with doctors and public health agencies able to use new tools of prevention, control and cure: vaccines, antiseptics, antisera, antitoxins, and above all antibiotics. These developments had an impact beyond communicable diseases, establishing the hegemony of the ontological conception of disease across medicine and encouraging researchers to look for the contagia of other types of disease, for example viral causes of cancer. New knowledge and new technologies gave doctors, public health officials and state agencies more powerful means to prevent, control and treat communicable diseases. Measures could be targeted at particular diseases and controversies over when and how to adopt quarantines largely disappeared, as policies were focused on individuals and their germs rather than on localities and communities. However, by the end of the twentieth century, there were fears that the predominance of reductionist approaches had been too great and that ecological understandings needed to be fostered.

Continuities with regard to contagion can be found principally in public and government responses to communicable diseases. As the panics in 2014 about the possible spread of Ebola to Europe and America demonstrate, fear of contagion remains potent. Fear of this disease was due, in part, to the realisation that medicine did not have a cure beyond the reliance on 'old' control methods of isolation and disinfection. Distrust of expert opinion has also been longstanding, renewed by erroneous reassurances in the 1990s about the non-contagiousness of bovine spongiform encephalopathy (BSE) and mixed messages about possible pandemics of bird flu, swine flu and severe acute respiratory syndrome (SARS). Tensions and conflicts between the public, the medical profession and state authorities persist in the twenty-first century over the extent to which the protection of population health legitimates the restriction of the freedoms of individuals and groups. Finally, no communicable disease has ever been experienced solely as physical disease. Those affected have always sought meanings for their individual suffering and the fate of their community. In the fifteenth century, solace and meaning were principally provided by religion. In subsequent centuries individuals and communities have also drawn increasingly on other cultural resources: political, social, and, of course, on modern biomedical science.

## Notes

1 M. Pelling, 'Contagion/germ theory/specificity', in W. F. Bynum and R. Porter (eds), *Companion Encyclopaedia of the History of Medicine, Volume 1*, London: Routledge, 1993, pp. 309–34.

2 V. Nutton, 'Did the Greeks have a word for it?', in D. Wujastyk and L. Conrad (eds), *Contagion: Perspectives from Pre-Modern Societies*, Aldershot: Ashgate, 2000, pp. 137–63; V. Nutton, 'The seeds of disease: an explanation of contagion and infection from the Greeks to the Renaissance', *Medical History* 27 (1983), pp. 1–34.

3 See Chapter 2 on humours and humoral theory by Jim Hankinson.

4 O. Temkin, 'Health and disease', in *The Double Face of Janus and Other Essays in the History of Medicine*, Baltimore, MD: Johns Hopkins University Press, 1977, pp. 395–407.

5 N. Pemberton and M. Worboys, *Rabies in Britain: Dogs, Disease and Culture, 1830–2000*, Basingstoke: Palgrave, 2012.

6 B. Wasi and M. Murphy, *Rabid: A Cultural History of the World's Most Diabolical Virus*, New York: Viking, 2012.

7 P. Baldwin, *Contagion and the State in Europe, 1830–1930*, Cambridge: Cambridge University Press, 1999; M. Harrison, *Contagion: How Commerce Spread Disease*, New Haven: Yale University

Press, 2012. Also see: B. H. Lerner, *Contagion and Confinement: Controlling Tuberculosis along the Skid Road*, Baltimore: Johns Hopkins University Press, 1998; A. Bashford and C. Hooker (eds), *Contagion: Historical and Cultural Studies*, London: Routledge, 2001; P. Wald, *Contagious: Cultures, Carriers, and the Outbreak Narrative*, Durham, NC: Duke University Press, 2008.

8  E. Ackerknecht, 'Anticontagionism between 1821 and 1867', *Bulletin of the History of Medicine* 22 (1948), pp. 562–93.

9  Harrison, *Contagion*, passim.

10  P. Mitchell, *Contagious Metaphor*, London: Bloomsbury, 2012.

11  International Movie Database (IMDb), 'Contagion' (http://www.imdb.com/title/tt1598778/ Accessed 22 September 2014); H. Swartout, 'How fear and ignorance is helping Ebola spread' *Time*, 26 Aug 2014 (http://time.com/3181835/ebola-virus-congo-west-africa/ Accessed 19 September 2014).

12  See Chapter 10 in this volume by Sam Cohn.

13  J. Arrizabalaga, J. Henderson and R. French, *The Great Pox: The French Disease in Renaissance Europe*, New Haven: Yale University Press, 1997.

14  Ibid., pp. 1–19

15  V. Nutton, 'The reception of Fracastoro's theory of contagion: The seed that fell among thorns?' *Osiris* 6 (1990), pp. 196–23.

16  Ibid., pp. 196–8.

17  J. Henderson, 'Historians and plagues in pre-industrial Italy over the longue durée', *History and Philosophy of the Life Sciences* 25 (2003), pp. 481–99.

18  B. T. Boehrer, 'Early modern syphilis', *Journal of the History of Sexuality* 1 (1990), pp. 197–214.

19  C. Lawrence, 'The history and historiography of surgery', in C. Lawrence (ed.), *Medical Theory, Surgical Practice*, New York: Routledge, 1992, pp. 1–47.

20  Arrizabalaga, *The Great Pox*, pp. 152–68 and 173–5.

21  Ibid., pp. 145–233 and 278–82.

22  A. G. Carmichael and A. M. Silverstein, 'Smallpox in Europe before the seventeenth century: Virulent killer or benign disease?', *Journal of the History of Medicine and Allied Sciences* 42 (1987), pp. 147–68.

23  J. Landers, *Death and the Metropolis: Studies in the Demographic History of London: 1670–1830*, Cambridge: Cambridge University Press, 1993.

24  D. R. Hopkins, *Princes and Peasants: Smallpox in History*, Chicago: University of Chicago Press, 1983.

25  G. Williams, *The Angel of Death: The Story of Smallpox*, Basingstoke, Palgrave, 2011.

26  S. Bhattacharya, M. Harrison and M. Worboys, *Fractured States: Smallpox, Public Health and Vaccination Policy in British India, 1800–1947*, New Delhi: Orient Longman, 2005.

27  Pelling, 'Contagion/germ theory/specificity', pp. 320–8.

28  S. Finger, *The Contagious City: The Politics of Public Health in Early Philadelphia*, Ithaca, NY: Cornell University Press, 2012.

29  For further discussion of the history of yellow fever, see the chapter in this volume by Monica Garcia.

30  M. Pelling, *Cholera, Fever and English Medicine, 1825–1865*, Oxford: Oxford University Press, 1976, pp. 18–22, 295–310, 157. M. S. Pernick, 'Contagion and culture', *American Literary History*, 14 (2002), pp. 858–65

31  C. Hamlin, *Cholera: The Biography*, Oxford: Oxford University Press, 2009.

32  C. Rosenberg, *The Cholera Years: The United States in the Years 1832, 1849 and 1866*, Chicago: Chicago University Press, 1962, pp. 40–64.

33  M. Sigsworth and M. Worboys, 'The public's view of public health in mid-Victorian Britain', *Urban History*, 21 (1994), pp. 237–50.

34  Rosenberg, *Cholera Years*, pp. 101–33.

35  J. Snow, *On the Mode of the Communication of Cholera*, London: John Churchill, 1849.

36  S. J. Snow, 'Commentary: Sutherland, Snow and water: The transmission of cholera in the nineteenth century', *International Journal of Epidemiology* 31 (2002), pp. 908–11.

37  J. Eyler, *Victorian Social Medicine: The Ideas and Methods of William Farr*, Baltimore, MD: Johns Hopkins University Press, 1979; W. H. Brock, *Justus von Liebig: The Chemical Gatekeeper*, Cambridge: Cambridge University Press, 2002, 183–214.

38  C. Hamlin, *More than Hot: A Short History of Fever*, Baltimore, MD: Johns Hopkins University Press, 2014.
39  K. Maglen, *The English System: Quarantine, Immigration and the Making of a Port Sanitary Zone*, Manchester: Manchester University Press, 2014.
40  J. R. Walkowitz, *Prostitution and Victorian Society: Women, Class, and the State*, Cambridge: Cambridge University Press, 1982; P. Levine, *Prostitution, Race and Politics: Policing Venereal Disease in the British Empire*, London: Routledge, 2003.
41  Walkowitz, *Prostitution*, pp. 184–214.
42  A. M. Silverstein and C. Ruggere, 'Dr. Arthur Conan Doyle and the case of congenital syphilis', *Perspectives in Biology and Medicine* 49 (2006), pp. 209–21.
43  Ministry of Agriculture Fisheries and Food, *Animal Health: A Centenary, 1865–1965*, London: HMSO, 1965; A. Hardy, 'Pioneers in the Victorian provinces: Veterinarians, public health and the urban animal economy', *Urban History* 29 (2002), pp. 372–87.
44  M. Worboys, *Spreading Germs: Disease Theories and Medical Practice in Britain, 1865–1900*, Cambridge: Cambridge University Press, 2000, pp. 51–60.
45  Ibid., pp. 73–90.
46  T. Romano, *Making Medicine Scientific: John Burdon Sanderson and the Culture of Victorian Science*, Baltimore, MD: John Hopkins University Press, 2002.
47  N. Tomes, *The Gospel of Germs: Men, Women, and the Microbe in American Life*, Cambridge, MA: Harvard University Press, 1999.
48  Worboys, *Spreading Germs*, pp. 124–49, 234–40.
49  G. M. Ayers, *England's First State Hospitals and the Metropolitan Asylums Board 1867–1930*, London: Wellcome Institute of the History of Medicine, 1971.
50  A. Hardy, *The Epidemic Streets: Infectious Diseases and the Rise of Preventive Medicine, 1856–1900*, Oxford: Oxford University Press, 1993.
51  J. V. Pickstone, *Medicine and Industrial Society*, Manchester: Manchester University Press, 1985, pp. 156–83.
52  C. Gradmann, 'Robert Koch and the invention of the carrier state: Tropical medicine, veterinary infections and epidemiology around 1900', *Studies in History and Philosophy of Science* Part C 41 (2010), pp. 232–40.
53  J. W. Leavitt, *Typhoid Mary: Captive to the Public's Health*, Boston, MA: Beacon Press, 1996.
54  Worboys, *Spreading Germs*, pp. 193–203.
55  Ibid., pp. 221–5.
56  H. Neale, 'The germ-theory of phthisis', *British Medical Journal* I (1885), p. 897.
57  S. Szreter, 'The importance of social intervention in Britain's mortality decline c.1850–1914: A re-interpretation of the role of public health', *Social History of Medicine* 1 (1988), pp. 1–37.
58  G. Williams, *Paralysed with Fear: The Story of Polio*, Basingstoke: Palgrave, 2013; H. Phillips and D. Killingray (eds), *The Spanish Influenza Pandemic of 1918–19*, London: Routledge, 2003.
59  N. Rogers, *Dirt and Disease: Polio before FDR*, New Brunswick, NJ: Rutgers University Press, 1992, pp. 10–11.
60  N. P. Johnson and J. Mueller, 'Updating the accounts: Global mortality of the 1918–1920 "Spanish" influenza pandemic', *Bulletin of the History of Medicine* 76 (2002), pp. 105–15.
61  M. Honigsbaum, *Living with Enza: The Forgotten Story of the Britain and the Great Flu Pandemic of 1918*, Basingstoke: Macmillan, 2009, pp. 3–17.
62  H. Phillips, 'The recent wave of "Spanish" flu historiography', *Social History of Medicine* 2014, doi: 10.1093/shm/hku066.
63  A. M. Brandt, 'Polio, politics, publicity, and duplicity: Ethical aspects in the development of the Salk Vaccine', *International Journal of Health Services* 8 (1978), pp. 257–70; H. Marks, 'The 1954 Salk poliomyelitis vaccine field trial', *Clinical Trials* 8 (2011), pp. 224–34; Williams, *Paralysed with Fear*, pp. 192–247.
64  M. Bresalier, 'Uses of a pandemic: Forging the identities of influenza and virus research in interwar Britain', *Social History of Medicine* 25 (2012), pp. 400–24.
65  A. Homei and M. Worboys, *Fungal Disease in Britain and the United States: Mycoses and Modernity, 1850–2000*, Basingstoke: Palgrave Macmillan, 2013, pp. 43–66.
66  R. Bud, *Penicillin: Triumph and Tragedy*, Oxford: Oxford University Press, 2006.

67 For further discussion of developments in medical bacteriology, see the chapter by Christoph Gradmann.
68 J. T. Macfarlane and M. Worboys, 'The changing management of acute bronchitis in Britain, 1940–1970: The impact of antibiotics', *Medical History* 52 (2008), pp. 47–72.
69 L. Bryder, *Below the Magic Mountain: The Social History of Tuberculosis in Twentieth Century Britain*, Oxford: Clarendon Press, 1988, pp. 253–7.
70 J. Anderson, F. Neary and J. V. Pickstone, *Surgeons, Manufacturers and Patients: A Transatlantic History of Total Hip Replacement*, Basingstoke: Palgrave, 2007.
71 R. M. Krause, *The Restless Tide: The Persistent Challenge of the Microbial World*, Washington, DC: National Foundation for Infectious Diseases, 1981.
72 J. Lederberg, R. E. Shope and S. C. Oaks (eds), *Emerging Infections: Microbial Threats to the United States*, Washington, DC: National Academy Press, 1992.
73 P. Farmer, 'Social inequalities and emerging infectious diseases', *Emerging Infectious Diseases* 2 (1996), pp. 59–69.
74 M. Gandy and A. Zumla (eds), *The Return of the White Plague: Global Poverty and the 'New' Tuberculosis*, London: Verso, 2003.
75 V. A. Harden, *AIDS at 30: A History*, Dulles, VA: Potomac Books, 2013.
76 A. S. Fauci, 'The acquired immune deficiency syndrome: The ever-broadening clinical spectrum', *Journal of the American Medical Association* 249 (1983), pp. 2375–6.
77 '"Moral Majority Report": HIV and AIDS 30 Years Ago', (https://hivaids.omeka.net/items/show/9 Accessed 15 January 2015).
78 P. Wallis, 'Debating a duty to treat: AIDS and the professional ethics of American medicine', *Bulletin of the History of Medicine* 85 (2011), 620–49.
79 J. Engel, *The Epidemic: A Global History of AIDS*, London: HarperCollins, 2006.
80 J. R. Youde, *AIDS, South Africa, and the Politics of Knowledge*, Farnham: Ashgate, 2013.
81 P. M. Sharp and B. H. Hahn, 'Origins of HIV and the AIDS pandemic', *Cold Spring Harbour Perspectives on Medicine*, 2011, 1(1), a006841, (doi: 10.1101/cshperspect.a006841 Accessed 21 January 2015). For further discussion of HIV-AIDS, see the chapter by Richard McKay.

## Select bibliography

P. Baldwin, *Contagion and the State in Europe, 1830–1930*, Cambridge: Cambridge University Press, 1999.

A. Bashford and C. Hooker (eds), *Contagion: Historical and Cultural Studies*, London: Routledge, 2001.

A. Hardy, *The Epidemic Streets: Infectious Diseases and the Rise of Preventive Medicine, 1856–1900*, Oxford: Oxford University Press, 1993.

M. Harrison, *Contagion: How Commerce Spread Disease*, New Haven: Yale University Press, 2012.

M. Pelling, 'Contagion/germ theory/specificity', in W. F. Bynum and R. Porter (eds), *Companion Encyclopaedia of the History of Medicine, Volume 1*, London: Routledge, 1993, pp. 309–34.

N. Tomes, *The Gospel of Germs: Men, Women, and the Microbe in American Life*, Cambridge, MA: Harvard University Press, 1999.

P. Wald, *Contagious: Cultures, Carriers, and the Outbreak Narrative*, Durham, NC: Duke University Press, 2008.

M. Worboys, *Spreading Germs: Disease Theories and Medical Practice*, Cambridge: Cambridge University Press, 2000.

D. Wujastyk and L. Conrad (eds), *Contagion: Perspectives from Pre-Modern Societies*, Aldershot: Ashgate, 2000.

# 6

# EMOTIONS AND MENTAL ILLNESS

*Elena Carrera*

The concept of mental illness began to gain prominence in Paris during the 1790s, when Philippe Pinel was working as physician at the hospice of Bicêtre, a state-funded institution which housed 4,000 inmates, including criminals, syphilitics, elderly paupers, and some 200 madmen. In 1794 Pinel presented a report to the Société d'Histoire Naturelle, arguing that some forms of mental alienation were curable and that it was essential for physicians to determine the cause of each patient's illness by making careful notes based on observations of their behaviour and individual interviews.[1] In promoting the use of close observation as a diagnostic method, he was following the authoritative example of Hippocrates, as he would note in the introduction to the *Traité Medico-Philosophique sur l'Aliénation mentale* (1801).[2] The *Traité* was soon translated into other European languages (German, 1801; Spanish, 1804; English, 1806; Italian, 1830), though Pinel's introduction was not always included in the translations.[3]

The main difference between Pinel's approach and that of the Hippocratic authors and their followers throughout the centuries is that, while they had sought to cure mental disturbance through changes in lifestyle and medical interventions that worked on the body, Pinel was more concerned with the forms of mental alienation that did not have obvious physical causes and thus required non-physical treatment. In a footnote to the introduction to the *Traité*, Pinel laments that Galen did not devote himself to the study of mental alienation, given the perspicacity he demonstrated in his diagnosis of the 'hidden moral suffering' (*affection morale cachée*) of the melancholic lady who was secretly in love with the actor Pylades.[4] Unable to explain her insomnia and agitation, Galen interviewed her slave and found that her suffering was caused by sorrow, and then observed that her pulse and the colour and expression of her face changed when the name of the actor was mentioned. This is an illustrative example of the observation and interview techniques that Pinel sought to introduce first at the Bicêtre, and then at the hospice of the Salpêtrière, where he worked as chief physician between 1795 and 1826. When he arrived there in 1795, and found some 7,000 ailing female paupers with no hope of being discharged, he set up an infirmary for those he identified as suffering from intermittent or periodical insanity.[5] Under his leadership, the Salpêtrière became a teaching and research centre, to which inquisitive medical students like Jean-Étienne Esquirol were attracted.[6]

Pinel went beyond his predecessors and contemporaries in raising public awareness about the role of emotional factors in mental health, and in advocating the understanding of the passions as part of the training of the medical staff offering state-funded therapeutic interventions.[7] In 1805, Esquirol published his medical thesis on the passions as causes, symptoms and cures of insanity, noting that Pinel was not alone in regarding the passions as the most common cause of insanity:

> Few authors have studied the relationship of mental alienation to the passions. Crichton offers precise ideas on the origins and development of the passions and their effect on the body. Professor Pinel agrees with him, regarding the passions as the most frequent cause of derangement of our intellectual faculties.[8]

At the time when he wrote and defended his thesis, Esquirol was running his own *maison de santé*, which has been seen as the first private psychiatric hospital in Paris.[9] In 1811 he became the administrator of the insane patients division of the Salpêtrière and in 1817 he began to lecture on mental illness to medical students.[10] Eight years later he was appointed chief physician of the large public asylum Maison Royale Des Aliénés de Charenton, in the outskirts of Paris. His treatise on madness, *De la folie* (1816), was republished in a larger volume, *Maladies mentales* (1838), which also included reports on the medico-legal issues related to mental alienation, statistical and hygienic reports, and 27 engravings.[11]

The four decades between the publication of Pinel's *Traité* (1800) and Esquirol's death in 1840 was a key period within the history of 'psychiatry' (a term first used by Johann Christian Reil in 1808), marked by the departure from a Lockean understanding of delusion as an intellectual error. Many of the patients treated by Pinel and Esquirol displayed extreme passions, such as fear, anger or pride, and often seemed to have deluded ideas about their own identity or other people's intentions, which could be explained as being rooted in such extreme passions. These physicians' emphasis on using non-physical methods for counteracting the non-physical causes of insanity had clear precedents in the Hippocratic approach to the non-natural (non-organic) factors of health and disease, promoted in the regimens of health (self-help manuals), which had enjoyed considerable popularity since the late Middle Ages and which continued to be published well into the nineteenth century.[12] The use of contrary passions had been recommended since Hippocratic times, following the principle of *contraria contrariis curantur*.[13] However, in contrast with the primarily prescriptive nature of existing writings by physicians and theologians on the uses of passions, Pinel and Esquirol provided ample evidence of the practical results of such methods.

This chapter examines Pinel's and Esquirol's views on the passions as causes and cures of insanity in the wider context of ideas about the curability of the insane. It will show how these two French physicians moved away from the physiological models which had underpinned dominant medical approaches to the passions and to insanity well into the eighteenth century, and focused instead on the forms of insanity which were unrelated to physical disease. The chapter will first consider their contributions to our understanding of the passions as causes of insanity, and then explore their views on the use of the passions as cures.

## The curable insane in public institutions

Pinel helped to transform public hospices into therapeutic institutions by proving that a significant proportion of the paupers who were furiously insane or who were too fearful to even want to eat could regain their sanity and become citizens able to work. He became known for allegedly overseeing the removal of the chains that had been used to restrain the furious inmates of the Bicêtre (even though this was probably the initiative of the non-medical superintendent, Jean-Baptiste Pussin). A similar reform movement had been started in 1785 in Florence by the 26-year-old physician Vincenzio Chiarugi, who by 1788 had succeeded in banning the use of chains in the overcrowded hospice of Santa Dorotea, and in 1793–94, published a three-volume text on the diagnosis, classification and treatment of mental illness, *Della pazzia*, in which he argued that mental institutions should promote therapy, rather than simply being used as custodial places.[14]

There had been earlier examples of hospitals, such as those founded in late medieval and early-modern Iberia (in Valencia in 1409, Cordova in 1419, Zaragoza in 1425, Seville ca. 1436, Toledo in 1483, Valladolid in 1489 and Granada in 1527), in which mad people had been treated through a regimen of rest, work, appropriate diet and pharmaceutical remedies, and had often recovered.[15] Pinel would indeed refer to the example of the hospital in Zaragoza, in which, according to a report published in Paris in 1791, many insane patients were engaging in manual work, doing jobs in the kitchen, dealing with the laundry, gatekeeping and farming.[16] As he argued in the *Traité*, such work had proved effective in helping the Zaragoza patients attain a solid cure, and it was remarkable that the noblemen in that institution, who disdained manual work and refused to do it, were denying themselves the opportunity to be cured of their 'derangements and delusions'.[17] Manual work was one of the 'moral' methods that Pinel promoted at the Salpêtrière, as we read in the 1845 English edition of Esquirol's *De la folie* (1816):

> Pinel recommends that an Establishment for the insane should have a farm connected with it, on which the patients can labor. The cultivation of the garden has succeeded happily in the cure of some insane persons. At the Salpêtrière, the best results follow the manual labor of the women in that Hospital. They are assembled in a large working room, where some engage in sewing or knitting; while others perform the service of the house, and cultivate the garden. This precious resource is wanting in the treatment of the rich of either sex. An imperfect substitute is furnished, in walks, music, reading, assemblages, etc. The habit of idleness among the wealthy counterbalances all the other advantages which this class enjoy for obtaining a cure.[18]

Esquirol did encourage walks in the garden, in line with longstanding views about the beneficial effect of looking at greenery, which would inform the construction of asylum and hospital gardens in the nineteenth and twentieth centuries.[19] He also acknowledged that working the land had a wide range of therapeutic effects (for instance, dissipating the paralysis of the tongue of a 47-year-old insane soldier).[20] Nonetheless, in promoting the use of farms and gardens 'to make patients work' (*pour faire travailler les malades*), Pinel's and Esquirol's primary aim was to change the patients' 'habits of idleness'.[21] They extended the traditional Hippocratic and Galenic approaches to non-organic factors of health and disease by placing the emphasis on changing 'habits of mind, feeling and action' as a way of attaining a cure.[22] They used the term 'moral' (in opposition to 'physical') to refer to mental, emotional and behavioural 'habits' (Lat. *mores*), as well as to any other non-physical factors, such as domestic troubles, reversals of fortune or political events.[23]

To some extent, Pinel's and Esquirol's projects were part of a much wider reform movement, which was inspired by Enlightenment thinking in a number of European contexts. While they received state support in the specific context of post-Revolutionary Paris, they also benefited from the wide dissemination of ideas across national borders. In London, William Battie had been a driving force in the foundation in 1751 of St Luke's Hospital, a public asylum, and had argued in his *Treatise on Madness* (1754) that management and confinement were more effective means of curing madness than medicine.[24] In Florence, Chiarugi drew on more than 50 medical authors, from antiquity, Germany, Switzerland, France, England, Scotland and Italy, and recommended the old traditional Hippocratic method of using contrary passions for some conditions, such as melancholy, which could be cured by promoting and encouraging hope.[25] Besides, he systematically dissected the brains of his deceased mental patients in the hope of finding the organic causes of their insanity: 'even if a substantial lesion [was] not evidently perceptible' it might have been 'sufficient to disturb the functions of the brain'.[26] In contrast with Chiarugi's organicist approach, Pinel suggested that in the dissections of corpses of lunatics performed by himself and others there was no evidence of brain lesion or cranial malformation. Turning against the dominant views of eighteenth-century anatomists, who had shown through their dissection of the corpses of lunatics that insanity was related to abnormality and degeneration in the brain substance, he claimed that it was most commonly rooted in 'an extreme exaltation of the passions', and thus could not be understood without knowledge of the 'medical history of the passions'.[27]

Benjamin Rush, who had initiated the reform of the care of the mentally ill at the Pennsylvania hospital in 1792, continued to use physical methods such as bleeding and treatment with mercury. In his *Medical Inquiries*, published in Philadelphia in 1812, he defended the view that 'the cause of madness is seated primarily in the blood-vessels of the brain, and it depends upon the same kind of morbid and irregular actions that continues other arterial diseases'.[28] Four years later, in *De la folie*, Esquirol explained the relationship between the brain and thought by drawing an analogy to the relationship between a muscle and its movement: while the brain was necessary in the production of ideas, these did not have an ontological existence in the brain.[29] He insisted on the need to distinguish between mental illness and other conditions caused by brain lesions (such as 'chronic inflammation of the meninges', cerebral haemorrhage, 'tubercles, cancers, and softening of the brain'), suggesting that 'organic lesions of the encephalon and its envelopes' had been observed in the brain dissections of deceased patients 'whose insanity was complicated with paralysis'.[30] He also argued that 'the causes of mental alienation, do not always act directly upon the brain, but upon organs more or less removed therefrom', noting that insanity was often complicated with brain lesions, just as it might be 'complicated with affections of the lungs, of the heart and intestines, and of the skin'.[31]

## The passions as causes of insanity

The view that extreme emotions or prolonged moods could cause diseases in the body and the mind had been recognized by medical authors since Hippocratic times.

As Robert Burton noted in *The Anatomy of Melancholy* (1.2.3.4), Hippocrates had seen sorrow as 'the mother and daughter of melancholy, her epitome, symptom, and chief cause'. Indeed, in the Hippocratic *Aphorisms* (23.6), prolonged sorrow and fear are presented as both causes and symptoms of melancholy. Drawing on Hippocrates, Galen had noted the importance of 'mental activity' or 'affections of the mind' as one of the key factors of health and disease, alongside other non-organic factors such as the air breathed, the type and quantity of food and drink taken, and the balance between sleep and wakefulness, rest and exercise, and retention and evacuation.[32] As he suggests in the *Commentary on Epidemics* (VI.5, 484–6), the types of mental activity that could cause (mental or physical) illness were excessive anger, fear of death, worry, grief and shame.[33]

In the influential *Lilium medicinae* written ca. 1305 by the Montpellier physician Bernard of Gordon, emotions and diet were still seen as the main causes of melancholia and madness.[34] The regimen of health written around that time by Arnald of Villanova gave self-help advice related to all six non-organic factors of health, including an appropriate diet, regular sleep and the need to avoid passions such as anger and fear, which could have a harmful impact on both the body and the mind:

> The passions and accidents of the soul can change or alter the body in powerful ways, and have a notable impact on the functioning of the understanding. Therefore, harming passions must be eschewed carefully and diligently, particularly anger and sadness. Anger makes all bodily parts hot and swollen. They make the heart so hot that they darken and disturb the functioning of reason. Therefore we must avoid all situations in which we might get angry, with the exception of wrongdoings.
>
> Sadness, by contrast, cools down the body and dries it. Therefore it makes the sad person drier and thinner. It makes the heart shrink, and the spirits thicker and darker. It weakens the understanding, hinders perception, confuses judgement and destroys memory.[35]

Among the regimens of health that proliferated in sixteenth-century Europe, Andrew Boorde's *Breviary of Helthe* (1547) explains that the 'impediment' of a 'lunatike person' may have organic causes, 'by nature and kynde, and than it is uncurable', or be the effect of an excessive passion or intellectual exertion.[36]

Views on the interaction between mind and body promoted in the regimens of health are also found in learned medical books well into the early-modern period, when diet continued to be seen as having a crucial impact not only on the body, but also on the mind. We see this, for instance, in Girolamo Mercuriale's *Responsorum et consultationum medicinalium* (1587–1620) and *Medicina Practica* (1602), in which he drew on a range of ancient, medieval and Renaissance authors, such as Galen, Hippocrates, Aretaeus, Avicenna, Averroes, Jacques Fernel and Girolamo Cardano. Using the standard contemporary phrase 'melancholia hypocondriaca' to refer to a condition characterized by a mixture of physical and mental symptoms, like flatulence and inexplicable sadness, Mercuriale noted how it was often caused by disorderly lifestyle factors, such as excessive eating. In line with the still prevalent Hippocratic–Galenic theories of the mind–body connection, Mercuriale explained that when food was not fully digested, it would produce noxious vapours which would rise up to the heart and the brain, producing fear, sadness and other *perturbationes animi* in people who either had a melancholic disposition or were turned melancholic by external circumstances such as misfortune.[37] In *Medicina Practica* he defined 'mania' as 'mental alienation' (*mentis alienatio*) or continuous ecstasy, without fever or inflammation, which could also affect speech, the imagination or memory; he distinguished it from melancholia, which was generally believed to affect only the imagination.[38]

In the mid-eighteenth century, Battie went beyond the emphasis placed by George Cheyne on nervous distempers in his influential *The English Malady* (1733) to distinguish 12 more remote causes of madness, which could affect all sectors of the population. These included external factors (such as sun stroke, fracture or concussion of the skull, and the intake of poisons, medicines and vinous spirits) and internal physical changes (such as growth of cranial bone), as well as 'tumultuous passions' (like joy and anger), 'quieter passions' (like love, grief and despair), the 'unwearied attention of the mind to one object', and 'gluttony' and 'idleness'.[39]

The passions continued to be seen as causes of both physical and mental illness. In the two-volume *Inquiry into the Nature and Origin of Mental Derangement* published in 1798, the Scottish physician Alexander Crichton noted that the passions 'produce constant effects on our corporeal frame, and change the state of our health, sometimes occasioning dreadful distempers, sometimes freeing us from them' and also 'produce beneficial and injurious effects on the faculties of the mind, sometimes exalting them, sometimes occasioning temporary derangement, and permanent ruin'.[40] His claims were in line with the views promoted in the traditional self-help regimens of health, though he also took account of recent theories of the imagination as the site capable of producing fear-arousing images, based on past experiences of fear.[41]

Crichton also suggested that the delusions of people who believed they were kings, lords or bishops might be caused by pleasurable passions such as vanity.[42] Besides providing physiological explanations of the causal links between passions and disordered mental states, he also noted the impact on the mind of social factors such as poverty or other people's emotions (like contempt or hatred):

> the notions which produce this passion [vanity] are constantly present to the mind, *a great train of imaginary happiness* is produced, that increases until the person is, as it were, absolutely intoxicated. As often as these thoughts occur, the blood rushes with impetuosity to the head; the sentient principle is secreted in preternatural quantity, and the excitement is at last so often renewed, and increases to such a degree, as to occasion an impetuous and permanent delirium. But when the expectations, and high desires, which pride and vanity naturally suggest, are blasted, when these passions are assailed by poverty, neglect, contempt, and hatred, and are unequal to the contest, they now and then terminate in despair, or in settled melancholy.[43]

For Crichton, delirium could be caused by strong passions (such as anger, grief, pride and love), as well as by mental exertion, fevers, poisoning, and intoxication (from spirits, wine, aether and opium). He also distinguished between hereditary and non-hereditary factors that predisposed one to mental disorders, noting in particular the debilitating effect of 'poor diet, bad drink, scrophula, over-fatigue of the body, excess of venery, self-pollution, excessive haemorrhages, and excessive discharges'.[44] While endorsing the Hippocratic emphasis on the potential impact of diet or the passions on bodily and mental health, he also displayed an interest in pride and vanity, and in the detrimental effects of reversals of fortune and poverty, which had become widespread concerns in the aftermath of the French Revolution.

Pinel summarized Crichton's views in his introduction to the *Traité*, noting, for instance, that the harmful effects of excessive anger included syncope, convulsions, and sudden death, as well as temporary madness.[45] He also referred to a number of medical authors cited by Crichton to support the idea that terror could, under certain conditions, cause violent spasms, convulsions, epilepsy, catalepsy, mania and death.[46] Echoing Crichton's account, he explained that terror (*terreur*) was more sudden and intense than fear, and had its own specific physiological manifestations: accelerated heartbeats; spasmodic contraction of the arteries, near the skin; paleness; 'a sudden dilation of the large blood vessels and of the heart'; a sudden interruption of respiration (as the muscles of the larynx go into spasm); trembling of the body and legs; loss of movement in the arms.[47]

As a teacher of internal medicine, Pinel recognized that some mental disorders had physical causes, as we know from the case histories recorded by his assistants in 1802.[48] He also drew on his observations of female patients at the Salpêtrière to warn that women were particularly vulnerable to insanity during their menstruation and after giving birth, when their anger, if habitual, might develop into furious delirium, or lead to other conditions such as stupor or dementia.[49] He nonetheless rejected speculative explanations based on the prevailing physiological model of fluids or vapours circulating in the nerves. Instead, he taught precise observation skills, which were at the basis of his rigorous method of classification of diseases, and showed a genuine interest in both understanding and treating insane patients.

Pinel's emphasis on the patients' experience, rather than on explaining their condition with reference to medical theory, was already evident in the footnote he added to his 1898 translation of Crichton's chapter on delirium. Expressing his disagreement with the Scottish doctor's physiological explanatory model of delirium as being caused by alterations in the quantity and quality of nervous fluid, Pinel emphasized instead the causative role of 'vivid emotions'.[50] In another footnote, he criticized Crichton for having based one of his arguments on a single case history and described his more reliable scientific approach, which involved systematic observations of 200 patients suffering from periodic insanity, their classification using the standard categories of idiocy, mania and melancholia, and detailed reports of the events that precipitated their condition, the 'premonitory signs of periodic bouts', and their symptoms.[51] In the *Traité*, however, Pinel explicitly endorsed Crichton's view that physicians should look at the passions as phenomena with natural causes that could be inquired into; he urged his medical colleagues to consider the human passions as an object of study which belonged completely to the discipline of medicine, proposing that these should be studied without any reference to notions of morality or immorality.[52]

In the clinical notes that Pinel wrote about the insane patients he was able to interview, he referred to the personal circumstances and events that might explain their insanity. Thus in the notes related to the 265 women he treated personally at the Salpêtrière between 1802 and 1805 we find precise references to passions as causes of insanity. Terror was the cause of the insanity of Dorothée Ducastel, a widow aged 45, who was brought to the hospice by the police, and who had experienced great fear for 15 days after dreaming of her dead husband telling her about God's forgiveness and announcing to her that she would die within a year.[53] This diagnosis seems to have been straightforward, given that fear was still the widow's dominant passion and that her symptoms and the content of her dream fitted into an existing medical category: 'religious melancholy'. By contrast, the 26-year-old Adélaide Faucheux, whose main symptom was her fear of being poisoned, was diagnosed as suffering from melancholia caused by jealousy after giving birth. It was only by listening to these women's personal stories that Pinel could go beyond his observations of their behaviour to establish which passions were at the root of their condition.

The most common cause of the insanity of women who were 'cured' by Pinel was domestic trouble, which comprised a wide range of family misfortunes and dissensions. Cécile Gonin was a 44-year-old widow diagnosed with 'delirium and desire to die' after she had thrown herself into a well, following a 15-month period of insanity caused by her husband's death. Josèphe Perrone, a 30-year-old seamstress diagnosed with 'delirium with fury', had suffered a bout of insanity when her husband joined the army, leaving her six weeks pregnant, and then every time she menstruated while still breastfeeding. The 50-year-old widow Thérèse Chevallier had become insane after losing her cat. Augustine Filliot, aged 56, suffered 'delirium with fury' and 'religious mania' caused not only by grief, but also by the deprivation she experienced in the six months she lived in the fields and woods, following her husband's arrest during the Revolution. Marguerite Langlois, aged 41, suffered from amenorrhea for six months after being maltreated by her husband, while 21-year-old Agathe Richer was diagnosed with 'delirium with fury' after being robbed of 800 livres, having also been maltreated by her father, who opposed her marriage plans.[54]

When Pinel argued in the *Traité* that intense passions (*affections morales très-vives*) such as excessive and misplaced ambition, religious fanaticism, deep sorrows, or unhappy love affairs had a significant impact on mental health, he was able to support his claims with statistical analyses of the clinical evidence he had gathered at the Bicêtre: of the 113 male insane patients from whom he had been able to obtain accurate information at the Bicêtre, 34 had been led to insanity by domestic trouble, 24 by obstacles which had prevented them from marrying the person they loved, 30 by the events of the Revolution, and 25 by a fanatic zeal or terror about the afterlife.[55] His arguments were also illustrated by a number of tables, one of them showing the causes of insanity of patients reported to be 'cured' (*guéris*) at the Bicêtre in 1792. These included eight cases of 'periodical fury with delirium' caused by unrequited love (a gardener aged 45), terror (two soldiers, aged 22 and 24), excessive ambition (two soldiers, aged 24 and 30), loss of money (a tailor aged 36), jealousy (a ferryman aged 28) and sorrow (a ploughman aged 68); four cases of acute mania, caused by sorrow (two tanners, aged 25 and 46), terror (a hairdresser aged 56) and excessive ambition (a soldier aged 25); and two cases of melancholia caused by sorrow (a tailor aged 36) and loss of money (a merchant aged 46).[56]

In referring to passions as causes of insanity in his printed works, Pinel went beyond standard lists provided by pre-modern physicians to consider specific events from a socio-historical context, such as the Revolution, and culturally bound states of mind such as 'enthusiastic patriotism'.[57] In both versions of the *Traité* he recounts the case of a tailor who, during the most 'effervescent' period of the Revolution, was heard making some remarks about the condemnation of Louis XVI and, as a consequence, was seen as unpatriotic by people in his neighbourhood. He went back home trembling, was unable to sleep and lost his appetite and the will to work; he began to experience constant frights, and became so obsessed with the idea that he had been condemned to the guillotine that he was taken to the Bicêtre.[58] This case illustrates well that the causes of insanity were not located only in the body, and that social environment was believed to have a strong influence in shaping people's passions and aggravating their impact on mental health.

Esquirol's medical thesis (1805) offered further statistical evidence on the impact of the passions on mental health, based on Pinel's report on the Salpêtrière hospice and on his own experience at his private *maison de santé*. At the Salpêtrière, Pinel had identified 611 cases of melancholia and mania, of which only 165 were attributed to physical causes, and 142 cases of dementia and idiocy, of which 36 had clear physical causes. Esquirol himself had identified 66 cases of melancholia or mania among his private patients, of which only 19 were attributed to physical causes, and 15 cases of dementia or idiocy, of which 11 had clear physical causes. He also agreed with Pinel's view that when melancholia and mania were of recent onset and had not been treated through physical methods, they were potentially curable, and that these should be distinguished from other conditions like idiocy and dementia, which were characterized by 'total damage of the intellectual faculties' and were incurable.[59]

Nine years later, in *De la folie*, Esquirol published new statistics based on evidence obtained at the Salpêtrière in 1811–12 and at his private clinic. These showed, for instance, that the insanity of 466 of the Salpêtrière patients was attributed to physical causes (including 105 to heredity, 60 to old age, 60 to apoplexy, 52 to post-partum complications, 14 to head injuries, 13 to fevers, 11 to the mother's convulsions during gestation and 8 to syphilis), while in 323 of the patients it was attributed to 'moral' causes: 105 to domestic sorrows, 77 to reversals of fortune, 46 to unrequited love, 38 to fright, 18 to jealousy and 16 to anger.[60] Claiming that pride, fear, fright, ambition, reversals of fortune and 'domestic trouble' were the most frequent moral causes of insanity, Esquirol also noted that religious fanaticism, while a major cause of madness in the past, was no longer significant (he had only seen one such case among the 337 patients admitted in his private clinic).[61] He also mentioned that he had seen young women who had become insane out of shame and sorrow after being raped.[62] He further suggested that the factors which influenced the frequency, character, duration and treatment of insanity included not only climate, seasons, age, sex, temperament, profession and lifestyle, but also 'laws, civilization, morals, and the political condition of people'.[63]

While agreeing with Locke's remark that the insane are 'like those who lay down false principles, from which they reason very justly, although their consequences are erroneous', Esquirol also pointed out that the wrong beliefs of the insane were underpinned by a dominant passion:

> Among the insane, some are stricken with terror, believe themselves ruined, tremble lest they shall become the victims of a conspiracy, fear death. Others are happy and gay; think only of the good which they enjoy, or of the benefits which they can dispense. They feel persuaded that they are elevated to the greatest dignity.[64]

From this perspective, observation and listening were crucial methods in establishing what particular passion or passions dominated each patient's mind.

## The passions as cures for insanity

Pre-modern Hippocratic–Galenic physicians saw the brain as the mind's instrument, and thus believed that pharmaceutical remedies (such as hellebore or poppy) and surgical remedies (like bloodletting) could help to produce changes in the way the mental faculties worked. In *A Treatise on Madness*, Battie suggested that when the causes of madness were 'tumultuous and spasmodic passions, such as joy and anger', nothing should be done, unless the patient was in danger of dying, since these passions and their muscular effects would normally subside. If the patient's life was at risk, the remedies he recommended were first 'depletion and diminution of maniacal pressure' (usually through bloodletting and purgatives), and then giving the patient poppy, known to be an efficacious narcotic. If, despite these remedies, the patient continued to be engrossed in any particular passion for an unusually long period, Battie left it to the physician's discretion to determine 'how far it may be advisable or safe to stifle it by a contrary passion'. He nonetheless warned that the patient's safety could not be guaranteed if such method was used, because the effect of the new passion was unpredictable: 'it is almost impossible by general reasoning to foretell what will be the effect of fear substituted in the room of anger, or sorrow immediately succeeding to joy'.[65]

In the introduction to the first edition of the *Traité*, Pinel noted that the most suitable approach in seeking to 'correct' the errors of patients suffering from mania was that of the fifth-century medical writer Caelius Aurelianus, who had suggested that the physician should avoid both excessive indulgence and repulsive harshness, and use an appropriate tone in communicating with those patients, mid-way between an 'imposing gravity' and the 'simple tone of genuine sensitivity'.[66] Like the Quakers at the York Retreat and other contemporaries of Pinel, Aurelianus had placed the emphasis on the physician's interactions with their patients, stressing the need for the physician to 'earn their respect and esteem by being frank and open, and by constantly eliciting their love and fear (*s'en faire constamment chérir et craindre*), a skill credited to certain Moderns whose actual source I note here'.[67]

Drawing on his observation of the ways in which Pussin handled patients at the Bicêtre and on his own clinical experience, Pinel praised the efficacy of methods such as 'intimidation without violence', 'evoking fear (*appareil de crainte*), opposing firmly the dominant ideas and obstinacy of some of the insane, and showing courage and determination without anger'.[68] Firmness was crucial, for instance, with patients who refused to eat, thereby making themselves physically weak and putting themselves at risk of becoming ill and dying. This was a situation encountered with some of the patients suffering from periodical or intermittent insanity, whom he diagnosed as having 'lesions of the functions of the understanding'; it was successfully dealt with by using threats to arouse fear sufficiently strong to counteract seemingly unfounded obsessive fears, such as the fear of divine punishment.[69] Fear, in Pinel's view, could dispel the patient's own obsessive ideas and anxieties by shaking his or her imagination.[70]

In his medical thesis (1805), Esquirol also recommended that repressive methods should be used 'either to shake the imagination vigorously, or to produce a feeling of fear which would tame and subject manic patients'.[71] He claimed that intimidation and the arousal of fear was an effective method to help patients calm down: 'a great display of force and power, a threatening appearance and props intended to inspire terror can put a stop to the most stubborn and dreadful resolutions.'[72] Drawing on Lorry's dictum *spasmum spasmo solvitur*, he argued that fear could produce a spasm through which the basis of the delusion might be shattered.[73] If used skilfully, 'physical and moral shocks or commotions (*des secousses*)' would have the effect of 'shaking and threatening the machine, so to speak', thereby restoring health.[74] He elaborated on these ideas in *De la folie*, where he suggested that instilling 'real fear' could be beneficial in displacing 'imaginary fear', and that it was crucial to ensure that the passions aroused by the physician during therapeutic exchanges with patients were stronger than the passions that controlled their thoughts and fed their delusions:

> We must oppose, and conquer the most obstinate resolutions, inspiring the patient with a passion, stronger than that which controls his reason, by substituting a real for an imaginary fear; now, secure his confidence, and raise his fallen courage by awaking hope in his breast. Each melancholic should be treated on principles resulting from a thorough acquaintance with the tendency of his mind, his character and habits, in order to subjugate the passion which, controlling his thoughts, maintains his delirium. . . . A sudden, strong, and unexpected emotion, a surprise, fear and terror, have sometimes been successful.[75]

As the North American translator of Esquirol's *Mental Maladies* noted, referring to the principle of curing by contraries: 'we apply it instinctively, in the treatment of those cases of mental disease, which are connected with depression of mind. We cheer up the desponding in heart; offer encouragement to the timid and doubting; and bid the weak be strong.' He also observed that it was not uncommon for physicians to feel compassion for patients suffering from depression because it affected their bodily health: 'so rare is it, in any of the forms of insanity connected with depression of mind, particularly in the early stages, to find the organic functions healthfully performed, that we can hardly fail to entertain a sympathy for the afflicted sufferer.'[76] In his view, the method of eliciting fear could be justified in some cases:

> One of the great advantages resulting from fear consists in this, that it leads the patient to exercise self-control, and also, in many cases, secures that submission which induces a ready compliance with those hygienic measures which are regarded as useful in restoring the bodily and mental health.[77]

Esquirol's translator warned, however, that a physician seeking to use fear as a healing method would need 'much tact, as well as prudence and discrimination, and much of the "milk of human kindness" in his nature, cultivated by education'.[78]

Pinel had explained in his much revised edition of the *Traité* in 1809 that a combination of kindness and frankness in talking to patients was the most appropriate way of eliciting patients' trust and accessing their thoughts: 'it takes several attempts, and a number of interviews, conducted with skill and with a good-natured and frank approach, to be able to penetrate their thoughts, dispel their doubts, and get rid of their apparent contradictions by drawing comparisons.'[79] This suggests that, even though interviewing patients required skill and perseverance, interviews themselves proved to be therapeutic for some of Pinel's patients.

Pinel also warned that physicians and attendants should avoid passions such as haughtiness, and strike the right balance between kindness and firmness, to prevent patients from responding with violent emotions (*émotions*): 'the principles followed in public and private establishments to manage them can also produce in them new ideas and emotions which they did not have before. If treated with haughtiness or with inappropriate harshness, they become furious, hateful and violent.'[80] It was only by treating the furious insane with kindness (rather than seeking to control them by hitting them or putting them in chains) that the pattern of their emotions and ideas could be observed: 'they can only be brought back to their own affections by being treated in a gentle and benevolent manner; it is only then that the accurate history of their derangements can be studied as an illness (*maladie*).'[81] Instead of simply applying to patients existing labels such as melancholia or mania, Pinel tried to establish their individual clinical histories to be able to determine how their specific forms of mental illness might be cured.

As Esquirol further argued, the physician required great dedication not only to ascertain what passions were at the root of the patients' disorders, but also to determine what passions to evoke in each of them to help them recover:

> In this assemblage of enemies, who know only how to shun, or injure each other; what application, what devotion to duty, what zeal are necessary, to unfold the cause, and seat of so many disorders; to restore to reason its perverted powers; to control so many diverse passions, to conciliate so many opposing interests; in fine, to restore man to himself! We must correct and restrain one; animate and sustain another; attract the attention of a third, touch the feelings of a fourth. One may be controlled by fear, another by mildness; all by hope.[82]

He even noted how physicians should actively and consciously use their voice, gestures, facial expressions, words and silences to elicit the patient's trust and confidence (*confiance* means both) and to 'produce an effect upon the mind or heart of the maniac'.[83] In Esquirol's view, physicians should ensure that they gave patients hope, courage and consolation, whether they treated mental or physical illnesses:

> This treatment besides, is not confined exclusively to mental maladies: it is applicable to all others. It is not enough to say to the sick, courage, courage, you will be better. A feeling heart must dictate these consoling words, that they may reach the mind and heart of him who suffers.[84]

Esquirol thus suggested that physicians should cultivate a 'feeling heart' (*l'accent du coeur*) to be able to connect with patients at an emotional level, and that this would help not only in regulating disordered passions and habits (by arousing contrary or stronger passions), but also in giving hope, which would aid the healing process. Instilling hope and confidence in patients was an important part of the therapeutic approach promoted in the Hippocratic corpus: 'sometimes simply in virtue of the patient's faith in the physician that a cure is effected.'[85] Fully endorsing this principle, Esquirol claimed that 'without confidence, there is no cure'.[86] He noted that, while the art of healing had traditionally been in the hands of the 'ministers of the altar', his contemporaries would now seek to place their faith in a reputable physician: 'his name, his consolations, his councils, are often more useful than his remedies, because his reputation commands confidence, and permits us not to doubt respecting a cure.'[87]

Trust and confidence were essential elements in Esquirol's therapeutic interventions, judging from the clinical histories included in *Maladies Mentales*. An example of his approach is his account of his first meeting with a retired general who had become suicidal and was committed to his care on 1 August 1817. After spending an hour in conversation with the patient, followed by a few minutes of silence during which he kept his eyes fixed on him, he addressed him in these terms:

> You wish to destroy yourself, and for want of some other means of effecting your purpose, you wish to keep your cravat. You shall not take your life. I will restore you to health, and return you to your family and happiness.

When the patient replied that there was no more happiness for him, Esquirol continued to seek to elicit his confidence and hope: 'General, I will restore you to happiness, and I wish to assure myself that you will make no attempt upon your life.'[88] He then sought to draw on the patient's sense of pride by giving him a choice between pledging his word of honour that he would not take his life, or having four attendants watching him in his room. The patient chose the pledge, as expected of a high-ranking army officer. When Esquirol visited him the following day, he congratulated and encouraged him further, still striving to gain his trust. This proved successful, since the patient opened up to him, telling him that he had tried to strangle himself more than 20 times, and that his word of honour had restrained him.

Esquirol combined his conversations with patients with traditional remedies, such as 'leeches to the anus, and foot baths, rendered stimulant by mustard', and walks in the garden. The general's first garden walk, suggested by Esquirol as a response to his confiding that he thought that his 'disease' was caused by his sedentary life after retiring from the army, gave him the opportunity to experience 'the most delightful emotions' (*les impressions les plus agréables*), and to see nature with a degree of pleasure that he had not enjoyed for a long time.[89] After six months in Esquirol's care, during which he was taken on trips and visited by relatives, the general was deemed to be cured in that he made no more attempts to take his life. Back in his home environment he was not suicidal, though he fell prey to such 'paroxysms of jealousy' that his wife was forced to move out and stay with her own family for some years.[90] This narrative suggests that Esquirol was able to instil trust and confidence in patients who were pathologically suspicious of others. It also supports his views about the impact of social environment and social interaction on mental illness: 'the cause of mental alienation often exists within the family circle . . . and the presence of the parents and friends of the patient, exasperates the evil. Sometimes, an excess of tenderness keeps up the disease.'[91]

In the ideal establishment for insane people which Esquirol described in *Mental Maladies*, hope and confidence would not simply be promoted through interactions between patients and the physician. Patients would become more hopeful about their own recovery if they saw other patients recovering and being discharged. Convalescent patients had a crucial role in consoling and encouraging the sick 'by their contentment, their advice and counsels'.[92] Nonetheless, Esquirol also noted that the task of promoting hope and confidence among patients could prove challenging:

> For this untiring devotion, an approving conscience must be our chief reward. For what can a physician hope, who is always considered wrong when he does not succeed, who rarely secures confidence when successful; and who is followed by prejudices, even in the good which has been obtained.[93]

As practitioners of moral therapeutics, Pinel and Esquirol maintained an unflagging faith in the benefits of their approach in the face of prejudice and criticism, while the majority of physicians continued to advocate physical remedies only for the treatment of insanity.

## Conclusion

Pinel is widely recognized for having laid the foundations of scientific psychiatry.[94] He has also been seen as one of the 'fearless and dispassionate pioneers of mental medicine' and as a 'pioneer in the medicalization of passions', which paved the way for the massive medicalization of emotional life in the twentieth and twenty-first centuries.[95] The claim that he was 'dispassionate' seems to be underscored by assumptions about the incompatibility of science and emotion which cannot go unchallenged, and certainly cannot be attributed to Pinel. The label 'dispassionate' may serve to convey the level-headedness with which Pinel and Esquirol dealt with their patients, but, as can be seen from the evidence discussed in this chapter, it does not capture their concern with engaging with patients at an emotional level.

In focusing on forms of insanity which did not have physical causes, did not always affect people's intellectual ability and could not be cured simply by reasoning, Pinel and Esquirol drew on a traditional (Hippocratic) hygienic understanding of non-physical factors in health and disease. Nonetheless, they paid closer attention than any of their predecessors to the role of the emotions as causes, symptoms and cures of mental disturbance. Rather than medicalizing the emotions, they saw them as objects of study in their own right, and as powerful therapeutic tools.

According to Esquirol's testimony, Pinel achieved remarkable success in eliciting the trust of his patients, who provided him with detailed accounts of their thoughts, memories and information regarding their family, their symptoms, and significant events, which helped him to construct clinical histories.[96] He also sought to give his therapeutic methods the 'character of true science' by ensuring that they were based on careful clinical observation and by using statistical analysis: 'the application of the calculus of probabilities'.[97] One of Pinel's and Esquirol's main contributions was to show how scientific methods could be used to keep records (and to evaluate the effectiveness) of their large-scale application of a therapeutic approach which took into account the age, sex, occupation, circumstances, crucial events and dominant passions of each of their insane patients deemed potentially curable: those whose symptoms were of recent onset and had not been treated through physical methods.

Pinel's and Esquirol's methods and statistical findings were widely disseminated abroad. As we see from the anonymous review of Esquirol's *Statistique*, published in 1838, two in three of Pinel's patients were diagnosed as suffering from a form of insanity produced by non-organic causes:

> Pinel found moral causes to operate in the proportion to physical, as 464 to 219, and the first question which he generally put to patients, who still preserved some intelligence, was 'Have you undergone any vexation or disappointment?' The reply was seldom in the negative.[98]

It would seem that the high proportion of 'moral' causes of insanity identified by Pinel might be explained as the result of his interviewing techniques, if he really asked such leading questions. Nonetheless, Pinel and Esquirol did not simply rely on interviews, but kept detailed notes of their clinical trials, and subjected them to quantitative analysis, thereby promoting a form of evidence-based medicine, which still prevails today.

While Pinel's statistics showed that two-thirds of his patients suffered from forms of mental illness produced by non-organic causes, Esquirol argued that, even though not all insane patients had something wrong with their brain, they all appeared to have disordered passions:

> This moral alienation is so constant, that it would appear to me to be an essential characteristic of mental alienation. There are insane persons whose delirium is scarcely noticeable; none whose passions and moral affections, are not disordered, perverted, or annihilated.[99]

This implied that the moral approach was relevant to all forms of insanity, whether to cure patients or simply to manage them.

The conscious efforts made by Pinel and Esquirol to elicit their patients' trust and confidence through their interviewing techniques helped not only to understand patients' ways of thinking, but also to regulate the passions which underpinned the patients' thoughts. Detailed clinical histories, such as that of the general treated by Esquirol in 1817, suggest that their interviewing techniques were somewhat successful in changing more harming patterns of thought and behaviour, even though they did not completely suppress other habits of thinking and feeling, which the patients had developed in response to their own social environment, outside the therapeutic institution.

Pinel is well known as one of the European pioneers who applied the humanitarian outlook of the Enlightenment to the care of the insane by banning chains and the use of violence by attendants. What is less known are his uses of fear and hope as ways of accessing the minds of patients who were unable to reason, an approach that he had drawn from the *Encyclopaedia Britannica*, as he noted in the first edition of his *Traité*.[100] Pinel believed in the efficacy of threats to treat some of his most obstinate patients. Such methods might seem questionable today, but in Pinel's day and nearly two decades after his death, when Esquirol's *Mental Maladies* appeared in English in Philadelphia, the use of fear with insane patients was seen as a potentially efficacious (though controversial) way of treating some patients.

Then, as now, health carers would be expected to show not only judiciousness, but also tact and kindness, 'cultivated by education'.[101] Then, as now, the education of medical professionals had to combine the science of healing, which Pinel and Esquirol helped to develop, with the art of observing and interacting with patients, which they had learned to practise, inspired by ancient practitioners such as Hippocrates, Galen and Caelius Aurelianus. In our age, when medical education needs to keep up with a rapid proliferation of technological devices for examining patients' brains and other internal organs, Pinel's and Esquirol's lessons on the importance of connecting with patients' minds at an emotional level would seem more enlightening and relevant than ever.

## Notes

1 The report is published in D. B. Weiner, 'Philippe Pinel's "Memoir on Madness" of December 11, 1794: A Fundamental Text of Modern Psychiatry', *American Journal of Psychiatry* 149, 1992, 725–32.

2 P. Pinel, *Traité médico-philosophique sur l'aliénation mentale*, Paris: Richard, Caille & Ravier, An IX (1800/1), p. xiv. On the influence of Hippocrates on Pinel, see J. M. Pigeaud, 'Le *rôle* des *passions* dans la pensée médicale de *Pinel* à *Moreau de Tours*', *History and Philosophy of the Life Sciences* 2, 1980, 123–40; *Aux Portes de la psychiatrie: Pinel, l'ancien et le moderne*, Paris: Aubier 2001, pp. 117, n. 5 and 188, n. 10.

3 I will cite from the 1806 English translation whenever possible, even though it contains a number of mistranslations and omits parts of Pinel's text, such as the Introduction (replaced with the translator's own Introduction); see Pinel, *A Treatise on Insanity*, trans. D. D. Davis, Sheffield: W. Todd, 1806, pp. i–lv. For an account of omissions and mistranslations in the English version, see D. B. Weiner, 'Betrayal! The 1806 English Translation of Pinel's *Traité médico-philosophique sur I 'aliénation mentale ou la manie'*, *Gesnerus* 57(1–2), 2000, 42–50.

4 *Traité*, p. xiii, n. 1.

5 See the report Esquirol presented to the Interior Minister in 1818 on the institutions for the insane in France and the ways in which their fate could be improved; *Des Établissements des aliénés en France et des moyens d'améliorer le sort de ces infortunés*, Paris: Huzard, 1819, p. 12.

6 Weiner, *Comprendre et soigner: Philippe Pinel (1745–1826) et la médecine de l'esprit*, Paris: Fayard, 1999, pp. 9 and 18.

7 J. Goldstein, *Console and Classify: The French Psychiatric Profession in the Nineteenth Century*, 2nd edn. Chicago: University of Chicago Press, 2001, p. 94; L. Charland, 'Science and Morals in the Affective Psychopathology of Philippe Pinel', *History of Psychiatry* 21, 2010, 38–53 (pp. 42–5); P. Huneman, 'Montpellier Vitalism and the Emergence of Alienism in France (1750–1800): The Case of the Passions', *Science in Context* 21, 2008, 615–47.

8 J. E. Esquirol, *Des passions considérées comme causes, symptômes et moyens curatifs de l'aliénation mentale*, Paris: Didot, 1805, p. 20.

9 Weiner, 'Esquirol's Patient Register: The First Private Psychiatric Hospital in Paris, 1802–1808', *Bulletin of the History of Medicine* 63, 1989, 110–20.

10 Esquirol, 'Introduction à l'étude des aliénations mentales', *Revue médicale française et étrangère* 8, 1822, 31–8, cited in Weiner, 'Mind and Body in the Clinic: Philippe Pinel, Alexander Crichton, Dominique Esquirol, and the Birth of Psychiatry', in G. S. Rousseau (ed.), *The Languages of the Psyche: Mind and Body in Enlightenment Thought*, Berkeley: University of California Press, 1990, pp. 331–402 (p. 357).

11 For the English translation, see Esquirol, *Mental Maladies: A Treatise on Insanity*, trans. E. K. Hunt, Philadelphia: Lea and Blanchard, 1845.

12 On the passions as non-natural factors of health and disease, see E. Carrera, 'Anger and the Mind-Body Connection in Medieval and Early Modern Medicine', in *Emotions and Health, 1200–1700*, Leiden/Boston: Brill, 2013, 95–146. On health regimes in the French Encyclopaedia, see W. Coleman, 'Health and Hygiene in the Encyclopedia', *Journal of the History of Medicine and Allied Sciences* 29, 1974, 399–442.

13  See, for instance, Felix Plater's advice in Felix Plater, Abdiah Cole, Nicholas Culpeper, *A Golden Practice of Physick*, London: Peter Cole, 1662, p. 37. See also S. Jackson, *Care of the Psyche: A History of Psychological Healing*, New Haven: Yale University Press, 1999, pp. 201–11.

14  G. Mora, 'Vincenzo Chiarugi (1759–1820) and His Psychiatric Reform in Florence in the Late 18th Century (on the occasion of the bi-centenary of his birth)', *Journal of the History of Medicine* 14, 1959, 424–33.

15  See for instance the Rules of the Cordova hospital, which were copied in 1753: *Ordenanzas*, 21 June 1473; Archivo de Protocolos, Córdoba, *Escrituras públicas*, fols 525–41, cited in G. García González, *Historia de la asistencia psiquiátrica en Córdoba*, Cordova: Imprenta Provincial, 1983, pp. 192–3 (192). See also L. García Ballester and J. García González, 'Nota sobre la asistencia a los locos y "desfallecidos de seso" en la Córdoba medieval: El Hospital de Jesucristo', *Asclepio* 30–31, 1979, 199–207.

16  The report was published by José Iberti in 1791; see J. Espinosa-Iborra, 'Un testimonio de la influencia de la psiquiatría española de la ilustración en la obra de Pinel: el informe de José Iberti acerca de la asistencia en el Manicomio de Zaragoza (1791)', *Asclepio* 16, 1964, 179–82.

17  Pinel, *Traité*, p. 226.

18  Esquirol, *Mental Maladies*, p. 53.

19  Ibid., p. 296. For examples of such therapeutic advice in the Middle Ages and the Renaissance, see P. E. Pormann (ed.), *Rufus of Ephesus: On Melancholy*, Tübingen: Mohr Siebeck, 2008, p. 102; J. L. Vives, *The Passions of the Soul: The Third Book of De Anima et Vita*, trans. Carlos G. Noreña, Lewiston, NY: Mellen, 1990, p. 74. For the later period, see Clare Hickman, *Therapeutic Landscapes: A History of English Hospital Gardens since 1800*, Manchester: Manchester University Press, 2013.

20  Esquirol, *Mental Maladies*, p. 90.

21  Esquirol, *Maladies mentales considerées sur les rapports médical, hygiénique et médico-légale*, Paris: Baillère, 1938, p. 72.

22  For the context of the phrase 'habits of mind, feeling and action', see Esquirol, *Mental Maladies*, p. 88.

23  The label 'domestic trouble' was used to refer to 'all the griefs, all oppositions, misfortunes and dissensions, that grow out of the family state'; ibid., pp. 46–7, 75.

24  W. Battie, *A Treatise on Madness*, London: J. Whiston and B. White, 1758. For the wider context, see R. Porter, *Mind-Forg'd Manacles: A History of Madness in England from the Restoration to the Regency*, Cambridge, MA: Harvard University Press, 1987, pp. 206–22.

25  V. Chiarugi, *On Insanity and Its Classification*, ed. and trans. G. Mora, Canton, MA: Science History Publications, 1987, pp. 137–8. See also L. Charland, 'The distinction between "Passion" and "Emotion". Vincenzo *Chiarugi*: a case study', *History of Psychiatry* 25(4), 2014, 477–84.

26  Cited in D. B. Weiner, 'The Madman in the Light of Reason. Enlightenment Psychiatry: Part II. Alienists, Treatises, and the Psychologic Approach in the Era of Pinel', in E. R. Wallace and J. Gach, eds, *History of Psychiatry and Medical Psychology: With an Epilogue on Psychiatry and the Mind-Body Relation*, New York: Springer, 2008, pp. 281–303 (p. 288). See also D. L. Gerard, 'Chiarugi and Pinel Considered: Soul's Brain/Person's Mind', *Journal of the History of the Behavioral Sciences* 33(4), 1997, 381–40.

27  Pinel made this point in response to what he saw as Condillac's failure to explain restlessness, desire and the passions; *Treatise*, pp. 83–4.

28  Benjamin Rush, *Medical Inquiries and Observations upon the Diseases of the Mind*, 3rd edn, Philadelphia: Crigg, 1827, p. 15.

29  Esquirol, *Mental Maladies*, p. 28.

30  Ibid., p. 70.

31  Ibid., pp. 54 and 56.

32  Galen, *The Art of Medicine*, in *Selected Works*, trans. P. N. Singer, Oxford: Oxford University Press, 1997, pp. 345–96 (p. 367).

33  *In Hippocratis Epidemiarum, VI.1–8*, ed. E. W. and F. Pfaff, Berlin: Academiae Litterarum, 1956, p. 483.

34  Gordonio, *Lilio de medicina*, ed. B. Dutton and M.N. Sánchez. 2 vols, Madrid: Arco/Libros, 1993, vol. I, p. 505.

35 Villanova, *El maravilloso regimiento*, trans. H. de Mondragón, Barcelona: Jaime Cendrat, 1606, p. 12; my translation.

36 A. Boorde, *The Breviary of Helthe*, London: Wylllyam Myddelton, 1547, fol. 85r.

37 G. Mercuriale, *Responsorum et consultationum medicinalium*, Book 2, Venice: Giolito, 1589, consultatio 103; book 3, Venice: J. de Franciscis, 1620, consultatio 108.

38 *Medicina Practica*, Frankfurt: J. T. Schönwetteri, 1602, p. 61.

39 Pinel, *Treatise on Insanity*, pp. 78–9.

40 A. Crichton, *An Inquiry Into the Nature and Origin of Mental Derangement*, 2 vols. London: T. Cadell Jr and W. Davies, 1798, vol. II, p. 99. See also Weiner, 'Mind and Body'; Louis C. Charland, 'Alexander Crichton on the Psychopathology of the Passions', *History of Psychiatry* 19, 2008, 275–96; Louis C. Charland, '"A moral line in the sand": Alexander Crichton and Philippe Pinel on the Psychopathology of the Passions', in Louis C. Charland and Peter Zachar (eds), *Fact and Value in Emotion*, Amsterdam/Philadelphia: John Benjamins, 2008, pp. 15–35.

41 Crichton, *Inquiry*, vol. I, p. 210.

42 Ibid., p. 182.

43 Ibid., pp. 168–9.

44 Ibid., p. 188.

45 Ibid., p. xxxiv.

46 Ibid., p. xxx.

47 Ibid.

48 See Weiner 'Mind and Body', p. 349.

49 *Traité* (1809), p. 26.

50 Pinel, 'Recherches sur les causes du délire, par A. Crichton', *Recueil périodique de littérature médicale étrangère* 1, 1798–99 (An VII), pp. 401–18 and 463–78 (p. 466, n. 1).

51 Ibid., p. 411, n. 1.

52 Pinel, *Traité*, p. xxii; Crichton, *Inquiry*, vol. II, pp. 98–9.

53 For further references to these clinical notes, see Weiner, *Comprendre et soigner*, pp. 227–31.

54 Ibid.

55 Pinel, *Traité*, p. 110. The reference to terror related to the afterlife is omitted in the translation; *Treatise*, p. 113.

56 Pinel, *Traité*, p. 250; *Treatise*, p. 240. A third of these patients were reported to have suffered relapses.

57 Pinel, *Treatise*, p. 15.

58 Pinel, *Traité* (1809), p. 349. Laure Murat offers a partial analysis of this case; L. Murat, *The Man Who Thought He Was Napoleon: Toward a Political History of Madness*, Chicago: University of Chicago Press, 2014, pp. 57–60.

59 Esquirol, *Passions*, p. 20

60 Esquirol, *Mental Maladies*, pp. 47 and 49.

61 Ibid., pp. 46–7.

62 Ibid., p. 51.

63 Ibid., p. 30.

64 Ibid., p. 12.

65 Battie, *Treatise*, p. 84.

66 Pinel, *Traité*, p. xi.

67 Ibid., p. xi. Weiner suggests this is a clear allusion to the founders of the York Retreat; see 'Mind and Body', p. 341. On the use of fear at the York Retreat, see L. Charland, 'Benevolent Theory: Moral Treatment at the York Retreat', *History of Psychiatry* 18, 2007, 61–80.

68 Pinel, *Traité*, p. 61. The English translation omits the idea of provoking fear: 'of intimidation, without severity; of oppression, without violence; and of triumph, without outrage'; *Treatise*, p. 63.

69 Pinel, *Treatise*, pp. 23–4.

70 The notion of shaking the patient's imagination is explicitly mentioned in the original French title of this section, 'ébranler l'imagination', translated into English as 'restrain upon the imagination'; *Traité*, p. 59; *Treatise*, p. 61.

71 Esquirol, *Passions*, p. 54

72  Ibid., p. 70.
73  Ibid., p. 54
74  Ibid., p. 70.
75  Esquirol, *Mental Maladies*, pp. 78–9.
76  Ibid., p. 334.
77  Ibid., p. 334–5.
78  Ibid., p. 335
79  Pinel, *Traité* (1809), p. 134.
80  Ibid., pp. 134–5.
81  Ibid.
82  Esquirol, *Mental Maladies*, p. 21.
83  Ibid., p. 404.
84  Ibid., p. 80.
85  Hippocrates, *Regimen*, II. See also *Precepts* 6, cited in Owsei Temkin, *Hippocrates in a World of Pagans and Christians*, Baltimore: Johns Hopkins University Press, 1995, p. 102.
86  Esquirol, *Mental Maladies*, p. 76.
87  Ibid., p. 80
88  Ibid., p. 296.
89  Ibid., p. 296 (cf. *Maladies Mentales*, p. 308).
90  Ibid., p. 298.
91  Ibid., p. 75.
92  Ibid., p. 77
93  Ibid., p. 21.
94  Petteri Pietikäinen, *Madness: A History*, Abingdon: Routledge, 2015, pp. 106–11, 159–63.
95  Ibid., pp. 107, 109.
96  See Esquirol's testimony: 'Observateur ingénieux et profound, habile à saisir les rapports, M. Pinel voit à travers les troubles de la raison, la pensée, les affections des aliénées, il révèle ce qu'il y a de plus mystérieux dans l'intelligence humaine', cited in Weiner, *Comprendre et soigner*, p. 22.
97  Pinel, *Résultats d'observations et construction des tables pour servir à determiner le degré de probabilité de la guérison des aliénés*, Paris: Baudouin, 1808; cited in Goldstein, *Console and Classify*, pp. 101–5.
98  Anonymous, 'Statistics of insanity in Europe (review article on *Esquirol's Statistique de la Maison Royale de Charenton, dans les Annales d'Hygiène Publique*)', *The Foreign Quarterly Review* 20, 1938, 39–54.
99  Esquirol, *Mental Maladies*, p. 16.
100  Pinel, *Treatise*, p. 103.
101  Ibid., p. 335.

## Select bibliography

Battie, W., *A Treatise on Madness*, London: J. Whiston and B. White, 1758.

Carrera, E., *Emotions and Health, 1200–1700*, Leiden/Boston: Brill, 2013.

Charland, L., 'Alexander Crichton on the Psychopathology of the Passions', *History of Psychiatry* 19, 2008, 275–96.

Charland, L., 'Benevolent Theory: Moral Treatment at the York Retreat', *History of Psychiatry* 18, 2007, 61–80.

Charland, L. 'Science and Morals in the Affective Psychopathology of Philippe Pinel', *History of Psychiatry* 21, 2010, 38–53.

Chiarugi, V., *On Insanity and Its Classification*, ed. and trans. G. Mora, Canton, MA: Science History Publications, 1987.

Crichton, A., *An Inquiry into the Nature and Origin of Mental Derangement*, 2 vols. London: T. Cadell Jr and W. Davies, 1798.

Esquirol, E., *Des passions considérées comme causes, symptômes et moyens curatifs de l'aliénation mentale.* Paris: Didot, 1805.

Esquirol, J. E., *Maladies mentales considerées sur les rapports médical, hygiénique et médico-légale.* Paris: Baillère, 1938./ *Mental Maladies: A Treatise on Insanity,* trans. E. K. Hunt, Philadelphia: Lea and Blanchard, 1845.

Goldstein, J., *Console and Classify: The French Psychiatric Profession in the Nineteenth Century,* 2nd edn., Chicago: University of Chicago Press, 2001.

Pinel, P., *Traité médico-philosophique sur l'aliénation mentale,* Paris: Richard, Caille & Ravier, Year IX (1800/1)/ *A Treatise on Insanity,* trans. D. D. Davis, Sheffield: W. Todd, 1806.

Porter, R., *Mind-Forg'd Manacles: A History of Madness in England from the Restoration to the Regency,* Cambridge: Harvard University Press, 1987.

Rousseau, G. S. (ed.), *The Languages of the Psyche: Mind and Body in Enlightenment Thought,* Berkeley: University of California Press, 1990.

Weiner, D. B. *Comprendre et soigner: Philippe Pinel (1745–1826) et la médicine de l'esprit,* Paris: Fayard, 1999.

# 7

# DEVIANCE AS DISEASE

## The medicalization of sex and crime

### *Jana Funke*

Behaviours that are 'deviant' in that they breach social standards have frequently been understood as forms of disease with underlying and potentially treatable medical causes. At the beginning of the twenty-first century, a number of 'strange', 'unusual' or 'undesirable' behaviours are classified as forms of disease: excessive drinking, smoking, eating, gambling or sexual activity are commonly labelled as addictions; hyperactivity in children and, increasingly, adults, is often viewed as ADHD (attention deficit hyperactivity disorder); grief and suicide can be seen as outcomes of depression and are thus linked to mental illness.[1] The relation between deviance and disease raises fundamental questions about the validity of disease categories and the constitution of medical knowledge and authority: how and why are certain behaviours framed as medical problems? What is at stake in labelling behaviours as forms of disease? Which actors, forces and structures are involved in legitimating claims that deviant behaviours should be understood in medical terms?

Scholars from across disciplines and fields, such as disability studies, feminist studies, history, literary studies, psychiatry, queer studies and sociology, have sought to address these questions. From a historical perspective, it has been argued that deviant behaviours that once fell outside the realm of medical expertize and were considered immoral, sinful or criminal, rather than pathological, have been understood increasingly as medical problems over the course of the late nineteenth and twentieth centuries.[2] The shifting relation between deviance and disease is at the heart of debates about medicalization, the reframing of non-medical aspects of life as medical problems. The ways in which deviant behaviours have been medicalized is often seen as indicative of the excessive expansion of medical authority over new areas of life. It is viewed as symptomatic of a proliferation of disease categories and has been described as 'overmedicalization' or 'disease mongering'.[3]

This chapter engages with these questions and debates, but also seeks to complicate linear and monolithic narratives of the medicalization of deviance.[4] It focuses, in particular, on the emergence of the intersecting fields of criminal anthropology, sexology and psychiatry from the second half of the nineteenth century onwards. It demonstrates that, throughout the past 150 years or so, the relation between criminal and sexually deviant behaviours and disease was debated critically inside and outside medical fields of knowledge. As such, the medicalization of deviance needs to be understood as a complex and unstable process that involves multiple actors and forces. Instead of illustrating a steady increase of medical authority, the medicalization of deviance has regularly inspired controversy about the expansion of medical authority and definitions of disease.

## Defining deviance and disease

Examining the medicalization of deviance brings to the fore the definitional insta-bility and uncertainty surrounding disease designations.[5] Deviance itself is a highly contested concept that has been understood as a statistical deviation from the norm or in terms of moral standards and principles. At least since the 1970s, social construc-tionist frameworks have frequently served to illuminate and examine critically the ways in which understandings of deviance are shaped or even constituted by social and cultural processes.[6] For example, sociologists like Howard Becker have argued that deviance cannot be measured objectively (either statistically or morally), but simply constitutes any form of behaviour that is labelled and recognized as deviant by society.[7]

Building on such constructionist perspectives, studies of the relation between deviance and disease have sought to draw attention to the fact that disease itself is an elusive category that defies straightforward definition in that it reflects and rein-forces social and cultural values. In the 1960s, sociologist Thomas Scheff developed a labelling perspective on mental illness to challenge the objectivity of the category of mental illness. He suggested that psychiatric classifications were used to label what he called 'residual rule-breaking', that is to say, behaviours that violate social norms.[8] At around the same time, psychiatrist Thomas Szasz's studies of mental illness and psychiatrist and historian David Musto's work on opiate addiction similarly began to highlight the social construction of specific disease categories.[9]

Historical approaches have offered a particularly powerful means to interrogate shifting definitions of disease. Social constructionism has exerted a lasting influence on social histories of medicine from the 1970s onwards. Resulting scholarship has drawn attention to the ways in which medical understandings of disease are embed-ded in and shaped by social and cultural contexts.[10] Sociological studies of deviance and disease have also embraced historical perspectives: in the early 1980s, Peter Conrad and Joseph W. Schneider developed an 'historical-social constructionist approach' to demonstrate that medicalized views of deviance needed to be under-stood as culturally and historically specific designations that can change over time.[11] One frequently cited example is masturbation, which was considered, by some, as a form of disease in the nineteenth century, but has since been mostly demedicalized and is even presented as healthy at times.[12] This indicates that categories used to classify behaviours as deviant or diseased do not reflect universal standards, but are historically contingent and changeable.

Insisting that disease classifications are shaped by social and cultural contexts has served to challenge positivist medical models that view human behaviour primarily or exclusively on the basis of biological causes. Some critics have drawn attention to the fact that disease categories have been applied to forms of deviant behaviour, such as alcoholism, homosexuality or hyperactivity, even in cases where a somatic or biolog-ical basis remains unclear or is possibly non-existent. The results have been (often radical) challenges, in particular, to psychiatric attempts to frame human behaviour on the basis of a medical disease model.[13] Other critics from within the anti-psychiatry movement and disability studies have challenged the application of medical models even with regard to conditions that might have an organic or physical dimension.[14] These voices have argued that all human experiences of illness and suffering are necessarily structured by social and cultural forces irrespective of causation. As such, discussions of the medicalization of deviance have served to test and contest the authority and legitimation of medical models of disease and its validity when applied to human behaviour. Such challenges have forcefully demonstrated, for instance, that medical disease categories do not always do justice to the psychosocial dimensions of human life. On the contrary, they locate disease within the individual, thus drawing attention away from the social conditions that can cause suffering.

Political questions concerning the power of medical authority and its control over human behaviour are central to these debates. In particular, the medicalization of deviance is seen as an expansion of medical jurisdiction and power to realms of behaviour that were previously governed by institutions like the church or the law. This, in turn, highlights how medical knowledge is linked to the exercise of social control.[15] Historical scholarship has demonstrated how disease categories served to reinforce social hierarchies based on categories of difference including race, class, gender and sexuality. 'Drapetomania', for instance, was used in the US in the mid-nineteenth century to describe a supposed mental illness causing black slaves to flee captivity.[16] Feminist historians and cultural and literary critics have shown that 'hysteria' often served as a catch-all disease to label as sick or unhealthy women who failed to observe gendered rules of conduct.[17] Similarly, at the beginning of the twentieth century, 'feeble-mindedness' was used as a disease category to justify the institutionalization of 'delinquent' young working-class women who were seen to transgress gender and sexual norms.[18] These examples demonstrate how disease designations could be used to enforce compliance with expected social roles and to uphold social values. Going beyond the emphasis on external control, Foucauldian scholars have also emphasized how medical knowledge can exert power by shaping the self-perception of individuals who internalize the social values expressed through disease designations.[19]

Overall, these diverse approaches, which can broadly be characterized as historical-social constructionist, have sought to challenge the status of medical disease categories as objective and neutral. Focusing on the medicalization of socially undesirable or suspect behaviours, in particular, has made it possible to draw attention to the ways in which disease designations are implicated in social and cultural values and power structures. However, historians and sociologists of medicine have also drawn attention to the limitations of radical constructionist approaches. As historian Charles Rosenberg points out, constructionist critics tend to focus on 'culturally resonant diagnoses – hysteria, neurasthenia, and homosexuality, for example – in which a biopathological mechanism is either unproven or unprovable'.[20] Sociologist Allan V. Horwitz has also argued that constructionist views of disease are flawed in that they can easily sidestep the question of whether certain forms of deviant behaviour are, in fact, linked to underlying biological causes. As a result, Horwitz suggests, social constructionist frameworks make it impossible to differentiate between behaviours that are socially designated as deviant and forms of disease that do have a biological basis.[21]

To address this problem, scholars have sought to develop hybrid approaches that combine constructionist sensibilities with an appreciation of the possible biological causes of behaviour. Like Horwitz, Jerome C. Wakefield maintains that many conditions that are currently regarded as mental illnesses are simply forms of deviant behaviour and should thus not be labelled as diseases.[22] Yet, Horwitz and Wakefield also believe that deviant behaviour can be indicative of mental disorders that exist beyond arbitrary social judgements in that they are based on biological dysfunctions. In cases of schizophrenia, bipolar disorder, or other psychoses, for instance, Horwitz suggests, socially disvalued behaviours can be symptomatic of internal dysfunctions and should be understood as forms of disease.[23] Horwitz's and Wakefield's work usefully highlights how constructionist views can flatten out differences between various conditions. At the same time, upholding the distinction between deviance as socially defined and disease as linked to biological causation also has serious intellectual and political implications. It can obscure the fact that biological explanations are inevitably shaped by social and cultural values, a point that has been made convincingly, especially by disability, feminist and queer scholars. Moreover, there is a risk of reinforcing unhelpful and potentially harmful distinctions between allegedly 'genuine' forms of disease that are biologically grounded and conditions that are perceived as psychological or psychosomatic.

Instead of attempting to resolve these ongoing disputes, the following sections trace how disease designations and their relation to deviant behaviours have been negotiated within the intersecting fields of criminal anthropology, sexology and psychiatry from the late nineteenth century onwards. Questions concerning the authority of medical disease frameworks, the validity of biopathological explanations of human behaviour, and the extent to which social values shape disease categories are not specific to late twentieth- and early twenty-first-century critiques of medicalization. On the contrary, such debates are at the heart of the process of medicalization itself.

## Crime and sex as disease

The relationship between deviance and disease has a long history that has been traced back to ancient Greece.[24] The interrelated ideas that deviant behaviour might be caused by disease, that disease might lead to deviant behaviour, and that deviance itself can be seen as a symptom of illness, are therefore not specific to the modern period. Still, much scholarship has focused on the medicalization of deviance over the past 200 years, and it has been argued that deviance was increasingly understood within a medical disease framework over the course of the nineteenth and twentieth centuries. One factor contributing to this development was the emergence of a causal concept of disease within nineteenth-century Western medical thought according to which diseases were best regarded as having specific natural causes. As K. Codell Carter has argued, the success of this etiological standpoint led to 'persistent efforts to expand the range of human abnormalities that are approached by the way of causes' and viewed as forms of disease.[25]

At the same time, the medicalization of deviance cannot be understood by focusing on isolated shifts within medical knowledge or by rehearsing teleological narratives of increasing medicalization. Critics of the concept of medicalization have argued that the power and authority of medicine have been overestimated and that 'medicalization should not be regarded as the sole, or possibly, even the major trend in deviance definition' in the modern period.[26] Non-medical explanations of deviant behaviours continued to intersect and exist in dialogue with medical views over the course of the nineteenth and twentieth centuries. It has also been cautioned that an emphasis on medicalization has tended to foreground medical control and power in the abstract without paying sufficient attention to the ways in which disease definitions were developed collaboratively by medical experts and other interest groups, including patients themselves.[27] Medicalized views of disease were not invented by medical professionals working in isolation, but were shaped and reshaped through processes of knowledge co-production, circulation and popularization that involved various medical and non-medical agents and groups.[28]

The complex dynamics of the medicalization of deviance can be illustrated by focusing on the history of psychiatry, criminal anthropology and sexology. These overlapping fields of knowledge emerged together at the end of the nineteenth century and shared a fascination with the etiological question of whether deviant forms of behaviour were due to biological causes and could thus be explained as natural forms of disease. This somatic turn towards a disease framework in late nineteenth-century psychiatry is often associated with the work of German psychiatrist Emil Kraepelin, whose *Psychiatrie* (1883) introduced the major diagnostic categories of 'dementia praecox' and 'depression', which were understood as forms of physical disease.[29] At the same time, sexologists and criminal anthropologists, many of whom were psychiatrists by training, investigated whether criminal and deviant sexual behaviours could equally be understood and classified as forms of physical disease with natural causes.

In addition to the rise of the causal disease concept, the tendency to interpret a wide range of human behaviours from a strictly biological point of view was also facilitated by widespread nineteenth-century theories of degeneration.[30] It was suggested that degenerative illnesses were characterized by a steady decrease of mental functioning from one generation to the next. Such narratives of decline were frequently embedded in an evolutionary framework; degeneration was seen as an atavistic return to the primitive origins of human development. This explanation was employed to argue that a very wide range of abnormal behaviours was based on 'hereditary taint'. These behaviours were seen as an outcome of underlying constitutional conditions that had been inherited and could be passed on to the next generation.

Italian psychiatrist Cesare Lombroso led the development of criminal anthropology by suggesting that crime was a degenerative disease.[31] In making this point, he built on deterministic phrenological and psychiatric arguments rehearsed earlier in the nineteenth century, which suggested that criminal behaviour was caused by abnormal properties of the brain and could have an hereditary basis.[32] In his book *L'uomo delinquente* (1876), translated into English as *Criminal Man*, Lombroso introduced the type of the 'born criminal', whose antisocial tendencies were grounded in his 'physical and psychic organisation, which differs essentially from that of normal individuals'.[33] The born criminal was seen as an individual suffering from 'moral insanity'.[34] Following Lombroso, proponents of the school of criminal anthropology suggested that crime could best be understood by researching the biological causes and origins of such behaviours and by identifying, studying and measuring the criminal as type.

Similarly, psychiatrists like Austro-German Richard von Krafft-Ebing or German Karl Westphal, who played a foundational role in shaping sexological investigations, examined whether sexual acts that were considered to be deviant (mainly because they were non-procreative) could be understood and classified as forms of physical disease.[35] Reflecting the broader psychiatric interest in neuropathology – the idea that mental disorders were situated in the nervous system and, particularly, the brain – these sexologists suggested that sexual abnormalities were caused by underlying nervous disturbances that were inborn and linked to degeneration.[36] A commonly rehearsed argument first developed by German jurist Karl Heinrich Ulrichs, for instance, was that same-sex desire was the cause of physical and psychological inversion: a male homosexual was born with a woman's brain in a man's body.[37] As numerous historians of sexuality have argued, these deterministic approaches, which viewed sexual behaviours as symptomatic of innate biological and psychological characteristics, facilitated the emergence of sexual types, such as the homosexual.[38] The 'born sexual deviant' can thus be considered as a counterpart to the 'born criminal'.[39]

This biologization of crime and sexual deviance as forms of degenerative disease placed questions of aetiology at the heart of modern understandings of criminal and sexual behaviour. Investigating the causes and origins of such behaviours was an important goal of criminal anthropology and sexology. Although biological factors were certainly seen as crucial, there were ongoing debates about environmental or social causes of deviance. Despite the emphasis on the born homosexual, for instance, sexologists continued to explore the role that environmental or social factors could play in causing same-sex behaviour. Possible factors ranged widely and included the climate, overexcitement, consumption of literature, seduction by an elder, or habits like masturbation. English sexologist Havelock Ellis and his co-author, classicist and Renaissance scholar John Addington Symonds, for instance, argued in *Sexual Inversion* (1896–7) that homosexuality was not always inborn or congenital, but could also be acquired, even if they believed such cases to be rare.[40] Similarly, Lombroso took into account environmental factors (prison life, contact with other criminals, abuse of alcohol, distress) that could encourage criminal activity in individuals who had a weaker predisposition (so-called 'criminaloids') or were not biologically pre-disposed towards crime at all (so-called 'juridical criminals').[41] Thus, the medical framing of deviant behaviours did not necessarily exclude consideration of non-biological explanations. On the contrary, medicalization of deviance in the fields of sexology and criminal anthropology inspired vital debate about the etiological significance of both biological and environmental or social factors. The emergence of a medical framework within which deviance could be understood did not signal an exclusive or narrow focus on biological causes and origins.

Sexologists and criminal anthropologists also questioned whether deviant behaviours were unchanging aspects of human life or whether they were dependent on shifting social values and judgements. Lombroso argued that 'pederasty' (used synonymously with homosexuality here) was a crime in the modern world, but would not have been considered as such in ancient Rome and Greece.[42] Ellis and Symonds made the same argument in *Sexual Inversion* to demonstrate that same-sex desires between men were not criminalized, but rather 'recognized and idealized' in the ancient world.[43] Similarly, Ellis rejected universal definitions of criminality in his 1880 study *The Criminal* (closely modelled on Lombroso's *Criminal Man*), adopting a relativistic stance instead: 'Criminality . . . consists in a failure to live up to the standard recognized as binding by the community.'[44] This indicates that an interest in the biological causes of deviant behaviours did not foreclose a consideration of the shifting social and cultural contexts in which such behaviours were understood and judged.

The argument that criminal or sexual behaviours were inborn forms of disease was frequently used to call for a change in social values and attitudes towards such behaviours and the individuals who displayed them. The notion that the born criminal or sexual deviant was biologically predisposed raised fundamental questions about guilt and culpability. If the individual did not choose to engage in deviant behaviours, but was driven towards crime or sexual transgression by a congenital condition, should he or she be held responsible and punished? Reframing deviant behaviours as forms of disease evoked the sense of a condition over which a person had little or no control. Lombroso rejected the classical criminological approach introduced by late-eighteenth-century Italian jurist Cesare Beccaria, who had insisted that the criminal could determine his own destiny through reason and free will and should therefore be punished for his or her behaviour. Lombroso argued instead that the born criminal's culpability was limited and that criminal anthropology had the 'object of curing, instead of punishing'.[45] Similarly, many sexologists including Krafft-Ebing, Ellis and German physician Magnus Hirschfeld argued that born homosexuals should neither be blamed nor punished for their behaviours and that male homosexuality should be decriminalized.[46]

These calls for decriminalization were made in the name of humanistic and liberal ideals, but the medicalization of criminal and sexual behaviours can also be seen as a key example of the drive to expand medical authority and control over areas of life that were traditionally governed by church or law. According to Szasz, for instance, Krafft-Ebing 'was not interested in liberating men and women . . . he was interested in supplanting the waning power of the church with the waxing power of medicine'.[47] This view is representative of other critical accounts of the rise of criminal anthropology and sexology that focus on the struggle to create 'ownership' over territories of intervention, power and social prestige. At the end of the nineteenth century, the newly emerging fields of criminal anthropology and sexology did strive for authority and legitimacy and their challenge to existing legal protocols formed part of this process. Yet, claims articulated by sexologists and criminal anthropologists were often contested and refuted.[48] As such, linear accounts of a straightforward rise of medical power that often underpin debates about medicalization need to be challenged. The fact that medical approaches did not simply replace legal or religious authority is illustrated by the fact that homosexuality was criminalized until 1967 in the UK and 1968–9 in Germany despite earlier efforts by sexologists like Ellis and Hirschfeld to challenge these laws, which had been newly instituted at the end of the nineteenth century.[49]

It has also been argued that the overtly humanistic rhetoric of medicalization can easily obscure the fact that describing deviant behaviours as forms of disease is simply a form of stigmatization under a different name. Casting crime or sexual deviance as disease continues to stigmatize and, more specifically, pathologize specific actions and desires. Individuals who exhibit such behaviours continue to be seen as deviant because of their alleged sickness.[50] Moreover, the disease framework opens up the possibility of treatment, cure and therapy and such interventions have been debated widely with regard to both crime and homosexuality from the nineteenth century onwards.[51] As such, a medical disease framework continues to carry moral weight and needs to be seen as a value judgement in its own right. It is also linked to the exertion of control since punishment under the law can easily be replaced by medical intervention.

While important, these criticisms of medicalization often rely on unhelpful polarizing distinctions between medical and non-medical understandings of human behaviour and run the risk of reinforcing reductive views of medical knowledge. It is easy to overlook the fact that disease frameworks have served different purposes and have also been put to ends that might be considered liberal or affirmative by some. Antagonistic views of medicalization also fail to do justice to the fact that disease frameworks, and the relation between disorder and pathology, in particular, were critically debated even within fields that have often been seen as forceful drivers of medicalization. Sexologists like Ellis or Hirschfeld were medically trained and worked within a medical framework to make the case that homosexuality was inborn and therefore neither sinful nor criminal. However, they also argued against the idea that homosexuals required treatment or therapy. Krafft-Ebing and Sigmund Freud, too, rejected as unnecessary emerging treatments that promised to 'cure' individuals who experienced same-sex desires.

Many sexologists also sought to distance themselves from the idea that homosexuality was always necessarily pathological or harmful. Ellis, for instance, compared homosexuality to the synesthetic sensation of 'coloured-hearing in which there is not so much defect, as an abnormality of nervous tracks producing new and involuntary combinations'.[52] Here, sexual deviance is seen as a form of disorder of the normal functioning of the organism and, thus, an abnormality. It is not, however, considered pathological since it is not harmful. In addition, sexologists turned to a range of evidence, including the case study and cross-cultural and cross-historical materials, to demonstrate that homosexuality could occur in healthy individuals.[53] This demonstrates that there were on-going attempts by sexologists to challenge pathologizing definitions of disease and to think critically about the uses of medical disease frameworks.

Finally, the medicalization of deviant behaviours that occurred at the end of the nineteenth century was not driven by medical professionals alone and must not be understood as a top-down process whereby medical classifications were imposed on 'passive victims of a medical juggernaut'.[54] As Harry Oosterhuis has convincingly argued, Krafft-Ebing's work on sexuality was crucially shaped and influenced by his own patients and the individuals who contacted him unsolicited after reading his work. By publishing case studies, often at length and in the individual's own voice, sexologists like Krafft-Ebing, Ellis and Hirschfeld, 'did not simply encourage medical treatment, restraint, and repression, but also offered a space in which sexual desire could be articulated' and in which medical understandings of sexual behaviours could be negotiated and shaped.[55] The medicalization of deviant behaviours was strongly influenced by demands made on behalf of different agents, including medical professionals, patients and other stakeholders. Moreover, individuals like Edward Carpenter, Hirschfeld, Symonds and Ulrichs, who themselves identified as homosexual, often laid claim to scientific and medical authority, and fundamentally shaped the field of sexology.[56] As such, it can be misleading to assume clear-cut distinctions between medical experts and the patients and subjects they studied, and it is important to acknowledge the different and often overlapping forms of expertize and credibility that constitute medical knowledge.

The following section takes up these issues, but shifts focus to the second half of the twentieth century when newly emerging psychiatric disease categories were intensely debated. This historical moment has often been seen as the zenith of 'medical imperialism', but as has already been shown, medicalization is not a straightforward process of medical expansion or domination. It involves multiple driving forces, agendas, and voices and produces critical debate about questions of definition both within and outside the medical realm. In the second half of the twentieth century, this is particularly evident when considering the tensions and controversies around the application of psychiatric disease categories to forms of behaviour, including sexual behaviours, which could be viewed as deviant or had been considered as such in the past.

## Diagnostic psychiatry and the medicalization of deviance

The 1980 publication of *DSM-III*, the third and much revised edition of the *Diagnostic and Statistical Manual* published by the American Psychiatric Association (APA), has widely been seen as a watershed moment in the history of Western psychiatry, indicating the return of a Kraepelian model of psychiatry. Kraepelin believed that mental illness could reliably be understood on the basis of observable symptoms and that these symptoms were caused by biological dysfunctions. Whereas Kraepelin had limited his diagnostic repertoire to two major disease categories, neo-Kraepelian psychiatry in the second half of the twentieth century produced an explosion of diagnostic categories, many of which were (and continue to be) used to label forms of deviant behaviour as specific kinds of disease. This is reflected in the proliferation of new and expanding disease categories with an increase from 106 disorders listed in the first *DSM* in 1952 to nearly 300 in *DSM-IV* in 1994.

From a historical perspective, this growth in diagnostic categories has been linked to a shift away from a Freudian model of dynamic psychiatry that had gained popularity in the first half of the twentieth century, particularly in the United States.[57] Freud and his followers argued that deviant behaviours, including those that had been classified as symptoms of mental illness or degeneration by psychiatrists like Kraepelin, often lacked underlying biological causes and could not easily be classified as specific forms of disease. Focusing mainly on psychosocial causes, the Freudian school of dynamic psychiatry blurred the distinction between the mentally ill and healthy and insisted that neurotic tendencies affected large sections of society and were not limited to a few isolated cases found in hospitals, asylums or prisons. In so doing, dynamic psychiatry expanded enormously the range of problems that fell within the realm of psychiatric expertize, but did not consider these problems as specific forms of disease. When diagnostic psychiatry gained authority towards the end of the twentieth century, Horwitz argues, it 'reclassified as specific diseases the huge realm of behaviour that dynamic psychiatry had already successfully defined as pathological'.[58] As a result, a very large number of heterogeneous human behaviours, experiences and emotions were classified as forms of disease with allegedly biological causes.

These changes in psychiatric thinking have often been seen as the outcome of a struggle to legitimate psychiatry. Several studies have suggested that the APA's shift towards a biomedical disease model in *DSM-III* served to defend psychiatry against the burgeoning anti-psychiatry movement of the 1960s and 1970s.[59] According to this view, embracing the idea that mental illnesses could be understood on the basis of a medical disease model, in the same way as physical illnesses, served to establish the scientific credentials of the discipline.

The expansion of psychiatric authority expressed in the proliferation of diagnostic categories has given rise to heightened controversy about whether diagnostic psychiatry can reliably differentiate between disease and deviance or whether it is at risk of biologizing and pathologizing forms of behaviour that are deemed undesirable or unacceptable because they offend social standards and values.[60] Critics of diagnostic psychiatry and of the *DSM*, in particular, have argued that many of the behaviours included in the manual are not indicators of discrete underlying diseases. They should rather be considered the result of personal distress, stressful social conditions or simply as forms of social deviance, which might be socially undesirable or offensive to some, but are not grounded in biological dysfunction.[61]

One area that has continued to inspire particular debate about the application and validity of disease frameworks in the second half of the twentieth century is the medicalization of allegedly deviant sexual behaviours. Arguments have centred on 'the difficult conceptual distinction between deviant sexual desires that are mental disorders and those that are normal variations in sexual preference (even if they are eccentric, repugnant, or illegal if acted upon)'.[62] Critics have pointed out that it is 'exceedingly difficult to eliminate historical and cultural factors from the assessment of unusual sexual interests'.[63] Moreover, there have been concerted political efforts on behalf of gay, lesbian, queer and feminist activist groups to challenge the medical framing of sexual behaviours on the grounds that the medicalization of sexuality reflects and reinforces social ideologies and hierarchies and is thus a form of social control.

The psychiatric profession itself has responded to such challenges and has also been a driver of change in its own right. The APA, for instance, has continuously revised the terminology and conceptual framework used to classify deviant sexual behaviours or 'paraphilias', a term that was introduced because it was seen as more neutral and less pejorative than alternatives like 'sexual perversion'.[64] Moreover, the sexual desires and acts that are included in the various editions of the *DSM* have shifted considerably over the past half-century. The most prominent example is homosexuality, which was formally removed from *DSM-II* in 1974.[65] This was partly the outcome of political lobbying on behalf of gay activist groups, such as the Gay Liberation Front, which drew inspiration from the anti-psychiatry movement of the 1960s. Resistance to the idea that homosexuality should be classified as a psychiatric disorder also came from within the APA with gay psychiatrists resisting the pathologization of homosexuality.[66]

The publication of *DSM-II* is often seen as a revolutionary moment and the demedicalization of homosexuality is used as a primary example to demonstrate that allegedly deviant behaviours that have been classified as forms of disease can be removed from a medical framework and can be reclaimed as alternative 'lifestyles' or 'orientations'. However, this argument overlooks the many ways in which homosexuality has continued to be medicalized. *DSM-II* removed the word 'homosexuality', but introduced the new category of 'sexual orientation disturbance', which applied only to those individuals who were unhappy with their homosexual desires and sought psychiatric help.[67] *DSM-IV* introduced the category of 'gender identity disorder' (GID), which, as psychologists and queer critics have argued, continued to medicalize gender-non-conforming behaviours that can be part of homosexual self-presentation, especially in children and young adults.[68] In the 2010s, homosexuality has also been rebranded as a form of 'addiction' requiring treatment. This move serves to justify the continued use of 'conversion' or 'reparative' therapies – psychotherapeutic attempts to convert homosexual into heterosexual desires – which have largely been rejected as unethical by major medical organizations, including the APA. Homosexuality has thus continued to be understood within a psychiatric framework even after its removal from *DSM-II.*

However, the continued medicalization of homosexuality in the second half of the twentieth century was not single-handedly driven by an allegedly reactionary psychiatric agenda, but involved processes of knowledge co-production and popularization. The AIDS epidemic, for example, has been linked to the re-medicalization and pathologization of male homosexuality, as the illness was, at least initially, framed as a 'gay disease'.[69] Yet, as Steven Epstein's sociological study of AIDS activism has shown, the AIDS crisis also led activists, many of whom were members of the gay community, to intervene in and drive medical research.[70] These lay people gained expertize in the fields of virology, immunology and epidemiology, successfully claimed scientific credibility, challenged anti-gay assumptions underpinning some research agendas, and ultimately produced medical knowledge in critical dialogue and collaboration with medical experts, healthcare providers, pharmaceutical and biotechnology companies, journalists and other agents. The AIDS crisis did result in the pathologization of male homosexuality and other forms of behaviour that were considered deviant, such as drug use or sex work. At the same time, diverse medical and non-medical experts participated in ensuing debates to negotiate critically the nature, causes and possible treatments of the disease and, in so doing, co-produced medical understandings of HIV/AIDS.[71]

Homosexuality has also come to be re-medicalized through genetic research, which illustrates further the role that popular understandings of scientific and medical research play in processes of medicalization.[72] From the early 1990s onwards, researchers have attempted to locate a genetic marker for homosexuality.[73] While these studies have been met with considerable scepticism, the notion of the 'gay gene' was and continues to be disseminated widely across popular culture. It has given rise to contradictory responses, which echo late-nineteenth-century debates about the medicalization of homosexuality. On the one hand, the possibility of genetic causation has been met with fears that homosexuals would be stigmatized and pathologized, leading the British press to caution against the possibility of genetic testing being used to abort selectively embryos with a genetic predisposition for homosexuality.[74] On the other hand, the promise of genetic proof of the inborn nature of homosexuality has been embraced by some gay and lesbian activists and their allies. Such groups reject the implication that biological difference equals pathology or dysfunction, but maintain that the 'gay gene' affirms an inborn sexual identity. They use genetic arguments to naturalize and defend homosexuality on the basis that sexual orientation is not a matter of choice or personal responsibility and should not be subject to therapeutic intervention. In this sense, there is a considerable demand for specific aspects of the medical disease framework on behalf of some members of the gay and lesbian community who negotiate and employ medical knowledge to serve personal and political purposes. Overall, the recent history of homosexuality shows that the removal of homosexuality from the *DSM* did not mark a watershed moment of demedicalization. Homosexual behaviour has continued to be medicalized and there are on-going debates about the potential biological causes of homosexuality among different groups and stakeholders.

The medicalization of other forms of deviant sexual behaviour also continues to inspire debate. In contrast to homosexuality, widely diverse sexual behaviours such as paedophilia, exhibitionism, voyeurism, sadism, masochism, or fetishism are still included in the *DSM* as paraphilias. *DSM-5* (2013) sought to confront some of the criticisms aimed at previous editions by introducing new terminology to increase diagnostic validity. It differentiates between paraphilias, which are to be seen as non-disordered sexual variations, and paraphilic disorders, which cause distress and harm to non-consenting victims.[75] This decision has been welcomed on the grounds that a large number of sexual behaviours, such as BDSM or kink, are no longer classified as mental disorders.

At the same time, the new terminology employed in *DSM-5* has also been exposed as conceptually weak and there are persistent doubts about the validity of diagnostic classification.[76] It is not clear, for instance, how the criterion of harmfulness can be used to distinguish reliably between disease and socially or legally defined deviance. First, the distress caused by sexual acts or desires cannot be separated from the social context and value system within which these acts or desires are experienced. Second, as Wakefield has argued, the concern about harm done to non-consenting victims should be a matter of criminal law rather than psychiatric diagnosis.[77] Physician Charles Moser and psychologist Peggy Kleinplatz make a similar point and suggest that demedicalizing illegal sexual behaviours like paedophilia by removing them from the *DSM* 'would focus attention on the criminal aspects of these acts, and not allow the perpetrators to claim mental illness as a defense or use it to mitigate responsibility for their crimes'.[78] This call effectively reverses the arguments made by late nineteenth-century criminal anthropologists in that it differentiates between criminal deviance and mental illness to increase the individual's personal responsibility.

## Conclusion

The medicalization of deviance continues to raise challenging questions about diagnostic validity and objectivity. The way in which allegedly deviant forms of behaviour, including sexual and criminal behaviours, have been linked to biopathological causes and framed in terms of disease draws attention to the social values that inevitably shape medical knowledge. The extension of medical authority over realms of human behaviour that have not always been considered as medical problems also calls into question the scope of medical jurisdiction. As such, the medicalization of deviance opens up important critical perspectives on medical knowledge. At the same time, it is crucial to acknowledge that medicalization needs to be understood as a complex process that is driven and influenced by different medical and non-medical experts and interest groups. Medical disease frameworks have been negotiated and used differently by various stakeholders and have served a number of purposes that go beyond the pathologization and stigmatization of deviant behaviour.

As this chapter has begun to show, the relation between deviance and disease has been debated widely across disciplines and fields and it is through cross-disciplinary and inter-disciplinary exchange that the complexities of medicalization can best be understood. Social histories of medicine, for instance, need to continue to turn to cultural history and literary studies to examine how understandings of deviance and disease were articulated, circulated and negotiated across different genres and audiences. Sociologists have developed useful methodological approaches to conceptualize and map processes of knowledge co-production that allow scholars to move beyond flawed views of medicalization as driven by medical experts alone. Critical race, disability, feminist and queer studies offer crucial tools to illuminate how social ideologies and hierarchies shape medical knowledge and raise awareness of the political struggles at stake in framing deviance as disease. Finally, nuanced accounts of the medicalization of deviance need to avoid antagonistic views of medicine and acknowledge that medical models of disease, and the ways in which these are applied to deviant forms of behaviour, have been considered carefully and critically by both medical and non-medical agents and groups. It is by integrating different disciplinary perspectives that scholars can achieve a fuller understanding of the cultural, medical, political and social factors that have shaped the relation between deviance and disease in the past and present.

## Notes

1  M. Ajzenstadt and B. E. Burtch, 'Medicalization and regulation of alcohol and alcoholism: The professions and disciplinary measures', *International Journal of Law and Psychiatry*, 1990, vol. 13, 127–47; G. L. Blackburn, 'Medicalizing obesity: Individual, economic, and medical consequences', *American Medical Association Journal of Ethics*, 2011, vol. 13, 890–95; A. V. Horwitz and J. C. Wakefield, *The Loss of Sadness: How Psychiatry Has Transformed Normal Sadness into Depressive Disorder*, Oxford: Oxford University Press, 2007; J. Rossol, 'The medicalization of deviance as an interactive achievement: The construction of compulsive gambling', *Symbolic Interactions*, 2001, vol. 24, 315–41; M. Smith, *Hyperactive: The Controversial History of ADHD*, London: Reaktion Books, 2012.
2  Foundational studies of the medicalization of deviance include: P. Conrad and J. W. Schneider, *Deviance and Medicalization: From Badness to Sickness*, Philadelphia: Temple University Press, 1992; I. Illich, *Medical Nemesis: The Expropriation of Health*, New York: Pantheon, 1976; K. S. Miller, *The Criminal Justice and Mental Health Systems*, Cambridge: Gunn & Hain, 1980; T. Szasz, *The Manufacture of Madness*, New York: Harper & Row, 1970.
3  The term 'disease mongering' was coined by L. Payer, *Disease-Mongers: How Doctors, Drug Companies, and Insurers Are Making You Feel Sick*, New York: John Wiley & Sons, 1992.
4  For nuanced accounts of medicalization, see: K. Ballard and M. A. Elston, 'Medicalization: A multi-dimensional concept', *Social Theory and Health*, 2005, vol. 3, 228–41; P. Conrad, *The Medicalization of Society: On the Transformation of Human Conditions into Treatable Disorders*, Baltimore: Johns Hopkins University Press, 2007; S. J. Williams and M. Calnan, 'The "limits" of medicalization: Modern medicine and the lay populace in "late modernity"', *Social Science and Medicine*, vol. 42, 1609–20.
5  Conrad, *Medicalization*, p. 3; italics in original.
6  For an overview of the diverse intellectual and political traditions feeding into social constructionism, see: P. Conrad and K. K. Barker, 'The social construction of illness: Key insights and policy implications', *Journal of Health and Social Behavior*, 2010, vol. 51, 67–79; and L. Jordanova, 'The social construction of medical knowledge', *Social History of Medicine*, 1995, vol. 8, 361–81.
7  H. S. Becker, *Outsiders: Studies in the Sociology of Deviance*, London: Free Press of Glencoe, 1963.
8  T. J. Scheff, *Being Mentally Ill: A Sociological Theory*, Chicago: Aldine, 1966.
9  D. Musto, *The American Disease: Origins of Narcotic Control*, New Haven: Yale University Press, 1973; T. Szasz, *The Myth of Mental Illness: Foundations of a Theory of Personal Conduct*, New York: Hoeber-Harper, 1961.

10 The social history of madness offers a good example. For an overview, see: R. Bivins and J. V. Pickstone (eds) *Medicine, Madness and Social History*, Basingstoke: Palgrave, 2007.
11 Conrad and Schneider, *Deviance*, p. 17.
12 Conrad, *Medicalization*, p. 97. For more on the history of masturbation, see: T. H. Engelhardt, 'The disease of masturbation: Values and the concept of disease', *Bulletin of the History of Medicine*, 1974, vol. 48, 234–48; T. Laqueur, *Solitary Sex: A Cultural History of Masturbation*, New York: Zone Books, 2003.
13 Key anti-psychiatric works include: D. Cooper, *Psychiatry and Anti-Psychiatry*, London: Paladin, 1971; E. Goffman, *Asylums: Essays on the Social Situation of Mental Patients and Other Inmates*, Garden City: Anchor Books, 1961; R. D. Laing, *The Divided Self: A Study of Sanity and Madness*, London: Tavistock Publications, 1960; Szasz, *Mental Illness*. For a more recent articulation of anti-psychiatric perspectives, see: P. Kinderman, *A Prescription for Psychiatry*, Basingstoke: Palgrave, 2014.
14 For a critical discussion of the social model in disability studies, see: T. Shakespeare and N. Watson, 'The social model of disability: An outdated ideology?', *Research in Social Science and Disability*, 2001, vol. 2, 9–28.
15 S. Cohen, *Visions of Social Control*, Cambridge: Polity Press, 1985.
16 K. White, *An Introduction to the Sociology of Health and Illness*, London: SAGE, 2002, pp. 41–2.
17 L. Appignanesi, *Mad, Bad and Sad: A History of Women and the Mind Doctors from 1800 to the Present*, London: Virago, 2009; E. Showalter, *The Female Malady: Women, Madness, and English Culture, 1830–1980*, London: Virago, 1987; A. Scull, *Hysteria: The Disturbing History*, Oxford: Oxford University Press, 2011.
18 M. A. Rembis, *Defining Deviance: Sex, Science, and Delinquent Girls, 1890–1960*, Urbana: University of Illinois Press, 2011.
19 For a discussion of Foucault's influence on debates about medicalization, see S. J. Williams, 'Sociological imperialism and the profession of medicine revisited: Where are we now?', *Sociology of Health and Illness*, 2001, vol. 23, 135–58.
20 C. E. Rosenberg, 'Framing disease: Illness, society and history', in C. E. Rosenberg and J. Golden (eds), *Framing Disease: Studies in Cultural History*, Newark: Rutgers University Press, p. xv.
21 A. V. Horwitz, *Creating Mental Illness*, Chicago: University of Chicago Press, 2002, pp. 9–10.
22 Horwitz, *Creating Mental Illness*; J. C. Wakefield, 'The concept of mental disorder: On the boundary between biological facts and social values', *American Psychologist*, 1992, vol. 47, 373–88. Wakefield's 'harmful dysfunction' model suggests that mental disorders should be understood as involving two distinct components: the failure of a mechanism to perform a function for which it was biologically designed and a (social) value judgement that the dysfunction is undesirable. For a critical discussion of this approach, see: D. Murphy and R. L. Woolfolk, 'The harmful dysfunction analysis of mental disorder', *Philosophy, Psychiatry, & Psychology*, 2000, vol. 7, 241–52.
23 Horwitz, *Creating Mental Illness*, p. 15.
24 See Conrad and Schneider, *Deviance*, for discussions of the 'long history' of the medicalization of deviance.
25 K. Codell Carter, *The Rise of Causal Concepts of Disease: Case Histories*, Aldershot: Ashgate, 2003, p. 2.
26 A. V. Horwitz, 'The medicalization of deviance', *Contemporary Sociology*, 1981, vol. 10, p. 752. For an early critique of 'medical imperialism', see: P. Strong, 'Sociological imperialism and the profession of medicine: A critical examination of the thesis of medical imperialism', *Social Science and Medicine*, 1979, vol. 13A, 199–215.
27 Rosenberg, 'Disease', p. xvi.
28 R. Cooter and S. Umfrey, 'Separate spheres and public places: Reflections on the history of science, popularization and science in popular culture', *History of Science*, 1994, vol. 32, 237–67; T. F. Gieryn, *Cultural Boundaries of Science: Credibility on the Line*, Chicago: University of Chicago Press, 1999; R. Porter (ed.), *The Popularization of Medicine: 1650–1850*, London: Routledge, 1992.
29 For an overview of Kraepelin's role in the development of descriptive psychiatry, see H. S. Decker, *The Making of DSM-III: A Diagnostic Manual's Conquest of American Psychiatry*, Oxford: Oxford University Press, 2013, pp. 35–52.

30  Degeneration theory was rooted in French psychopathology, especially the writings of Valentin Magnan and Bénédict Morel, but was soon taken up as a broader framework of explanation in scientific, literary, philosophical and political circles. See: J. E. Chamberlin and S. L. Gilman (eds), *Degeneration: The Myth of Progress*, New York: Columbia University Press, 1985.

31  For more on Lombroso, see: C. Beccalossi, *Female Sexual Inversion: Same-Sex Desires in Italian and British Sexology, c. 1870–1920*, Basingstoke: Palgrave, 2011; C. Beccalossi, 'Sexual deviancies, disease, and crime in Cesare Lombroso and the "Italian school" of criminal anthropology', in R. Peckham (ed.), *Disease and Crime: A History of Social Pathologies and the New Politics of Health*, New York: Routledge, 2014, pp. 40–55. For more on the history of criminal anthropology more generally, see: P. Becker and R. F. Wetzell, *Criminals and Their Scientists: The History of Criminology in International Perspective*, Washington: German Historical Institute, 2006; N. Rafter, *Creating Born Criminals*, Urbana: University of Illinois Press, 1997.

32  For an overview of these earlier biological explanations, see: N. Rafter, *The Criminal Brain: Understanding Biological Theories of Crime*, New York: New York University Press, 2008.

33  C. Lombroso, *Criminal Man*, New York: Knickerbocker Press, 1911, p. 5.

34  Ibid., p. 52.

35  For useful introductions to the history of sexology, see: L. Bland and L. Doan (eds), *Sexology in Culture: Labelling Bodies and Desires*, Chicago: University of Chicago Press, 1998; V. Sigusch, *Geschichte der Sexualwissenschaft*, Frankfurt: Campus, 2008; C. Waters, 'Sexology', in H. Cook and M. Houlbrook (eds), *Palgrave Advances in the Modern History of Sexuality*, Basingstoke: Palgrave, 2006, pp. 41–63. For more on Krafft-Ebing, see: H. Oosterhuis, *Stepchildren of Nature: Krafft-Ebing, Psychiatry, and the Making of Sexual Identity*, Chicago: University of Chicago Press, 2000.

36  H. Oosterhuis, 'Richard von Krafft-Ebing's "step-children of nature": Psychiatry and the making of homosexual identity', in V. A. Rosario (ed.), *Science and Homosexualities*, New York: Routledge, p. 71.

37  Importantly, Ulrichs rejected the idea that sexual inversion was a sign of degeneration or pathology. See: H. Kennedy, 'Karl Heinrich Ulrichs, first theorist of homosexuality', in V. A. Rosario (ed.), *Science and Homosexualities*, New York: Routledge, pp. 26–45.

38  This point was influentially argued in M. Foucault, *The History of Sexuality: The Will to Knowledge*, New York: Vintage, 1976. See also A. I. Davidson, *The Emergence of Sexuality: Historical Epistemology and the Formation of Concepts*, Cambridge: Harvard University Press, 2001.

39  Beccalossi, 'Deviancies', p. 51.

40  H. Ellis and J. A. Symonds, *Sexual Inversion*, Basingstoke: Palgrave, 2007, p. 128.

41  Lombroso, *Criminal*, pp. 100–21.

42  Beccalossi, *Inversion*, p. 123.

43  Ellis and Symonds, *Sexual Inversion*, p. 113. For more on the reception of ancient Greek ideas, see: J. Bristow, 'Symonds' history, Ellis' heredity: *Sexual Inversion*', in L. Bland and L. Doan (eds), *Sexology in Culture: Labelling Bodies and Desires*, Chicago: University of Chicago Press, 1998, pp. 79–99; J. Funke, '"We cannot be Greek now": Age difference, corruption of youth and the making of *Sexual Inversion*', *English Studies*, 2013, vol. 94, 139–53.

44  H. Ellis, *The Criminal*, New York: Scribner and Welford, 1890, p. 206.

45  Ibid., p. 5.

46  See: J. D. Steakley, *The Homosexual Emancipation Movement in Germany*, New York: Arno Press, 1975.

47  T. Szasz, *Sex by Prescription*, Garden City: Anchor Press, pp. 19–20.

48  S. Brady, *Masculinity and Male Homosexuality in Britain, 1986–1913*, Basingstoke: Palgrave, 2005, pp. 119–56; M. Cook, *London and the Culture of Homosexuality, 1885–1914*, Cambridge: Cambridge University Press, 2003, pp. 76–8.

49  Paragraph 175 was introduced as part of the German Criminal Code in 1871. The Labouchère Amendment in Britain was passed in 1885.

50  On the patient as deviant, see T. Parsons, *The Social System*, New York: Free Press, 1951.

51  T. Dickinson, *'Curing Queers': Mental Nurses and Their Patients, 1935–74*, Manchester: Manchester University Press, 2015.

52  Ellis and Symonds, *Inversion*, p. 204.
53  I. Crozier, 'Introduction: Havelock Ellis, John Addington Symonds and the construction of *Sexual Inversion*', in I. Crozier (ed.) *Sexual Inversion*, Basingstoke: Palgrave, 2007, pp. 1–86; K. Fisher and J. Funke, 'British sexual science beyond the medical: Cross-disciplinary, cross-historical, and cross-cultural translation', in H. Bauer (ed.), *Sexology and Translation: Cultural and Scientific Encounters Across the Modern World*, Philadelphia: Temple University Press, 2015, pp. 95–114.
54  Oosterhuis, 'Step-children', p. 70.
55  Ibid., p. 83. For more on Ellis' and Hirschfeld's use of case studies, see: I. Crozier, 'Havelock Ellis, eonism and the patient's discourse; or, writing a book about sex', *History of Psychiatry*, 2000, vol. 11, 124–54; J. Funke, 'Narrating uncertain sex: The case of Karl M.[artha] Baer', in B. Davies and J. Funke (eds), *Sex, Gender and Time in Fiction and Culture*, Basingstoke: Palgrave, 2011, pp. 132–53.
56  Fisher and Funke, 'British sexual science'.
57  Decker, *DSM-III*; Horwitz, *Creating Mental Illness*, pp. 1–2.
58  Horwitz, *Creating Mental Illness*, p. 2. For more on neo-Kraepelian psychiatry (and Kraepelin's legacy), see E. J. Engstrom and M. M. Weber, 'Making Kraepelin history: A great instauration?', *History of Psychiatry*, 2007, vol. 18, 267–73.
59  Decker, *DSM-III*, p. 3–33.
60  For some recent discussions of this question, see: M. Abouelleil and R. Bingham, 'Can psychiatry distinguish social deviance from mental disorder?', *Philosophy, Psychiatry, & Psychology*, 2014, vol. 21, pp. 243–55; D. Bolton, *What is Mental Disorder?*, Oxford: Oxford University Press, 2008; J. C. Wakefield, 'What makes a mental disorder mental?', *Philosophy, Psychiatry, & Psychology*, 2006, vol. 13, pp. 123–31.
61  Horwitz, *Creating Mental Illness*, pp. 14–5.
62  J. C. Wakefield, 'DSM-5 proposed diagnostic criteria for sexual paraphilias: Tensions between diagnostic validity and forensic utility', *International Journal of Law and Psychiatry*, 2011, vol. 34, p. 195.
63  C. Moser and P. Kleinplatz, 'DSM-IV-TR and paraphilias', *Journal of Psychology & Human Sexuality*, 2006, vol. 17, p. 92. For other criticisms of the medicalization of sex, see: T. Cacchioni and L. Tiefer, 'Why medicalization? Introduction to the special issue on the medicalization of sex', *Journal of Sex Research*, 2012; vol. 49, 307–10; J. R. Fishman and L. Mamo, 'What's in a disorder: A cultural analysis of medical and pharmaceutical constructions of male and female sexual dysfunction', *Women & Therapy*, 2001, vol. 24, 179–93; G. Rubin, 'Thinking sex: Notes for a radical theory of the politics of sexuality', in C. S. Vance (ed.), *Pleasure and Danger: Exploring Female Sexuality*, London: Pandora Press, 1992, 267–319.
64  *DSM-I* (1952) and *DSM-II* (1968) included the umbrella term 'sexual deviation', which was replaced in *DSM-III* with 'paraphilia'. This was seen as preferable 'because it correctly emphasizes that the deviation (para) [lies] in that to which the individual is attracted (philia)', APA cited in H. S. Decker, *DSM-III*, n. 369. The term 'paraphilia' was coined by Austrian psychologist Wilhelm Stekel in the 1920s and popularized by Johns Hopkins psychologist John Money in the 1960s. For more on the history of the term 'paraphilia' and its use by Money, see: L. Downing, I. Morland and N. Sullivan, *Fuckology*, Chicago: University of Chicago Press, 2014, pp. 41–68.
65  Homosexuality had previously been included as a 'sociopathic personality disturbance' in *DSM-I* and as a 'personality disorder and non-psychotic mental disorder' in *DSM-II*. It was also included in the manuals of the World Health Organization at the time. The history of the medicalization of homosexuality has been discussed, for instance, by Conrad and Schneider, *Deviance*, pp. 172–214; J. Terry, *An American Obsession: Science, Medicine, and Homosexuality in Modern Society*, Chicago: University of Chicago Press, 1999; V. A. Rosario (ed.), *Science and Homosexualities*, New York: Routledge, 1997.
66  H. S. Decker, *DSM-III*, pp. 31–3 and 208–17. It has been suggested that the removal of homosexuality was mainly a political decision on behalf of the APA. See: R. Bayer, *Homosexuality and American Psychiatry: The Politics of Diagnosis*, New York: Basic Books, 1981.
67  Similarly, *DSM-III* used the term 'ego-dystonic sexual orientation' to describe an attraction that causes anxiety and the desire to change, because it is at odds with the individual's idealized self-image. This category was removed from the revised edition *DSM-III-R* in 1987.

68  E. Kosofsky Sedgwick, 'How to bring your kids up gay', *Social Text*, 1991, vol. 29, pp. 18–27; I. Wilson, C. Griffin and B. Wren, 'The validity of the diagnosis of gender identity disorder (child and adolescent criteria)', *Clinical Child Psychology and Psychiatry*, 2001, vol. 7, pp. 335–51.

69  P. Kayal, *Bearing Witness: Gay Men's Health Crisis and the Politics of AIDS*, Boulder: Westview, 1993, p. 197; S. Epstein, 'Moral contagion and the medicalizing of gay identity', *Research in Law, Deviance, and Social Control*, 1988, vol. 9, pp. 3–36.

70  S. Epstein, *Impure Science: Aids, Activism, and the Politics of Knowledge*, Berkeley: University of California Press, 1996.

71  Post-traumatic stress disorder (PTSD) and Alzheimer's disease offer further examples of diseases that were medicalized through critical dialogue involving social interest groups: P. Fox, 'From senility to Alzheimer's disease: The rise of the Alzheimer's disease movement', *Milbank Quarterly*, 1989, vol. 67, pp. 57–101; W. J. Scott, 'PTSD in DSM-III: A case of the politics of diagnosis and disease', *Social Problems*, 1990, vol. 37, pp. 294–310.

72  For more on the remedicalization of homosexuality, see: Conrad, *Medicalization*, pp. 70–96. For critical discussions of the 'gay gene' rhetoric, see: P. Conrad and S. Markens, 'Constructing the "gay gene" in the news: Optimism and skepticism in the US and British press', *Health*, vol. 5, pp. 373–400; K. O'Riordan, 'The life of the gay gene: From hypothetical genetic marker to social reality', *Journal of Sex Research*, 2012, vol. 49, pp. 362–68.

73  Studies either report distinct areas of the brain related to homosexuality or a high concordance of twins with homosexual attraction: D. Hamer and P. Copeland, *The Science of Sexual Desire: The Search for the Gay Gene and the Biology of Behaviour*, New York: Simon & Schuster, 1994; Simon LeVay, *The Sexual Brain*, Cambridge, MA: MIT Press, 1993; G. Wilson and Q. Rahman, *Born Gay: The Psychobiology of Sex Orientation*, London: Peter Owen, 2008.

74  P. Conrad and S. Markens, 'Constructing the "gay gene"', p. 386.

75  APA, *Diagnostic and Statistical Manual of Mental Disorders: 5th Edition*, Washington: APA, 2013.

76  Wakefield, 'Paraphilias'.

77  Ibid.

78  Moser and Kleinplatz, 'DSM-IV-TR', p. 105.

## Select bibliography

T. Cacchioni and L. Tiefer (eds), Special Issue on the Medicalization of Sex, *Journal of Sex Research*, 2012, vol. 49.

A. E. Clarke, J. R. Fishman, J. R. Fosket, L. Mamo and J. K. Shim, 'Biomedicalization: Technoscientific transformations of health, illness, and U.S. biomedicine', *American Sociological Review*, 2003, vol. 68, 161–94.

P. Conrad, *The Medicalization of Society: On the Transformation of Human Conditions into Treatable Disorders*, Baltimore: Johns Hopkins University Press, 2007.

H. S. Decker, *The Making of DSM-III: A Diagnostic Manual's Conquest of American Psychiatry*, Oxford: Oxford University Press, 2013.

A. V. Horwitz, *Creating Mental Illness*, Chicago: University of Chicago Press, 2002.

R. Peckham (ed.), *Disease and Crime: A History of Social Pathologies and the New Politics of Health*, New York: Routledge, 2014.

N. Rafter, *Creating Born Criminals*, Urbana: University of Illinois Press, 1997.

C. E. Rosenberg and J. Golden (eds), *Framing Disease: Studies in Cultural History*, Newark: Rutgers University Press, 1992.

T. Szasz, *The Medicalization of Life*, Syracuse: Syracuse University Press, 2007.

J. Terry, *An American Obsession: Science, Medicine, and Homosexuality in Modern Society*, Chicago: University of Chicago Press, 1999.

# Part II

# PATTERNS

# 8

# PANDEMICS

## *Mark Harrison*

The last few decades have seen numerous scares over the spread of disease – not only 'classic' pandemic diseases such as influenza but many previously unknown ones such as MERS (Middle Eastern Respiratory Syndrome), SARS (Severe Acute Respiratory Syndrome) and Ebola. Respected figures in public health have warned that we are entering a new 'Age of Pandemics' in which globalization, environmental degradation and climate change will combine to produce a maelstrom of infection.[1] But while there is widespread agreement about the potential for pandemics to develop in our globalized world, the response to these threats has been far from consistent. In some cases, it has been tardy; in others, heavy-handed. The competitive pressures generated by globalization have even induced some states to turn pandemics to their advantage, using them to justify measures that evade international law or compromise civil liberties.[2] Pandemics have thus become powerful tools in national and international politics, susceptible to being manipulated for political, economic and professional gain.

In these respects, our present 'Age of Pandemics' bears some resemblance to earlier times. The most obvious analogy is with the nineteenth century, when diseases such as cholera and plague began to circulate rapidly in a world newly connected by railways and steam-navigation. Imperial powers initially responded to this threat by trying to contain these infections, creating a sanitary barrier against places and peoples deemed generative of disease. The situation today is rather different. Certain regions such as tropical Africa retain their stigma but it is evident that pandemics may arise anywhere. The recent 'pandemic' of H1N1 'Swine Flu' had its origins in North America, while BSE (Bovine Spongiform Encephalopathy) and its human equivalent, variant CJD (Creutzfeldt-Jakob Disease) originated in the United Kingdom. Another crucial difference from the imperial era lies in our attitudes to infectious disease. From the late nineteenth century, confidence in the ability of governments to combat infectious diseases mounted and, by the middle of the twentieth century, it seemed that they might even be eradicated. But with the exception of a handful of diseases such as polio or leprosy, such ambitions are now more rhetorical than real. Rather, pandemic diseases tend to be conceived of as risks to be mitigated, accepting the near impossibility of their elimination.

Risks are relative truths, which are, in turn, responses to uncertainty. They are products of knowledge and cannot be overcome by greater knowledge. In a public health context, the implication is that we cannot fully control the factors that contribute to disease. Risk analyses aim to determine the probability of diseases occurring

129

in certain circumstances and their likely impact if risks are not mitigated. They differ from earlier attempts to describe disease statistically because their intention is to anticipate or, in effect, to manufacture future catastrophes.[3] These imagined events can acquire a life of their own, generating research, policy and commercial opportunities. But the influence of risk upon public health – and of the uncertainties it is supposed to reduce – is poorly understood. In this chapter, I shall examine the culture of risk in relation to what are rather loosely termed 'pandemics', as well as the idea of security with which both are intertwined.

## Pandemics in a world before risk

Until the last century, the present usage of the term 'pandemic' was rare. Although the word has an ancient lineage, it was generally used to refer to social *mores*, particularly so-called vulgar forms of love ('pandemic love'). Nor, on the rare occasions when it did pertain to disease, did the term 'pandemic' refer only to maladies considered contagious, as it does now. Such was the case with yellow fever in the eighteenth century. Although some people did regard this malady as contagious, most insisted that it resulted from accelerated putrefaction in hot climates.[4] In this sense, usage of the term pandemic followed those ancient authors who ascribed such outbreaks to excesses of heat or humidity.[5] As yellow fever was confined largely to West Africa and the Caribbean, it could easily be considered a climatic disease. But yellow fever was known occasionally to move out of its normal abodes, doing so spectacularly in the 1790s, when outbreaks in the Caribbean were followed by epidemics along the Eastern seaboard of North America.[6] By 1800, the disease had also reached the Mediterranean shores of Europe, probably for the first time.[7] In view of this expansion, more people came to regard yellow fever as a contagious disease. But it was an opinion that remained controversial and the term 'pandemic' was still seldom applied.

Although many contemporaries viewed these outbreaks as random events, the 1790s marked the beginning of a great epidemiological upheaval – the most important since the sixteenth century. The coming decades saw the transformation of cholera into an epidemic disease and the resurgence of many others, including plague and animal diseases such as rinderpest.[8] All circulated the globe with a rapidity that was previously unimaginable. But unlike earlier periods of turmoil – the First Plague Pandemic of the sixth to eighth centuries,[9] the Black Death of the fourteenth century,[10] and the Columbian Exchange of the sixteenth century[11] – there is no label that one can readily attach to the collective horrors of the 1800s. Indeed, the epidemiological significance of the nineteenth century has been generally overlooked.[12] One reason for this is that the impact of epidemic disease was generally less catastrophic in demographic terms. Whereas the Black Death removed between a third and a half of the population of Europe in the fourteenth century, no country experienced such a drastic decline in the nineteenth. Mortality could often be high but it tended to be localized and of little long-term consequence compared with endemic diseases such as malaria.[13] Historians of this period have also tended to concentrate on single epidemics or countries, their interest being chiefly to analyse social tensions which surfaced at periods of crisis.[14] While some scholars have considered infectious diseases on a larger canvas,[15] we still have little sense of the relationships between the century's numerous pulses of pandemic and panzootic disease.

The Victorian 'Age of Pandemics' was unprecedented in that it was the first to affect all inhabited continents. Similarly, diseases did not spread in any particular direction, as they had done in the past. Previously, plague had travelled predominantly from East to West, while most other diseases spread from the Old World to the New – syphilis being the sole, probable exception. In the 1800s, however, plant and animal diseases originated in the Americas as well as in Europe and Asia. New diseases – like cholera – appeared and spread simultaneously in many directions. Many contemporaries also believed that there was a close relationship between these disparate events: some attributing them to divine intervention, others to new and expanding networks of trade. Despite growing awareness of the scale of this epidemiological upheaval, the term 'pandemic' was rarely used during the 1800s and the term 'panzootic' was wholly unknown. Nor, on the rare occasions when a 'pandemic' was referred to, did it imply that the disease was contagious. In the 1860s, for instance, the British Army doctor Robert Lawson used historical data to show the influence of what he termed 'pandemic waves' – mysterious geomagnetic forces which were said to account for global fluctuations in a range of epidemic diseases. According to this theory, the wave would interact with local conditions – social, hygienic and meteorological – to produce a variety of ailments.[16]

This theory attracted attention for a while but its adherents had largely disappeared by the 1870s.[17] By that time, the waterborne theory of cholera had taken root and the notion of pandemic waves seemed, to many, to be wilfully obscure. However, the wave theory was a product of several decades in which medical statistics had been collected in numerous locations around the world. Colonial, military and maritime practitioners played an important part in this and were a major force in the new science of epidemiology; a discipline that enabled disease outbreaks to be understood statistically, and their course to be plotted and depicted graphically. As Robert Peckham points out in this volume, this endeavour was often frustrated by microbial movements that appeared random and inexplicable. Nevertheless, epidemiological studies allowed the scale, and to some extent the movement, of pandemics to be comprehended by both the medical profession and the lay public. British newspapers such as *The Times* reported the annual lectures of the presidents of the Epidemiological Society of London, for example.[18] Yet some people – including medical professionals – continued to regard cholera, plague and yellow fever as purely local phenomena: as products of particular places or climates. In this respect, there was a close analogy between war and epidemics. Both were viewed as likely to emanate from places ('seats') which were inherently unstable or dangerous.[19] For commercial and geopolitical reasons, the British were more inclined towards this localized view of disease. At the other end of the spectrum were the French, who tended to stress the communicability of diseases such as cholera – and the need for quarantine – partly in order to disrupt British interests.[20]

By the early 1900s, there was more agreement about the role of human transmission in cholera and yellow fever – diseases which had preoccupied the imperial powers. This became possible, in part, because of the identification of the bacterium causing cholera in 1884 and confirmation of the mosquito vector theory of yellow fever in 1900. Although there was still some dispute about both diseases, there was increasing agreement on the need to regulate the movement of human carriers and animal vectors.[21] Emerging states, such as Meiji Japan, strove to implement quarantine systems of their own, while cholera and yellow fever sometimes provided a pretext for intervention by imperial powers.[22] Some of these initiatives disrupted commercial and other forms of traffic but increasingly accurate epidemiological data (the result of more reliable and timely reporting) enabled preventative measures to be targeted more precisely. Inspection of suspect cases and fumigation of ships began to replace quarantine, thereby reducing damage to trade. Economic imperatives likewise encouraged many states to seek the standardization of protective measures.[23] This culminated in binding international regulations and the foundation of the first international health organizations at the beginning of the twentieth century.[24]

Improvements in international arrangements were accompanied or, more precisely, enabled by sanitary reforms. Many attributed the diminished frequency of epidemics in Western nations not to quarantine but to environmental improvements. Having plagued many nations in the middle years of the century, cholera was increasingly confined to poorer parts of the world, which possessed little or no sanitary infrastructure.[25] It was much the same with yellow fever. Following the terrible epidemic in the USA in 1878 (the worst in the nation's history), there was a concerted effort to improve the sanitary state of port cities and to centralize control measures. Outbreaks became less severe and more easily contained.[26] Yellow fever thus came to be seen as a disease of less developed nations, many of which accepted the assistance of new philanthropic bodies such as the Rockefeller Foundation.[27]

The decreasing threat from cholera and yellow fever meant that these diseases played relatively little part in the genesis of the modern concept of the pandemic. That dubious honour goes to two other infections. The first of these was the so-called Russian Flu of 1889–91 – a disease which claimed victims from all social classes. In this sense, it was very different from cholera, which affected the lower orders predominantly. Its dissemination was also harder to predict, unlike cholera and yellow fever, which normally followed established pathways of commerce and pilgrimage.[28] The second of these diseases was plague. Spreading from southern China in the 1890s, the re-emergence of this ancient malady – long confined to endemic areas in China and the Middle East – shocked many because of its associations with the medieval past.[29] Ironically, plague had been released from its Asian captivity by the forces of modernity; by new trade routes (especially those established by the British) and new technologies such as the steamship and the railway. The pandemic was experienced primarily within the world's great sea-ports, the major exceptions being India and Manchuria, where the disease caused enormous mortality and disturbance inland. In the latter cases, its movement was closely associated with transportation by rail.[30]

More than ever before, the rampages of disease were discussed in popular media, which meant that *fin de siècle* pandemics were very much social phenomena – the products of an increasingly global public sphere. For much of the nineteenth century, the literate elites of Europe and the Americas had read about the spread of disease but some time had generally elapsed between disease outbreaks and reports in newspapers, especially those far distant from events. Pandemics were experienced as disjointed outbreaks and information about them was not always widely or systematically disseminated. One reason for this was that the vast majority of people continued to receive information about epidemics by word of mouth, often in the form of rumour. This was still largely true at the end of the century but information was becoming easier to obtain. Many countries – including some of the European colonies – had seen a rise in literacy due to initiatives in elementary education and there was a corresponding increase in popular news media. Access to newspapers was no longer confined to the elite. Many of the publications aimed at the masses contained regular reports on outbreaks of disease and the apparent course taken by what were increasingly termed 'pandemics'. These reports were sometimes accompanied by photographs depicting the peoples and places affected by plague and the often drastic measures used to contain it, such as mass quarantine and destruction of property. This probably heightened the sense of dread that many readers would have felt but also enabled an idea of the pandemic to crystallize in the public imagination. Pandemics thus became more definite entities, with presumed points of origin, sequential development and termination. But as well as existing in particular locations, pandemics inhabited a new kind of space: a relational space or *space of flows* constituted by new technologies of communication and transportation.[31]

The increasing immediacy of news coverage was enabled by the expansion of the electrical telegraph. Telegraphic intelligence was also invaluable for port-health officials and others who were charged with monitoring the spread of disease and taking pre-emptive action. But the media rarely subjected reports of disease to critical scrutiny. Information normally came in bite-sized chunks, with little or no analysis. Thus, for the general public, the principal effect of telegraphic reporting was to convey a sense of rapid movement and imminent threat. The result was often panic and, more insidiously, a generalized anxiety which may have been a precondition for the culture of risk which would emerge over the coming century.[32] It may, at any rate, help to explain the large spike in the modern usage of the term 'pandemic', a word which conveyed the magnitude of global catastrophes – real and imagined. Database tools such as Google Books Ngrams Viewer, *The Times* Digital Archive, 1785–2009 and ProQuest's *The Times of India* Online (1861–) reveal very few references to 'pandemics' before the mid-nineteenth century. Usage of the term rose at the end of the century subsequent to the Russian influenza and plague but it was not until a much more deadly pandemic occurred in the form of the 1918–19 influenza that the term became commonplace, even in medical circles.[33] It was in the aftermath of this cataclysmic event (which claimed around 50 million lives) that virologists, epidemiologists and others began to digest mortality data and detect patterns which some turned into models. The idea that pandemics (at any rate, those of influenza) followed predictable courses took hold and became an important element of pandemic prevention and disaster planning.[34]

## Security and uncertainty

Later in this chapter, I will show how the concept of the pandemic became steadily more important in national and international public health. Before that, we need to consider the history of two other concepts with which it would become entangled – security and risk. Let us begin with the former. According to some scholars (chiefly in the field of international relations) security is a foundational and enduring principle in public health.[35] But the principle of security is more elusive than one might expect, for there is little agreement about what it actually means. For some, security amounts to no more than the protection of nation states and their citizens from clearly identifiable enemies – external and internal. At the other extreme, there are rather vague formulations of 'human' and 'global' security, which refer to protection from what might be termed 'global risks' such as climate change. Security is also recognized by some political theorists as having a subjective or inter-subjective element and is therefore conceived differently in different states as well as over time, as is the balance between considerations of security and those of liberty.[36] For these reasons it is difficult to identify a principle of security operating persistently and unambiguously in the field of public health. Health security has several dimensions, too. Firstly, there is 'biosecurity' within state borders. This attempts to neutralize the hazards that may arise from activities such as farming, food processing and retailing, as well as those arising from the environment more generally.[37] Secondly, there is security at the border: the sanitary policing of immigration, trade and so forth. Thirdly, there is the principle of intervention in other states in order to protect the health or interests of the intervening nation. While the first two notions of security have been evident in varying forms for centuries, the latter is a more recent idea and distinguishes some recent attempts at pandemic prevention from earlier ones.

But security, however one chooses to define it, is rarely the sole determinant of public health. From its origins in the Renaissance, public health has been animated to varying degrees by the idea of the 'common good' and continues to embody notions of social justice and collective responsibility.[38] Even when it comes to secular action designed to defend the integrity and stability of nation states, the modern principle of security fails to encompass the numerous ways in which public health has been used in statecraft. Arrangements made for quarantine, for example, were intended not merely to keep disease at bay but for social control. Sanitary arrangements also came to figure strongly in diplomacy and one of the most important factors governing the use of quarantine over the last three or four centuries has been the need to meet the expectations of *foreign* governments – not considerations of domestic security.[39] Sanitary measures have also evolved from an exclusive focus on nation states to the policing of international networks – a process in which global and regional bodies have assumed increasing importance.[40]

If security cannot be regarded as a transcendent principle in public health, what of that other characteristic feature of contemporary pandemic prevention – risk? Again, there are substantial differences between modern attitudes and those of barely a century ago. What we would now refer to as 'risk assessments' did play some part in determining the safety of medical procedures or in actuarial work, but they did not feature prominently in other areas of public health. The dangers presented by pandemic or epidemic disease were referred to in less precise terms; for example, as 'threats', 'dangers' or 'menaces'.[41] After World War I, risk analysis became more common as a result of publications like Frank H. Knight's *Risk, Uncertainty and Profit* (1921). Knight argued that uncertainty could never be eradicated from any walk of life but that it could be reduced, in so far as this was compatible with other political objectives.[42] By the middle of the twentieth century, the idea that uncertainty could be managed through the calculation of probabilities became widespread and integral to some aspects of medicine, for example in determining the safety of new drugs or inoculations like BCG against tuberculosis.[43] But it was not yet implicated in the evolving notion of the pandemic, even though the management of uncertainty was becoming a feature of international health.

There are intimations of this in the work of bodies such as the League of Nations (founded in 1919), whose Health Organization gathered epidemiological intelligence and orchestrated action to prevent pandemics. Pandemic prevention was the principal objective of the League's Health Organization and also the chief reason for its refugee relief activities in Eastern Europe and the Balkans/Middle East after the end of World War I.[44] One can discern in the work of the League and its partners an attempt to anticipate the catastrophe of 'pandemic' disease and to do so with an alliance of actors – state and non-state – which is familiar to us today. The international sanitary conferences of the preceding decades resulted in the standardization of national sanitary laws and established offices that collected epidemiological data. The League's Health Organization augmented these initiatives but its anti-epidemic activities were different because they were cooperative interventions, often conducted in association with voluntary bodies such as the Red Cross and Save the Children.[45] Although they were designed to serve humanitarian ends, the main purpose of these interventions was to protect the interests of member states. In both senses they may be considered an important precursor to recent ventures in the name of 'global health'. But while conditions in some countries were said to pose a threat to other nations, the language of risk was rarely used. Some distance therefore separates assessments of pandemic threats over the last few decades and those of earlier times.

## Remaking pandemics as risk phenomena

The relationship between pandemics, security and risk became more explicit during World War II, with the development of what came to be known as 'medical intelligence'. When the US Army prepared to enter the war, a Medical Intelligence section was created within the Office of the Surgeon General of the US Army and such intelligence became an integral part of operational planning.[46] Similar arrangements were made in the armies of other combatant nations. Although the main concern of such investigations was the potential impact of local diseases on military populations, the military and civil authorities were also concerned about the potential for wartime and post-war conditions to bring about a major pandemic, as in 1918–19. In Germany, Japan and other occupied territories, as well as in newly liberated countries like Korea, information about disease was collected systematically in detailed reports. The intelligence and medical communities began to work more closely together and their reports considered the risk factors existing in different locations with a view to determining and mitigating the threats posed by disease to Allied troops and social stability.[47] These concerns reflected the need to maintain order in the ideological (and sometimes literal) front-line between communism and capitalism.

Pandemic disease threatened homeland security, too. Even before the war had ended, the USA had begun to consider the implications of another pandemic on the scale of 1918–19. To assist this process, the US Army established a Commission on Influenza in 1941, while work began on the development of a vaccine. The Commission survived the war and, in 1957, it attempted a risk assessment to determine the likelihood of an influenza pandemic emerging from the Far East. A similar enterprise was conducted in 1968 but had no discernible impact on policy. At this time, such assessments were rare in the epidemiology of epidemic diseases, although they were already familiar in the epidemiology of chronic diseases, where the notion of the 'risk factor' was generally accepted.[48] As far as influenza was concerned, the epidemiological data suggested that influenza A virus pandemics occurred on average every 10–11 years and the assumption that this would be repeated led to an unprecedented mobilization of resources following an outbreak of H1N1 at Fort Dix, New Jersey, in 1976. This outbreak spurred a rapid risk assessment and a decision was made to purchase large quantities of influenza vaccine. A national vaccination pro- gramme was implemented (championed by President Ford) and 40 million people were immunized before the US public health authorities were forced to acknowledge that the predicted pandemic had failed to materialize. The Ford administration was heavily criticized for this decision, which, apart from wasting resources, produced serious side-effects (such as Guillain-Barré Syndrome) in some of those vaccinated.[49]

Unsurprisingly, these events were not followed by a rash of similar examples. But risk analysis began to figure more prominently in pandemic control after the emer- gence of HIV/AIDS in the early 1980s. As Allan Brandt has argued, AIDS produced important changes in what was formerly termed international health, giving birth to a new field – global health. According to Brandt, this was distinguished by a union between clinical and preventive medicine and a greater emphasis on human rights.[50] But while this slow-burning pandemic brought a new sense of global solidarity in mat- ters of health, risk was becoming more prominent in the analysis of infectious disease, with the emergence of new concepts such as 'risk groups' and 'risk behaviour' in relation to HIV. In Western countries, this sometimes translated into coercive action but generally found expression in public health campaigns based on education and self-regulation.[51] In the latter sense, the response to AIDS was in accord with what some branded the 'New Public Health'.[52] However, risk assessments also formed the basis of a new vision of international health, in which risk was more closely aligned with national security agendas.

Pandemics had long been regarded as threats to the stability and wealth of states but for some years they had not featured prominently as a security concern. The advocates of 'securitization' now sought to extricate public health from the liberal, internationalist paradigm that they believed to have been dominant since the foun- dation of the WHO in 1948.[53] This shift in thinking can been seen in the US National Intelligence Council briefing of 2001, which listed HIV/AIDS and other diseases as direct threats to the interests of US citizens, especially in countries in which the US had important economic and strategic interests.[54] But there was nothing uniquely American about this new doctrine, for it originated in the work of the British intelli- gence services. Nor was it simply a consequence of AIDS. Although AIDS was by far and away the most serious medical threat to the stability of countries in sub-Saharan Africa, it was one among many diseases which were appearing or reappearing as signif- icant threats to humanity. Publication databases show a massive increase in references to 'pandemics' at this time. Paramount in most people's minds was the heightened risk posed by globalization and the advent of cheap mass air travel. Closely allied to these fears was climate change, which allowed many pathogens and vectors to expand their range. Following the fall of the Soviet Union, there were also concerns that that nation's expertize or its stockpile of biological weapons would fall into the hands of terrorists or states that were likely to use them.

These complex and interrelated threats were increasingly described in the lan- guage of risk, which reflected growing uncertainty about science as well as the future of the planet. Ever since fatalistic attitudes to disease began to give way to a belief in the capacity of human intervention, many had speculated about its mastery. By the middle of the twentieth century these dreams seemed to be becoming reality, as a battery of new inoculations eradicated diseases from the developed world and then, in the case of smallpox, from the world as a whole. This coincided with the advent of many powerful antibiotics and new insecticides like DDT, capable of killing the vectors of diseases such as malaria.[55] The prospect of victory over many infectious diseases seemed imminent. But soon after the creation of the first antibiotics, drug-resistance began to be noted and continued to spread thereafter.[56] The destruction of disease vectors with DDT also proved more difficult than expected because of insecticide resistance and heavy expenditure. At the same time, many began to criticize its harmful effects on wildlife.[57] Concern was also expressed over antibiotic residues in food and diseases arising from over-medication.[58] The confidence that many had previously placed in science was evidently evaporating, even before AIDS demonstrated the remarkable persistence and vitality of humanity's microbial foes.

Probability rather than certainty reflected the new and potentially frightening possibilities of a world in which science no longer ruled the roost. New concepts such as 'emerging' and 'reemerging' infections encapsulated these fearful possibilities and subtly encoded many assumptions about the risks presented to the developed world by marginal populations and failed or poorly developed states.[59] Films such as *Outbreak* (1995) and best-selling books such as Richard Preston's *Hot Zone* (1994) reflected and contributed to public disquiet. In these and similar works, pandemic disease came to express other concerns, among them climate change, environmental degradation and globalization – not to mention terrorism.[60] But these disparate anxieties became so entangled as to become indistinguishable and with this entanglement came a narrowing of focus. Thorny and complex problems of poverty and climate change were compressed or sidelined and pandemics came to be seen primarily as problems of circulation, amenable to control by technological intervention at key strategic points.

This was the essence of most of the calls for surveillance and health 'securitization' in the 1990s and 2000s. Superficially, these demands resembled the impulses of Victorian times, when imperial powers attempted to create a sanitary buffer between themselves and localities they regarded as pathogenic.[61] But the analogy between the globalized world and the Victorian era fails to capture the complexity of our times. Power is more widely dispersed today than it was in the nineteenth or early twentieth centuries and the precautionary attitudes exhibited by the USA or European countries may now be found equally in China, Indonesia or Singapore – perhaps more so. The movement of pandemic disease is also less predictable than at the height of the imperial era, when it moved from country to country, predominantly by railway or ship. While the distribution of disease by air can easily be modelled, the multiplicity of possible transmission routes and the difficulties of detection are infinitely greater.

## Pandemics and globalization

If contemporary pandemic control is different from old style sanitary imperialism, what can we say about it? The obvious place to start is with the SARS pandemic of 2003, which set the agenda for pandemic control in the coming decade. Although it killed relatively few people by contrast with most other pandemic diseases (8,422 recorded cases and 916 deaths), the uncertainty surrounding SARS caused great alarm and threatened for a time to destabilize the global economy. Coming in the wake of the terrorist attacks of September 2001, the response to SARS was infused with the rhetoric of the War on Terror. Nations that failed to observe the new protocols of 'germ governance' (the requirement of governments to determine and mitigate risks) were seen as little better than the 'pariah states' which sponsored terrorism.[62] There was a sense that the world had become a more dangerous place, in which nations would struggle to protect their sovereignty and their borders.

The response to SARS invariably took the form of quarantine and isolation; violation of which, in some countries, was severely punished. While there was general acceptance that strict measures were desirable, some governments were criticized for unnecessary harshness. In Taipei, for example, the city's homeless population was rounded up because of fears they were likely to spread the disease.[63] There, and in several other countries, healthy people were placed in quarantine alongside the infected.[64] These reactions were driven by fear – not simply fear of infection or social disorder but of a failure to meet international expectations: states which appeared weak or negligent faced censure and the loss of business and investment.[65] Governments and commentators took stock of the potential losses from a more severe pandemic and began to issue dire warnings about imagined future catastrophes. Most continued to emphasize the need for 'early intervention and risk assessment'.[66]

The new mood was encapsulated in the World Health Organization's International Health Regulations (IHR), formulated in 2005, coming into force in 2007. The IHR require WHO's 194 member countries to notify the WHO of any public health emergency that may constitute a matter of international concern. States also have an obligation to maintain health surveillance at their borders. Having received information from member states, the WHO decides whether or not to declare an international public health emergency.[67] The main aim of these regulations is to control, but not disrupt, the global flow of people, animals and commodities. Pandemic protection is thus in one sense extra-territorial but national systems of surveillance, particularly quarantine and screening at airports, are also vital to the functioning of the global network which has been assembled to govern the circulation of people and commodities.

This ordering of public health conceives of disease primarily as a problem of mobility within a global system. As the sociologist Sven Orpitz has argued, it is a 'thoroughly post-humanist affair', in which the social person reverts to an organic entity – like plants or cattle – to be screened for infection.[68] Pandemics, similarly, have been extricated from the social, industrial and environmental contexts in which they arise. One beneficial aspect of this 'de-socialization' is that disease controls should, in theory, be free from racial or other forms of prejudice. In practice, this is not always the case. The recent Ebola epidemics produced several instances of disengagement from African countries (for example, by airlines), discouragement of travel from West Africa, and in the North Korean case, a complete ban on non-business or diplomatic travel.[69] Nor have the IHR – which are advisory rather than legally binding – been able to prevent serious economic disruption. During the H1N1 'pandemic' in 2009, for example, many countries imposed a ban on pork from North America despite a declaration from the WHO that such produce was safe. This prompted allegations of sanitary protectionism – a refrain that has become increasingly familiar since the dismantling of formal tariff barriers and the creation of the World Trade Organization in 1995.[70]

The WHO's declaration of a pandemic in 2009 revealed the term's remarkable elasticity. Although the WHO lacked a precise definition of what a pandemic was, it had formulated what it later referred to as a 'definition-description', the essential features of which were that the term should be applied only to infectious diseases that had spread over two or more continents. In 2009, however, one of the previous criteria was omitted. A 'pandemic' disease no longer needed to be 'especially destructive of human life' in order to merit that description – this proved to be the case with Swine Flu. After it became apparent that H1N1 was no deadlier than ordinary 'seasonal' influenza, the credibility of the WHO and national public health agencies (which often made exaggerated predictions of mortality) was called into question. There were concerns, too, over links between WHO advisors and pharmaceutical industries, and widespread criticism of a response that many regarded as disproportionately aggressive.[71]

The 2009 Swine Flu 'pandemic' illustrates not only the plasticity but also the potency of the term. As Charles Rosenberg once observed, the simple act of naming an 'epidemic' elicits primordial fears of death and chaos.[72] The 'emotional urgency' generated by such a designation is magnified many times in the case of pandemics – a word that has become synonymous with global catastrophe. But rather than bringing states together to face a global threat, the announcement of a pandemic seems as likely to divide them. In 2009 many states used the laxity of the IHR – and escape clauses in international trade law – to gain competitive advantage over others, imposing import restrictions much stricter than those that would normally be permitted. The only way of avoiding the commercial chaos that can so easily arise from such emergencies is to seek agreements similar to that reached by the Asia Pacific Economic Cooperation (APEC) prior to the outbreak of SARS.[73] Anticipating such an eventuality, APEC determined that member states would not use disease as a pretext for gaining advantages in investment or trade. Without such agreements, risk assessments are incapable of providing a platform for united action in the face of a public health crisis. Science has never been, and is never likely to become, a neutral arbiter.

Nevertheless, the value of risk assessments in pandemic prevention remains largely unquestioned. Virtually all countries now possess 'risk registers' in which pandemics figure prominently. The private sector is also heavily involved. For example, the UK-based company Maplecroft provides an Influenza Pandemic Risk Index, which 'enables governments, intergovernmental organizations and business to identify potential risks to populations and supply chains'. Countries are ranked and classified according to different risk levels along three indices: the risk of the emergence of new strains, the risk of their spread, and the capacity of states to contain an epidemic. This entails the analysis of cultural factors influencing behaviour, environmental factors, and the perceived adequacy of regulations.[74] Agencies responsible for domestic and global health have also responded to the pervasive culture of risk by making its analysis central to their attempts to control pandemic disease. The US Centers for Disease Control (CDC), for example, have recently developed an Influenza Risk Assessment Tool (IRAT) to evaluate the human pandemic potential of emergent strains of influenza A virus in animal populations. IRAT uses 10 criteria grouped around three themes: the properties of the virus; attributes of the risk population; and the ecology and epidemiology of the virus. This analysis takes account of both the risk of a new strain emerging and of its potential impact on human health. However, the CDC explains that IRAT is not a predictive tool and that the risks it identifies are not precisely quantified.[75] The WHO, similarly, has placed risk analysis at the centre of its recently published guidance on pandemic influenza. The aim behind this approach is to enable countries to respond more flexibly than in 2009, when governments were unsure how to deal with a disease less severe than anticipated.[76]

Risk assessments are generally valued because they appear to provide an objective and flexible basis for intervention in public health but some critics insist that risks are merely social constructions that enable interests to be advanced beneath a veil of objectivity. Despite some degree of convergence between these contrasting viewpoints, a synthesis remains elusive.[77] If or until this is achieved, the most important questions are: Who is to conduct risk assessments? Who should decide on which risks should be mitigated? And in what ways should they be mitigated? Technocratic and participatory models of risk assessment each have their advocates, in which experts are either free to make such assessments or are subject to some degree of deliberation.[78] Equally, if oversight is regarded as necessary, it is far from clear who should exercise it.

Presently, it would seem that both the assessment of risk and its mitigation are skewed in particular directions. There is a distinct preference for surveillance and containment, with calls for 'early-warning systems' that will detect and track the emergence of new strains of disease, the aim being to create a kind of 'global immune system'.[79] Another (unintended) consequence of imagining pandemics as risk phenomena is that it normalizes them, leading us to accept such events as rare but inevitable by-products of modern life. Recent discussions about pandemic disease – particularly influenza – have been characterized by a 'not if but when' attitude, although the 'when' seems never to be determined. Despite the existence of an industry of pandemic prophecy, the virus, as Carlo Carduff has noted, is always one step ahead. Indeed, prophecy seems to thrive on uncertainty. Experts of varying stripes ask the public to accept on faith the 'inevitability' of the next viral cataclysm despite never knowing when or even whether that event will occur.[80] And, if rational calculation can never master the microbe, it would seem that all that can be done is to prepare for the consequences; hence the current orientation towards disaster management.

In these respects, pandemic prevention stands in marked contrast from other areas of public health, especially those for which a new approach or 'Fifth Wave' is being openly advocated. Focusing largely on chronic disease, advocates of the Fifth Wave recommend a more holistic approach in which health is closely related to issues such as social inequality and environmental sustainability.[81] This new thinking has made little headway in the dryly securitized world of pandemic prevention but there are some intimations of change. In a 2012 article in *PLOSMedicine*, Bogick *et al.* propose that pandemic prevention be seen as an aspect of international development, with attention directed to health-care infrastructure and the alteration of pathogen dynamics between humans and their environments, particularly their relationships with other animals.[82] The recent epidemic of Ebola shows the necessity of such an approach but it shows, too, that it has yet to be taken seriously. Of the factors which enabled the disease to spread widely in West Africa, the absence of health-care infrastructure was paramount. Similarly, this and previous outbreaks had their origin in high-risk practices – the consumption of bushmeat – which have become entrenched through ignorance and poverty.

## Conclusion

In this chapter I have argued that 'the pandemic' is a relatively recent idea and that it covers an enormous range of possibilities, from potentially cataclysmic mortality on the scale of 1918–19 to a disease like SARS, which claimed fewer than a thousand lives. This imprecision represents a weakness in some respects but it is also tactically useful. Labelling a disease 'pandemic' endows it with a potency lacked by others, even those considered to be epidemic. When a pandemic is announced – and particularly, as in the case of Swine Flu, when this is accompanied by intense media speculation – it generates fear and with fear comes a craving for security, or a particular idea of security at least. There is also an expectation that states – as well as other public and private bodies – should take robust measures to avoid liability. This creates numerous opportunities, including the power to garner resources.[83] Over the last few years, these have flowed chiefly in the direction of medical and pharmaceutical research, border security, and epidemiological surveillance. These are all clearly important but the commercial, social and environmental causes of pandemics – particularly the factors driving disease emergence – are receiving far less attention.

These biases emerge at the level of risk assessment and, in a more obvious way, when it comes to mitigation. Presently, the processes by which these decisions are taken are opaque. What is clear, however, is that risk, security and pandemics are bound together in a powerful nexus. Just as pandemics tend to evoke a very particular kind of security response, risk assessments have become increasingly selective, conditioned by dominant conceptions of security. There is a circularity and exclusivity in these relationships, which appear to be animated not by ideals of human welfare but by the interests of powerful corporations and nation-states. Thus, responses to international public health emergencies have been either robust or inadequate depending on the extent to which these interests appear to be threatened. In the case of SARS – which affected relatively few people but posed a serious threat to international commerce – the response of most countries was decisive. In the case of the recent Ebola epidemics, which had a catastrophic effect on the afflicted countries, there was a strong desire on the part of many people and businesses to disengage from infected regions, while national governments and global bodies were initially reluctant to intervene. Eventually, nascent ideas of global responsibility – articulated most forcefully outside the governmental sector – and probably a belated awareness of the security and economic implications of state failure prompted a change of policy. But the response to Ebola – like that of some nation states to the last pandemic of Swine Flu – shows a worrying lack of coordination and, in the latter case, a tendency to exploit public health emergencies for commercial gain. The culture of risk arguably contributes to this fragmentation by encouraging an ethics of liability as opposed to one of collective responsibility. If so, this is particularly problematic for those involved in global health because it rubs abrasively against the core ideals of the movement – equality and human rights.

## Notes

1 Nathan Wolfe, *The Viral Storm: The Dawn of a New Age of Pandemics*, London: Allen Lane, 2011; Larry Brilliant, 'The Age of Pandemics', *Wall Street Journal*, 2 May 2009, http://online. wsj.com/article/SB12412196574048983.html; accessed 16/05/15.

2  Ulrich Beck, *World at Risk*, Cambridge: Polity, 2009, pp. 175–6.

3  Ulrich Beck, *Risk Society: Towards a New Modernity*, trans. M. Ritter, London: Sage, 2000.

4  E.g. Benjamin Moseley, *A Treatise on Sugar, with Miscellaneous Medical Observations. Second Edition, with considerable Additions*, London: John Nichols, 1800; Noah Webster, *A Brief History of Epidemics and Pestilential Diseases*, vol. 1, Hartford: Hudson & Goodwin, 1799, pp. 11–12.

5  See, for example, Ammiamus Marcellinus, *The Roman History of Ammianus Marcellinus*, trans. C.D. Yonge, London: Henry G. Bohn, 1853, Book XIX, p. 122.

6  J. Worth Estes and B.G. Smith (eds), *A Melancholy Scene of Devastation: The Public Response to the 1793 Philadelphia Yellow Fever Epidemic*, Philadelphia: College of Physicians of Philadelphia, 1997; Sean P. Taylor, '"We live in the midst of death": Yellow Fever, Moral Economy and Public Health in Philadelphia, 1793–1805', Northern Illinois University PhD thesis, 2001; J.H. Powell, *Bring Out Your Dead: The Great Plague of Yellow Fever in Philadelphia in 1793*, Philadelphia: Philadelphia University Press, 1949.

7  Mercedes Pascual Artiaga, *Fam, malatia I mort: Alacant I la fibre groga de l'any 1804*, Simat de la Valldigna: La Xara, 2000; M. Cabal, 'Medidas adoptadas por la Junta de Sanidad del Principado ante la posible invasión peninsular de la fiebre amarilla existente en Cádiz y Real Isla de León en el siglo XVII al XIX', *Boletín del Instituo de Estudios Asturias*, 42, 1988, 409–28; William Coleman, *Yellow Fever in the North: The Methods of Early Epidemiology*, Madison: University of Wisconsin Press, 1987.

8  Karen Brown and Daniel Gilfoyle (eds), *Healing the Herds: Disease, Livestock Economies, and the Globalization of Veterinary Medicine*, Athens, OH: Ohio University Press, 2010; Pule Phoofolo, 'The Rinderpest Epidemics in Late Nineteenth-Century Southern Africa', *Past & Present*, 138, 1993, 112–43; Terrie M. Romane, 'The Cattle Plague of 1865 and the Reception of the "Germ Theory" in Mid-Victorian Britain', *Journal of the History of Medicine and Allied Sciences*, 52, 1997, 51–80. See also the chapters in this volume by Robert Peckham and Akihito Suzuki.

9  William Rosen, *Justinian's Flea: Plague, Empire and the Birth of Europe*, London: Viking, 2007; Lester K. Little (ed.), *Plague and the End of Antiquity: The Pandemic of 541–750*, Cambridge: Cambridge University Press, 2007.

10  Ole J. Benedictow, *The Black Death 1346–1353: The Complete History*, Woodbridge: The Boydell Press, 2004; Stuart J. Borsch, *The Black Death in Egypt and England: A Comparative Study*, Austin: University of Texas Press, 2005; Michael J. Dols, *The Black Death in the Middle East*, Princeton: Princeton University Press, 1977.

11  Alfred W. Crosby, *The Columbian Exchange: Biological and Cultural Consequences of 1492*, Westport: Greenwood Press, 1972; Alfred W. Crosby, *Ecological Imperialism: The Biological Expansion of Europe, 900–1900*, Cambridge: Cambridge University Press, 1986; William H. McNeill, *Plagues and Peoples*, Garden City, NY: Anchor Press, 1976; Noble David Cook, *Born to Die: Disease and New World Conquest, 1492–1650*, Cambridge: Cambridge University Press, 1998.

12  Emmanuel Le Roy Ladurie's influential piece, 'A Concept: The Unification of the Globe by Disease', deals only with the three centuries after the initial outbreak of the 'Black Death'. See his *Mind and Method of the Historian*, Brighton: Harvester, 1981, pp. 28–83.

13  Mark Harrison, 'Disease and World History from 1750', in J.M. McNeill and K. Pomeranz (eds), *The Cambridge World History: Volume 7: Production, Destruction and Connection, 1750–Present*, Cambridge and New York: Cambridge University Press, 2015, pp. 237–58.

14  E.g. Frank Snowden, *Naples in the Time of Cholera 1884–1911*, New York: Cambridge University Press, 1995; Catherine J. Kudlick, *Cholera in Post-Revolutionary Paris: A Cultural History*, Berkeley: University of California Press, 1996; Mariola Espinosa, *Epidemic Invasions: Yellow Fever and the Limits of Cuban Independence, 1878–1930*, Chicago: University of Chicago Press, 2009.

15  Myron Echenberg, *Africa in the Time of Cholera: A History of Pandemics from 1817 to the Present*, New York: Cambridge University Press, 2011; Christopher Hamlin, *Cholera: The Biography*, Oxford: Oxford University Press, 2009.

16  Robert Lawson, 'Observations on the Influence of Pandemic Causes in the Production of Disease', *Transactions of the Epidemiological Society*, 2, 1862–3, 88–100.

17  Charles A. Gordon, *Notes on the Hygiene of Cholera*, Madras: Gantz Bros., 1877, p. 218.
18  See: 'Epidemic Diseases', *The Times*, 12 November 1875, p. 5; 'Epidemic Diseases', *The Times*, 24 November 1876, p. 4; *The Times*, 10 November 1887, p. 9.
19  *Ninth Annual Report of the Local Government Board. Supplement containing Report and Papers submitted by the Medical Officer on the Recent Progress of the Leventine Plague and on Quarantine in the Red Sea*, London: George E. Eyre and William Spottiswoode, 1881, 'The Seats of Plague', map facing p. 2; 'The Seat of War', *North Otago Times*, 2 May 1877, p. 2; 'The Seat of War in Europe', *Reynolds Newspaper*, 19 August 1877, p. 3.
20  Mark Harrison, 'Quarantine, Pilgrimage, and Colonial Trade: India 1866–1900', *Indian Economic and Social History Review*, 29, 1992, 117–44.
21  Peter Baldwin, *Contagion and the State in Europe 1830–1930*, New York: Cambridge University Press, 1999, pp. 123–243.
22  Jeongran Kim 'The Borderline of "Empire"': Japanese Maritime Quarantine in Busan c.1876–1910', *Medical History* 57, 2013, 226–48; Espinosa, *Epidemic Invasions*.
23  Valeska Huber, 'The Unification of the Globe by Disease? The International Sanitary Conferences on Cholera, 1851–1894', *The Historical Journal*, 49, 2006, 453–76; Valeska Huber, *Channelling Mobilities: Migration and Globalisation in the Suez Canal Region and Beyond*, Cambridge: Cambridge University Press, 2013. See also the chapters in this volume by Akihito Suzuki and Monica Garciá.
24  Sylvia Chiffoleau, *Genèse de la santé publique international: De la peste d'Orient à l'OMS*, Rennes: Presses Universitaire de Rennes, 2012; Céline Paillette, 'De l'Organisation d'hygiène de la SDN à l'OMS: mondialisation et regionalism européen dans le domaine de la santé, 1919–1954', *Bulletin de l'Institut Pierre Renouvin*, 32, 2010, 238–53; Marcos Cueto, *The Value of Health: A History of the Pan American Health Organization*, Washington, DC: Pan American Health Organization, 2007.
25  Baldwin, *Contagion and the State in Europe*, ch. 3; Richard Evans, *Death in Hamburg: Society and Politics in the Cholera Years, 1830–1910*, Oxford: Oxford University Press, 1987; Charles E. Rosenberg, *The Cholera Years: The United States in 1832, 1849, and 1866*, Chicago: University of Chicago Press, 2009. On other regions, see: Echenberg, *Africa in the Time of Cholera*; David Arnold, *Colonizing the Body: State Medicine and Epidemic Disease in Nineteenth-Century India*, Berkeley: University of California Press, 1993, ch. 4.
26  Margaret Humphreys, *Yellow Fever and the South*, Baltimore: Johns Hopkins University Press, 1992.
27  Marcos Cueto and Steven Palmer, *Medicine and Public Health in Latin America: A History*, Cambridge: Cambridge University Press, 2015, pp. 106–56.
28  Mark Honigsbaum, 'The Great Dread: Cultural and Psychological Impacts and Responses to the "Russian" Influenza in the United Kingdom, 1889–1893', *Social History of Medicine*, 23, 2010, 299–319; Mark Honigsbaum, *A History of the Great Influenza Pandemics: Death, Panic and Hysteria, 1830–1920*, London: I.B. Tauris, 2013.
29  Carol Benedict, *Bubonic Plague in Nineteenth-Century China*, Stanford: Stanford University Press, 1996; and the chapter in this volume by Sam Cohn.
30  William C. Summers, *The Great Manchurian Plague of 1910–1911: The Geopolitics of an Epidemic Disease*, New Haven: Yale University Press, 2012; Myron Echenberg, *Plague Ports: The Global Urban Impact of Bubonic Plague 1894–1901*, New York: New York University Press, 2007; Mark Gamsa, 'The Epidemic of Plague in Manchuria 1910–1911', *Past & Present*, 190, 2006, 147–84; Rajnarayan Chandavarkar, 'Plague Panic and Epidemic Politics in India, 1896–1914', in T. Ranger and P. Slack (eds), *Epidemics and Ideas*, Cambridge: Cambridge University Press, 1992, pp. 203–40; Arnold, *Colonizing the Body*, ch. 5.
31  Roland Wenzlhuemer, 'Globalization, Communication and the Concept of Space in Global History', *Historische Sozialforschung*, 35, 2010, 19–47.
32  Joanna Bourke, *Fear: A Cultural History*, London: Virago, 2005, p. 351. More generally, see Robert Peckham (ed.), *Empires of Panic: Epidemics and Colonial Anxieties*, Hong Kong: Hong Kong University Press, 2014.
33  Howard Phillips and David Killingray (eds), *The Spanish Influenza Pandemic of 1918–19*, London: Routledge, 2003.

34 Mark Honigsbaum, 'The Great Dread: Influenza in the United Kingdom in Peace and War, 1889–1919', University of London PhD thesis, 2011.

35 Andrew Price-Smith, *Contagion and Chaos: Disease, Ecology and National Security in the Era of Globalization*, Cambridge, MA: MIT Press, 2009.

36 Ian Loader and Neil Walker, *Civilizing Security*, Cambridge: Cambridge University Press, 2007.

37 Stephen J. Collier and Andrew Lakoff, 'The Problem of Securing Health', in A. Lakoff and S.J. Collier (eds), *Biosecurity Interventions: Global Health and Security in Question*, New York: Columbia University Press, 2008, pp. 7–28.

38 Dorothy Porter, *Health, Civilization and the State: A History of Public Health from Ancient to Modern Times*, London: Routledge, 1999.

39 Mark Harrison, *Contagion: How Commerce Has Spread Disease*, London and New Haven: Yale University Press, 2013.

40 Alison Bashford (ed.), *Medicine at the Border: Disease, Globalization and Security: 1850 to the Present*, Basingstoke: Palgrave, 2006.

41 E.g. *The Times*, 'The Threatened Epidemic of Cholera', 18 August 1865, p. 10.

42 Frank H. Knight, *Risk, Uncertainty and Profit*, Boston and New York: Houghton Mifflin Co., 1921, pp. 347–8.

43 Thomas Schlich, 'Risk and Medical Innovation: A Historical Perspective', in T. Schlich and U. Troehler (eds), *The Risks of Medical Innovation: Risk Perception and Assessment in Historical Context*, Abingdon: Routledge, 2006, pp. 1–5.

44 Iris Borowy, *Coming to Terms with World Health: The League of Nations Health Organisation*, Berlin: Peter Lang Verlag, 2009; Alison Bashford, 'Global Biopolitics and the History of World Health', *History of the Human Sciences*, 19, 2006, 67–88; Paul Weindling (ed.), *International Health Organizations and Movements, 1918–1939*, Cambridge: Cambridge University Press, 1995.

45 For example, International Red Cross Missions to Greece and Syria, MS 66 and MS 75–6, ICRC Archives, Geneva.

46 http://en.wikipedia.org/wiki/National_Center_for_Medical_Intelligence.

47 Jessica Reinisch, *The Perils of Peace: The Public Health Crisis in Occupied Germany*, Oxford: Oxford University Press, 2013; 'Medical Notes on Korea', EF 37-1/A, 16-1, 1951 BUMED General Correspondence, 1947–51, RG 52 Record of the [US] Bureau of Medicine and Surgery, National Library, Seoul.

48 Schlich, 'Risk and Medical Innovation', pp. 5–6; Luc Berlivet, '"Association or Causation?": The Debate on the Scientific Status of Risk Factor Epidemiology, 1947–c.1965', *Clio Medica*, 75, 2005, 39–74.

49 George Dehner, *Influenza: A Century of Science and Public Health Response*, Pittsburgh: University of Pittsburgh Press, 2012; Harvey Fineberg V and Richard E. Neustadt, *The Epidemic That Never Was: Policy-Making and the Swine Flu Scare*, New York: Vintage Books, 1983.

50 Allan Brandt, 'How AIDS invented Global Health', *New England Journal of Medicine*, 368, 2013, 2149–52.

51 Peter Baldwin, *Disease and Democracy: The Industrialized World Faces AIDS*, Berkeley: University of California Press, 2005; Virginia Berridge, *AIDS in the UK: The Making of Policy, 1981–1994*, Oxford: Oxford University Press, 1996.

52 Alan Petersen and Deborah Lupton, *The New Public Health: Self and Health in an Age of Risk*, London: Sage, 1996.

53 Price-Smith, *Contagion*.

54 Gwyn Prins, 'AIDS and Global Security', *International Affairs*, 80, 2004, 931–52.

55 Nancy Leys Stepan, *Eradication: Ridding the World of Diseases Forever?*, Ithaca, NY: Cornell University Press, 2011.

56 Scott H. Podolsky, *The Antibiotic Era: Reform, Resistance, and the Pursuit of Rational Therapeutics*, Baltimore: Johns Hopkins University Press, 2013.

57 David Kinkela, *DDT and the American Century: Global Health, Environmental Politics, and the Pesticide that Changed the World*, Chapel Hill: University of North Carolina Press, 2011.

58 Ivan Illich, *Medical Nemesis: The Expropriation of Health*, London: Pantheon Books, 1982.

59 Institute of Medicine, *Emerging Infections: Medical Threats to Health in the United States*, Washington, DC: National Academies Press, 1992; S. Harris Ali and R. Keil (eds), *Networked Disease: Emerging Infections in the Global City*, Oxford: Wiley-Blackwell, 2008.

60 Richard Preston, *The Hot Zone: A Terrifying True Story*, New York: Random House, 1994; Laurie Garrett, *The Coming Plague: Newly Emerging Diseases in a World Out of Balance*, London: Penguin, 1995.

61 Nicholas B. King, 'Security, Disease, Commerce: Postcolonial Ideologies of Global Health', *Social Studies of Science*, 32, 2002, 763–89.

62 David Gratzer, 'SARS 101', *National Review*, 19 May 2003. On 'germ governance', see David P. Fidler, 'Germs, Governance, and Global Public Health in the Wake of SARS', *Journal of Clinical Investigation*, 113, 2004, 799–804.

63 Barbara Demick, 'Taiwan Takes No Chances on SARS', *Los Angeles Times*, 15 May 2003, http://articles.latimes-com/2003/may/15/world/fg-taisars15, accessed 15/09/13.

64 Tim Brookes, *Behind the Mask: How the World Survived SARS, the First Epidemic of the Twenty-First Century*, Washington, DC: American Public Health Association, 2005; C. Low ed., *At the Epicentre: Hong Kong and the SARS Outbreak*, Hong Kong: Hong Kong University Press, 2004.

65 Harrison, *Contagion*, pp. 260–3.

66 E.g. Michael T. Osterholm, 'Getting Prepared', *Foreign Affairs*, 84, 2005, p. 36.

67 World Health Organization, *International Health Regulations (2005)*, 2nd edn, Geneva: WHO, 2005.

68 Sven Orpitz, 'Regulating Epidemic Space: The Nomos of Global Circulation', *Journal of International Relations and Development*, 20 February 2015, doi: 10.1057/jird.2014.30.

69 Business and diplomatic travellers had to endure a 21-day period of quarantine.

70 Harrison, *Contagion*, pp. 271–5.

71 Peter Doshi, 'The Elusive Definition of Pandemic Influenza', *Bulletin of the World Health Organisation*, 89, 2011, 532–38, doi: 10.2471/BLT.11.086173, accessed 13/04/15.

72 Charles E. Rosenberg, 'What is an Epidemic? AIDS in Historical Perspective', in *Explaining Epidemics and Other Studies in the History of Medicine*, New York: Cambridge University Press, 1992, pp. 278–92.

73 Harrison, *Contagion*, p. 264.

74 http://maplecroft.com/about/news/influenza_pandemic_risk_index.html, accessed 10/09/13.

75 http://www.cdc.gov/flue/pandemic-resources/tools/risk-assessment.htm, accessed 12/04/15.

76 WHO, 'Pandemic Influenza Risk Management: WHO Interim Guidance', p. 2, http://www.who.int/influenza/preparedness/pandemic/GIP_PandemicInfluenzaRiskManagement InterimGuidance_Jun2013.pdf, accessed 1/07/15.

77 Martin Kusch, 'Towards a Political Philosophy of Risk: Experts and Publics in Deliberative Democracy', in T. Lewens ed., *Risk: Philosophical Perspectives*, London: Routledge, 2007, pp. 131–52; Sheila Jasanoff, 'Bridging the Two Cultures of Risk Analysis', *Risk Analysis*, 13, 1993, 123–9.

78 Tim Lewens, 'Introduction', in Lewens, ed., *Risk*, p. 12.

79 Jeffrey K. Taubenberger and David M. Morens, 'Influenza: The Once and Future Pandemic', *Public Health Reports*, 125, 2010, Supplement 3, 16–26; Wolfe, *Viral Storm*, pp. 165–7.

80 Carlo Carduff, 'Pandemic Prophecy, or How to Have Faith in Reason', *Current Anthropology*, 55, 2014, 296–315.

81 P. Hanlon, S. Carlisle, M. Hannah, D. Reilly, A. Lyon, 'Making the case for a "fifth wave" in public health', *Public Health*, 125, 2001, 30–36; Sally C. Davies, Eleanor Winpenny, Sarah Ball, Tom Fowler, Jennifer Rubin, Ellen Nolte, 'For debate: a new wave in public health improvement', www.thelancet.com, published online 3 April 2014 http://dx.doi.org/10.1016/S0140-6736(13)62341-7, accessed 19/05/15.

82 Tiffany L. Bogick, Rumi Chunara, David Scales, Emily Chan, Laura C. Pinheiro, Aleksei A. Chumra, Peter Daszak, John S. Brownstein, 'Preventing Pandemics Via International Development: A Systems Approach', *PLOSMedicine*, 11 December 2012, doi: 10.1371/journal.pmed.1001354, accessed 19/05/15.

83 Philip Alcabes, *Dread: How Fear and Fantasy have Fueled Epidemics from the Black Death to Avian Flu*, New York: Public Affairs, 2009.

## Select bibliography

Alcabes, Philip, Dread: *How Fear and Fantasy Have Fueled Epidemics from the Black Death to Avian Flu*, New York: Public Affairs, 2009.

Bashford, Alison, ed., *Medicine at the Border: Disease, Globalization and Security: 1850 to the Present*, Basingstoke: Palgrave, 2006.

Carduff, Carlo, 'Pandemic Prophecy, or How to Have Faith in Reason', *Current Anthropology*, 55, 2014, 296–315.

Doshi, Peter, 'The Elusive Definition of Pandemic Influenza', Bulletin of the World Health Organisation, 89, 2011, 532–38, doi: 10.2471/BLT.11.086173, accessed 13/04/15.

Echenberg, Myron, *Plague Ports: The Global Urban Impact of Bubonic Plague 1894–1901*, New York: New York University Press, 2007.

Fineberg, Harvey V. and Neustadt, Richard E., *The Epidemic That Never Was: Policy-Making and the Swine Flu Scare*, New York: Vintage Books, 1983.

Harrison, Mark, *Contagion: How Commerce Has Spread Disease*, London and New Haven: Yale University Press, 2012.

Harrison, Mark, 'Disease and World History from 1750', in J.M. McNeill and K. Pomeranz (eds), *The Cambridge World History. Volume 7: Production, Destruction and Connection, 1750–Present*, Cambridge and New York: Cambridge University Press, 2015, 237–58.

Honigsbaum, Mark, *A History of the Great Influenza Pandemics: Death, Panic and Hysteria, 1830–1920*, London: I.B. Tauris, 2013.

Huber, Valeska, 'The Unification of the Globe by Disease? The International Sanitary Conferences on Cholera, 1851–1894', *The Historical Journal*, 49, 2006, 453–76.

McNeill, William H., *Plagues and Peoples*, Garden City: Anchor Press, 1976.

Phillips, Howard and Killingray, David (eds), *The Spanish Influenza Pandemic of 1918–19*, London: Routledge, 2003.

Prins, Gwyn, 'AIDS and Global Security', *International Affairs*, 80, 2004, 931–52.

Summers, William C., *The Great Manchurian Plague of 1910–1911: The Geopolitics of an Epidemic Disease*, New Haven: Yale University Press, 2012.

# 9

# PATTERNS OF ANIMAL DISEASE

*Abigail Woods*

Animal disease is a key shaper and product of human, animal and environmental history. Its emergence and spread can be traced to the ecological relationships between animals and their environments, and to the ways in which humans have used and manipulated them to better serve human ends. Disease not only impacted on animal health and well-being. Heavy human dependence on animals for food, income, transport, companionship, military strength, cultural capital and the creation of scientific knowledge, meant that it also had profound ramifications for human society, politics, economics, health, science and nutrition.

These disease dynamics and their historical significance are increasingly recognised by historians. While short, descriptive accounts of animal disease have long featured in histories of war, agriculture, colonialism, politics and economics, the last 10 to 15 years have witnessed a considerable expansion in dedicated, critical historical literature. This chapter will review works published in the English language. Building on earlier analyses, it opens with some general historiographical reflections on the scope of the field and its sources, themes and approaches.[1] It then offers a summary of the current state of knowledge before concluding with some suggestions for future lines of enquiry.

It is perhaps inevitable that when studying patterns of animal disease, historians have focused on those that left the most prominent historical records. These records are largely text-based. Many were created in response to highly dramatic or problematic disease events that affected animals valued by humans and amenable to human surveillance and control. Governments and scientific institutions produced particularly voluminous records, but their use has resulted in a rather uneven historical picture, which privileges the relatively small number of high-profile diseases that inspired scientific and policy responses in Western and colonial settings during the eighteenth, nineteenth and early twentieth centuries.

Insights into different diseases, times and places are offered by other historical source materials that historians are only just beginning to tap. Texts devoted to the understanding and cure of sick animals, such as manuscripts, books, the records of veterinary practices, drug recipes and advertisements, offer oblique glimpses into the types of diseases that affected animals from the medieval period to the present day.[2] Oral histories offer a counterbalance to recent official and scientific narratives by revealing the disease perceptions and experiences of animal keepers and healers.[3] Novel perspectives on the identities and impacts of medieval animal diseases are offered by inter-disciplinary analyses that utilise paleopathology, molecular clock analysis and climatic data alongside textual sources.[4]

For medieval historians, the paucity of textual sources favours a European, inter-regional approach to livestock disease.[5] For the opposite reason, modern historians tend to use the colonial or nation state as their dominant frame of analysis. This is in spite of the propensity of infectious diseases to cross national borders, and the late-nineteenth- and twentieth-century growth of international organisations for their control.[6] Like the administrations they study, modern historians frame animal disease as a problem of political economy, public health or (for Colonial Africa) land use. They usually examine diseases singly, in biographical fashion, focusing particularly on the scientists who investigated them, and the officials and veterinarians involved in the development and implementation of vertical control measures. Animal keepers feature largely as the targets or opponents of these policies. Their contributions to disease control, and those made by privately funded scientists, industrial corporations and expert advisors, are historically neglected, along with the experiences of the animal victims. The UK, USA and South Africa are particularly well studied. There are some analyses of disease patterns in Canada, Western Europe, Sub-Saharan Africa, Australia and New Zealand, but little is known about Asia, Latin America, Eastern Europe and the Arab world.

The animal disease represented most frequently in the historical literature is cattle plague or rinderpest, an extremely fatal and contagious disease that swept repeatedly across the globe before its elimination in 2011. Other highly visible disease outbreaks amongst domestic animals, and diseases that spread to humans via meat and milk consumption, are also well described, to the neglect of less dramatic disease events and those which fell beyond the purview of the state. Cows dominate the literature. Horse diseases are touched upon in some recent accounts but have attracted less dedicated attention than one might expect given their significance to transport and the military.[7] Sheep and pigs feature infrequently and, despite a burgeoning literature on the history of dogs, analysis of dog diseases is largely confined to rabies.[8] Cat and bird diseases are virtually absent from the historiography, and wildlife feature only when their health impacted on human concerns.[9] Although the health of laboratory animals was recognised as a problem by early-twentieth-century scientists, the diseases they suffered have not been studied.[10]

Distinctive approaches to animal diseases can be identified within different historical sub-fields. Medical historians tend to follow Charles Rosenberg in approaching animal disease as a social and biological phenomenon.[11] They examine how the manifestations, interpretations of, and responses to disease were moulded both by its epidemiological and clinical characteristics and by wider social, economic and political milieux. By contrast, medieval historians are particularly concerned with reaching retrospective diagnoses, often employing inter-disciplinary methodologies to achieve this goal.[12] Environmental and economic historians often 'black box' disease conceptions and focus instead upon the spread of contagion and its effects. Environmental historians also study the ecological relationships between disease and its environments, but their considerations rarely extend to Western contexts and indoor animal environments. Some historians use animal disease to mount a critique of colonialism. They argue that devastating disease events were precipitated by colonial rule, and managed in ways that advanced colonial interests at the expense of indigenous peoples.[13] Other colonial historians challenge this stance by claiming that there was no simple dichotomy between indigenous and settler interests, knowledges and practices, and that all could benefit from disease control.[14]

For heuristic purposes, the following summary will divide animal diseases into four categories: epizootics, zoonoses, other infectious diseases and non-communicable diseases. These categories are widely recognised today. They also existed loosely in the past, although their exact definitions and labels have changed over time.[15] They are rooted partly in the epidemiological characteristics of disease: epizootics are animal epidemics, and zoonoses spread between humans and animals. These categories are also social constructs that reflect the framing and management of particular diseases at specific points in time. Epizootics that have potentially devastating effects on animal health and zoonoses that threaten human health are more likely to be perceived as public problems in need of state and scientific intervention than lower-profile infections and non-communicable diseases. Diseases can also move between categories as social, political, economic and scientific developments reshape their understanding and measures adopted for their control.[16]

## Epizootic diseases

Owing to their high visibility and often devastating effects, epizootic diseases have been recorded since antiquity.[17] While their contagious nature was widely recognised, prior to the late nineteenth century they were also believed to generate spontaneously owing to the influence of the atmosphere and the conditions in which animals were kept.[18] This notion was eventually dispelled by the germ theory, the efficacy of quarantine regulations, and the discovery, particularly in tropical regions at the turn of the twentieth century, that some diseases had insect and tick vectors whose distribution influenced disease geography.[19]

At least eight major outbreaks of cattle epizootics occurred in early post-classical Europe, although their identity is difficult to discern.[20] There was also a severe horse epizootic, retrospectively diagnosed as Eastern Equine Encephalomyelitis, which killed 90 per cent of Charlemagne's heavy war-horses in 791 AD and prevented him from waging war as he did in virtually every other year of his reign.[21] Like subsequent epizootics, these were spread by, and impacted on trade, human migration, and military campaigns. Extreme weather and climatic anomalies may also have played a role by creating food shortages that undermined disease resistance and prompted migration.[22] During the nineteenth century, the development of railways and steamships, the rise of free trade, colonial conquest and commerce, and the growth of urban populations, resulted in more frequent, long-distance animal movements that intensified epizootic disease spread.[23] Expanding cities were supported by ever-increasing horse populations whose susceptibility to influenza epizootics brought transport to a virtual standstill.[24] Rising demand for milk stimulated the development of overcrowded, insanitary urban dairies which were known as 'hotbeds' of disease.[25] The development of settler agriculture at the Cape, and the dramatic expansion of sheep grazing there and in New Zealand and Australia, impacted on livestock stocking densities, land use patterns, and the distribution of insect vectors and game reservoirs of disease, all of which contributed to epizootic disease outbreaks.[26]

The most historically significant epizootic disease (although it never became established in the Americas, Australia or New Zealand) was the highly contagious, fatal cattle plague or rinderpest.[27] Earlier claims that this was responsible for devastating European epizootics in 569–70 and 986–88 AD have been challenged by molecular clock analyses, which demonstrate that rinderpest and measles had not yet evolved into separate pathogens. It has been proposed recently that these epizootics were caused by the now-extinct ancestral virus.[28] Rinderpest was probably the cause of a fourteenth-century, pan-European cattle epizootic. Since cows in medieval society not only supplied meat and milk but also fertiliser and draught power, this 'Great Bovine Pestilence' impacted substantially on human nutrition, and potentially increased human vulnerability to subsequent outbreaks of bubonic plague.[29]

While cattle plague epizootics of the fifteenth, sixteenth and seventeeth centuries have not been subjected to critical historical analysis, those of the eighteenth and nineteenth centuries are well documented. Successive waves of the disease hit eighteenth-century Europe, killing an estimated 200 million cattle. Appuhn argues that its spread was precipitated by the development of new markets for beef cattle in Western Europe which were supplied by the growth of cattle ranching on the Hungarian plains.[30] The devastating effects of cattle plague stimulated some of the first, organised responses from the state and medical profession. According to Wilkinson, they also precipitated the late-eighteenth-century creation of the veterinary profession, though recent scholarship has shed doubt on this claim.[31]

Inspired by responses to the Black Death, cattle plague controls aimed to quarantine infected herds, slaughter sick animals and restrict the livestock trade. The evolution and outcome of these measures have been studied for Britain, France, the Netherlands and parts of Germany. Authors conclude that their effectiveness was impeded by public evasion, opposition, and the weakness of the state.[32] An alternative, empirical measure, inoculation, was attempted but failed to displace the so-called 'stamping out' method.[33] Around 1800, rinderpest disappeared from Western Europe but remained endemic in Russia and parts of Eastern Europe. During the 1860s it re-invaded. To enable stamping out, governments created permanent veterinary departments which granted vets a state-sanctioned role in contagious animal disease control. In Britain, there were extensive scientific investigations and futile attempts at cure. Stamping out proved unpopular initially and achieved widespread acceptance only after it had eliminated disease from the nation.[34]

Rinderpest also entered India in the 1860s and spread quickly through South Asia. It was introduced into Africa by the 1888 Italian invasion of Ethiopia, and swept south to reach South Africa in 1896.[35] In parts of Africa, where cattle were a 'means of production and reproduction, not only of labour power, but of society itself',[36] death rates exceeded 90 per cent, depriving indigenous peoples of fuel, fertiliser, food, clothing, traction and currency. The collapse of transport and the livestock economy prompted the growth of state intervention and veterinary services. Restrictions on livestock trade were sometimes enforced by military troops and provoked uprisings by indigenous peoples in certain colonial contexts. At the Cape, the impracticality of stamping out, combined with developments in Western bacteriology, resulted in the application of a new form of inoculation. Developed by Robert Koch, improved upon by local vets and taken up by supportive Anglophone farmers, it changed the course of the rinderpest epidemic and became compulsory in Rhodesia in 1898.[37] A similar method was adopted in India. By then, the disease had been endemic for over three decades, but the Indian government had made no attempt to control it. According to Mishra, this neglect reflected the irrelevance of cattle to the colonial economy and officials' disregard for the plight of ordinary Indians.[38] However, the propensity of cow slaughter to provoke religious and political unrest was probably also a factor.[39]

The effects of cattle plague were compounded by other epizootic disease outbreaks. In South Africa, the disease known as horse sickness destroyed around 40 per cent of horses in 1854–5. Cavalry regiments were struck down during military campaigns of the 1870s and the South African wars of 1880–81 and 1899–1902. Housing and transhumance (trekking with animals) offered some protection from the midges that were later discovered to spread the disease. Vaccines were developed in the 1930s.[40] Horses and cattle were also killed by nagana (animal trypanosomiasis), which Zulu pastoralists tried to avoid through game slaughter, bush clearance and transhumance. Their ideas informed investigations performed by David Bruce in 1894–7, which confirmed the role of tsetse fly vectors and game reservoirs. Rinderpest unexpectedly reduced the prevalence of cattle nagana by killing game, but subsequent efforts to preserve game increased the risk, leading to conflicts between preservationists and settler farmers. Different colonies adopted different strategies for nagana control, ranging from fly catching or trapping to bush clearance, game culling, compartmentalisation of the landscape, and the control of human, livestock and game movements. From the mid-1940s these were largely superseded by DDT spraying, which proved highly effective.[41]

Another unanticipated result of late-nineteenth-century rinderpest was the appearance in South Rhodesia of the highly fatal, tick-borne, East Coast Fever (ECF), which was imported with replacement cattle from Tanzania. Spreading into North Rhodesia and the Transvaal, it crippled the mining industry, which depended on ox transport. Koch was summoned, but his method of inoculation did not work. Instead, governments adopted livestock dipping and quarantine, and constructed fences to prevent stock moving between 'clean' and 'dirty' areas.[42] Like other such regulations, these measures interfered with trade and pasture use, prompting conspiracy theories and inspiring rebellion in parts of the Transkei in 1914.[43]

Ticks were also implicated in the most devastating animal disease to affect late-nineteenth-century USA: tick-borne fever. Endemic in Mexico and the American South, it made annual incursions into northern areas, killing nearly all infected cattle. Efforts to bar southern animals from particular states were not particularly effective and gave way to dipping in the 1890s, after the tick's role was discovered.[44] Dipping was also used to kill the mite responsible for sheep scab, which caused severe itching and the costly deterioration of wool. In the New World, this disease was a product of the Columbian exchange.[45] Its appearance in 1780s Australia threatened the expansion of sheep farming and led, from the 1830s, to the first Australian animal health laws.[46] During the 1870s, Natal and Cape Colony adopted similar regulations, requiring the isolation and dipping of sheep.[47]

Livestock in Africa suffered from other serious epizootics such as blue tongue, red water, heart water, liver fluke, and contagious bovine pleuro-pneumonia (CBPP, or 'lung-sickness'), which spread globally in the mid-nineteenth century, along with the highly contagious but generally non-fatal foot and mouth disease (FMD). CBPP was introduced into the Cape in 1853 by Dutch cattle and killed around 20 per cent of the cattle population. In the midst of colonial conflict and social change, it gave rise to a Xhosa prophesy that the active killing of cattle would cause the dead to rise, white man to perish, and new cattle to issue from the earth. Although chiefs promoted this policy in 1856, the prophesy was not fulfilled and enormous hardship ensued.[48] Colonialists sometimes tried to control CBPP by inoculation, as devised by Belgian physician, Louis Willems, in the 1840s, but although it produced some immunity it could also transfer infection.[49] Stamping out was adopted at the Cape in 1881 and in the USA in 1884, with eradication declared just eight years later.[50] Britain also opted for stamping out,[51] but other countries used inoculation to control CBPP, either alone (as in New Zealand, where it disappeared within a decade of its 1863 introduction from Australia),[52] in combination with stamping out (in Australia),[53] or as a means of reducing incidence to a point at which stamping out became possible. By 1900, CBPP had been controlled if not eliminated from many Western countries.[54]

In Britain, stamping out was applied to FMD from 1869, and extended in 1878 to swine fever, a frequently fatal disease of pigs. As these were familiar, endemic problems with low mortality (FMD) and variable symptoms (swine fever), many stockowners questioned whether the benefits outweighed the costs.[55] However in 1886, FMD was eliminated and thereby transformed into a dreaded, alien plague. It reappeared frequently over the next 80 years, causing occasional, devastating epidemics. In early-twentieth-century Germany, France, Holland and Italy, serum was the preferred method of FMD and swine fever control. Effective vaccines became available mid-century and enabled some countries to progress to stamping out.[56] However, FMD remained endemic in parts of Africa, Latin America and Asia. An epidemic in Mexico, 1946–52, led to US assistance in stamping out, and generated new appointments of state veterinarians, and the improvement of veterinary education and research.[57] In 1992, with disease at a low ebb, FMD vaccination was halted throughout the EU in favour of stamping out. This left the region vulnerable to a devastating epidemic, which struck in 2001 following a global resurgence of FMD.[58]

When selecting epizootic control policies, governments bore in mind the likely costs and benefits, chances of success, and public response. Stamping out aimed to eradicate disease from regions or nations, while disease control was the object of inoculation, serum treatment, vaccination, dipping and *cordons sanitaires*. The latter measures were imposed with varying degrees of compulsion and depended on effective biological products, which were generally developed within state laboratories such as Onderstepoort Veterinary Institute in South Africa.[59] Stamping out was primarily adopted by rich nations that were geographically remote from epicentres of infection and had low to moderate disease incidence. It required well-defined, easily policed borders, well-resourced veterinary services, and compliant publics who were prepared to report suspect cases and abide by controls. It appealed especially to importing nations that were both vulnerable to disease invasion and capable of imposing sanitary standards upon their trading partners. These partners frequently challenged these measures. They claimed – not without reason – that sanitary regulations were applied for political reasons or to protect domestic producers.[60] During the early twentieth century there were attempts to resolve such conflicts through an internationally agreed system of trade controls, but it was decades before a workable system came into operation under the Office International des Epizooties.[61]

The costs and benefits of control policies were not distributed equitably. Some parties had more power than others to define the disease problem and the manner of its solution. For example, British FMD control policy was shaped by elite breeders and privileged their interests over grass roots producers, whose protests resurfaced in every major epizootic up to and including 2001.[62] Likewise, Olmstead and Rhodes note the uneven impacts of US policy, and the resistance generated, but conclude that state actions were justified given the overarching benefits of disease control.[63] The actions of colonial states often involved incursions into traditional husbandry practices. In the Cape, official attempts to control disease through livestock movement restrictions actually undermined traditional methods of disease avoidance though transhumance.[64] For Phoofolo, such controls were part of an on-going colonial strategy to marginalize Africans and subjugate them to colonial rule.[65] Others have tempered this claim. Brown and Gilfoyle argue that while measures were intended to promote white settler agriculture, Africans also benefited from the diminution of disease.[66] Similarly, Waller claims that policies initially intended to support Kenyan settler elites were subsequently redirected for the benefit of Africans, in response to new scientific understandings and colonial development priorities.[67]

Myxomatosis, a new world disease of rabbits spread by blood sucking insects, is a rare example of a modern epizootic that was allowed to run its course. Although the victims – wild rabbits – had some utility to humans as food and fur, they were more widely regarded as crop-consuming pests. The disease was introduced into New South Wales in 1950 with a view to their destruction. It broke out in continental Europe in 1952, then entered Britain, where farmers were complicit in its spread. Despite the devastating effect on rabbit populations, governments declined to act.[68]

## Zoonotic diseases

One of the earliest recognised zoonoses was cowpox. Dairy farmers had long been aware of its capacity to protect humans against smallpox infection, but it was not until Edward Jenner demonstrated this fact in 1796 and published his findings two years later that vaccination became a medical practice, made compulsory by many governments during the nineteenth century.[69] Cowpox was unusual in benefiting human health. Rabies and glanders were fatal, though sporadic. Their capacity to spread from dogs and horses respectively was known by the early nineteenth century, although other origins were proposed. Legislation to control glanders was made more effective by the 1892 discovery of mallein, a diagnostic product that could identify infected but asymptomatic horses. Produced by government laboratories, and applied by civilian and military officials under compulsory test and slaughter policies, it resulted in the eradication of glanders from most of Europe and North America by WWII.[70]

The horrific symptoms of rabies provoked disproportionate fear and panic, leading occasionally to the mass public slaughter of dogs. In the late nineteenth century, several countries passed legislation in the face of owner resistance for the muzzling, quarantine and destruction of dogs. Vaccines were applied from the 1930s.[71] In Southern Africa, where rabies was a predominantly rural and often unreported disease that circulated and spread through livestock and wildlife, there were unsuccessful twentieth-century attempts to eradicate the main animal vectors: meerkats and jackals.[72] Similar methods were applied, with similar outcomes, to the rodent vectors of bubonic plague, whose role was identified at the turn of the twentieth century in the context of a devastating pandemic.[73]

Other zoonotic diseases rose to prominence in the mid to late nineteenth century, influenced by the same factors that contributed to epizootic disease spread, and the rising production and consumption of diseased meat and milk.[74] Concurrently, the adoption of germ theories, the growth of epidemiological, pathological and bacteriological research, and the assumption of new state responsibilities for human and animal health, resulted in the identification of new epizootic diseases that were subjected to novel and frequently controversial forms of state intervention.[75]

One such disease was anthrax or 'splenic fever', a sporadic but potentially devastating disease of horses, sheep and cattle that was associated with particular soils. During the 1870s and 1880s, scientists discovered that it had the same bacterial cause as two diseases associated with the expanding textile industry in Western Europe and the United States: 'woolsorters disease', a fatal pneumonia, and 'malignant pustule', a skin disease. It transpired that the growth of the global wool trade was exposing Western wool workers to anthrax spores contained in the fleeces of Asian and South African sheep. Anthrax generated a range of responses: disinfection of fleeces and the factory environment, the special burial of animal carcasses, and the use of serum and vaccines. It later resurfaced as a biological weapon for use against animals and humans.[76]

The role of meat in disease transmission was first elucidated for the pork-borne parasitic disease, trichinosis, which could cause death in humans. It was identified in human muscle tissue in 1835, and its life-cycle elucidated in 1850–70 by Rudoph Virchow and others. The increasing identification of human deaths prompted various German states to establish public slaughterhouses, where meat was subjected to microscopic inspection for trichinosis and to general veterinary inspection for the detection of other zoonotic and epizootic diseases.[77] This system was taken up by some other European countries and led, from 1879, to restrictions on the importation of suspect pork from the USA. Despite the protests of American officials, Germany lifted its restrictions only in 1891, following the passage of American legislation to require the microscopic inspection of pork for export.[78]

The late-nineteenth-century development of meat inspection was also driven by fears surrounding bovine tuberculosis (bTB) and its transmissibility to humans. These fears were confirmed in 1882 when Koch announced that TB in humans and animals had the same bacterial cause. Control was problematic because bTB was often prevalent but only became clinically evident in its advanced stages. Butchers, vets and doctors laid rival claims to expertise in the identification and handling of diseased carcasses. Difficulties were compounded by Koch's controversial 1901 announcement that the diseases were not, after all, identical.[79]

By then, attention was turning to bTB transmission via milk. Its role in the spread of human typhoid, scarlet fever and diphtheria had already been postulated by British public health doctors in the 1870s and 1880s, in the face of strong opposition from veterinary surgeons and dairy farmers.[80] Fears of bTB added impetus to efforts to improve the sanitary status of milk. However, effective action was frequently impeded by conflict between interested parties, the disconnect between regulatory regimes that had evolved to tackle either human or animal disease, the physical distance between sites of milk consumption and production, the sheer scale of the problem, difficulties in enforcing regulations, conflict over the costs and benefits of milk pasteurisation, and the outbreak of WWI.[81]

Two methods emerged in the late nineteenth century for the control of bTB in cows. German veterinarian, Robert von Ostertag, advocated the clinical identification and slaughter of advanced cases. Danish veterinarian, Bernhard Bang aimed to identify (then isolate or slaughter) infected cows through injections of tuberculin, a diagnostic substance whose effects were often contested. Vaccines were subsequently developed, notably by the Pasteur Institute in inter-war France, where their favourable effects were used to justify the extension of the BCG vaccination to children.[82] Government attempts at bTB control typically began with the removal of clinical cases and progressed to the use of tuberculin, initially voluntarily and then on a compulsory basis. Regional measures often preceded national campaigns.[83] The same approach was later applied, in conjunction with vaccination, to the eradication of brucellosis. This disease caused contagious abortion in cows and was discovered, in the 1920s, to spread via milk to cause undulant fever in humans.[84]

The timing and progress of public bTB campaigns were influenced by disease incidence, public attitudes, and governments' willingness to bear the costs. Action was initiated by the Finnish government in 1898. In the USA, where less than 5 per cent of dairy cows were infected, state campaigns began in the 1900s and a federal campaign in 1917. Denmark and the Netherlands followed during the inter-war period. In Britain, where 40 per cent of dairy cows were infected, piecemeal inter-war interventions were superseded in the 1950s by a full-scale eradication campaign. By then, the compulsory pasteurisation of milk had effectively abolished the threat to human health.[85] Fears of trade restrictions imposed by countries that aimed to eliminate bTB drove the 1970s adoption of national eradication schemes in Northern Ireland and Australia.[86] Many of these campaigns were highly successful. However in Britain and New Zealand, early progress was later overturned. Wildlife disease reservoirs – badgers and possums respectively – were held to blame. Possum control proved relatively straightforward, but in Britain the situation is still unresolved owing to politicisation, disputes over scientific evidence and expertise, and conflicting cultural attitudes to the badger.[87]

Co-ordinated efforts to address zoonotic diseases in developing countries began in 1948, with the foundation of a Veterinary Public Health Unit under the World Health Organization. Working closely with the Food and Agriculture Organization, its programmes were key vehicles for improving human health and nutrition through improved meat hygiene and zoonotic disease control.[88] Meanwhile, in the West, new zoonotic disease threats were identified. Species of malaria thought to be specific to monkeys were found to transmit to humans,[89] and the discovery that pigeons could harbour psittacosis fuelled campaigns to remove them from cities.[90] There was an increased risk of food poisoning from inadequately cooked meat and eggs. This arose partly from more intensive farming methods that encouraged the spread of salmonella and campylobacter. Slaughterhouse practices were also to blame. Variable standards of hygiene resulted in the cross-contamination of carcasses, while germs went undetected by traditional, macroscopic inspection methods.[91]

Towards the end of the twentieth century, scientists traced several emerging human infections to animals. They discovered that HIV/AIDS had developed from non-human African primates, and SARS from civets. In 1996, they linked a new variant of the fatal human brain disease, CJD, to the consumption of meat from cows suffering from BSE. Also known as mad cow disease, BSE had appeared in Britain a decade earlier. Although the British government took steps to reduce the disease risk to humans, its earlier assurance that meat was safe, and the devastating impact of vCJD on its young victims, generated a crisis of trust in science and the state. Zoonotic disease concerns subsequently shifted to swine and avian influenza, but early-twenty-first-century fears of a major human pandemic have not as yet been realized.[92]

## Other infectious and non-communicable diseases

Despite the historical attention awarded to epizootic and zoonotic diseases, the vast majority of diseases experienced by animals fell outside these categories. Relatively little is known about their histories, particularly in the pre-modern era. They included infections now identified as influenza in cats and distemper, parvo virus and kennel cough in dogs, together with lameness, infertility, and respiratory, gastro-intestinal and parasitic diseases in horses and farmed livestock. There were also many non-communicable diseases, including injuries, lameness, infertility, diseases associated with feeding, and chronic conditions such as cancer. Some affected individuals, others populations, and many had complex aetiologies. Their effects ranged from death to symptomatic illness to sub-clinical reductions in performance. Responsibility for their identification and management fell to animal keepers and their expert advisors. While nineteenth- and twentieth-century governments supported some research into economically important diseases of livestock, they only intervened in disease control under exceptional circumstances, as in WWII, when food shortages led the British government to subsidise practising veterinary surgeons in the control of bovine mastitis and infertility.[93]

The incidence, perception and impacts of these diseases were shaped by the ways in which humans used, managed and valued animals. Paleopathological analysis and the records kept by animal healers reveal the prevalence of wounds and musculo-skeletal problems in horses. This reflects their use as power sources, while the attention paid to breeding difficulties in livestock illustrates how humans relied upon them for meat, milk and profit.[94] Dog owners, alarmed by the death and suffering caused by the infectious disease distemper, stimulated inter-war British research into the disease and helped to test the vaccine which resulted.[95] During the later twentieth century, efforts to understand and manage chronic diseases in pets, such as Feline Urological Syndrome, reveal the growth of humanitarian and consumerist attitudes to these animals. Awarded a similar status to family members, pets became part of a new 'economy of love' that encouraged the circulation of surgical techniques between human and veterinary surgery for the management of their orthopaedic conditions.[96]

For grazing animals, pasture and its management had important impacts on health. In New Zealand, the poisonous plant, tutu, was known since the pre-colonial period to cause significant losses, estimated for the mid-nineteenth century at 25–75 per cent of sheep flocks.[97] Mineral-deficient soils and poisonous plants were identified as causes of disease by the first colonial veterinarians at the Cape. By the early twentieth century, poisonous plants made horse-rearing impossible in parts of South Africa, and by 1920, they killed more livestock than infectious disease. In earlier periods, African pastoralists and settlers had used transhumance to avoid affected areas, but this was impeded by land privatisation and increased stocking densities.[98] Pastures could also harbour infection, as in colonial New Zealand, where pastoralists encountered the prevalent and costly problem of foot rot in sheep, and responded by breeding new types of sheep that were less susceptible to infection.[99] Known since the eighteenth century, the sheep disease scrapie was also associated with certain pastures, although infection and heredity were also suggested as causes. Twentieth-century scientists demonstrated its communicability, and hypothesised the involvement of an unusual disease agent, the prion, which was subsequently implicated in BSE.[100]

Housed livestock experienced a different set of diseases, whose emergence was associated with the mid- to late-twentieth-century shift towards intensive husbandry regimes. Intensification also enabled greater surveillance of animal bodies that made the effects of disease more visible. Meanwhile, farmers' narrowing profit margins and, from the 1960s, the emergence of animal welfare agendas, led to the increasing problematisation of disease. This context favoured the growth of disease research, veterinary services, and the development and use of new drugs, notably antibiotics, but these did not always achieve the desired ends.[101] For example, in the case of mastitis in dairy cows, controls developed as a result of scientific research simply enabled farmers to pursue more intensive forms of production, which led to the unanticipated emergence of new forms of the disease.[102]

## Conclusion

This chapter offers an overview of a field of historical enquiry that is advancing rapidly in scope and intellectual ambition. Epizootic and zoonotic disease history is now a well-established genre, underpinned by a substantial body of literature. Authors have probed the emergence, spread, and impacts of these diseases; the linked development of state veterinary services and scientific research; and the origins and effects of regulations for their control and the controversies that often ensued. New perspectives are emerging which push beyond existing, state-centred narratives, to examine diseases in previously overlooked parts of the world; their relationships with land use patterns in non-Western contexts; and their identities and impacts within medieval societies. There is still potential to extend these enquiries in space and time to produce genuinely global histories of animal disease. This will require attention to the under-studied early-modern era and post-WWII decades; to regional and international disease impacts and responses; and to epizootic and zoonotic diseases that fell beyond the purview of the state.

However, perhaps the greatest priority for future scholarship is to shift the focus away from the zoonoses and epizootics, towards other infectious and non-communicable animal diseases. Despite a slowly developing trend in this direction, these diseases are still neglected by historians. This is surprising when one considers that such diseases were not only numerous but also difficult to prevent or eliminate. Compared to zoonoses or epizootics – which were exceptional events – they were encountered frequently and had a more substantial impact on the lives of animals and their keepers. Yet because historians prefer to use easily accessible archives, and have been more interested in scientists and the state than animals and their keepers, zoonotic and epizootic diseases continue to dominate historical scholarship.

Extending the sphere of analysis to everyday diseases offers exciting possibilities for rewriting the script of animal disease history. It brings wildlife, birds, pets, laboratory, and zoo animals into the picture, as well as their habitats and the people who cared for them, and who advised upon and investigated their health. It shifts the scale of analysis from the nation state or livestock economy to the farm, stable, household, firm or laboratory. In these settings, diseases were not simply economic or public health problems but also threats to ecosystems, production systems, communities, scientific research, animal well-being, and the human–animal bond. Examining their histories using the variety of source materials outlined in the introduction will provide new insights into the contexts that gave rise to disease and into the ways in which animal keepers understood and responded to it. Foregrounding these individuals, and approaching them as disease experts rather than subjects, shapers and opponents of government policy, will also assist in the long-overdue production of an animal health history 'from below'.

There is also scope for integrating investigations into animal disease history with other types of disciplinary enquiry. Medieval historians already draw upon archaeological findings and scientific insights to help interpret their often fragmentary documentary evidence. According to Newfield, 'it is via interdisciplinarity that our understanding of past non-human animal health and disease . . . will improve'.[103] While this approach may appeal less to modern historians, the integration of historical analyses with social scientific and scientific perspectives offers constructive opportunities to situate current disease patterns within a longer historical trajectory, to identify the contributing factors, and to learn from past attempts to understand and control them. In this way, historians can make their insights relevant to a world in which animal disease continues to threaten human health and nutrition, animal welfare and the environment.

# Notes

1 K. Brown and D. Gilfoyle, 'Introduction', in K. Brown and D. Gilfoyle (eds), *Healing the Herds: Disease, Livestock Economies and the Globalization of Veterinary Medicine*, Athens: Ohio University Press, 2010, pp. 1–18; S. Mishra, 'Veterinary history comes of age', Introduction to special virtual issue, *Social History of Medicine*, 2014, http://www.oxfordjournals.org/our_journals/sochis/veterinaryhistory.html; T. Newfield, 'Domesticates, disease and climate in early post-classical Europe: The cattle plague of c940 and its environmental context', *Post-Classical Archaeologies* 5, 2015, 95–126.

2 A. Woods and S. Matthews, '"Little, if at all, removed from the illiterate farrier or cow-leech": The English veterinary surgeon, c. 1860–85, and the campaign for veterinary reform', *Medical History* 54, 2010, 29–54; L. Hill Curth, *'A Plaine and Easie Waie to Remedie a Horse': Equine Medicine in Early Modern England*, Leiden: Brill, 2013; H. A. Shehada, *Mamluks and Animals: Veterinary Medicine in Medieval Islam*, Leiden: Brill, 2013.

3 W. Beinart and K. Brown, *African Local Knowledge and Livestock Health*, Woodbridge: Boydell and Brewer, 2013. See also: the George Ewart Evans collection of interviews, http://sounds.bl.uk/Oral-history/George-Ewart-Evans-collection; oral history initiatives underway at the University of Minnesota, USA, http://blog.lib.umn.edu/ahc-ohp/ahc-oral-history-project/college-of-veterinary-medicine/; Cornell University College of Veterinary Medicine, USA, http://www.vet.cornell.edu/Legacy/PartI.cfm/; and Royal College of Veterinary Surgeons, UK, http://knowledge.rcvs.org.uk/grants/awards-made/collaborations/capturing-life-in-practice/.

4 R. Thomas, 'Non-human paleopathology', in J. Buikstra and C. Roberts (eds), *The Global History of Paleopathology*, Oxford: Oxford University Press, 2012, pp. 652–64; B. Upex and K. Dobney, 'More than just mad cows: Exploring human–animal relationships through animal paleopathology', in A. Grauer (ed.), *A Companion to Paleopathology*, Chichester: Wiley Blackwell, 2012, pp. 191–213; S. De Witte and P. Slavin, 'Between famine and death:

England on the eve of the Black Death – Evidence from paleoepidemiology and manorial accounts', *Journal of Interdisciplinary History* xliv, 2013, 37–60; Newfield, 'Domesticates'.

5  Newfield, 'Domesticates'.

6  P. Zylberman, 'Making food safety an issue: Internationalized food politics and French public health from the 1870s to the present', *Medical History* 48, 2004, 1–28; C. Knab, 'Infectious rats and dangerous cows: Transnational perspectives on animal diseases in the first half of the twentieth century', *Contemporary European History* 20, 2011, 281–306; A. Woods and M. Bresalier, 'One health, many histories', *Veterinary Record* 174, 2014, 650–54.

7  Curth, *A Plaine and Easie Waie*, G. Winton, *Theirs Not to Reason Why: Horsing the British Army 1875–1925*, Solihull: Helion, 2013.

8  H. Ritvo, 'Pride and pedigree: The evolution of the Victorian dog fancy', *Victorian Studies* 29, 1986, 227–53; K. Kete, *The Beast in the Boudoir: Pet-Keeping in Nineteenth-Century Paris*, London: University of California Press, 1994; K. Grier, *Pets in America: A History*, Chapel Hill: University of North Carolina Press, 2006; N. Pemberton and M. Worboys, *Mad Dogs and Englishmen: Rabies in Britain, 1830–2000*, Basingstoke: Palgrave Macmillan, 2007.

9  S. Jones, 'Framing animal disease: Housecats with Feline Urological Syndrome, their owners, and their doctors', *Journal of the History of Medicine and Allied Sciences* 52, 1997, 202–35; A. Cassidy, 'Vermin, victims and disease: UK framings of badgers in and beyond the bovine TB controversy', *Sociologia Ruralis* 52, 2012, 192–214.

10  R. Kirk, '"Wanted – standard guinea pigs": Standardisation and the experimental animal market in Britain ca. 1919–1947', *Studies in History and Philosophy of Biological and Biomedical Sciences* 39, 2008, 280–91.

11  C. Rosenberg, 'Disease in history: Frames and framers', *The Milbank Quarterly* 67, Supplement 1, 1989, 1–15.

12  T. Newfield, 'Human-bovine plagues in the Early Middle Ages', *Journal of Interdisciplinary History* XLVI, 2015, 1–38.

13  P. Phoofolo, 'Epidemics and revolutions: The rinderpest epidemic in late nineteenth-century southern Africa', *Past and Present* 138, 1993, 112–43.

14  R. Waller, '"Clean" and "dirty": Cattle disease and control policy in Colonial Kenya, 1900–1940', *Journal of African History* 45, 2004, 45–80.

15  In the nineteenth century, the term 'infectious' had a relatively narrow definition, extending only to diseases that were directly or indirectly communicable. The term 'zoonosis' referred primarily to animal diseases, and only secondarily to human diseases acquired from an animal. The labelling of diseases transmissible from animals to man as zoonoses occurred in the context of 1950s World Health Organisation discussions on their control. There remains some ambiguity around the scientific definitions of 'zoonoses' and 'infectious diseases'. The origins of the term 'epizootic' have not been investigated, though its application to dramatic, rapidly spreading livestock diseases dates from at least the late eighteenth century: R. Fiennes, *Zoonoses of Primates*, London: Weidenfeld and Nicolson, 1967, pp. 2–6; F. Condrau and M. Worboys, 'Second opinions: Epidemics and infections in nineteenth-century Britain', *Social History of Medicine* 20, 2007, 147–58.

16  J. McEldowney, W. Grant and G. Medley, *The Regulation of Animal Health and Welfare: Science, Law and Policy*, London: Routledge, 2013.

17  G. Fleming, *Animal Plagues: Their History, Nature and Prevention*, London: Chapman and Hall, 1871, pp. 1–32.

18  W. Beinart, 'Vets, viruses and environmentalism: The Cape in the 1870s and 1880s', *Paideuma* 43, 1997, 227–52; A. Woods, *A Manufactured Plague: The History of Foot and Mouth Disease in Britain*, London: Earthscan, 2004.

19  M. Worboys, 'Germ theories of disease and British veterinary medicine, 1860–1890', *Medical History* 35, 1991, 308–27; M. Harrison, *Contagion: How Commerce Has Spread Disease*, New Haven: Yale University Press, pp. 211–46.

20  Newfield, 'Human-bovine plagues'.

21  C. Gilmour, 'The 791 equine pestilence and its impact on Charlemagne's army', *Journal of Medieval Military History* 3, 2005, 23–45.

22  Newfield, 'Human-bovine plagues'; Harrison, *Contagion*.

23  Harrison, *Contagion*.

24  C. McShane and J. Tarr, *The Horse in the City: Living Machines in the Nineteenth Century*, Baltimore: Johns Hopkins University Press, 2007; A. Greene, *Horses at Work: Harnessing Power in Industrial America*, London: Harvard University Press, 2008.

25  A. Woods, 'From practical men to scientific experts: British veterinary surgeons and the development of government scientific expertise, c1878–1919', *History of Science* li, 2013, 457–80.

26  Beinart, 'Vets, viruses and environmentalism'; J. Fisher, 'Technical and institutional innovation in nineteenth century Australian pastoralism: The eradication of psoroptic mange in Australia', *Journal of the Royal Australian Historical Society* 84, 1998, 38–55; R. Peden, 'Sheep breeding in colonial Canterbury (New Zealand): A practical response to the challenges of disease and economic change, 1850–1914', in Brown and Gilfoyle (eds), *Healing the Herds*, pp. 215–31.

27  C. Spinage, *Cattle Plague: A History*, London: Kluwer Academic, 2003.

28  Newfield, 'Human-bovine plagues'.

29  T. Newfield, 'A Cattle panzootic in early fourteenth-century Europe', *Agricultural History Review* 57, 2009, 155–90; P. Slavin, 'The Great Bovine Pestilence and its economic and environmental consequences in England and Wales, 1318–50', *Economic History Review* 65, 2012, 1239–66; S. De Witte and P. Slavin, 'Between famine and death: England on the eve of the Black Death – Evidence from paleoepidemiology and manorial accounts', *Journal of Interdisciplinary History* xliv, 2013, 37–60. See also the chapter in this volume by Sam Cohn.

30  K. Appuhn, 'Ecologies of beef: Eighteenth-century epizootics and the environmental history of Early Modern Europe', *Environmental History* 15, 2010, 268–87.

31  L. Wilkinson, *Animals and Disease: An Introduction to the History of Comparative Medicine*, Cambridge: Cambridge University Press, 1992, pp. 35–64.

32  C. Hannaway, 'The Societe Royale de Medicine and epidemics in the ancient regime', *Bulletin of the History of Medicine* 46, 1972, 257–73; J. Broad, 'Cattle Plague in Eighteenth-Century England', *Agricultural History Review* 31, 1983, 104–15; Wilkinson, *Animals and Disease*; P. Koolmees, 'Epizootic diseases in the Netherlands', in Brown and Gilfoyle (eds), *Healing the Herds*, pp. 19–41; D. Huenniger, 'Policing epizootics: Legislation and administration of cattle plague in eighteenth century northern Germany as continuous crisis management', in Brown and Gilfoyle (eds), *Healing the Herds*, pp. 76–91; D. Brantz, '"Risky business": Disease, disaster and the unintended consequences of epizootics in 18th and 19th century France and Germany', *Environment and History* 17, 2011, 35–52.

33  C. Huygelen, 'The immunization of cattle against rinderpest in eighteenth-century Europe', *Medical History* 41, 1997, 182–96.

34  J. Fisher, 'The economic effects of cattle disease in Britain and its containment, 1850–1900', *Agricultural History* 52, 1980, 278–94; Worboys, 'Germ theories'; J. Fisher, 'British physicians, medical science and the cattle plague, 1865–6', *Bulletin of the History of Medicine* 67, 1993, 651–99; T. Romano, 'The cattle plague of 1865 and the reception of the germ theory', *Journal of the History of Medicine* 52, 1997, 51–80.

35  T. Ofcansky, 'The 1889–97 rinderpest epidemic and the rise of British and German colonialism in eastern and southern Africa', *Journal of African Studies* 8, 1981, 31–8; C. Ballard, 'The repercussions of rinderpest: cattle plague and peasant decline in colonial Natal', *International Journal of African Historical Studies* 19, 1986, 421–50; G. Campbell, 'Disease, cattle, and slaves: The development of trade Between Natal and Madagascar, 1875–1904', *African Economic History* 19, 1990–91, 105–33; H. Weiss, '"Dying cattle": Some remarks on the impact of cattle epizootics in the Central Sudan during the nineteenth century', *African Economic History* 26, 1998, 173–99; D. Doeppers, 'Fighting Rinderpest in the Philippines, 1886–1941' in Brown and Gilfoyle (eds), *Healing the Herds*, pp. 108–28; M. Barwegan, 'For better or worse? The impact of the veterinarian service on the development of agricultural society in Java (Indonesia) in the nineteenth century', in Brown and Gilfoyle (eds), *Healing the Herds*, pp. 92–107; S. Mishra, 'Beasts, murrains, and the British Raj: Reassessing colonial medicine in India from the veterinary perspective, 1860–1900', *Bulletin of the History of Medicine* 85, 2011, 587–619.

36  Phoofolo, 'Epidemics and revolutions', 118.

37 D. Gilfoyle, 'Veterinary research and the African rinderpest epizootic: The Cape Colony, 1896–1898', *Journal of Southern African Studies* 29, 2003, 133–54; E. Gargallo, 'A question of game or cattle? The fight against trypanosomiasis in Southern Rhodesia (1898–1914)', *Journal of Southern African Studies* 35, 2009, 737–53.

38 Mishra, 'Beasts, Murrains'.

39 P. Chakrabarti, 'Beasts of burden: Animals and laboratory research in Colonial India', *History of Science* 48, 2010, 125–52.

40 D. Gilfoyle, 'Veterinary immunology as colonial science: Method and quantification in the investigation of horsesickness in South Africa, c1905–45', *Journal of the History of Medicine* 61, 2005, 26–65; K. Brown, 'Frontiers of disease: human desire and environmental realities in the rearing of horses in nineteenth and twentieth-century South Africa', *African Historical Review* 40, 2008, 30–57.

41 C. Mavhunga and M. Spierenburg, 'A finger on the pulse of the fly: Hidden voices of colonial anti-tsetse science on the Rhodesian and Mozambican borderlands, 1945–1956', *South African Historical Journal* 58, 2007, 117–41; K. Brown, 'From Ubombo to Mkhuzi: Disease, colonial science and the control of nagana (livestock trypanosomosis) in Zululand, South Africa, c1894–1953', *Journal of the History of Medicine and Allied Sciences* 63, 2008, 285–322; Gargallo, 'A Question of game or cattle'.

42 P. Cranefield, *Science and Empire: East Coast Fever in Rhodesia and the Transvaal*, Cambridge: Cambridge University Press, 1991.

43 C. Bundy, '"We don't want your rain, we won't dip": Popular opposition, collaboration and social control in the anti-dipping movement, c1908–16', in W. Beinart and C. Bundy (eds), *Hidden Struggles in Rural South Africa*, London: Currey, 1987, pp. 191–221.

44 A. Olmstead and P. Rhodes, *Arresting Contagion: Science, Policy, and Conflicts over Animal Disease Control*, Cambridge: Harvard University Press, 2015, pp. 251–77.

45 A. Crosby, *The Columbian Exchange: Biological and Cultural Consequences of 1492*, Westport: Greenwood, 1972.

46 Fisher, 'Technical and institutional innovation'.

47 W. Beinart, 'Transhumance, animal diseases and environment in the Cape, South Africa', *South African Historical Journal* 58, 2007, 17–41.

48 J. Peires, 'The central beliefs of the Xhosa cattle-killing', *Journal of African History* 28, 1987, 43–63.

49 C. Andreas, 'The spread and impact of the lungsickness epizootic of 1853–57 in the Cape Colony and the Xhosa chiefdoms', *South African Historical Journal* 53, 2005, 50–72.

50 C. Hutson, 'Texas fever in Kansas, 1866–1930', *Agricultural History* 68, 1994, 74–104; A. Olmstead, 'The first line of defense: Inventing the infrastructure to combat animal diseases', *Journal of Economic History* 69, 2009, 327–57; Beinart, 'Transhumance'.

51 Woods, 'From practical men'.

52 J. Fisher, 'The origins, spread and disappearance of contagious bovine pleuro-pneumonia in New Zealand', *Australian Veterinary Journal* 84, 2006, 439–44.

53 L. Newton, 'CBPP in Australia: Some historic highlights from entry to eradication', *Australian Veterinary Journal* 69, 1992, 306–17.

54 J. Fisher, 'To kill or not to kill: The eradication of contagious pleuro-pneumonia in Western Europe', *Medical History* 47, 2003, 314–31.

55 Woods, *Manufactured Plague*; Woods, 'From practical men'.

56 O. Stalheim, 'The hog cholera battle and veterinary professionalism', *Agricultural History* 62, 1988, 116–21; Woods, *Manufactured Plague*.

57 L. A. Lomnitz and L. Mayer, 'Veterinary medicine and animal husbandry in Mexico: From empiricism to science and technology', *Minerva* 32, 1994, 144–57.

58 Woods, *Manufactured Plague*.

59 K. Brown, 'Tropical medicine and animal diseases: Onderstepoort and the development of veterinary science in South Africa 1908–1950', *Journal of Southern African Studies* 31, 2005, 513–29.

60 J. Kastner, 'Scientific conviction amidst scientific controversy in the transatlantic livestock and meat trade', *Endeavour* 29, 2005, 78–83; K. Asdal, 'Making space with medicine', *Scandinavian Journal of History* 31, 2006, 255–69; Woods, 'From practical men'.

61 Knab, 'Infectious rats'.

62 Woods, *Manufactured Plague*.

63 Olmstead and Rhodes, *Arresting Contagion*.

64 Beinart, 'Transhumance'.

65 Phoofolo, 'Epidemics and revolutions'.

66 Brown and Gilfoyle, 'Introduction'.

67 Waller, '"Clean" and "dirty"'.

68 P. Bartrip, *Myxomatosis: A History of Pest Control and the Rabbit*, Palgrave, London, 2008.

69 H. Bazin, *Vaccination: A History from Lady Montagu to Genetic Engineering*, John Libbey Eurotext, 2011.

70 L. Wilkinson, 'Glanders: Medicine and veterinary medicine in common pursuit of a contagious disease', *Medical History* 25, 1981, 363–84; J. Derbyshire, 'The eradication of glanders from Canada', *Canadian Veterinary Journal* 43, 2002, 722–6; Mishra, 'Beasts, murrains'.

71 L. Van Sittert, 'Class and canicide in Little Bess: The 1893 Port Elizabeth rabies epidemic', *South African Historical Journal* 48, 2003, 207–34; Pemberton and Worboys, *Mad Dogs*; P. Teigen, 'Legislating fear and the public health in Gilded Age Massachusetts', *Journal of the History of Medicine and Allied Sciences* 62, 2007, 141–70.

72 K. Brown, *Mad Dogs and Meerkats: A History of Resurgent Rabies in Southern Africa*, Athens: Ohio University Press, 2011.

73 S Jones, 'Plague's Third Pandemic: A History of Disease Ecology', Annual lecture in the history of health and medicine, Kings College London, 2015.

74 P. Koolmees, 'Veterinary inspection and food hygiene in the twentieth century', in D. Smith and J. Phillips (eds), *Food, Science, Policy and Regulation in the Twentieth Century*, London: Routledge, 2000, pp. 53–68.

75 A. Hardy, 'Animals, disease and man: Making connections', *Perspectives in Biology and Medicine* 46, 2003, 200–15. See also the chapter in this volume by Christoph Gradmann.

76 D. Gilfoyle, 'Anthrax in South Africa: Economics, experiment, and the mass vaccination of animals, c1910–45', *Medical History* 50, 2006, 465–90; M. Cassier, 'Producing, controlling and stabilising Pasteur's anthrax vaccine: Creating a new industry and a health market', *Science in Context* 21, 2008, 253–78; S. Jones, *Death in a Small Package: A Short History of Anthrax*, Baltimore: Johns Hopkins University Press, 2010; J. Stark, *The Making of Modern Anthrax, 1875–1920: Uniting Local, National and Global Histories of Disease*, London: Pickering and Chatto, 2013.

77 D. Brantz, 'Animal bodies, human health, and the reform of slaughterhouses in 19th century Berlin', in P. Y. Lee (ed.), *Meat, Modernity and the Rise of the Slaughterhouse*, London: University of New Hampshire, 2008, pp. 71–85; Tatsuya Mitsuda, 'The development of veterinary medicine in Germany, circa 1770–1930', unpublished manuscript, 2014.

78 J. Gignilliat, 'Pigs, politics and protection: The European boycott of American pork, 1879–1891', *Agricultural History* 35, 1961, 3–12; J. Cassedy, 'Applied microscopy and American pork diplomacy: Charles Wardell Stiles in Germany 1898–1899', *Isis* 62, 1971, 4–20.

79 K. Waddington, *The Bovine Scourge: Meat, Tuberculosis and Public Health, 1850–1914*, Woodbridge: Boydell Press, 2006; D. Berdah, 'Meat, public health and veterinary expertise: Eating beef from tuberculous bovines in France, late nineteenth century', unpublished manuscript, 2013.

80 L. Wilson, 'The historical riddle of milk-borne scarlet fever', *Bulletin of the History of Medicine* 60, 1986, 321–42; J. Steere-Williams, 'The perfect food and the filth disease: Milk-borne typhoid and epidemiological practice in late Victorian Britain', *Journal of the History of Medicine and Allied Sciences* 65, 2010, 514–45.

81 P. Atkins, 'The pasteurisation of England: The science, culture and health implications of milk processing, 1900–1950' in D. Smith and J. Phillips (eds), *Food, Science, Policy, and Regulation in the Twentieth Century*, London: Routledge, 2000, pp. 37–51; M. French and J. Phillips, *Cheated not Poisoned? Food Regulation in the United Kingdom, 1875–1938*, Manchester: Manchester University Press, 2000, pp. 158–84; B. Orland, 'Cow's milk and human disease: Bovine tuberculosis and the difficulties involved in combating animal diseases', *Food and History* 1, 2003, 179–202; S. Jones, *Valuing Animals: Veterinarians and Their Patients in Modern America*, Baltimore: Johns Hopkins University Press, 2003, pp. 63–90; S. Jones, 'Mapping

a zoonotic disease: Anglo-American efforts to control bovine tuberculosis before World War I', *Osiris* 19, 2004, 133-48; G. Colclough, '"Filthy vessels": Milk safety and attempts to restrict the spread of bovine tuberculosis in Queensland', *Health & History* 12, 2010, 6–26.

82 D. Pritchard, 'A century of bovine tuberculosis 1888–1988: Conquest and controversy', *Journal of Comparative Pathology* 99, 1988, 357–99; D. Berdah, 'Between human medicine and veterinary medicine: The fight against bovine tuberculosis in France and in the United Kingdom, 1920–60', unpublished draft, 2013.

83 K. Waddington, 'To stamp out "so terrible a malady": Bovine tuberculosis and tuberculin testing in Britain, 1890–1939', *Medical History* 48, 2004, 29–48; Olmstead and Rhodes, *Arresting Contagion*, pp. 278–301.

84 E. Madden, *Brucellosis: A History of the Disease and Its Eradication from Cattle in Great Britain*, London: MAFF, 1984; R. Kaplan, 'The politics of the herd: Big government and libertarian imperative in United States brucellosis eradication, 1945–1985', *Western Humanities Review*, 2015, forthcoming.

85 P. Atkins, 'Milk consumption and tuberculosis in Britain, 1850–1950' in A. Fenton (ed.), *Order and Disorder: The Health Implications of Eating and Drinking in the Nineteenth and Twentieth Centuries*, East Linton: Tuckwell Press, 2000, pp. 83–95; B. Orland, 'Cow's milk'; Olmstead and Rhodes, *Arresting Contagion*.

86 D. Cousins and J. Roberts, 'Australia's campaign to eradicate bovine tuberculosis: The battle for freedom and beyond', *Tuberculosis* 81, 2001, 5–15; P. Robinson, 'A history of bovine tuberculosis eradication policy in Northern Ireland', *Journal of Epidemiology and Infection*, 2015, forthcoming.

87 W. Grant, 'Intractable policy failure: The case of bovine TB and badgers', *British Journal of Politics and International Relations* 11, 2009, 557–73; A. Cassidy, 'Vermin'; G. Enticott, 'Biosecurity and the bioeconomy. The case of disease regulation in the UK and New Zealand', in A. Morley and T. Marsden (eds), *Researching Sustainable Food: Building the New Sustainability Paradigm*, Earthscan: London, 2014, pp. 122–42.

88 Woods and Bresalier, 'One Health'.

89 R. Mason Dentinger, 'Patterns of infection and patterns of evolution: How a malaria parasite brought "monkeys and man" closer together in the 1960s', *Journal of the History of Biology*, 2016, forthcoming.

90 C. Jerolmack, 'How pigeons became rats: The cultural-spatial logic of problem animals', *Social Problems* 55, 2008, 72–94.

91 A. Hardy, *Salmonella Infections, Networks of Knowledge, and Public Health in Britain, 1880–1975*, Oxford: Oxford University Press, 2014; Koolmees, 'Veterinary inspection'.

92 H. Ritvo, 'Animal planet', *Environmental History* 9, 2004, 204–20. On pandemics, see the chapter in this volume by Mark Harrison.

93 A. Woods, 'The farm as clinic: Veterinary expertise and the transformation of dairy farming, 1930–50', *Studies in History and Philosophy of Biological and Biomedical Sciences* 38, 2007, 462–87.

94 Jones, *Valuing Animals*; M. MacKay, 'The rise of a medical specialty: The medicalisation of elite equine care c.1680–c.1800', unpublished PhD thesis, University of York, 2009; A. Gardiner, 'Small animal practice in British veterinary medicine', unpublished PhD thesis, University of Manchester, 2010; A. Woods and S. Matthews, '"Little, if at all, removed from the illiterate farrier or cow-leech": The English veterinary surgeon, c. 1860–85, and the campaign for veterinary reform', *Medical History* 54, 2010, 29–54; Upex and Dobney 'More than just mad cows'.

95 M. Bresalier and M. Worboys, '"Saving the lives of our dogs": The development of canine distemper vaccine in interwar Britain', *British Journal for the History of Science* 47, 2014, 305–34.

96 S. Jones, 'Framing'; A. Gardiner, 'The animal as surgical patient: A historical perspective in the 20th Century', *History and Philosophy of the Life Sciences* 31, 2009, 355–76; M. Schlünder and T. Schlich, 'The emergence of "implant-pets" and "bone-sheep": Animals as new biomedical objects in orthopedic Surgery (1960s–2010)', *History and Philosophy of the Life Sciences* 31, 2009, 433–66; C. Degeling, 'Negotiating value: Comparing human and animal fracture care in industrial societies', *Science, Technology, & Human Values* 34, 2009, 77–101.

97 N. Clayton, '"Poorly co-ordinated structures for public science that failed us in the past"? Applying science to agriculture: A New Zealand case study', *Agricultural History* 82, 2008, 445–67.

98 Beinart, 'Transhumance'; K. Brown, 'Poisonous plants, pastoral knowledge and perceptions of environmental change in South Africa, c. 1880–1940', *Environment and History* 13, 2007, 307–32.

99 R Peden, 'Sheep breeding'.

100 M. Schwartz, *How the Cows Turned Mad*, London: University of California Press, 2003; K. Kim, *The Social Construction of Disease: From Scrapie to Prion*, London: Routledge, 2006.

101 Jones, *Valuing Animals*, pp. 91–114; M. Finlay, 'Hogs, antibiotics and the industrial environments of postwar agriculture', in S. Schrepfer and P. Scranton (eds), *Industrializing Organisms*, London: Routledge, 2004, pp. 237–60; K. Smith-Howard, 'Antibiotics and agricultural change: Purifying milk and protecting health in the postwar era', *Agricultural History* 85, 2010, 327–51; A. Woods, 'Is prevention better than cure? The rise and fall of veterinary preventive medicine, c1950–80', *Social History of Medicine* 26, 2013, 113–31.

102 A. Woods, 'Science, disease and dairy production in Britain, c1927–80', *Agricultural History Review* 62, 2014, 294–314.

103 Newfield, 'Domesticates', 117.

## Select bibliography

Beinart, W., 'Transhumance, animal diseases and environment in the Cape, South Africa', *South African Historical Journal* 58, 2007, 17–41.

Brown, K. and Gilfoyle, D. (eds), *Healing the Herds: Disease, Livestock Economies and the Globalization of Veterinary Medicine*, Athens: Ohio University Press, 2010.

Cranefield, P., *Science and Empire: East Coast Fever in Rhodesia and the Transvaal*, Cambridge: Cambridge University Press, 1991.

Fisher, J., 'To kill or not to kill: The eradication of contagious pleuro-pneumonia in Western Europe', *Medical History* 47, 2003, 314–31.

Hardy, A., *Salmonella Infections, Networks of Knowledge, and Public Health in Britain, 1880–1975*, Oxford: Oxford University Press, 2014.

Jones, S., *Death in a Small Package: A Short History of Anthrax*, Baltimore: Johns Hopkins University Press, 2010.

Newfield, T., 'Domesticates, disease and climate in early post-classical Europe: The cattle plague of c940 and its environmental context', *Post-Classical Archaeologies* 5, 2015, 95–126.

Olmstead, A. and Rhodes, P., *Arresting Contagion: Science, Policy, and Conflicts over Animal Disease Control*, Cambridge: Harvard University Press, 2015.

Orland, B., 'Cow's milk and human disease: Bovine tuberculosis and the difficulties involved in combating animal diseases', *Food and History* 1, 2003, 179–202.

Pemberton, N. and Worboys, M., *Mad Dogs and Englishmen: Rabies in Britain, 1830–2000*, Basingstoke: Palgrave Macmillan, 2007.

Slavin, P., 'The Great Bovine Pestilence and its economic and environmental consequences in England and Wales, 1318–50', *Economic History Review* 65, 2012, 1239–66.

Waddington, K., *The Bovine Scourge: Meat, Tuberculosis and Public Health, 1850–1914*, Woodbridge: Boydell Press, 2006.

Waller, R., '"Clean" and "dirty": Cattle disease and control policy in Colonial Kenya, 1900–14', *Journal of African History* 45, 2004, 45–80.

Woods, A., 'Science, disease and dairy production in Britain, c1927–80', *Agricultural History Review* 62, 2014, 294–314.

Woods, A., *A Manufactured Plague: The History of Foot and Mouth Disease in Britain*, Earthscan: London, 2004.

# 10

# PATTERNS OF PLAGUE IN LATE MEDIEVAL AND EARLY-MODERN EUROPE

*Samuel Cohn, Jr.*

Infectious diseases are more than their pathogenic agents alone. Despite the general nosology or scientific naming of diseases after their pathogens (*Vibrio cholera, Yersinia pestis*, etc.), diseases are due to relationships between viruses, bacteria, fungi, or protozoa and their human hosts. These relationships can often vary radically over time and place, depending on environmental and genetic changes in both pathogens and hosts.[1] As a consequence, changing signs and symptoms, rates of mortality, morbidity, lethality, and changes such as the poor becoming the ones overwhelmingly afflicted can transform ailments beyond recognition. For instance, during the first half of the sixteenth century, physicians and other commentators saw the venereal disease that we now generally label as syphilis changing fundamentally. In his study of contagious diseases of 1546, Girolamo Fracastoro maintained that the prominent characteristic that had given the disease one of its most common names – the Great Pox – had largely disappeared over the past 20 years: now pustules appear 'in very few cases'. He added that the horrible pains with sleepless nights so often described by doctors and evoked in the poetry of those afflicted 'had somewhat subsided'. In their place, new symptoms surfaced such as the disappearance of eyebrows and hair on top 'falling out', making 'men look ridiculous'.[2]

Because of these changes in the signs and symptoms of diseases, double Nobel-prize-winner, Frank Macfarlane Burnet, cautioned historians over 60 years ago about identifying diseases in past times and advised that epidemiological patterns might prove a better guide than a disease's clinical features. Yet he realised that epidemiological patterns could also change as with the tendency of populations to build up immunity and the disease to become more prevalent in children than adults. Before the development of antibiotics in the second quarter of the twentieth century, the acquisition of a population's herd immunity to a particular disease could run in cycles. Declining mortalities and widening gaps between epidemics of a disease lowered a population's innate immunity to that disease – a recurrent pattern seen over the 500-year history of the plague during the 'Second Pandemic'.[3]

This chapter will investigate patterns and presentations of plague in Europe from the Black Death, 1347–51, to successive waves of this disease as late as the early nineteenth century when it flared briefly in isolated places such as Corfu, Malta, and Noja (Noicattaro) near Bari in 1814–15.[4] Detailed written records from municipalities and other states in Europe provide an opportunity to study the plague of the 'Second Pandemic' variety over a longer period of time than for any other infectious disease. These patterns of plague regard not only the disease's clinical and epidemiological aspects but also changes in society, politics and mentalities. I shall argue that long before the 'Laboratory Revolution' of the late nineteenth century, action within the human realm altered the patterns of the plague and possibly the pathogen itself. The chapter begins by investigating clinical aspects of plague, then the epidemiological, and at the end the social, political, and psychological dimensions of the disease. In so doing, it seeks to raise new questions which historians, together with those in the biological sciences, need to confront in the future.

## Signs and clinical diagnosis

As far as the signs and symptoms of historical plague go, the patterns differed from diseases such as early-modern syphilis or tuberculosis: few striking changes occurred from the period of its immediate shocks on the bodies of the previously unexposed in 1347–51 to its last appearances in the early nineteenth century. Throughout this history the plague's tell-tale signs remained relatively constant. Chronicles, doctors, and storytellers such as Giovanni Boccaccio described not only buboes in the three principal lymph nodes – the groin, armpits, and the cervical region around the neck and below the ears as 'big as an apple and others the size of an egg';[5] they also described other skin disorders including carbuncles, rashes and other smaller bumps forming outside these main lymph regions and in places rarely, if ever, seen with *Yersinia pestis* during the 'Third Pandemic' from its appearance in Hong Kong in 1893–4 to the present – that is, on backs, eyelids, penises, and up noses. Most prominent of the skin disorders that formed outside the three main lymph regions were black and blue spots, marks, or tokens called by various names – *macchie nere o lividi, lenticulae, puncticulae, morbilli* – that could cover entire bodies. From the earliest descriptions of plague in Europe at Messina, Sicily, in the autumn of 1347, these signs were described as mingled with the buboes and carbuncles and often on the same bodies, and physicians and chronicles pointed to them as the plague's most deadly marks. The fifteenth-century merchant chronicler of Florence, Giovanni Morelli, described these 'tiny balls . . . on the skin in various places' as more deadly than the larger swellings. From the former, patients had 'little chance of any cure'.[6] Later, the Capuchin friar Paolo Bellintani, officiating as head of *lazzaretti* (pest houses), first at Milan in 1576–7, then at Brescia a year later, and finally at Marseille in the early 1580s, described his experiences based on thousands of plague patients: while those with buboes and other pestilential signs had some chance of survival, those afflicted with the black spots, he reported, were doomed.[7] Of the 10 signs he catalogued for recognising plague, only these 'led to certain death'. During the same plague wave at Genoa, the surgeon Luchino Boerio concurred: the smaller spots were the most deadly, and from them, patients could perish within 24 hours.[8] At Noja in Italy's last plague (1815) before the 'Third Pandemic', the Neapolitan physician Vitangelo Morea continued to describe the presence of these deadly black and brown spots that accompanied larger buboes and carbuncles.[9]

The appearance of the deadly tokens during plague were not peculiar to the Italian peninsula. For the plague of 1348–9, the Oxfordshire chronicler Geoffrey le Baker described the same combination of large 'ulcers' principally in the three lymph nodes and the small black 'pustuli' that covered entire bodies. As with the Italians, he claimed the spots were the more deadly of the two: from them 'hardly any survived'.[10] In a Welsh poem written during the second plague of 1362, Llywelyn Fychan lamented the deaths of his four children. A medley of deadly signs appear: first the larger boils – 'bitter head of an odious onion'; 'a swelling under the armpit, grievous sore lump,/white knob, poisonous misfortune'; then the smaller pustules – 'a little boil which spares no one'; 'Inflamed burning of brittle coal fragments'; 'A shower of peas giving rise to affliction,/messenger of swift black death'; 'berries, it is painful that they should be on fair skin'.[11]

As in Italy, so too in England and elsewhere, these plague signs persisted over the long term. During London's last plague of 1665–6, Dr Nathaniel Hodges described the clinical condition of those afflicted by citing numerous cases from his practice during the epidemic. Of the eight sections of his 224-page report, Section V: 'Of the Manifest Signs of the Late Pestilence' was the longest, comprising nearly a quarter of his tract. Buboes formed in the groin, armpits, and behind the ears and varied in number from one to cases in which 'all the glands were tumified' with them.[12] The smaller carbuncles 'about the Bigness of a Millet Seed' could spread across bodies and form in places such as women's breasts and on fingers but rarely were fatal.[13] On the other hand, 'the pestilential Spots', called *Petechie* or *tokens* by the 'Common People', formed in the 'Neck, Breast, and Back', were 'most commonly very numerous', and 'in some . . . so thick, as to cover the whole skin', constituting 'certain Characters of Death imprinted in many places'. They were 'Pledges of Death' from which individual patients would pass from perfect health to death in a matter of hours.[14]

In France, the physician of three popes who survived four waves of plague from 1348 to the 1380s, Raymond Chalin de Vinario, described the same combination of larger tumours in the lymph nodes and the black spots across the body.[15] These black spots were signs of plague and endured to the end of France's 'Second Pandemic'. In searching out plague cases and plague deaths in the parishes of Marseille during the plague of 1720, the English physician Richard Mead advised that 'diligent men' should replace 'ignorant old women' and should be instructed on the plague signs, 'particularly with livid spots, bubo's, or carbuncles, to give notice thereof to the Council of Health who should immediately send skilful physicians to examine the suspected bodies'.[16] Finally, during the last major plague in the Baltic countries, from 1709 to 1713, diaries, doctors' reports, and their questionnaires on the criteria for identifying plague pointed to the same combination of buboes and spots, even though this plague erupted during frigid winter months, when the Baltic seas had frozen over, blockading harbours and creating famine conditions.[17] In addition, these reports also describe the frequent formation of buboes and carbuncles outside the principal lymph nodes: in vaginas, anuses, navels and the corners of eyes.[18]

These clinical observations find confirmation in the remarkable *Libri dei Morti* of Milan, in which university-trained physicians described plague deaths over the course of six plague years from 1452 to 1524. In addition to describing the course of the disease, the number of days from first signs to death, physicians recorded the number and precise positions of buboes and carbuncles on the bodies of the afflicted, and whether black, blue, or (very infrequently) red pustules formed. Across these years these spots appeared in about a third of 6,993 cases the physicians identified as plague. On average those with the spots died in less than three days – slightly more quickly than those with buboes or carbuncles alone,[19] but not with the spectacular speed of those afflicted with septicaemic plague during the 'Third Pandemic', when untreated, death comes within eight hours.

This combination of signs did not signify two separate diseases occurring simultaneously. In the first instance, no disease flared during the Middle Ages or early-modern period when these spots appeared alone without the larger carbuncles or buboes and which also killed great numbers with such quickness and such terrible rates of lethality. Second, even by the fourteenth century contemporaries could distinguish between plague and other diseases with similar skin disorders, whether smallpox or a disease in Italy called *Pondi*. For instance in 1390, two diseases spread through the countryside of Florence, both during the height of summer, both with similar black pustules. Some thought the two were the same. However, given their divergent epidemiological characteristics, the Florentine chronicler of the Minerbetti family argued convincingly that two different diseases were spreading concurrently: one moved through the countryside slowly and was a slow killer, taking weeks before any of its victims succumbed, while the other possessed the usual characteristics of plague: it raced through the countryside and killed its victims in a matter of days, not weeks.[20] Similarly, an English chronicle described a plague in 1369 accompanied by another disease, 'an illness that men callen "the pokkes"'. Despite the similarity of their signs, he distinguished between the two: whereas the 'pokkes' (probably smallpox) was slow to spread or kill their victims, the 'pestilens' killed quickly.[21] Others such as Dr Hodges during the London plague of 1665 pointed to differences in the 'punctures' between the pestilential 'tokens' and the small spots caused from flea bites or scurvy.[22] In 1477–8 Milan was afflicted by an epidemic of *morbilli* that spread over bodies, killing over a thousand in little over a year. The health board's doctors, however, did not consider this disease plague. First, these *morbilli* were predominantly red and not black or blue and neither carbuncles nor buboes accompanied them; second, their diagnosis relied on epidemiology: the average period of illness for those with the *morbilli* in 1477–8 survived for more than two weeks (14.7 days on average) with less than 8 per cent dying within a week. By contrast, less than 2 per cent of those who died from plague survived beyond a week.[23]

If the pathogen of the plague that spread worldwide from 1894 was the same as that of the Black Death disease that persisted in parts of Europe to the second decade of the nineteenth century, then something radical, and yet uncharted, happened between c. 1820 and 1894. After a half millennium of stasis, the plague's signs had become transformed. From 1894 to the present, in 95 per cent of plague cases only a single bubo appears on the victim's body, and overwhelmingly this one was in the groin. By contrast, during the 'Second Pandemic', multiple buboes and carbuncles formed in the three principal lymph regions and often others strayed to other bodily parts, and sometimes in places hardly ever seen with bubonic plagues of the 'Third Pandemic'.[24] Most strikingly, black or blue pustules that spread across bodies had become an extremely rare occurrence post-Yersin's discovery. In over 3,000 clinical reports of plague victims recorded from hospitals at the turn of the twentieth century in the Presidency of Mumbai only two patients may have shown these signs, but even these were described as 'black blisters', and not spots, bumps, or pustules. With another two sets of clinical reports, one compiled from two areas of China in the 1920s comprising 9,500 cases, and another from two hospitals in the Presidency of Mumbai from 1896 to 1908, comprising 13,600, not a single clinical report described the black spots or pustules.[25] Such signs have, however, occurred, elsewhere as in Chile in 1903, Ecuador in 1908, and Peru in the early twentieth century and were called 'small-pox plague'. Yet, when these signs appeared, they signalled the exact opposite prognosis from the deadly *morbilli* of late medieval or early-modern Plague. Instead of being plague's most deadly 'tokens', they were the signs of 'benign plague', when few, if any, died.[26]

## Symptoms

While the signs of plague remained unusually stable for an infectious disease over the long term from 1347 to the early nineteenth century, this was not the case with the disease's symptoms. Yet on this score I know of no recent historians or health scientists to comment on the change. However, contemporaries as early as the mid-sixteenth century were aware of it and puzzled by it. In 1348 plague descriptions emphasised the Black Death's two or three forms. Most famously, the pope's physician at Avignon, Guy de Chauliac, described the Black Death striking Avignon first in January, when it was characterised by continuous fever and the spitting of blood; it was highly contagious and victims died in three days. By March, Guy claimed the disease had changed to a bubonic form, which remained contagious and deadly but its victims took four to five days to perish.[27] Louis Sanctus, Matteo Villani, Tommaso del Garbo, the anonymous comments in a Necrology (dead list) of Cividale del Friuli, Johannes della Penna, the Irish friar John Clyn,[28] and many others pointed to the pneumonic and bubonic forms of the disease in 1348, but did not see these as clustering into two distinctive seasons: a winter pulmonary phase followed by a spring and summer bubonic one. During the third plague at Lucca in 1373, the physician Ser Iacopo di Coluccino kept a register (*cedola*) of visits to his plague patients. Four of the seven were stricken with pneumonic plague characterised by the coughing and spitting of blood. Interspersed with them were three plague patients with the characteristic buboes but without any pulmonary symptoms. There was no seasonal separation. The colder months had not brought on the pulmonary form. His observations stretched from May to November without any cases during the winter months.[29] Other sources such as letters from merchants at Avignon in 1398–9 corresponding with Marco di Francesco Datini's bank at Prato describe plague patients suffering from both forms of plague simultaneously and taking four days or more to die.[30] Along with Guy de Chauliac and the doctor from Lucca, others reported that those afflicted with pneumonic plague invariably died (quickly), but the norm was still two days or more,[31] and not with the lightning speed of modern pneumonic plague, when victims die within 24 hours if not treated immediately with antibiotics.

My point, however, is different. By the beginning of the fifteenth century, contemporary reports of pneumonic plague – descriptions of coughing, vomiting blood, or other pulmonary complications – vanished almost completely. A few such as the merchant chronicler Giovanni Morelli continued to refer to these plague symptoms. He did so, however, while chronicling the Black Death of 1348 and, as he admitted, his information came from Boccaccio, not from his own experiences of plague in early-fifteenth-century Florence.[32] The latest of such descriptions of spitting or coughing of blood during a medieval or Renaissance plague that I have sighted come from outside Western Europe at a plague in Novgorod in 1424.[33]

Such knowledge of these earlier fourteenth-century plagues continued into the sixteenth century, when large sections of physicians' plague tracts were devoted to recognising the signs and symptoms of plague, not only for the guidance of their fellow physicians but for the public, who were obliged to report plague cases in their homes and parishes on pain of serious penalties, including death.[34] While the Genoese surgeon Boerio listed eight signs, the head of lazzaretti in three cities, Bellintani, cited 10, and Palermo's protomedico, Giovan Filippo Ingrassia, listed 52.[35] Clinical criteria focused on the skin disorders – the larger buboes in the principal lymph nodes, the smaller 'carbunculi' in and outside these regions, and the smaller, more numerous and more deadly *morbilli, herpes, petecchie*, 'et simili mali' across entire bodies.[36] Yet, despite their detailed attention to these clinical features, none described the coughing or spitting of blood as present in the plagues of their day, even though they were aware of the writings of Guy de Chauliac and others who had described them for the Black Death. One such tract published in 1566 by a physician from Trento, then practising at Frankfort-am-Main, was completely dedicated to pneumonic plague 'with the spitting of blood'.[37] But, unlike the vast majority of plague tracts of late medieval and early-modern Europe, it was purely historical and mentioned no plagues since the fourteenth century that were pneumonic. A decade later, the chair of medicine at Padua, Alessandro Massaria, revisited descriptions of pneumonic plague in 1348 and 1362 but confirmed that in his day plague had changed and admitted that he could not explain why the coughing and spitting of blood described by fourteenth-century writers were no longer present.[38]

Physicians' clinical descriptions of plague deaths in Milan's *Libri dei Morti*, corroborate Massaria's judgements. In these records physicians classified 806 deaths (of 6,993 plague cases) as plague even though death came so quickly – in a day or less – that the usual skin disorders, buboes, carbuncles, or pustules, had no time to form. Yet even with these, Milan's physicians reported none who coughed or spat blood.[39] Instead, the Milanese physicians found consistently a trio of symptoms with plague – continuous fever, headaches, and vomiting. In hundreds of plague tracts across Europe into the eighteenth century as well as in the Milanese death records, physicians made clear that this vomiting was intestinal, with the bloating of stomachs, nausea, and diarrhoea, and had not derived from complications affecting the lungs.[40]

These features of plague in the early-modern period were not peculiar to Italy or to the warm climates of the Mediterranean. After plague at Novgorod in 1424 and before the Third Pandemic, I know of only one source to mention pneumonic symptoms of plague without specifying that they pertained to plagues of the fourteenth century. A plague tract by Dr Gilbert Skeyne (who in 1568 became James VI's physician) mentions the 'spitting of blude' in a list of plague characteristics but without specifying any plague and relying heavily on foreign sources.[41] The next chapter ('Signs of deth in pestilential personis'), also lists characteristics of plague but now mentions no pulmonary signs and instead makes clear that vomiting concerned 'dolore of the intestynis'.[42] Not even in the frigid conditions of the Baltic plagues in 1709–13 were pneumonic characteristics described. Instead, detailed diaries, chronicles, and physicians' reports from numerous villages, towns, and cities in Finland, Sweden, Denmark, and Northern Germany described the skin disorders of bubonic plague in great detail stretching through these unusually cold winters. These were the same buboes, carbuncles, and black spots seen with plagues of hot Mediterranean summers and elsewhere.[43] Did the general disappearance of pneumonic plague from Western Europe around 1400 or before depend on an adaptation between the pathogen and the host, or was it the consequence of a genetic mutation? Perhaps later genetic analysis with specifically dated plague pits might be able to unravel this enigma.

## Seasonality and epidemiology

Plague presents further patterns over the long term of the 'Second Pandemic' that were not as stable as its signs or as relatively straightforward as its one-time switch in symptoms. Like the patterns of signs and symptoms described above, these features also fail to match well with plague of the 'Third Pandemic' from 1894 to the present. First, the seasonality of medieval and early-modern plagues illustrates wide variability but also distinctive patterns. In 1348–9, plague could persist through the winter months as at London, where surviving last wills and testaments from the Court of Hustings show plague deaths mounting from December 1348 to April 1349.[44] Unlike Guy de Chauliac's description of the Black Death in Avignon, no source points to any change in the character of the disease in England from a pneumonic to a bubonic plague that may have corresponded with changing temperatures. Similarly, as we have seen, the last plague wave in Scandinavia and northern Europe in 1709–13 flared during exceptionally frigid winters even for these climes and yet remained bubonic. The remarkable flexibility in the seasonality of the Black Death disease over the long term of the 'Second Pandemic' fails to correspond with the close relationship between the fertility cycle of fleas and plague outbreaks during the 'Third Pandemic', especially when these outbreaks were predominantly of the bubonic form.

Nonetheless, plagues of late-medieval and early-modern Europe assumed seasonal patterns. In his mid-fifteenth-century *Commentaries*, Pope Pius II called plague a 'summer contagion'.[45] Evidence from last wills and testaments, necrologies of friars and religious confraternities, and the earliest burial records of entire cities (first Arezzo in 1373;[46] then Florence in 1398), confirm Pius's generalisation, at least for regions of the Mediterranean. Almost without fail the disease there began to climb sharply in May and reached its peak at the hottest points in the calendar, either in June or July, and by September had virtually disappeared, even if, as the early-fifteenth-century Giovanni Morelli claimed, cases might smoulder unreported through the winter months.[47] These patterns and their expectations contributed to long-lasting consequences for the habits and culture of urban residents in Renaissance cities, along with architectural changes in the countryside. After 1348, summers signalled a seasonal migration, at least for elites, who escaped cities for the bucolic life of the countryside. Increasingly, they built country villas, extolled the virtues of antique Roman leisure on country estates and attitudes of *noblesse oblige* over supposedly loyal country retainers and peasants as seen in Leon Battista Alberti's *Della famiglia* or Giovanni Rucellai's *Zibaldone*.[48] To be sure, Florentine patricians possessed country estates before the Black Death as with a branch of Florence's banking family, the Bardi, who had built castles in the mountainous districts north of Florence at Mangona and Vernio, but these were military fortifications, not the villas of the Renaissance, and had been constructed more or less for year-round residency and not to escape the summer's environmental dangers in cities.[49] Moreover, land transactions recorded in the notarial archives of Florence show that this Black Death impetus extended to urban artisans and even the occasional skilled worker in the wool industry. Increasingly, they too invested in plots of land and cottages beyond city walls, to where they could escape during the summer months.[50]

This summer pattern of plague in the Mediterranean is found beyond the Italian peninsula,[51] but diverges sharply from the seasonality of the plagues in the twentieth century, dependent on reservoirs of rats and flea vectors. When bubonic plague appeared in places such as Taranto (Puglia) in 1945, it flared between September and November, and peaked in October, that is three to four months later than the medieval plagues had reached their peaks. Given the seasonal cycle of fleas and the Mediterranean rat flea (*N. fasciatus*) in particular, mid-summer would have been the least likely moment for an outbreak of plague: June and July mark the lowest point in these insects' populations because of high average daily temperatures and lower humidity. But, curiously, in colder northern cities in Europe, where published last wills and testaments or other death records such as necrologies and burial records survive – Besançon, Caen, Arras, Lille, Douai, Cambrai, Lübeck, Lüneberg, Hamburg, and Braunschweig – another pattern can be detected that differed markedly from the Mediterranean pattern: mortalities here peaked in the cooler months of these cooler climes, that is, in October or November.[52] Why colder weather should have favoured late medieval-Renaissance plague in the north has yet to be explained. But as in the south, these northern peaks of late autumn, even early winter, fail to correspond with the heights in the rat flea's fertility cycle, especially given that by the fifteenth century 'the Little Ice Age' was already forming in northern Europe (though not yet in the Mediterranean), causing temperatures to drop significantly lower in these regions than they are today.[53]

Late-medieval and early-modern plague in Europe presents other patterns equally enigmatic. From 1347 to 1351 the Black Death appears to have been as disastrous in large cities and towns as in small villages or even in isolated and sparsely populated mountain districts, as William Rees discovered almost a century ago for hamlets in the region of Snowdonia. Moreover, the mountain outpost of Mangona in the Alpi Fiorentini was the only commune in the territory of Florence to receive special compensation from the city because of the ravages of the Black Death in 1348.[54] As far as the statistics allow, certain villages such as ones on the estates of Crowland Abbey,[55] or around St. Flour (Auvergne) in the mountains of the Massif Central, experienced mortalities that exceeded 70 per cent,[56] and were as high as any mortality rates in crowded densely packed European cities of 1348. But in the countryside, the mortality rates could vary dramatically from one village to the next, as in Cambridgeshire in 1348, for instance, between 5 and 70 per cent. As Matteo Villani commented, plague in Friuli and Slovenia in 1358 'struck in the way of a hail storm, leaving one area intact while destroying another'.[57] For the second plague wave in Central Italy of 1362, he changed his metaphor but emphasised the same pattern. In the mountainous region of the Casentino, northeast of Florence, plague struck certain hamlets but skipped others, 'similar to early thin clouds through which rays of sunlight appear, casting light here but not there'.[58] Here, the reader might find parallels with the seemingly random character of the spread of plague in the late nineteenth and twentieth centuries, which was dependent more on the pathways of rodents than of people. These similarities, however, end with the pervasiveness with which the Black Death spread through most of Europe in a little over three years. As George Christakos and his team of epidemiologists have shown, the Black Death covered space in a given period of time by two orders of magnitude greater than had any plague anywhere since Yersin's discovery in 1894, despite the latter's advantages of transmission by railway, steamship, and motorised vehicles.[59]

The scattered facts of high mortalities in certain villages may also lead the reader to conclude that the geographical mortality patterns of late-medieval and early-modern plague correspond with those the epidemiologist Major Greenwood discovered for plagues in India during the early twentieth century: unlike diseases that spread person-to-person as with measles or influenza, plague of the 'Third Pandemic' has exhibited an inverse relationship between the size of a community and rates of mortality.[60] Plague rates of mortality in villages of the Punjab during the early twentieth century greatly exceeded those in Mumbai or Lucknow: for the former they could reach as high as 30 per cent, while for the latter, they never exceeded the 2.7 per cent scored at Mumbai in 1903. No such patterns, however, are seen with plague rates of mortality in Europe, either during the Black Death of 1347–51 or for later plagues through the early-modern period. The population losses in Europe's large cities such as London, Barcelona,[61] Tournai, Ghent, and many others topped 50, even 60 per cent, in 1348. The urban population of Florence may even have fallen by as much as three-quarters during six months of plague in 1348.[62]

The return of plague in the seventeenth century could be as devastating. In the plague of 1630–1 Milan lost over half of its inhabitants, and in Italy's last major plague of 1656–7, Genoa and Naples lost between 60 per cent and two-thirds of their populations.[63] By contrast, from the fifteenth to the seventeenth century, the relationship was the opposite of that which Greenwood argued for twentieth-century plague: in market towns and country hamlets plague deaths fell, as can be seen for Tuscany during the fifteenth century, when the city of Florence's population stagnated, while those of its surrounding countryside (*contado*) gathered pace, despite inward migration to cities. Scholars have even claimed that plague became 'mainly an urban event' during the early-modern period and not one of small places or of the countryside.[64] As Guido Alfani has shown, early-modern plague played a key role in reducing Italian urbanisation rates'.[65] Even when mortalities climbed in villages and penetrated through the countryside in various regions of Italy with the plagues of 1630–3 and 1656–7, still in places such as the diocese of Salerno, where rates of mortality can be measured and compared, urban parishes experienced the higher rates,[66] and within the countryside rates of mortality tended to vary positively with population density.[67]

Why plagues from at least the fifteenth century should have become more virulent in cities is yet another unanswered question and in some respects is counter-intuitive. The major investments and breakthroughs in sanitation and public health in early-modern Europe – innovations in hospitalisation, systems of quarantine and surveillance, nursing and food subsidies for plague victims, their families, and the poor – centred on cities, not on the countryside.[68] As historians such as E. L. Sabine argued in the 1930s for medieval London,[69] and Carole Rawcliffe has now demonstrated for medieval cities across England, the Black Death was a stimulus for sanitary and environmental reform with investments in public latrines, sewage, systems of piped water, controls on butchering and garbage disposal.[70] In addition, the halving or more of urban populations, combined with notions of miasma and fears that smells, trash, squalor, and urban concentration sparked plague and other diseases,[71] encouraged major changes in urban landscapes. Cities such as Florence became more open as elites destroyed the tangles of twisted medieval streets to broaden boulevards and public squares and to build palaces and gardens as models of the new 'ideal' cities of the Renaissance. Yet, despite these reforms and sanitary changes, cities such as Florence, hard-hit by plague until the early sixteenth century, increasingly had to rely on excess numbers from the countryside to replenish their urban populations, which failed to return to pre-Black Death levels until the second half of the nineteenth century.

## Mentalities and policies

Patterns of plague had other momentous effects on mentalities. With few exceptions, plague mortality rates declined steadily and steeply from 1348 to the mid-fifteenth century. As Raymond Chalin charted: 'In 1348, two-thirds of the population were afflicted, and almost all died; in 1361, half contracted the disease, and very few survived; in 1371, only one-tenth were sick, and many survived; in 1382, only one-twentieth became sick, and almost all survived'.[72] No doubt, Chalin's trend-line was overly optimistic, especially for 1382, but as seen from age pyramids constructed from the Florentine tax record, the Catasto of 1427,[73] and from last wills and testaments in Tuscany and Umbria of the first half of the fifteenth century, plague mortalities had settled down to 2 to 5 per cent of the population (and not the catastrophic fourteenth-century levels of 20 to 70 per cent), that is, until the disastrous plague of 1449–52 and that of 1478–80 in Italian cities such as Florence and Milan.[74] Reflecting further the symbiosis between plague and its human hosts, the Black Death became increasingly a childhood disease: the second plague of 1361–3, which chroniclers across Europe called the plague of children, was no one-off exception. According to burial records at Siena (that divided plague deaths into children and adults) and at least one chronicler – the Pisan Ranieri Sardo – the disease sharply and steadily became one of children between 1348 and 1383: the proportion of victims under twelve rose from a third in 1348 to half in 1361, to two-thirds in 1374–5, and to 88 per cent in 1383.[75]

Correspondingly, the rapid changes in the body's adaption to its new pathogen sparked an abrupt shift in attitudes: no 'l'histoire immobile' was at play here.[76] With the first onslaught of plague, 1347–51, religious leaders, merchant chroniclers, and physicians looked to the heavens and stars to explain the apocalyptic levels of mortality, which they proclaimed had never before been seen even in Biblical times. Their explanations centred on sin or cosmic transformations well above the human sphere. Either way, secular human action, whether through governmental dicta or medical intervention, was viewed with scepticism, even hostility. As the Parisian doctor Simon de Couvin declared in 1350, the Black Death had confounded all doctors: 'the art of Hippocrates was lost' and 'no medicine was more efficacious than flight'.[77] Those outside the medical profession were more damning. For Boccaccio, 'neither the advice of physicians nor any medicine was of any value'.[78] Siena's principal chronicler, Agnolo di Tura del Grasso, was forced to bury his five sons with his own hands in the summer of 1348 and claimed that 'people' realised 'any [medicine] tried only brought a quicker death'.[79] His Florentine contemporary, Matteo Villani, concurred: 'doctors from all over the world . . . had no remedy . . . they visited the sick only to make money'.

This pessimism did not, however, continue. As early as the second plague, physicians were interpreting changes in the body's resistance as the consequence of their own medical procedures and interventions. By the fourth plague in the early 1380s the physician of three popes at Avignon reflected on the plague trends and praised his generation of physicians for the steep declines in mortality and morbidity. Amongst others, a Venetian physician at the end of the fourteenth century joined the chorus of self and professional adulation, claiming to have cured a hundred plague patients in his practice. Around the same time a Portuguese physician made similar claims: in the 'last big plague', by his methods he had cured 'an infinite number'. Other doctors cured themselves (or so they believed) and wished to make their experiments and experiences available to fellow doctors, their citizens, 'and to the world'.[80] At the end of the fourteenth century, Dr Stephanus of Padua described himself and his wife afflicted with plague – four days 'of horrendous fevers and the detestable signs'. But with his regime of remedies he cured both of them and afterwards (so he claimed) 'many other citizens of Padua'. Emboldened by his 'successes', he boasted to have 'triumphed over the plague', and wrote down his recipes to benefit the citizens of Padua, to whom he dedicated his tract.[81] The remarkable growth of this virtually new form of medical writing is testimony to the confidence, even hubris, of this new post-Black Death generation of physicians. According to the medical historian Arturo Castiglioni, by the fifteenth century, the plague tract had become 'a new genre of popular literature' and thousands of them continue to survive in rare book rooms across the globe.[82]

Doctors were not the only ones to advertise medical successes. By the early fifteenth century, the Florentine merchant Giovanni Morelli advised his children and future progeny to seek out and observe 'diligently the remedies of valiant doctors':

> [In 1348] no one had any remedy or cure; the plague was so great and fierce that none could help in any way, and for these reasons people died without any relief. Yet today because of the present plague and many others that we have lived through, there are now cures . . . Doctors' advice and their preventive rules provide an arm to defend against this disease's poison.[83]

Such a change in confidence regarding the power of human intervention to alter the natural world was not limited to Italy. As early as the late 1350s, John of Burgundy boasted that his advice on plague treatment came from 'long experience' and had given his generation of physicians a new seal of authority: they now surpassed Arabic and ancient authorities, 'who had never witnessed a plague of the duration and magnitude of the plagues in his day'. By the third plague, Johannes Jacobi of Montpellier went further: 'since the plague had invaded us so frequently, we have much to say from experience, while the ancients could say little'.[84] With the next plague in 1382 Raymond Chalin was less respectful: the ancients did not understand the causes of plagues or know how to deal with them . . . they had left everything in confusion'.[85]

With the rise in plague mortalities in the second half of the fifteenth century, Renaissance confidence began to wane. Again, patterns of plague conditioned patterns of mentality. For more than a century, medical writing on plague shifted from the vernacular back to Latin and once again became steeped in erudite debates filled with citations that bowed to the authority of Arabic and ancient authors. Boasting from experience and experimentation, with claims of curing 'infinite numbers', had now disappeared.[86] But with plague engulfing Italy from coastal towns in southern Sicily to the mountain passes north of Trento in 1574, plague writing shifted again and their publication increased exponentially. Now, physicians and even some theologians challenged more vigorously than ever before the old Galenic principles and theological causes of plague. In their place, writers pointed to poverty, poor housing, insanitary sources of water and the absence of plumbing as the culprits, and called on the state to intervene.[87]

Did conscious human endeavours to control plague, in turn, contribute to shaping the contours of plague? During the second half of the fourteenth century, gaps between successive strikes of plague were roughly the same throughout Europe, one every 10 to 14 years. Then, first at Ragusa, the gaps between successive plagues widened. Between 1482 and 1526, no plagues entered Ragusa's gates, and the city experienced its last plague of the Second Pandemic in 1533. Corresponding with these patterns were political and bureaucratic innovations within the city of Ragusa: not only did it create the first plague quarantine (actually, a trentina) in 1377; less well known, it created the first health office in 1392, first permanent health office in 1397, and throughout the fifteenth century strengthened its systems of disease surveillance, as Zlata Blažina Tomič and Vedsna Blažina have recently shown.[88] In the sixteenth century, plague gaps widened in Italy, where new forms and procedures of quarantine and plague control were also implemented. To be sure, such elongations might reflect changes in the adaptation of pathogens to their human hosts, changes in environment, or possibly changes in the pathogens alone. But that these changes should reflect sharp differences in political as opposed to ecological borders suggests that conscious human factors were at play. Indeed, they correlate with differences in developments of public health care policies and infrastructure. For Europe, the Italians led the way. Early initiatives included the plague legislation at Pistoia in 1348, quarantine procedures in Milan in 1374, or lazaretti at Venice in 1422. With the Italian plague of 1574–7, changes in policies and practice became more pervasive through the peninsula. They included developments of health boards backed by city-state and territorial authority, enforcing martial law, building and improving new plague hospitals that more rigorously enforced the isolation of plague patients, places and procedures for quarantining and cleansing goods, new organisation for street cleaning and the emptying of latrines, the invention of health passports, the posting of guards at city gates and mountain passes, and new systems of gathering information on plague outbreaks with spy networks that by the end of the sixteenth century extended to intelligence gathering by Italian cities across Europe and into Africa and Asia. Italian governments and physicians were the first to address what they began to see as the fundamental causes of plague – filth, excrement, bad housing, bad plumbing, and poverty – and they made concrete proposals for improving quarantine, the cleaning of latrines, and the implementation of better plumbing.[89] These new approaches left their mark. While plague remained more or less endemic in France, England, Scandinavia, and German-speaking regions into the seventeenth century, with epidemic outbreaks every generation and in places more often,[90] the gaps separating successive epidemics in Italy lengthened, especially during the sixteenth century to its last major plagues in 1656–7. Between 1540 and 1666 devastating plagues struck nine times in London, which had yet to adopt the Italian systems of gate and border guards and quarantine,[91] and in Denmark (which did not institute state-wide plague regulations until 1625) 11 major plagues hit between 1550 and 1660.[92] In addition to numerous plagues, which appear to have sprung internally within the regions of central Europe, a further 11 plagues from outside penetrated its ports and spread through the region between 1560 and 1660.[93]

By contrast, after 1524 only two plagues afflicted Milan and Venice, and only one outbreak struck Florence. Before 1576, plague had not struck Palermo for a hundred years,[94] and afterwards, only one occurred, its finale, in 1624. No plagues hit Genoa, Rome and many other major cities between 1579 and 1656, while for Milan and Venice, the devastating one of 1630 was their last one.[95] The Italian peninsula as a whole was the first region of Europe, Asia, or Africa, beyond a single city-state, to rid itself of the major plagues of the late-medieval and early-modern variant.[96] From the late sixteenth to the mid-seventeenth century, public officials and physicians in other regions of Europe began translating the great outpouring of public-health-minded tracts produced by Italian physicians and other intellectuals during the plague of 1576–9,[97] and soon afterwards began systematically to adopt Italian anti-plague systems as in Central Europe after 1640. As a result, a pattern that replicated the Italian experience of a century earlier began to take shape with a dramatic decline in the frequency of plagues but with at least one last catastrophic strike as had happened in parts of Italy: Stockholm, Copenhagen, Danzig, and places in East Prussia paid the price of the absence of plague in their regions with one last devastating plague in 1709–13, as did Marseille, its region and other towns in southern France a decade later, when their *grande finale* after a long absence reaped between half and two-thirds of their populations.[98]

## Conclusion

The patterns of late-medieval and early-modern plague, including variations in signs and symptoms, trends in adaptation between pathogens and hosts, and shifts in seasonality, show some parallels with other infectious diseases as well as enigmas that remain to be studied. Some of these puzzles, but not all of them, relate to the abrupt, and yet to be explained, changes in the character of plagues during the late Middle Ages and early-modern period and those of the 'Third Pandemic', such as differences in speeds of transmission, levels and nature of contagion, seasonality, and signs and symptoms. These findings invite new questions and should encourage new cooperation between historians and those involved in the biological sciences – bio-archaeologists, microbiologists, and geneticists who analyse traces of ancient DNA. How do we explain, for instance, the transformation of the plague around 1400 from a disease that had combined bubonic and pneumonic forms of plague as evinced by the violent spitting of blood to ones in which all pneumonic symptoms had disappeared? Was it simply a matter of a gradual adaptation between pathogen and host or did a significant mutation in the pathogen suddenly occur? The same question could be asked of the shift in plague during the seventeenth century in Italy, when the disease recovered its capacity to spread rapidly through rural areas, to kill across social classes, and to reap levels of mortality in cities such as Milan, Naples, and Genoa comparable to what they had been with the Black Death in 1348. In mapping various patterns of plague over the long term of the 'Second Pandemic', this chapter has also attempted to raise other points currently of little concern to geneticists and microbiologists, such as how patterns of plague changed social practices and mentalities in late-medieval and early-modern Europe and, in turn, how conscious human endeavours well before the 'laboratory revolution' could to some extent control this major scourge of late-medieval and early-modern societies.

## Notes

1 I thank Lawrence Weaver for an early reading of this chapter. For discussions of conflicting notions of diseases, see: O. Temkin, 'The Scientific Approach to Disease: Specific Entity and Individual Sickness', in A. Crombie (ed.), *Scientific Change: Historical Studies in the Intellectual, Social and Technical Conditions for Scientific Discovery*, London: Heinemann, 1963, pp. 629–47; C. Rosenberg, 'What Is Disease? In Memory of Owsei Temkin', *Bulletin of the History of Medicine*, 2003, vol. 77, 491–505; C. Rosenberg, 'Epilogue: Airs, Waters, Places. A Status Report', *Bulletin of the History of Medicine*, 2012, vol. 86, 661–70; C. Rosenberg and Janet Golden (eds), *Framing Disease: Studies in Cultural History*, New Brunswick: Rutgers University Press, 1992; and M. Worboys, *Spreading Germs: Disease Theories and Medical Practice in Britain, 1865–1900*, Cambridge: Cambridge University Press, 2000, esp., 278–86.
2 G. Fracastoro, *De Contagione et Contagiosis Morbis et eorum Curatione, Libri III*, ed. and trans. W. Wright, New York: Putnam, 1930, pp. 138–9.
3 F. Macfarlane Burnet, *Natural History of Infectious Disease*, 3rd edn, Cambridge: Cambridge University Press, 1962, pp. 5–6 and 296. The first pandemic, sometimes called the Justinianic plague, lasted from 541 AD to circa 750 AD. The Black Death of 1347 to 1352 set off 'the Second Pandemic', which reappeared in numerous waves into the nineteenth century; and 'the Third Pandemic' is usually dated to the disease reaching Hong Kong in 1893 and continuing to the present.
4 V. Morea, *Storia della peste di Noja*, Naples: A Trani, 1817. For a still later outbreak of the plague of the stamp of the 'Second Pandemic', see M. and J. Peset, *Muerta en España (Política y sociedad entre la peste y el colera)*, Madrid: Seminarios y Ediciones, 1972, p. 115, for a plague in Mallorca in 1820.
5 G. Boccaccio, *Tutte le opere di Giovanni Boccaccio. Vol. 4: Decameron*, ed. V. Branca, Milan: Mondadori, 1976, p. 10.
6 Giovanni di Pagolo Morelli, *Ricordi*, in V. Branca (ed.), *Mercanti Scrittori: Ricordi nella Firenze tra Medioevo e Rinascimento*, Milan: Rusconi, 1986, p. 207.
7 P. Bellintani, *Dialogo della peste*, ed. E. Pacagnini, Milan: Libri Scheiwiller, 2001, pp. 127–30; on Bellintani, see S. Cohn, *Cultures of Plague: Medical Thinking at the End of the Renaissance*, Oxford: Oxford University Press, 2010, pp. 64, 236, 267, 289.
8 Cohn, *Cultures of Plague*, pp. 50–1; L. Boerio, *Trattato delli bubonic, e carboni pestilenti con loro cause, segni, e curationi*, Genoa: Giuseppe Pavoni, 1630; first published, 1579, p. 56.
9 Morea, *Storia della peste*, p. 5.
10 *Chronicon Galfridi le Baker de Swynebroke*, ed. E. Thompson, Oxford: Clarendon, 1889, p. 100.
11 *Galar Y Beirdd: Marwnadau Plant/ Poets' Grief: Medieval Welsh Elegies for Children*, ed. and trans. D. Johnston, Cardiff: Tafol, 1993, pp. 56–8. The chronicler of the second plague in England, John of Reading, also described these black spots across the bodies of plague victims: *Chronica Johannis de Reading et Anonymi Cantuariensis*, ed. J. Tait, Manchester: University of Manchester Press, 1914, p. 212.
12 N. Hodges, *Loimologia: Or, an Historical Account of the Plague in London in 1665*, London: E. Bell, 1721, p. 117.
13 Ibid., pp. 119 and 124–7.
14 Ibid., pp. 128–36. This tri-partite division of plague signs persisted beyond Europe in Egypt and other places in the Middle East and was categorised by 'levels of intensity'. The first level – those with buboes – were the easiest to cure, while the third – those with 'les pétéchins' that could spread 'indiscriminately to all parts of the body' – was the most dangerous: A. Brayer, *Neuf années a Constantinople, observations sur . . . la peste*, 2 vols. Paris: Bellizard, 1836, II, pp. 23 and 31–3.
15 J. Hecker, *The Black Death in the Fourteenth Century*, trans. B. Babington, London, 1833, p. 26.

16  R. Mead, *A Discourse on the Plague*, 9th edn, London: A. Miller, 1744, pp. 110 and 152–3, where he instructed on how to distinguish between the pustules of smallpox and those of plague.

17  K. Frandsen, *The Last Plague in the Baltic Region 1709–1713*, Copenhagen: Museum Tusculanum Press, 2010, pp. 43, 152–3, 163.

18  Ibid., pp. 44, 152, 153, 163,193.

19  For a more detailed analysis of these records, see Cohn, *Cultures of Plague*, ch. 2, especially, pp. 61–5.

20  *Cronica volgare di anonimo fiorentino dall'anno 1385 al 1409 gia` attribuita a Piero di Giovanni Minerbetti*, ed. E. Bellondi, Rerum Italicarum Scriptores, vol. 27, part 2, Citta di Castello: E. Lapi, 1915–18, p. 110.

21  For this and other similar cases of diagnosis during the fourteenth and fifteenth century, see S. Cohn, *The Black Death Transformed: Disease and Culture in Early Renaissance Europe*, London: Edward Arnold, 2002, pp. 136–7.

22  Hodges, *Loimologia*, pp. 81–2, 128.

23  Cohn, *Cultures of Plague*, pp. 62–3.

24  For more details on these clinical characteristics and comparisons from reports in India and China in the late nineteenth and early twentieth centuries, see Cohn, *Cultures of Plague*, ch. 2.

25  Ibid., pp. 61–2.

26  A. Macchiavello, 'Chapter 47: Plague', in R. Gradwohl, L. Soto, and O. Felsenfeld (eds), *Clinical Tropical Medicine*, London: H. Kimpton, 1951, pp. 444–76, especially p. 460.

27  Guy de Chauliac, *Inventarium sive chirugia magna* I, ed. M. McVaugh, Leiden: Brill, 1997, pp. 117–18.

28  Cohn, *The Black Death Transformed*, pp. 59, 83–7, 90–5.

29  *Il Memoriale di Iacopo di Coluccino Bonavia Medico Lucchese (1373–1416)*, in P. Calamari (ed.), *Studi di Filologia Italiana*, XXIV, Florence, 1966, 55–428, p. 397.

30  J. Hayez, 'Quelques témoignages sur les épidémies à Avignon, 2e moitié XIV siècle' (unpublished article).

31  In addition to Guy de Chauliac, see the tract of Louis Sanctus in 'La Peste en Avignon (1348) décrite par un témoin oculaire', Andries Welkenhuysen (ed.), in R. Lievens, E. Van Mingroot and W. Verbeke (eds), *Pascua Mediaevalia: Studies voor Prof. J. M. De Smet*, Leuven: Universitaire Pers Leuven, 1983, 452–92, pp. 465–9; and the Egyptian chronicler Maqrīzī, *Al-sulūk li-ma 'rifat duwal al-mulūk*, trans. G. Wiet, in 'La grande pest noire en Syrie et en Egypte', *Etudes d'orientalisme dédiées à la mémoire de Lévi-Provençal*, Paris: G.-P. Maisonneuve et Larose, 1962, vol. 1, 367–80, p. 370.

32  Morelli, *Ricordi*, p. 207.

33  *The Chronicle of Novgorod 1061–1471*, trans. R. Michell and N. Forbes, Camden Society, 3rd Series, vol. 15, London: J. B. Nichols, 1914, p. 145.

34  For Sicily during the plague of 1576–7, for instance, see *Bando et ordinationi fatte per . . . città di Palermo*, Palermo: Giovan Mattheo Mayda, 1575, 3v–4r.

35  G. Ingrassia, *Informatione del pestifero et contagioso morbo . . .*, Palermo: Giovan Mattheo Mayda, 1576; reprinted Milan: FrancoAngeli, 2005.

36  Cohn, *Cultures of Plague*, pp. 50–2.

37  A. Gallo, *Fascis de Peste, Peripneumonia pestilentiali cum sputo sanguinis . . .*, Brescia: Io. Baptistae Bozolae, 1566.

38  A. Massaria, *De Peste: Libri duo*, Venice: Altobellum Salicatium, 1579, 31v.

39  Pleurisy was mentioned in 3 of 6,993 cases, all in children between one and six years old during the 1485 plague – Cohn, *Cultures of Plague*, p. 67.

40  For descriptions of these symptoms, ibid., pp. 65–72.

41  Gilbert Skeyne, *Ane Breve Descriptiovn of the Pest (1568)*, in *Tracts by Dr Gilber Skeyne, Medicinar to His Majesty*, Edinburgh: The Bantyne Club, 1860, pp. 1–46, p. 14. Also see Charles Creighton, *A History of Epidemics in Britain*, ed. D.E.C. Eversley, London: Frank Cass, 1965, I, p. 364.

42  Skeyne, *Ane Breve Descriptiovn*, p. 14. In a recent conversation, Lars Walløe, Professor Emeritus of pathology, Oslo, informed me that he has discovered documents describing pneumonic symptoms on an island in the Baltic during an eighteenth-century plague.

43  Frandsen, *The Last Plague in the Baltic Region.*

44  See Cohn, *The Black Death Transformed,* p. 184.

45  A. Piccolomini (Pope Pius II), *I Commentarii,* ed. L. Totaro, Milan: Adelphi Edizioni, 1984, vol. 1, p. 1615.

46  L'Archivio della fraternità dei Laici di Arezzo, Libri dei morti, no. 882.

47  Morelli, *Ricordi.*

48  Leon Battista Alberti, *I libri della famiglia,* ed. R. Romano and A. Tenenti, Turin, Einaudi, 1994; *Giovanni Rucellai ed il suo Zibaldone,* ed. A. Perosa, London: The Warburg Institute, 1960.

49  See *I Capitoli del Comune di Firenze,* ed. C. Guasti, Florence, Cellini, 1865, vol. 1, pp. 107–8; Bardi e Mangona (Vendita di Mangona) on January 15, 1340 (Florentine style); and Emanuele Repetti, *Dizionario geografico, fisico, storico della Toscana,* Florence: A. Tofani, 1833–46, vol. 3, pp. 42–7. For other magnate families with principal residences in the country-side before the Black Death, see Cohn, *Creating the Florentine State: Peasants and Rebellion, 1348–1434,* Cambridge: Cambridge University Press, 1999.

50  These conclusions are based on thousands of contracts, mostly rents and land transactions, redacted by the Mazzetti family of notaries from 1348 to 1426, who worked within a ten-mile radius of Florence's city walls – Cohn, *Creating the Florentine State,* pp. 18, 102–3, 108.

51  Cohn, *The Black Death Transformed,* p. 88, for places such as Millau (Aveyron) in southern France, Barcelona and Valencia.

52  See ibid., pp. 178–87.

53  H. Lamb, *Climate, History and the Modern World,* 2nd edn, London: Routledge, 1995, p. 207. By contrast, these colder conditions caused by 'blocking anticyclones' had affected only northern Europe until the sixteenth century. In the Mediterranean, temperatures would have remained about +1°C warmer than temperatures in 1970. In his through-going attack on the so-called human flea (*Pulex irritans*) as the vector of plagues during the 'Second Pandemic' and his argument that *N. fasciatus* instead must have been the vector, F. Audoin-Rouzeau, *Les Chemins de la Peste- Le rat, la puce et l'homme,* Rennes: Presses universitaires de Rennes, 2003, does not consider the wide seasonal discrepancies between the outbreaks of plague, c. 1347 to 1820, in Europe and the fertility cycle of this (or any other) flea.

54  Cohn, *Creating the Florentine State,* p. 70.

55  F. Page, *The Estates of Crowland Abbey: A Study in Manorial Organization,* Cambridge: Cambridge University Press, 1934, pp. 120–5.

56  H. Dubois, 'La dépression: XVIe et XVe siècles', in J. Dupâquier, *Histoire de la population française,* Paris: Presses Universitaires de France, 1988, vol. I, 313–66, p. 321.

57  *Matteo Villani, Cronica con continuazione di Filippo Villani,* ed. G. Porta, 2 vols., Parma: Fondazione Pietro Bembo, 1995, II, p. 301.

58  Ibid., pp. 301 and 585–6.

59  G. Christakos, et al., *Interdisciplinary Public Health Reasoning and Epidemic Modelling: The Case of Black Death,* Berlin: Springer, 2005.

60  O. Benedictow, *The Black Death, 1346–1353: The Complete History,* Woodbridge: Boydell, 2004, pp. 281–96; and my criticism of his statistics in *The New England Journal of Medicine,* 352, 2005, 1054–5.

61  R. Gyug, 'The Effects and Extent of the Black Death of 1348: New Evidence for Clerical Mortality in Barcelona', *Medieval Studies,* 1983, vol. 45, 385–98; idem, *The Diocese of Barcelona during the Black Death: Register Notule Comunium (1348–49),* Toronto: University of Toronto Press, 1994.

62  From tax records immediately before the Black Death and in 1351, I have estimated that the death toll in Florence in 1348 may have been as high as 75 per cent.

63  Among other places, see C. Cipolla, *Public Health and the Medical Profession in the Renaissance,* Cambridge: Cambridge University Press, 1976, p. 56, who estimates that this plague wiped out even more at Genoa, 55,000 of a population of 73,000.

64  G. Alfani, 'Plague in Seventeenth Century Europe and the Decline of Italy: An Epidemiological Hypothesis', *European Review of Economic History,* vol. 17, 2013, 408–30, esp. pp. 414 and 424.

65  Ibid., p. 424.

66 C. Corsini and G. Delille, 'La peste de 1656 dans le diocèse de Salerne: quelques résultats et problèmes', in H. Charbonneau and A. Larose (eds), *The Great Mortalities: Methodological Studies of Demographic Crises in the Past*, Liège: Ordina, 1979, 51–9, p. 55.

67 Alfani, 'Plague in Seventeenth Century Europe', pp. 420–21.

68 R. Palmer, 'The Control of Plague in Venice and Northern Italy, 1348–1600', PhD thesis: University of Kent at Canterbury, 1978; Cohn, *Cultures of Plague*; and J. Crawshaw, *Plague Hospitals: Public Health for the City in Early Modern Venice*, Farnham: Ashgate, 2012.

69 E. Sabine, 'Butchering in Medieval London', *Speculum*, vol. 8, 1933, pp. 335–53; E. Sabine, 'Latrines and Cesspools of Mediaeval London', *Speculum*, vol. 9, 1934, pp. 303–21; and E. Sabine, 'City Cleaning in Mediaeval London', *Speculum*, vol. 12, 1937, pp. 19–43.

70 C. Rawcliffe, *Communal Health in Late Medieval English Towns and Cities*, Woodbridge: Boydell, 2013.

71 J. Henderson, 'The Black Death in Florence: Medical and Communal Responses', in S. Basset (ed.), *Death in Towns: Urban Response to the Dying and the Dead, 100–1600*, Leicester: Leicester University Press, 1992, pp. 136–50; J. Henderson, '"La schifezza, madre della corruzione": Peste e società nella Firenze della prima età moderna, 1630–1631', *Medicina & Storia*, 2001, vol. 2, 23–56.

72 Ibid., p. 191.

73 D. Herlihy and C. Klapisch-Zuber, *Les Toscans et leurs familles: Une étude du Catasto de 1427*, Paris: École des hautes études en sciences sociales, 1978.

74 The notable exceptions were the plagues of 1399–1401 and 1437–8.

75 On these sources, see Cohn, *The Plague Transformed*, pp. 212–15.

76 On the assumptions that changes in mentality were so slow as to be unrecognisable before the Enlightenment, see E. Le Roy Ladurie, 'L'histoire immobile', *Annales E.S.C.*, vol. 29, 1974, 673–92.

77 'Opuscule relalif à la peste de 1348 composé par un contemporain', ed. E. Littre, *Bibliothèque d'École des chartes*, 1841, vol. 2, pp. 201–43.

78 Boccaccio, *Decameron*, p. 11.

79 *Cronaca Senese attribuita ad Agnolo di Tura del Grasso*, in A. Lisini and F. Iacometti (eds), *Cronache senesi*, Rerum Italicarum Sciptores, vol. 15, pt. 6, Bologna: Nicola Zanichelli, 1933–5, p. 555.

80 'Ex libro Doinysii Secundi Colle', in H. Haeser (ed.), *Geschichte der epidemischen Krankheiten in Lehrbuch der Geschichte der Medizin und der epidemischen Krankheiten*, II (Jena, 1865), p. 169.

81 'Ein Paduaner Pestkonsilium von Dr. Stephanus de Doctoribus', *Sudhoff Archiv für Geschichte der Medezin* [hereafter, *Sudhoff*], 1913, Bd 6, p. 356.

82 A. Castiglioni, 'Ugo Benzi da Siena ed il "Trattato utilissimo circa la conservazione della sanitate"', *Rivista di Storia Critica delle Scienze Mediche e Naturali*, 1921, vol. 12, p. 75.

83 Morelli, *Ricordi*, pp. 209–10.

84 'Pestshcrift des Johannes Jacobi', *Sudhoff*, Bd 5, 1911, p. 58.

85 'Das Pestwerkchen des Raymundus Chalin de Vinario', *Sudhoff*, Bd 17, 1925, pp. 38–9.

86 Cohn, *Cultures of Plague*, concentrates on a third shift in plague patterns and mentalities with the peninsula-wide threat of plague in 1575–8 and, except for Venice and Brescia, the successful defence against it. A fourth shift came with the disastrous plagues of 1656–7 – G. Alfani, 'Plague in Seventeenth Century Europe and the Decline of Italy: An Epidemiological Hypothesis', *in European Review of Economic History*, vol. 17, 2013, pp. 408–30.

87 Cohn, *Cultures of Plague*, especially chapters 1, 6, 7, and 8.

88 Z. Tomič and V. Blažina, *Expelling the Plague: The Health Office and the Implementation of Quarantine in Dubrovnik 1377–1533*, Montreal: McGill-Queen's University Press, 2015.

89 These changes in physicians' attitudes and advice on plague is the principal theme of Cohn, *Cultures of Plague*. Despite the absence of political unity from Palermo to Turin, Florence to Venice, a cultural and intellectual unity based largely on training and contacts made at the University of Padua knitted together medical thought through the peninsula; ibid., pp. 32–8.

90 See, for instance, P. Slack, *The Impact of Plague in Tudor and Stuart England*, Oxford: Clarendon, 1990, pp. 68 and 151: serious epidemics swept through London and parts of England from Antwerp and Amsterdam and coincided with outbreaks in Germany and the Low Countries in 1498, 1535, 1563, 1589, 1603, 1625, and 1636.

91  Ibid., p. 151.
92  P. Christensen, 'Appearance and Disappearance of the Plague: Still a Puzzle?' in *Living with the Black Death*, L. Bisgaard and L. Søndergaard (eds), Odense, 2009, pp. 17 and 19.
93  E. Eckert, *The Structure of Plagues and Pestilences in Early Modern Europe: Central Europe, 1560–1640*, Basel, 1996; and E. Eckert, *The Retreat of Plague from Central Europe, 1640–1720: A Geomedical Approach, Bulletin of the History of Medicine*, 74, 2000, pp. 23–4.
94  Ingrassia, *Informatione*, p. 30.
95  For these figures and the bibliography on the 1656–7 Italian plague wave, see I. Fusco, *Peste, demografia e fiscalità nel Regno di Napoli del XVII secolo*, Milan: FrancoAgnelli, 2007 and E. Sonnino, 'Cronache della peste a Rome: Notizie dal Ghetto e lettere di Girolamo Gastaldi (1656–1657), in *Roma moderna e contemporanea*, 14, 2006, pp. 35–8.
96  See: P. Christensen, 'Appearance and Disappearance of the Plague'; P. Slack, 'The Disappearance of Plague: An Alternative View', *Economic History Review*, n.s. 33, 1981, pp. 469–76; and G. Restifo, *Le ultime piaghe: le pesti nel mediterraneo (1720–1820)*, Milan: Selene, 1994. Sicily's last plague was at Messina in 1743 but was contained largely to this one city and its suburbs.
97  Cohn, *Cultures of Plague*, pp. 259–61.
98  J.-N. Biraben, *Les hommes et la peste en France et dans les pays européens et méditerranéens*, 2 vols, Paris: Mouton, 1975–6, II, pp. 230–2.

## Select bibliography

Alfani, G., 'Plague in Seventeenth Century Europe and the Decline of Italy: An Epidemiological Hypothesis', *European Review of Economic History*, vol. 17, 2013, 408–30.

Audoin-Rouzeau, F., *Les Chemins de la Peste – Le rat, la puce et l'homme*, Rennes: Presses universitaires de Rennes, 2003.

Biraben, J.-N., *Les hommes et la peste en France et dans les pays européens et méditerranéens*, 2 vols, Paris: Mouton, 1975–6.

Brayer, A., *Neuf années a Constantinople, observations sur . . . la peste*, 2 vols. Paris: Bellizard, 1836.

Carmichael, A., *Plague and the Poor in Renaissance Florence*, Cambridge: Cambridge University Press, 1986.

Christensen, P., 'Appearance and Disappearance of the Plague: Still a Puzzle?' in *Living with the Black Death*, L. Bisgaard and L. Søndergaard (eds), Odense, 2009.

Cohn, S., *The Black Death Transformed: Disease and Culture in Early Renaissance Europe*, London: Edward Arnold, 2002.

Cohn, S., *Cultures of Plague: Medical Thinking at the End of the Renaissance*, Oxford: Oxford University Press, 2010.

Christakos, G., Olea, R. A., Serre, M. L., Yu, H.-L. and Wang, L.-L., *Interdisciplinary Public Health Reasoning and Epidemic Modelling: The Case of Black Death*, Berlin: Springer, 2005.

Frandsen, K., *The Last Plague in the Baltic Region 1709–1713*, Copenhagen: Museum Tusculanum Press, 2010.

Henderson, J., 'The Black Death in Florence: Medical and Communal Responses', in S. Basset (ed.), *Death in Towns: Urban Response to the Dying and the Dead, 100–1600*, Leicester: Leicester University Press, 1992, pp. 136–50.

Hodges, N., *Loimologia: Or, an Historical Account of the Plague in London in 1665*, London: E. Bell, 1721.

Macfarlane Burnet, F., *Natural History of Infectious Disease*, 3rd edn, Cambridge: Cambridge University Press, 1962.

Rawcliffe, C., *Communal Health in Late Medieval English Towns and Cities*, Woodbridge: Boydell, 2013.

Slack, P., *The Impact of Plague in Tudor and Stuart England*, Oxford: Clarendon, 1990.

Tomič, Z. and V. Blažina, *Expelling the Plague: The Health Office and the Implementation of Quarantine in Dubrovnik 1377–1533*, Montreal: McGill-Queen's University Press, 2015.

Worboys, M. *Spreading Germs: Disease Theories and Medical Practice in Britain, 1865–1900*, Cambridge: Cambridge University Press, 2000.

# 11

# SYMPTOMS OF EMPIRE

## Cholera in Southeast Asia, 1820–1850

*Robert Peckham*

Dr. Stewart referred to the report of the disease which prevailed in the
Milbank Penitentiary, at a time when the country as little dreamed of a
visit from cholera, as from the emperor of China . . .

('Westminster Medical Society: Discussion on
the Cholera', *Lancet*, 17 March 1832, vol. 1, no. 446, 869–72 (870))

This chapter explores the interrelationship between empire and infectious disease
during the close of the second era of imperial expansion in South and Southeast
Asia.[1] In 1817, an outbreak of cholera – an acute diarrheal infection now understood to be caused by ingestion of the bacterium *Vibrio cholerae* – diffused from
Bengal across the world, in what has since been termed the first Asiatic cholera
pandemic (1817–24).[2] The global dispersal of cholera coincided with an expansion
of British influence in Southeast Asia following the Napoleonic Wars, the erosion
of French influence in the Subcontinent, and a retrenchment of Dutch trading
activity.

In 1811, the British invaded Dutch Java, strengthening their hold over the East
Indian Archipelago (Indonesia). Although Java was relinquished in 1816, British influence in the Malay Peninsula was formally recognized with the Anglo-Dutch Treaty of
1824 – concluded the same year that a British expeditionary force seized control
of Rangoon from the Burmese Konbaung dynasty during the First Burma War. As
a result of the Anglo-Dutch Treaty, Britain was ceded Malacca (Melaka), while the
Dutch recognized British claims to Singapore, the East India Company (EIC) trading post established in 1819 by Thomas Stamford Raffles. In 1826, Malacca, Dinding
(Manjung), Penang or Prince of Wales Island, and Singapore were subsumed in the
'Straits Territories'.[3] Meanwhile, a Siamese invasion of the Sultanate of Kedah in 1821
led to Lord Hasting's appointment of John Crawfurd as head of an embassy to the
court of King Rama II, with a view to forging better relations between the Kingdom
of Siam and the EIC. Further east, the British continued to push for concessions in
China. In 1816, Lord Amherst was dispatched on a mission to seek further trading
privileges from the Qing dynasty.[4] The progress of disease from Bengal in 1817 was
thus intertwined in complex ways with efforts to extend and consolidate British commercial interests across Asia, and in particular with the promotion of the opium trade.

It is, of course, a commonplace to claim that infectious disease spreads along the
expanding networks of empire. Trade and disease are coeval processes.[5] Epidemics of

cholera in the nineteenth century called attention to the multiplying webs of communication between Britain and the colonies, tracing out 'the contact lines of empire'.[6] As Alan Bewell has remarked,

> the British understanding of and response to cholera cannot be explained without reference to its global context. Cholera crossed many of the boundaries – cultural, geographical, and climatic – that were thought to exist between Britain and its colonial possessions, and by so doing it challenged those boundaries and led to their reconceptualization. It changed how the British saw themselves and their place in the colonial world.[7]

While there is a copious literature exploring the ways in which cholera was 'Asianized' in the West and the fears triggered by the specter of an 'Oriental scourge',[8] the emphasis in contemporary scholarship on Asia has tended to be exclusively on India and, in particular, on tracking the disease's inexorable westwards march from its purported 'home' in the deltas of the Ganges and Brahmaputra rivers. There has been good reason for such a focus, of course, given the scale of mortality from cholera during the nineteenth century. Although the figures remain debated, an estimated 15 million died of cholera in India between 1817 and 1865.[9]

For all the emphasis in recent imperial historiography on networks and interconnectedness, however, the contemporary story of cholera – its 'biography' as Christopher Hamlin would have it – remains a strangely one-dimensional affair, with Southeast Asia and China invariably relegated to a footnote, or absent altogether.[10] In the words of the eminent Irish surgeon Robert James Graves, '*There is a popular idea current, that [cholera's] course was westward; such was the case in Europe, but in most of Asia it was eastward*' [original italics].[11] As Kerrie MacPherson has observed of cholera in China:

> Although watched with keen interest wherever it has prevailed since 1817, its westward diffusions into a susceptible Europe (1830) have attracted the most attention, obscuring its eastward advances. This bias partly explains why authorities disagree on the chronology of the six pandemics that swept the world, and in the case of Chinese researchers even the number of pandemics between 1817 and 1925.[12]

Most recent histories of cholera follow in the steps of nineteenth-century accounts in their sketchy description of the pandemic's progress across Asia. China, Korea and Japan are invariably referenced as sites of distant infection. Their inclusion in a litany of 'exotic' place-names is largely gestural, serving to underline the disease's unprecedented global reach, with minimal discussion of specific cholera episodes or the impact that they may have had on local communities. China tends to be conjured as a symbolic geographical marker; a vanishing point where disease disperses into the hazy vastness of an uncharted landmass. The very nebulousness of 'China' functions as a framing device for bringing a sharper clarity to cholera's unexpected eruption at 'home'. As *The Lancet* declared in a discussion on cholera in 1832, the idea that Britain might experience a visitation of this Asiatic pestilence seemed to many contemporaries as absurd an idea as a sojourn in England by the emperor of China.[13]

There is thus an imperative to reappraise cholera's global history, moving beyond the token inclusion of Southeast Asia in essentially 'Western'-oriented works, to a more thorough and thoughtful analysis of how cholera was experienced and understood, and what impacts it may have had on the Asian regions it affected. In this chapter, my scope is necessarily more limited. The aim in what follows is to explore how British commentators – surgeons, traders and predominantly medical missionaries – conceptualized Southeast Asia and China in relation to cholera: from the first pandemic through the 1830s to the Opium War of 1839–42, which saw the demise of the 'Canton system', the establishment of Hong Kong as a British colony, and the creation of strategic Western-dominated treaty ports along the Chinese coast at Shanghai, Ningpo (Ningbo), Foochow (Fuzhou) and Amoy (Xiamen).

My purpose is to explore how British efforts to establish strategic port-cities across Southeast Asia, adopting more aggressive policies of 'opening up' China to the opium trade, might be related to shifts in how the diffusion of cholera was understood. There is a significant change in how British commentators wrote and thought about cholera in China from the 1840s, when more systematic attempts were made to situate the disease's distinctively local manifestations within a broader global understanding of the disease's etiology and modes of transmission. In Southeast Asia, this period also saw a more overt form of British imperialism characterized by the rise of commercial firms after the Charter Act of 1833 brought an end to the EIC's trading monopoly in China.[14] Cholera, as a local but transnational phenomenon, I suggest, has historically been 'pegged' to imperialism; the historical context of this 'pegging' – the fixing of an equivalence between 'choleric' and 'imperial' imaginaries – forms the principal focus of the chapter.

How did British residents and travellers across Southeast Asia during the first two cholera pandemics view themselves in relation to extending imperial networks? In addressing this question, I seek to show how efforts to make sense of cholera through plotting its incidence not only helped to define the identity of a disease, but also served to delineate irregular networks of British interest across Asia from India to the Straits Settlements, the East Indian Archipelago, Canton (Guangzhou) and beyond. The trade routes opened up by opium and tea, and bolstered by the movement of troops, were understood to be pathways for the circulation of cholera. From the eighteenth century, the EIC had monopolized the sale of opium in India, licensing private traders to export it to China: by 1830, over 30,000 chests were being shipped, rising to 40,000 in 1838.[15] Empire, the opium trade and global capitalism were profoundly entangled in Southeast Asia.[16] Colonial budgets and labor markets remained dependent on the revenue generated by opium until the end of the century.[17] Singapore provides a particularly striking example of opium's importance. Explicitly conceived at its foundation in 1819 as a 'fulcrum for the support of [British] Eastern and China trade' and as a strategic locale for promoting a 'nearer link' to China, the city's development to hub-port was predicated on the import and export of the drug.[18] 'For a full century', observes Carl Trocki, 'Singapore was "Opium Central: Southeast Asia."'[19]

Despite recurrent epidemic episodes, it was not until the mid-century that the disease became a focus of policy and an object of more systematic enquiry. As MacPherson has observed, in the new treaty ports such as Shanghai, established after 1842, cholera assumed 'pride of place' in medical investigations: 'no other disease from then until the 1880s [when the cholera bacillus was discovered] generally so terrified Europeans.'[20] From the 1840s, China's southern and central coasts were increasingly viewed as entryways for the diffusion of cholera to North China, Korea and Japan, where the disease (*korori/korera*) may have been introduced through Nagasaki or Shimonoseki in 1822.[21] More concerted efforts were made to track the disease's transmission with a view to implementing preventative strategies. Although reports on the cholera in Southeast Asia do not pervade the British archives in the 1820s and 1830s, what evidence survives does, nonetheless, provide a critical counterpoint to more familiar views of cholera's Indian 'origins' and its westward march, thereby offering a means of recovering an often ignored facet of cholera's global history. At the same time, re-considering the pandemic's eastward trajectory may provide a fresh perspective on an earlier and much less theorized period of imperial history.[22]

## Shadow imaginaries: empire and disease

The emphasis, here, is on the visualization of cholera; on ways of seeing; on empire and disease first and foremost understood as spatial phenomena. Visualization, in this context, designates not only the pictorial representation of information – for example in maps or the inclusion of images in disease reports – but also what the art historian W. J. T. Mitchell has termed 'textual pictures', that is, representations of space across a range of media, from newspaper accounts to official documents and scientific literature.[23]

Many British observers in Asia sought to understand cholera by inferring comparisons between places on the basis of the disease's movements in time and space, with China as one nodal point in a ramifying global network of disease. A recurrent theme in early nineteenth-century cholera reports is the importance attached to the spatialization of infection. The maps that accompanied such reports drew on a strand of eighteenth-century medical cartography that endeavored to map the progression of diseases between places within a locality.[24] Cholera reports linked the disease to the murky 'tanks and ponds of the Gangetic Delta' and to the 'swampy surfaces of the Sunderbunds [which] were converted into apparent spiracles of poison'.[25] Yet if cholera was attributed to insalubrious configurations of place and atmosphere, contagionist ideas gained increasing ground. It was noted that disease outbreaks coincided with the movement of human populations. Analyses of cholera's diffusion illustrated 'with singular clearness, the constancy with which the disease followed the track of ships, armies, pilgrims, caravans, and individuals, from one country to another'.[26] Idioms of trade, war and flight were applied to the disease itself: cholera was 'exported', 'attacked' and 'migrated'. Enumerating the 'geographical itinerary' of the disease as it moved between villages and towns in Bengal 'would almost form an itinerary gazetteer of the province'.[27]

The physician James Jameson prefaced his influential *Report on the Epidemick Cholera Morbus* in Bengal between 1817 and 1819 with a map drawn up using data furnished by responses to a questionnaire sent out by the Bengal Medical Board, which strove to gather 'the collective experience of a large number of individuals'. The object of the enquiry was explicitly framed, with the deployment of visual tropes, as the bringing to light of 'hidden causes'. Tracking the path, flow and current of the 'pestilential virus' would help to 'elucidate the laws, by which its progress was regulated' and in so doing plot 'a correct narrative of the circumstances which marked the origin and progress of a malady, perhaps more destructive in its effects, and more extensive in its influence, than any other recorded in the annals of this country'.[28] The emphasis in Jameson's black-and-white index map 'shewing the places chiefly visited by the Epidemick', was on British garrisons where outbreaks of cholera had been reported. [29] Despite the arguments presented in the main body of the *Report* – where the British are exonerated of blame for the infection's dispersal and the emphasis is on meteorological factors including the 'distempered' weather – cholera locations and colonial placements become synonymous (see Figure 11.1). As Tom Koch has recently observed, Jameson's map suggests 'the commonality of British military and political outposts as a single field of epidemic infection distinct from any occurring in the local populations of Indian cities and towns'.[30]

A similar relationship between colonial locales and the progression of disease is underscored in other maps produced in the 1830s, such as Frederick Corbyn's 1832 index map of India, which documents cholera outbreaks in British regimental stations across the country.[31] While India is depicted as the epicenter of cholera in many such maps, others aimed to show the disease's progressive diffusion along global pathways. Thus, the British physician Alexander Turnbull Christie in his *Treatise on Epidemic Cholera* noted the disease's arrival in Malacca (1819), Siam (1820), Canton, Whampoa (Huangpu) and Macao (1820), 'where the alarm produced by it among the Chinese merchants was so great as to occasion a serious interruption to commerce' (see Figure 11. 2).[32] 'From the Bay of Bengal', wrote the Indian army surgeon

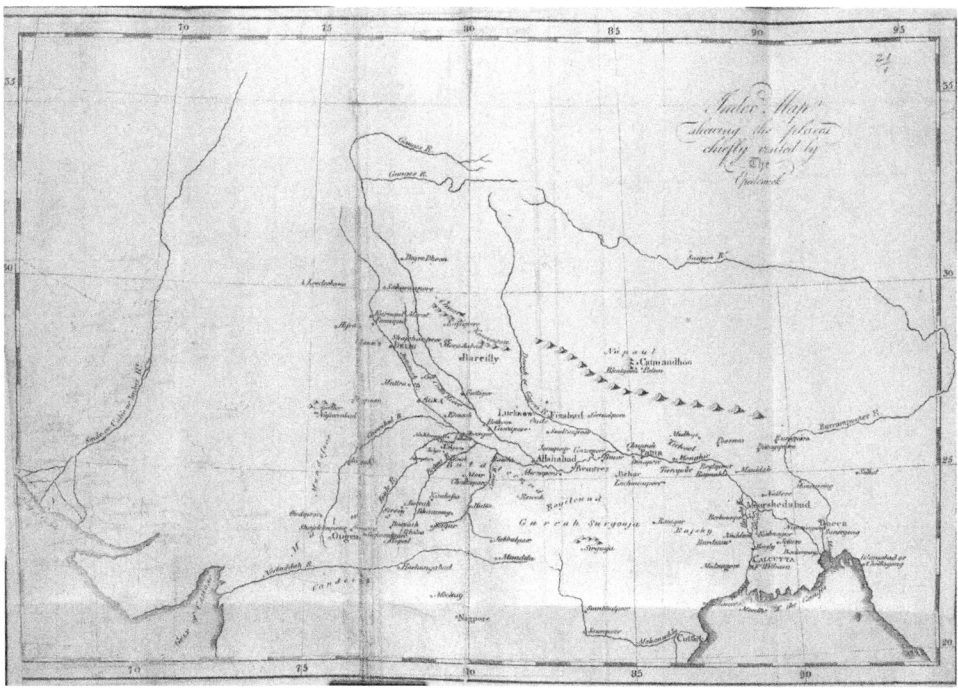

*Figure 11.1* 'Index Map Shewing the Places Chiefly Visited by the Epidemick.' From James Jameson, *Report on the Epidemick Cholera Morbus: As it Visited the Territories Subject to the Presidency of Bengal, in the Years 1817, 1818, and 1819*, Calcutta: Government Gazette Press [A. G. Balfour], 1820. © British Library.

James Kennedy in 1831, 'the cholera extended eastward along the coast of the Asiatic continent, and through the islands of the Indian Ocean, to the farther boundaries of China' with Canton 'invaded' in the fall of 1820 and Peking (Beijing) suffering from contagion the following year.[33] 'In the continent of Eastern Asia', remarked William Macmichael in an article on the *cholera spasmodica* in 1832, 'the cholera also followed the great media of communication between mankind': 'It has mastered every variety of climate, surmounted every natural barrier, conquered every people.'[34]

Cholera was construed as a condition that could be educed from a collection of disparate symptoms. Robert Strange, for example, in his 'Observations on the Cholera Morbus of India', sought to argue for a continuity of disease across Southeast Asia, even while recognizing differences in its local manifestation. As a Company surgeon on board the East Indiaman the *Charles Grant*, Strange had sailed from India to China (1825–7). During the course of the journey, the ship had experienced two visitations of a debilitating disease, which Strange diagnosed as cholera: one as the ship lay at anchor on the Hooghly River in West Bengal and one at Whampoa, an anchorage approximately 10 miles downstream from Canton. Despite the differences between the symptoms in the two cases, Strange argued that they were both 'cholera', suggesting that the disease could take various forms and was produced either by 'infectious miasma arising from putrid stagnant marshes, or by the contagion arising from the bodies infected by the disease'.[35]

*Figure 11.2* 'Map of the Countries Visited by the Epidemic Cholera From the Year 1817 to the Year 1830, inclusive.' From Alexander Turnball Christie, *A Treatise on the Epidemic Cholera; Containing its Histories, Symptoms, Autopsy, Etiology, Causes and Treatment*, London: J. and C. Adlard, 1833. © British Library.

A lengthy editorial on the progress of the 'blue' cholera in *The Lancet* in 1831 similarly emphasized the process by which cholera's 'new' identity (in contradistinction to the 'old' cholera) was to be constructed through the amalgamation of its symptoms.[36] This was, in part, a visual process, the aim being to bring the disease 'within the immediate range of our own grasp and vision' in order to ensure that the new disease entity remained 'unincumbered [*sic*] by the clouds with which far-distant objects are so frequently invested'.[37] A map was included, along with instructions for the reader: 'In the examination of the map it should be observed, that places which the malady has visited, are marked with a dot, and surrounded by a black circle'.[38] Despite the map's purported aim to show 'the progress of the cholera in Asia, Europe, and Africa', the focus is squarely on Europe, with Asia extending no further than the east coast of India in the Bay of Bengal. An inset map emphasizes the area around Calcutta and Mysore as the 'home' of the disease. While the map is concerned chiefly with the disease's 'uninterrupted tour from the Gangetic Delta to the river Wear', the emphasis in the body of the text is on the worldwide spread of cholera – on its 'geographical history'.[39] 'Pestilence, though the offspring of one locality, soon extends its desolation to many', the author disclaimed.[40] Emphasis was placed on the disease's 'reiterated persecution' as the author tracked 'the migration of the cholera over the gulfs and arms of the ocean which wash the littoral boundaries of the Indian peninsula' to Ceylon (Sri Lanka), Malacca, Macao, Canton and across China.[41] This is an epidemic mapping that delineates an expanding empire and the consolidating junctures of Britain's global trade and military interventions, in particular the penetration of Asia. EIC troops and commerce had brought the cholera to Burma where it was manifest during the first Anglo-Burmese War (1824–6); from Burma it moved to Singapore and eastwards to Canton, Wenchow (Wenzhou), Ningpo and the Yangtze Valley, with outbreaks in Peking and central and northern China between 1822 and 1824.

Raffles himself tracked the cholera as it moved from Bengal along the rim of the Indian Ocean, threatening British possessions there. In a letter to his friend the English Orientalist William Marsden from Calcutta dated October 1818, he remarked that 'the heat has been extremely oppressive, and the whole of India very sickly – it is computed that not less than two millions have fallen a sacrifice to what is here called the *cholera morbus*'.[42] In February 1820, Raffles observed in another letter, penned at sea en route for Sumatra: 'The cholera morbus has lately committed dreadful ravages at Acheen, Penang, and Quedah: it is now raging at Malacca, and I have great apprehensions for Singapore.'[43] And from Bencoolen (Bengkulu), where he had been appointed Governor-General, Raffles wrote in July 1821: 'Java, I am concerned to say, is suffering under all the miseries of the exterminating cholera; the deaths average eight hundred a-day, and from ten minutes to four hours is the usual period of illness.'[44]

Empire becomes analogous to an emergent disease, which is discernable through an assemblage of symptoms and 'hot spots' on the map. Indeed, a preoccupation with maps and mapping is evident throughout Raffles' letters. Thus, in a letter to his cousin written on a journey to Calcutta in 1819: 'If you refer to the map, and observe the commanding position of Singapore, situated at the extremity of the Malay Peninsula, you will at once see what a field is opened for our operations.'[45] Like disease, British influence in the region is thought of as an aggregation of hubs in an extended political community, however loosely conjoined. Imperial 'bridging' involves an equivalent process of visualization and, specifically, the mapping of connections that reveal the contours of otherwise invisible entities: disease and the often ambiguous, quasi-colonial emplacements which constitute emergent empire. In summary, disease and empire function as shadow imaginaries: the one implicitly outlining the form of the other.

As Matthew Edney has argued, geography played a crucial role in the development of the British Empire, particularly in the EIC's subjugation of South Asia: from James Rennell's survey of Bengal in the eighteenth century to Sir George Everest's completion of the first phase of the Great Trigonometric Survey of India in 1843.[46] In this chapter, however, the focus is not on nineteenth-century cartographic technologies as tools of governance or technologies of power, but on the ways in which disease maps and 'textual pictures' acted as a spur to a particular kind of imperial 'cartographic imagination'.[47] The dispersal of cholera from India to East Asia after 1817 prompted British commentators to reflect on the likely origins of a virulent and apparently portable disease. In mapping cholera's arbitrary but relentless progress, other networks became visible.

Put somewhat differently, the mapping of cholera embedded within it an idea of evolving empire. We might, in fact, substitute 'empire' for 'disease' in Koch's formulation that

> to understand disease and its history we need to think about *seeing* at every scale. It is in the *seeing* – of the animalcule, the parts of the infected body, and the shared sets of symptoms evidenced across maps of the city, nation, and world that the unknown is made real, its public nature asserted.[48]

Empire, too, involved different scales of *seeing* and the aggregation of different parts into a public nature. In seeking to understand cholera, British commentators found themselves compelled to consider the continuities and discontinuities of British interests across Asia and globally: disease was a trigger for deliberating on the origins, limits and future of Britain's global influence at a moment when trade links were intensifying and services expanding with pre-industrial manufacturing.[49] While proliferating networks provided increasing opportunities for commerce, they also suggested the heightened possibility of future ruptures since, as A. G. Hopkins has observed, the proto-globalization of the early nineteenth century promoted fragmentation even as it facilitated unification.[50]

## China and the world

By 1820, the year in which the first cholera pandemic was widely reported to have reached China from Bengal, there was an established Western presence in the Qing Empire. For part of each year, foreign traders acting through Chinese merchant families were permitted to trade in a designated area outside the port city of Canton on the Pearl River Delta. By the early nineteenth century, the British and Americans had come to dominate this China trade.[51] The nature of the British presence in China has been much debated, particularly after 1842. Some historians have claimed it constituted a species of 'informal empire', whereby indirect political means were exerted to support British commercial interests.[52] Others have maintained that although

Britain did have significant influence on China, this was not enough to substantiate the notion of 'informal empire', but perhaps at most equated to a condition of 'semi-colonialism'.[53] As Ruth Rogaski has observed, China was 'at once not a colony, yet still the site of multiple colonialisms'.[54]

In 1819, under the pseudonym 'Dusty Traveller', the British missionary Robert Morrison had published a slim volume in Chinese entitled *A Brief Account of Things That I Have Seen and Heard During a Voyage Westwards Around the World*.[55] The book, which was printed in Canton, purported to be a Chinese traveller's account of a trip to Europe through India, returning by way of America. The aim was to 'open' Chinese minds to the historical and geographical interdependencies of the world. Maps of the world – of China, Asia and Europe – also accompanied the English Congregationalist missionary Walter Medhurst's *Geographical Catechism* published in Malacca the same year.[56]

While officials of the EIC, such as Raffles, were mapping Southeast Asia, British residents in China, such as Morrison and Medhurst, were similarly concerned with the place of China, and specifically of Canton, within the broader British world. This is evident from the first edition of the *Canton Miscellany*, a journal founded by Medhurst in 1831. In the introduction to the first issue, the editor concedes that China marks the 'Eastern extremity of the Earth at nearly the furthest point of removal from the favored seats of Science and Civilization'.[57] For citizens of post-Revolutionary Europe ('the agitating vortex' of the world), the countries of the East appear to 'pass in transient review before their eyes as shadows in the distance, and rarely become subjects even of temporary observation and discussion'. Indeed, 'in proportion as the scene of an event is distant the interest, which is calculated to excite becomes languid'.[58] Yet the construction of far-flung China as a vanishing point is qualified by an appreciation of the proliferating connections that link Britain and China, with the 'Ocean itself' promoting 'intercourse between the most remote Regions'. In a metaphor that uncannily prefigures contemporary theorizing of empire as a 'web', the author imagines British industry – and in particular the manufacture of cloth – in terms of enmeshment, as a 'thread, which if brought in a continuous line might connect the empires'.[59]

Global interconnections and China's increasingly networked relationship with the world were brought to the fore by events in 1819. The *Indo-Chinese Gleaner*, a quarterly missionary journal published from Malacca with an original monthly circulation of 500 rising to 1,000, reported that cholera had swept 'along the continent and shores of India' to reach the Malay peninsula.[60] In a letter to Morrison in Canton, the journal's editor and fellow missionary, William Milne, noted:

> The cholera morbus has visited Malacca. On the 2nd instant, sixteen persons died: two funerals have just passed our door to-day in course of the last two hours. Seven funerals passed our door the other day; Klings and Malays have chiefly suffered yet. Two of our domestics have died of the cholera, and one more has been very ill. Our lives are in his hand; living and dying may we be his.[61]

Cholera's arrival in Malacca in 1819 had coincided with the disembarkation of troops, although it was claimed that Chulia and Malay communities were first affected, along with the city's Chinese inhabitants.[62] By 1820, 'the circle of this dreadful scourge' had enlarged, with infection spreading 'through the whole of the vast Indo-Chinese countries'.[63] In May, it caused perhaps as many as 30,000 deaths in the British trading post of Penang (the Governor of the settlement, John Alexander Bannerman, had succumbed to cholera the preceding August). In Java, the Dutch authorities recorded 1,255 deaths from the disease within 11 days in Semarang and 778 deaths in Batavia.[64] In a letter to a friend in England in 1820, Milne noted the unrelenting advance of infection through the 'colonies' of Java, Singapore and Penang: 'We have lately been visited with the Cholera Morbus in these countries, which has carried off multitudes.'[65] In Bangkok, in May 1822, Crawfurd noted a recrudescence of the disease, which had committed 'dreadful ravages' two years previously,[66] killing an estimated 100,000 people.[67]

On 9 October 1820, Manila in the Philippines witnessed 'one of those terrific outbursts of barbarian despair which have more than once signalized the progress of this pestilence'.[68] An angry mob believing that the disease was a foreign plot to poison the local population attacked and killed foreigners including a number of British sailors and Chinese. Among those murdered was 25-year-old Captain David Nicoll of the EIC ship the *Merope*, which some claimed had introduced the epidemic disease to Manila from Calcutta.[69] In a dispatch to W. A. Chibley, Secretary to the Government at Penang, J. W. Campbell, commander of *HMS Dauntless*, observed that on his arrival in Manila, 'I perceived even before I landed that some dreadful catastrophe had marked its progress with desolation and had produced stagnation in the commercial operations on the River and in the Port.'[70] Writing to a relative from Canton in December 1820, Morrison noted:

> There has been a very shocking massacre of from thirty to forty Europeans of different nations of Europe, and of about eighty Chinese, at Manilla. The perpetrators of this cruel act were the native Manilla people. The pretext was a supposition that foreigners had introduced the disease called cholera morbus, which had prevailed extensively, and was very fatal.[71]

Meanwhile, in Canton and surrounding areas of the Pearl River, 'thousands were dying'.[72] Between 1820 and 1823, some 10,000 alone were said to have perished in Ningpo, a seaport in the eastern province of Zhejiang.[73] Rumors attributed the Jiaqing emperor's death at the beginning of September 1820 – which led to a struggle for succession – to the cholera.[74] As the *Morning Post* in London reported,

> To add to the distraction of the Chinese empire from these causes, we learn that the *cholera morbis*, that fatal epidemic, had found its way hither from Bengal, and was producing the most fatal ravages, the inhabitants dying by thousands. With every precaution, its effects had been severely felt even among the crews of the British ships at Canton.[75]

An imperial edict the following year postponed the imperial examinations on the grounds of the public danger posed by an outbreak of 'epidemical disease'.[76] So high was the mortality in Peking in 1821 and 1822, according to one report, 'that the government was obliged to furnish coffins, and other funereal apparatus, for the use of the lower classes'.[77]

Writing from Canton in 1824, Dr. John Livingstone, a surgeon with the EIC, observed that the cholera as it was manifest in China appeared to be an entirely new phenomenon and a disease of exceptional virulence. Unusually, he plotted its course overland. 'It seems to have appeared first in Tartary afterwards in the N.W. of China', he wrote, tracking its ostensible trajectory southwards from Central Asia 'by irregular leaps'.[78] Credited as the first person who 'systematically brought medical aid within reach of the Chinese',[79] Livingstone had first-hand experience of cholera's virulence. He had been a pall-bearer at the funeral of Morrison's first wife Mary who had died from the disease in Macau in June 1821.[80]

From the 1820s, then, cholera provided a critical context for Anglo-Chinese relations, flaring into periodic epidemics, particularly in the 1830s with the onset of the second cholera pandemic (1829–51). In May 1835, for example, the *Chinese Repository* reported: 'Many cases of sickness and death have occurred in Canton and its vicinity during the last two or three months: some of these, so far as we can ascertain, are evidently cases of the epidemic of malignant cholera.'[81]

## Cholera and the Opium War

In 1800, a Qing edict had banned opium imports and domestic production of the drug in China. This was followed by the outlawing of opium smoking in 1813.[82] British traders, however, continued to evade Chinese authorities and by the mid-1820s the Chinese economy was experiencing the consequences in an outflow of silver which raised prices, triggering social unrest.[83] The end of the EIC monopoly of the China Trade in 1833 exacerbated the situation with an influx of foreign traders and rising opium sales. As the Qing sought to clamp down more concertedly on the opium trade, British commercial interests clamored for military intervention.[84]

In 1840, Indian troops had assembled in Calcutta and Madras to embark for China. Cholera was then widespread in Bengal and the expeditionary force re-imported the disease to the Straits Settlements in April 1840, and thence to China, where an initial epidemic broke out soon after troops had landed on the island of Chushan (Zhoushan) off the coast below Shanghai in July.[85] As the Yokohama-based American physician Duane B. Simmons later noted in his history of cholera: 'The Government of India dispatched a native army to China in the interest of the opium trade. This force carried with it the seeds of cholera, which not only arrived at Peking, but followed the track of the caravans westward to Russia' (see Figure 11.3).[86]

Writing of the epidemic in Malacca, the physician Thomas Oxley remarked that since the cholera had visited the settlement in 1821 and 1826 'no symptoms of the complaint have been observed; indeed the remembrance of cholera had nearly faded from all memories when the present epidemic (1840) sprung up amongst us'.[87] The *Chinese Repository* commented that 'large numbers of native inhabitants' were being carried off, as well as British residents, including the Reverend John Evans, Principal of Malacca's Anglo-Chinese College.[88] The Superintendent Surgeon of the Straits, William Montgomerie, reported to the Medical Board 'the remarkable circumstance that epidemic cholera broke out, early in the year, along the sea-shore towns bordering the Straits, and slowly advanced from the south to Malacca'.[89]

According to Captain Arthur Cunynghame, aide-de-camp to Major-General Lord Saltoun, troops freshly arrived in the field of action who had been 'so long cooped up within the narrow limits, of their overcrowded ships' were particularly susceptible, while 'severe sickness' was brought on by sun exposure, 'want of rest and a too free

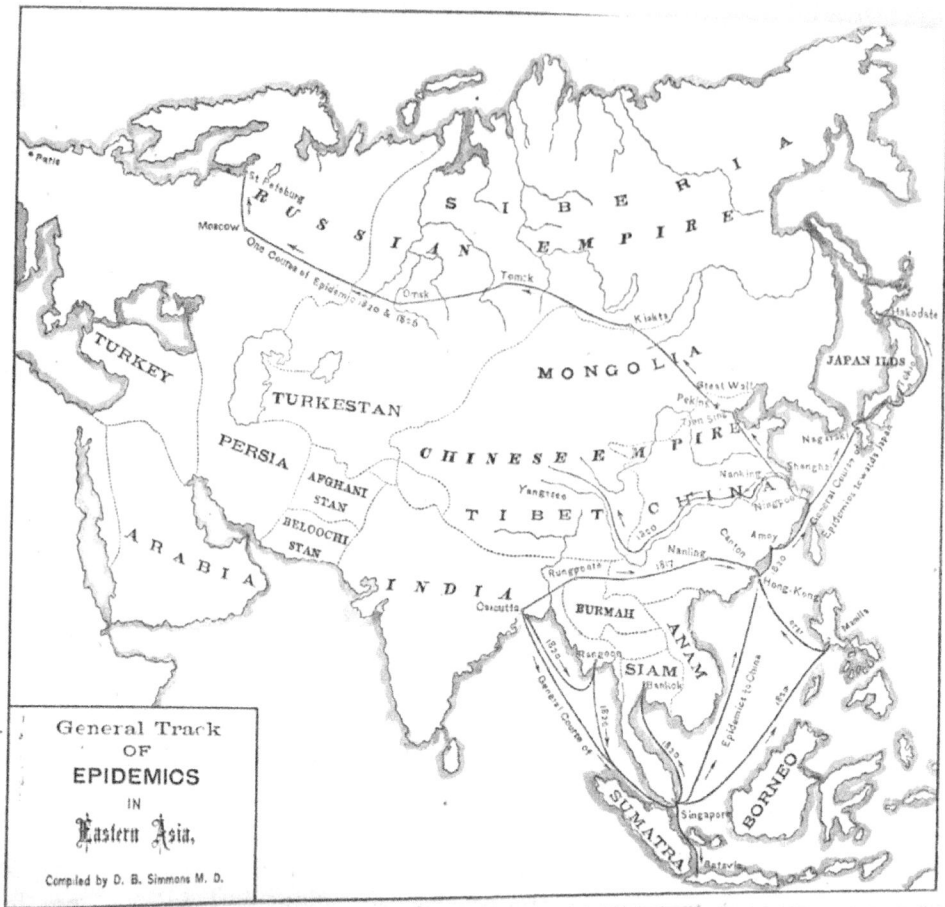

*Figure 11.3* 'General Track of Epidemics in Eastern Asia.' From D. B. Simmons, 'Cholera
Epidemics in Japan', in *China: Imperial Maritime Customs, Series 2, Medical Reports for
the Half-Year Ended 30th September 1879*, no. 18, Shanghai: Statistical Department of
the Inspectorate General, 1880, p. vii.

indulgence in unripe fruits and raw vegetables, which their officers could not persuade them from indulging in'.[90] As Sir Hugh Gough declared, following the capture
of Chinkiang (Zhenjiang) on the Yangtze River in July 1842: 'I regret to say that,
notwithstanding every precaution, I have lost several officers and men by cholera.'[91]
Although the cholera 'had already made its appearance amongst the troops',[92] the
prospect of a full-blown cholera epidemic amongst the native population was viewed
as 'a greater calamity than the destruction of all the cities on the coast'.[93] The surgeon
Charles Macnamara subsequently noted that 'the English Government unknowingly
inflicted on the unfortunate inhabitants of the Celestial Empire one of the most
frightful visitations of disease to which any nation was ever subjected'.[94]

Realization of cholera's global dissemination along the pathways of empire, and
in particular the linkage made between cholera, opium and the movement of troops,

fuelled an increasing interest in fixing the cause and likely transmission routes of disease. On the one hand, disease was viewed as the result of an imperial or colonial *modus operandi*, a conviction that confirmed the view held by many Chinese commentators who apportioned blame for cholera outbreaks squarely on foreigners.[95] James French, appointed Superintending Surgeon of the forces serving in China in 1841, for example, concluded that from inquiries made among missionaries and other inhabitants, the disease was not endemic and did not exist before the arrival of British troops in Chushan.[96]

On the other hand, the disease was viewed as endemic. The dangers posed by infectious disease to British soldiers and sailors in the China Station were dramatically underscored by the naval physician Alexander Bryson in his *Statistical Reports on the Health of the Navy for the Years 1837–43*.[97] In 1842, William C. Milne – the son of the missionary William Milne who had reported to Morrison on the cholera from Malacca in 1819 – had been requested to write a report on the disease by British naval and military authorities.[98] Handed a list of queries, which guided his investigation in Tinghai (Dinghai) and Ningpo, Milne concluded, in contrast to French, 'that Asiatic cholera has already, and not unfrequently [*sic*] nor slightly, but often and with great severity, visited China, in common with almost every other nation under heaven'.[99] Milne's report was based on an evaluation of existing Chinese names for 'cholera' such as *huoluan*, a word meaning 'sudden [intestinal] turmoil', as well as on interviews with Chinese practitioners.[100]

The literature on the cholera produced in China suggests, then, a shift in knowledge; a more thoughtful engagement with Chinese medical sources; and an appreciation that understanding Chinese history and customs was a prerequisite for a new kind of Western presence. After the Treaty of Nanking in August 1842, the Canton system was overturned and self-governed foreign settlements were established in five treaty ports. In these extraterritorial enclaves, anxieties about the threat of endemic cholera (which medical reports showed was rampant), influenced by metropolitan sanitary concerns, became a spur to ambitious municipal works – the building of roads, hospitals and waterworks – which precipitated, as Rogaski has argued, a form of 'hygienic modernity'.[101] Epidemics of cholera were to recur in the second half of the century, during the Second Opium War (1856–60) and the Taiping Rebellion (1850–64), when the disease again reached China along the networks of the British Empire from India to Hong Kong. By the 1880s, however, British residents in the colonial and quasi-colonial outposts of Southeast Asia and China were concerned with a different kind and scale of mapping.[102] The emphasis was on preserving health through hygienic vigilance and the segregation of Western from potentially infective Asian bodies. Increasingly informed by bacteriological science and epidemiology, sanitarians drew up reports replete with maps, architectural surveys and plans for public works. Cholera now revealed not only the proliferating pathways that empire fostered, but also – and above all – the necessity for colonial interventions to close off dangerous circulations.

## Conclusion

This chapter has sought to show how debates about cholera's identity were interconnected with changing views of Britain's place in the world, as the mercantilism of the early nineteenth century shifted to a more aggressive search for markets and trade in the East, and as the Canton system gave way to more permanent self-governed foreign settlements in China. Rethinking British imperialism from Southeast Asia and China, specifically in relation to the first two cholera pandemics, provides a fresh perspective on the anxieties and expectations which shaped and informed the new global imperialism after 1830. In a sense, the aim in this chapter has been to understand the intensifying networks of empire in Asia in terms of the 'things' that these networks transported: including cholera and opium. As Arjun Appadurai has suggested, commodities accrue 'social lives' as they move across different 'regimes of value'. They are enmeshed in complex economic, political and cultural processes.[103] Rethinking the changing configurations of empire through mapping the circulations of cholera thus becomes a way of instantiating uneven patterns of global connectivity, the legacy of which continues to shape the contemporary world of emerging infections across Southeast Asia.

At the same time, the chapter has suggested that reappraising empire through the lens of epidemic disease may help to produce a more critical global history – one that pays attention to the friction between local circumstances and transnational relations.[104] While the British Empire invariably appears on maps as an unvariegated color-coded space, viewed through the prism of cholera its contours appear much less clearly delineated, often tapering into ambivalent zones of influence and partial-sovereignty. 'Territorial control was, in many places, an incidental aim of imperial expansion', Lauren Benton has observed. Empires, she notes, 'did not cover space evenly but composed a fabric that was full of holes, stitched together out of pieces, a tangle of strings'. In plotting the often obscure pathways of cholera, early-nineteenth-century British soldiers, traders and missionaries coincidentally made visible the heterogeneous fabric of an emergent empire, characterized by 'narrow bands, or corridors, enclaves and irregular zones'.[105]

In short, examining the history of cholera from the multi-vantage of scattered ports and missions across Southeast Asia and China may help to challenge conventional narratives of imperial expansion. As opposed to the steady aggregation of territories and the smooth extension of sovereignty, we begin to discern empire as an entanglement of networks that produced often highly irregular and disjointed spaces. Indeed, descriptions from the 1820s and 1830s of cholera's 'itinerary', reflect a tension between attempts to trace the disease's systematic 'progress' and a recognition of cholera's often inexplicable movements. *The Lancet*, for example, noted how the disease spread across Bengal 'like radii from a common centre' and the aim was to map its trajectory in relation to specific 'points'. Yet this quasi-mathematical endeavor to locate the coordinates of disease is undermined by its irregular spread: cholera marches and countermarches and appears unconstrained by the 'particular lines and divisions of the country'.[106]

## Notes

1 C. A. Bayly has characterized this epoch as 'the first age of global imperialism'; see his arguments in 'The First Age of Global Imperialism, c. 1760–1830', *Journal of Imperial and Commonwealth History*, 1998, vol. 26, no. 2, 28–47.

2 On the six cholera pandemics, see R. Pollitzer, *Cholera*, Geneva: World Health Organization, 1959, pp. 11–50. There is some disagreement about the periodization of these pandemics. See William Johnston, 'Epidemics Past and Science Present: An Approach to Cholera in Nineteenth-Century Japan', *Harvard Asia Quarterly*, 2012, vol. 14, no. 4, 28–35.

3 Penang was the capital of the Straits Settlements until 1832, when the seat of government was transferred to Singapore. The Settlements became a British Crown colony in 1867.

4 A mission which ended in failure; see Jonathan D. Spence, *The Search for Modern China*, rev. edn., New York: W. W. Norton, 2013 [1990], p. 147.

5 Mark Harrison, *Contagion: How Commerce Has Spread Disease*, New Haven, CT: Yale University Press, 2012; Erwin H. Ackerknecht, *History and Geography of the Most Important Diseases*, New York: Hafner, 1965, pp. 24–9. For an account of how infectious diseases have 'functioned as cautionary metaphors for concerns about the dangers of commerce, understood in the broadest sense, between the East and the West', see Jo Robertson, 'In Search of *M. Leprae*: Medicine, Public Debate, Politics and the Leprosy Commission to India', in Leigh Dale and Helen Gilbert (eds), *Economies of Representation, 1790–2000: Colonialism and Commerce*, Aldershot: Ashgate, 2007, pp. 41–57 (41).

6 Alan Bewell, *Romanticism and Colonial Disease*, Baltimore, MD: Johns Hopkins University Press, 1999, p. 247; on 'webs' of empire, see Tony Ballantyne and Antoinette Burton, 'Introduction: Bodies, Empires, and World Histories', in Tony Ballantyne and Antoinette Burton (eds), *Bodies in Contact: Rethinking Colonial Encounters in World History*, Durham, NC: Duke University Press, 2005, pp. 1–18 (3).

7 Bewell, *Romanticism and Colonial Disease*, p. 244.

8 Christopher Hamlin, *Cholera: The Biography*, Oxford: Oxford University Press, 2009, pp. 39–46. The notion of an 'Oriental scourge' was commonplace; see, for example, James Mouat, 'On Cholera Morbus', *Transactions of the Medical and Physical Society of Calcutta*, vol. 4, Calcutta: Thacker and Co., 1829, pp. 265–307 (266). Cholera was also associated with China and likened to a 'yellow horde'; see Alexandre Moreau de Jonnès, *Rapport au conseil supérieur de santé sur le choléra-morbus pestilentiel*, Paris: Cosson, 1831, pp. 340–1.

9 David Arnold, *Colonizing the Body: State Medicine and Epidemic Disease in Nineteenth-Century India*, Berkeley, CA: University of California Press, 1993, p. 161; 'Cholera and Colonialism in British India', *Past and Present*, 1986, vol. 113, no. 1, 118–51 (120–21); 'Cholera Mortality in British India, 1817–1947', in Tim Dyson (ed.), *India's Historical Demography: Studies in Famine, Disease, and Society*, London: Curzon, 1989, pp. 261–83.

10 Hamlin, *Cholera*.

11 Robert James Graves, *Clinical Lectures on the Practice of Medicine*, Dublin: Fannin and Co., 1864, p. 298.

12 Kerrie L. MacPherson, 'Cholera in China, 1820–1930: An Aspect of the Internationalization of Infectious Disease', in Mark Elvin and Ts'ui-jung Liu (eds), *Sediments of Time: Environment and Society in Chinese History*, Cambridge: Cambridge University Press, 1998, pp. 487–519 (488).

13 'Westminster Medical Society: Discussion on the Cholera', *Lancet*, 17 March 1832, vol. 1, no. 446, 869–72 (870).

14 'Imperialism' might be defined, here, 'as the complex of intentions and material forces which predispose states to an incursion, or attempted incursions, into the sovereignty of other states'; see Bayly, 'The First Age of Global Imperialism', p. 28.

15 Spence, *The Search for Modern China*, p. 149.

16 Carl A. Trocki, *Opium, Empire and the Global Political Economy: A Study of the Asian Opium Trade, 1750–1950*, London: Routledge, 1999.

17 For a recent study exploring the interconnections between the articulation of a rationale for opium regulation and the legitimation of colonial rule in colonial Burma after 1826, see Ashley Wright, *Opium and Empire in Southeast Asia: Regulating Consumption in British Burma*, Basingstoke: Palgrave Macmillan, 2013.

18 'East Indies: Singapore', *Times* [London], 7 September 1819, 2.

19 Carl A. Trocki, *Opium and Empire: Chinese Society in Colonial Singapore, 1800–1910*, Ithaca, NY: Cornell University Press, 1990, p. 50.

20 Kerrie L. MacPherson, *A Wilderness of Marshes: The Origins of Public Health in Shanghai, 1843–1893*, Hong Kong: Oxford University Press, 1987, p. 29.

21 Ann Bowman Jannetta, *Epidemics and Mortality in Early Modern Japan*, Princeton, NJ: Princeton University Press, 1987, pp. 157–9.

22 Bayly, 'The First Age of Global Imperialism'.

23  W. J. T. Mitchell, *Picture Theory: Essays on Verbal and Visual Representation*, Chicago, IL: University of Chicago Press, 1994.
24  Lloyd G. Stevenson, 'Putting Disease on the Map: The Early Use of Spot Maps in the Study of Yellow Fever', *Journal of the History of Medicine and Allied Sciences*, 1965, vol. 20, no. 3, 226–61.
25  Anon, 'History of the Rise, Progress, Ravages, &c. of the Blue Cholera of India', *Lancet*, 19 November 1831, vol. 1, no. 429, 241–84 (242).
26  Ibid., p. 261. The causes contributing to the emergence of this 'new' virulent form of cholera have been much debated by historians. A recent argument has been made linking the pandemic to the eruption of Tambora on the island of Sumbawa in Indonesia in April 1815 and the impact that this had on the disease ecology of the Bay of Bengal; see Gillen D'Arcy Wood, *Tambora: The Eruption that Changed the World*, Princeton, NJ: Princeton University Press, 2014. This argument rehearses many views held by commentators in Europe in the 1830s, who attributed the cholera to natural disasters, including volcanic eruptions; see François Delaporte, *Disease and Civilization: The Cholera in Paris, 1832*; translated by Arthur Goldhammer, Cambridge, MA: MIT Press, 1986, p. 99.
27  Anon, 'History of the Rise, Progress, Ravages, &c. of the Blue Cholera of India', pp. 267, 243.
28  James Jameson, *Report on the Epidemick Cholera Morbus: As It Visited the Territories Subject to the Presidency of Bengal, in the Years 1817, 1818, and 1819*, Calcutta: Government Gazette Press [A. G. Balfour], 1820, pp. i–iii. Of the 238 persons to whom the questionnaire was sent, 124 replies were received, although only 100 provided substantive information (viii).
29  Ibid. On the links between the military and cholera in India, see Arnold, 'Cholera and Colonialism', pp. 126–9.
30  Tom Koch, *Disease Maps: Epidemics on the Ground*, Chicago, IL: University of Chicago Press, 2011, p. 97.
31  Frederick Corbyn, *A Treatise on the Epidemic Cholera, as It Has Prevailed in India*, Calcutta: W. Thacker and Co., 1832.
32  Alexander Turnbull Christie, *A Treatise on the Epidemic Cholera; Containing Its History, Symptoms, Autopsy, Etiology, Causes, and Treatment*, London: J. and C. Adlard, 1833, p. 19.
33  James Kennedy, *The History of the Contagious Cholera: With Facts Explanatory of Its Origin and Laws, and of a Rational Method of Cure*, London: J. Cochrane and Co., 1831, p. 200.
34  William Macmichael, 'The Cholera', *Quarterly Review*, 1832, vol. 46, 170–212 (170, 183).
35  Robert Strange, 'Observations on the Cholera Morbus of India', *Lancet*, 20 August 1831, vol. 2, no. 416, 644–9 (648).
36  On the 'old' and 'new' choleras, see Hamlin, *Cholera*.
37  Anon, 'History of the Rise, Progress, Ravages, &c. of the Blue Cholera of India', p. 242.
38  Ibid., p. 284.
39  Ibid., p. 252.
40  Ibid., p. 241.
41  Ibid., pp. 241, 245.
42  Sophia Raffles, *Memoir of the Life and Public Services of Sir Thomas Stamford Raffles, Particularly in the Government of Java 1811–1816, Bencoolen and Its Dependencies, 1817–1824*, 2 vols, London: James Uncan, 1835, II, p. 5.
43  Ibid., p. 88.
44  Ibid., p. 194.
45  Ibid., p. 67; on mapping, see also, pp. 15, 16, 18, 51, 84, 95, 298, 302, 320, 328, 334, 342, 401.
46  Matthew H. Edney, *Mapping an Empire: The Geographical Construction of British India, 1765–1843*, Chicago, IL: University of Chicago Press, 1997.
47  Tony Ballantyne, 'Empire, Knowledge, and Culture: From Proto-Globalization to Modern Globalization', in A. G. Hopkins (ed.), *Globalization in World History*, London: Pimlico, 2002, pp. 115–40 (122–3).
48  Koch, *Disease Maps*, p. 4.
49  See, here, the arguments made about proto-globalization in C. A. Bayly, *The Birth of the Modern World, 1780–1914: Global Connections and Comparisons*, Malden, MA: Blackwell, 2004.
50  A. G. Hopkins, 'Introduction: Globalization – An Agenda for Historians', in Hopkins (ed.), *Globalization in World History*, pp. 1–10 (3).

51 On the 'Canton system', a term taken to describe the particular trade arrangements between foreigners and the Qing from 1700 to 1842, see Paul A. Van Dyke, *The Canton Trade: Life and Enterprise on the China Coast, 1700–1845*, Hong Kong: Hong Kong University Press, 2005.

52 See, for example, J. Gallagher and R. Robinson, 'The Imperialism of Free Trade', *Economic History Review*, 1953, vol. 6, no. 1, 1–15; and E. S. Wehrle, *Britain, China and the Anti-Missionary Riots, 1891–1900*, Minneapolis: University of Minnesota Press, 1966.

53 For critiques of 'informal empire', see B. Dean, 'British Informal Empire: The Case of China', *Journal of Commonwealth and Comparative Politics*, 1976, vol. 14, no. 1, 64–81; Wang Gungwu, *Anglo-Chinese Encounters since 1800: War, Trade, Science and Governance*, Cambridge: Cambridge University Press, 2003; on 'semi-colonialism', see J. Osterhammel, 'Britain and China, 1842–1914', in Andrew Porter (ed.), *The Oxford History of the British Empire: The Nineteenth Century*, vol. 3, Oxford: Oxford University Press, 2009 [1999], pp. 146–69; on the inadequacy of the terms 'imperialism' and 'informal empire' to describe the British presence in China, see Ulrike Hillemann, *Asian Empire and British Knowledge: China and the Networks of British Imperial Expansion*, Basingstoke: Palgrave Macmillan, 2009, p. 172.

54 Ruth Rogaski, *Hygienic Modernity: Meanings of Health and Disease in Treaty-Port China*, Berkeley, CA: University of California Press, 2004, p. 3.

55 Chenyou jushi, *Xiyou diqiu wenjian lüezhuan* (1819).

56 *Dili biantong lüezhuan* (1819).

57 Anon, 'Introduction', *Canton Miscellany*, no. 1, China: Published by the Editors, 1831, i–xii (ii)

58 Ibid.

59 The editors also added that 'the tie . . . would be a fragile and insecure one'. Ibid., pp. i–iii.

60 'Cholera Morbus', *Indo-Chinese Gleaner*, January 1820, no. 11, 247–54 (247); see 'Memoir of the Late Rev. William Milne, DD. Missionary to the Chinese, &c.', *Evangelical Magazine and Missionary Chronicle*, vol. 1, London: Francis Westley, April–May 1823, pp. 133–9, 177–81 (179).

61 Eliza Morrison, *Memoirs of the Life and Labours of Robert Morrison*, 2 vols, London: Longman, Orme, Brown, Green and Longmans, 1839, II, p. 14. Morrison had, in fact, enrolled in a course for missionaries at St. Bartholomew's Hospital, London, where he acquired rudimentary medical knowledge.

62 Thomas John Newbold, *Political and Statistical Account of the British Settlements in the Straits of Malacca*, 2 vols, London: John Murray, 1839, I, pp. 117–18.

63 'The Spasmodic of Cholera Morbus', *Asiatic Journal and Monthly Register for British and Foreign India, China, and Australasia*, vol. 6, London: Parbury, Allen, and Co., September–December 1831, pp. 325–36 (329).

64 Peter Boomgaard, 'Morbidity and Mortality in Java, 1820–1880: The Evidence of the Colonial Reports', in Norman G. Owen (ed.) *Death and Disease in Southeast Asia: Explorations in Social, Medical, and Demographic History*, Singapore: Oxford University Press, 1987, pp. 48–69 (50).

65 'Memoir of the Late Rev. William Milne', p. 180.

66 John Crawfurd, *Journal of an Embassy from the Governor-General of India to the Courts of Siam and Cochin China*, 2 vols, 2nd edn, London: Henry Colburn and Richard Bentley, 1830, I, p. 227.

67 See B. J. Terwiel, 'Asiatic Cholera in Siam: Its First Occurrence and the 1820 Epidemic', in Owen (ed.), *Death and Disease in Southeast Asia*, pp. 142–61.

68 Anon, 'History of the Rise, Progress, Ravages, &c. of the Blue Cholera of India', p. 245.

69 There is a memorial plaque on the east wall of the churchyard at Kirriemuir, Angus, dedicated to: 'Capt. DAVID NICOLL, late commander of the ship *Merope* of Calcutta, who was killed in the massacre at Manilla, 9th Oct. 1820, aged 25 years.' See Andrew Jervise, *Epitaphs and Inscriptions from Burial Grounds and Old Buildings in the North East of Scotland*, 2 vols, Edinburgh: David Douglas, 1879, II, p. 361.

70 'Papers Concerning the Philippines and Penang', British Library, (India Office Records) IOR/H/77, 349-56. See also, *The Calcutta Annual Register, for the Year 1821*, Calcutta: Government Gazette Press, 1823, pp. 256–7.

71 Morrison, *Memoirs of the Life and Labours of Robert Morrison*, p. 37; on cholera in early-nineteenth-century Manila, see Ken De Bevoise, *Agents of Apocalypse: Epidemic Disease in the Colonial Philippines*, Princeton, NJ: Princeton University Press, 1995, pp. 28–9; José P. Bantug, *A Short History of Medicine in the Philippines*, Quezon City: Colegio Médico-Farmacéutico de Filipinas, 1953, pp. 25–37.

72 'The Spasmodic Cholera Morbus', p. 329. A history of cholera in China remains to be written, but see Wu Lien-teh (Wu Liande), J. W. H. Chun, R. Pollitzer and C. Y. Yu, *Cholera: A Manual for the Medical Profession in China*, Shanghai: National Quarantine Service, 1934.

73 John Dudgeon, *The Diseases of China: Their Causes, Conditions, and Prevalence, Contrasted with those of Europe*, Glasgow: Dunn & Wright, 1877, p. 45.

74 Although see William C. Milne, 'Notices of the Asiatic Cholera in China', *Chinese Repository*, vol. 12, Canton: Printed for the Proprietors, 1843, pp. 485–9 (466). Milne's research suggested that the emperor had, in fact, died of 'a stroke, apoplexy of paralysis'.

75 'The Death of the Emperor of China', *Morning Post* [London], 19 March 1821, 3.

76 Reported in the *Asiatic Journal and Monthly Register for British India and its Dependencies*, vol. 14, London: Kingsbury, Parbury, & Allen, July–December 1822, p. 568.

77 Bisset Hawkins, *History of the Epidemic Spasmodic Cholera of Russia*, London: John Murray, 1831, p. 178.

78 J. Livingstone, 'Observations on Epidemic Cholera as It Appeared in China', *Transactions of the Medical and Physical Society of Calcutta*, vol. 1, Calcutta: Thacker and Co., 1825, pp. 202–10 (205).

79 William Warder Cadbury and Mary Hoxie Jones, *At the Point of a Lancet: One Hundred Years of the Canton Hospital, 1835–1935*, Shanghai: Kelly and Walsh, 1935, p. 13.

80 Livingstone had taken up a position in Macau in 1808 and later helped Morrison establish a public dispensary for Chinese patients.

81 *Chinese Repository*, vol. 4, Canton: Printed for the Proprietors, 1836, p. 48.

82 Spence, *The Search for Modern China*, p. 130.

83 Ibid., p. 148.

84 Ibid., p. 149.

85 Pollitzer, *Cholera*, p. 26.

86 D. B. Simmons, 'Cholera Epidemics in Japan', in *China: Imperial Maritime Customs, Series 2, Medical Reports for the Half-Year Ended 30th September 1879*, no. 18, Shanghai: Statistical Department of the Inspectorate General, 1880, pp. 1–30 (2–3).

87 Quoted in Charles Macnamara, *A History of Asiatic Cholera*, London: Macmillan and Co., 1876, p. 142.

88 *Chinese Repository*, vol. 10, Canton: Printed for the Proprietors, 1841, p. 54.

89 Macnamara, *A History of Asiatic Cholera*, p. 142. Montgomerie died of cholera in India; he was the brother of Major General Sir Patrick Montgomerie (1793–1872) of the Madras Artillery who served in China; see P. F. Pearson, *People of Early Singapore*, London: University of London Press, 1955, p. 12.

90 Arthur Cunynghame, *An Aide-de-Camp's Recollections of Service in China, a Residence in Hong-Kong, and Visits to Other Islands in the Chinese Seas*, 2 vols, London: Saunders and Otely, 1844, I, pp. 98, 111.

91 'Peace and Treaty with China', *Foreign and Colonial Quarterly Review*, January–April 1843, vol. 1, 301–33 (313).

92 Ibid., p. 318.

93 Quoted in David McLean, 'Surgeons of the Opium War: The Navy on the China Coast, 1840–42', *English Historical Review*, 2006, vol. 121, no. 491, 487–504 (493, 498).

94 Macnamara, *A History of Asiatic Cholera*, p. 143.

95 MacPherson, *A Wilderness of Marshes*, p. 29.

96 Ibid., pp. 143–4.

97 Ibid., p. 19; see, also, Macnamara, *A History of Asiatic Cholera*, p. 143.

98 Milne, 'Notices of the Asiatic Cholera in China'.

99 Ibid., p. 486.

100 On the meaning of *huoluan*, see Marta E. Hanson, *Speaking of Epidemics in Chinese Medicine: Disease and the Geographic Imagination in Late Imperial China*, Abingdon: Routledge, 2011, pp. 134-7; on *huoluan* and the question of cholera's endemicity in China, see also Wu et al., *Cholera: A Manual for the Medical Profession in China*, pp. 7–16.

101 Rogaski, *Hygienic Modernity*.

102 In 1885, the presence of the *Vibrio cholerae* had been confirmed in Shanghai; see Wu et al., *Cholera: A Manual for the Medical Profession in China*, p. xi.

103 Arjun Appadurai, 'Introduction: Commodities and the Politics of Value', in Arjun Appadurai (ed.), *The Social Life of Things: Commodities in Cultural Perspective*, Cambridge: Cambridge University Press, 1988, pp. 3–63.
104 See Sarah Hodges, 'The Global Menace', *Social History of Medicine*, 2012, vol. 25, no. 3, 719–28.
105 Lauren Benton, *A Search for Sovereignty: Law and Geography in European Empires, 1400–1900*, Cambridge: Cambridge University Press, 2010, pp. 1–39.
106 Anon, 'History of the Rise, Progress, Ravages, &c. of the Blue Cholera of India', p. 243.

## Select bibliography

Arnold, David. 'The Indian Ocean as a Disease Zone, 1500–1950', *South Asia: Journal of South Asian Studies*, 1991, vol. 14, no. 2, 1–21.

Bayly, C. A. *Imperial Meridian: The British Empire and the World, 1780–1830*, London: Longman, 1989.

Benton, Lauren. *A Search for Sovereignty: Law and Geography in European Empires, 1400–1900*, Cambridge: Cambridge University Press, 2010.

Hamlin, Christopher. *Cholera: The Biography*, Oxford: Oxford University Press, 2009.

Harrison, Mark. *Contagion: How Commerce Has Spread Disease*, New Haven, CT: Yale University Press, 2012.

Hillemann, Ulrike. *Asian Empire and British Knowledge: China and the Networks of British Imperial Expansion*, Basingstoke: Palgrave Macmillan, 2009.

Koch, Tom. *Disease Maps: Epidemics on the Ground*, Chicago, IL: University of Chicago Press, 2011.

MacPherson, Kerrie L. 'Cholera in China, 1820–1930: An Aspect of the Internationalization of Infectious Disease', in Mark Elvin and Ts'ui-jung Liu (eds), *Sediments of Time: Environment and Society in Chinese History*, Cambridge: Cambridge University Press, 1998, pp. 487–519.

Owen, Norman G., (ed.). *Death and Disease in Southeast Asia: Explorations in Social, Medical, and Demographic History*, Singapore: Oxford University Press, 1987.

Stockwell, A. J. 'British Expansion and Rule in South-East Asia', in Andrew Porter (ed.), *The Oxford History of the British Empire: The Nineteenth Century*, vol. 3, Oxford: Oxford University Press, 2009 [1999], pp. 371–94.

# 12

# DISEASE, GEOGRAPHY AND THE MARKET

## Epidemics of cholera in Tokyo in the late nineteenth century

*Akihito Suzuki*

Historians of disease often follow the concept of 'framing disease' put forward by Charles Rosenberg in 1992, which incorporates both the lived reality of the disease and the socio-cultural framework in order to understand the illness.[1] Since many diseases are based on environmental elements, histories of disease have routinely included the examination of the agency of environment with both natural and man-made aspects, promoting what Christopher Sellers has called 'a revival of Hippocratic ways of thinking'.[2] The integration of physical, environmental, and socio-cultural aspects of disease into historical studies demands different methodological approaches and diverse historiographical elements from various disciplines: medical, biological and environmental sciences for the study of the physical realities of disease and its occurrence; and intellectual, cultural, and social histories for the analysis of human reactions to the disease in the past. Perspectives and analytical tools taken from these disciplines are combined and integrated by medical historians to reconstruct pictures of diseases in the past.

Historians need first to examine how diseases and human surroundings interacted with human bodies in the past, and then investigate how those interactions were imagined, understood, and controlled. Human communities, with their natural and man-made environmental conditions, have been differently affected by diseases and the analyses of how different environments shaped the content and extent of a disease have revealed important insights into environmental determinants in the past. At the same time, the study of why different societies assumed different reactions to the disease can shed light on the socio-cultural frameworks through which diseases in the past were dealt with.

When we examine the history of cholera and other infectious diseases, the categories of human reactions can be divided into strategies that centred on isolation and quarantine and others that were aimed at sanitizing places.[3] Indeed, analysis of the contrasting choices of prophylactic strategies against cholera by nineteenth-century European states has been a major field of historiographical development, with the classic work of Erwin Ackerknecht, who identified contagionism with the authoritarian states of Russia and German states and anti-contagionism with the liberal states of

Britain and France, and Peter Baldwin's new classic, which has convincingly refuted Ackerknecht's simple equation and presented more complex pictures.[4] It should be noted, however, that Ackerknecht, Baldwin, and many others have examined the political aspects of the human reactions to cholera and other infectious diseases, and largely neglected the economic side of the problem, particularly the question of consumerism and material culture. People at risk of suffering from infectious diseases were not just political agents but also consumers who did act through the market and buy certain commodities such as food, drugs, and other items which were regarded as prophylactic measures according to their beliefs. Although the significance of commercial measures was trivialized, ridiculed, and criticized by Rudolf Virchow, these measures nevertheless present an important aspect of the history of human reactions against diseases and suggest that historical analysis should integrate both political and economic aspects of societies in the past.[5] Integration is particularly important when one examines a society characterized by uneven developments in politics and economy. Instead of examining typical cases of politically 'backward' and economically poor countries or politically advanced and economically rich states, countries with uneven states of politics and economy reveal interesting aspects of disease, human reactions, and society.

Japan in the late nineteenth century presents such an 'uneven' case, as a result of its recent departure from a feudal societal system and the high development of market economy for a couple of centuries.[6] This chapter attempts to combine historiographies of environment, contagionism, and consumeristic prophylaxes to describe a picture in which natural and man-made circumstances, politics, and economy together formulated Japan's experience of diseases. Using the epidemics of cholera in Tokyo in the late nineteenth century as an example, this chapter first examines how the disease spread through the urban environment of Tokyo, considering both natural and autonomous elements and man-made and historically constructed societal factors.[7] I have tried to highlight the role played by the complex geography of the area of Tokyo and the history of civic engineering and the segregation of residences of different classes. As for human understanding and reactions to the disease, this chapter emphasizes the role of market and consumerism in the making of public health in modern Japan, rather than the political and administrative enforcement of isolation and segregation. Since Japan and some other countries in East Asia had not developed a system of public health comparable to those in Europe until its encounter with Western countries in the nineteenth century, historians need to reconstruct a different path (or several different paths) through which Japan and other East Asian countries achieved their health transition. The influence of the Western model of isolation and segregation obviously played a large role in the making of the hygienic modernity of Japan, but so did models that had *not* been imported from the West. The influence of Shintoism, a religion which is indigenous in Japan and has a strong orientation toward purity of the body and the environment, is one factor, and the formation of the attitude toward public health through consumerism is another.[8] The first section will present a topographical analysis of cholera epidemics in Tokyo, and the second section will offer an overview of issues related to the isolation of patients of cholera. The third section will discuss the role of consumer behaviour in the making of public health in Japan through episodes taken from the epidemics of cholera in Tokyo in the late nineteenth century.[9]

## Cholera and the topography of Tokyo

Cholera first visited Japan in 1822, during its first pandemic which started in Bengal in 1817.[10] This early visitation is hardly surprising: Japan was one of the nodes of the flourishing trading sphere which included India, Southeast Asia and China, with an increasingly large role being played by the United Kingdom, other European powers and the United States. Although Japan at that time strictly regulated foreign trade, its links with the trading zone of China, Korea and the Eastern half of the Indian Ocean were nevertheless strong.[11] The disease entered the country from either Tsushima or Nagasaki, both officially approved ports for foreign trade. The outbreak was geographically limited to the southwestern part of Japan. Although Osaka, the second largest city in Japan at that time, was severely hit, Edo, the capital and the largest city, with a population of about one million, was spared from the disease.

The second epidemic was in 1858, the year when the Tokugawa Shogunate signed a humiliating unequal treaty with the USA and subsequently with other European powers.[12] In July, the US Navy's *Mississippi* brought the disease from the coastal cities of China to Nagasaki. In the port city, more than 800 people immediately perished. Cholera quickly moved eastward along the major highway. The disease was rampant in Osaka in September and October, reputedly occasioning more than 10,000 deaths. Edo was ravaged around the same time, resulting in around 30,000 deaths in about two months. The disease waned in Edo in late October, only to be rekindled in the next year in several cities.[13]

The two epidemics of cholera in the Tokugawa period were characterized by limited involvement of the Shogunate or the feudal lordships of domains, apart from distributing medicines or issuing pamphlets on cure and prevention. Local studies reveal that each village devised its own way to fight against the epidemic: village officials often collected information and travelled widely in search of effective magical-religious talismans.[14] The lack of interest of public authorities, particularly the Tokugawa Shogunate, and the contrasting active local government fit in with the general picture of medical development in early-modern Japan, where local governments and private agencies played large roles in the promotion of medicine.[15]

Cholera returned to Japan for the third time in 1877, when the new Meiji government faced the Seinan War, the largest rebellion in the southwestern corner of the country. For the next two decades, cholera was almost semi-endemic in Japan, with particularly large numbers of cases in 1879 and 1886, both exceeding 100,000 deaths.[16] The epidemic of cholera in 1886 was the largest outbreak of the disease after the Meiji Restoration. In Tokyo, the first case was reported on 9 July 1886, and the outbreak continued until the end of October. The Statistical Book of Tokyo counted 10,813 cases reported in the 15 wards of the city of Tokyo, and the database created from a newspaper's daily reports of the number of patients in 1,500 streets allows close examination of the patterns of the incidence.[17]

Table 12.1 shows the number of patients, the incidence rate, the size of population, and the population density (number of residents per square kilometre) of the 15 wards of Tokyo, which are arranged in order of the incidence rate. The contrast between the wards is striking. Geographically speaking, the disease hit much more severely the central and eastern part of the city, which faced the sea or were on the River Sumida. The western and northwestern wards situated inland, on the other

*Table 12.1* The number of patients, the incidence rate of cholera, the size of the population, and the population density of the 15 wards of Tokyo

| | Ward | Number of patients | Incidence (per 100,000) | Number of residents | Population density (per kilometre) |
|---|---|---|---|---|---|
| 1 | Nihonbashi | 2,363 | 1,534 | 153,996 | 55.5 |
| 2 | Kanda | 1,621 | 1,315 | 123,241 | 42.1 |
| 3 | Fukagawa | 808 | 1,118 | 72,278 | 16.7 |
| 4 | Kyobashi | 1,594 | 933 | 170,816 | 48.2 |
| 5 | Honjo | 723 | 842 | 85,868 | 19.9 |
| 6 | Asakusa | 953 | 801 | 119,042 | 30.9 |
| 7 | Shiba | 836 | 749 | 111,681 | 18.1 |
| 8 | Hongo | 456 | 733 | 62,205 | 16.1 |
| 9 | Shitaya | 493 | 628 | 78,462 | 23.1 |
| 10 | Kojimachi | 273 | 532 | 51,303 | 6.4 |
| 11 | Koishikawa | 197 | 482 | 40,840 | 7.4 |
| 12 | Ushigome | 163 | 374 | 43,550 | 9.1 |
| 13 | Yotsuya | 115 | 374 | 30,733 | 19.9 |
| 14 | Azabu | 151 | 370 | 40,814 | 11.0 |
| 15 | Akasaka | 67 | 253 | 26,526 | 5.9 |

*Source:* Data taken from cases reported in Yomiuri Newspaper, July–December 1886.

hand, suffered much smaller numbers of patients and much lower incidence rates. This pattern of regional contrast among the wards of Tokyo was consistent in other outbreaks of cholera in the 1880s and 1890s.

One of the major factors that contributed to the pattern was the topography of the city of Tokyo, both in its natural and man-made aspects. The region of Tokyo has a complex geological structure, which had been adapted through a series of civic engineering projects since the late sixteenth century to accommodate the very large capital of Edo. Geologists and historians have revealed that the area of Tokyo had three different types of land: Musashino Plateau, Tokyo Lowland, and the reclaimed land.[18] The Musashino Plateau was originally an old alluvial fan created by a river about 140,000 years ago, on which volcanic ashes from Mt. Hakone and Mt. Fuji came down 50,000–80,000 years ago to form layers of loams. The Tokyo Lowland was a delta which lay in the east formed by rivers, much later than the formation of the Musashino Plateau. The plateau was eroded by rivers and seas, which resulted in a complex structure of ridges and valleys. The bottom of the valley sometimes became marshy wetland, while the remaining part of the plateau formed complicated hills. The area of Tokyo thus became a low wetland in the east, high plateau in the west, and complicated hills and valleys in between through natural forces which had operated for thousands of years.

This geo-complexity was further altered through the power of humans from the late sixteenth century, when Ieyasu Tokugawa (1543–1616), one of the great warriors and later the first Shogun, started to build Edo as the capital.[19] The marshlands at the centre-east, which intersected with shallow sea, were reclaimed to become the heart of politics, commerce, and transportation. Redirecting rivers, building canals, and establishing dockside markets, the Tokugawa Shogunate built up an impressive city which was by the eighteenth century to harbour a population of one million. Providing this population living on a complex topography with water to drink was a major problem. People in the hill areas could avail themselves of water from the wells.

The lowland area was provided with potable water through waterworks. In 1630, the Kanda waterworks were completed, drawing water from a lake for about 22 kilometres to Edo through pipes made of stone and wood. In 1653, the Tamagawa waterworks were finished, drawing water from Tama River for 42 kilometres to the southern part of Edo. The marshy areas in the east were most dispossessed in terms of the provision of water. Since neither wells nor waterworks were available, these areas were served by water-sellers, who received a licence from the government to take water from Kanda waterworks and bring it to the marshy residential areas for sale, which was frequently depicted in *ukiyoe*-prints.[20] Three types of water supply thus developed in Edo: well-water for high-lying areas, piped water for much of low-lying areas, and the service of water-sellers for the marshy area. The residential segregation of *samurai* (warrior class) and *chōnin* (commoners or non-warrior classes) during the Tokugawa period generally overlapped with the topography: high-lying areas were the places for samurai while low-lying areas and marshlands were for commoners.

The intersection of topography, civil engineering, and societal aspects which had been created during the early-modern period provided Tokyo in the late nineteenth century with an important background for the epidemiology of cholera. The disadvantages of the low-lying and marshy parts of Tokyo are clearly shown by the fact that the seven waterfront wards all recorded high incidence rates. Streets with high incidence of cholera (>2,000) were situated in the low-lying Shitamachi area close to the sea or the river, with only one exception in the north, which was a short-lived red-light district. If one compares the difference between the plateau and low-lying land in a single ward, the disadvantage of low-lying topography becomes clearer. A survey in 1877 of the quality of piped water in Tokyo conducted by Robert William Atkinson, then Professor of Chemistry at the University of Tokyo, shows a remarkable difference in water taken at various places in Tokyo.[21] Atkinson and two Japanese assistants took samples of water from various wells and conduits for chemical analysis for particles of dirt, chlorine, ammonia, and nitrogenous compounds. Well-water was cleaner than piped water. Indeed, with the exceptions of wells in low-lying Fukagawa and Honjo, samples of well-water in Tokyo were better than piped water in London. As for the differences in piped water in various places, they found that waterworks supplied water of better quality in the areas close to the sources of the conduit and that the quality of the water deteriorated as the distance from the starting point increased. The water was worse in areas around the lowest end of the conduit. Although this was not a result based on bacteriological analysis, residents in the plateau area drank less contaminated water from the wells, and residents in the lower stream of the conduit drank more contaminated water.

Perhaps more important was the disposal of the waste water. The human wastes of night soil were collected by farmers for fertilizer regardless of the area. Waste water from washing clothes was disposed of via gutters which led to rivers and to the sea. The contamination of drinking water through the wells and the damaged wooden pipes was one of the major concerns during the epidemics of cholera, and some incidences of cholera were traced to the contamination of that nature. The likelihood of contamination differed according to the length of time that waste water stayed in the gutter. In the high, plateau-like areas with steep hills, gravity carried waste water more quickly out of the area. In the flat and low-lying land, by contrast, waste water stayed longer in the area and had more chance to contaminate wells and pipes.[22] The difference between the nature of soils in the plateau and the low-lying land might have contributed to the probability of contamination. The soil of the plateau was loam or volcanic ash, which was porous and did not hold water for a long period. The low-lying land, on the other hand, had on its surface peat-like soil, which is dense and keeps waste water longer in the ground.

Another danger of the lowland was exposure through transportation along canals. Epidemics of cholera in various locales taught Japanese public health officers that boatmen were likely both to be infected by cholera and to infect the locale they visited. In his popular work of hygiene published in 1912, Shibasaburō Kitazato (1852–1931) wrote succinctly: 'Who mediate cholera? Boatmen do.'[23] The development of canals in Edo and other major cities was remarkable in the seventeenth and eighteenth centuries, and the database reveals that streets around the canals of Tokyo showed a significantly higher incidence of cholera.[24] In Kanda, streets with a high incidence rate were concentrated in the eastern part of the ward and they were all on the low-lying land, criss-crossed with canals, and dock-markets along the canals were the centres of diffusion. The dockside market was a juncture of waterborne and land transportation, as well as a place for exchange.[25] During the Tokugawa period, goods from across Japan were delivered to Edo and landed at the dockside for sale. Each dockside market specialized in a certain kind of goods, the most famous being the fish market which had been at Nihonbashi until the Great Kanto Earthquake in 1923. By the early Meiji period, there were 65 dockside markets, many of which were situated in the central low-lying land. The eastern half of Kanda was indeed surrounded by major dockside markets, around which high-incidence streets were situated. The markets were where people assembled and met other people. When the major means of transportation involved ships, the dockside markets were places where infection was most likely to be introduced from outside and caught by local people.

Another reason for the danger of dockside markets was the fact that dock labourers carried goods landed from the ships. In major ports such as Kobe, Yokohama or Nagasaki, longshoremen or dock labourers had long been recognized as the first to catch cholera and infect the rest of the population. The 60 dock-markets in Tokyo needed a huge number of carriers to move goods around. Contemporaries noted that dock-markets attracted poor labourers, who made unstable livings and formed slums in cheap rented houses close to the market.[26] Contemporary observers found that the hygienic state of these slums was deplorable and feared that they would become the hub of infection of cholera. Slum clearance thus became routine in the context of preventive measures against cholera in Japan in the 1880s. Dock-markets were likely to become gathering places for poor labourers who had no choice other than to live in unhygienic conditions.

Epidemics of cholera in Tokyo in the late nineteenth century were thus conditioned by the natural, the traditional and the modern. They depended on the natural topography of the environment of the region, because of the great role played by the time-old geological and geographical aspects of the area. The construction of the city from the seventeenth century, such as the reclamation of land and the building of water supply, canals and dockside markets, provided the background of the epidemiology of cholera in the late nineteenth century. Cholera in Tokyo was also influenced by the modern, in the sense that the diffusion of cholera outside India started in the early nineteenth century and continued throughout that century.

## Cholera and public health

The Meiji government established its administrative rules for public health from the late 1870s to the late 1890s, the period when recurring attacks of cholera were the major threats.[27] In 1877, the Home Ministry (to which the Sanitary Bureau belonged) drafted a set of rules, 'Guides to the Prevention of Cholera', the first national regulations on cholera prevention. Facing the fierce epidemic in 1879, the Ministry developed the rules into another set, 'Provisional Rules for the Prevention of Cholera'. In 1880, these were enlarged into 'Rules for the Prevention of Infectious Diseases', which stated fairly detailed regulations to fight against cholera and five other infectious diseases: typhoid, dysentery, diphtheria, typhus, and smallpox. Subsequently, numerous amendments and additions were made, which were crystallized in 1897 into the 'Law for the Prevention of Infectious Diseases', the first national codification of public health measures in modern Japan. Between 1877 and 1897, cholera repeatedly ravaged the country, and the new Meiji government struggled to create a framework of public health measures and to establish national and local organizations for that purpose.

In their attempts to create an effective public health strategy, the government was eager to learn from the West how to combat this disease and quickly incorporated measures based upon Western medical science and public health. In the 1870s and early 1880s, the Sanitary Bureau utilized the service of foreign doctors who were employed by the government, as well as Japanese doctors who had a smattering of Western medicine. Erwin von Baeltz (1849–1913), who had studied under Carl Wunderlich and became a professor of medicine at the University of Tokyo, was among the most prominent of the former. The advice of foreign doctors was largely in line with the miasmatic theory, and strong emphasis was laid on cleaning smelly dirt. At the same time, quarantine and the isolation of patients were vigorously pursued.[28] In 1888, Tadanori Ishiguro (1845–1941), surgeon-general of the army, was sent by the government to see Robert Koch in Berlin to ask how to combat cholera in Japan. Later, those who had studied medicine under Koch and other prominent German professors were actively engaged in public health measures.[29] Shibasaburō Kitazato was the most eminent of those coteries of German-trained doctors who became the leading figures in public health in Japan. Those German-trained Japanese medical scientists quickly trained younger students in Japan, both at the University of Tokyo and the Institute for the Research of Contagious Diseases established in Tokyo in 1892. By the late 1890s, bacteriological research in Japan was sufficiently sophisticated to produce its own vaccine and to discover different strains of cholera bacillus. Due to the acrimonious rivalry between doctors and scientists based at the University of Tokyo and those at the Institute for the Research of Contagious Diseases, the vaccine and the strains generated fierce controversies.[30] Despite those controversies, modern Japan established the basic principles, such as disinfection, cleanliness, quarantine, and isolation, that had not changed from the first establishment of state policies in 1877.

Devising policies was one thing; implementing them was quite another, however. Central government policy encountered considerable challenges and resistance.[31] Especially difficult was the isolation of patients in hospitals. The institution of hospitalization did not exist in Japanese society in the early-modern period, and sending patients away from home was definitely alien to the traditional pattern of cure and care of sick patients in Japan, where familial care was emphasized and legitimated through Confucian ideology. People were unaccustomed, or even antagonistic, to the practice of sending the sick to hospitals away from home. The high death rate of patients sent to isolation hospitals and the wretched conditions of cheap and makeshift buildings further increased distrust and hostility among people. The new government's unpopular measures in the early 1870s, such as the Conscription Law (1873) and the introduction of a police force, further increased people's distrust of hospitals, a new institution introduced by the government from the West.

Isolation hospitals were thus feared and hated. Rumours ran that doctors disembowelled patients alive and sold the livers as medicine, overlapping the popular image of cannibalistic demons from the other world. Particularly during the epidemic of cholera in 1879, there were 50 incidences of popular riots against government measures. In Niigata, about 1,000 peasants gathered in the manner of a traditional peasants' uprising and demanded the closure of isolation hospitals. When their demand was not heard, they resorted to violence, killing several officials of local government and looting rich merchants' houses.[32] In Chiba in the same year, a doctor who worked for the local isolation hospital was pursued, beaten, and killed.[33] He had been extremely unpopular because of his former practice of digging up corpses for the purpose of anatomical study. The practice of isolation was thus an issue of confrontation between the government and the common people: cholera riots contributed to a conflict between the modern and the traditional, between the culture of the ruling class and that of common people. The situation in Japan in the 1870s was somewhat comparable to that in the West in the nineteenth century, when tension between the policies that originated from medicine and the custom of common people was high.

This antagonistic dichotomy between tradition and the modern or between the classes should not be overemphasized, however. The schism between the modernizing elite and the traditional common people was not the major framework of the Japanese response to cholera. There were numerous signs of compromise and adaptation on the part of both the government and the populace. Central and local governments took pains to soften stern measures.[34] Isolating patients in their own home instead of hospitals was soon allowed or permitted, but the practice of domestic quarantine was soon found too cumbersome and of little use, and its enforcement considerably diminished. Common people often actively supported the government's policies against cholera. Donation of money and disinfectant medicine to local offices was widely practised. Brothel houses voluntarily proposed to build their own isolation hospitals, and donation of money from prostitutes was routinely reported in the press.[35] Theatres were also quick to disinfect and clean their premises.[36] Local elites chose to enter the isolation hospital in order to set an example for common people.[37] Both the government and the common people regarded policies as a mixture of enforcement and adaptation.

## Cholera, regimen and consumerism

In contrast to the practice of isolation, which was regarded as something new imported from the West, the significant locus of the merging of the traditional and the modern was the marketplace in the Japanese context. This marketplace of health was the social space where continuity rather than discontinuity was obvious and the presence of both elite and common people was evident. Through examining the response of common people to the epidemic of cholera during the nineteenth century, one can construct the medical history of modern Japan from the viewpoint of the social history of 'health for sale', conceived by Roy Porter.[38] Epidemics of cholera, as in many other countries, were a crucible of modernization of medicine in general and public health in particular. More important than the isolation of patients in the Japanese context of the making of modern hygiene was dietary regimen, which stretched over the Tokugawa Period and the Meiji Period. It was also practised across diverse social classes. Significantly, continuity and social inclusiveness were achieved through market and consumer society. The market of food and the knowledge to choose proper food for the prevention of diseases created a social space for hygienic citizenship.

During the Tokugawa period, more than one hundred books on general regimen (yōjō) were published, among which *Yōjō-kun* (1713) by Kaibara Ekken (1630–1714) was the most famous. These works on regimen were widely read and popularized through circulating libraries.[39] Choice of food or dietary regimen had been a crucial part of the prevention of cholera in Japan since the early nineteenth century. When Japanese doctors first encountered cholera in 1822, they quickly combined the Western and Chinese concepts of the disease. They learned from Dutch sources that the disease which hit them at that time was called 'Asiatic cholera' by Western doctors. They also found that Chinese medicine was helpful in understanding the disease, identifying Asiatic cholera with *kakuran*, a condition discussed in classic texts of Chinese medicine.[40] Cholera and *kakuran* shared epidemic seasons (which was the summer) and key symptoms – violent diarrhoea and vomiting, coldness of the extremities, cramps of the legs, the agony of the patient, and the rapid succession of death. In terms of the diagnosis, cholera was not a strange or peculiar disease for Japanese doctors, who found somewhat familiar phases of *kakuran* understood in the traditional medicine in East Asia.

This identification profoundly influenced the subsequent medical discourse and people's responses. Both learned discourse about cholera and popular measures against the disease were formulated with the aetiology of *kakuran* in mind and practised with the dietary regimen for digestive troubles. *Kakuran* in Chinese medicine had long been regarded as caused by the combination of eating immoderately and cooling one's stomach. Japanese medicine in the early-modern period developed this idea and added indigestion as another factor: when the food taken stayed too long in the stomach and turned putrid, the putrid matter would become poisonous and harm the stomach, causing violent diarrhoea or vomiting.[41] The process was called *shokushō*, meaning alimentary harm. There were many reasons for food staying too long in the stomach: most typical were taking too much food and eating particular kinds of food which were hard to digest. Eating food that was already becoming putrid had a similar effect. The situation of one's stomach was also important. When the stomach was deficient in vital heat, it lacked the power to digest food and stagnation and *shokushō* would follow. The aetiology of *Kakuran* was framed around the stagnation of food in the stomach, due to the types of food and the heat of the stomach.

The prevention of cholera was integrated into dietary regimen and somatic hygiene developed from the concept of *kakuran*: in order to prevent cholera, one should avoid the stagnation of food in the stomach and follow a special dietary regimen.

Interestingly, the basic rules that had a clear resonance with Chinese medicine were formulated by Pompe van Meerdervoort (1829–1908), a Dutch military surgeon who was invited to teach medicine in Nagasaki. During the epidemic of cholera in 1858, Pompe (as he was called in Japan) learned from his Japanese students that the disease, or one with very similar symptoms, was called *kakuran* in Chinese and Japanese medicine.[42] Although Pompe thought that cholera was more contagious than *kakuran*, his subsequent rules for the prevention of cholera for the city of Nagasaki clearly had *kakuran* in mind. The Dutch doctor notified the municipal governor that people should avoid cucumber, watermelon, apricot, and unripe plum and that they should not spend the night in a naked state. Later, the governor added sardines, mackerels, tuna, octopus, and others to the list of foods to be avoided. The rules fitted well with the aetiology and prophylaxis of *kakuran*, through its focus on digestion by way of the selection of food and on regimen by keeping the heat of the stomach. Pompe may well have found that these precepts made sense also in the Western medical system. The cucumber and the melon, which had long been regarded as 'cold' and possibly harmful food in the Galenic system of dietary regimen, were regularly invoked as one of the causes of cholera in nineteenth-century Europe and North America.[43] Dietary regimen provided a common ground for Western and Chinese medicine in the early and mid-nineteenth century. It continued well into the Meiji period: indeed, it was preached with intensified ardour. The Home Ministry's *Korera Yobō Yukai* [Instructions for the Prevention of Cholera] (1876) continued the dietary

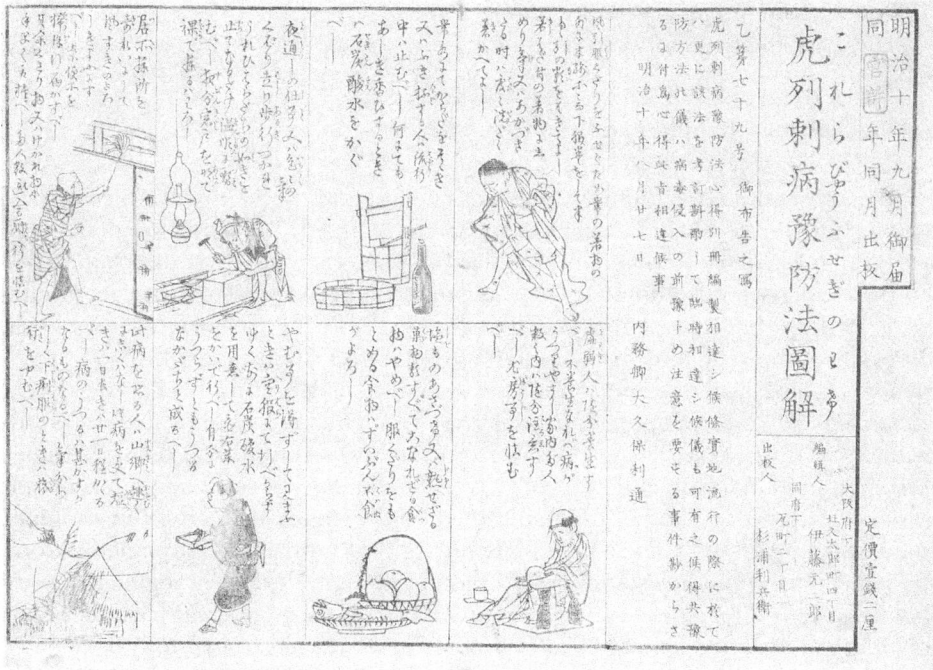

*Figure 12.1* Korerabyo fusegi no zukai (Illustrated Guide for the Prevention of Cholera), 1877. Courtesy of the Naito Museum of Drugs.

rules, advising people not to eat bad fish, shellfish, oysters, and prawns, as well as unripe or overripe fruits. The seventh on the list advised people to put on a belly-warmer when asleep and not to sleep naked. Elite doctors who studied medicine in Western countries (mainly Germany) regularly included those rules of regimen for the prevention of cholera, which had unmistakable resonance with the aetiology of *kakuran* and *shokushō*.

The emphasis on diet, foods to be avoided, and keeping one's stomach warm cut across social classes and more popular advice manuals shared the same strategy. Newspapers regularly reported incidences of cholera allegedly caught by eating particular food items. A broadsheet entitled '[An] Illustrated Guide to the Prevention of Cholera' issued in 1877 for the populace told its readers not to expose one's stomach to cold air, and to avoid indigestible food, as well as preaching cleanliness, temperance, and suitable rest (see Figure 12.1). The broadsheet issued in 1886 listed foods to eat and not to eat in the style of a sumo league table, recommending eggs, soles, and eels and discouraging octopus, crabs, and cucumbers. In order to help common people memorize the rules, two verses were composed, printed, and distributed in 1879. They are about food, regimen, and the stomach, as well as about cleanliness and miasma:

> Eat and drink moderately
> Avoid things that are smelly
> Don't catch cold at stuffy night
> Keep away from any crowded site
> Put on clothes that are clean
> These are the rules for your hygiene
>
> Greasy food, seafood, green fruit, and sushi
> Noodles, and dumplings do you harm, you see?[44]

The dietary regimen for the prevention of cholera and the theory of dietary pathogenesis showed remarkable tenacity in the late nineteenth and early twentieth centuries, both in learned and popular discourses on cholera. It also straddled boundaries between indigenous/traditional and Western/modern medicine, as mentioned above. Perhaps because of this, it was supported both by the progressive and the conservative, the elite and the common people. Most importantly, the dietary regimen was hailed as important by progressive-minded Westernizers. The newspaper *Yomiuri Shinbun*, for example, embraced Western medicine and preached preventive measures against cholera based on Western medical science. It also showed unrestrained contempt for practitioners of Chinese medicine, maintaining that 'roots and barks' were ineffectual and outmoded remedies. The newspaper's hostility to 'superstitious' healing methods such as amulets and religious rituals was particularly strong. The paper was, nonetheless, adamant in maintaining that dietary regimen was the most important. The newspaper even launched an attack on the emphasis on germs, isolation, and disinfection. Not that the newspaper was out of touch with the latest developments in bacteriology. On the contrary, it closely followed the discoveries of French and German medical scientists. In particular, it extensively covered Robert Koch's discovery of cholera bacillus in Calcutta, his triumphant return to Berlin and his receiving an honour from the German emperor. Nonetheless, this newspaper insisted that eating improper food resulting in the disturbance of the stomach was the chief cause of cholera. In an editorial which ran for two days, the paper made a foray into the contested terrain of the aetiology of cholera.[45] Although it sounded somewhat apologetic in not respecting some expert opinions, the editorial adopted the familiar 'seed and soil' model in the aetiology of the disease and laid very strong emphasis on the soil, namely the health of the stomach.[46] Devising its own metaphor of oil and fire, it insisted that without the accumulation of combustible material, a spark should not cause fire: the cholera bacillus identified by Koch was compared to a spark, and the food that became putrid due to an inactive stomach was the combustible material. On the basis of this metaphor, the editorial maintained that the stagnation of putrid matter in the stomach was a necessary cause of cholera. Thus the 'seed and soil' model was an important theoretical apparatus that secured continuity with the indigenous preventive measures of dietary regimen.

Dietary regimen was largely concerned with which food to buy, at least for residents of large cities of Tokugawa Japan. With the development of water-borne transportation and the establishment of Edo as a huge centre of consumption, common people in Edo started to consume diverse kinds of food. Sushi and tempura, now the two most internationally famous components of the Japanese cuisine, were sold on the street of Edo for artisans and labourers in the eighteenth and nineteenth centuries. Since food became something over which people could exercise choice as consumers, dietary regimen was closely linked with the consumer culture of food in early-modern Japan.

These instructions were not just preached, but at least some of them were followed. Earlier records of epidemics often contained which particular food was avoided or sought after. When certain items were alleged harmful and others beneficial, and when a large number of people followed the advice, the prices of those food items were affected. From around the late seventeenth century, chronicles recorded the fluctuations of prices of particular foods during an epidemic almost as a matter of routine.[47] This was because dietary regimen was a major part of the preventive measure against epidemics of cholera and other infectious diseases such as smallpox and measles. One humorous print depicted how vendors of forbidden foods such as fish, sushi, soba and others were taking revenge on the disease of measles (see Figure 12.2). A *Chronicle of Edo* noted large fluctuations in the prices of various food items during the cholera epidemic of 1858:

> Vendors of fish became very small in number, because fish would turn out to be fatal when eaten. Accordingly, fishermen and fishmongers suffered heavy loss. So did restaurants and bistros. Sardines were thought to be especially poisonous, and few people bought them even when they were fresh. On the other hand, prices of eggs and vegetables rose.[48]

During the cholera epidemics of the 1870s and 1880s, similar variations in food prices according to the rules of dietary regimen occurred. Sudden shifts in demand and the prices of particular food were regularly reported in the press. In Kyoto in 1878, matsutake mushroom, a delicacy usually much loved by the Japanese, was reputed to have caused cases of cholera. Its price suffered a heavy slump immediately. The next

213

*Figure 12.2* Tosei zatsugo ryukō machin fusegi (Contemporary miscellany on the battle of the epidemic of measles), c.1860. Courtesy of the Naito Museum of Drugs.

year, fishmongers of Kyoto were at a loss for what to do with their octopus, which nobody ate lest they should catch cholera.[49] In Tokyo in 1879, the prices of Chinese melons suffered a heavy downturn. Also in Tokyo in 1882, stalls that sold ice lollies diminished from 108 to 79 due to the cholera epidemic in the summer.[50] On the other hand, eels and loaches were reputed to be good and their prices soared in 1884, although some cases were attributed to eating those kinds of fish.[51] In the outbreak of cholera in 1886, *Yomiuri Shinbun* conducted a survey of the prices of various food items in Tokyo. On 26 June the newspaper published an article which listed the ups and downs of the sales and prices of various food items.[52] Items that recorded good sales and high prices were eggs, poultry, beef, dried bonito, grilled eels, vegetables, pickled radish, milk, starch gruel, and dry confectionaries. The food items whose sales slumped included: raw fish, salted fish, tempura, sushi, shellfish (which suffered the heaviest slump), and tōfu. In Yokohama in the same year, stalls selling iced waters, fishmongers, tempura-bars, soba-noodle bars and fruit shops had no customers, while poultry, eggs, eels, and Western food were in high demand.[53] In 1886, farmers in Chiba who brought peaches for sale to Edo found that the price had gone down so much that they could not pay the cost of transportation.[54] Likewise, farmers of the agricultural hinterland of Tokyo found that bringing and selling Chinese melon to cholera-struck Tokyo did not pay.[55]

The connection between epidemics and buying trends was such that some merchants would exploit it. A producer of pickles in Odawara reputedly made a fortune during the cholera epidemic in 1858. Learning this, a merchant speculated on pickles and prepared a huge stock, but, alas, pickles this time did not become fashionable and he suffered a heavy loss. Such practices had a long history and had been a well-established part of people's life since the early Tokuagawa period. In Edo in 1699, the city was hit by an epidemic of an unidentified disease called *korori*. During the epidemic, prices of pickled plum and fruit of nandina (*Nandina domesticat*) soared, due to the reputed preventive qualities of these foods. It was, however, later found that a grocer invented the theory. He had had a large stock of pickled plum imported from Osaka and he found that the supply of plums would be short this year. Intending to exploit this situation, he tried to beguile people into buying the food. In the end, however, his unethical business was revealed and he was severely punished.[56]

Such stories of unethical exploitation of the shifting market of dietary commodities suggest that there were public concerns around the market, dietary regimen, and public health during the time of epidemics. People changed their diet in response to epidemics and rules of dietary regimen, which were not just 'private' preventive measures. Indeed, it was repeatedly claimed to be one of the core public duties of an individual in the time of epidemics. The dietary regimen straddled individual well-being and public welfare. The dual nature was put into sharp relief during cholera epidemics, because of the highly contagious nature of cholera and the 'seed and soil' theory in which it was conceptualized. Indulgence in a desire to eat and drink would bring cholera not just to the individual, but also to his or her family members, neighbours, fellow villagers and citizens. Gluttony of an individual would cause stagnation of undigested food in his or her stomach, cause cholera in him or her, and then spread the disease. An editorial of *Yomiuri Shinbun* was outraged at the selfish indulgence of a handful of people: 'despite their knowledge that certain foods were harmful, they ate thirty peaches, drank six glasses of iced water, and devoured tuna'.[57] Bad food items were often delicacies eaten for pleasure rather than for subsistence – sushi, tempura, soba noodles were (and still are) pleasure food, so to speak. The pleasure of cooling one's body on stuffy and humid summer nights was also frowned upon, since it would invite cholera to the individual and spread the disease to other people.

Ogata Masanori, professor of hygiene at the University of Tokyo, succinctly summed up in his popular lecture on cholera: 'those who indulge in immoderate eating and drinking are manufacturers of cholera'.[58] Giving up those temporary pleasures of the body and the senses was to protect the health of both the individual in question and the community to which they belonged.

People's behaviour in terms of the choice and consumption of food was thus an integral part of their citizenship in the hygienic community of modern society. The food market acted as a social space that created conditions for hygienic citizenship.[59] Although we have ample reasons to believe that dietary regimen was practised by many people across diverse social sectors, not all of them followed the rules. In other words, the sphere of food consumption driven by the rules of dietary regimen was not comprehensive: a significant minority stayed outside this culture of health-oriented food consumption. Some city dwellers did not participate in the dietary regimen mediated by the food market. Many urban poor stayed outside the culture of preventing cholera through changing their food. Some consumers tried to exploit the low prices of food that was deemed harmful: *Yomiuri Shinbun* noted that a man who ate many Chinese melons when their prices went down due to its reputed pathogenic quality died from cholera: he was, in the view of the newspaper, duly punished for his greed and indulgence.[60]

Urban slums presented more serious problems. In large cities in early Meiji Japan, urban slums, whose residents suffered from chronic destitution, mushroomed. In the mid-1890s, journalists and social investigators started to visit those slums and publish what they saw in lurid and sensationalistic language. Journalists such as Matsubara Iwagorō and Yokoyama Gennosuke depicted the almost sub-human conditions of those who lived in urban squalor.[61] Suzuki Umeshiro's report on Nago-chō, Osaka's most destitute slum, included detailed and fascinating observations of people's attitude to cholera, since the reporter was staying there when cholera broke out in Osaka.[62] Umeshiro found that residents of Nago-chō had absolutely no qualms about eating foods that were deemed harmful. Fishmongers sold disgusting fish, such as bony scraps or half-rotten fish discarded by other fishmongers as unsuitable for respectable customers. Observing people eating such food, the reporter wrote: 'every item sold in the shop was a powerful cholera-causing material in its own right'. Expressing the theory of dietary pathogenesis of cholera, the reporter also claimed that the rapid diffusion of cholera in this area was primarily due to the voracious consumption of half-rotten food.

From the viewpoint of slum dwellers, eating proper food was far beyond their means: their income was so low that they could not buy rice and they collected half-rotten discarded food to survive. One of Nago-chō's informants protested against the charge that their dietary habits were propagating cholera: 'Rich people blame us for eating improper food and thus diffusing cholera to society. When we try to buy proper food, we find that we cannot make ends meet unless we engage ourselves with illegal activities.'[63] Although there is some doubt over the authenticity of the informant's words, Suzuki pointed out the crux of the problem: if eating properly was a requisite of hygienic citizenship, the urban poor, who could not buy proper food, faced the hard choice of being either a criminal or a cholera-spreader. The vision of hygienic citizenship through the regimen under the marketplace excluded the poor sector of society.

## Conclusion

This chapter has examined the history of cholera in terms of its impact on humans through the environment and people's understandings of the disease. Through the analysis of cholera epidemics in Tokyo in the late nineteenth century, I have tried to emphasize the importance of the historical construction of the environment and ideas of prevention. Although Japan was one of the first and arguably most successful non-Western countries which modernized and Westernized its medical and public health provisions, its path was far from a story of progressive modernization and Westernization. The pattern of modernization was markedly different from one social sphere to another, and this chapter has highlighted stark differences between the sphere of the policy of the state and other public authorities, on the one hand, and the sphere of individual consumption of food in the marketplace, on the other. Japan's modernization of the state public health machinery represented a sharp break around the Meiji Restoration, while the commercialization of health-seeking behaviour that had developed much earlier in Edo and other large cities showed remarkable continuity in models for the prevention of the disease. Commodification of health was flexible, or even protean, absorbing traditional *yōjō*, Western medicine, elite discourse and popular culture. This chapter highlights the importance of studying the role of the marketplace as a meeting point of tradition and modernity. Fernand Braudel wrote that the 'clamour of the market-place has no difficulty in reaching our ears'.[64] Perhaps it is time for medical historians to listen more carefully to the clamour of the marketplace in order to grasp the complex processes in the modernizations of medicine.

## Notes

1 C. Rosenberg, 'Introduction: Framing Disease: Illness, Society and History', in C. Rosenberg and J. Golden (eds), *Framing Disease: Studies in Cultural History*, New Brunswick: Rutgers University Press, 1992, xiii–xxvi; J.C. Burnham, *What Is Medical History?* Cambridge: Polity, 2005, 55–79.

2 C. Sellers, 'Health, Work, and Environment: A Hippocratic Turn in Medical History', in M. Jackson (ed.), *Oxford Handbook of the History of Medicine*, Oxford: Oxford University Press, 2011, 450–68. For the dual nature of the history of disease, see also P. Slack, *The Impact of Plague in Tudor and Stuart England*, Oxford: Clarendon Press, 1985.

3 For an introductory and interesting account of the history of cholera, see C. Hamlin, *Cholera: The Biography*, Oxford: Oxford University Press, 2009.

4 E.H. Ackerknecht, 'Anticontagionism between 1821 and 1867', *Bulletin of the History of Medicine*, 22, 1948, 562–93; P. Baldwin, *Contagion and the State in Europe 1830–1930*, Cambridge: Cambridge University Press, 1999.

5 H. Waitzkin, 'One and Half Centuries of Forgetting and Rediscovering Virchow's Lasting Contributions to Social Medicine', *Social Medicine*, 1, 2006, 5–10.

6 A. Gordon, *A Modern History of Japan: From Tokugawa Times to the Present*, Oxford: Oxford University Press, 2003; M.B. Jansen, *The Making of Modern Japan*, Cambridge, MA: Harvard University Press, 2000.

7 English works on the history of cholera in Japan are still few in number. An important exception is the works by William Johnstone, See, for example, http://socialsciences.blogs. wesleyan.edu/files/2012/11/Johnston-Cholera-Paper-for-Interdisciplinary-Group2.pdf. Accessed 8 September 2015.

8 Social and cultural history of Japanese hygiene through the prism of Shintoism remains to be written. For a work integrating mainly Western ideas of hygiene, culture and society, see V. Smith, *Clean: A History of Personal Hygiene and Purity*, Oxford: Oxford University Press, 2007.

9 Some parts of the third section overlap with the argument provided in A. Suzuki and M. Suzuki, 'Cholera, Consumer, and Citizenship: Modernization of Medicine in Japan', in Hormoz Ebrahimnejad (ed.), *The Development of Modern Medicine in Non-Western Countries: Historical Perspectives*, London: Routledge, 2009, 184–203.

10 For basic information on the history of cholera in Japan, see S. Yamamoto, *Nihon Korera-shi* [*History of Cholera in Japan*], Tokyo: University of Tokyo Press, 1982.

11 For revised understanding of the exclusion policy in early-modern Japan, see Jansen, *Making of Modern Japan*, pp. 91–5.

12 Yamamoto, [*History of Cholera in Japan*], pp. 14–26.

13 Although people reacted with horror, there were no signs of mass flight from Edo and other cities, which represents a sharp contrast with the mass flight observed in European and American cities hit by epidemics of plague during the early-modern period: Slack, *The Impact of Plague*. For a mass flight in New York in the epidemic of cholera in 1832, see C. Rosenberg, *The Cholera Years: The United States in 1832, 1849, and 1866*, Chicago: University of Chicago Press, 1987.

14 Studies of cholera in the Tokugawa Period are now vast. An interesting discussion of local policies combined with popular religion is found in S. Takahashi, *Bakumatsu Orugi* [*Cholera and the World-Turned-Upside-Down at the Close of the Tokugawa Period*], Tokyo: Asahi Shinbunsha, 2005.

15 See: T. Aoki, *Edo Jidai no Igaku* [*Medicine during the Edo Era*], Tokyo: Yoshikawa Kôbunkan, 2012; R. Umihara, *Kinsei Iryô no Shakaishi* [*A Social History of Early Modern Medicine in Japan*], Tokyo: Yoshikawa Kôbunkan, 2007.

16 Yamamoto, [*History of Cholera in Japan*], pp. 27–95.

17 The database has been created from the daily news of *Yomiuri Shinbun*, which reported the number of cases for 1,500 streets of the city of Tokyo. It has about 140,000 records about daily numbers of patients of the streets. Although the analysis presented here does not include detailed discussion of street-based analysis of cholera in Tokyo, some further discussion in that direction is now being prepared.

18 For the geography of the area of Edo and Tokyo, see the work of Masao Suzuki, *Edo ha Kôshite Tsukurareta* [*The Making of Edo*], Tokyo: Chikuma Shobō, 2000.

19 Suzuki, [*The Making of Edo*], pp. 132–76.

20 For examples of water-sellers in *ukiyoe*, see the collection at Waseda University, http://www.enpaku.waseda.ac.jp/db/enpakunishik/results-big.php?shiryo_no=500-0901. Accessed 1 September 2015.

21 R. Atkinson, M. Kuhara and M. Miyazaki, *Tokyo-fuka Yōsui Shikensetsu* [*Examination of Water in Tokyo-fu*], 1881.

22 Suzuki, [*The Making of Edo*], pp. 15–29.

23 S. Kitazato, *Korera Yobôhô* [*Prevention of Cholera*], Tokyo: Katei no Eisei-sha, 1912.

24 For the building of canals and the development of dock-markets, see Suzuki, [*Making of Edo*], pp. 132–76 and 301–8.

25 N. Kawana, *Kashi* [*The Dockmarket*], Tokyo: Hôsei University Press, 2007.

26 Such opinions were expressed routinely. See, for example, City of Tokyo, *Taisho Juichi-nen Korera-byô Ryukoshi* [*Outbreak of Cholera in Taisho 11*], 1925.

27 Yamamoto, [*The History of Cholera in Japan*], pp. 249–355.

28 For works on local public health reforms, see: K. Hidehiko, *Nihon no Iryô Gyōsei* [*Medical Administration of Japan*], Tokyo: Keio University Press, 1999; Y. Baba, 'Sanshinpōki no Toshigyōsei' ['Urban Administration of the Era of Three New Local Legislations'], *Historia*, no. 141, 1993, 48–66; K. Ozaki, '1879 Nen Korera to Chihō Eisei-seisaku no Tenkan' ['Cholera of 1879 and the Transformation of Local Hygienic Administration'], *Nihonshi Kenkyū*, no. 418, 1997, 23–50.

29 For a brilliant analysis of the German connection, see H. E. Kim, *Doctors of Empire: Medical and Cultural Encounters between Imperial Germany and Meiji Japan*, Toronto: University of Toronto Press, 2014.

30 For the history of the Institute of the Research into Contagious Diseases, see T. Odaka, *Densenbyō Kenkyūjo* [*Institute of the Research into Contagious Disease: A History*], Tokyo: Gakkai Shuppan Centre, 1992.

31 Local studies of cholera riots are now numerous. See, among others: S. Obinata, 'Korera Sōjō wo Meguru Minshū to Kokka' ['People and the State over Cholera Riots'], in Minshūshi Kenkyūkai (ed.), *Minshūshi no Kadai to Hōkō* [*Problems and Directions of Popular History*], Tokyo: San-ichi Shobō, 1978, 235–52; H. Sugiyama, 'Oboegaki: Bunmei Kaikaki no Hayariyamai to Minshūishiki' ['Notes on Epidemics and Popular Consciousness'], *Machidashi Jiyūminken Shiryōkan*, no. 2, 1988, 19–59.

32 Obinata, ['People and the State over Cholera Riots']. See also M. Nakano, 'Meiji Juninen Niigata Korera Sojo' ['Cholera Riots in Niigata in 1879'], *Chihoshi Kenkyu*, no. 149, 1977.

33 Numano Genshō, *Korera-i Genshō* [*Gensho: A Cholera Doctor*], Tokyo: Kyōei Shobō, 1978.

34 Yamamoto, [*History of Cholera in Japan*], pp. 407–584.

35 For hygienic cooperation of brothel houses and prostitutes, see *Yomiuri Shinbun* [Yomiuri Newspaper, hereafter YN] 1879/7/27; 1879/8/16; 1886/8/7; 1886/8/17; 1886/8/19; 1886/9/21; 1886/9/25; 1886/9/29.

36 YN 1879/8/9; 1886/8/12.

37 YN 1879/8/28.

38 R. Porter, *Health for Sale: Quackery in England 1660–1850*, Manchester: Manchester University Press, 1989.

39 For books on regimen from the Tokugawa period, see T. Takizawa, *Kenkō Bunkaron* [*Health and Culture*], Tokyo: Taishūkan, 1998, pp. 18–42.

40 M. Ōtsuki, 'Bunsei Jingo Tenkō Banki Kon Kakuran Ryōranbyo Zakki' ['Kakuran of the Year of Bunsei Jingo'], *Chugai Iji Sinpō*, no. 1131, 1928, 45–9; no. 1132, 1928, 106–7; no. 1133, 1928, 162–4; no. 1134, 1928, 216–18.

41 For a perceptive discussion of alimentary harm in early-modern Japan, see K. Daidōji, 'Edo no Shokushō' ['Alimentary Harm in the Edo Period'], in A. Suzuki and H. Ishizuka (eds), *Shokuji no Gihō* [*Technology of Eating*], Tokyo: Keio University Press, 2005, 147–67.

42 On Pompe van Meerderwoort, see T. Miyanaga, *Pompe: Nihon Kindai Igaku no Chichi* [*Pompe: The Father of Modern Japanese Medicine*], Tokyo: Chikuma Shobō, 1985.

43 During the epidemic of cholera in England in 1832, an article in *The Edinburgh Medical and Surgical Journal* wrote that 'repletion and indigestion should be guarded against; all raw vegetables, acescent, unwholesome food and drink avoided', and an article in *The Foreign Review* said to avoid 'exposure to cold, to chills, to the night dew, to wet and moisture; the use of cold fluids, and of cold, flatulent and unripe fruit'. See: R.J. Morris, *Cholera 1832*, New York: Holmes & Meier Publisher, 1976, p. 175; Rosenberg, *The Cholera Years*, p. 30.

44 YN 1879/9/13.

45 YN 1885/9/2; 1885/9/4.

46 On the social implications of seed and soil theory, see M. Worboys, *Spreading Germs: Disease Theories and Medical Practice in Britain, 1865–1900*, Cambridge: Cambridge University Press, 2000.

47 Kanei Kengo (ed.), *Bukō Nenpyō* [*A Chronicle of Edo*], 3 vols, Tokyo: Chikuma Gakugei Bunko, 2003–4.

48 [*A Chronicle of Edo*], vol. 3, pp. 103–4.

49 YN 1879/6/12.

50 YN 1879/7/22; 1882/7/18; 1882/7/23.

51 YN 1884/8/28.

52 YN 1886/6/23; 1886/6/26.

53 YN 1886/7/6.

54 YN 1887/7/30.

55 YN 1886/7/30.

56 A. Sōhaku et al., *Bōbyō Sōsetsu* [*Collected Treatises on Cholera*], Tokyo: for Kitazawa Ihachi, 1879.

57 YN 1886/7/25.

58 YN 1890/7/21.

59 For the concept of hygienic citizenship, see, for example: D. Armstrong, 'Public Health Spaces and the Fabrication of Identity', *Sociology*, no. 27, 1993, 393–410; D. Lupton, *The Imperative of Health: Public Health and the Regulated Body*, London: Sage Publications, 1995;

and R. Rogaski, *Hygienic Modernity: Meanings of Health and Disease in Treaty-Port China*, Berkeley: University of California Press, 2004.

60 YN 1879/7/22.

61 I. Matsubara, *Saiankoku no Tokyo* [*The Darkest Tokyo*], Tokyo: Iwanami Shoten, 1988; G. Yokoyama, *Nihon no Kasō-shakai* [*The Lower Societies of Japan*], Tokyo: Iwanami Shoten, 1949. These and similar works are discussed in K. Jun'ichiro, *Tokyo no Kasō-shakai* [*The Lower Societies of Tokyo*], Tokyo: Chikuma Shobō, 2000.

62 U. Suzuki, 'Osaka Nago-cho Hinmin-shakai no Jikkyo Kiryaku', ['A Reportage of the Society of the Destitute of Nago-cho of Osaka'], in N. Nishida, *Toshi Kasō-shakai* [*Urban Underclasses*], Tokyo: Seikatsusha, 1949, pp. 213–62.

63 Suzuki, ['A Reportage of the Society'], pp. 196–7.

64 Fernand Braudel, *Civilization and Capitalism 15th–18th Century: Volume Three. The Perspectives of the World*, translated by Sian Reynolds, London: HarperCollins Publishers, 1984, p. 25.

## Select bibliography

Chris Aldous and Akihito Suzuki, *Reforming Public Health in Occupied Japan, 1945–52: Alien Prescriptions?*, London: Routledge, 2012.

Alexander Bay, *Beriberi in Modern Japan: The Making of a National Disease*, Rochester, NY: University of Rochester Press, 2012.

Susan Burns, 'Making Illness Identity: Writing "Leprosy Literature" in Modern Japan', *Japan Review*, 16 (2004), 191–211.

Ann Jannetta, *Epidemics and Mortality in Early Modern Japan*, Princeton: Princeton University Press, 1987.

Ann Jannetta, *The Vaccinators: Smallpox, Medical Knowledge, and the "Opening" of Japan*, Stanford: Stanford University Press, 2007.

William Johnston, *The Modern Epidemic: A History of Tuberculosis in Japan*, Cambridge, MA: Harvard University Press, 1995.

Michael Shi-Yung Liu, *Prescribing Colonization: The Role of Medical Practice and Policy in Japan-Ruled Taiwan*, Ann Arbor, MI: Association for Asian Studies, 2009.

Akihito Suzuki, 'Measles and the Transformation of the Spatio-Temporal Structure of Modern Japan', *Economic History Review*, 62 (2009), 828–56.

Brett Walker, *The Conquest of Ainu Lands: Ecology and Culture in Japanese Expansion, 1590–1800*, Berkeley: University of California Press, 2001.

Brett Walker, *Toxic Archipelago: A History of Industrial Disease in Japan*, Seattle, WA: University of Washington Press, 2011.

# 13

# HISTORIES AND NARRATIVES OF YELLOW FEVER IN LATIN AMERICA

*Mónica García*

This chapter debates the historiographical perspectives that have dominated historical narratives of yellow fever in Latin America. Almost invariably, historians have assumed that past accounts of yellow fever refer to the same yellow fever that historians know and understand according to modern medicine. We identify the development of medical bacteriology as the watershed between old ideas and the new path that led to current notions of the fever – as caused by a microorganism, a virus. But no matter how we intend to historicize medical notions about the fever, most of us have actually avoided questions about what did 'vômito preto', black vomit, yellow fever or periodic fevers of the yellow fever variety refer to, according to contemporaries' interpretations and worldviews. The historiography of yellow fever in Latin America conveys the assumption that when historians find these terms in documents produced by doctors and policy-makers at any moment or location, those expressions refer invariably to one and the same yellow fever: acute viral haemorrhagic disease transmitted by infected mosquitoes and distinguishable thanks to salient symptoms such as jaundice, black vomit and fever. The unintended consequence of this apparently unproblematic decision is that we ignore one of the fundamental questions related to studying diseases historically: the contentious character of medical knowledge, hence of the things designated by it. The aim of this chapter is to analyse the historiographical approach to yellow fever, particularly the one that takes for granted current medical knowledge and performs retrospective diagnosis – an approach that could be named presentism in the historiography of yellow fever.[1] The varied narratives that have been produced from such a standpoint are summarized here. Finally, a brief exposition of some of the perspectives and themes that could help us to circumvent such presentism is discussed in the last part of this chapter.

## Historiographical perspectives

It is common for historians of yellow fever to start their narratives from the secure point of scientific understanding of the fever according to current medical knowledge – what we consider the 'true' yellow fever. We reify what we think is the actual yellow fever – that is, we treat yellow fever as if it were a non-transmutable reality. After we are convinced that we are dealing with a clear-cut natural object, we feel safe to devote ourselves to the political, institutional, economic, cultural or environmental factors involved in the efforts to understand and tackle yellow fever at any particular time and locale. As a consequence, we believe that the yellow fever that we find in past accounts is the true yellow fever as dictated by nature, the same fever as ours. Any difference between past accounts of the fever and our yellow fever is imputed to institutions, policies, economies and culture, all of these considered to be dependent on human action.

In the 1980s, Charles Rosenberg's influential definition of disease reinforced the distinction between nature and culture in histories of yellow fever.[2] According to Rosenberg's definition:

> Disease is at once a biological event, a generation-specific repertoire of verbal constructs reflecting medicine's intellectual and institutional history, an aspect of and potential legitimation for public policy, a potentially defining element of social role, a sanction for cultural norms, and a structuring element in doctor/patient interactions. In some ways disease does not exist until we have agreed that it does – by perceiving, naming, and responding to it.[3]

Historians of disease in Latin America who have claimed to work from a socio-constructivist approach, have enthusiastically taken up Rosenberg's definition as a way to justify the study of the social, cultural and political dimensions of disease.[4] However, as Adrian Wilson has pointed out, Rosenberg's essentialism that fixes the biological basis of diseases might have hindered the socio-constructivist impulse of the 1980s which looked at medicine as a social practice.[5] Socio-constructivists created a space for the historicization of medical knowledge and facts.[6] The fact that historians of yellow fever continue to consider contemporary medical knowledge to be the truth and yardstick to write the fever's history only confirms the success of Rosenberg's definition. It is also a sign of the fact that these historians seldom refer to the insights of fields such as history of science and the sociology of scientific knowledge (SSK) which have taken the socio-constructivist approach further.

Although sharing with historians of science and the SSK in general the goal of unveiling the ways natural knowledge is produced,[7] historians of medicine and public health have been very slow or reluctant to take up the insights of these fields. Historical, sociological and anthropological reconstructions of how scientists work have extensively shown that natural and social facts are the result of processes involving varied strategies and material arrangements. These works emphasize the material basis of scientific endeavours and the practices involved in the co-production of the natural and the social. Historians of medicine seem to have been very slow in adopting these understandings, as Michael Worboys denounced a few years ago.[8] Partly because they have been more interested in learning lessons from the past,[9] or because they deal with health, charged as it is with ethical compromises,[10] it may have been more difficult for historians of medicine to fully assume the consequences of historicizing medical knowledge and facts. However, it may be also possible that this reluctance is due to the fact that these historians, particularly in the Anglo-American context, have not been willing to abandon either the security of the 'real thing' – the yellow fever certified by the community of contemporary physicians and researchers – or the acknowledgment that comes with intellectual proximity to these certifiers.

Ilana Löwy, a biologist and historian of science, is one of the few historians of yellow fever in Latin America who has tackled the history using some of the insights of SSK. In her work on yellow fever in Brazil between 1880 and 1950, she explores the science around yellow fever which involved international circulation of knowledge, varied practices and specific policies.[11] Löwy acknowledges that objects such as the virus causing yellow fever or what we identify as the disease itself are the result of numerous mediations between society and nature, the result of a human activity. However, Löwy warns us that this does not mean that yellow fever is a pure expert construction. Maladies have an existence outside our experience and diverse human societies develop symbolic and practical tools to deal with them: diseases are bio-cultural phenomena, 'a mixture of human-made elements'.[12] Yellow fever would fall into the category of diseases that, besides patients' experiences, is perceived thanks to methods that make them visible. So, Löwy argues, the history of the fever is inseparable from the history of these methods. That is why she locates one of the turning points in the history of yellow fever in the 1930s, when our current medical understanding of the fever – as caused by a well-known virus – found its first foothold in the technologies that made it visible. It was from then on that we had at our disposal techniques to identify the virus directly – reproducing the fever in animals or indirectly – via antibodies. Before the 1930s, Löwy continues, identification of the fever was only via clinical experience, sometimes supported by some pathological findings in the liver of the diseased and in epidemiological indexes.

In accounts of the fever before the 1930s, argues Löwy, one cannot discard that other pathologies were included which were different from that of the virologists. Indeed, she reminds us that the symptoms of the 'true yellow fever' – high fever, jaundice, black vomit – are not specific to the fever; patients with malaria, leptospirosis, dengue, or typhoid fever could also portray them.[13] Thus, when we refer to people suffering from yellow fever, we have, according to Löwy, to locate and explain what are the criteria used to define this pathology: non-specialized definitions, doctors' definitions, or laboratory analysis. Nevertheless, even if we do not have the technology to make a retrospective diagnostic, she argues, 'this problem is not an important issue when what we study is not the fever itself'.[14] 'For example,' Löwy continues, 'when we find colonial physicians descriptions of the epidemics, it is not important if the subjects suffer from leptospirosis, malaria, typhoid fever or liver inflammation.'[15] It is notable that Löwy highlights the role of technologies in producing medical understanding of the fever. However, by separating the study of the fever itself or the 'true yellow' fever from its historical study she is implicitly separating nature from culture.

As this brief analysis shows, histories of yellow fever in Latin America produced over the last three decades were written and continue to be written reifying what historians think yellow fever is. This is so even if historians acknowledge, as does Löwy, the social and material mediations that have been part of its history. In the next section, I summarize the narratives around which historians have constructed the history of yellow fever in the region from this historiographical perspective, that is, when historians take current medical knowledge at face value. Histories of yellow fever in Latin America have been produced by social historians, historians of medicine and public health and environmental historians who have constructed two main narratives around the fever: (1) that yellow fever is an historical actor that has shaped history; (2) that the history of yellow fever is intertwined with power relations including the racialization of public health discourse, the role of centres and peripheries in science, and US public health colonialism. All these narratives have as a point of departure the notion of yellow fever as defined by current medical knowledge.

## Yellow fever as an historical actor

John Robert McNeill's and Mariola Espinosa's work on yellow fever in the Great Caribbean and Cuba respectively, are good examples of how historians have explicitly considered yellow fever – as defined by current medical knowledge – to be an important actor in shaping and explaining history.[16] McNeill, an environmental historian, has been interested in how ecological changes that affected the development of yellow fever and malaria shaped empire, war and revolution in the Greater Caribbean between 1620 and 1914. Interestingly, as the author claims, his work provides a perspective that takes into account nature – viruses, plasmodia, mosquitoes, monkeys, swamps – as well as humankind in making political history.[17] Clearly, McNeill's standpoint in his consideration of yellow fever and malaria is 'nature' as defined by current medical knowledge, with the novelty that McNeill attributes to these natural actors an active role in shaping history and vice versa. This standpoint makes it possible for him to argue that, for example, the ecological change resulting from the establishment of plantation economies in the Colonial Caribbean since the 1640s would have improved breeding and feeding conditions for mosquito species involved in the transmission of yellow fever and malaria. The rationale of McNeill's account of yellow fever is that sugar production would have brought with it deforestation which would have diminished insectivorous birds, hence reducing predation of mosquito eggs and larvae. Keeping water for dry seasons in kegs, barrels, buckets, as well as water collected in clay pots used in the production of semi-refined sugar, would have provided breeding grounds for mosquitoes.[18] As for the virus, McNeill argues that already existing reservoirs in the bigger islands of Cuba, Jamaica, Hispaniola and in South and Central America were supposedly imported from Africa with the slave trade,[19] and must have been built upon with further cases of the virus imported with the African monkey migration in ships of slaves that crossed the Atlantic in the seventeenth and eighteenth centuries.[20] 'If true,' continues McNeill, 'all this could help to explain how local populations developed immunities to yellow fever . . . It suggests that especially on the big islands and the mainland, yellow fever could be endemic in the countryside and on the plantations, not merely in the cities.'[21]

The presentism in McNeill's work with regards to medical knowledge thus explains not only the occurrence of yellow fever in maritime ports (Atlantic trade) and on the mainland, but also why foreigners appear defenceless in contrast to resistant locals. Differential immunity across the Atlantic would have been used to political advantage. Until the 1770s, McNeill argues, immunity would have helped the Spanish to protect their empire from British and French invaders. Some leaders of Latin American independence movements of the nineteenth century would also have recognized the differential effects of yellow fever among locals and foreigners and adjusted their war plans with immunity in mind.[22]

The fact that McNeill includes yellow fever as a factor that could explain historical outcomes is indeed an important call for historians in general, despite McNeill's presentism with regards to medical knowledge. But this presentism makes him judge past knowledge as true or false according to the rationale dictated by modern medicine.

He describes medical ideas and preventive and therapeutic practices surrounding the fever in the Great Caribbean, including both the European tradition (Hippocratic and Galenic medicine), and Afro-Caribbean medicine. However, McNeill makes judgements about the usefulness of this knowledge and these therapeutics according to his understanding of the fever. Of course, these judgements make little contribution to understanding the people who supported these practices. Both 'conventional wisdom' and certain practices appeared, at least in McNeill's eyes, to be truly helpful in keeping the fever at bay, even if those measures were produced for other purposes.[23]

Using current medical knowledge as the yardstick to investigate the history of yellow fever, Mariola Espinosa also confers to this ailment some sort of historical agency. Espinosa explores the role of yellow fever in the USA–Cuba relations during the decades surrounding the independence of the island from the Spanish crown (1878–1939). Her argument is that the yellow fever virus had a crucial and long-lasting impact on the relationship between Cuba and America.[24] Yellow fever had repeatedly spread from Havana to the US South in the late nineteenth century, affecting not only the southern economy but also the entire country. Thus, yellow fever became a source of apprehension to the US government long before the United States declared war on Spain in 1898. Particularly after the economy of the Mississippi Valley was seriously affected during the 1878 epidemic, the idea that the United States should acquire Cuba so as to ensure the health of the South became an inexorable conclusion and Cuba was invaded in 1898 to end the threat of yellow fever.[25]

Espinosa shows that the US war on yellow fever that started with the invasion of Cuba and continued until 1909 was not principally aimed at protecting the occupying force or improving the health of Cubans but was an attempt to eliminate the source of the infection. For this reason, Cubans resisted this colonial intervention in public health, acknowledging that keeping the island free of yellow fever was primarily a benefit to the United States. Espinosa highlights how Cubans struggled against the US characterization of Cuba as inherently dirty and disease-ridden and fought to make it known that the United States was dependent on Cuba for public health rather than the reverse. As part of these claims, Cubans refused to accept US myth-making regarding the scientific victory over yellow fever and insisted on acknowledging and celebrating the important role of their compatriot, Dr Carlos J. Finlay, in the story.[26]

Considering yellow fever as an historical actor that should be taken into account in the production of historical narratives and explanations is an important insight for historians in general. As the work of McNeill and Espinosa shows, historians can undoubtedly see in past descriptions of yellow fever how the virus, the mosquito and the illness itself unfolded, even if contemporaries were blind to the fever's 'real nature' – that of current medical knowledge. This latter presentist approach has also informed the second kind of historical narrative that has dominated the historiography of yellow fever in Latin America, namely the politics of yellow fever.

## The politics of yellow fever

The bulk of the historiography revolves around the idea that yellow fever is intertwined with power relations. Historians have thus constructed narratives that shed light on the multiple ways in which yellow fever created opportunities to set differences among people and between societies, in terms of (a) the racialization of public health discourse; (b) the role of centres and peripheries in science; and (c) US public health colonialism.

Sydney Chalhoub's account of yellow fever in Brazil during the second half of the nineteenth century intends to illustrate how yellow fever was involved in racialized arguments about Latin American inhabitants.[27] Chalhoub shows how medical explanations of Rio's yellow fever epidemic in the early 1850s and then in the early 1870s moved between medical thinking and political and racial ideology. During these epidemics Brazilian doctors dealt with the problem of why Africans and Afro-Brazilians suffered comparatively less from the disease than local whites and, especially, European immigrants. Brazilians, argues Chalhoub, did not follow the racial argument according to which Africans were naturally fit to work in the difficult conditions of warm climates, neither did they follow the line of thinking which emphasized the detrimental conditions of the Brazilian environment for foreigners. Contrary to American and European explanations, which oscillated between these two arguments, Brazilians explained the differential susceptibility between locals and foreigners in terms of general notions of environment and acclimatization. To them yellow fever was produced under poor sanitary conditions featuring filthy marshes and animal and vegetable material in decomposition. People who had long been exposed to such conditions – native Brazilians from the city of Rio – tended to fare better during an epidemic than those who were still becoming acquainted with such an environment, namely immigrants recently arrived from Europe.

In the 1870s, continues Chalhoub, this environmental language acquired new political and racial significance. Yellow fever had become the major public health challenge in Brazil and one of the main obstacles to the project of Brazilian planters: European immigrants, the main victims of yellow fever, were those who would cover possible losses in the workforce from the emancipation of slaves that eventually happened in 1888. Yellow fever was perceived to be a hindrance to the anticipated smooth transition from slavery to free labour through the use of European immigrants. The destruction of immigrant settlements in Rio was considered necessary because they were associated with immigrants dying of yellow fever and this association could discourage potential European immigrants from taking their chances in Brazil. Yellow fever was perceived as an obstacle to the country's progress and civilization. Thus, concludes Chalhoub, Brazilian medical thought and sanitary policies were deeply informed by a specific racial ideology: they had become active components in the making of the whitening ideal.

The racialization of the health discourse in late-nineteenth-century Brazil as presented by Chalhoub hints at the complexity that should be considered when dealing with nineteenth-century accounts of yellow fever. This kind of discourse was also present in Colombia as is shown in the final section of this chapter. Chalhoub's account of yellow fever, however, presumes that yellow fever is the fever of contemporary medicine; epidemics of yellow fever since the 1870s would have thus been involved in the labour crises of Brazil contributing to shape racial ideologies. Nature and culture, although related, are kept separated.

Historians working from a presentist perspective have also seen in the history of yellow fever an opportunity to make claims about how science developed in Latin America. These claims can be summarized in terms of the centre and periphery approach. Since the 1980s, one of the frames that has been used to explain how science developed in regions like Latin America is the diffusionist model according to which science is produced in some centres (European or American) and eventually spreads to or acclimatizes in the peripheries.[28] Without contending this colonial model, historians of science in Latin America have tried to show how science does not arrive to fill an existing vacuum and have even looked for cases of scientific success in the periphery.[29] The controversy around the establishment of the mosquito as the vector of yellow fever between 1878 and 1900 is a case in point: some contemporaries and historians argue that the Cuban, Carlos Finlay, who in 1878 hypothesized that mosquitoes *Aedes aegypti* transmitted yellow fever, was the actual author of this discovery while others argue that this discovery should be attributed to the US Yellow Fever Commission appointed to Cuba which took up and demonstrated Finlay's hypothesis 20 years later, in 1900. Indeed, according to Espinosa, Cuban contemporaries discredited the narrative of the scientific triumph over yellow fever that was prevalent in the United States and that excluded Finlay and granted all credit to US scientists. The act of vindicating the glory of Finlay, even while acknowledging his participation with the US Commission, had a political purpose: to vindicate an independent Cuba.[30] For Cubans then and today, Carlos Finlay should be credited with the discovery of the mosquito as the vector of the fever, and they blame the US Yellow Fever Commission for having used Finlay's theory, instruments and preparations to steal the authorship of the discovery.[31]

Historians of medicine and science in Latin America, on the other hand, have been concerned with the place of Latin America in the production of universal science. Debates around the discovery of the mosquito transmitting yellow fever are insightful. Among historians of yellow fever it is conventional to argue that Carlos Finlay did correctly hypothesize that the mosquito transmitted yellow fever, but also that the US scientists of the Yellow Fever Commission confirmed Finlay's hypothesis by performing convincing experiments two decades later.[32] According to this version Finlay, while having correctly denied that filth and fomites would transmit yellow fever, would have however failed to demonstrate his mosquito hypothesis since he had not realized that this insect does not become infectious immediately after biting a yellow fever patient.[33] By combining the theory of Finlay with the findings of a US sanitarian Henry R. Carter, who established that 10 to 12 days elapsed between an initial infection of a previously disease-free location and subsequent secondary cases, the US Yellow Fever Commission was able to perform experiments convincingly proving Finlay's hypothesis.[34] In this account, scientific failure on the side of Finlay would explain the lack of recognition for the work of the Cuban doctor for 20 years and also the position he shares with US scientists in the discovery of the mosquito as transmitter of yellow fever.

Other historians have tried to give more credit to Finlay and wonder not what was wrong with Finlay's science but why it took two decades for his mosquito hypothesis to be taken up by the scientific community. Even further, they have asked why it took the US Yellow Fever Commision only two months to confirm Finlay's hypothesis, despite the fact that they committed 'many of the same errors which Finlay had been accused of making'.[35] The 20 years' delay has been explained partly because Americans condemned scientific work in Latin America. George Miller Sternberg, the army medical officer who had visited several Latin American countries in the 1880s in his quest for the bacteriological cause of the fever, condemned the work of Brazilian and Mexican doctors who had claimed to have found not only the germ causing yellow fever but also a preventive method. Despite Sternberg's contact with Finlay, it would not be until the acceptance of insect vectors in disease and the US occupation of Cuba that interest and resources were brought together so as to make possible serious consideration of Finlay's idea.[36]

Francoise Delaporte, on the other hand, has claimed that Finlay was not original in his idea of the mosquito as key in the transmission of yellow fever. Finlay was indebted to Scottish physician Patrick Manson for the idea of the mosquito – even if Finlay did not recognize this or mention Manson – from which he formed the notion of the insect as agent of transmission. What explains the success of the US Yellow Fever Commission is Ronald Ross's hypothesis that the mosquito serves as intermediate host. Ross was key in taking into account the period during which the yellow fever germ incubates in the insect's body for a successful experimental case. The 20 years' limbo of Finlay's hypothesis was not due to intentional or unintentional denial from Americans, but this was simply the time that it took to unravel the mechanism of malarial infection.[37]

The historical epistemology of Francoise Delaporte and efforts by historians to judge the correctness of the experiments performed by Finlay and the US Yellow Fever Commission imply some sort of compromise on the part of historians with regard to what is considered to be true science. Instead, historians should better historicize science itself. In the last section of this chapter some of the questions that historians of science and the SSK have addressed are explored in order to point to ways that could help us to take up such an endeavour. For the moment it is sufficient to say that historians dealing with the discovery of the mosquito have helped us to understand the politics of this insect since they have clearly described how once *Aedes aegypti* was established as the vector of yellow fever it became paramount for the US colonialist public health project in Latin America. Eradicating this insect from maritime settlements became the cornerstone of US campaigns against yellow fever in Latin America in the first half of the twentieth century, hence fostering US economic expansion and intervention in the region.

The politics of the *Aedes aegypti* is indeed evident in the work of the philanthropic US institution, the Rockefeller Foundation (RF). Historians have described the role of the RF in promoting public health and in supporting public health education and public health services in Latin America between the 1910s and the 1940s, in line with RF's aims of 'advancing developing economies, promoting international goodwill, improving productivity, and preparing the state and professionals for modern development'.[38] Historians have also denounced the RF's colonial approach to Latin America as well as its role in fostering US economic and cultural expansion in the region. From tempering anti-US sentiments in post-revolutionary Mexico,[39] to subjecting people to open-ended experiments with anti-hookworm treatments,[40] and ignoring local initiatives and needs,[41] according to these historians the RF worked with the premise that Latin American societies were uncivilized. Historians have criticized the RF's narrow-minded approach to public health that ignored local health and social problems of much greater magnitude or concern.[42]

The work of the RF on yellow fever was at the heart of imperial expansion of the US in the region. Indeed, although eradicating yellow fever was not the RF's primary goal, the call of the US military after 1914 encouraged the RF's involvement in yellow fever campaigns. The reason was the opening of the Panamá Canal that year. After successfully eliminating yellow fever from Cuba in 1900 by destroying the mosquito, the US's next concern was Panamá. Once Panamá separated from Colombia in 1903 with the involvement of the US, Americans were able to take up the construction of the canal with the help of the US officials who had sanitized Cuba. These actions set the basis for preventing the Caribbean sources of yellow fever reinfecting the southern US, as contemporaries feared with the opening of the canal in 1914.[43]

The involvement of the RF started with the Yellow Fever Commission sent to South America in 1916. This commission set the basis for future campaigns against the fever which hinged around the idea that only few endemic centres of the disease, the so-called key centres (large coastal cities), served as seedbeds for infection. The RF thus concentrated eradication efforts on cities that harboured, or were suspected foci of, endemic yellow fever in maritime settlements in Central America, Ecuador, Guatemala, Perú, México and Brazil during the 1910s and 1920s with some success. Brazilians had managed to eradicate the fever with anti-mosquito measures in Rio de Janeiro from 1903. The method used by the RF was the use of yellow fever brigades in the destruction of the mosquitos and larvae, as well as house inspections, the depositing into water of small fish receptacles that would eat mosquitoes' larvae, and the use of petroleum in ponds and marshes where the female mosquitoes lay their eggs.

Historians have shown how the RF's work on yellow fever was shaped not only by US economic interests but also by local politics. In Mexico, the RF's yellow fever campaign helped to stabilize and legitimize Mexico as a state and created the basis for future institutional developments in post-revolutionary Mexico.[44] Veracruz was one of the most important US enclaves in Latin America with the monopoly of the oil industry. The US invasion in 1914 of Veracruz shows how threatening the Mexican revolution was (1910–29) to US investments. Anti-US sentiments, as well as clashes between local sanitarians who had worked in the control of yellow fever before the RF arrived, were common. In a tense environment, the RF started its campaign against yellow fever in 1921. This campaign alongside other health measures helped anti-US sentiments to be turned around. In other regions where the US also had economic interests, such as in Yucatán, fears of separatist movements and bolshevism prompted support from the national government for the RF campaign.[45]

In Brazil, the RF began to work in 1923 in the north-eastern part of the country where yellow fever threatened migration and commerce. The campaign has been described as being implemented on the line between persuasion and coercion.[46] By 1928 the RF claimed that yellow fever was close to being eradicated. However, this statement was contested by Brazilian doctors who argued that cases of yellow fever had widely increased in the interior of the country, in the north and in Minas Gerais.[47] This was the case also in Colombia, where RF officials who visited Colombia in 1916 ignored the work of local doctors who in 1907 – and later in 1923 – described cases of yellow fever in the mainland that would eventually be known as the sylvan yellow fever.[48] The disease's unexpected re-emergence in Rio de Janeiro in late 1928 ended up convincing RF doctors that forests were and had been sources of yellow fever for decades, resulting in a massive two-decade-long yellow fever campaign. This campaign developed in close cooperation with Getulio Vargas' regime (1930–45) and eventually extended into rural areas. The RF controlled the epidemiological surveys via viscerotomy (analysis of livers from people who died of yellow fever), the systematic elimination of *Aedes aegypti*, and from 1937 the production and distribution of the vaccine developed by RF laboratories. The last big yellow fever epidemic was in Rio in 1928–29. Later sporadic cases and the sylvan yellow fever were controlled with vaccination.[49]

## Themes and methods

One of the reasons that could explain presentism in the historiography of yellow fever is the fact that historians have not been interested in the insights of the history of science and of the SSK developed since the 1980s. Even if medical knowledge has not been the object of enquiry of these historians, it is possible to argue that by ignoring the contingent character of scientific knowledge, hence of the things designated by it, these historians have overlooked the concerns of the people they study that were connected with yellow fever as an historical actor or even with the politics of the fever. Before illustrating these points let me first consider what it is that SSK and history of science invite us to see.

SSK encourages us to be impartial with respect to truth or falsity, or success or failure, in knowledge claims and to be symmetrical with regard to the causal explanations about why certain knowledge came to be judged as true or false.[50] The presumption is that explanations for the cognitive decisions of scientists reside neither in natural reality nor in the logical structures of individual cognition, but must be found in the contingent judgement involved in scientific work.[51] This means that the content of knowledge is radically under-determined by the forms and structures of natural reality; that the application of scientific concepts to the natural world is ultimately a matter of inductive judgement, not deductive logic; and that the meaning and use of such concepts is therefore capable of changing over time.

As the vast historiography of science of the last three decades shows, contingency in the history of science is related to several factors. First, there are certain worldviews within which what is considered scientific and valid is agreed on by groups supporting these worldviews – the 'styles of knowledge' of Ludwik Fleck.[52] Second, there are literary technologies, such as public demonstrations, debates in newspapers, scientific congresses, textbooks and peer review, that make it possible for scientists to gain the trust of both experts and lay people and as a result gain legitimacy.[53] Third, the practices through which scientists translate the objects they work with into portable inscriptions – charts, numbers, tables, curves, etc. – circulate and allow scientists to make claims of universality.[54] Fourth, tacit knowledge is involved in performing experiments according to specific traditions.[55] Finally, there are complex networks of people and objects involved in scientific work, of which scientists are just one component.[56] One can hardly see in the historiography of yellow fever in Latin America a consideration of any of these points, either in the subject that has attracted historians' attention – the discovery of the mosquito – or in the subsequent efforts in understanding the epidemiology of the fever, either in the research involving the discovery of the virus in the 1920s, or in the production of the vaccine in the late 1930s. If historians take some of these proposals seriously, some of the themes that have been only hinted at within the presentist historiography of yellow fever arise as truly relevant historical problems. Furthermore, new objects of historical enquiry emerge. These new topics could complement and deepen the narratives that have dominated the historiography of yellow fever in the region, but from a non-presentist view of the fever with regard to medical knowledge and facts. Let me take the case of the nosology of fevers in nineteenth-century medicine, the climatic and geographic determinism of diseases, and the debates surrounding mainland yellow fever.

In 1981, William Bynum called attention to the fact that there were not many historical works dealing with fevers, even though the bulk of medical literature did deal with them, at least before the bacteriological revolution.[57] Indeed, if we look at textbooks from the mid-nineteenth century, we find that fevers, defined by heat and acceleration of pulse, were among the few diseases around which pathologists organized pathology (inflammation, poisoning, haemorrhage and diseases specific to organs and tissues were among the few others). This classification of diseases reveals the tension between eighteenth-century nosology and anatomo-pathology. Indeed, the so-called *essential* fevers were considered to be a group of diseases on their own via a negative definition: they were fevers not associated with any inflammation and whose organic basis was unknown. Thus, because of the lack of a lesion that would explain the fevers, doctors used a system of classification of fevers according to symptoms. This system revolved around genres, species and varieties. The criteria for sorting out fevers in such a hierarchical scheme were variations of the fever in time as well as associated symptoms such as headache, haemorrhages, diarrhoea, eruption on the skin. For example, in the third edition of the French textbook of pathology used in Europe and Latin America, the *Traité élémentaire et pratique de pathologie interne* by Augustin Grisolle,[58] fevers were divided into five genera – continuous, eruptive, intermittent, remittent, and hectic fevers – depending on the intervals at which the fever happened and on the associated symptoms. Hectic fevers were those fevers which were the expression of a hidden disease that would eventually become visible to the doctor. In the continuous group Grisolle located several species: yellow fever or typhus of America; typhoid fever; typhus of Europe; Eastern typhus or Plague; bilious fever of the warm countries, which was supposed to be the most common fever in the southern regions of the US; and inflammatory fevers. Descriptions of yellow fever were based on information from epidemics in the US, in Europe and from the French working in the French Antilles.

Grisolle characterized yellow fever by both jaundice and black vomit and classified it as typical of warm countries of the Americas, some parts of Africa and southern Europe. Foreigners were considered to be more susceptible to acquiring the fever than natives. In this frame of thought, the boundaries between yellow fever and the bilious fever of warm countries, for example, could be difficult to draw and it may be the case that a general practitioner would see yellow fever coexist with intermittent pernicious fevers, making diagnosis difficult. To make things even more complicated, doctors could also witness the transformation of intermittent pernicious fevers into remittent fevers and then into continuous fevers with the symptoms of yellow fever (jaundice and black vomit in nineteenth-century terms).

In Colombia, on the other hand, mid-nineteenth-century doctors grouped fevers into two genres: continuous and periodic fevers. In contrast to Grisolle, whose textbook they knew, Colombian doctors grouped together eruptive and continuous fevers, believing that the eruptions of the skin in smallpox and of the intestine in typhoid fever were part of the same pathological process. They also identified the yellow fever variety as belonging to the intermittent pernicious fever group and not to the continuous group as their French counterpart did. Yellow fever was grouped with the periodic fevers, probably as a way of emphasizing the miasmatic origin of these fevers which doctors associated with agricultural production in the low warm lands. This association allowed them to claim the local origin of intermittent fevers of the yellow fever variety and reject any idea in favour of the imported hypothesis. Colombian doctors followed the widely known French anticontagionist, Nicolas Chervin, in this argument. Chervin had studied the epidemics of the Caribbean and the US between 1820 and 1822 and the Gibraltar epidemic of 1828. He was adamant with regards to the miasmatic and local origin of the fever. Colombians found Chervin's argument persuasive since it allowed them to highlight their scientific expertise and superiority when compared to Europeans. Indeed, with regards to pathologies locally produced, such as fevers, Colombian doctors claimed that because they were in direct contact with these ailments, they knew them better than their European counterparts.[59]

The transformation of nineteenth-century continuous or discontinuous fevers into what are today known as specific fevers (typhoid fever, typhus, malaria or yellow fever) has scarcely been the subject of historical analysis, at least one that escapes teleological assumptions with respect to medical knowledge. Leonard Wilson and Dale C. Smith have shown how the continuous fevers, typhoid and typhus, reached an anatomo-pathological identity by mid-nineteenth century in Europe and America;[60] but very few historians have been interested in how periodic fevers became yellow fever and malaria. The attention of historians to epidemics in Europe up until 1857 and to those happening in North America up until the 1878 epidemic has focused on the contagionist/non-contagionist debate and on public health policies surrounding epidemics, always from a presentist perspective.[61]

In Colombia, the transformation of periodic fevers into yellow fever and malaria took more than two decades, from the moment the first account of the family of fevers was produced in 1859 until yellow fever became framed as a distinct disease deserving to be studied in its own right around 1887. Until 1886, the blurred boundaries between the species and varieties of periodic fevers, as well as the miasmatic theory according to which these fevers were locally produced by putrefaction of organic matter, formed a framework that suited doctors' arguments in favour of building a national medicine. Colombian doctors claimed that studying local pathologies, such as the periodic fevers of the warm climates of which Europeans, unlike them, had no first-hand contact, would be the way to truly know the nature of such pathologies.[62] These arguments had changed by 1887, when a hands-on practice, the preventive inoculation of a microorganism against yellow fever, triggered a debate among Colombian medical elites that culminated with the acceptance of yellow fever as a specific disease caused by a germ yet to be established.[63] The rhetoric of building a national medicine was replaced by claims around doctors becoming members of the universal land of science via bacteriology in a process that involved the consolidation of new ways of knowing among Colombian scientific elites.[64]

As this example shows, the decision not to take for granted what yellow fever means and to be impartial and symmetrical in the ways we approach yellow fever historically, makes it possible to see that for nineteenth-century medicine yellow fever was a variety within a fluid family of fevers which therefore could be transformed from other continuous or periodic fevers, depending on the medical community historians are referring to, in this case the French and the Colombian respectively.

Historical explanations about doctors' claims with regard to the place of cases of fever in any of the genera, species and varieties of the family of fevers, but most importantly historical explanations about why doctors decided whether the variety of yellow fever was imported or originated locally, should be looked for not in the disease itself – as we know it – but in the historical contingency of knowledge and of the things designated by it. From this perspective, it is possible to argue that nineteenth-century yellow fever was not a disease on its own. Of course, from the presentist view it is possible to argue that yellow fever was a distinct disease if we take jaundice and black vomit as the true markers of the fever.[65] These two symptoms were indeed the symptomatic traits by which mid-nineteenth-century medical authors such as Grisolle separated yellow fever from other fevers, but the fever was not confined to that in the cosmology of these doctors. By giving attention to the frame and worldview to which yellow fever belonged in the nineteenth century, we can also notice that climate and geography were considered to be part of its identity in many ways.

Arguments about the local origin of disease were coupled – at least in Colombia – with ideas about the climatic determinism of disease and people, which rekindled Hippocratic notions about the role of climate in producing disease.[66] Doctors linked medical geography, plant geography and transformist ideas in order to explain the distribution of diseases and people's susceptibilities to falling ill through a racialized discourse about people's capabilities.[67] According to views about the inferiority of natives and African people, inherited from colonial times, Colombian medical elites of the mid-nineteenth century considered the so-called blacks to be more resistant to diseases of warm climates in lowlands, such as periodic fevers, thanks to centuries of acclimatization. Natives of the Andes Mountains and Spanish descendants would be more adapted to temperate European-like climates of the highlands and would be therefore more susceptible to illnesses typical of the lowlands. The alleged susceptibility to periodic fevers of people coming from the highlands down to the areas of tobacco production in the lowlands alongside the Magdalena River during the 1840s and 1850s would have reinforced this argument for climatic and racial determinism of people and disease. Furthermore, mirroring the social division between the old colonial castes known as whites, blacks and natives – or the so-called races in the new republic (1810) – elite doctors extended this determinism to explain labour divisions: whites would be more suitable for intellectual work, whereas blacks would be fit to work under the burning heat of the lowlands.[68]

Given the pre-eminence of the climatological and geographical notions with which Europeans and Latin American elites understood differences between the old and new world after Spanish rule ended in Latin America,[69] it is worthwhile to wonder whether Latin American medical communities also used the geographic determinism that pervaded nineteenth-century worldviews in considerations of diseases and of fevers in particular. Needless to say, this world of geographic determinism has escaped the majority of historians of yellow fever in Latin America who have worked from a perspective of yellow fever in contemporary terms.

The reader could argue that since the infectious character of yellow fever was agreed on by around 1900 (a microorganism transmitted by a mosquito), there is no need to be aware of earlier views of the fever since from that moment the fever described by twentieth-century doctors was the same as ours. Following this rationale, one would easily accept that Fred Soper, working for the RF in Brazil, established for the first time in 1935 the sylvan cycle of yellow fever, that is, that yellow fever was not confined to the maritime settlements. Again, by taking for granted our knowledge in relation to the fever and the idea that science develops in centres that spread to the peripheries – the USA and Latin America for that matter – we have overlooked the fact that long before the RF conducted research in Latin America and Africa, Latin American physicians were debating the possibility of yellow fever – or the variety yellow fever of the periodic fevers – unfolding inland, alongside rivers, without even linking it to importation, and in rural areas far away from maritime settlements. Indeed, Jaime Benchimol has suggested that the nineteenth-century epidemics of yellow fever occurring in mainland areas of Brazil had an impact on widely held conceptions of yellow fever.[70] Benchimol argues that these epidemics might have helped hygienists, clinicians and bacteriologists to defend the idea of the fever being a specific disease of the inter-tropical regions. Unfortunately, it seems that nobody has taken up the analysis of these debates.

In Colombia, debates about periodic fevers typical of the low warm lands – including the variety of yellow fever – intensified during the 1860s and 1880s when epidemics of periodic fevers occurred alongside rivers and also up to the highlands. Doctors struggled to identify the nature of such fevers, which prompted doubts as to whether it was the yellow fever variety or not.[71] Even after a specific yellow fever had been defined as an infectious disease and the idea of transmission by the mosquito was accepted, twentieth-century Colombian physicians diagnosed the 'jungle yellow fever' around the 1910s in villages surrounded by forests, far away from maritime settlements. This work was ignored by the RF officials from 1916 and until the 1930s, and by historians until recently.[72]

## Epilogue

The pervasiveness of presentism in the historiography of yellow fever in Latin America and the consequent selection of historical themes that resonate with historians' views of the fever and of science could explain how themes such as those mentioned above have been neglected. Indeed, there are more histories about the discovery of the mosquito or about the debates surrounding the bacteriological cause of yellow fever during the 1880s in Latin America[73] than the political nature of the nineteenth-century nosology of fevers, the Latin American doctors' debates with regards to the mainland yellow fever variety, the particular practices that have shaped medical knowledge about the fever, the circulation of people, knowledge and objects in the region, and the climatic and geographic identity of the fever.

Historians have described how the climatological worldview and ideas about the geographical influences in the populations, nature, bodies and diseases, have informed visions of the post-colonial new world. There are few histories of diseases in Latin America – presentist or not – that point to the fact that neo-Hippocratic ideas and medical geography, with their implicit climatic determinism, were alive until the beginning of the twentieth century.[74] This dimension has escaped historians of yellow fever in particular because of their decision to consider nature a fixed object according to contemporary medicine. Historians of yellow fever in Latin America have so far not found it problematic to take for granted modern science, partly because they invest their work also with the authority that our societies grant to medicine and surely because they have neglected the insights of fields such as the history of science and SSK studies.

## Acknowledgements

I am grateful to Stefan Pohl-Valero, Franz Hensel and Mark Jackson for discussions on this chapter.

## Notes

1  As pointed out by Mark Jackson in the introduction to this volume, one of the challenges that historians of medicine face is the use of language. The way yellow fever is used in this chapter refers to what historians have claimed to historicize on behalf of either nature or of a particular scientific community.
2  I. Löwy, *Virus, Moustiques et Modernité. La fièvre jaune au Brésil entre science et politique*, Paris: Éditions des archives contemporaines, 2001, p. 19.
3  C. Rosenberg, 'Framing Disease: Illness, Society and History', *Explaining Epidemics and Other Studies in the History of Medicine*, Cambridge: Cambridge University Press, 1992, p. 305.
4  D. Armus, *Between Malaria and AIDS: History and Disease in Modern Latin America*, Durham, NC: Duke University Press, 2003, p. 1, and D. Obregón, *Batallas contra la lepra: Estado, Medicina y Ciencia en Colombia*, Medellín: Eafit, 2002, p. 27.
5  A. Wilson, 'On the History of Disease-Concepts: The Case of Pleurisy', *History of Science*, 2000, 38, 271–318.
6  L. Jordanova, 'The Social Construction of Medical Knowledge', *Social History of Medicine*, 1995, vol. 8, 3, 361–81 and P. Wright and A. Treacher, *The Problem of Medical Knowledge: Examining the Social Construction of Medicine*, Edinburgh: Edinburgh University Press, 1982.
7  J. H. Warner 'The History of Science and the Sciences of Medicine', *Osiris*, 1995, vol. 2, 10, 164–93, 165.
8  M. Worboys, 'Practice and the Science of Medicine in the Nineteenth Century', *Isis*, 2011, 102, 109–15, 110.
9  M. Jackson 'Introduction', *The Oxford Handbook of the History of Medicine*, Oxford: Oxford University Press, 2011, 4–5.
10 S. Müller-Wille, 'History of Science and Medicine' in M. Jackson (ed.), *The Oxford Handbook of the History of Medicine*, Oxford: Oxford University Press, 2011, 1–2.
11 Löwy, *Virus, Moustiques et Modernité*.
12 Ibid., 19.
13 Ibid., 23.
14 Ibid.
15 Ibid.
16 J. R. McNeill, *Mosquito Empires: Ecology and War in the Greater Caribbean, 1620–1914*, Cambridge: Cambridge University Press, 2010 and M. Espinosa, *Epidemic Invasions: Yellow Fever and the Limits of Cuban Independence, 1878–1939*, Chicago/London: University of Chicago Press, 2008.
17 McNeill, *Mosquito Empires*, 2.
18 Ibid., 48.
19 Some scholars have questioned the common assumption that yellow fever was imported from Africa to the Americas and the supposedly natural immunity of Africans or 'black' people to yellow fever. See S. Watts, 'Yellow Fever Immunities in West Africa and the Americas in the Age of Slavery and Beyond: A Reappraisal', *Journal of Social History*, 2001, vol. 34, 4, 955–67.
20 McNeill, *Mosquito Empires*, 49–50.
21 Ibid., 50.
22 Ibid., 303.
23 Ibid., 69–72.

24 Espinosa, *Epidemic Invasions*, 3.
25 Ibid., 27–9.
26 Ibid., 109–11.
27 S. Chalhoub, 'Yellow Fever and Race in Nineteenth Century Rio de Janeiro', *Journal of Latin American Studies*, 1993, vol. 25, 3, 441–63.
28 G. Basalla, 'The Spread of Western Science', *Science*, 1967, vol. 156, 3775, 611–22 and A. Lafuente, E. Alberto and M. L. Ortega (eds), *Mundialización de la ciencia y cultura nacional*, Madrid: Doce Calles, 1993.
29 J. G. Peard, *Race, Place, and Medicine: The Idea of the Tropics in Nineteenth-Century Brazil*, Durham, NC: Duke University Press, 2000.
30 Espinosa, *Epidemic Invasions*, 113–15.
31 See G. Delgado, 'Trascendencia de la obra científica del doctor Carlos J. Finlay en el 170 aniversario de su nacimiento', *Revista Cubana de Medicina Tropical*, 2004, vol. 56, 1, 6–12.
32 Espinosa, *Epidemic Invasions*, p. 56 and A-E. Birn, *Marriage of Convenience. Rockefeller International Health and Revolutionary Mexico*, Rochester, NY: University of Rochester Press, 2006, 49; P. Sutter, 'El control de los zancudos en Panamá: los entomólogos y el cambio ambiental durante la construcción del Canal', *Historia Crítica*, 2005, 30, 67–90, 71; and C. Alcalá, 'De miasmas a mosquitos: el pensamiento medico sobre la fiebre amarilla en Yucatán, 1890–1920', *Historia, Ciências, Saúde-Manguinhos*, 2012, vol. 19, 1, 72–5.
33 Espinosa, *Epidemic Invasions*, 56.
34 Ibid., 108 and Löwy, *Virus, Moustiques et Modernité*, 61.
35 N. Stepan, 'The Interplay between Socio-Economic Factors and Medical Science: Yellow Fever Research, Cuba and the United States', *Social Studies of Science*, 1978, vol. 8, 4, 397–423, 402.
36 Stepan, 'The Interplay between Socio-Economic Factors', 407–9.
37 F. Delaporte, *The History of Yellow Fever: An Essay on the Birth of Tropical Medicine*, Cambridge, MA: MIT Press, 1991, 8.
38 Birn, *Marriage of Convenience*, 25.
39 A. Solórzano, 'The Rockefeller Foundation in Revolutionary Mexico: Yellow Fever in Yucatan and Veracruz' in Marcos Cueto (ed.), *Missionaries of Science: The Rockefeller Foundation in Latin America*, Bloomington: Indiana University Press, 1994, 53–69.
40 S. Palmer, 'Toward Responsibility in International Health: Death Following Treatment in Rockefeller Hookworm Campaigns, 1914–1934', *Medical History*, 2010, vol. 54, 2, 149–70.
41 E. Quevedo, E. C. Manosalva, M. Tafur, J. Bedoya, G. Matiz and E. Morales, 'Knowledge and Power: The Asymmetry of Interests of Colombian and Rockefeller Doctors in the Construction of the Concept of "Jungle Yellow Fever", 1907–1938', *Canadian Bulletin of Medical History*, 2008, vol. 25, 1, 71–109; P. Mejía, 'De ratones, vacunas y hombres: el programa de fiebre amarilla de la Fundación Rockefeller en Colombia, 1932–1948', *Dynamis*, 2004, 24, 119–55; and Espinosa, *Epidemic Invasions*.
42 Birn, *Marriage of Convenience*, 25; Mejía, 'De ratones, vacunas y hombres', 120.
43 Espinosa, *Epidemic Invasions*, 120; M. Cueto, *Missionaries of Science: The Rockefeller Foundation in Latin America*, Bloomington: Indiana University Press, 1994, xii; and Birn, *Marriage of Convenience*, 28.
44 Birn, *Marriage of Convenience*.
45 Solórzano, 'The Rockefeller Foundation in Revolutionary Mexico'.
46 Löwy, *Virus, Moustiques et Modernité*, 139.
47 Löwy, *Virus, Moustiques et Modernité*, 144–5.
48 Quevedo et al., 'Knowledge and Power'.
49 Löwy, *Virus, Moustiques et Modernité*, 165 and The World Health Organization, 'Present Status of Yellow Fever: Memorandum from a PAHO Meeting', *Bulletin of the World Health Organization*, 1986, vol. 64, 4, 511–24.
50 D. Bloor, *Knowledge and Social Imagery*, Chicago/London: University of Chicago Press, 1991, 7.
51 B. Barnes, D. Bloor and J. Henry, *Scientific Knowledge: A Sociological Analysis*, London: Althlone, 1996, 54–56 and B. Barnes, 'On the Conventional Character of Knowledge and Cognition', *Philosophy of the Social Sciences*, 1981, vol. 11, 3, 303–33, 309.

52  L. Fleck, *The Genesis and Development of a Scientific Fact*, Chicago: University of Chicago Press, 1979; first printed in German (1935).
53  S. Shapin, 'Pump and Circumstance: Robert Boyle's Literary Technology', *Social Studies of Science*, 1984, vol. 14, 4, 481–520; and J. A. Secord, *Victorian Sensation: The Extraordinary Publication, Reception, and Secret Authorship of Vestiges of the Natural History of Creation*, Chicago/London: University of Chicago Press, 2001.
54  B. Latour, *Laboratory Life: The Social Construction of Scientific Facts*, Los Angeles: Sage, 1979; *Science in Action: How to Follow Scientists and Engineers through Society*, Cambridge, MA: Harvard University Press, 1987; and *The Pasteurization of France*, Cambridge, MA: Harvard University Press, 1988.
55  H. Collins, *Tacit and Explicit Knowledge*, Chicago/London: University of Chicago Press, 2010.
56  Latour, *Laboratory Life*.
57  W. F. Bynum, 'Cullen and the Study of Fevers in Britain, 1760–1820' in Theories of Fever from Antiquity to the Enlightenment, *Medical History*, 1981, Supp. 1, 135–47, 145.
58  A. Grisolle, *Traité élémentaire et pratique de pathologie interne*, vol. I, París: Victor Masson, Libraire-éditeur, 1848, 3rd edn.
59  C. M. García, 'Las "fiebres del Magdalena": medicina y sociedad en la construcción de una noción médica colombiana, 1859–1886', *Historia, Ciências, Saúde-Manguinhos*, 2007, vol. 14, 1, 63–89.
60  L. G. Wilson, 'Fevers and Science in Early Nineteenth Century Medicine', *Journal of the History of Medicine and Allied Sciences*, 1978, 33, 386–407; D. C. Smith, 'Gerhard's Distinction between Typhoid and Typhus and Its Reception in America, 1833–1860', *Bulletin of the History of Medicine*, 1980, vol. 54, 3, 368–85; and D. C. Smith, 'The Rise and Fall of Typhomalarial Fever', *Journal of the History of Medicine and Allied Science*, 1982, vol. 37, 182–220 and 287–321.
61  W. Coleman, *Yellow Fever in the North: The Methods of Early Epidemiology*, Madison: University of Wisconsin Press, 1987 and M. Humphreys, *Yellow Fever and the South*, New Brunswick, NJ: Rutgers University Press, 1992.
62  García, 'Las 'fiebres del Magdalena'.
63  M. Garcia, 'Producing Knowledge about Tropical Fevers in the Andes: Preventive Inoculations and Yellow Fever in Colombia, 1880–1890', *Social History of Medicine*, 2012, vol. 25, 4, 830–47.
64  M. García and S. Pohl-Valero, 'Styles of Knowledge Production in Colombia, 1850–1920', *Science in Context* (forthcoming).
65  N. L. Stepan, *Picturing Tropical Medicine*, London: Reaktion Books, 2001, 162–3.
66  For an analysis with regards to the persistence of ideas of the influence of climate in health and disease see the Special Issue: Modern Airs, Waters, and Places of the *Bulletin of the History of Medicine*, 2012, vol. 86, 4.
67  M. Garcia, 'Clima, enfermedad y raza en la medicina colombiana del siglo XIX', in G. Hochman, S. Palmer and M. S. Di Lisia (eds.), *Patologías de la Patria. Enfermedades, enfermos y Nación en América Latina*, Buenos Aires: Lugar Editorial, 2012, 59–74.
68  Garcia, 'Clima, enfermedad y raza'.
69  J. Cañizares, 'Entre el ocio y la feminización tropical: ciencia, élites y estado-nación en Latinoamérica, siglo XIX', *Asclepio*, 1998, vol. 1, 2, 11–31; Stepan, *Picturing Tropical Medicine*, 149–63.
70  J. Benchimol, *Dos mosquitos aos micróbios. Febre amarela e a revolução pasteuriana no Brasil*, Rio de Janeiro: Fundação Oswaldo Cruz/Editora, 1999, 15.
71  Garcia, 'Producing Knowledge about Tropical Fevers'.
72  Quevedo, 'Knowledge and Power'.
73  Benchimol, *Dos mosquitos aos micróbios*; Garcia, 'Producing Knowledge about Tropical Fevers'; S. Lozano, 'Importation et destin de la première théorie des germs au Mexique: développement des premières recherches sur la fièvre jaune dans les années 1880', *História, Ciências, Saúde-Manguinhos*, 2008, vol. 15, 2, 451–71; and M. Warner, 'Hunting the Yellow Fever Germ: The Principle and Practice of Etiological Proof in Late Nineteenth-Century America', *Bulletin of the History of Medicine*, 1985, vol. 59, 361–82.

74 Stepan, *Picturing Tropical Medicine*, M. Cueto, 'Nationalism, Carrión's Disease and Medical Geography in the Peruvian Andes', *History and Philosophy of the Life Sciences*, 2003, vol. 25, 319–35; García, 'Las 'fiebres del Magdalena'.

## Select bibliography

W. F. Bynum, 'Cullen and the Study of Fevers in Britain, 1760–1820' in Theories of fever from Antiquity to the Enlightenment, *Medical History*, 1981, Suppl 1., 135–47.

W. Coleman, *Yellow Fever in the North: The Methods of Early Epidemiology*, Madison: University of Wisconsin Press, 1987.

S. Chalhoub, 'Yellow Fever and Race in Nineteenth Century Rio de Janeiro', *Journal of Latin American Studies*, 1993, vol. 25, 3, 441–63.

F. Delaporte, *The History of Yellow Fever: An Essay on the Birth of Tropical Medicine*, Cambridge, MA: MIT Press, 1991.

M. Espinosa, *Epidemic Invasions. Yellow Fever and the Limits of Cuban Independence, 1878–1939*, Chicago/London: University of Chicago Press, 2008.

C. M. García, 'Las "fiebres del Magdalena": medicina y sociedad en la construcción de una noción médica colombiana, 1859–1886', *Historia, Ciências, Saúde-Manguinhos*, 2007, vol. 14, 1, 63–89.

M. García, 'Producing Knowledge about Tropical Fevers in the Andes: Preventive Inoculations and Yellow Fever in Colombia, 1880–1890', *Social History of Medicine*, 2012, vol. 25, 4, 830–47.

S. Lozano, 'Importation et destin de la première théorie desgerms au Mexique: développement des premières recherches sur la fièvre jaune dans les années 1880', *História, Ciências, Saúde-Manguinhos*, 2008, vol. 15, 2, 451–71.

J. R. McNeill, *Mosquito Empires: Ecology and War in the Greater Caribbean, 1620–1914*, Cambridge: Cambridge University Press, 2010.

J. G. Peard, *Race, Place, and Medicine: The Idea of the Tropics in Nineteenth-Century Brazil*, Durham, NC: Duke University Press, 2000.

E. Quevedo, C. Manosalva, M. Tafur, J. Bedoya, G. Matiz and E. Morales, 'Knowledge and Power: The Asymmetry of Interests of Colombian and Rockefeller Doctors in the Construction of the Concept of "Jungle Yellow Fever", 1907–1938', *Canadian Bulletin of Medical History*, 2008, vol. 25, 1, 71–109.

A. Solórzano, 'The Rockefeller Foundation in Revolutionary Mexico: Yellow Fever in Yucatan and Veracruz' in M. Cueto (ed.), *Missionaries of Science: The Rockefeller Foundation in Latin America*, Bloomington: Indiana University Press, 1994, pp. 53–69.

N. Stepan, 'The Interplay between Socio-Economic Factors and Medical Science: Yellow Fever Research, Cuba and the United States', *Social Studies of Science*, 1978, vol. 8, 4, 397–423.

N. Stepan, *Picturing Tropical Medicine*, London: Reaktion Books, 2001.

M. Warner, 'Hunting the Yellow Fever Germ: The Principle and Practice of Etiological Proof in Late Nineteenth-Century America', *Bulletin of the History of Medicine*, 1985, 59, 361–82.

L. G. Wilson, 'Fevers and Science in Early Nineteenth Century Medicine', *Journal of the History of Medicine and Allied Sciences*, 1978, 33, 386–407.

# 14

# RACE, DISEASE AND PUBLIC HEALTH

## Perceptions of Māori health

*Katrina Ford*

In the first decade of the twentieth century, Dr Peter Buck, also known by his Māori name of Te Rangi Hiroa, was concerned about the devastating impact of tuberculosis on Māori communities in New Zealand. Arguing for urgent measures to prevent the spread of the disease, Buck rejected the idea that Māori constituted 'new soil' for the tubercle bacillus.[1] 'Virgin soil' theories were prominent in early-twentieth-century discussions about tuberculosis amongst 'primitive' peoples, and could discourage public health efforts to combat the disease in indigenous communities.[2] To counter fatalism about Māori immunological weakness, Buck insisted that tuberculosis had existed amongst pre-European Māori, its effects controlled by their active, outdoor and sanitary mode of life. Buck believed sanitary reform could therefore arrest the impact of the disease: 'If we can prove that phthisis is not one of the many gifts of the pakeha [European], we can at all events concentrate our attention upon local conditions without the haunting fear of the "luxuriant growth in new soil" theory.'[3]

Buck's argument indicates how conceptions of the relationship between race and disease could have profound significance for public health practice. The nineteenth century saw the rise of modern epidemiological science, and the investigation and prevention of epidemic disease.[4] The analysis of disease in populations encouraged scientists to seek explanations as to why some groups of people were more susceptible to certain diseases than others. The nineteenth and early twentieth centuries were also an age of racial science, as scientists explained human diversity in terms of theories that organised peoples of the world into racial hierarchies.[5] Apparent variations in susceptibility to disease were among the most obvious and intriguing of the differences between races, and were part of the edifice of ideas now known as 'scientific racism'. The relationship between race and disease was a preoccupation of many doctors, public health officials, epidemiologists and scientists until at least the middle decades of the twentieth century. The words of Dr Arthur Thomson, a mid-nineteenth-century physician and scientist, in his account of Māori society, would have resonated with many scientists during this era: 'Intimately connected with the history of a people are the diseases which kill them, for although all mankind are born to die, the diseases which produce this result are dissimilar in different races.'[6]

Historians are also interested in examining the relationships between race and disease. Practitioners of the social history of medicine, with its understanding of medicine as bodies of knowledge linked to socio-political contexts, analyse the embedding of racial ideas within medical discourse, and the role of medical science in creating and affirming racial inequalities.[7] Analysis of tropical medicine in the late nineteenth

239

century has demonstrated how integral racial and imperial ideologies were to the creation of this medical sub-discipline.[8] Post-colonial scholars have been particularly interested in the entwining of race and disease in colonial discourses, justifying and enabling colonial policies of racial exclusion and segregation.[9] While the specific nature of these configurations varied from place to place, there are also common themes across places: as Christian W. McMillen has noted with regard to tuberculosis among Native Americans, medical theories about disease susceptibility 'transcended national and racial boundaries'.[10] In diverse spaces and places, historians have shown that ideas about race and disease were key components of socio-political hierarchies in the nineteenth and twentieth centuries.

This chapter demonstrates these themes through an examination of the configurations of race, disease and power in New Zealand in the late nineteenth and early twentieth centuries. In the British settler colony of New Zealand, health and race were foundational concepts. New Zealand was a 'racialized state', a product of an Empire that was 'increasingly ruled and organized through discourses and practices of race'.[11] New Zealand was also 'a healthy country', supposedly the British colony with the healthiest climate and the most favourable living conditions.[12] In the nineteenth century, immigration companies used mortality and morbidity statistics showing New Zealand to be healthier than any of the other British colonies to entice migrants.[13] New Zealand was the colony where the worst traits of the old society were abandoned and the very best stock of the British race was gathered, to create a new society and a new people – 'Better Britons'.[14] In the late nineteenth century, changes in New Zealand's political landscape saw the rise of the Liberal party, whose leading politicians sought to create a nation based on imperial ties and a homogenous 'British' racial identity.[15] New Zealand's burgeoning self-image was closely tied to health and racial values.

## Race and science

New Zealand's founding myths have been dependent upon a narrative of supposedly favourable race relations between Europeans and the indigenous inhabitants, known as Māori from the mid-nineteenth century.[16] The British Empire annexed New Zealand in 1840, during a period when paternalistic humanitarianism influenced the British Colonial Office.[17] There was a desire to avoid the previous mistakes and atrocities that occurred in dealings with indigenous peoples in other places, particularly in the Australian colonies. The popular opinion that the indigenous people of New Zealand were a 'superior type of native' supported these aims.[18] Some observers believed Māori demonstrated levels of intelligence and adaptability that would make it possible for them to become 'Brown Britons'.[19] Official discussions about the future of Māori in settler society were framed by ideas about racial amalgamation, the eventual absorption of the Māori into the European population.[20]

As was the case in other colonial settings, the arrival of Europeans precipitated a demographic decline for Māori.[21] From around 1800, there were reports of waves of epidemics causing substantial loss of life. Influenza, measles, whooping cough, bacillary dysentery, typhoid, tuberculosis and syphilis were among the diseases that decimated Māori communities in the nineteenth century. The scattered nature of settlement may have helped to prevent the development of major pandemics, at least until the mid-nineteenth century.[22] Some historians have suggested that nineteenth-century observers, whose perceptions were shaped by previous colonial experiences and assumptions about biologically inferior native bodies, exaggerated the degree of depopulation and 'fatal impact'.[23] Uncertainties about the Māori population at first contact with Europeans in the late eighteenth century, with estimates ranging from 86,000 to 175,000, also make it difficult to assess the extent of depopulation. However, even the lower end of these estimates indicates that incorporation into global networks of microbial exchanges was a disaster for Māori. The Māori population dropped to around 42,000 in 1896, before beginning to recover.[24]

Explanations for Māori susceptibility to disease reveal the mutually constitutive nature of concepts of race and disease. Discourses of 'the Dying Maori' were a prominent feature in nineteenth- and early-twentieth-century public discussions; indeed, 'the Dying Maori' persisted well into the twentieth century, long after demographic data indicated a steady recovery in the Māori population.[25] Reflecting the influence of evolutionary ideas in Victorian culture, many commentators regarded the looming extinction of Māori as an expression of the natural order of existence, whereby superior races replaced weak and inferior peoples.[26] Some argued that Māori had been dying out even before the arrival of Europeans. In one infamous account, Dr Alfred Newman, a prominent physician, businessman and politician, claimed that 'the Maoris were a disappearing race before we came here', due to them being a 'thoroughly worn out' race, whose approaching death 'has been hastened by the struggles with a newer and fresher race'.[27] In 1884, Dr Walter Buller, President of the Wellington Philosophical Society and celebrated ornithologist, discussed the inevitability of Māori extinction, suggesting to his audience: 'Our plain duty as good, compassionate colonists is to smooth down their dying pillow. Then history will have nothing to reproach us with.'[28] Explanations of Māori susceptibility to disease sometimes referred to inherent physiological weaknesses; it was a common observation, for example, that Māori had 'weak chests', making them prone to respiratory diseases, especially phthisis.[29] Such arguments had a large element of self-justification, absolving Europeans of responsibility for Māori demographic decline by reference to supposed physical weaknesses.[30] The 'Dying Maori' was a convenient trope for settlers who looked forward to a time when Pakeha could exert sovereignty without opposition from an inconvenient indigenous population. As the missionary James Buller noted of the 'popular opinion' that Māori were dying out: 'With many, I fear, the "wish is father to the thought'.'[31]

However, views which rooted Māori susceptibility to disease in a biologically inferior body were but one strand in the complex conglomeration of ideas that was racial science in New Zealand. James Hector, Director of the Colonial Museum and unofficial leader of the scientific community in New Zealand, was critical of the argument that Māori decline was due to 'an inherent tendency to decay', which he decried as an attempt 'to excuse ourselves by any other natural law but that of might'.[32] Instead, he laid the fault for Māori ill-health at the foot of colonisation, which had left Māori adrift between their old ways and the modern world: 'we have destroyed their social organization and not replaced it with our own'.[33] Other commentators voiced similar opinions; Māori had been a strong and healthy race, but the disintegration of their society through the impact of European arrival had created an environment where disease thrived.[34] These arguments indicate the various shades and dimensions of racial thought in New Zealand in the nineteenth and early twentieth centuries.

Rather than being inherently biological, scientific racism could also take on a cultural guise. Theories of racial immunity or susceptibility to disease were composed of a complex mixture of biological and cultural factors.[35] Many commentators regarded what Warwick Anderson has termed 'the cultural liabilities of race' as explanation enough for variations in disease.[36] Rather than emphasising innate biological factors, the habits and customs of inferior races explained why they succumbed to disease. These cultural differences often took on an essential and fixed dimension within racial discourses, 'as if culture was "in the genes".'[37] In New Zealand, 'the cultural liabilities of race' were a dominant feature of Pakeha explanations of Māori ill-health in the nineteenth and early twentieth centuries. Public officials and numerous self-appointed experts emphasised the unhealthy conditions of Māori life. Inappropriate clothing, poorly ventilated *whare* [houses], bad diet, insistence on living in unhealthy localities, and lack of attention to cleanliness and basic sanitation were the main factors thought to explain Māori depopulation.[38] The gathering of people at *hui* [meetings] and *tangi* [funerals] also came in for particular criticism from European observers. In 1881, Major Gilbert Mair, Native Officer for the Auckland and Waikato region, claimed that the annual meetings of the Waikato tribes 'generate disease from the over-crowding and over-feeding and poverty that follows'.[39] Another official in 1886 blamed ill-health in his district on the fact that locals were spending too much time and energy on 'political meetings, discussions, and plottings'.[40] Officials rarely mentioned the social and economic devastation many Māori communities experienced due to war and land confiscation as reasons for ill-health. While some recognised a connection between Māori poverty and sickness, they usually attributed poverty to lack of industry, laziness and an inherent inability to plan for the future. Māori were therefore responsible for their own demise. After listing the numerous sanitary transgressions observed in the local Māori community, one government official lamented in 1884: 'What can one hope to do with people who are so foolish as to persist in a course which can only be described as suicidal in the extreme?'[41]

Others with an interest in Māori health believed that while the situation was dire, the cause was not hopeless. James Pope, Inspector of Native Schools, believed Māori could 'escape extermination' if they rejected the unhealthy aspects of their culture and emulated the healthy ways of Europeans.[42] This perspective formed the core of Pope's 1884 book, *Health for the Maori*, written for the instruction of Māori children in Native Schools. The provision of pure air and clean water were the most significant remedies needed to improve Māori health, reflecting the sanitarian approaches of British public health.[43] Pope's analysis combined these sanitary understandings with Victorian middle-class morality. He blamed drinking, improper dancing, and irregular lifestyles for Māori ill-health.[44] The substitution of *tangi*, *hui* and a communal, subsistence way of life with European-style funerals, picnics and tea parties, individualised land holding and regular work was also crucial to Māori health reform. Pope's ideas about Māori health had an important influence on a generation of reformers committed to saving the race through the modernisation of Māori communities.[45]

## Public health and Māori health reform

The beginning of the twentieth century signalled a new era for public health in New Zealand.[46] With the establishment of the Department of Public Health in 1901, public health officials sought to impress upon the public an awareness of the individual's duty in maintaining public health. Infectious diseases could undermine the robust settler; people should not take the healthy country for granted, but needed to actively defend it through the performance of hygienic rituals. Self-discipline and the internalisation of hygienic imperatives were vital. Constant vigilance was required to maintain public health and ensure the progress and prosperity of the nation.

This new model of preventative medicine provided impetus for the reform of sanitary conditions and hygienic behaviour in Māori communities. Māori health reform was a particular interest of Dr James Malcolm Mason, the Chief Health Officer.[47] In 1901, Mason appointed Dr Maui Pomare to the position of Native Health Officer, with the task of transforming sanitary conditions in Māori communities. Pomare was an ex-pupil of Te Aute College, the elite Anglican Māori college, and had graduated from the American Medical Missionary College in Chicago in 1899, becoming the first qualified physician of Māori descent.[48] Pomare emphasised the need for Māori to engage with modern concepts of hygiene if the race was to survive, and he criticised those aspects of Māori culture he believed were detrimental to health. Yet he also believed Māori could not be forced into changes that they did not understand or agree with. Therefore, the key to effecting transformation in Māori communities was to adapt modern methods of sanitation to Māori knowledge and traditions.[49] In 1905, Dr Peter Buck, a graduate of Otago Medical School, joined Pomare in the Department. Like Pomare, Buck was of European and Māori ancestry, had been educated at Te Aute College and was a firm advocate of the need for Māori to engage with the Pakeha world to survive.[50] His approach to Māori health reform was also grounded in the belief that change could only be successful if Māori were themselves convinced of the principles of modern hygiene.[51] He embarked upon an intensive study of Māori language and traditions to make his arguments for sanitary reform more persuasive.[52] This inquiry laid the basis for Buck's later career in anthropology.

Pomare and Buck's emphasis upon Māori engagement with modern sanitary science was reflected in another development in Māori health reform in the early twentieth century, the Maori Councils Act of 1900.[53] This legislation was driven by James Carroll, a Liberal politician of mixed Māori and European descent, and the 'Young Maori Party', a new generation of Māori leadership, which included Pomare and Buck.[54] The Maori Councils Act set up a system of Māori local bodies, with responsibility for limited governance in their communities. Village Committees, or Komiti Marae, were to oversee the administration of important health measures such as sanitation and water supplies in each *kainga* [settlement]. Pomare worked closely with the Councils in his first years as Native Health Officer, trying to ensure his messages of sanitary reform were understood and implemented.[55]

Along with the Māori Councils, the Department of Public Health appointed a cadre of native sanitary inspectors to assist Pomare and Buck.[56] Unlike their European equivalents, these were not men with formal qualifications in sanitary science, but leaders whose communities would respect their opinions and authority. As the historian Raeburn Lange has suggested, their knowledge of, and commitment to, the sanitary principles they were supposed to be espousing might not have been complete, but the influence and experience these men had in their communities made them significant figures in the campaign for Māori health reform.[57]

For Buck and Pomare, the knowledge of disease offered by Western medical science was a powerful tool that needed to be in the hands of Māori. A well-known example of Pomare's public health education methods illustrates the centrality of germ discourses in his campaign. At a *hui* in the Hawkes Bay region, Pomare was attempting to bring his message of sanitary reform to the people:

> he began by condemning the water supply which was generally used for drinking purposes at the meeting. He was about to be bundled out of the meeting house, when he pleaded with the people to bring him a bucket of water from their drinking well . . . he fixed a microscope on top of the bucket and invited the people to see for themselves what they were drinking . . . Those who looked in the bucket, scattered around, spitting on the ground and cried out saying *E hika ma e; Kei te Kai tatau I te ngarara*. (O people; We are eating live creatures.)[58]

This story, which was widely reported in newspapers at the time, was a dramatisation of the revelatory power of modern science.[59] Pomare's performance with bucket and microscope exemplified his approach to Māori health reform; the adaptation of modern scientific methods and concepts to a Māori context.

Sanitary reform in Māori communities involved criticisms of some of the most fundamental institutions and practices of Māori culture. Public health officials now condemned Māori communal life because it encouraged the exchange of germs of disease.[60] The practice of *whanau* sleeping together in single-roomed *whare*, and the gathering of people at *hui* and *tangi* were constructed as sites for the growth and spread of disease germs. In a paper given by Pomare in 1904 at a Te Aute College Students' Association Conference, bacteriological knowledge and Pomare's rhetoric repackaged and strengthened nineteenth-century criticisms of Māori customs:

> Let us trace one of these germs [of tuberculosis] in one of its journeyings. Here is a Maori pa [fortified village], with everybody in it free and happy. A tangi party arrives . . . and among them is an old crone who is coughing and spitting about the floor . . . Each expectoration . . . has millions in it. Our lassies have to turn out at all hours, and in all conditions of weather, to get kai [food] for the visitors . . . These poor girls have been unconsciously preparing soil for a crop of germs, and as the old crone's sputa dries . . . sure enough a crowd of witnesses arise in the form of germ-laden dust, to testify to the evils of tangis, and crowding of whares. Poor Hine breathes in the death-charged dust as she sweeps merrily along, but soon, alas! They chant the death wail over her, as she is gathered to her forefathers, before her three score and ten years.[61]

Moving from the vague criticisms of the late nineteenth century that such gatherings led to disease, to a more precise exposition on how the tubercle bacillus claimed its victims at *tangi* and *hui* helped provide a scientific foundation for criticisms of Māori cultural practices as dangerous and uncivilised.

Māori resistance to isolating the victims of infectious diseases brought the sternest condemnation from public officials. Isolation was not a new public health technique, but it was intensified by the development of germ theories of disease, as bacteriological science offered new knowledge about how infectious diseases were spread.[62] Officials highlighted the refusal of Māori to relinquish customs of visiting the sick and prolonged physical contact with corpses during mourning as major transgressions. Both Native School teachers and Native Health nurses discussed the difficulty of introducing concepts of isolation into communities where communal responses to disease and death were prevalent.[63] The teacher at Otamauru Native School observed: 'Many Maoris have a distinct prejudice against isolation. It appears to be a question of etiquette that the healthy should visit the sick regardless of the nature of the disease or the welfare of the patient.'[64] Nurse Amelia Bagley, who helped to establish the Native Health Nursing Service, referred to problems implementing isolation to demonstrate the difficulties of nursing among Māori:

> half of these people are almost uncivilised. You know there were six deaths from this fever. They had a tangi over one for days, and with the children, the Korow people slept with the corpse. So you cannot expect me to work with them as I might white folk.[65]

Often, a nurse's first priority on entering a Māori community was to attempt to introduce boundaries between the sick and the healthy:

> in the patient's room I found six women, as many children (their ages ranging from a month to seven years), and as many men as could crowd in, while the overflow filled the doorway . . . My first business was to clear the patient's room.[66]

Public health officials complained generally about the difficulty of persuading people to comply with isolation practices, but in the case of Māori communities, laxity in isolation was regarded as evidence of the inferior and illogical nature of the Māori mind. Descriptions of Māori behaviour in cases of infectious disease were often redolent with the disgust and anxiety many European observers felt when confronted with Māori communal customs. For the teacher at Kawhia Native School, the failure of the community to practice isolation was an indication of the failings of the Māori race as a whole:

> Last night when I went down to Maketu I saw a sight which would have opened the eyes of those who think Maoris are amenable to reason. I have in the past few years been trying to show them the danger of keeping corpses too long, of persons going near them or indeed near persons suffering from typhoid but it seems to have been without avail. In a bed on the floor were lying in a space of about 5 ft wide, the dead body of Poutu, Pouaka (suffering from typhoid . . .) Ngahete the father and Martha just recovered from the fever. At the foot of the bed lay the coffin. There were several others in the whare.[67]

Such cultural criticisms closely mirrored middle-class public health officials' disapproval of the customs of working-class and immigrant communities in Europe and North America.[68] For these officials, the failure to isolate cases of infectious disease not only indicated a failure in health and sanitation, but also represented an inherently immoral and degraded communal lifestyle, when demarcations between the healthy and the ill, between young and old, between sleeping, living and eating, and indeed, between life and death, were not recognised or respected. In the context of a colonial society, Māori bore the brunt of fears about the dangers of disease and disorder which might result from such transgressions.

## The 'Māori menace' to public health

Despite an initially optimistic outlook regarding Māori health reform, public health officials demonstrated increasing frustration and impatience towards Māori by the second decade of the twentieth century. Pomare and Buck left the Public Health Department for political careers, and other officials lacked their understanding of the barriers to sanitary reform in Māori communities.[69] These frustrations reflected an important change in the wider significance of infectious diseases in Māori communities. Public health officials increasingly represented Māori as a threat to European health in the early twentieth century.[70] This was a shift from the nineteenth century, when European commentary on unhealthy Māori customs had focused mainly on the threat to Māori health. Claims that specific outbreaks of infectious disease among Europeans had originated in Māori communities appeared in the reports of the Department of Public Health as early as 1902.[71]

From 1911, identification of Māori as the source of epidemics of infectious disease became more strident. In official discourses, Māori were repeatedly described as 'a menace' to public health.[72] In comments published in newspapers, the Auckland District Health Officer (DHO) and Government Bacteriologist, Dr Robert Makgill, described the danger from epidemics among Māori and claimed 'under present conditions these people form a constant menace to the rest of the population'.[73] Likewise, Dr Herbert Chesson, Wellington DHO, identified Māori transgressions of the most basic hygienic behaviour as a source of danger to Māori and Europeans:

> Privies are rarely found, and the habits of these people in depositing their excreta about the outskirts of the *pas* [fortified villages] and settlements constitute a serious menace to the public health not only of the Māori race, but of the Europeans near whom they live.[74]

Fears about the spread of disease from Māori to Europeans were part of a context of expanding European settlement in many parts of the North Island at the turn of the century, as the impetus of New Zealand's increasing involvement in the global dairy trade 'opened up' remote areas to agricultural development.[75] Hester Maclean, the Assistant Inspector of Hospitals, explicitly referred to the 'serious menace to the growing European population' in isolated districts, as settlement exposed Pakeha to the unsanitary conditions in which Māori lived.[76] Rural Māori communities were problematic because they represented both an economic and a sanitary obstacle to the civilising imperatives of settler society.

The construction of typhoid as a rural Māori problem in the early twentieth century demonstrates the interaction of racial anxieties with aetiological concepts and new public health methods. Public health officials most often invoked the Māori menace in connection with the presence of typhoid in Māori communities. Their concerns reflected changes in the explanations of typhoid epidemics. While older explanations tended to attribute outbreaks of enteric fever to the general presence of filth in the environment, the new approaches of public health regarded typhoid as the result of infection with a specific micro-organism. Typhoid was still associated with filth, but there was more focus on identifying the exact sources of typhoid bacilli. This approach made public health officials and sanitary inspectors more aware of the presence of typhoid in Māori communities.

The decline in rates of typhoid within European communities in the early twentieth century also highlighted the Māori association with the disease. As local councils constructed the drainage and sewerage infrastructure that reduced the possibility of infection, typhoid gradually ceased to be endemic in the main cities. Public health officials came to regard the presence of typhoid as an embarrassing indictment on a community's sanitary standards, evidence of primitive and uncivilised behaviour.[77] The historian Anne Hardy observed that connections between typhoid and notions of civilisation, 'and by association, barbarism', were common in early-twentieth-century discussions of the disease.[78] These associations made typhoid outbreaks within rural Māori communities more obvious and more troubling. In 1911, in a heated exchange of memos with the Under Secretary for Native Affairs, Makgill repeatedly raised the danger of typhoid spreading from Māori communities to Europeans:

> I only know that Natives are allowed to die right and left without registration, notification, or supervision. They are a danger to their white neighbours, and it seems to me an extraordinary thing that any part of the population should be living and dying under these conditions.[79]

Makgill's anger was directed both at what he perceived as the neglect of Māori sanitary conditions by the Native Department, and the effect this neglect might have on the Pakeha population.[80] Māori typhoid was an affront to a civilised society.

The identification of Māori communities as hotbeds of typhoid became so prevalent that any account of illness amongst Māori could be magnified into a typhoid outbreak. The *Evening Post* reported in 1901 that Dr Pomare visited a settlement near Napier following reports of an 'alarming outbreak of typhoid', but found nothing worse than 'the prevailing epidemic of influenza'.[81] In 1908, the Department instructed Pomare to visit settlements on the Whanganui River following reports of an epidemic of typhoid fever among Māori, but he telegraphed back to Wellington: 'Absolutely no foundation for alarming reports.'[82] Some Māori were sensitive to the negative impact such unfounded reports could have on perceptions of Māori. Discussing an incident where a European had mistakenly reported an outbreak of typhoid in a village, Māori Sanitary Inspector Riapo Puhipi complained such 'false reports . . . give a bad name to the Maoris'.[83] The claim that an outbreak of typhoid in 1913 in the Waiapu district near Gisborne was due to a recent large *hui* in the area provoked a strong reaction from locals and public officials; Apirana Ngata, the local Māori Member of the House of Representatives (MHR), described the accusation as 'a libel'.[84]

Officials and observers usually blamed typhoid outbreaks on Māori unwillingness to adhere to basic standards of hygiene and sanitation. This is not to say that public health officials were uncritical of European habits. But the public health discourses of Māori sanitary shortcomings presented Europeans as the unwitting victims of Māori habits. According to public health officials, Māori polluted water sources with excrement and household filth, which were then used by unsuspecting Europeans; Māori living in filthy conditions sold oysters contaminated with the typhoid bacillus to Europeans; flies from Māori villages carried pathogenic germs to the food of Europeans; and Europeans drank infected milk from dirty Māori dairy farms.[85] Many of these sources of infection only became apparent or acquired greater significance with the development of germ theories of disease. Bacteriological knowledge had multiplied the ways in which the germs of disease were known to gain access to the human body. Even the most vigilant citizen was at risk from the presence of a race of people who were unwilling or unable to adhere to the basic precepts of modern hygiene. Unlike Pomare and Buck, Makgill was unconvinced of the ability of Māori to respond appropriately to health education and to uphold sanitary laws.[86] He complained that 'the native soon wearies of well-doing' and therefore needed 'constant supervision'.[87] In a particularly irate moment in 1911, he suggested to the Chief Health Officer, Dr Thomas Valintine, that the only answer to the Māori menace might be to follow the American example and segregate Māori into native reservations.[88] In his view, the native mind was unable to maintain the self-discipline and vigilance that was necessary to uphold the public health. Such views depended upon racial constructions of natives as irresponsible, lazy and lacking in self-discipline, all qualities inimical to the requirements of modern public health.[89]

The intersection of racial discourses and bacteriological ideas was particularly apparent in discussion of Māori typhoid carriers.[90] In the context of the Philippines, Warwick Anderson noted the increasing identification of germ carriage with the native body in the early twentieth century, a combination perceived as particularly dangerous because of the unhygienic racial customs and habits of the Filipinos.[91] Similar associations were evident in racial discourse on typhoid in New Zealand. In 1913, an editorial in the *Poverty Bay Herald* called on the Department of Public Health to address the problem of identifying and controlling the undiagnosed '"typhoid" Māori', who was responsible for spreading the disease far and wide amongst 'the Natives, the half-castes and the Europeans'.[92] This may have been a play on words from the infamous typhoid carrier 'Typhoid Mary', whose arrest and incarceration in the United States was receiving worldwide public attention at this time; the author called on the public health authorities to isolate typhoid carriers 'as is now being done in American cities'. According to the article, Māori typhoid carriers were more dangerous, because the germs they carried were more virulent, whereas in Europeans, 'the virus becomes attenuated' due to successive generations of early and continued treatment. Thus, the newspaper used popular interpretations of bacteriological concepts to construct a pathological native body.

This racially specific construction of the typhoid carrier was also encouraged by pronouncements from the Department of Public Health. In 1914, the *Evening Post* newspaper carried a report of an interview with Makgill, in which he advised against the formation of a contingent of Māori volunteers for the war. According to Makgill, the presence of typhoid carriers amongst Māori volunteers would almost inevitably result in an outbreak.[93] When questioned why this was not also a problem in the main expeditionary force, Makgill argued that the problem of typhoid carriers was particular to Māori because the 'habits' of 'white troops . . . were not so conducive to the spread of infection'.[94] According to the *Auckland Star*, Makgill warned 'there would be an immediate danger to all who came into contact with Maori volunteers travelling from any of the Auckland settlements'.[95] Makgill's arguments racialised the carrier problem. Māori bodies were far more likely than Europeans to carry the germs of typhoid, and Māori habits made the carrier state in Māori far more dangerous to the public health. Māori hygienic shortcomings meant that, in Makgill's eyes, they were ineligible for the one of the greatest responsibilities of the male citizen in the early-twentieth-century state – military service.

These changing scientific understandings of disease and fears of racial contamination contributed to shifts in public health strategies towards Māori. Public health officials became more willing to use the coercive power of the state to compel Māori to comply with sanitary ideals.[96] Public health measures targeting Māori came to the fore most dramatically during the 1913 smallpox epidemic. Apart from a couple of small outbreaks in the late nineteenth and early twentieth centuries, New Zealand had remained free of smallpox, and the 1913 epidemic was the first time the disease had impacted upon Māori communities. Smallpox probably entered the country with an American Mormon missionary proselytising amongst Māori and the Māori communities in the districts immediately north and south of Auckland were worst affected. The epidemic was a 'racializing moment', a point at which the construction and contestation of racial difference in New Zealand can be examined.[97] Travel restrictions and quarantine measures aimed specifically at Māori reflected the widespread belief that they were responsible for the outbreak. Under Section 18 of the 1908 Public Health Act, the Public Health Department issued proclamations forbidding Māori and half-caste Māori from travelling from infected areas, unless they possessed a certificate from a public vaccinator or doctor testifying to a successful vaccination.[98] The railways, trams, taxis and shipping companies could not carry Māori unless they had a certificate and the Borough Councils of Onehunga and Cambridge voted to prevent Māori entering these towns.[99]

Many people expressed the desire for more extreme methods of segregation and control. In Parliament, some members suggested more stringent measures were required. These ranged from prohibiting all Māori, even those vaccinated, from travelling on railways, to the suggestion that all Māori travelling on public transport should be fumigated, based on the belief that even those who were vaccinated could still 'carry germs about their persons'.[100] Some newspaper commentary on the outbreak expressed virulent racial antagonism: 'In filthy and insanitary Māori villages the disease roots itself, thrives and extends. Diseased Māoris wander over the country, stray into the towns, work on the wharves, loiter on the streets without any action being taken until the disease is widespread.'[101] There were complaints that Māori children in districts far from the areas affected by smallpox were being excluded from public schools because of fears they would infect European children.[102] The fear and loathing so evident in public discussion is a long distance from the benevolent paternalism that characterised most public comment on Māori health problems in the late nineteenth century. In the interim, fears about contamination had become much more prevalent, encouraged by public health policies on the control of infectious disease, and public pronouncements about the danger of infection from Māori communities.

Public authorities rigorously enforced the restrictions, especially those relating to rail travel, against all Māori, regardless of their individual circumstances and status. Some people contested this attempt to create a strict dichotomy between Māori and Pakeha in the interests of public health. In Parliament, complaints were raised by some MHRs, who argued that half-castes in their districts who lived European, that is presumably sanitary, lifestyles were not a public health threat, and were suffering 'indignities' under the regulations. In one case, a man 'who had been living for some years as a European, with a European wife and who employed several Europeans on his farm', had attempted to travel into Gisborne with his employees, but was refused permission to enter the train.[103] These complaints reflected a belief that half-castes who conformed to European expectations of the assimilated native deserved exemption from discriminatory regulations.[104] The stipulation that the regulations applied to half-castes but not quarter-castes could lead to disputes over racial identity. For example, Alfred A. Yates telegraphed William Herries, the Minister of Railways, claiming he was a quarter-caste and should be permitted to travel by train.[105] Local officials disputed this, claiming that the 'general opinion' was that Yates was a half-caste, not a quarter-caste.[106] The historian James Bennett has observed that in New Zealand, racial classifications such as half-caste were used in a very loose sense compared to other parts of the colonial world.[107] While this may have been the case generally, in the 'racial moment' of the smallpox epidemic, public health strategies attempted to strictly enforce boundaries between Māori and Europeans, and degrees of blood mixture had significance for personal freedom of movement. Enforcing the boundary between contamination and purity could erase the negotiated identities that lay between Māori and Pakeha.[108]

Opposition to the quarantine regulations from some Māori interpreted the measures as evidence of the marginalisation of Māori from social and political power. A letter to the *Evening Post*, signed 'Heio (a half caste Maori)', described the Department of Health's actions as evidence of the decline of Māori power and status relative to the Pakeha since the wars of the nineteenth century: 'In those days we were the power in the land and were all in all to you. Today the position is reversed, and it would seem that we must henceforth be viewed as something more akin to animals.'[109] In a letter to the Railways Department, the Reverend Himepiri Munro complained that a station guard forced him off a train: 'I need not enlarge the humiliating position in which we were placed, being put off the train like a lot of sheep before a midday travelling public.'[110] Others complained that the regulations encouraged racial discrimination and segregation, raising the spectre of 'the color line' being drawn between Māori and Pakeha.[111] In a letter to the editor of the *Poverty Bay Herald*, Rawiri Karaha highlighted the segregationist intent of the regulations and linked them to a hardening of racial attitudes in the country: 'there is a growing population in our land who are inclined to favour the oppression of the Maori'. He argued that the regulations seemed to be at odds with the government's professed desire for Māori and Pakeha to be treated equally: 'I had thought that a Government so solicitous for the equal political grading of the two races would be the last to connive at a regulation whose burden is the separation of the two races.'[112] For these people, the construction of Māori as dangerous disease carriers had troubling implications for the future of relations between Māori and Pakeha in New Zealand.

## Conclusion

Alison Bashford's assertion that 'the pursuit of "health" has been central to modern identity formation' rings particularly true in the case of New Zealand.[113] More than any other of the British colonies, New Zealand's identity was entwined with its reputation as 'a healthy country'. Therefore, Māori, as a supposedly unhealthy and insanitary race, did more than just threaten the public health; they threatened the very identity of the nation as it was being defined by politicians, public officials and social commentators in the late nineteenth and early twentieth centuries. Racial difference was a challenge to the health and homogeneity of Pakeha society.

Public health attitudes to the problem of Māori ill-health indicate some of the more divisive impulses beneath the assimilationist agenda that dominated the New Zealand state's racial policies in the early twentieth century. It has been said that in Pakeha constructions of the native race, the 'dying Māori' was replaced by the 'whitening Māori', who was amenable to Europeanisation and therefore full participation in New Zealand society.[114] Pomare's microscope and bucket of water symbolised the Māori ability to understand the gift of modern medical science. However, behind this lurked the 'Typhoid Māori', roving around the countryside saturated with the germs of diseases, whose inherent inability to learn and maintain hygienic behaviour placed them outside the boundaries of civilised society. Europeans interpreted Māori determination to continue to practise social customs that transgressed hygienic mores as evidence of their essentially uncivilised nature. Bacteriological conceptions of disease, with their menacing emphasis on hidden means of infection and their accent upon the human body as the most potent source of pathogens, amplified European uncertainty about the place of Māori in modern New Zealand society. Configurations of the relationship between race and disease thus had important consequences for public health policies and practices, as well as wider significance for the constitution of racial hierarchies in New Zealand society.

## Notes

1 Public Health, *Appendices to the Journals of the House of Representatives* (AJHR), 1908, H-31, p. 131; P. H. Buck, 'Medicine Amongst the Maoris in Ancient and Modern Times: A Thesis for the Degree of Doctor of Medicine (N.Z.) by "Abound"', 1910, p. 72.
2 M. Worboys, 'Tuberculosis and Race in Britain and its Empire, 1900–50', in W. Ernst and B. Harris (eds), *Race, Science and Medicine, 1700–1960*, London: Routledge, 1999, p. 145; C. W. McMillen, '"The Red Man and the White Plague": Rethinking Race, Tuberculosis, and American Indians, ca.1890–1950', *Bulletin of the History of Medicine*, 82, 3, 2008, 608–45, 613.
3 Public Health, AJHR, H-31, 1908, p. 131.
4 L. Wilkinson, 'Epidemiology', in W.F. Bynum and R. Porter (eds), *Companion Encyclopedia of the History of Medicine*, New York: Routledge, 1993, pp. 1272–3.
5 N. Stepan, *Idea of Race in Science: Great Britain 1800–1960*, London: Macmillan, 1982.
6 A.S. Thomson, *The Story of New Zealand*, Vol. 1, London, 1859, p. 211.
7 This literature is extensive. See, for example, D. Arnold, *Colonizing the Body: State Medicine and Epidemic Disease in Nineteenth Century India*, Berkeley: University of California Press, 1993; M. Harrison, *Climates and Constitutions: Health, Race, Environment and British Imperialism in India 1600–1850*, New Delhi: Oxford University Press, 1999; A. M. Kraut, *Silent Travellers: Germs, Genes, and the "Immigrant Menace"*, New York: Basic Books, 1994; K. Ott, *Fevered Lives: Tuberculosis in American Culture since 1870*, Cambridge, MA: Harvard University Press, 1996; R. M. Packard, *White Plague, Black Labor: Tuberculosis and the Political Economy*

*of Health and Disease in South Africa*, Berkeley: University of California Press, 1989; N. Shah, *Contagious Divides: Epidemics and Race in San Francisco's Chinatown*, Berkeley: University of California Press, 2001; J. Brown, 'Purity and Danger in Colour: Notes on Germ Theory and the Semantics of Segregation, 1895–1915', in J. Gaudilliere and I. Löwy, (eds), *Heredity and Infection: The History of Disease Transmission*, London: Routledge, 2001, pp. 101–32; H. Deacon, 'Racism and Medical Science in South Africa's Cape Colony in the Mid-to-Late Nineteenth Century', *Osiris*, 2nd Series, 15, 2001, 190–206; and essays in D. Arnold (ed.), *Imperial Medicine and Indigenous Society*, Manchester: Manchester University Press, 1989; Ernst and Harris, *Race, Science and Medicine*; and in R. MacLeod and M. Lewis (eds), *Disease, Medicine, and Empire: Perspectives on Western Medicine and the Experience of European Expansion*, London: Routledge, 1988.

8 W. Anderson, 'Immunities of Empire: Race, Disease and the New Tropical Medicine', *Bulletin of the History of Medicine*, 70, 1, 1996, 94–118; D. Arnold (ed.), *Warm Climates and Western Medicine: The Emergence of Tropical Medicine, 1500–1900*, Amsterdam: Rodopi, 1996.

9 W. Anderson, *The Cultivation of Whiteness: Science, Health and Racial Destiny in Australia*, Melbourne: Melbourne University Press, 2002, and *Colonial Pathologies: American Tropical Medicine, Race and Hygiene in the Philippines*, Durham, NC: Duke University Press, 2006; A. Bashford, *Imperial Hygiene: A Critical History of Colonialism, Nationalism and Public Health*, New York: Palgrave Macmillan, 2004; L. Manderson, *Sickness and the State: Health and Illness in Colonial Malaya, 1870–1940*, Cambridge: Cambridge University Press, 1996; M. Vaughan, *Curing Their Ills: Colonial Power and African Illness*, Cambridge: Polity Press, 1991.

10 McMillen, '"The Red Man and the White Plague"', p. 613.

11 D.I. Salesa, *Racial Crossings: Race, Intermarriage and the Victorian British Empire*, Oxford: Oxford University Press, 2011, p. 17.

12 J. Belich, *Making Peoples: A History of the New Zealanders from Polynesian Settlement to the End of the Nineteenth Century*, Auckland: Penguin Press, 1996, pp. 289–300; M. Fairburn, *The Ideal Society and Its Enemies: The Foundations of Modern New Zealand Society 1850–1900*, Auckland: Auckland University Press, 1989; R. Grant, 'New Zealand "Naturally": Ernst Dieffenbach, 'Environmental Determinism and the Mid Nineteenth-Century British Colonization of New Zealand', *New Zealand Journal of History*, 37, 1, 2003, 22–37, 25–7; T. D. Selesa, '"The Power of the Physician": Doctors and the "Dying Maori" in Early Colonial New Zealand', *Health and History*, 2001, 3, 13–40, 26–30.

13 L. Bryder, 'Introduction', in L. Bryder (ed.), *A Healthy Country: Essays on the Social History of Medicine in New Zealand*, Wellington: Bridget Williams Books, 1991, p. 4; M. Nicolson, 'Medicine and Racial Politics: Changing Images of the New Zealand Maori in the Nineteenth Century', in Arnold, *Imperial Medicine*, p. 83.

14 J. Belich, *Paradise Reforged: A History of the New Zealanders from the 1880s to the Year 2000*, Auckland: Allen Lane, 2001, Part 1.

15 Ibid., pp. 216–32.

16 Ibid., pp. 189–90.

17 A. Lester, 'British Settler Discourse and the Circuits of Empire', *History Workshop Journal*, 54, 1, Autumn, 2002, 25–9; R. McGregor, 'Degrees of Fatalism: Discourses on Racial Extinction in Australia and New Zealand', in P. Grimshaw and R. McGregor (eds), *Collisions of Cultures and Identities: Settler and Indigenous People*, Melbourne: University of Melbourne, 2007, pp. 246–7; Nicolson, 'Medicine and Racial Politics', pp. 75–6; A. Ward, *A Show of Justice: Racial 'Amalgamation' in Nineteenth-Century New Zealand*, 2nd edn., Auckland: Auckland University Press, 1995, pp. 34–6; Salesa, *Racial Crossings*, ch. 1.

18 Belich, *Paradise Reforged*, pp. 206–7; Grant, 'New Zealand "Naturally"', p. 26; R. Lange, *May the People Live: A History of Maori Health Development 1900–1920*, Auckland: Auckland University Press, 1999, p. 60; Pat Moloney, 'Savagery and Civilization: Early Victorian Notions', *New Zealand Journal of History*, 35, 2, 2001, 153–76, 161–3; Nicolson, 'Medicine and Racial Politics', p. 68.

19 Belich, *Paradise Reforged*, pp. 209–10.

20 M.P.K. Sorrenson, 'Maori and Pakeha', in G. Rice (ed.), *The Oxford History of New Zealand*, 2nd edn., Auckland: Oxford University Press, 1992, pp. 142–3; Ward, *A Show of Justice*. For discussion of state policies of assimilation in the early twentieth century, see R. S. Hill, *State*

*Authority, Indigenous Autonomy: Crown–Māori Relations in New Zealand/Aotearoa 1900–1950,* Wellington: Victoria University Press, 2004.

21  I. Pool, *Te Iwi Māori: A New Zealand Population Past, Present and Projected,* Auckland: Auckland University Press, 1991, pp. 42–6; A.W. Crosby, *Ecological Imperialism: The Biological Expansion of Europe, 900–1900,* 2nd edn., Cambridge: Cambridge University Press, 2004, ch. 10.

22  Pool, *Te Iwi Māori,* p. 46.

23  Belich, *Making Peoples,* pp. 173–8.

24  Ibid., p. 78; Pool, *Te Iwi Maori,* ch. 5.

25  D. A. Dow, *Maori Health and Government Policy 1840–1940,* Wellington: Victoria University Press, 1999, p. 84; Lange, *May the People Live,* pp. 53–62; Nicolson, 'Medicine and Racial Politics', pp. 83–92.

26  J. Belich, *The New Zealand Wars and the Victorian Interpretation of Racial Conflict,* Auckland: Auckland University Press, 1986; J. Stenhouse, 'Darwinism in New Zealand, 1859–1900', in R. L. Numbers and J. Stenhouse (eds), *Disseminating Darwinism: The Role of Place, Race, Religion, and Gender,* Cambridge: 1999, pp. 61–89.

27  A.K. Newman, 'A Study of the Causes Leading to the Extinction of the Maori', *Transactions and Proceedings of the New Zealand Institute,* 14, 1882, 459–77. For more detailed discussion of Newman's paper, and its place and influence within New Zealand history, see J. Stenhouse, '"A disappearing race before we came here": Doctor Alfred Kingcome Newman, the Dying Maori, and Victorian Scientific Racism', *New Zealand Journal of History,* 30, 2, 1996, 124–40.

28  'Wellington Philosophical Society, Address by the President', *Transactions and Proceedings of the New Zealand Institute,* 17, 1884, 556. Buller's now infamous 'dying pillow' comment was a paraphrase of a remark made by the Wellington politician Dr Isaac Featherston several years earlier.

29  See, for example, Newman, 'A Study of the Causes', p. 475; W. P. Reeves, *The Long White Cloud,* 4th edn., London, 1898, p. 60.

30  Lange, *May the People Live,* pp. 59–60.

31  J. Buller, *Forty Years in New Zealand,* London: Hodder and Stoughton, 1878, p. 164.

32  Proceedings of the Wellington Philosophical Institute, Sixth Meeting, *Transactions and Proceedings of the New Zealand Institute,* 14, 1881, 539.

33  Ibid. See also J. Stenhouse, '"A disappearing race before we came here"', pp. 124–40.

34  F. von Hochstetter, *New Zealand: Its Physical Geography, Geology and Natural History,* Stuttgart, 1867; Thomson, *Story of New Zealand,* p. 216; Archdeacon Walsh, 'The Passing of the Maori: An Inquiry into the Principal Causes of the Decay of the Race', *Transactions and Proceedings of the New Zealand Institute,* 40, 1907, 154–75.

35  W. Ernst, 'Introduction', in Ernst and Harris, *Race, Science and Medicine,* pp. 5–6.

36  Anderson, 'Immunities of Empire', pp. 109–13.

37  Ernst, 'Introduction', p. 6.

38  See for example, the reports from Native Department officials, which repeated these criticisms year after year. Reports from Officers in Native Districts, *Appendices to the Journals of the House of Representatives* (AJHR), 1880, G-4, pp. 2–3, 6; AJHR, 1881, G-3, pp. 2, 6; AJHR, 1881, G-8, p. 15; AJHR, 1883, G-1, pp. 6–7; AJHR, 1883, G-1A, pp. 5, 10; AJHR, 1884, G-1, pp. 1–2, 4, 5, 6, 18, 20; AJHR, 1885, G-2, pp. 1–2, 6, 7, 8–9, 12–13; AJHR, 1885, G-2A, pp. 1–2, 3, 5, 8–9, 12–13; AJHR, 1886, G-1; AJHR, 1888, G-1, p. 1; AJHR, 1892, G-3, p. 1.

39  Native Districts, AJHR, 1881, G-8, p. 15.

40  Native Districts, AJHR, 1886, G-1, p. 3.

41  Native Districts, AJHR, 1884, G-1, p. 2.

42  J.H. Pope, *Health for the Maori: A Manual for Use in Native Schools,* Wellington, 1884, p. 4. The Native Schools system was set up 1877 with the aim of civilizing Māori, with a particular emphasis upon teaching children English and inculcating them with 'European' ideas of hygiene. Native Schools served Māori communities in remote areas, although the European children of settlers also attended if there was no other school in the area. Lange, *May the People Live,* pp. 77–82; J. Simon and L. T. Smith, *A Civilising Mission? Perceptions and Representations of the New Zealand Native Schools System,* Auckland: Auckland University Press, 2001.

43  Pope, *Health for the Maori,* pp. 39–51.

44  Ibid., pp. 76–80, 91–102.

45  Lange, *May the People Live*, pp. 78–9.

46  D. A. Dow, *Safeguarding the Public Health: A History of the New Zealand Department of Health*, Wellington, 1995, ch. 2; K. Ford, 'The Tyranny of the Microbe: Microbial Mentalities in New Zealand, c.1880–1915', PhD Thesis, University of Auckland, 2013, ch. 1.

47  Dow, *Māori Health*, p. 93.

48  Pomare was of European, Ngati Mutunga and Ngati Toa descent. See J. F. Cody, *Man of Two Worlds: Sir Maui Pomare*, Wellington, 1953; Graham Butterworth, 'Pomare, Maui Wiremu Piti Naera', from the Dictionary of New Zealand Biography, Te Ara – the Encyclopedia of New Zealand, updated 30 October 2012, http://www.TeAra.govt.nz/en/biographies/3p30/pomare-maui-wiremu-piti-naera (accessed 25 November 2012).

49  Lange, *May the People Live*, pp. 157–9.

50  Buck was of Ngati Mutunga descent. J. B. Condliffe, *Te Rangi Hiroa: The Life of Sir Peter Buck*, Christchurch: Whitcombe and Tombs, 1971; M. P. K. Sorrenson, 'Buck, Peter Henry – Biography', from the Dictionary of New Zealand Biography. Te Ara – the Encyclopedia of New Zealand, updated 4 July 12, http://www.TeAra.govt.nz/en/biographies/3b54/1 (accessed 12 November 2012).

51  Lange, *May the People Live*, pp. 163–6.

52  Condliffe, *Te Rangi Hiroa*, pp. 92–3.

53  Lange, *May the People Live*, pp. 140–6; Dow, *Maori Health*, pp. 99–102; See also Hill, *State Authority*, pp. 50–64.

54  Belich, *Paradise Reforged*, pp. 200–6; Hill, *State Authority*, pp. 43–7.

55  Public Health, AJHR, 1903, H-31, pp. 66–9.

56  Ibid., p. vi.

57  Lange, *May the People Live*, pp. 205–16, provides a comprehensive account of the backgrounds and duties of the Native Sanitary Inspectors.

58  T.H. Mitira, *Takitimu: A History of the Ngati Kahungunu People*, Wellington: A.H. and A.W. Reed, 1944, p. 222. Italics in original.

59  'All Sorts of People', *New Zealand Free Lance*, 22 February 1902, p. 3.

60  Public Health, AJHR, 1902, H-31, p. 70; AJHR, 1904, H-31, p. 64; AJHR 1906, H-31, p. 74; AJHR, 1907, H-31, p. 18; AJHR, 1908, H-31, p. 92; AJHR, 1912, H-31, p. 77.

61  'Health for the Maori', *Poverty Bay Herald*, 25 January 1904, p. 4.

62  A. Bashford and C. Strange, 'Isolation and Exclusion in the Modern World: An Introductory Essay', in A. Bashford and C. Strange (eds), *Isolation: Places and Practices of Exclusion*, London: Routledge, 2003, p. 7.

63  The Department of Public Health officially established the Native Health Nursing Scheme to attend to the health needs of remote Māori communities in 1911, although the Department had sent nurses to work in these areas prior to this. Lange, *May the People Live*, pp. 169–70; and Dow, *Māori Health*, pp. 130–5.

64  Maori Schools – Policy – Closure of Maori Schools Owing to Epidemics, 1902–1921, BAAA 1001 104b 44/1/32 5, Archives New Zealand – Auckland (ANZA).

65  'The Health of the Maoris: Measures Being Taken', *Kai Tiaki*, IV, 3, July 1911, 109–10.

66  'Three Weeks in a Pah', *Kai Tiaki*, April 1913, 74.

67  Maori Schools Kawhia Log Book, 1899–1902, 5 June 1896, BAAA 1003 1l, ANZA.

68  D.S. Barnes, *The Great Stink of Paris and the Nineteenth-Century Struggle against Filth and Germs*, Baltimore: Johns Hopkins University, 2006, pp. 161–6; A. Hardy, *The Epidemic Streets: Infectious Disease and the Rise of Preventive Medicine, 1856–1900*, Oxford: Clarendon Press, 1993, pp. 270–3; Kraut, *Silent Travellers*, pp. 110–12. F. Cooper and A.L. Stoler, 'Between Metropole and Colony: Rethinking a Research Agenda', in F. Cooper and A.L. Stoler (eds), *Tensions of Empire: Colonial Cultures in a Bourgeois World*, Berkeley: University of California Press, 1997, p. 9, have commented on the resonances between European class politics and colonial racial policies.

69  Buck became Member of the House of Representatives for Northern Maori in 1909, Pomare Member of the House of Representatives for Western Maori in 1911. Buck's political career was short-lived however, and he later became Director of the Maori Hygiene Division of the Department of Health, before pursuing a full-time career in anthropology overseas.

70  Lange, *May the People Live*, pp. 184, 232; Dow, *Māori Health*, pp.142, 144.

71  Public Health, AJHR, 1902, H-31, p. 27.
72  For particular references to Māori as a 'menace', see Public Health, AJHR, 1911, H-31, p. 50; AJHR, 1912, H-31, p. 77; AJHR, 1913, H-31, pp. 12–13, 66; Makgill to Under Secretary Native Affairs, 21 January 1911, MA 21 20, Administrative Papers relating to Medical Care of Maoris, 1906–1919, Archives New Zealand, Wellington (ANZW). For more general discussions of Māori as a threat to European health, see Public Health, AJHR, 1911, H-31, p. 3; AJHR, 1911, H-31, p. 182; AJHR, 1912, H-31, pp. 3, 65–6; J.H. Crawshaw, 'Some Remarks on the Treatment of Tuberculosis amongst the Maoris at Tuahiwi Park', *New Zealand Medical Journal*, XIII, 56, October 1914, 348.
73  Public Health, AJHR, 1913, H-31, p. 66. Extracts from this report were reprinted as '"A Menace": Epidemics Among the Maori', *Evening Post*, 5 February 1914, p. 3.
74  Public Health, AJHR, 1912, H-31, p. 77.
75  Belich, *Paradise Reforged*, pp. 70–1; T. Brooking, 'Economic Transformation', in Rice, *Oxford History of New Zealand*, p. 234. The term 'opening up' was common in the discourses of settlement.
76  Public Health, AJHR, 1913, H-31, pp. 12–13.
77  See Public Health, AJHR, H-31, 1902, pp. 10, 26, and Public Health, AJHR, H-31, 1904, p. 3, for references to the high number of typhoid cases in the Auckland health district as evidence of the backward state of sanitary affairs in the region.
78  A. Hardy, '"Straight Back to Barbarism": Antityphoid Inoculation and the Great War, 1914', *Bulletin of the History of Medicine*, 74, 2, 2000, 265–90, 269.
79  Makgill to Under Secretary Native Affairs, 28 January 1911, MA 21 20, Administrative Papers relating to Medical Care of Maoris, 1906–1919, Archives New Zealand – Wellington.
80  For the wrangling between the two Departments over the responsibility for Māori health, see Dow, *Maori Health*, pp. 95–9.
81  *Evening Post*, 16 September 1901, p. 4.
82  *Evening Post*, 1 May 1908, p. 8.
83  Public Health, AJHR, 1908, H-31, p. 135.
84  'Typhoid Epidemic', *Poverty Bay Herald*, 9 April 1913, p. 3; *Poverty Bay Herald*, 10 April 1913, p. 2; 'Typhoid in the Waiapu', *Poverty Bay Herald*, 11 April 1913, p. 2.
85  Public Health, AJHR, 1911, H-31, p. 33; AJHR, 1912, H-31, p. 77; AJHR, 1907, H-31, pp. 18, 22; AJHR, 1903, H-31, p. 7.
86  Public Health, AJHR, 1911, H-31, p. 50.
87  Ibid.; Makgill to Under Secretary for Native Affairs, 31 January 1911, MA 21 20, Administrative Papers relating to Medical Care of Maoris, 1906–1919, ANZW.
88  Makgill to Valintine, 14 February 1911, quoted in Lange, *May the People Live*, p. 184.
89  See Bashford, *Imperial Hygiene*, pp. 103–4, for the construction of 'native' inability to 'perform health and hygiene' in the Australian context.
90  A carrier is someone who carries virulent micro-organisms, but who shows no symptoms. For the development of this concept in medicine, see J.W. Leavitt, *Typhoid Mary: Captive to the Public's Health*, Boston: Beacon Press, 1996.
91  Anderson, *Colonial Pathologies*, pp. 91–2.
92  'Necessary Precautions', *Poverty Bay Herald*, 21 April 1913, p. 2.
93  Hardy, '"Straight Back to Barbarism"', 277–8, discusses British fears about typhoid carriers in the army leading up to the First World War, after the disastrous example of the Boer War, where the disease had wreaked havoc amongst British forces.
94  'Maori Contingent', *Evening Post*, 18 September 1914, p. 2.
95  'Danger of Typhoid if Maoris are Mobilised', *Auckland Star*, 18 September 1914, p. 6.
96  Ford, 'Microbial Mentalities', pp. 335–7.
97  A. Holland, 'Introduction', in A. Holland and B. Brookes (eds), *Rethinking the Racial Moment: Essays on the Colonial Encounter*, Newcastle upon Tyne: Cambridge Scholars, 2011, pp. 1–20.
98  *New Zealand Gazette*, 1913, II, p. 2184.
99  Ibid., pp. 2183–4; AJHR, 1914, H-31, p. 52.
100  *New Zealand Parliamentary Debates* (NZPD), 1913, 162, pp. 357, 458; NZPD, 1913, 163, p. 443.
101  'Suspected Smallpox', *Evening Post*, 10 July 1913, p. 3.

102 'Outlook Brighter – Epidemic in the North – Maori Health Improving', *Evening Post*, 17 July 1913, p. 8; 'Smallpox', *Evening Post*, 25 July 1913, p. 8; 'Maori Health – Complaints Against Restrictions – Color Line Suggested', *Poverty Bay Herald*, 26 September 1913, p. 4; NZPD, 1913, 162, p. 724; NZPD, 1913, 163, p. 443; NZPD, 1913, 165, pp. 358.
103 NZPD, 1913, 163, p. 443; NZPD, 1913, 165, p. 574.
104 Salesa, *Racial Crossings*, p. 235, observed that according to the policy of racial amalgamation, half-castes were entitled to privileges and protections, but this was conditional on their removal from native categories and allegiances, and was subject to their alignment with colonial institutions.
105 Alfred A. Yates to Mr Herries Minister of Railways, n.d., R 3 W2334 13 1913/3819, Precautions against Smallpox and Restrictions in Travel, 1913–1920, ANZW.
106 A. Steven, Station Master to District Traffic Manager Railways Auckland, 18 September 1913, Precautions against Smallpox and Restrictions in Travel, 1913–1920, R 3 W2334 13 1913/3819, ANZW.
107 J. Bennett, 'Maori as Honorary Members of the White Tribe', *Journal of Imperial and Commonwealth History*, 29, 3, 2001, 36–7.
108 H. Fischer-Tiné and S. Gerhmann, 'Introduction: Empires, Boundaries, and the Production of Difference', in H. Fischer-Tiné and S. Gerhmann (eds), *Empires and Boundaries: Rethinking Race, Class, and Gender in Colonial Settings*, New York: Routledge, 2009, p. 5, discuss the negotiated nature of racial status in colonial settings.
109 'Racial Distinction and Smallpox – To the Editor', *Evening Post*, 17 July 1913, p. 8.
110 H. Munro to Traffic Manager, 1 September 1913, Precautions Against Smallpox and Restrictions in Travel, 1913–1920, R 3 W2334 13 1913/3819, ANZW.
111 See: 'The Color Line – Letter to the Editor', *Poverty Bay Herald*, 22 July 1913, p. 4; 'The Color Line', *Poverty Bay Herald*, 22 July 1913, p. 6; 'Smallpox', *Evening Post*, 25 July 1913, p. 8; 'Maori Health – Complaints Against Restrictions – Color Line Suggested', *Poverty Bay Herald*, 26 September 1913, p. 4; NZPD, 1913, 163, p. 443.
112 'The Color Line – Letter to the Editor', *Poverty Bay Herald*, 22 July 1913, p. 4.
113 Bashford, *Imperial Hygiene*, p. 4.
114 Belich, *Paradise Reforged*, pp. 206–7; Hill, *State Authority*, p. 19.

## Select bibliography

Anderson, W. 'Immunities of Empire: Race, Disease and the New Tropical Medicine', *Bulletin of the History of Medicine*, 70, 1, 1996, 94–118.
Anderson, W. *The Cultivation of Whiteness: Science, Health and Racial Destiny in Australia*, Melbourne: Melbourne University Press, 2002.
Anderson, W. *Colonial Pathologies: American Tropical Medicine, Race and Hygiene in the Philippines*, Durham, NC: Duke University Press, 2006.
Arnold, D. (ed.). *Imperial Medicine and Indigenous Society*, Manchester: Manchester University Press, 1989.
Bashford, A. *Imperial Hygiene: A Critical History of Colonialism, Nationalism and Public Health*, New York: Palgrave Macmillan, 2004.
Cooper, F. and Stoler, A. L. (eds). *Tensions of Empire: Colonial Cultures in a Bourgeois World*, Berkeley: University of California Press, 1997.
Dow, D. A. *Maori Health and Government Policy 1840–1940*, Wellington: Victoria University Press, 1999.
Ernst W. and Harris, B. (eds). *Race, Science and Medicine, 1700–1960*, London: Routledge, 1999.
Fischer-Tiné, H., and Gerhmann, S. (eds), *Empires and Boundaries: Rethinking Race, Class, and Gender in Colonial Settings*, New York: Routledge, 2009.
Holland, A. and Brookes, B. (eds). *Rethinking the Racial Moment: Essays on the Colonial Encounter*, Newcastle upon Tyne: Cambridge Scholars, 2011.
Lange, R. *May the People Live: A History of Maori Health Development 1900–1920*, Auckland: Auckland University Press, 1999.
Stepan, N. *The Idea of Race in Science: Great Britain 1800–1960*, London: Macmillan, 1982.

# 15

# RE-WRITING THE 'ENGLISH DISEASE'

## Migration, ethnicity and 'tropical rickets'

*Roberta Bivins*

Since the 1980s, historians of medicine have paid increasing attention both to chronic illness and to the reciprocal impacts of medicine and empire. Here, rickets in Britain will serve as a lens through which to examine – and integrate – these analytical strands. I will demonstrate that the geographies of chronicity and empire are far from distinct, despite the historiographical tendency to situate studies of the 'rise of chronicity' in the developed world and to locate 'colonial' (and increasingly 'post-colonial') medicine in developing nations. Thus, if the image of rickets in 1900 was that of a bow-legged London urchin, by the 1960s and 1970s, it was a markedly bowed and unmistakeably dark-skinned infant or a knock-kneed British Asian schoolgirl. In between, Vienna's starving children and the rickety youth in Abram Games' famously banned 'Your Britain: Fight for it now' World War II propaganda poster served as innocent faces of deprivation. Indeed, it is a matter of some historical irony that in the early 2000s, the public face of rickets became that of sun-screened and socially networked children of affluence; the more vulnerable housebound elderly remained below the threshold of public awareness, just as they had for the entire twentieth century. Shaping and reflecting these changing pictures, the fall and rise of rickets in mid- and late-twentieth-century Britain exposes the interplay between epidemiological, biochemical, molecular and social models of disease in this period. As the tools of the clinician and the public health officer shifted from household surveys, 'tactus eruditus', and the clinical gaze to x-ray imaging, biochemical, and then molecular analysis, 'rickets' changed from a disease of gross deformity and patent malnutrition to one of subtle signs, asymptomatic deficiency, and 'risk'. But if the symptoms and meanings of rickets have been fluid since the nineteenth century, the disease itself has consistently retained political significance and emotive power.

From the mid-seventeenth to the late twentieth century, the identity of the bone-deforming childhood disease rickets was largely stable. Although rickets was omnipresent and well recognised across northern Europe, it was most closely associated with Britain. Defined by English physician Daniel Whistler in his 1645 Leyden medical dissertation as 'morbo puerili Anglorum' – 'the English [children's] disease' – rickets' close ties to the British Isles have been variously associated with Britain's far northern latitudes and dark winters, its cloudy climate, and the murky skies produced by early and thoroughgoing urban industrialisation.[1] Indeed, environmental conditions have long been seen as a key causal factor in rickets. Nonetheless, by the late nineteenth century, clinicians and researchers had advanced other putative causes, and there was significant debate about whether the disease was environmental, dietary, hereditary or infectious in origin.[2] While its causes remained obscure well into the twentieth century, two cures for rickets were firmly established and widely known by the mid-1920s: cod liver oil and skin exposure to strong natural or artificial ultraviolet light.[3]

Today, rickets and the related condition of osteomalacia in adolescents and adults are understood as metabolic disorders of bone mineralisation caused either by deficiency of vitamin D, calcium or phosphorus, or by specific congenital impairments of metabolism. 'Vitamin D' is a group of secosteroid hormones – most importantly $D_2$ (ergocalcipherol), and $D_3$ (cholecalciferol) – that humans can ingest from fortified food and a narrow range of natural foods or synthesise via the exposure of cholesterol in the skin to ultraviolet irradiation.[4] By far the majority of rickets and osteomalacia cases globally result from dietary malnutrition or insufficient sun exposure. Medical consensus about the biochemistry of rickets and osteomalacia and its appropriate treatment emerged and solidified over the course of the twentieth century. However, expert agreement about the proximate causes of the diseases did not make them any less controversial. Instead, the intimate links between vitamin D deficiency (and its clinical sequelae, including rickets and osteomalacia), diet, behaviour, and environment rendered these conditions a lightning rod for political debates about the respective roles played by the social determinants of health and individual or community behavioural choices in promoting or undermining well-being. It is the abiding tension between models of rickets that emphasise individual choices and culturally sanctioned practices (from modest dress to the use of high-powered sunscreen) and those that stress the limitations on choice and burdens to health imposed by poverty and impoverished environments that make the disease so interesting. A final aspect of the rickets story, and one that attracted particular attention in the era of postcolonial migration, is the role contentiously attributed to the confounding factor of skin pigmentation, or 'race'. By tracing responses to, and interpretations of, rickets in Britain over the course of the twentieth century, this chapter will locate the condition at the nexus between changing models of public health and individual responsibility and imperial and post-imperial identities.

## 'The English Disease': rickets in 'slumdom', 1900–39

Research on the basic constituents and metabolic processes involved in nutrition rapidly expanded in the early twentieth century. Diagnostic tools, practices and technologies shifted, creating new and sometimes controversial understandings of 'disease'. As these changes took hold, rickets became a favourite case study for researchers on both sides of an increasingly heated debate about the relative importance of environment and diet in producing physical health.[5] Both the prevention and the diagnosis of rickets became enmeshed in a deeply political struggle over the respective roles and responsibilities of the state, society and individuals in promoting health for all. Did infants and young children develop rickets because of parental (almost universally understood as maternal) ignorance or neglect of their dietary needs; because poverty impeded the ability of families to meet their nutritional needs; or because the social groups most likely to be affected by rickets were also those trapped in unhealthy environments? These were scientific as well as social questions. Leading biochemists, and particularly those committed to the 'newer knowledge of nutrition', with its emphasis on 'vitamins' and other as-yet undefined food factors and metabolic processes, insisted that diet alone could explain the presence or absence of rickets. While this left them politically unencumbered (the responsibility for poor diet could be blamed either on society's failure to redress poverty or on individual failures of character or education), it distanced them from the battle to improve living conditions more generally. By contrast, some clinical teams and practitioners remained convinced that rickets resulted from environmental factors, specifically 'defective hygienic surroundings': the inadequate, dark and airless housing of the slums, and their residents' lack of access to open spaces, exercise, light and fresh air. In this view, only intervention by local or central authorities would solve the desperate inequalities of which rickets was the medical symptom. Summarising the state of play in 1928, one expert complained of 'two schools of thought – by the one rickets is regarded as a primary dietetic disease, and by the other as due to defective hygiene'; the 'controversy between the protagonists of these theories' had become 'acrimonious'.[6] Further complicating the picture, even if florid rickets, the rickets of the clinical gaze and experienced touch, was gradually disappearing from Britain's cities by the late 1920s and early 1930s,[7] new and more sensitive technologies – in particular the x-ray – identified the disease at an earlier stage and thus rendered it 'curable' when diagnosed early. The emergence of 'radiological rickets' amplified pressure to eradicate the condition, but at the same time raised (largely in the political sphere) what would become recurrent questions about whether 'mild cases' should be classified as 'rickets' at all.[8]

Controversies about its specific causation and signs notwithstanding, early-twentieth-century medical consensus certainly held rickets to be a preventable disease of temperate 'slumdom' (Figure 15.1). This shared view underpinned the significance and status of the disease for medical and political pundits alike. As a *Lancet* editorial observed in 1940, the persistence of rickets was undoubtedly 'galling' to researchers who had already explained 'the whole story of rickets' and its cure. Their frustration at the 'mild cases' still all too common even among the carefully parented children presented to welfare clinics in Britain's poorer areas contributed to the condition's continued visibility long after its physical signs had become almost undetectable even to the expert eye.[9] Writing as total war gripped Britain, the editors made their own stance clear: 'economics rather than lack of the application of scientific knowledge' explained the persistence of 'mild rickets'. In other words, the mere presence of rickets, even 'mild' rickets, was a reproof not to the medical profession but to society as a whole.

During the interwar period too, some authors began to describe rickets specifically as a disease of civilisation. Thus in the 1920s, an Assistant Medical Officer of Health in the northern manufacturing town of Huddersfield confidently reported that 'confining conditions of modern civilisation' were 'at least predisposing factors in the development of the disease'.[10] Late in the same decade, prominent Birmingham physician Leonard Parsons publicly reinforced this claim. Describing the appearance of rickets in environments ranging from urban slums to indigenous villages on the Labrador peninsula, Parsons was convinced that 'civilisation' produced dangerously pathogenic conditions. Citing 'preservation of the natural habitat . . . and diet' as the most important preventive factors, Parsons identified 'the urbanisation of society, the migration into cities and darkened dwellings' with the rise in rickets. Rickets, he concluded, was 'associated with the progress of civilisation, and in particular, with the rise of industrialism and the decline of breast-feeding'.[11]

*Figure 15.1* 'The Great Goal: Good Health'. This puzzle, commissioned by the Infant's Hospital, St Vincent Square London, c. 1930–39 and produced by the Chad Valley Company Ltd., Birmingham, illustrates understandings of rickets as a health penalty of 'slumdom'. Courtesy of the Science Museum, London, Wellcome Images.

Initially observers based their identification of rickets with civilisation on the apparent absence of rickets among even the most poorly nourished children of the colonial world, provided they retained their 'native' ways of life. That their immunity was not rooted in any distinctive biological attributes – that it was not, in other words, 'racial' – seemed clear in light of their vulnerability to rickets when exposed to the same pathological urban environments experienced by their northern peers. Margaret Ferguson, commissioned to explore the aetiology of the disease by Britain's

National Health Insurance Joint Committee in 1918, brought this fact to the attention of her superiors, themselves closely associated with the 'Glasgow School' and its emphasis on environmental causes. '[U]nknown when savage races live under natural conditions', she wrote, rickets became 'exceedingly rife when these peoples dwell in civilised countries . . . The same is a striking feature here [i.e. in Glasgow]. I do not recall ever seeing a non-rachitic half-caste negro child.' Reinforcing the association of rickets with poverty as well as environment, Ferguson added: 'It is notorious that these native races in civilised countries inhabit the worst quarters of our towns.'[12]

As the priorities and approaches of both metropolitan medical research and imperial medical observation shifted, however, the relationship between rickets and 'civilisation' in the colonial world took on a new inflection. If the pathogenic effects of civilisation in Europe and North America resulted from industrial urbanisation and an artificial diet, closer investigation of nutrition and health in the colonies revealed an apparent link between rickets and culture. Both Worboys and Arnold have pointed out that colonial malnutrition – like other chronic conditions similarly associated with burning questions of economic efficiency or racial politics – was 'discovered' by medical researchers in the interwar period, though as Arnold demonstrates, civil servants and medical officers had been addressing its prevention and effects, at least in India, throughout the nineteenth century.[13] Early research on rickets in India appeared to confirm the Glasgow School's heavy emphasis on the importance of environment, rather than diet, in the aetiology of the condition. In the tropics, however, the relationship between culture and 'environment' also became manifest:

> in India . . . while low caste children did not develop rickets, the disease was common amongst high caste children who lived under the 'purdah' system. Late rickets was also common amongst 'purdah' women in spite of the fact that their diet and that of their children contained far more fat than did that of the low caste women.[14]

Contrastingly, for Walter Fletcher, Secretary of the Medical Research Committee and himself a biochemist, India was a prime location for observational studies aimed at identifying the sometimes-elusive contributions of food to health, and to prove their primacy over environmental and social factors. In 1929, he extolled the 'great field for work in India by trained investigators and especially with regard to nutrition as related to osteomalacia and rickets'.[15] Exemplifying contemporary ideas about the uses and territory of 'colonial science', he added enthusiastically that such work would in turn generate 'new problems for more primary work by investigators here'.

Crucially, whichever aetiological model researchers supported with their colonial research, their interpretations included assumptions about the negative health impacts of 'native' religious beliefs. Tropical rickets and osteomalacia, environmentally inexplicable in Indian sunshine and nutritionally puzzling among well-fed prosperous Indians, could bolster either theory if blamed on (non-Christian) religious dogma which doomed the 'natives' to darkened rooms or nutritionally unsound dietaries. As this chapter will argue below, wartime and post-war research combining colonial observation with metropolitan biochemistry integrated these two models successfully to locate rickets' aetiology (at least for postcolonial populations) in 'culture'.

## The 'English disease' no more? Rickets, equity and eradication, 1933–63

The continued prevalence of rickets in Britain, and claims that its incidence was rising unheeded by a heartless state, gave the disease political saliency in the interwar years. In debate after debate, rickets symbolised the wider failings of the British state (and especially its Conservative legislators) to respond humanely to want and to the effects of enduring and unrelieved unemployment on workers, their families, and the nation's children. Debating the 1933 Unemployment Bill (intended to reduce the deficit in the National Insurance scheme by slashing payments to the unemployed), medical MPs criticised the 'starvation or semi-starvation' of the unemployed in Britain and repeatedly cited reports from local Medical Officers of Health that rickets in particular was on the rise.[16] In a 1936 debate on malnutrition, for example, Jarrow MP Ellen Wilkinson cited increasing rates of rickets – medical investigators had recently found 83 to 87 per cent of children in some areas affected by the condition – as a sign of the spread of malnutrition more generally:

> Take the question of rickets . . . Rickets is admittedly a disease due to malnutrition, bad or insufficient food, or lack of sunlight, and it is, therefore, a poverty disease . . . We on this side object to the fact that good health is becoming a class question.[17]

Although they disputed the figures indicating its rising incidence, opponents posed no challenge to this characterisation of rickets as a 'poverty disease'.[18] By failing to address malnutrition, one MP resoundingly claimed, Britain risked 'growing a race of pygmies'.[19] As the shadow of conflict in Europe moved ever closer, the persistence of such dysgenic and potentially disruptive social conditions came to be seen as a threat to national stability and efficiency.[20]

Interwar anxieties about the meaning, as well as the physical effects, of rickets – its status as an entirely preventable and easily treatable 'disease of the slums', and thus as a metric of failures in social equity – meant that its prevention loomed large in the minds of Britain's wartime food planners. They feared that any growth in the incidence of rickets (along with other signs of malnutrition, including scurvy) among British children due to the exigencies of war would be a body blow to national morale. Consequently, the Food Policy and Jameson Committees, the Ministry of Health, the Ministry of Food and other key actors incorporated its prevention into their advice about the national diet under rationing.[21] In particular, margarine (long associated with the poor, large and improvident families who were also regarded as most likely to suffer from rickets)[22] was mandatorily fortified with vitamin D and never rationed. So too were the infant foods available to new mothers. At the same time, groups deemed particularly vulnerable could access free supplemental foods like cod liver oil and orange juice, while schoolchildren under the age of 14 benefited from a number of programmes, notably the provision of milk and meals in schools.

Medical professionals scrutinised the impact of rationing on child health throughout the war. Contemporary evaluations were mixed. In 1941–2, there were conflicting reports on the weight and growth of schoolchildren and the incidence of deficiency diseases among them. By the end of 1942, however, with the rationing regime bedded in, researchers and clinicians were increasingly confident that British children could be protected at least from the desperate malnutrition that had affected their European peers after World War I.[23] Admittedly, all was not entirely rosy. In 1944, the journal *Public Health* declared rickets 'an important index' of the overall effects of dietary restriction on Britain's children.[24] And despite the state's best efforts, the British Paediatric Association's 1944 national survey – which revealingly looked for rickets and rickets alone – examined 5,283 children both clinically and radiologically and found that while rickets had not increased, 'the maximum measure of freedom from rickets has not been attained'.[25]

Notwithstanding these doubts and the grim persistence of rationing well into the 1950s, a consensus gradually emerged that the wartime food planners had achieved their aim, at least in relation to rickets and perhaps child health more generally. 'The first large-scale application of the science of nutrition to the population of the United Kingdom' had been a success.[26] Between 1938 and 1945, rates of child malnutrition fell and – despite a small rise in London – rickets in particular was generally reported as declining. Even sceptics accepted that the 'sweeping statements that rickets nowadays can only be learnt about in text-books' were 'in some measure true'. However, in crediting wartime 'propaganda and education' for the reduction, rather than the state provision of free or reduced cost supplements and protective foods, they foreshadowed the decline of direct interventions in population nutrition in favour of increasingly targeted education campaigns.[27]

By the 1950s, and particularly after rationing finally ended, victory over rickets was repeatedly declared. Sir Henry Dale, chairman of the Wellcome Trust, claimed that rickets, 'once the common childhood trouble to be seen in . . . every hospital in the country', had been 'practically wiped out'.[28] Similarly, *The Guardian* newspaper cheered, 'the rickets cases and other *symptoms of the slums* . . . have disappeared'.[29] These celebrations would be short-lived, but for those interested in studying rickets as a tool for understanding metabolism, attention returned to Britain's colonies and tropical research centres.

## 'Tropical rickets' and medical postcoloniality

In the late interwar period, long-established colonial researcher Dagmar Curjel Wilson entered into a new collaboration, this time with pioneering biochemist and dietary researcher Elsie May Widdowson. Wilson and Widdowson undertook an ambitious four-year (1936–40) study of diets and nutrition in India, sampling populations from across its geographic regions and ethnic groups. This research would come to have a marked impact on domestic British health policy. In 1942, the researchers published their results as 'A Comparative Nutritional Survey of Various Indian Communities'. It combined detailed observations of India's 'natural experiments in nutrition' and family dietary studies with equally meticulous biochemical analyses performed in London on material gathered in the Indian field.[30]

In Wilson and Widdowson's account, the regions of India were defined – in line with established research traditions – not by religion, politics, culture or ethnicity, but by the staple grains produced and consumed within them.[31] The diverse populations involved, however, were introduced specifically through a discussion of their assumed faiths' 'dietetic precepts' and general food customs. Implicit is the assumption that religion would play a significant role in dietary choice and nutritional status. Wilson and Widdowson looked closely for the effects of religious dietary proscriptions within communities which otherwise depended on the same staple cereal and local crops, and listed religion first among the 'complicating factors' they considered in interpreting their raw data. Religious 'taboos', they reasoned, 'may be expected to have some influence on the composition of the diet'.

Such views would later be conveyed from colonial settings back 'home' with researchers returning to Britain in the wake of decolonisation. Yet even in India, these expectations were confounded. The team found little evidence in the field that religious factors significantly affected diet in the majority of Indian contexts. Instead, through observation and biochemical analysis, they (re)discovered the overwhelming importance of economics in determining consumption in India, just as in Britain. 'In practice,' they concluded, 'religion does not appear to be the factor of first importance.'[32] Nonetheless, implicit criticism of the impacts of religion on dietary choice persisted in their final report.[33]

While retreating from assumptions that religious beliefs played a *primary* role in dietary choice for most Indians, Wilson and Widdowson made no claims for individual choice or agency. In fact, they had chosen their methodology, the study of family groups, because the individual 'native' was perceived to be largely inaccessible: 'Among native communities in India and other parts of the world . . . an individual method of study is extremely difficult and tedious, though not impossible to carry out.'[34] Their published results repeatedly stressed this inability to study or define the individual and his or her diet in the colonial setting, foreshadowing the later tendency for ex-tropical nutrition researchers in Britain to focus likewise on 'Asian' familial and community patterns, rather than individual choices or preferences.

On the other hand, Wilson and Widdowson engaged with – and disputed – one of the abiding homogenising tendencies in Indian nutritional research: the habit of making broad generalisations about entire populations based on their staple cereal.[35] As David Arnold has demonstrated in his study of beriberi in India, the conflation of diet, culture and race was an established part of colonial nutrition work in the period spanning the two world wars. Researchers condemned rice-based and vegetarian diets, while associating the use of wheat, meat and dairy products not only with superior health, but with 'racial' physical and mental superiority in general.[36] In contrast with the early-twentieth-century studies described above, 'race' was a crucial analytical category for Wilson and Widdowson, and the authors clearly regarded it as a contributory factor to differences between ethnic groups.

> There are a number of factors, racial, developmental, social and hygienic, which play an important part in determining the growth of the child . . . The difference between the Hindu and Muslim children may be racial rather than dietetic in origin. Muslims are often the descendants of invaders of tall stature who have come from the north.[37]

Implicitly, diet alone could not explain the observed differences between India's populations.

Consideration specifically of rickets and osteomalacia formed only a part of this study. However, Wilson's long experience in studying both (she published, either independently or collaboratively, some 15 studies of these conditions between 1929 and 1932) brought considerable data to bear on the question of their relative prevalence and origins. Here, in contrast to the rest of the study, religion was given a central role. Although alert to compounding factors including poor housing stock, Widdowson and Wilson's conclusions allowed for no doubt that religiously mandated seclusion and modest dress were the most important factors in the generation of vitamin D deficiency in India: 'It must be concluded . . . that the confinement of girls indoors in North India is the chief single factor responsible for the occurrence of rickets. In South India the abundance of sunshine which girls as well as boys enjoy more than makes up for their poorer diets.'[38]

The presence of serious vitamin D deficiency in India, its relationship to religion, and the role of the built environment all would attract attention to Wilson and Widdowson's research from nutrition workers in the Ministry of Health/Department of Health and Social Services (DHSS) – but not until the 1960s and 1970s, when metropolitan British medical attention was focused closely on rickets and osteomalacia in the 'tropical' South Asian bodies which had suddenly become 'local' in the metropole. Not coincidentally, this was also the period in which, as one actor recalled, 'the approach to nutrition in Britain . . . depended in practice on ex-patriot scientists from the Third World'.[39]

In the intervening years, much domestic British research looked inward, searching out the signs and symptoms of an increasingly subtle disease, exploring the metabolic machinery that underpinned those signs and symptoms, and investigating the diets of Britain's own urban indigenes.[40] In contrast, Britain's tropical colonies offered more than vital 'clinical material' to British metabolic and nutritional research.[41] They served as living laboratories within which partisans on each side of the debates that transfixed British nutritional science and policy – debates about whether research and interventions in nutrition should focus on aetiology and metabolic biochemistry, or on social and environmental factors – could test their convictions in practice. Research workers deployed both structural and technical approaches to the problems of malnutrition and hunger on colonial and postcolonial populations.[42] Many acknowledged and celebrated the 'invaluable role' that 'the third world' played 'as a training ground for British nutritional science'.[43] Moreover, based on their pioneering work, whether clinical, epidemiological or biochemical, such researchers often achieved dominant positions in Britain's domestic medical hierarchy on their return from colonial settings in East and West Africa, India, Malaya and the British Caribbean.[44] Ex-colonial scientists and medical officers came to direct the Medical Research Council's highly influential Dunn Nutrition Unit, the Rowett Research Institute in Aberdeen, and key laboratories and teaching hospital research units across the UK.

The role of the colonies and successor nations as sites of research did not end with formal empire, since the Medical Research Council continued to fund clinical and especially tropical medicine research facilities across the New Commonwealth.[45] Nor was the impact of such research felt only, or even primarily, in the tropics; researchers returning from the colonies (and post-colonies) described their experiences as 'priceless for the development of diet-orientated health programmes in the United Kingdom'. Their tropical experiences were enduringly transformative of their own research interests, but also – particularly as they rose through the biomedical ranks – of the institutions to which they returned. As R. G. Whitehead (Director of the Dunn

Nutrition Unit and an ex-Ugandan medical researcher) put it, 'the basic investigative philosophies that have become central to nutritional science in general' evolved in a specifically colonial context.[46] Just as debates taking place in British biomedicine between the wars were played out in colonial medical research and service, so debates that emerged from late colonial and international health would find expression in postcolonial British medical research and practice.

## The return of rickets, 1963–71

In 1963, less than a decade after the end of rationing – and in the wake of steady, if always controversial, reductions in welfare feeding – a new report emerged from research teams based in London and Glasgow. Rickets was back. Given the highly politicised nature of the disease, such reports provoked disquiet.[47] 'Inquiries are being made in Glasgow and London into the incidence of rickets which was thought to have disappeared with rising prosperity', reported *The Guardian*, before complaining that rickets was 'on the increase, but . . . no regular statistics are kept'. The only consolation the paper could provide its readers came in the form of speculation that these cases of rickets, unlike those of the interwar period, 'might not necessarily be caused by poverty and neglect', but instead by the 'misguided devotion' of breast-feeding 'immigrant mothers'.[48] Each of these claims hints at the underlying political and scientific debates.

As in the interwar period, medical consensus on who suffered from rickets and why the disease had reappeared was scarce. Researchers in Glasgow and Britain's industrial Midlands expressed scandalised concern that the condition affected the children of large, poor families in the majority community and urged a return to the interventionist preventive policies of the wartime and immediately post-war era. Publicly, the Ministry of Health ridiculed such claims; privately, its medical officers commissioned urgent investigations into the nutritional status of Britain's children. In the interim, rickets was instead presented as an 'immigrant' disease. Ministry officials speculated that New Commonwealth migrants, whose arrival had already provoked considerable attention in public health circles, might have imported malnourishment along with their other health deficits. In 1964, Minister for Health Anthony Barber (and other government members of parliament) suggested that their rickets might be the result of dietary ignorance, or a failure to rapidly adapt to the British climate and mode of life.[49] In 1965, a *Guardian* reporter invoked such hypotheses to propose a new aetiology for the once-English disease, rooted in failed adaptation to grey skies and modern nutrition alike. Rickets was, Nesta Roberts argued, one of the 'disorders of transplantation'. After being 'almost completely eliminated' from Britain, she continued, 'rickets appeared again with people who came from a sunny climate and who had never needed cod liver oil . . . they had no idea of the uses of welfare foods'.[50]

Yet while the association of rickets with increased immigration was politically attractive to those eager to challenge any sense that Britain was losing ground, it was far from uncontested. Rickets became a vehicle for heated criticisms not only of a mean-spirited and penny-pinching state, but of the British diet in general, and indeed of the failure of the Welfare State to guarantee the modern, egalitarian society it had once promised as recompense for pre-war inequality and wartime privations:

The reality is that our standards of nutrition are somewhat worse now than they were in the austerity of 1950 . . . (And our standards are generally some way below those of America and Russia.) Rickets, the most emotion-charged of all the symptoms of deprivation, is still found in Britain – mainly but not wholly in immigrant families. More nebulously, nutritionists are worried at the accumulating evidence that children in large families are smaller than only children; that health and physique are somehow related to social class; and that in many respects the health gap between rich and poor seems to be widening.[51]

It was, *The Times* suggested, 'the old story of poverty, social incompetence and poor health marching together'.[52] While disputing this abiding association with poverty, even DHSS experts reluctantly acknowledged the *social* (but not the medical) significance of rickets. Principal Medical Officer W.T.C. Berry, head of the Department's Nutrition Unit, admitted in 1968 that rickets 'excited great emotion' as 'a sign of social regression'.[53]

Berry himself did not share this view; indeed he rejected it ferociously in favour of arguments that rickets was an evanescent symptom of mass migration, and of imported bodies still marked by the poverty of their countries of origin. As a former colonial medical researcher first in Nyasaland (Malawi) and then in The Gambia, Berry had personally participated in unrewarding technocratic campaigns to research and treat 'malnutrition' in the form of various hypothesised vitamin deficiencies, while ignoring the wider context of desperate 'undernutrition'. The inefficacy of the New Nutrition as a 'panacea' for Africa's ills was for him an abiding and bitter memory, and one that may have significantly influenced his resistance to supplementation and fortification interventions in relation to what he saw as largely latent rickets in a British context of dietary abundance.[54] In fact, the Nutrition Unit employed a number of ex-colonial researchers in senior positions. Among Unit medical staff like Berry and Senior Medical Officer ex-Ugandan doctor Sylvia Darke, attitudes fostered in Britain's tropical colonies – about what constituted a problem of nutrition, who was 'vulnerable', and what forms of intervention and engagement were appropriate to particular (homogenised) populations – drove research and policy interventions.[55] In addition, as in colonial tropical medicine, assumptions about the diets, customs, beliefs and aptitudes of 'Asians' and other New Commonwealth groups profoundly influenced Departmental responses to the unwelcome return of rickets.

Yet despite his scepticism, Berry was unable to completely resist public and professional pressure, evidenced in a series of regional internal reports, scientific papers, medical editorials and popular news reports, to respond to the reappearance of rickets.[56] In the late 1960s, Berry confirmed the Department's discovery of 'overt rickets' in 'small pockets, *either socioeconomic or racial*, which might respond well to fairly simple modifications of our existing fortification system'.[57] The presence and then persistence of such a familiar and easily resolved deficiency disease in Britain was shocking to all. To members of the Ministry who had begun their careers in colonial settings, the political ramifications of such a manifest failure in the state's duty of care to its population would have been well known. As one researcher recalled, in the late colonial era 'the concept of a protein deficiency in a British Protectorate was deemed politically objectionable. Infections were unavoidable . . . vitamin deficiencies were new and thus not having planned for them was defendable; but something as basic as a lack of protein was quite unacceptable!'[58] Decades after the widespread recognition of vitamin deficiencies as causes of ill-health, in a developed and proudly 'modern' nation, and during a time of comparative abundance, how much more unacceptable might deficiency rickets become?

Media reporting and parliamentary debates demonstrate exactly this point: they presented the return and persistence of rickets in Britain as scandalous, and demanded – unsuccessfully – immediate action. Initially, both political and media outrage focused strongly on fears that rickets was once again an indicator of economic inequality, revitalised by reductions in welfare feeding and increased means-testing in its distribution. However, with national surveys producing contradictory evidence about rickets among Britain's white working-class communities, the DHSS denied an indigenous problem and directed attention towards Britain's newcomers. In this, they were supported by some sectors of the medical community. Surveying its members in 1962, for example, the British Paediatric Association sought to establish the impacts of immigration and was keen to separate the nutritional status of migrants from that of the majority community. In their view, '[l]arge scale immigration . . . of families from the Commonwealth' explicitly 'caused' the seeming increase in nutritional disorders; they therefore urged their members to be attentive to the race and birthplace of each case of malnutrition reported.[59] Through such surveys, and through the efforts of clinicians and specialist metabolic researchers located in areas of high migration, abundant evidence confirmed high rates of rickets and osteomalacia among children and adolescents of South Asian descent.

## Colonising communities: 'Asian rickets' and British medicine, 1971–81

Initially, the close association between post-war rickets and Britain's growing population of recent migrants from South Asia and East Africa and their British-born descendants offered attractive cover to both cost-cutting politicians and Britain's central health authorities.[60] Most of the affected families were (at least in theory) eligible to receive welfare foods and dietary supplements even under the new, restrictive regime; thus welfare cuts could not be blamed for the rise in rickets. At the same time, their failure to access or use the benefits and medical services to which they were entitled could readily be ascribed to cultural or linguistic differences, rather than flaws in the services themselves. The link between Asian ethnicity and rickets was accepted and even fostered by medical professionals. By 1973, it had generated a new nomenclature for the once 'English disease': 'Asian rickets'.[61] But the visibility of the affected population, the growing interest its members provoked among elite researchers eager to unravel the mysteries of human metabolism, and – perhaps most importantly – the persistence of claims that rickets and osteomalacia were in fact far more widespread eventually forced the Department's hand.

Finally prodded into action, the DHSS and its expert Committee on the Medical Aspects of Food [COMA] echoed colonial precedents in initially prioritising methods of assessment and surveillance of rickets over its immediate prevention. Treatment was, in any case, a matter for local health authorities and practitioners, who also provided much public health education. Paralleling colonial nutrition research, the Committee focused first on establishing a solid foundation in the basic sciences.

This approach also met with some support among the wider medical profession; a 1971 leading article in *The Lancet* documented the challenges faced by practitioners attempting to diagnose early rickets. Radiography, once the gold standard, was now regarded as insufficiently sensitive to early signs of sub-clinical deficiency – now known as 'biochemical rickets'. It was also subject to technical artefact and professional disagreement, and potentially risky to the child-patient. Like clinical examination before it, radiographic diagnosis had become 'subjective'. Blood tests, on the other hand, were subject to wide variation between individuals and even with automation had proven difficult to standardise. 'Normal variation' in levels of key serum constituents (serum alkaline phosphatase [SAP], in particular, the 'sheet-anchor of diagnosis' for biochemical rickets) was known to exist – but its parameters were undefined for any given population, and levels did not vary consistently with treatment, rendering its utility as a diagnostic sign dubious.[62] As *The Lancet*'s editors pointed out, there might also be 'racial differences' in enzyme levels. They concluded pointedly that 'Rickets matters', not just to 'a few coloured children in temperate climates', but to 'any infant' deprived of sunlight or vitamin supplementation, and exhorted the profession on two fronts. First, they urged continued clinical examination to screen all children; second, they called for further investigation of SAP as a diagnostic sign.[63]

When in October 1972 COMA's Panel on Child Nutrition tackled 'Asian rickets' by establishing four expert working groups (clinical, biochemical, epidemiological, and dietary) to address 'various aspects of the problem of rickets and osteomalacia in the immigrant population', rather than any direct interventions, they were acting in line with at least some segments of professional opinion. Charged with developing standardised methodologies for determining the incidence and severity of disease, and only secondarily its causes, the panels were led by internationally recognised expert researchers.[64] COMA invited the head of the University of Manchester's cutting-edge metabolic unit, S. W. Stanbury, to chair the clinical working group alongside paediatrician Professor C. E. Stroud of the Department of Child Health, King's College Hospital, University of London.

An ex-colonial researcher and an active practitioner in a community increasingly marked by the presence of immigrant and ethnic communities, Stroud was a strong advocate for action. He had already published research on the problem of rickets among New Commonwealth migrants, including those with origins in the Mediterranean, Caribbean and Africa, as well as South Asia.[65] However, his preference for speedy intervention was stymied by Stanbury's primary commitment to basic research. Indeed, while Stanbury bragged enthusiastically about the research benefits to be had from working closely with Manchester's growing Asian community ('local co-operation is so good that I have little if any doubt about [their] willing availability'), he also insisted that for his unit the DHSS work 'could only be done if a by-product of our clinical examinations were the acquisition of biochemical information relevant to our personal research interests'.[66]

Three months later, the leaders of the working groups framed a plan of action. Their first goal was the standardisation of measures for normal biochemistry; this was to involve not only determining the best method for assaying samples biochemically, but also establishing 'normal' levels of the various indicators (calcium, phosphate, alkaline phosphatase) in human serum. For this, comparisons with 'a control group of Caucasians' were required – an uncontested assertion that offers convincing evidence of two key underlying assumptions held by the group: that British 'Caucasians' were the norm from which other groups might deviate; and that – evidence from Glasgow notwithstanding – rickets and osteomalacia were not generally present among the majority population. Such assumptions were commonplace: in contemporary medical literature, 'healthy Caucasians' were often specified as controls, while in some studies, even apparently healthy Asians could only be 'symptomless'.[67] The final component of this study was a longitudinal study of 250 of the recently arrived Ugandan Asian refugees, from whom blood had been opportunistically gathered, to assess any 'changes in their condition', testing the hypothesis that rickets might be an imported ailment rather than a deficiency arising from life in Britain.[68] Thus the bodies and availability of colonial and post-colonial populations remained vital to basic research in this field.

A subsequent Panel meeting in April 1974 aired the possibility of one additional study, suggested by LSHTM professor John Waterlow, the recently returned founder of Jamaica's MRC-funded Tropical Metabolism Research Unit. Waterlow proposed a parallel study in 'populations from which the immigrants come, e.g. Pakistan, or in countries where rickets is common', and offered to put forward a proposal to the Tropical Medicine Research Board. This suggestion was encouraged and approved – neatly joining the colonial and postcolonial strands of tropical medicine.[69] However, such a study had in fact already been performed and published by Stanbury's group in 1973, comparing the blood chemistry of 119 residents of Ludhiana (in the Punjab) with their previous results for a similar group in Rochdale. Leaning heavily on Wilson and Widdowson, Stanbury noted that all existing explanations of rickets among Britain's Asian immigrants had been shown to be unlikely explanations of rickets in India itself. Stanbury's group asserted that 'Indian and Pakistani people in Britain tend to maintain their social and dietary customs'.[70] Therefore, they argued, if diet or traditional behaviours caused rickets and osteomalacia, they should find significant evidence of the disease in Ludhiana. However, they found only one case, and concluded that 'Asian rickets' was in fact the English disease reborn: a product of 'inadequate solar exposure' alongside low intake of vitamin D.

In closing, the authors returned to Wilson and Widdowson's vision of India as a place of 'natural experiments in nutrition'. They wrote:

> This opportunity is now shared with Europe, and the hard-earned experience of Indian physicians and nutritionists must be remembered. We also suggest that more might be learned from the history of vitamin D deficiency and rickets in Britain than from the pursuit of improbable causes of Asian rickets and osteomalacia.[71]

Such lessons from history, however, were clearly not ones that the DHSS wanted either to learn or to teach. Not only did history offer clear evidence of a different standard of response to 'Asian rickets' than to 'the English disease', but it might restore an unwelcome focus either on living conditions in British Asian communities, or on the politically sensitive subject of the physical attributes of race – in particular skin pigmentation.[72] Far preferable were studies that might produce evidence that 'Asian rickets' resulted from 'Asian' behaviours (including a voluntary failure to gain 'solar exposure'), which could in theory be changed through education and assimilation alone.

While the Working Groups continued their deliberate progress in assessing 'whether rickets constituted a public health problem', clinicians and local health authorities in areas with large Asian communities fumed at DHSS inaction.[73] In July 1974, for example, a Birmingham consultant clinician, W. Trevor Cooke wrote in some annoyance to the DHSS:

> There are many of us in this country who are not quite so complacent about the [rickets] situation as the Ministry appears to be. It is an open secret that for many years . . . the Ministry were openly critical of the findings in Glasgow suggesting the incidence of rickets in the immigrant community there.[74]

He also pointed out that the Ministry had focused its attention almost exclusively on infants and children under five – the group most likely to be protected even by the weakened fortification regime. This focus also reflected the impact of 'population' approaches and clinical priorities imported back to Britain by returning colonial medical officers and tropical researchers accustomed to seeing infants and their mothers as the most vulnerable to nutritional deficit.

The popular press, too, began to ramp up their coverage of 'Asian rickets', producing headlines like 'Rickets Danger for Asians in Britain', 'Asians "Face Danger Over Their Diet"', and – perhaps most revealingly – 'Asians Advised to Spread Margarine' in response to calls from the British Nutrition Foundation and Community Relations Commission for increased health education about rickets for Asian communities.[75] In part, press interest stemmed from the continued political salience of rickets as a marker of poverty and inequality, an issue effectively spotlighted by Labour parliamentarians in reaction to Conservative Education Secretary Margaret Thatcher's 1971 withdrawal of free milk that had been provided for school children since WWII.[76] However, it also reflected growing attention to questions of 'race relations' and racial equality in the public services and British society at large. By 1973, and for the remainder of the decade, the relationship between rickets and 'racial', rather than socio-economic, inequality would increasingly drive policy responses to the condition within Whitehall and Westminster.

As this transition in the signification of rickets took hold, officials in the Department of Health found themselves caught between two contradictory discourses. On one hand, they and their political masters envisioned a far narrower sphere of action for central government, and particularly the central health authorities, in managing the public health. Only those problems that affected the entire nation called for a central response; matters affecting any smaller segment of the population could be addressed at the local level. Thus, defending the Department's resistance to direct nutritional interventions – including the fortification of chapatti flour as advocated by key metabolic researchers and clinicians to (once again) eradicate rickets – Sylvia Darke explained in 1977 that:

> Nutritional problems can be dealt with either by changes in national policy or locally by area health authorities. Alterations in national policy are in general reserved for problems which affect the national health and which can only be solved by Government action . . . It is as well to remind ourselves that the public health means the health of *56 000 000* people and also that the effects of changes in nutritional policy may not reveal themselves for a generation.[77]

Because of these potentially long-ranging effects, she argued, any central action required 'sound evidence'; 'intervention' in particular required this, and demanded a 'well-informed' public able to make a 'wise choice'. By these definitions and under these constraints, direct national action to end rickets was of course beyond the scope of 'the public health': 'the problem of rickets' was 'confined chiefly to Asians in urban areas'.[78]

On the other hand, successive Race Relations Acts since the mid-1960s, and especially the Race Relations Act of 1976, had raised expectations of government action and support in maintaining and improving 'race relations' and in ensuring equality in the public services. The 1976 Act specifically mentioned health care as an area in which indirect discrimination, as well as direct discrimination, was illegal, and established a public duty for local and health authorities to ensure equal access. Moreover, by establishing the Commission for Racial Equality and endowing it with some investigative powers, the Act also opened new channels for community action, and for the new Community Health Councils [CHCs] established in 1974 to press for change. By the mid-1970s, the CHCs included growing numbers of ethnic minority members, particularly in urban areas with high rates of primary and secondary migration – exactly the areas in which both 'Asian rickets' and the influential specialists who researched and treated it were to be found.

One such specialist, A. S. Truswell, included the prevention of rickets and osteomalacia among his proposed nutrition goals for Britain. 'The "English disease" is back with us', Truswell proclaimed. '[P]ossible measures' to eliminate it included:

> (a) health education, encouraging Asians to expose their skins to sunlight, eat more margarine and use welfare vitamin D, (b) enrichment of chapatti flour, (c) enrichment of milk (as in the USA), (d) enrichment of butter. None of these is without difficulties and objections, any will require money and there is a lingering fear of hypercalcaemia.

Yet while these challenges were universal, he observed, rickets 'seems to be more prevalent here than across the Channel or across the Atlantic'. Its eradication was a useful nutritional goal for Britain precisely 'because it is theoretically achievable'. Truswell certainly persuaded his expert audience; in a subsequent straw poll of 105 ballots, they accepted rickets and osteomalacia prevention as one of the nation's top 15 health priorities.[79]

While individual members of the Unit were frustrated with their slow progress and seeming inability to return rickets to the history books in the mid-1970s, they and their peers across Whitehall nevertheless resisted calls for fortification. Again, their arguments often echoed colonial assumptions: Asians might, they assumed, object to fortified foods on religious grounds. It might even cause 'political, racial, and religious problems'.[80] However, under pressure from a wide range of organisations, from the British Medical Association to the Community Relations Commission and the Confederation of Indian Organisations, by the late 1970s they were eager to be seen to be doing something about the rickets problem.[81]

Even the Nutrition Unit was therefore surprised when, in 1978, its own specialist COMA panels rejected any idea of fortification. Crucially, they argued for a new understanding of rickets' renewed but selective prevalence in the UK:

> There is evidence that, far from being the 'English disease', rickets and oste-
> omalacia among young women are now largely diseases of the tropics and
> sub tropics . . . Asian child immigrants sometimes arrive in this country with
> active rickets, and . . . as the Asian people adjust to conditions of life in Great
> Britain the incidence of vitamin D deficiency is declining.[82]

'Asian rickets' was redefined not only as 'tropical' but – like other diseases associated
with migration, including tuberculosis – as imported. In the face of subsequent, and
often indignant professional debate and heated public criticism, COMA (and consequently the DHSS) was obdurate.[83] There would be no fortification of chapatti flour
or other additional foods; instead, the 'Asian people' must be educated to 'adjust'. Yet
they were not allowed the budget required for effective mass education. Describing
the Department's efforts to educate via pamphlets as 'pathetic', Sylvia Darke pleaded
with her superiors:

> I think at night of all the Asian children with rickets who could be cured
> or their illness prevented so easily. Is it my fault? To some extent we have
> failed. In 1972, our hunch was education but . . . [we] cannot easily launch a
> campaign. What is needed is TV and radio time in short bursts at intervals to
> repeat and repeat the message. We have not access to radio and TV!![84]

Darke would not get access; in line with the Department's vision of 'public health',
she was told that the target was too small to merit such action on a national scale.

Ironically, television would eventually catalyse a shift in the Department's approach
to 'Asian rickets'. Damning criticism of DHSS inaction on Granada Television's popular
'World in Action' programme in 1979 prompted the newly appointed (and medically
qualified) Minister of Health Gerard Vaughan to invite 'community leaders' into the
policy-making process.[85] Vaughan was determined to 'show that the Department was
concerned' about Britain's ethnic minorities.[86] The resulting DHSS/Save the Children
'Stop Rickets' campaign was designed to engage long-ignored communities with the
NHS through cost-effective education-only campaigning. Its disproportionate focus
on an easily prevented and cured condition also served as an effective distraction from
larger health concerns that affected Britain's Asian ethnic minorities.

## Conclusion

The identity of rickets changed repeatedly over the course of the twentieth century.
From a nearly inevitable disease of slumdom, it became a marker of inequality and a
highly politicised sign of social deprivation. Still later, the absence of 'clinical' (that
is, frank and visually apparent) rickets became a sign of post-war egalitarian and scientific modernity, while any return of the disease marked 'social regression'. Worried
civil servants and politicians re-inscribed such rickets as a 'tropical' disease (imported,
rather than acquired by the new British populations among which it was most prevalent), while frantically resisting claims that a new 'epidemic' of 'biochemical rickets'
invisibly afflicted the children of the indigenous poor. In politics, the press and professional journals, 'tropical rickets' was an evanescent 'disease of transplantation',
bound to yield to assimilation and accessible medical care. 'Biochemical rickets',

on the other hand, might not exist at all. When 'tropical' rickets persisted, its 'immigrant' identity transformed into an ethnic one. 'Asian rickets' replaced the 'English disease', and approaches and anxieties shaped by colonial medicine were re-applied to postcolonial populations in Britain.

Each of these shifts in the identity of rickets – each chapter of its twentieth-century biography – reflected and provoked changes in diagnostic technologies, medical and public health practices, and the meaning of the disease for sufferers, communities and policy-makers. The emergence of 'Asian rickets', for example, initially offered valuable protection to politicians eager to screen themselves and their policies from the political fallout of rickets' return. In an era of increasing attention to 'race relations' and with the help of new legislation and regulatory bodies (specifically, the Race Relations Act of 1976 and the CHCs and CRE), it also opened the door for 'Asian' communities and equality activists to demand a national response to this racialised condition, albeit at the price of drawing attention away from wider health disparities. And at the same time, 'Asian rickets' continued to obscure another at-risk population, the housebound elderly of all ethnic origins. Despite strong evidence that they experienced high rates of treatable osteomalacia, this group was once again ignored by public health campaigners, demonstrating the potentially damaging impacts of politicised – and racialised – disease identities on the assessment, measurement and protection of the public health.

# Notes

1 Whistler was closely followed by other commentators and researchers. See G. T. Smerdon, 'Daniel Whistler and the English Disease: A Translation and Biographical Note', *Journal of the History of Medicine and the Allied Sciences* 5, 1950, 397–415; Francis Glisson, *A Treatise of the Rickets: Being a Disease Common to Children*, translated and edited by N. Culpeper, London: P. Cole, 1651; 'Francis Glisson (1597–1677) and the "discovery" of rickets', *Archives of Disease in Childhood. Fetal and Neonatal Edition*, 78, 1998, 154; Edwin Clarke, 'Whistler and Glisson on Rickets', *Bulletin of the History of Medicine* 36, 1962, 45–61.

2 See Karl Pearson, *Darwinism, Medical Progress, and Eugenics: The Cavendish Lecture, 1912. An Address to the Medical Profession*, London: Dulau and Co., 1912.

3 Rima Apple, *Vitamania: Vitamins in American Culture*, New Brunswick, NJ: Rutgers University Press, 1996, pp. 33–53; Harriette Chick, 'Study of Rickets in Vienna 1919–1922', *Medical History* 20, 1976, 41–5; Kumaravel Rajakumar, 'Vitamin D, Cod-Liver Oil, Sunlight, and Rickets: A Historical Perspective', *Pediatrics* 112, 2003, 132–5.

4 Anthony W. Norman, 'From Vitamin D to Hormone D: Fundamentals of the Vitamin D Endocrine System Essential for Good Health', *The American Journal of Clinical Nutrition* 88, 2008, 491S–9S.

5 David F. Smith and Malcolm Nicholson, 'Chemical Physiology Versus Biochemistry, the Clinic Versus the Laboratory: The Glaswegian Opposition to Edward Mellanby's Theory of Rickets', *Proceedings of the Royal College of Physicians Edinburgh* 19, 1989, 51–60. On the historical politics of hunger itself, see James Vernon, *Hunger: A Modern History*, Cambridge, MA: Harvard University Press, 2007.

6 Leonard Parsons, 'Ingleby Lectures: On Some Recent Advances in Our Knowledge of Rickets and Allied Diseases', *The Lancet* 212, 8 September 1928, 433–8.

7 See the grumblings of Wilfred Sheldon, 'Observations on Rickets', *The Lancet* 225, 19 January 1935, 134.

8 See David Smith, 'Nutrition Science and the Two World Wars', in David Smith (ed.), *Nutrition in Britain: Science, Scientists and Politics in the Twentieth Century*, London: Routledge, 1997, pp. 142–65, pp. 151–2.

9  'Rickets', *The Lancet* 235, 13 January 1940, 84–5, 84.

10  Amy Hodgson, 'Vitamin Deficiency and Factors in Metabolism: Relative to the Development of Rickets', *The Lancet* 198, 5 November 1921, 945–9, 945.

11  Parsons, 'Ingleby Lectures', p. 437. In fact, prolonged breast feeding would later be recognised as a risk factor for infantile rickets.

12  Margaret Ferguson, 'A Study of Social and Economic Factors in the Causation of Rickets', London: Medical Research Council, National Health Insurance, 1918, p. 58.

13  Michael Worboys, 'The Discovery of Colonial Malnutrition between the Wars', in David Arnold (ed.), *Imperial Medicine and Indigenous Societies*, Manchester: Manchester University Press, 1988, pp. 208–25; David Arnold, 'The "Discovery" of Malnutrition and Diet in Colonial India', *Indian Economic Social History Review* 31, 1994, 1–26, and on the nineteenth century, 7–10; David Arnold, 'British India and the "Beriberi Problem", 1798–1942', *Medical History* 54, 2010, 295–314.

14  Parsons, 'Ingelby Lectures', 435.

15  The National Archives (TNA) FD 1/1974 Walter Fletcher to George Newman, 13 November 1929 (emphasis added). See also Joan Austoker, 'Walter Morley Fletcher and the Origins of a Basic Biomedical Research Policy', in Joan Austoker and Linda Bryder (eds), *Historical Perspectives on the Role of the MRC: Essays in the History of the Medical Research Council of the United Kingdom and Its Predecessor, the Medical Research Committee, 1913–1953*, Oxford: Oxford University Press, 1989, pp. 23–33.

16  'Unemployment Bill', House of Commons official report (Hansard), 6 December 1933, vol. 283 cc. 1499–623. Opponents censured the Bill for ignoring the role of 'industrial capitalism' in creating 'the victims of the unemployment' [col. 1499]. On rickets as a marker, see Dr Alfred Salter's speech, col. 1567–8.

17  'Malnutrition', House of Commons official report (Hansard), 8 July 1936, vol. 314 cc. 1229–349 at col. 1266.

18  'Malnutrition', Hansard, 8 July 1936.

19  David Adams, 'Malnutrition', Hansard, 8 July 1936, at col. 1306.

20  For an actor's summary, see Dorothy Hollingsworth, 'Food and Nutrition Policies in Wartime', in Elsie May Widdowson and John C Mathers (eds), *The Contribution of Nutrition to Human and Animal Health*, Cambridge: Cambridge University Press, 1992, pp. 329–39.

21  Smith, 'Nutrition Science', pp. 142–65.

22  Alysa Levene, 'The Meanings of Margarine in England: Class, Consumption and Material Culture from 1918 to 1953', *Contemporary British History* 28, 2014, 145–75.

23  'Effect of War-Time Food on Children', *The Lancet* 240, 11 July 1942, 41–2; C. Fraser Brockington, 'Effects of War-Time Nutrition on Children', *Public Health* 55, 1941, 175–8; Margaret O'Brien, 'Diets in War-Time Nurseries', *Public Health* 56, 1942, 128–30; Dagmar Curjel Wilson, 'Nutrition of Rural Children in War-Time', *The Lancet* 238, 4 October 1941, 405.

24  'Child Nutrition in War-Time', *Public Health* 57, 1943, 108.

25  'Child Nutrition in War-Time', 108. The report appeared as 'The Incidence of Rickets in War-Time', *Archives of Disease in Childhood* 19, 1944, 43–67.

26  'Children's Health in War-Time Food and Mortality', *The Lancet* 251, 31 January 1948, 188–9, at 188. See also Hollingsworth, 'Food and Nutrition Policies in Wartime', 336–7.

27  Elenora Simpson, 'The Future of Child Welfare Clinics', *Public Health* 66, 1952, 156–8.

28  'Medical Advances "Not due to Flash of Genius": Sir Henry Dale Confident of Progress', *Manchester Guardian*, 8 October 1957, 3.

29  'Flat Feet not Rickets, now: Change in Clinic Patients Through the Years', *Manchester Guardian*, 2 July 1959, 4.

30  Dagmar Curjel Wilson and Elsie May Widdowson, 'A Comparative Nutritional Survey of Various Indian Communities', *Indian Medical Research Memoirs* No. 34, Supplementary Series to the *Indian Journal of Medical Research*, Calcutta: Indian Research Fund Association, 1942, p. 1. Reflecting 1940s Colonial Welfare and Development Act, the study was intended to illustrate the 'major defects of diet', and thus to identify related failings in traditional food production practices for future correction. p. 7.

31  Arnold, 'The "Discovery" of Malnutrition', 11–14.

32 Wilson and Widdowson, 'A Comparative Nutritional Survey', p. 19.

33 Ibid., p. 52.

34 Ibid., p. 5.

35 Arnold, '"Beriberi Problem"'.

36 Ibid., 311–12.

37 Wilson and Widdowson, 'A Comparative Nutritional Survey', p. 50.

38 Ibid., p. 80.

39 W. P. T. James and Ann Ralph, 'Translating Nutrition Knowledge into Policy' in Widdowson and Mathers, *Contribution of Nutrition*, pp. 340–50 at p. 345.

40 Celia Petty, 'Primary Research and Public Health: The Prioritization of Nutrition Research in Inter-war Britain', in Austoker and Bryder, *Historical Perspectives on the Role of the MRC*, pp. 83–108; Smith, *Nutrition in Britain*; Harmke Kamminga and Andrew Cunningham (eds), *The Science and Culture of Nutrition, 1840–1940*, Amsterdam: Rodopi, 1995.

41 Roberta Bivins, '"The English Disease" or "Asian Rickets": Medical Responses to Post-Colonial Immigration', *Bulletin of the History of Medicine* 81, 2007, 533–68; Worboys, 'Discovery of Colonial Malnutrition', 221; Anne Hardy, 'Beriberi, Vitamin B1 and World Food Policy, 1925–1970', *Medical History* 39, 1995, 61–77.

42 For contemporary reference to India in particular as a 'nutrition worker's paradise', see Arnold, '"Discovery" of Malnutrition in India', 17–18.

43 R. G. Whitehead, 'Kwashiorkor in Uganda', in Widdowson and Mathers, *Contribution of Nutrition*, 303–13, at 303.

44 See the accounts of Roger Whitehead, John Waterlow, and Erica Wheeler, and the critical perspective offered by William P. T. James and Ann Ralph in Widdowson and Mathers (eds), *Contribution of Nutrition*.

45 Sabine Clarke, 'The Research Council System and the Politics of Medical and Agricultural Research for the British Colonial Empire, 1940–52', *Medical History* 57, 2013, 338–58.

46 Whitehead, 'Kwashiorkor', p. 303

47 See Roberta Bivins, 'Ideology and Disease Identity: The Politics of Rickets, 1929–1982', *Medical Humanities* 40, 2013, pp. 3–10.

48 'Increase of Rickets in Two Cities', *The Guardian*, 25 September 1963, p. 16.

49 'Welfare Foods', House of Commons official report (Hansard), 23 March 1964, vol. 692 cc. 28–9; Michael Noble, 'Rickets', House of Commons official report (Hansard), 29 July 1964, vol. 699 cc. 1404–6 at col. 1405.

50 Nesta Roberts, 'Britain has always been multi racial', *The Guardian*, 10 August 1965, p. 4.

51 John Barry 'There is Now Considerable Evidence that Malnutrition May Occur in Britain on a Far Larger Scale than Anyone Has Realised', *Sunday Times*, 6 August 1967, 3; Bivins, 'Ideology and Disease Identity'.

52 Barry, 'There is Now Considerable Evidence', p. 3.

53 W. T. C. Berry, 'Nutritional Aspects of Food Policy', *Proceedings of the Nutrition Society* 27, 1968, 1–8, at p. 3.

54 W. T. C. Berry, *Before the Wind of Change*, Halesworth, Suffolk: Halesworth Press, 1983, pp. 40–57.

55 W. P. T. James and Ann Ralph, 'Translating Nutrition Knowledge into Policy', in Widdowson and Mathers (eds), *Contribution of Nutrition*, pp. 340–50 at p. 345. See Roberta Bivins, 'Coming "Home" to (post)Colonial Medicine: Treating Tropical Bodies in Post-War Britain', *Social History of Medicine* 26, 2013, 1–20.

56 E.g. TNA MH55/2336 'Scottish Health Services Council Standing Medical Advisory Committee: Incidence of Rickets and Diet of Pre-School Children', August 1962; G. C. Arneil, J. C. Crosbie, M. D. Dalhousie, 'Infantile Rickets Returns to Glasgow', *The Lancet* 282, 31 August 1963, 423–5; P. F. Benson, C. E. Stroud, N. J. Mitchell et al., 'Rickets in Immigrant Children in London', *British Medical Journal* [*BMJ*] 1, 11 May 1963, 1054–6; 'Rickets in British Children', *BMJ* 1, 2 July 1966, 558–9; Barry, 'There Is Now Considerable Evidence'.

57 Berry, 'Nutritional Aspects', 3. Emphasis added.

58 Whitehead, 'Kwashiorkor in Uganda', 305.

59 TNA MH55/2336. Draft Letter and Survey Form BPA. May 1962 and 1962 n.d.

60 See Bivins, 'Ideology and Disease Identity' for discussion.
61 J. A. Ford, 'Proceedings: Aetiology of Asian Rickets and Osteomalacia in the United Kingdom', *Archive of Diseases in Childhood* 48, 1973, 827–8; S. W. Stanbury, P. Torkington, G. A. Lumb, P. H. Adams, P. de Silva, and C. M. Taylor, 'Asian Rickets and Osteomalacia: Patterns of Parathyroid Response in Vitamin D Deficiency', *Proceedings of the Nutrition Society* 34, 1975, 111–17; I. Robertson, A. Kelman, and M. G. Dunnigan, 'Chapatty Intake, Vitamin D Status and Asian Rickets', *BMJ* 1, 1977, 229–30; M. G. Dunnigan and I. Robertson, 'Residence in Britain as a Risk Factor for Asian Rickets and Osteomalacia', *The Lancet* 315, 5 April 1980, 770; M. G. Dunnigan, W. B. McIntosh, G. R. Sutherland, et al., 'Policy for Prevention of Asian Rickets in Britain: A Preliminary Assessment of the Glasgow Rickets Campaign', *BMJ* 282, 31 January 1981, 357–60; 'Asian Rickets in Britain', *The Lancet* 318, 22 August 1981, 402.
62 'Diagnosis of Nutritional Rickets', *The Lancet* 298, 3 July 1971, 28–9.
63 Ibid., 29.
64 On this COMA subgroup (called the COMA Panel on Child Nutrition – Rickets and osteomalacia') see TNA MH 148/623.
65 Bivins, 'The "English Disease" or "Asian Rickets" Medical Responses to Post-Colonial Immigration', *Bulletin of the History of Medicine* 81, 2007, 533–68, 548–56. Stroud promoted an wide view of the profession's role in maintaining and improving the health of migrant populations – C. E. Stroud, 'The New Environment', *Postgraduate Medical Journal* 41, 1965, 599–602.
66 TNA MH 148/623S. W. Stanbury to J. M. L. Stephen, 6 March 1973. See also Bivins, 'The "English Disease" or "Asian Rickets"'.
67 M. A. Preece, W. B. McIntosh, S. Tomlinson, J. A. Ford, M. G. Dunnigan, J. L. H. O'Riordan, 'Vitamin-D Deficiency Rickets among Asian Immigrants to Britain', *The Lancet* 301, 23 April 1973, 907–10 at 908.
68 TNA MH148/623 Panel on Child Nutrition, Minutes of a Meeting of the Chairmen of Working Groups, 18 December 1972.
69 TNA MH148/623 Minutes, Panel on Child Nutrition Working Party on Rickets and Osteomalacia Meeting of Chairmen of Working Parties, 9 April 1974.
70 P. Hodgkin, P. M. Hine, G. H. Kay, G. A. Lumb, S. W. Stanbury, 'Vitamin D Deficiency in Asians at Home and in Britain', *The Lancet* 302, 28 July 1973, 167–72 at 167.
71 Ibid., 171.
72 Stanbury indignantly rejected subsequent press interpretations of his research as indicating the effects of skin pigmentation, claiming that his paper had been 'perverted'. S. W. Stanbury, 'Vitamin-D Deficiency in Asians', *The Lancet* 302, 25 August 1973, 446.
73 TNA MH148/623 Minutes, Meeting of Chairmen of Working Parties, 9 April 1974.
74 TNA MH 148/623 Letter, Cooke to Knight, 26 July 1974.
75 TNA MH 148/623 Clippings File. The articles appeared in 1974 in *The Times, The Telegraph*, and *The Guardian* respectively.
76 'Education (Milk) Bill', House of Commons official report (Hansard), 14 June 1971, cc. 42–167.
77 Sylvia J. Darke, 'Monitoring the Nutritional Status of the UK Population', *Proceedings of the Nutrition Society* 36, 1977, 235–40 at 240. Emphasis original.
78 Darke, 'Monitoring The Nutritional Status', 239.
79 A. S. Truswell, 'The need for change in food habits from a medical viewpoint', *Proceedings of the Nutrition Society* 36, 1977, 307–16 at 314, 315.
80 TNA MH 148/623 H. M. Goodall, 'Background notes and draft answer', 7 May 1976.
81 TNA MH 148/624, 'Health Education and Diet for Ugandan Asians: Meeting with Community Relations Commission 19 October, 1976'; TNA MH148/624 S. J. Darke to K. Nagda, 22 December 1977; TNA MH 148/624 P. J. Everett [Committee for Community Medicine, British Medical Association] to R. P. Pole [Public and Environmental Health Division, DHSS], 16 June 1978.
82 TNA MH 148/624 'Report of the Chairman of the Committee on Medical Aspects of Food Policy in December 1978'.
83 For details, see TNA MH148/624.
84 TNA MH148/624 Memo, Darke to Dr. A. Yarrow, 13 December 1978.

85 MH 160/1447 'Asians: Rickets – Meeting with Dr Vaughan, Minister of State (Health) 31 July, 1979'.
86 Ibid.

## Select bibliography

Apple, Rima, *Vitamania: Vitamins in American Culture*, New Brunswick, NJ: Rutgers University Press, 1996.

Arnold, David, 'The "Discovery" of Malnutrition and Diet in Colonial India', *Indian Economic Social History Review* 31, 1994, 1–26.

Barona, Josep L., *From Hunger to Malnutrition: The Political Economy of Scientific Knowledge in Europe, 1818–1960*, New York: I.E. Peter Lang, 2012.

Bivins, Roberta, 'The "English Disease" or "Asian Rickets" Medical Responses to Post-Colonial Immigration', *Bulletin of the History of Medicine* 81, 2007, 533–68.

Bivins, Roberta. 'Ideology and Disease Identity: The Politics of Rickets, 1929–1982', *Medical Humanities* 40, 2013, 3–10.

Kamminga, Harmke and Andrew Cunningham (eds), *The Science and Culture of Nutrition, 1840–1940*, Amsterdam: Rodopi, 1995.

Smith, David F. (ed.), *Nutrition in Britain: Science, Scientists and Politics in the Twentieth Century*, London: Routledge, 1997.

Smith, David F. and Jim Phillips, (eds), *Food, Science, Policy and Regulation in the Twentieth Century: International and Comparative Perspectives*, London: Routledge, 2000.

Vernon, James, *Hunger: A Modern History*, Cambridge, MA: Harvard University Press, 2007.

Ward, John D. and Christian Warren, *Silent Victories: The History and Practice of Public Health in Twentieth Century America*, Oxford: Oxford University Press, 2006.

Warren, Christian. 'The Gardener in the Machine: Biotechnological Adaptation for Life Indoors' in Virginia Berridge and Martin Gorsky (eds), *Environment, Health, and History*, London: Palgrave, 2011, 206–23.

Webster, Charles. 'Healthy or Hungry Thirties', *History Workshop Journal*, 13, 1982: 110–29.

Widdowson, Elsie May and John C. Mathers (eds), *The Contribution of Nutrition to Human and Animal Health*, Cambridge: Cambridge University Press, 1992.

Worboys, Michael. 'The Discovery of Colonial Malnutrition between the Wars' in David Arnold (ed.), *Imperial Medicine and Indigenous Societies*, Manchester: Manchester University Press, 1988, 208–25.

# 16

# SOCIAL GEOGRAPHIES OF SICKNESS AND HEALTH IN CONTEMPORARY PARIS

## Toward a human ecology of mortality in the 2003 heat wave disaster[1]

*Richard C. Keller*

Descriptions of the catastrophe are uniform, cast in a language of extremity and exception. In the first two weeks of August 2003, France was struck by 'an unprecedented heat wave' of 'exceptional intensity, duration, and geographic extent,' one of a 'magnitude' and 'severity' unmatched in more than a century. A 'murderous heat wave' that served as a capstone to an already 'particularly murderous summer,' the *canicule* of August 2003 wrought 'dramatic health consequences' on the French population in its death toll of nearly 15,000.[2] Such language rightly points to a meteorological crisis. The degree of this disaster is perhaps most tellingly measured by the absence of any significant reduction of mortality in the aftermath of the heat wave, which indicates that most of the deaths were directly attributable to heat, rather than to a 'harvesting effect' of anticipated deaths in an already weak population.[3]

Yet such descriptions are misleading in their attribution of this crisis to an acute natural disaster. They are uniform not only in their language, but also in the sense of *inevitability* that they convey. By this logic the collision of extreme temperatures and an aging population sets the stage for an ineluctable tragedy. Moreover, given demonstrable increases in global temperatures and disturbing forecasting models that predict more frequent heat waves of even greater duration and intensity in the coming decades, climatologists also argue that such situations are getting worse, with potentially egregious health consequences.[4] As convincing as these models are, they privilege temperature over human society as the decisive variable in shaping mortality. Several factors point to the need for pushing beyond these explanations of the proximate causes of heat mortality. Critical demographic disparities and a micro-geography of death indicate patterns of mortality that are at least as ascribable to the social ecology of urban France as to the more ostensibly 'natural' causes of disaster. Because of the importance of social phenomena in shaping the catastrophe of the *canicule*, the crisis deserves the attention of historians attuned to the making of a dangerous social landscape in the course of the late twentieth century.

Climatologists tend to examine weather and nature on a worldwide scale: 'global' warming is merely one example of this phenomenon. Yet despite a pervasive discourse on the relentless pace of globalization, we reside largely in local environments, however subject they are to wider global influences. Much as environmental historians have done for years – and as medical geographers and urban reformers began doing in the eighteenth and nineteenth centuries – epidemiologists and climatologists have thus begun to call for a 'downscaling' of investigations of climate change to local contexts and conditions as well as analyses of the relationship between human society and the environment.[5] While they tend to mean by this a series of quantitative assessments of local environments, scholarship on the social dimensions of environment and health suggest that an emphasis on the local demands careful qualitative analysis as well.[6]

Disasters such as industrial accidents, heat waves, droughts, floods, famines, and tsunamis find their origins in global phenomena, but their effects depend almost entirely on the local human environments with which they collide. As Amartya Sen and Mike Davis have each argued, droughts lead to famine only in specific political environments. Tragedies such as the Bhopal explosion originate in multinational capitalism and the outsourcing of chemical production, but the explosion's effects are exacerbated by the fragile social landscape of an unregulated industrial environment. A flood wall spares Mark Twain's home from the swelling of the Mississippi but reinforces the river's volume and intensity, thereby worsening conditions for poor neighbourhoods downstream. Most relevant to this study, sociologist Eric Klinenberg's brilliant analysis of the Chicago heat wave of 1995 described how patterns of social isolation, fears of urban crime, and ineffective public responses led heat to kill hundreds of elderly African Americans in the city, but very few whites in the suburbs.[7] Such events reveal human vulnerabilities to extreme crisis. Yet as these events disproportionately affect the *most* vulnerable members of a given society, they are assimilated into the fabric of human experience as acute incidences of chronic suffering.[8] Human suffering is often more a consequence of social divisions, economic policy, and the failure of political will than the whimsical, if tragic, powers of climate.

Taking the French heat wave of 2003 as its focus, this chapter argues that there is an important place for historians in this project. Drawing on recent social studies of disaster, I argue that the heat wave disaster exploited a fragile social ecology with deep historical roots in the social geography of sickness and health of modern Paris. What follows aims at a social outline of human vulnerability by pointing to the curious intersection of phenomena that are apparently discrete, yet whose combination greatly exacerbated the health effects of deadly temperatures. Describing in turn the rise of a social geography of pathology in Paris in the nineteenth century, the social isolation of the elderly, and the emergence of particular cityscapes that exacerbated vulnerability during the disaster, the chapter indicates the possibilities for a historically situated examination of epidemiological disparities in the contemporary city.

## Medical and social geographies of the city

The relationship between health and place is an ancient one. In the Western tradition, it dates at least to Hippocrates, who in *Airs, Waters, Places* describes a delicate equilibrium between environment and health. By the early-modern period, medical geography was an entrenched specialization that was tightly enmeshed with the expansion of European empires. Military physicians such as James Lind wrote about healthy and unhealthy landscapes for Europeans, characterizing the tropics as a lethal environment due to the 'numerous swarms of flies, gnats, and other insects, which attend putrid air'; their 'thick, noisome fogs'; and the fact that in these zones 'a corpse becomes intolerably offensive, in less than six hours.' Lind noted that 'in such places, during excessive heats, and great calms, it is not altogether uncommon, especially for such Europeans who are of a gross habit of body, to be seized at once with the most alarming and fatal symptoms of what is called the yellow fever.'[9] Such readings of the connections between latitude, health, and race were deeply characteristic of the late eighteenth century and persisted well after the so-called bacteriological revolution of the late nineteenth century.[10]

By the early nineteenth century, however, the rapid urbanization of Europe facilitated the emergence of a new form of medical geography, one that focused on the social disparities of sickness and health that characterized distinct regions of cities. Social reformers such as the British hygienist and social reformer Edwin Chadwick chronicled the horrific living conditions of the English working classes, borrowing from the rhetoric of Thomas Malthus to attribute the high rates of disease among workers and paupers to both immorality and the failings of the English poor laws. Focusing on the emerging industrial cities of England, Friedrich Engels, by contrast, attributed these conditions to the inherent inequalities of capitalism, arguing that an economic geography of class – marked by pollution and disease – predisposed vulnerable English workers to poor health outcomes.

In France, figures such as Louis-René Villermé and Alexandre Parent-Duchâtelet pioneered the study of morbidity and mortality in the early nineteenth century. In the 1820s, Villermé engaged in a study of all-cause mortality that generated an economic geography of life and death in Paris. He drew on a newly available data set collected by Paris's nascent bureau of vital statistics, and sought to determine why certain neighbourhoods experienced high mortality and others experienced low mortality.[11] Exploring deaths over a five-year period, Villermé examined the physical geography of the city, seeking to determine the roles played by factors such as proximity to the river, wind direction, elevation, water quality, and population density in mortality statistics. Yet careful study indicated that these factors played no role, leaving only economic inequality as the critical variable, pointing toward a political economy of mortality in early-nineteenth-century Paris.[12] With the arrival of cholera in Paris in the 1830s, Villermé's work assumed a new importance. Combined with the studies of Parent-Duchâtelet and other urban hygienists, Villermé's investigations of the links between poverty and mortality indicated the social, rather than moral or miasmatic, foundations of the cholera epidemic.

Such work contributed to the wholesale reconstruction of Paris in the mid-nineteenth century. In 1853, the Baron George-Eugène Haussmann, then Prefect of the Seine, launched a two-decade reimagination of the city that remains largely intact at present. Public health concerns were paramount in the project.[13] Paris's population more than doubled in the mid-nineteenth century, growing from some 800,000 in the 1820s to more than two million in the 1870s. This congestion placed unsustainable pressure on the city's housing, water, and drainage infrastructure, and created a perfect ecology for diseases such as cholera, tuberculosis, typhus, and other killers.[14] Hygienist concerns and rhetoric were deeply entwined in the building project, which, as the geographer Matthew Gandy has argued, employed bodily metaphors to describe a functioning city as a 'holistic' body.[15] The Bois de Vincennes and the Bois de Boulogne, at the city's eastern and western extremes respectively, were the lungs of Paris, while its new fresh water systems and its main traffic arteries were its circulatory systems; the new sewers, among the most critical of these systems, were the city's bowels. Haussmann largely succeeded in his goal: cholera never returned to the city in a serious way (although tuberculosis remained a powerful scourge) and Edwin Chadwick purportedly said to Napoleon III, 'may it be said of you that you found Paris stinking and left it sweet.'[16]

The remaking of Paris recreated the city's social as well as its medical geography. An extensive historiography outlines the ways in which Haussmannization destroyed a medieval city and erected a modern one in its place. In particular, critics have signaled the ways in which Haussmann's remaking of the city amounted to an assault on poverty and the poor. The principal targets for renewal were exactly the slums whose staggering mortality levels Villermé had noted in the 1820s. Most of these neighbourhoods were destroyed in their entirety. New construction, by contrast, catered to an entrenched bourgeoisie: a Paris of luxury and leisure was born out of the destruction of the city's poorest and most marginalized districts. The displacement involved some 300,000 people, forced from miserable, squalid housing into homelessness.[17] The remaking of the city amounted to a marginalization of poverty through the establishment of a bourgeois city centre that was ringed by a proletarian 'red belt' in the outer arrondissements and, eventually, the Paris banlieue, a geography of class that has remained essentially in place to the present.[18]

## Social vulnerability in the contemporary city

Where nineteenth-century hygienists pioneered a social geography approach to urban epidemiology, contemporary scholars have refined these techniques, developing quantitative indices and qualitative evaluations of vulnerability in the modern environment. Among the best examples of these approaches is the work of Susan Cutter. A geographer who studies natural disasters, Cutter has developed a Social Vulnerability Index with the aim of providing a metric for determining vulnerability to natural hazards in a given environment. The index includes factors such as age, personal wealth, density of the built environment, condition of housing stock, tenancy rates, ethnicity, the occupational makeup of a given region, and infrastructure dependence among other data points, which provide for a comparative framework for gauging broad social vulnerability. For example, disproportionately elderly communities may be more vulnerable to natural hazards because of higher rates of disability that coincide with old age: such factors may well inhibit evacuation in the face of an oncoming disaster. Likewise, high poverty or tenancy rates often coincide with lower rates of insurance carriage, decreasing a community's resilience in the face of meteorological hazards.[19]

These methods characterized the approach of epidemiologists and demographers to the post-mortem analysis of the 2003 heat wave in France, and in large French cities in particular. When measured in terms of mortality, the heat wave remains the worst meteorological disaster in French history to date. Yet investigations into the heat wave's mortality disparities quickly made clear that the disaster was a social as well as a meteorological catastrophe. The August heat wave was in fact the third heat wave to sweep through France and across Europe in less than six weeks. The first struck in late June, and the second in mid-July, each pushing daily high temperatures near

40 degrees Celsius for extended periods. Combined with weeks of little or no rainfall, such temperatures were devastating: news reports between late June and early July placed heat, rural drought, and their effects for livestock among their top stories.

Yet the August heat wave was of a different magnitude in its meteorological and social effects. Temperatures spiked throughout the country beginning on 2 August. Between 4 and 12 August, Paris witnessed mean high temperatures that surpassed 100° Fahrenheit (38.1° Celsius), with daily lows that never dipped below 73° F (23.3° C). Temperatures in 15 percent of France's weather stations surpassed 104° F (40° C) during the period, hitting 108° F in Orange on 11 and 12 August. On the same days, Paris's daily lows broke all records, dipping only to 78° F (25.5° C).[20] Ozone levels in Paris were also the highest in years, making for a noxious mix of enervating temperatures and foul pollution.

Bodies living and dead streamed into emergency rooms throughout France in droves beginning on 4 August, mounting steadily in the following weeks. In Paris, Patrick Pelloux, the director of France's emergency room physicians' union, called attention to the crisis situation on 9 August in a report in Le Parisien, backed by an interview with the television station TF1 on the following day in which he noted dozens of unanticipated deaths in his hospital alone.[21] Emergency services and the government were caught off guard. Thousands of hospital beds had been closed as a cost-saving measure, while physicians and nursing staff were cut to the bone due to the tradition of August vacations. Meanwhile, the government itself was away. President Jacques Chirac was attending the G-8 summit in Canada, while Prime Minister Jean-Pierre Raffarin and Minister of Health Jean-François Mattéi, both on vacation, were extremely slow to respond. Both conceded to television interviews on site during their vacations, but only returned to Paris to address the crisis on 13 August, after the heat had already broken. On 14 August Raffarin launched the 'plan blanc' in Paris, and later throughout France, opening many of the closed hospital beds, but far too late to ameliorate the situation. Chirac only addressed the nation concerning the heat wave on 21 August, nearly three weeks after the crisis began.

Epidemiological reports revealed startling and steadily rising numbers of excess deaths – that is, numbers of deaths that surpassed daily averages for the five previous years.[22] Beginning with 400 excess deaths on 4 August, the toll reached 3,900 by 8 August and 14,800 by 20 August, well after temperatures had dipped significantly; on the single day of 12 August, France experienced nearly 2,200 unanticipated deaths.[23] Funeral services and morgues were completely overwhelmed; the city of Paris was forced to resort to storing the bodies in refrigerated food warehouses at Rungis on the outskirts of the city. In the aftermath, teams of genealogists donated their services pro bono to attempt to find the next of kin for hundreds of unclaimed bodies.[24]

Heat stroke was the most commonly noted cause of death – a pathology characterized by the body's core temperature surpassing 105° F at the time of death; reported cases included some whose core temperatures had soared to 109° F before death. More important, however, is that despite its widespread discomfort, such intense heat struck unequally. Certain populations are at abnormally high risk for heat stroke, as the death toll indicates. Those over 75 experienced a mortality rate 70 per cent above normal, and constituted over 80 per cent of total excess mortality during the heat wave.[25] Much of this risk is a result of natural failures of thermoregulation in aging bodies, in which bodies cease to dissipate heat efficiently. Signal processing is another factor: messages of dehydration begin to move much more slowly from the body to the brain as one ages, meaning that the elderly often take in less fluid than the body requires. Mental disorders also play an important role in elevating risk for all populations, regardless of age. Many of the mentally ill suffer from the same failures of signal processing as the elderly, greatly exacerbating risk for many older psychiatric patients. Schizophrenics taking neuroleptic drugs such as Thorazine or Haloperidol also suffer from exacerbated dehydration related to their medications, again elevating risk.

Yet a number of social factors – many conforming closely to Cutter's Social Vulnerability Index – confound these more strictly biological or 'natural' functions of risk. Age was the most obvious of these. With some 80 percent of the heat wave's nearly 15,000 victims aged 75 or older, politicians and the media instantly seized on the social isolation that appeared to surround the victims of the heat wave, attributing the catastrophe to a failure in family and community solidarity through descriptions of the elderly as living on the margins of society. The story of Adèle Angèle offers an important example. At age 92, Adèle had never married and had no family, and lived alone in the banlieue apartment she had occupied since it was built in the 1960s. Her neighbours unanimously attested to her good condition despite her age. The clerk at the supermarket 600 metres from her home noted that she was 'a bit stooped,' but always 'alert.' A neighbour – who departed on vacation at the end of July – claims that she checked on Adèle before the onset of the heat, and that Adèle refused all offers of help, even an alert system in her home in case of emergency. Yet on 15 August, the fire department discovered her collapsed on the floor, a chair overturned, a half-eaten meal and an emergency telephone number on the table before her. No one knows exactly when she died, although it was clearly several days before she was discovered: the reporting officer noted that the heat certainly 'doesn't help preserve the body.' Three hours later, the mortician arrived and sealed her body in a plastic bag and a coffin, promising to return shortly to remove her from the apartment. The besieged funeral service in fact arrived 12 days later to remove the body.[26]

The case points to the problematic nature of a discourse of isolation and its role in the state's response to the crisis. Witnesses were eager to point out that Adèle had repeatedly refused offers of help: she insisted upon doing her own daily marketing and living without domestic assistance. By extension her isolation appears self-imposed: a failure of neither social solidarity nor the state. One thus finds a concomitant pattern of denial and the shifting of blame from the agencies of the state onto the victim. Distancing the health ministry from responsibility, Jean-François Mattéi insisted that his staff 'acted properly, and in a timely fashion' to the emerging crisis, and was quick to note during a press conference that 'half of these deaths took place *outside* the hospital.'[27] A similar elision was at work in denials that deaths were heat-related at all, that is, the ascription of excess mortality not to heat but to the heat's 'harvesting effect' that accelerated the deaths of those already with one foot in the grave.[28]

Such claims are typical yet tragic. They help city and state authorities to distance themselves from acute crises, much as ascription of natural disaster to an 'act of God' provides cover for infrastructural or political problems that place populations at heightened risk.[29] Yet they also effectively write the elderly poor out of existence as fully human individuals. If their isolation is self-imposed, if their deaths are inevitable and merely hastened by extreme temperatures, if their deaths took place 'outside the hospital,' it is as if they have opted out of the social contract. By this logic their aging, poverty, and isolation constitute transgressions against citizenship. Their lives of self-imposed social abandonment, or, as one of Adèle's neighbours claimed, lives that are 'invisible,' are somehow less than fully human. Public statements by the state reinforced this dehumanization of the elderly. When asked about the possibility of implementing an effective heat warning system for vulnerable populations, Mattéi claimed: 'You know, the elderly, as they don't have very good memories, often from one moment to another, so the preventive messages that we could air . . . well, they'd forget them the same day!'[30]

As a global phenomenon, aging has preoccupied anthropologists in particular, who have demonstrated the fascinating ways in which the cultural dimensions of aging constitute an important field of inquiry into the intersections of biology and culture.[31] In France in particular, for the past several decades social scientists, government functionaries, and physicians have called attention to the significant aging of the French population. Accompanying this awareness has been the growth of a rich literature surrounding the politics and pragmatics of aging. Advice manuals for the aged and their families, a social scientific and journalistic literature on nursing homes, and polemical essays on styles of aging have fueled a vigorous debate over how one grows old in contemporary France. Historians have by contrast had relatively little to say about the phenomenon of aging in the modern West. Yet recent works point to important questions that force a reexamination of social knowledge about the elderly and the *canicule*.

Mattéi's characterization of forgetful seniors and the idea of a decrepit elderly population living in a state of marginalization find their origin in a crisis of representation of agedness that has marked French politics and media since the early 1960s. As sociologist Isabelle Mallon has argued, images of the elderly have 'oscillated between admiration and pity.'[32] On the one hand is an image promoted by advertisers, developers of retirement communities, and the leisure industry, catering to the desires of a new demographic: the idea of a fit, tanned class of successful early retirees, enjoying the first phases of grandparenthood and deserved leisure. Yet such an image contrasts powerfully with the figure of the impoverished, vulnerable, but also parasitic senior, a victim of poor economic circumstances and a failing body. The former notion is an artifact of postwar prosperity, which witnessed the development of the new figure of the youthful senior.[33] The latter, by contrast, has been far more prevalent in twentieth-century political discourse. As Hervé Le Bras has argued, the pervasiveness of this notion is in part a product of language: in contrast to the English term 'aging', which connotes a natural chronological process in a subject of any age, the French term 'vieillissement,' or 'getting old,' signifies a process of decay specific to the end of life. 'Vieillesse,' or old age, is always a 'degradation.'[34] It is a step toward death, rather than a process of maturation.

This association between the aging of the population and national decadence itself was a frequent iteration of the interwar period. A concern about a crisis in France's population in the early twentieth century – a realization of a birth gap between France and Germany in particular – was given a fresh impetus by the devastating death toll of the First World War.[35] Demographers seeking to explain the phenomenon of French depopulation pointed not only to lower birth rates, but also to the aging of the French population as a critical factor in the development of a national vulnerability. As a result, the elderly became the objects of significant blame. For nationalist demographers and politicians, by growing older and withdrawing from productive activity, they became a drain on a society desperate for economic recovery from the war; as parents of too few children, they bore responsibility for France's depopulation. A significantly gendered language informed this invective: combined with women's longer life expectancies, the war's heavy toll on men meant that by the end of the next world war the face of French aging was female. In the aftermath of the Occupation, the elderly had become an icon of national weakness. 'The conditions of modern warfare,' one demographer argued in 1948, demanded armies comprised of 'the young, in full possession of their physical means.'[36] As two of his colleagues concurred, an 'invasion' by the elderly preceded the invasion of the Germans; thus 'the terrible failure of 1940, as much moral as it was material, should be linked in part to this sclerosis.'[37]

This stigmatization of the elderly points to a fascinating genealogy about the relegation of the elderly to a status on the margins of citizenship. Far from a romanticized notion of an appreciation of senescent wisdom that some historians have seen in the pre-war era,[38] it suggests a critical set of epistemological links between spheres of production, reproduction, and citizenship. In a 1976 lecture at the Collège de France, Foucault argued that biopolitics – or government through the deployment of discourses of public health, hygiene, and population – first seized on aging as an important site of intervention beginning in the modern era. For Foucault, the rise of industrial capitalism initiated a deep concern for the ineluctability of the process of aging and its inevitable extraction of subjects from 'the field of capacity, of activity.'[39] Aging and its social knowledge constituted a dividing practice that measured inclusion and exclusion by reifying social phenomena as biological expressions; discourses of aging thus operated in a manner analogous to those surrounding insanity, medical pathology, and sexuality in the same period. Yet whereas Foucault's investigations of other dividing practices have spawned enormous scholarship, historians have been relatively slow to engage with aging, leaving critical questions about the culture of aging in France under-examined. When, specifically, did aging become a significant concern of the state? How has aging changed in the transition from industrial to post-industrial society? How has isolation become the dominant discourse about the elderly in France? What cultural factors have contributed to its development, and how have they shaped the patterns of bare life that have come to dominate representations of the elderly?

The consumer orientation of French society in the post-war era witnessed a tightening of the link between aging and economic productivity. The early 1960s marked the first explicit engagement of the French state with the question of aging, via an official report by the Prime Minister's Commission on the Problems of Old Age, issued in 1962.[40] Named for its author, Pierre Laroque, the report demonstrates an important shift in the notion of national decadence and old age: it records a transition from the sentiment that the elderly weakened the nation to the idea that they signaled economic and political danger to the Fifth Republic. Noting that 'of all the countries in the world,' France had the 'highest proportion' of people over 60, the Laroque report connects to earlier fears by linking the aging of the population both to the 'insufficiency of the French birthrate' and to a dystopian future. The report seized particularly on the threats that aging posed for the nation at this dynamic moment in France's development, characterized by economic and social modernization, as 'politically and psychologically, aging means conservatism, attachment to old habits, the failure of mobility, and an inability to adapt to the evolution of the current world.'[41]

The Laroque report vacillates between a sympathetic analysis of the real difficulties faced by the elderly, and a language of objectifying difference portraying the elderly as a self-centred drain on social resources. The report recognizes that the 'misery and suffering of the elderly population' are 'no longer characteristic of a civilized nation,' and actively 'condemns' any 'segregation of the elderly population,' seeking instead social reintegration of the aged into the fabric of modern life.[42] Yet it also depicts that population as itself resisting this integration. Whereas the traditional family structure of the past assured the 'maintenance' of the elderly, such arrangements were 'no longer compatible with the state of mind [of the aged] or the conditions of modern life.'[43] Rejecting life with their descendents, the elderly appear to prefer the difficulties of managing a household to the 'psychological obstacles of everyday life' as dependents of their juniors.[44]

The report thus initiated a core element of discussions of aging in the post-war era: the notion of the self-imposed isolation of the elderly. The central emphasis of the report echoes Foucault's outline of a biopolitics of aging, as its key recommendation is the 'encouragement' that the elderly remain 'active' members of the labour and social markets.[45] Toward this end the obligation of the state is to provide pensions for those who cannot work after the age of 60, and for all after 65, while also ensuring the positions of those over 50 in the labour market. Yet as scholars of aging in France have noted, the link between aging and economic production placed the elderly in a classic bind. The framing of aging in terms of production has forced an alignment between retirees and labour syndicalism, producing powerful tensions around an inherent conflict of interest that sets the protection and job security of elderly workers against the interests of younger workers in an era often marked by staggering unemployment.[46]

This notion of the elderly as a burden has persisted into the present, in constant tension with the image of a youthful, socially integrated senior population enjoying a so-called 'third age of life.' In contrast with this 'Papy boom' is what journalist Gérard Badou has described as the 'realities of growing old,' which have produced a 'new third estate.'[47] The French have entered a moment in which many can expect to reach their mid-eighties. Yet, Badou notes, the collective enfeeblement of an aging population (with 500,000 cases of Alzheimer's disease in France alone) has diverted the science and health industries from productive endeavours and the advancement of human health into a losing battle against aging. The elderly, for Badou, have become the 'gluttons' of the health-care sector, all in the interest of preserving life at the limits of human existence.

This condition is exemplified by the plight of those in French nursing homes and hospices, the site of some 20 per cent of France's heat wave deaths in 2003. A growing literature has pointed to the problems of retirement homes and hospices as 'the privileged space of dependent aging.'[48] Despite the Laroque commission's efforts to avoid the 'segregation' of the elderly, the nursing home has become what anthropologist João Biehl, speaking in a different context, has called a 'zone of social abandonment,' or a de facto space of bare life.[49] The nursing home is an entryway into death's waiting room: 'to be confined here is already to die a little.'[50] Private sector homes have come to constitute a 'market' for the 'trafficking of the elderly,' according to one journalist: repeated investigations have pointed to an epidemic of unqualified personnel, dramatic overcrowding, dreadful hygiene, and illegal drug-dispensing, including the use of neuroleptic drugs as 'pharmacological straitjackets' in the private nursing home industry.[51] In the public sector, for the poorest of the elderly, these homes are, according to one critic, 'neither camp nor prison. In a sense, they are worse.' Here 'one dies of the deep sadness of death, without tragedy . . . Here, nothing but the end, nothing but the end.'[52] As a product of the fiscal belt-tightening of the 1990s, cuts in social services have produced dramatic staff shortages, dissatisfied and unqualified personnel, and often egregious living conditions. Their residents are 'citizens at the end of their rights,' their personnel dehumanized through a 'daily habituation to decrepitude and death.'[53] Placement thus relegates one to abandonment of the 'good life' in an Aristotelian sense – the political life, the life of citizenship and the individual – for the bare life of invisibility, an act that amounts to an abandonment of personhood and the right to the good death as an individual that contains meaning for family and community. The heat wave was thus less a scandal in its own right than a revelation of scandalous conditions of the vulnerability of the aged – a vulnerability resulting not from a failed community, but from a failed commitment to the elderly by the state.[54]

## The heat wave as urban pathology

In addition to the dramatic age disparities of the death toll, a second important disparity merits historians' close attention. A deep divide marked mortality between urban and rural populations, with cities bearing the brunt of excess deaths. Whereas rural communes and small cities experienced death rates between 40 and 50 per cent higher than normal, the Paris region witnessed rates 140 per cent higher than expected. Epidemiologists have ascribed much of this difference to two phenomena: the existence of stronger social ties in rural settings than in alienating cities, and what climatologists call the 'urban heat island effect.' The latter is the more significant in most analyses of the *canicule*. Land-use patterns have powerful effects on heat concentrations. A preponderance of reflective surfaces, vehicle exhaust, ozone concentrations, the effect of tall buildings on stifling breezes, large concentrations of heat-producing sources, and other factors particular to urban environments create important temperature differences across landscapes, at times as much as 10° F between downtown regions and surrounding rural areas.

The heat island effect raises important questions about the relationship between nature and culture that have been at the heart of a growing literature in environmental history, and specifically the environmental history of France.[55] Such questions include, most prominently, what about a disaster is natural if its origins are found in human actions, including human effects on global climate change and the effects of architecture and land use on dangerous urban microclimates? More thoroughly in the domain of social history, however, it is clear that the effect of the temperatures created by heat islands on mortality is open to question: temperatures were higher in much of the South of France for more sustained periods than in Paris, with significantly lower mortality rates. Much of this may be attributed to the habituation of those in the Midi to higher temperatures and an experienced adaptation to heat waves (although a 1983 heat wave in Marseille was devastating to that population). Yet other elements appear to play equally important roles in shaping the urban epidemiology of heat deaths.

Although heat islands have a powerful impact on the microgeography of climate, their role is deeply confounded with a range of forms and disorders of urban sociability and habitation. In its 'definitive report' on the *canicule* the Assemblée Nationale signaled the 'disastrous' condition of many Paris apartments as a major variable in mortality patterns.[56] Although it emphasized the heat island effect, the Institut de Veille Sanitaire's extensive study of urban heat mortality confirmed, at least implicitly, the importance of concentrated poverty in its ranking of key risk factors for death among those living at home during the crisis.[57] Those who were at greatest risk for death during the heat wave were: those identified 'socio-professionally' as 'working-class or other,' the lowest income category considered, who were 264 per cent likelier to die than those with professional backgrounds; those living in older buildings without elevators (40 per cent likelier to die); those living on the highest floors of apartment buildings, directly beneath zinc and aluminum roofs (133 per cent); those living without private bath facilities (147 per cent); and those living without televisions or radios (100 per cent).[58]

Each of these problems contributed both directly and indirectly to mortality. As the Institut's study pointed out, those without televisions had more limited access to reports about the heat wave and health warnings; those without baths had little recourse to washing as a survival strategy; those on the highest floors suffered the most punishing heat. But each is also a symptom of poverty and marginalization in a consuming society. The study, however, ignored the explicit role of poverty, as well as the concentration of these factors in certain neighbourhoods and the effects of entire marginalized zones on mortality. Yet individual habits and urban space affect one another reciprocally. Marginal, 'difficult,' or 'sensitive' neighbourhoods – such as the 'outlaw zones' of the *banlieue* – encourage the seclusion of their residents, a sensibility that reinforces the desertion and desolation of the streets. According to a range of studies, such phenomena disproportionately affect seniors, who find themselves at heightened vulnerability to crime as their bodies begin to fail.[59] Behaviour during the heat wave that is typical of those who live in these difficult circumstances affected risk significantly, according to the study. Those who never left their apartments were nearly four times as likely to die as those who left daily, while those who participated in no 'social, religious, cultural, or leisure activities' outside the home were over six times as likely to die as those who did.[60] Because they ignore the social environment from which these cases isolate themselves, such figures misleadingly suggest that 'self-imposed' forms of isolation were the key element in determining vulnerability.

The contemporary city's microgeography of mortality has at least as much to do with local forms and disorders of urban sociability and habitation as it does with the city itself as a pathological mechanism. The key question is perhaps not *whether* cities show higher mortality rates than other environments, but instead *where* cities show concentrations of mortality. What particular environments and neighbourhood structures coincide with higher risk of death during heat waves and other disasters? Such an analysis allows for a finer accounting of how to evaluate risk and vulnerability in urban communities by potentially exposing more specific factors that influence health outcomes. These factors might include building styles, neighbourhood layouts, crime rates, concentrations of relative wealth and poverty, access to transportation, and proximity to hospitals and other health-care services among others.

Among the most fascinating studies of the spatial dimensions of vulnerability during the disaster are those generated by the Atelier Parisien de Santé Publique. The Paris mayor's office hired the Atelier's staff to conduct a number of official studies of the distribution of mortality in the city.[61] Their studies plot excess mortality throughout the city, indicating important intersections of neighbourhood and risk, with attention to factors such as socioeconomic status of local households and the median age of the local population.[62] Such projects indicate that land-use patterns and neighbourhood config- urations may well have profound impact on mortality rates. For example, the highest concentrations of mortality are grouped in the southern arrondissements (the twelfth, thirteenth, fourteenth, and fifteenth). Yet there are other important correlations of mor- tality with the cityscape. The architectural styles of different neighbourhoods appear to correlate with differential mortality rates. The thirteenth and fifteenth arrondissements show high excess mortality; these are also the districts of the city with the highest con- centrations of planned high-rise commercial and residential structures that date to the 1960s and 1970s. Such structures offer a textbook context for studying the urban heat island effect. They are concrete, steel, and glass towers with no shade or green space, heated further by underground parking garages. They also point to the significance of a social ecology of place, a critical site of inquiry for urban environmental historians, who have highlighted the ways in which the modern city marks an intersection of tech- nology, nature, and society.[63] These critics and others have pointed to the complexities that are built into the city as a technological system, one that exacerbates vulnerability for certain populations due to its effects on sociability and culture.[64]

These high-rise complexes represent the dream of mid-twentieth-century urban planners who saw in them the capacity to construct futuristic neighbourhoods, which organized residential building towers around esplanades that contained businesses and retail outlets.[65] The towers' residents could take advantage of elevators to retreat from the hustle-and-bustle of city life, while maintaining easy pedestrian access to supermarkets, restaurants, and places of employment. The architect Le Corbusier was among the chief proponents of such complexes. Describing them as 'machines for living,' Le Corbusier argued that a technologically inspired architecture utilizing concrete and steel held the promise to bring light and air, those luxuries of the bour- geoisie, into working-class living environments.

Yet while these structures appeared utopian at the moment of their invention, they have proven deeply problematic in the decades since their inception. Many critics have argued that Le Corbusier's 'machines for living' have now become technologies of alienation.[66] Although they inserted a new community form into given areas, they also powerfully disrupted existing community structures. Among the best examples is that outlined in Henri Coing's 1966 study of the demolition of 'Îlot 4' in the thirteenth arrondissement. The 'ilots insalubres' were zones within Paris that city officials had targeted for urban renewal for decades. For city planners, they were bounded regions of filth, poverty and disease; for local residents, they were *quartiers*. The mass clear- ing of Îlot 4 for the construction of the Olympiades complex near the Place d'Italie entailed a complete fragmentation of the existing community. Much like the process of Haussmannization, the replacement of former buildings with the new complex displaced the community, who could no longer afford the higher rents imposed by the new construction.[67] But another phenomenon also attended the recomposition of the neighbourhood. The construction precipitated steep economic declines for existing businesses. Small, locally owned cafés, bakeries, and shops were forced out by the development, replaced by large, impersonal supermarkets and restaurants. In Îlot 4, the number of cafés and bakeries declined by 90 per cent; in the Rue Nationale alone, 48 cafés were replaced by one.[68]

As Coing noted in the mid-1960s, the hundred-odd cafés in Îlot 4 served the neighbourhood's distinct ethnic and social communities, offering a living space for those who suffered daily indignities in tiny, overcrowded and unhealthy apartments. They helped residents 'find a truly human way of living in an inhuman place.'[69] While the redevelopment of the neighbourhood dramatically cleaned up the area, it also fragmented an existing community. The new agglomerations engender insularity. With many essential services located on these campuses, residents tend not to leave – further reinforcing their isolation – or to reject the services and shops in the area out of nostalgia for the locally owned businesses that formerly characterized the neighbourhood, and patronize something reminiscent of these outside their new community.[70] Any venture to the outside world entails a long wait for an elevator, followed by an often-circuitous route down to the sidewalk. Children thus play on the concrete esplanades rather than in green parks, and residents walk their dogs across concrete esplanades instead of local sidewalks. The closing of the centres of social life in a marginal region such as Îlot 4 assumes a new importance in light of the staggering mortality rate in the neighbourhood: this segment of the thirteenth was among the worst affected during the heat wave. Considering that social isolation was a critical factor that exacerbated mortality risk during the heat wave – isolation inspired by age, mental illness, disability, or addiction – an urban organization that reinforces isolation merits close scrutiny as a risk factor in its own right.

There has been a great deal of scholarship on the tandem issues of insecurity and urban form in the Paris banlieue.[71] But it is critical to recognize that these kinds of complexes inflicted an identical urbanism on some of Paris's most marginal neighbourhoods – especially near the Place d'Italie, the Front de Seine or 'Beaugrenelle' in the fifteenth, the Rue de Flandre in the nineteenth and the Rue de Belleville in the twentieth. *Habitations à loyer modéré*, or HLMs – the self-contained *cités* of the banlieue that some have blamed for the riots of fall 2005 as well as other forms of urban insecurity[72] – have also overtaken these neighbourhoods in Paris *intra-muros*. They constitute a sort of banlieue in the city, complete with new forms of mobility and sociability that are more characteristic of the peri-urban region than of life in the centre city. They fractured pedestrian life, for example, and prompted a retreat from public life into the insularity – or isolation – of the home.[73]

In addition to the everyday insecurities that these complexes impose on their communities, many of these structures place their residents at particularly high risk during heat waves. Few of them are effectively ventilated, for example, and their windows only open a few inches – a design intended to prevent suicides. The reduced water pressure experienced during heat waves is particularly acute on higher floors of these structures. One resident on the seventeenth floor of a high-rise on the Place d'Italie, who had just brought a newborn home from the hospital when the heat wave struck, described the intensity of the heat as literally dizzying – she found herself walking into walls in delirium. She also told me that filling her tub took more than an hour due to the low water pressure, and that by the time the water was deep enough to cool herself and her baby, it had warmed beyond effectiveness.[74] Power outages in these buildings render their elevators useless, trapping residents in place, while those who do get out are forced to walk across elevated esplanades in the blazing sunlight rather than shaded sidewalks. It is unsurprising that one of the highest concentrations of heat wave mortality appears in the fifteenth arrondissement bordering the Seine, a district that consists of an agglomeration of such buildings, while the more traditionally designed neighbourhood of Auteuil directly across the river in the sixteenth experienced among the lowest death rates during the crisis. To be sure, major differences in the socioeconomic level of these communities compounds their architectural organization, but community layout surely contributes to the vulnerability of the one and the resilience of the other.

## Conclusions

The heat wave of 2003 has been framed chiefly as a natural disaster. The term itself is largely meaningless: natural phenomena only become 'disasters' through their effects on human life and society.[75] Moreover, the social ecologies that natural disasters expose through their destructive force invariably reveal the important element played by human agency in the shaping of these impacts. This is especially the case with France's deadly heat wave. The heat itself is – arguably – an essential 'natural' element in the disaster. Yet the disparities of mortality left in the heat wave's wake point to the critical role of patterns of modern urban sociability in this intersection of nature and culture.

This chapter admittedly raises far more questions than it answers. Although it deals with the contemporary, the social nature of the problems it outlines suggests a need for a historical perspective on the present. Charles Rosenberg argued in his classic work on cholera that epidemics operate as a social 'sampling technique.'[76] By virtue of their extremity, they draw attention to the violence of everyday life that remains concealed by a veneer of civility in conditions of relative normality.[77] Historians have a role to play in analysing the origins of such conditions and how they framed the social ecology of disaster in the summer of 2003. A critical history of aging in France that accounts for the marginalization of the elderly to a zone of social abandonment is a necessary first step toward an understanding of the heat wave and toward the production of a social history of risk, as is a close examination of the ways in which particular urban forms have exacerbated vulnerability through their effects on sociability. Through its intersections of nature and culture, France during the heat wave presented a disease ecology as particular as a malaria environment, and one whose circumstances merit the same scrutiny at the human and social level as at a more formally medical one.

## Notes

1  For more extensive discussions of the problems outlined in this essay, see: Richard C. Keller, *Fatal Isolation: The Devastating Paris Heat Wave of 2003* (Chicago: University of Chicago Press, 2015); and Richard C. Keller, 'Place matters: mortality, space, and urban form in the 2003 heat wave disaster,' *French Historical Studies* 36, 2013: 299–330.

2 Quotations, in order, are from Joël Belmin, 'Les conséquences de la vague de chaleur d'août 2003 sur la mortalité des personnes âgées: Un premier bilan,' *La Presse médicale* 32 (18 October 2003): 1591–4, on p. 1591; Françoise Lalande et al., *Mission d'expertise et d'évaluation du système de santé pendant la canicule 2003* (unpublished report, 2003); Denis Hémon and Eric Jougla, *Estimation de la surmortalité et principales caractéristiques épidémiologiques* (Paris: INSERM, 2003); Denis Hémon and Eric Jougla, *Surmortalité liée à la canicule d'août 2003: Rapport remis au ministre de la santé et de la protection sociale* (Paris: INSERM, 2004); I. Grémy et al., 'Conséquences sanitaires de la canicule d'août 2003 en Ile-de-France: Premier bilan,' *Revue d'épidémiologie et de santé publique* 52 (2004): 93–108, on p. 93; Gilles Brücker, 'Impact sanitaire de la vague de chaleur d'août 2003: premiers résultats et travaux à mener,' *BEH* 45–6 (2003): 217; France 2, 'Canicule: ce que disent les hommes de foi,' *Islam* (airdate 26 October 2003); David Schnall and Jean Brami, 'Décès à domicile dus à la canicule: enquête sur les décès survenus en août 2003 dans le 19ᵉ arrondissement de Paris,' *La revue du practicien—médecine générale* 18, no. 662/663 (27 September 2004): 1007–11, at 1007.

3 Hémon and Jougla, *Surmortalité liée à la canicule d'août 2003*, 42.

4 European Environment Agency, *Impacts of Europe's Changing Climate: An Indicator-Based Assessment* (Luxembourg: Office for Official Publications of the European Communities, 2004); Christina Koppe et al., *Heat-Waves: Risks and Responses* (Copenhagen: World Health Organization, 2004); Gerald A. Meehl and Claudia Tebaldi, 'More intense, more frequent, and longer lasting heat waves in the 21st century,' *Science* 305, no. 5686 (2004): 994–7; Christoph Schär et al., 'The role of increasing temperature variability in European summer heatwaves,' *Nature* 427 (2004): 332–6.

5 For two examples of environmental histories that interrogate human intersections with landscapes, see: William Cronon, 'Introduction: in search of nature,' in *Uncommon Ground: Toward Reinventing Nature*, ed. William Cronon (New York: W.W. Norton, 1995), pp. 23–56; and Mike Davis, *Ecology of Fear: Los Angeles and the Imagination of Disaster* (New York: Vintage, 1998). For recent epidemiological and climatological studies on locality and human impact on climate change, see Peter A. Stott et al., 'Human contribution to the European heat wave of 2003,' *Nature* 432 (2004): 610–14. On medical geography and its history, see Mark Harrison, *Climates and Constitutions: Health, Race, Environment and British Imperialism in India, 1600–1850* (New York: Oxford University Press, 1997).

6 See Gregg Mitman, Michelle Murphy, and Christopher Sellers (eds), *Landscapes of Exposure: Knowledge and Illness in Modern Environments*, Osiris 19 (Chicago: University of Chicago Press, 2004).

7 Amartya Sen, *Poverty and Famines: An Essay on Entitlement and Deprivation* (Oxford: Oxford University Press, 1981); Mike Davis, *Late Victorian Holocausts: The El Niño Famines and the Making of the Third World* (London: Verso, 2002); Kim Fortun, *Advocacy after Bhopal: Environmentalism, Disaster, New Global Orders* (Chicago: University of Chicago Press, 2001); Ted Steinberg, *Acts of God: The Unnatural History of Natural Disaster in America* (New York: Oxford University Press, 2000); Eric Klinenberg, *Heat Wave: A Social Autopsy of Disaster in Chicago* (Chicago: University of Chicago Press).

8 Kai Erikson, *A New Species of Trouble: Explorations in Disaster, Trauma, and Community* (New York: Norton, 1994); Nancy Scheper-Hughes, *Death without Weeping: The Violence of Everyday Life in Brazil* (Berkeley: University of California Press, 1992).

9 James Lind, *An Essay on Diseases Incidental to Europeans in Hot Climates with the Method of Preventing Their Fatal Consequences* (London, 1771), pp. 138–9.

10 See: James Ranald Martin, *The Influence of Tropical Climates on European Constitutions* (London: J. Churchill, 1856); Philip Curtin, 'The promise and the terror of a tropical environment,' in *The Image of Africa: British Ideas and Action, 1780–1850* (Madison: University of Wisconsin Press, 1964); and Harrison, *Climates and Constitutions*.

11 Louis-René Villermé, 'De la mortalité dans les divers quartiers de la ville de Paris,' *Annales d'hygiène publique et de médecine légale* 3 (1830): 294–341; Joshua Cole, *The Power of Large Numbers: Population, Politics, and Gender in Nineteenth-Century France* (Ithaca: Cornell University Press, 2000); David Barnes, *The Making of a Social Disease: Tuberculosis in Nineteenth-Century France* (Berkeley: University of California Press, 1995); and David Barnes, *The Great*

*Stink of Paris and the Nineteenth-Century Struggle against Filth and Germs* (Baltimore, MD: Johns Hopkins University Press, 2006).

12 William Coleman, *Death is a Social Disease: Public Health and Political Economy in Early Industrial France* (Madison: University of Wisconsin Press, 1982), pp. 149–71; Cole, *The Power of Large Numbers*, p. 69; Barnes, *The Making of a Social Disease*, pp. 32–3.

13 See Barnes, *The Great Stink of Paris*, p. 50; David P. Jordan, *Transforming Paris: The Life and Labors of Baron Haussman* (New York: Free Press, 1995), esp. pp. 48, 96–7, 270–4.

14 Jordan, *Transforming Paris* and 'Haussmann and Haussmannization: the legacy for Paris,' *French Historical Studies* 27, no. 1 (2004): 87–113; Pinkney, *Napoleon III*; Barnes, *The Great Stink of Paris*; T. J. Clark, *The Painting of Modern Life: Paris in the Art of Manet and His Followers* (Princeton: Princeton University Press, 1985); David Harvey, *Consciousness and the Urban Experience: Studies in the History and Theory of Capitalist Urbanization* (Baltimore, MD: Johns Hopkins University Press, 1985).

15 Matthew Gandy, 'The Paris sewers and the rationalization of urban space,' *Transactions of the Institute of British Geographers*, NS 24 (1999): 23–44.

16 David Pinkney, *Napoleon III and the Rebuilding of Paris* (Princeton: Princeton University Press, 1971), p. 127.

17 Marshall Berman, *All That Is Solid Melts into Air: The Experience of Modernity* (New York: Penguin, 1982).

18 Tyler Stovall, *The Rise of the Paris Red Belt* (Berkeley: University of California Press, 1990); Tyler Stovell, 'From red belt to black belt: race, class, and urban marginality in twentieth-century Paris,' in *The Color of Liberty: Histories of Race in France*, ed. by Sue Peabody and Tyler Stovall (Durham, NC: Duke University Press, 2003), pp. 351–70; Berman, *All That Is Solid Melts into Air*; and Clark, *The Painting of Modern Life*.

19 Susan L. Cutter, Bryan J. Boruff, and W. Lynn Shirley, 'Social vulnerability to environmental hazards,' *Social Science Quarterly* 84, no. 2 (2003): 242–61.

20 All figures are from Grémy et al., 'Conséquences sanitaires de la canicule.'

21 See Patrick Pelloux, *Urgentiste* (Paris: Fayard, 2004); *TF1 20 heures* (airdate 10 August 2003).

22 For an analysis of mortality reporting during this and similar disasters, see Carine Vassy, Richard C. Keller, and Robert Dingwall, *Enregistrer les morts, identifier les surmortalités: Une comparaison Angleterre, Etats-Unis et France* (Rennes: Presses de l'EHESP, 2010).

23 Hémon and Jougla, *Estimation de la surmortalité*.

24 See France 3, 'Généalogistes: Décès canicule,' *19 20: Edition nationale* (airdate 23 September 2003).

25 All figures from Hémon and Jougla, *Estimation de la surmortalité*.

26 France 3, *Vieillir ensemble? Les dossiers de France 3* (airdate 23 September 2003).

27 *TF1 20 heures* (airdate 14 August 2003).

28 See the interview with Dr Pierre Carli, director of SAMU in Paris, who argued that 'it is extremely difficult to know whether the deaths we are observing at present are solely the result of heat and hyperthermia. It is of course clear that heat has an aggravating effect on people who are aged and already sick' – *TF1 20 heures* (airdate 10 August 2003).

29 Klinenberg, *Heat Wave*; Steinberg, *Acts of God*.

30 France 2, *20 heures le journal* (airdate 14 August 2003).

31 Julie Livingston, 'Pregnant children and half-dead adults: modern living and the quickening life cycle in Botswana,' *Bulletin of the History of Medicine* 77 (2003): 133–62, on 135; Margaret Lock, *Encounters with Aging: Mythologies of Menopause in Japan and North America* (Berkeley: University of California Press, 1993); Lawrence Cohen, *No Aging in India: Alzheimer's, the Bad Family, and Other Modern Things* (Berkeley: University of California Press, 1998).

32 Isabelle Mallon, *Vivre en maison de retraite: Le dernier chez-soi* (Rennes: Presse Universitaire de Rennes, 2004), 7.

33 For examples, see Maximilienne Levet and Chantal Pelletier, *Papy boom* (Paris: Grasset, 1988) and Robert Rochefort, *Vive le papy-boom* (Paris: Odile Jacob, 2000). On the creation of a 'new age' in the twentieth century, see Martine Segalen, 'Les changements familiaux depuis le début du XX$^e$ siècle,' in *Histoire de la population française*, ed. Jacques Dupâquier et al. (4 vols; Paris: Presses Universitaires de France, 1988), IV: 499–541.

34 Hervé Le Bras, *Marianne et les lapins: L'obsession démographique* (Paris: Hachette, 1993), p. 123.

35 On natalism and concerns over population in the interwar period, see Mary Louise Roberts, *Civilization without Sexes: Reconstructing Gender in Postwar France, 1917–1929* (Chicago: University of Chicago Press, 1993); also Cheryl Koos, 'Gender, anti-individualism, and nationalism: The alliance nationale and the pronatalist backlash against the femme moderne, 1933–1940,' *French Historical Studies* 19, no. 3 (1996): 699–723.

36 Jean Daric, *Vieillissement de la population et prolongation de la vie active* (Paris: Presses Universitaires de France, 1948; Institut National d'Etudes Démographiques, Cahier 7), p. 42.

37 Cited in Elise Feller, 'Les femmes et le vieillissement dans la France du premier XXe siècle,' *Clio* 7 (1998): 199–222, on 201 and in Le Bras, *Marianne et les lapins*, 118. See also Patrice Bourdelais, *L'âge de la vieillesse* (Paris: Odile Jacob, 1993), esp. pp. 90–154.

38 Segalen, 'Les changements familiaux,' 516.

39 Michel Foucault, *'Society Must Be Defended': Lectures at the Collège de France, 1975–1976*, trans. David Macey (New York: Picador, 2003), p. 244.

40 Premier Ministre, 'Haut comité consultatif de la population et de la famille,' *Politique de la vieillesse: Rapport de la Commission d'étude des problèmes de la vieillesse* (Paris: La Documentation française, n.d. [1962]). Hereafter 'Laroque Report.'

41 Ibid., 3–4.

42 Ibid., 259, 262.

43 Ibid., 262, 110–11.

44 Ibid., 111.

45 Ibid., 263.

46 Elise Feller, 'L'entrée en politique d'un groupe d'âge: La lutte des pensionnés de l'Etat dans l'entre-deux-guerres et la construction d'un 'modèle français' de retraite,' *Mouvement Social* 190 (2000): 33–59; also Dominique Argoud and Anne-Marie Guillemard, 'The politics of old age in France,' in *The Politics of Old Age in Europe*, ed. Alan Walker and Gerhard Naegele (Buckingham: Open University Press, 1999), pp. 83–92.

47 Gérard Badou, *Les nouveaux vieux* (Paris: Le Pré aux clercs, 1989).

48 Mallon, *Vivre en maison de retraite*, p. 7.

49 Biehl, 'Vita.'

50 France 3, *Les dossiers de France 3: Vieillir ensemble?* (airdate 23 September 2003).

51 Jean-François Lacan, *Scandales dans les maisons de retraite* (Paris: Albin Michel, 2002), pp. 48–50.

52 Cited in Badou, *Les nouveaux vieux*, p. 249.

53 Lacan, *Scandales dans les maisons de retraite*, p. 106.

54 Martine Bungener, 'Canicule estivale: La triple vulnérabilité des personnes âgées,' *Mouvements* 32 (2004): 75–82; and interview with Bungener in Villejuif, 11 January 2005.

55 Cronon, 'Introduction: in search of nature'; Davis, *Ecology of Fear* and *Late Victorian Holocausts*; Michael Bess, *The Light-Green Society: Ecology and Technological Modernity in France, 1960–2000* (Chicago: University of Chicago Press, 2003); Caroline Ford, 'Landscape and environment in French historical and geographical thought: new directions,' *French Historical Studies* 24, (2001): 125–34; Sara Pritchard, 'Reconstructing the Rhône: the cultural politics of nature and nation in contemporary France, 1945–1997,' *French Historical Studies* 27 (2004): 765–99.

56 Assemblée Nationale de France, *Rapport fait au nom de la Commission d'enquête sur les conséquences sanitaires et sociales de la canicule* (Paris: Assemblée nationale, 2004): http://www.assemblee-nat.fr/12/rap-enq/r1455-t1.asp, p. 145.

57 Institut de Veille Sanitaire, *Etude des facteurs de risque de décès des personnes âgées résidant à domicile durant la vague de chaleur d'août 2003* (Paris: INVS, 2004): http://www.invs.sante.fr/publications/2004/chaleur2003_170904/rapport_chaleur2003.pdf.

58 Ibid., pp. 51–2, 56, 60.

59 Klinenberg, *Heat Wave*.

60 InVS, *Etude des facteurs de risque de décès*, 52.

61 F. Canouï-Poitrine., E. Cadot, A. Spira, 'Excess deaths during the August 2003 heat wave in Paris, France,' *Revue d'épidémiologie et de santé publique* 54 (2005): 127–35; see also Emmanuelle Cadot and Alfred Spira, 'Canicule et surmortalité à Paris en août 2003, le poid des facteurs socio-économiques,' *Espaces, populations, sociétés* 2–3 (2006): 239–49.

62 In this case, the excess mortality rate is determined by subtracting the 'expected' mortality rate (based on an average of the number of deaths experienced in August for several years prior to the heat wave) from the actual number of deaths experienced in August 2003. The result yielded more than a thousand 'excess' deaths attributable to the heat wave in Paris. As some of these deaths occurred in hospitals, the APSP's studies focused on those who died at home: including hospital deaths would exaggerate the death rate in arrondissements with hospitals. Moreover, the rate of those dying at home was far higher than that of those dying in hospitals: while in any given August, some 200 Parisians die in their homes, in 2003 roughly 900 died at home.

63 Christine Meisner Rosen and Joel Arthur Tarr, 'The importance of an urban perspective in environmental history,' *Journal of Urban History* 20, no. 3 (1994): 299–310; Martin V. Melosi, *Garbage in the Cities: Refuse, Reform, and the Environment* (College Station: Texas A&M Press, 1981); Jeffrey K. Stine and Joel A. Tarr, 'At the intersection of histories: technology and the environment,' *Technology and Culture* 39, no. 4 (1998): 601–40.

64 Manuel Castells, *The City and the Grassroots: A Cross-Cultural Theory of Urban Social Movements* (Berkeley: University of California Press, 1983); David Harvey, *Social Justice and the City* (Oxford: Blackwell, 1988); Martin V. Melosi, 'Cities, technological systems, and the environment,' *Environmental History Review* 14, no. 1–2 (1990): 45–64, and 'The place of the city in environmental history,' *Environmental History Review* 17, no. 1 (1993): 1–24.

65 Paul Clerc, *Grandes ensembles, banlieues nouvelles: Enquête démographique et psycho-sociologique* (Paris: Centre de Recherche d'Urbanisme, INED, 1967); Virginie Picon-Lefebvre, Paris-ville moderne: Maine-Montparnasse et la Défense, 1950–1975 (Paris: Norma, 2003).

66 David Harvey, *The Condition of Postmodernity* (Oxford: Blackwell, 1990); Jane Jacobs, *The Death and Life of Great American Cities* (New York: Vintage, 1961).

67 Robert Franc, *Le scandale de Paris* (Paris: Grasset, 1971), pp. 201–30.

68 Henri Coing, *Rénovation urbaine et changement social* (Paris: Editions Ouvrières, 1966); Norma Evenson, *Paris: A Century of Change, 1878–1978* (New Haven: Yale University Press, 1979), esp. pp. 255–63. These estimates are of course problematic, as economic instability was a critical element determining the siting of these structures in the first place.

69 Coing, quoted in Evenson, *Paris*, 261.

70 See the sociologist Marie-Christine Pouchelle's account, *Vivre dans un grand ensemble: 'Les cimentiers'* (Paris: Epi, 1974).

71 For a particularly subtle analysis, see Eric Fassin and Didier Fassin, *Question sociale, question raciale?* (Paris: Editions la Découverte, 2006). Other references – mostly significantly less nuanced than Fassin and Fassin – include Danièle Authier, 'Représentations sociales du sentiment d'insécurité chez des personnes âgées résidant dans des quartiers opposés par leur taux de criminalité' (med. thesis; Lyon: Institut Universitaire Alexandre Lacassagne, 1986); Alain Bauer and Xavier Raufer, *Violences et insécurité urbaines* (Paris: Presses Universitaires de France, 1998); and France Estrosi, 'L'intéraction avant tout,' in *Le sentiment d'insécurité: Les seniors face à la délinquance* (Paris: Editions Taitbout, 1994), pp. 15–18; and Sebastian Roché, *Sociologie politique de l'insécurité: Violences urbaines, inégalités et globalisation* (Paris: Presses Universitaires de France, 1998).

72 See Christopher Caldwell, 'Revolting high rises: were the French riots produced by modern architecture?' *The New York Times Magazine* (27 November 2005): 28–9.

73 David Jordan has argued that they exceeded the worst of Haussmann's razings; see 'Haussmann and Haussmannization,' pp. 106–8. See also Richard Cobb, *Paris and Elsewhere* (New York: John Murray, 1999).

74 Interview with a tower resident, Place d'Italie, thirteenth arrondissement, June 2007.

75 Cronon, 'Introduction: in search of nature'; Davis, *Ecology of Fear*; Erikson, *A New Species of Trouble*; Steinberg, *Acts of God*.

76 Charles Rosenberg, *The Cholera Years: The United States in 1832, 1849, and 1866* (Chicago: University of Chicago Press, 1987 [1962]), p. 4.

77 Klinenberg, *Heat Wave*, p. 23.

## Select bibliography

Barnes, David. *The Making of a Social Disease: Tuberculosis in Nineteenth-Century France.* Berkeley: University of California Press, 1995.

Bungener, Martine. 'Canicule estivale: La triple vulnérabilité des personnes âgées.' *Mouvements,* 32, 2004: 75–82.

Canouï-Poitrine. F., E. Cadot, A. Spira, 'Excess deaths during the August 2003 heat wave in Paris, France.' *Revue d'épidémiologie et de santé publique,* 54, 2005: 127–35.

Coleman, William. *Death Is a Social Disease: Public Health and Political Economy in Early Industrial France.* Madison: University of Wisconsin Press, 1982.

Cutter, Susan L., Bryan J. Boruff, and W. Lynn Shirley. 2003, 'Social vulnerability to environmental hazards.' *Social Science Quarterly,* 84, 2003: 242–61.

Erikson, Kai. *A New Species of Trouble: Explorations in Disaster, Trauma, and Community.* New York: Norton, 1994.

Fortun, Kim. *Advocacy after Bhopal: Environmentalism, Disaster, New Global Orders.* Chicago: University of Chicago Press, 2001.

Hémon, Denis, and Eric Jougla. Surmortalité liée à la canicule d'août 2003: Rapport remis au ministre de la santé et de la protection sociale. Paris: INSERM, 2004.

Keller, Richard C. 'Place matters: mortality, space, and urban form in the 2003 heat wave disaster.' *French Historical Studies,* 36, 2013: 299–330.

Keller, Richard C. *Fatal Isolation: The Devastating Paris Heat Wave of 2003.* Chicago: University of Chicago Press, 2015.

Klinenberg, Eric. *Heat Wave: A Social Autopsy of Disaster in Chicago.* Chicago: University of Chicago Press, 2002.

Rosenberg, Charles. *The Cholera Years: The United States in 1832, 1849, and 1866.* Chicago: University of Chicago Press, 1987.

Steinberg, Ted. *Acts of God: The Unnatural History of Natural Disaster in America.* New York: Oxford University Press, 2000.

Villermé, Louis-René. 'De la mortalité dans les divers quartiers de la ville de Paris.' *Annales d'hygiène publique et de médecine légale,* 3, 1830: 294–341.

# Part III

# TECHNOLOGIES

# 17

# DISABILITY AND PROSTHETICS IN EIGHTEENTH- AND EARLY NINETEENTH-CENTURY ENGLAND

*David M. Turner*

On 5 May 1749, the press announced the arrival in London of 'the famous Sieur Rocquet', a Parisian surgeon, who offered a variety of medical services to the wealthy. According to an account published in the *Scots Magazine*, he charged 1l. 1s for 'cutting off a thigh (leg included)', or 10s 6d for removing a leg below the knee. To remove an arm close to the shoulder (including wrist, hand, fingers and thumb) would cost another 1l. 1s, while a single hand, foot, thumb, toe or finger could be removed for 5 shillings. To the dismembered, Rocquet sold 'wholesale or retale [*sic*] all sorts of legs, arms, eyes, noses, or teeth, made in the genteelest manner', as worn by 'persons of rank in France'.[1]

Fifty years later, the wearing of prosthetic limbs seemed to have become ubiquitous. In an account published in 1799, the French traveller Jacques-Henri Meister remarked that 'I have no where seen more wooden legs, or persons who have lost an arm' than on the streets of London. Here there were more prostheses-wearing amputees than any of the 'great cities' of Europe. The proliferation of amputees may be assumed to have been the consequence of Britain's prolonged involvement in overseas warfare, first in America and most recently in Europe, but intriguingly among those prosthesis-wearing 'cripples' that drew Meister's eye on the streets of Georgian London were 'many females'. Truly this was, in the opinion of Meister's translator, an age of 'extraordinary lopping of limbs, and flourishing state of the manufacture of wooden legs'.[2]

A prosthesis is an external appliance used either to restore the body's function or appearance. While the word 'prosthesis' had entered the English language in the mid-sixteenth century, it was not until the eighteenth century that it acquired its more modern meaning of replacing a missing part of the body with an artificial one.[3] Design and innovation in prosthetic and assistive devices has been seen as part of the advance of 'technological medicine' which, according to Stanley Reiser, has transformed approaches to the diseased body since the nineteenth century, marking advances in rehabilitative medicine that stand alongside transformations in diagnostic tools, from the stethoscope to fMRI scanners.[4] The use and manufacturing of prostheses has expanded significantly in Britain and the United States since the mid-nineteenth century. Between 1846 and 1873, Stephen Mihm has shown, Americans submitted some 167 patents for prosthetic devices.[5] An international trade in industrially manufactured artificial limbs expanded significantly in the aftermath of the American

Civil War.[6] On both sides of the Atlantic, a variety of prosthetic devices, including artificial eyes, eardrums and limbs, were increasingly advertised to the public in ways that emphasised their value in restoring function and the appearance of normality.[7] Catering in particular to injured servicemen and victims of industrial accidents, the history of modern prosthetics reflects a number of important themes in the history of disability, product design and medicine, ranging from the growing role of the state and other agencies in providing support for the injured (particularly in the aftermath of the First World War) to the relationship between commercial and medical interests in designing products for the maimed or diseased.[8] In the twentieth century, prostheses and technology transformed the experiences of patients suffering from various acute and chronic illnesses. In addition to the increasing availability of artificial limbs and hearing aids, the development of total hip replacement between the wars improved the experiences of arthritis patients, enabling them to live longer and more active lives.[9]

However, while the use of medical technology is often seen as a feature of modernity, Meister's observations from the late eighteenth century remind us that the use of prosthetics is nothing new. Prosthetic devices to assist the maimed can be traced back to ancient times and by the sixteenth century Parisian manufacturers had invented a wide variety of products including false teeth, eyes 'counterfeited and enameled' in such a way as to retain the 'brightness or gemmy decency of the natural Eye', and even an 'artificial yard' for men who had lost their penises to injury or venereal disease.[10] By the late eighteenth century, inventors were developing increasingly sophisticated 'technologies of the body', from rupture trusses to postural devices, which used new material such as cast steel and innovations in springs and gears to increase functionality.[11] The late eighteenth and early nineteenth century was a critical period in the development of modern prostheses, yet is often overlooked by historians who have focused instead on the rapid developments taking place later in the nineteenth century and beyond. During this period, prosthetics were becoming increasingly commercialised, but manufacturers retained the 'craft' status that had long characterised makers of medical technologies.[12] Prostheses were visible evidence of the increasing power of technology to restore the broken bodies of the diseased and maimed, but also carried a range of cultural meanings that reflected on the status and gender identity of the wearer. This chapter examines the ways in which prostheses were used to treat a variety of diseases and injuries and analyses the ways in which representations of prosthesis-wearing amputees shaped cultural responses to disability.

This work raises bigger questions about the relationship between histories of disability and histories of disease, and about the place of amputees in these histories. Modern disability scholars and activists have often sought to distance experiences of people with disabilities from patients suffering from ill-health, arguing that people with impairments are disabled not by physical restrictions on their capabilities but by the ways in which built environments and political systems are geared towards able-bodied norms.[13] Consequently, as Beth Linker has argued, in order to avoid viewing people with impairments as medicalised 'patients', scholars of disability have tended to avoid examining the experiences of the 'unhealthy disabled' – the long-term sick and ill – and paid more attention to people with sensory impairments, such as the deaf and blind, or to amputees as the 'predictably impaired' – people who do not need recurrent medical attention. In this way, the history of disability (focused on the barriers to citizenship facing people with disabilities in the past) has distanced itself from the history of medicine (concentrating on the history of diseases and the progress of medical intervention).[14] Nevertheless, although they may be conceptually different categories, 'disease' and 'disability' overlap in various ways. Many diseases can cause permanent impairments or physical deformities, whereas people with severe mobility impairments may be vulnerable to infectious diseases, particularly in the lungs and bladder.[15]

In the eighteenth century, doctors assumed that the 'deformed' body was necessarily an unhealthy one. As the physician William Buchan wrote in his popular *Domestic Medicine* (1769), a 'deformed body is not just disagreeable to the eye, but by a bad figure both the animal and the vital functions must be impeded, and, of course, health impaired'.[16] In this period amputation and prosthetics were viewed in a social, cultural and medical context in which bodily abnormality was subsumed into the broad category of 'disease'. For instance, the Royal Infirmary in Edinburgh, which published annual figures of patients treated for various conditions in the mid-eighteenth century, listed 'amputations' as part of the 'diseases' with which patients were 'afflicted', making no distinction between operations carried out in cases of trauma and those performed on the diseased.[17] Eighteenth-century hospitals more generally saw their role as tending to the 'sick and lame', not differentiating between patients suffering from disease and those incapacitated through physical deformity.[18] This chapter begins by examining the causes and context of amputation. It goes on to explore developments in prosthetic technology as a response to sickness and injury. Finally, it discusses the variety of cultural meanings associated with prosthetics in the past. The result is a case study that speaks to wider questions about how we view disease, disability and their representations in the past.

## Amputation and its discontents

The most common prostheses were those intended to supply the loss of limb, either through congenital defect or amputation.[19] In the past, as today, the most common use of amputation applied to the lower limbs.[20] In the eighteenth century, amputations were used in response to a series of medical circumstances.[21] The Prussian army surgeon Johann Ulrich Bilguer in his 1764 treatise on amputation described six common causes where dismemberment was 'universally thought necessary', including: mortification reaching the bone; 'a limb so hurt, that a mortification is highly probable'; trauma such as 'violent contusion of the flesh which at the same time has shattered the bone'; amputation as a preventative measure to stop life-threatening haemorrhage; 'incurable caries [rottenness] of the bones'; and cancer.[22] In some cases, amputation was carried out in response to a progressive worsening of health whose exact causes were difficult to pinpoint. For instance, the *Derby Mercury* reported in 1727 the case of a 69-year-old man for whom amputation was proposed after mortification spread down his right leg. The mortification was 'occason'd by an ill State of Health he had labour'd under for a considerable Time, which at last fix'd itself in the Foot, attended with great Pain, Swelling and Inflammation'.[23] However, most popular accounts of amputation, found among the plentiful lists of 'casualties' printed in eighteenth-century newspapers, described amputations that were the consequence of traumatic accidents or injuries.[24] For example, one Mr Swain of Lad Lane in the City of London was reported to have undergone an amputation in July 1754 after an accident in which he was thrown out of a chaise, breaking one of his legs and 'continues so dangerously ill, that 'tis feared he cannot recover'.[25] In 1758, Henry Therond of Trinity College, Cambridge, fell from his horse 'whereby he broke his leg in such a terrible Manner that both the Bones forced their way through his boot' so that the leg 'was obliged to be cut off'. Once again, there were 'but small Hopes of his Recovery'.[26]

When the surgeon Percivall Pott described limb amputation in 1779 as 'an operation terrible to bear' and 'horrid to see', he expressed a view that was commonplace in an age before anaesthetics.[27] Amputation was often described as a 'melancholy' procedure; patients were said to '*submit* to an amputation', indicating resignation to their fate.[28] Nevertheless, the belief that there were certain circumstances where its use was unavoidable made amputation, in Pott's view, an act of 'humanity', which reflected the rational judgement of medical professionals.[29] But in the second half of the eighteenth century the procedure of amputation was a matter of keen debate, raising questions not just about the necessity and safety of the operation, but also about the authority and competence of surgeons and the extent of their power over the patient's body. Amputation might proceed from a humane wish to save life by removing a diseased or shattered irrecoverable part, but it highlighted the role of medical intervention in causing disablement, leaving those who survived the procedure in a 'mutilated imperfect state'.[30] 'Most people are shocked at the mention of any amputation, or at the sight of a poor creature who has lost an hand, an arm, a foot or leg, wretchedly crawling along on crutches or a wooden leg', noted Bilguer, 'and consider the total privation of a limb, as a much greater misfortune than when it is preserved, though perhaps unshapely, and uncapable of performing several of its primitive functions'.[31]

One frequent complaint was that amputation, rather than being offered as a drastic final attempt to save life, was often proposed too readily without considering other options. In a particularly tragic case, a correspondent to the *Oxford Journal* in 1753 described how on a journey to Suffolk he had encountered a girl with 'two Wooden Legs'. When he asked about the 'Manner of her becoming lame' he was told that as a result of a numbing in both her legs, local surgeons had recommended amputation 'for fear of the worst Consequences of a Mortification'. However, they had sought a further opinion from a 'surgeon of great skill in the Town' before they proceeded, who had advised them to wait 'in Hope of seeing the Limbs take a favourable Turn'. Yet the girl's father, 'being either delirious or fearful of the Mortification', took action himself, and 'chopped off both his Daughter's Legs with a Hatchet'.[32]

In this story, the surgeons were represented as being rational and measured in their judgement – in contrast to the father's terrible impetuosity. However, in many cases the lack of skill or heedlessness of surgeons was blamed for untimely dismemberment. Vendors of quack remedies were quick to criticise surgeons and offer their own products or services as a more effective alternative to amputation. For example, a 1750 advertisement for 'Jackson's Tincture' included a testimony from William Ore, a shoemaker of Drury Lane, who suffered a mortification in his foot as a result of 'the bad Management of the Surgeon' who treated a cut in one of his toes, such that doctors deemed him 'incurable' and told him he must undergo amputation of his leg, but after the application of Jackson's medicine his foot regained 'its natural Warmth and Colour'.[33] The cause of the great number of amputees in London in

1799 according to Jacques-Henri Meister was not foreign warfare but the 'number of hospitals opened to receive the poor and indigent, maimed by accidental fractures, or lamed through disease', and the practices of 'surgeons of the first celebrity for operation' whose 'deficiency of skill' in tending wounds meant that they found it 'easier to amputate than to set a fractured limb'.[34]

Amputation was therefore the focal point for a broader critique of the role and conduct of medical professionals in an age of increasing professionalisation of surgery and expansion of hospital provision. By 1800 there were 44 voluntary hospitals for the 'sick and lame' in England and Wales.[35] As Mary Fissell and others have shown, the development of hospital medicine – including an expanding educative role of hospitals in spreading surgical expertise – increased the authority of doctors in the diagnosis and treatment of sickness, resulting in a gradual 'deskilling' of patients as interpreters of their own medical conditions.[36] Yet for some eighteenth-century commentators the growing authority of hospital surgeons manifested itself in greater 'inhumanity' towards patients. Hospital medicine's most powerful eighteenth-century critic, William Nolan, published *An Essay on Humanity: Or a View of Abuses in Hospitals with a Plan for Correcting them* in 1786 in which he identified several aspects of institutional care whose 'inhumanity is in diametric opposition to the very name and nature of a hospital'. Surgeons, he argued, were too often guilty of 'harsh language and apparently unfeeling treatment', manifested above all in the

*Figure 17.1* Thomas Rowlandson, *Amputation* (London: S. Fores, 1793). Courtesy of the Wellcome Library, London.

practice of amputation which was too often recommended 'without any previous consultation, whether a possibility exists of effecting a cure'. Indiscriminate amputation, he argued, was an act of 'cruelty and [in]justice' that caused intolerable pains to the mind and body of the patient.[37]

Nolan's criticisms were fiercely refuted by those who claimed 'internal knowledge' of the hospital system, and in theory at least most hospitals urged a cautious approach to dismemberment, calling for a broad consensus among the surgeons and physicians before the operation took place.[38] During the second half of the eighteenth century there was a proliferation of new methods and remedies proposed to make the operation less necessary or to remove some of the 'many Inconveniences' that patients suffered.[39] Bilguer's concerns about the safety and necessity of amputation led him to recommend more effective measures to prevent dismemberment becoming necessary, for example by applying Peruvian bark (quinine) to wounds to prevent mortification from spreading. He prescribed a course of treatment that involved making repeated linear incisions down to living tissues in gangrenous limbs, followed by at least one daily dressing to encourage the separation of dead tissues. Few patients, however, may have wished for such prolonged and painful treatment.[40] The use of agaric of oak to prevent bleeding after amputation was greeted enthusiastically in the press after it was proposed by a French country surgeon in 1751. This, it was claimed, offered hope of survival and relief for patients by easing them of a 'great deal of pain'.[41] The quality and positioning of the amputation was also crucial for the patient's future prospects of mobility. Consequently, some surgeons proposed new methods of amputation that would reduce distress to the patient by leaving neat stumps more amenable to the fitting of prosthetics.[42]

However, the belief that surgeons acted out of a callous disregard for the feelings or well-being of patients persisted and was reinforced by reports that military surgeons received a payment of 5 l. for every limb they removed.[43] Critics of amputation supported their arguments with examples of cases where lay people had successfully resisted medical advice. Nolan cited the case of a patient at London's St Bartholomew's hospital who refused to have his 'violent[ly] inflamed' left arm amputated by an eminent surgeon. The patient 'very rationally' considered that if medicine was prescribed for 'correcting and purifying the blood, it would probably abate the inflammation and on that event's taking place, external applications might operate more effectively'. This being done, the patient was discharged within a month, his arm 'perfectly healed'. In Nolan's view, this was a 'living monument of the cruelty and injustice of having recourse to amputation, unless in cases where the symptoms of mortification made it indispensably necessary'.[44] In 1794 the case of a collier who recovered from a compound fracture after he had 'refused to be removed to the county infirmary or submit to an amputation', provided 'striking proof of the necessity there is for great deliberation in cases where amputation may be thought necessary'.[45] Meister's translator added the example of a gentleman 'well known to many persons in the west end' of London who survived after fracturing his leg and thigh from falling out of a window. Because it was feared that he would die if surgeons attempted to amputate his limb he was 'left to his fate', but he 'happily recovered' and 'with the help of a crutch-stick and a high shoe, is often met taking the exercise of walking'.[46] Although evidence of resistance is anecdotal, opposition to amputation questioned the limits of medical authority over the bodies of the maimed and called for patients to be treated with humanity and respect.

## Prosthetics: technological approaches to disease and disability

For those who survived dismemberment, recovery could be a long and painful process that required personal fortitude and sometimes frequent medical intervention. One of the most vivid accounts was left by Lieutenant George Spearing whose left leg was taken off by surgeons below the knee on 2 May 1770 following an accident in which he had fallen down a disused mine shaft outside Glasgow. He described constant bleeding in the days following his operation with his principal artery having to be 'sewn up four different times before the blood was stopped'. After that:

> I suffered much for two or three days, not daring to take a wink of sleep; for the moment I shut my eyes, my stump (though constantly held by the nerve) would take such convulsive movements, that I really think a stab to the heart could not be attended with greater pain. My blood too was become so very poor and thin, that it absolutely drained through the wound near a fortnight after my leg was cut off. I lay for 18 days and nights in one position, not daring to move, lest the ligature should again give way; but I could endure it no longer, and ventured to turn myself in bed contrary to the advice of my surgeon, which I happily effected, and never felt greater pleasure in my life.

Spearing's rehabilitation took over nine months, during which he learnt to adapt to the loss of a limb and to walk using a prosthetic.[47]

The challenge of restoring the function and, to an extent, the appearance of lost limbs was taken up by a variety of artisans and craftsmen in the early-modern period, including barbers, locksmiths, clockmakers and wheelwrights.[48] Already by the late sixteenth century in France, manufacturers were experimenting with a variety of designs to restore broken bodies. Ambroise Paré described the innovations of a smith living in Paris known as Le Petit Lorrain who had devised a variety of products 'not onely profitable for the necessity of the body, but also for the decency and comeliness thereof'. These included various devices made from iron, including a hand that used cogs and gears to allow the fingers to flex; an arm with a spring-jointed elbow; and realistic-looking artificial legs. These sophisticated devices contrasted with 'the form of a Wooden leg made for poor men', a simple peg leg that attached to the amputee's stump using leather straps.[49] French and Swiss craftsmen continued to develop increasingly sophisticated designs during the seventeenth and eighteenth centuries, in part stimulated by interest in automata.[50] Some manufacturers gained an international reputation. Mid-eighteenth-century London newspapers marvelled at reports of the inventions of Monsieur Laurent, a Flemish 'Engineer at Bouchain . . . well known for his Talents in Mechanick Operation' who 'invented an artificial Arm, which imitates all the Motions of a natural one'. In 1760 it was reported that thanks to his skills, a soldier whose arms had been blown off by a cannonball leaving 'but five Inches remaining of the left one' was now able to 'eat and drink, take Snuff and write' with his artificial arm. This 'ingenious invention' was presented to the King and Queen of France and to L'Academie des Sciences.[51]

As well as being objects of admiration, innovations in prostheses were the subject of much sardonic comment in the eighteenth-century press. Newspapers satirised the sale of life-like artificial legs and arms as symbols of foppish affectation, a luxurious accoutrement to the body as subject to the whims of fashion as the latest hair or clothing style. A mock advertisement published in the *Public Advertiser* in April 1772 for 'Jonathan Lightfoot' proclaimed the invention of 'a cork-leg of so admirable a construction, that it not only answers all the Purposes of natural ones, but is free from all their defects', so that 'such Gentlemen and Ladies . . . as choose to be fitted with handsome limbs, may have their old ones very artificially taken off'.[52] A satire marking the benefits of the commercial treaty signed with France in 1786 announced that consumers would now be able to purchase an array of 'false hips, false rumps . . . false nails, false fingers, false hair, and every other falsity that can beautify and adorn the English, so as to make them as amiable and as elegant as the French'.[53] However, after the outbreak of war with France in 1793, newspaper advertisements for prosthetics remodelled artificial limbs as symbols of patriotic valour, providing support and 'comfort' to those who had sacrificed their limbs in battle. Advertisements for the 'patent artificial leg' sold by the surgeon Thomas Ranby Reid, variously presented it as a 'DISCOVERY very interesting to MUTILATED PERSONS', as providing 'CONSOLATION TO OUR BRAVE, BUT UNFORTUNATE DEFENDERS', or simply as being dedicated 'To the ARMY and NAVY'.[54]

In contrast to the functional peg leg, advertisers marketed products that boasted technological sophistication and superior materials so as to ease the comfort of wearers by being both 'light' and 'durable', and emphasised the 'utility' of their products in 'reliev[ing] the afflictions of human nature'.[55] Most striking of all was the 'cork leg' crafted by James Potts of Chelsea for the Marquis of Anglesey, which was remarkable for its lifelike qualities. It was, according to an account published in 1815, 'exactly the shape of the natural leg, and so ingeniously contrived with elastic strings and joints, as to play at the ankle and knee with the same facility, and accommodate itself to every movement in sitting down and walking'.[56] While Anglesey's leg would eventually provide a model for mass-produced prosthetics in the expanding Anglo-American limb industry of the mid-Victorian period, the accent of English newspaper advertisements for prosthetic limbs around the turn of the eighteenth into the nineteenth century was on the individuality of the wearer.[57] Advertisements contained testimonies from satisfied customers, sought to foster personal relationships with clients through offering personal interviews and consultations, and responded to potential concerns about the comfort or appearance of their products.[58] Prostheses were more than medical appliances designed to replace missing parts. They were advertised in ways that spoke to the polite aspirations of wearers, rewarding them for their taste and trust in the efficacy of certain designs and claiming that products would not just restore function, but would also enhance the appearance of the body – 'as well an object of admiration as the means of an inestimable blessing' as Ranby Reid's 1794 advertisement put it.[59]

No matter how much advertisers emphasised the 'ease' and 'convenience' of their products, amputees faced a complicated and sometimes lengthy process of adaptation in which they came to terms with their prostheses. Accidents involving wearers of wooden legs were documented in newspapers and coroners' reports, testifying to the manifold dangers faced by amputees in Georgian England. In June 1731, a 'poor man with one leg, that used to go on Errands at the Earl of Pomfret's House in Hannover Square, coming down stairs, his wooden leg slipt, so that he fell from the top to the Bottom, and bruised himself in so violent manner that it thought he cannot live'.[60] Amputees in the past (as today) required greater energy to walk than non-amputees.

Figure 17.2 *A Broken Leg, or the Carpenter the Best Surgeon* (London: Laurie and Whittle, 1800). Courtesy of the Wellcome Library, London.

Artificial limbs slowed down the wearer, making them vulnerable to traffic accidents and easy prey for robbers. In July 1765, the *Public Advertiser* reported that a 'poor Man' had been knocked over by a chariot in Great George Street, Westminster, an accident caused by 'his having a wooden leg, which prevented his getting out of the way'.[61] In September that year, a master stay-maker who 'wears a wooden leg' was set upon by two footpads who robbed him of his watch, half a guinea and some silver as he travelled home at night, and to prevent pursuit took away his prosthetic limb which was found the following morning thrown into a ditch.[62] The urban environment was ill-adapted to those with mobility impairments. Meister noted that London's streets contained many 'small pitfalls' to trap the unwary pedestrian.[63] Artificial legs of all kinds were liable to break, sometimes with tragic consequences for wearers. In September 1766, newspapers reported that 'a poor Boy, with a wooden leg, lately discharged from an Hospital' fell down in Fleet Street after his wooden leg broke, 'and shattered his Thigh in a terrible Manner, so that he was carried again to the same Charity'.[64]

Indeed, for all its professed 'natural' appearance and superior movement, some amputees rejected the more realistic-looking 'cork' leg for being uncomfortable and liable to breakage. Major Thomas Austin, who lost his leg fighting in the Low Countries in 1810, recalled that in spite of 'employing at great expense one of those London manufacturers of mechanical contrivances', the 'common wooden' leg worked best. Although a 'cork' leg enabled him to walk 'exceedingly well in a room, or on smooth ground', it was unsuited to rougher terrain, which caused the springs to give way.[65] In some cases, amputees went further by inventing their own devices, drawing on their personal experiences of daily accommodations and adaptations of everyday objects that were needed to get by. The most fascinating example is the series of inventions documented by Captain George Webb Derenzy in *Enchiridion: Or, a Hand for the One Handed* (1822). Derenzy lost his right arm at the battle of Vitoria (1813), and aimed his inventions at his fellow veterans. They included devices to aid washing and shaving, inventions to aid dressing such as boot hooks to allow a 'one-handed person to draw on his boots with more expedition than any attendant could do it for him', and inventions to help eating and to enable activities such as writing and holding playing cards.[66] The articles were manufactured to his specifications by J. Millikin, Surgical Instrument Maker in the Strand. The values underlying his designs, he stated, were 'Cleanliness, Economy, and above all, INDEPENDENCE' and they were intended to allow a maimed person to dispense with the assistance of a servant or a friend 'to supply to him those minute arrangements of neatness and economy, which the modes and refinement of social life render indispensable to personal comfort and appearance'.[67] For Derenzy, assistive technology was a tool for maintaining manliness and authority – being able to wash, dress and eat independently were not just practical necessities, they were also crucial to avoiding infantile dependency on others and for ensuring that the amputee was not excluded from social life. Devices such as a one-handed penknife and holder to enable a pen nib to be sharpened ensured that the impaired man was able to take control over his household and business affairs. Practical products such as these, as much as elaborate artificial limbs, were important in enabling the impaired to avoid full disability. Whereas most people, Derenzy argued, took their hands for granted, once a hand is lost an individual is 'made sensible, in a thousand painful instances of daily and inevitable occurrence of the full use he had for two'.[68]

## Representations of amputees and artificial limbs

Beyond the advertisements of specialist manufacturers, prosthetic limbs had a wider cultural visibility in eighteenth- and early-nineteenth-century England. Prostheses, particularly wooden legs, were freighted with cultural meaning. Simon Dickie has argued that in eighteenth-century culture the man with a wooden leg was 'almost a master trope for testing the possibilities and limitations of sympathy'.[69] Many accounts presented wooden legs as symbols of poverty. Wooden legs were, as a newspaper pointed out in 1765, 'badges of begging'.[70] According to an account published in 1766, '[t]he wooden leg proclaims that the man has had a real loss, and pleads in his favour', a symbol of want powerful enough even to 'draw a halfpenny out of the pocket of a miser', who might normally think that 'a beggar wants nothing but a whipping'.[71] Part of the sympathetic appeal of the wooden leg lay in its association with heroic sacrifice on the battlefield. 'An old, tattered, military coat, and a wooden leg, always soften my heart to pity, and dispose me to acts of benevolence', wrote the narrator of a story, 'The Old Soldier', published in the *Lady's Monthly Museum* in 1799.[72] Time and again in eighteenth-century stories, characters with prostheses were asked by others to 'give the history' of their wooden legs, so that audiences could indulge their compassion and patriotism.[73] Yet the capacity for such stories to be faked, or for healthy limbs to be trussed up in order to present a more pitiable spectacle of suffering, acted as a persistent source of doubt and a barrier to full sympathy. A typical example of deception was that reported in the *Morning Herald and Daily Advertiser* in December 1784, which described an intoxicated beggar 'who had apparently lost a leg and an arm', getting into a quarrel with a man in America Square in London's parish of Crutched Friars. One bystander 'suspecting, from some circumstances, that the fellow was a cheat, kicked away his wooden leg, which was immediately replaced by a sound one, which had been concealed under the skirts of his clothes'. The beggar then 'stripped off his coat and waistcoat, produced another arm as sound as his new-found leg, and offered to box any of the spectators'.[74]

Amputees were not necessarily regarded as passive figures, automatically rendered dependent or 'disabled' by their impairments. *Read's Weekly Journal* reported from Dublin that 'one Maxwell, a Cripple with two wooden legs', was to be whipped through the city in January 1735 'for assaulting a Coachman and beating him in a severe manner'. Encountering a one-legged stranger might elicit feelings of fear as much as compassion or admiration. As a group of tradesmen were crossing the fields in Islington one Monday evening in August 1765, they were

> stopp'd by a fellow with a wooden leg; who swore as he knew them to be tailors, he would have the clipping of their pockets, or out with their body linings; which struck them with so great a panic, that though there were five in company, not any of them had the power to secure the fellow.

They ran away.[75] A witness to the 1816 Parliamentary Select Committee on the State of Mendicity in London described one beggar, a soldier with one leg, as a more 'violent and desperate character' than any in the metropolis, recently tried at the Westminster Sessions for 'assaults on the servants at houses where he has applied' for food or money.[76]

Overwhelmingly, representations of prosthesis-wearing amputees focused attention on the effects of disease or impairment on the *male* body. Amputation posed important questions for masculine identity. Dismemberment shattered bodily integrity, while the convulsive, involuntary motion of stumps threatened an ideal of masculinity based on self-control.[77] The nervous sensitivity of stumps was a source of vulnerability in a period where delicacy of nerves was increasingly seen as a characteristic of female bodies.[78] If disease or trauma threatened loss of control or emasculation, the artificial limb offered the hope of regaining it, concealing what was soft, sensitive and wounded, with something hard and honorific.[79]

In the patriotic atmosphere of the late eighteenth and early nineteenth centuries, prosthetic legs became symbolic of a British manliness that had been tested on the international stage. In an essay published in 1780 extolling the 'noble and truly generous character' of the British sailor, it was proudly noted that 'to a British tar' there was 'no reward like a wooden leg, and no retreat so honourable as Greenwich Hospital', the institution for treating injured naval veterans founded in 1692.[80] Similarly, an account of the wooden-legged naval heroes, Admirals Bowyer and Parsley, published in 1794, noted that a wooden leg was

> a truer and more noble mark of honour, than any which it can be in the power of King and Parliament to bestow – The Testimony of vigorous and successful exertions in the service of their country; and of the danger from which they have rescued her by their bravery and intrepidity.[81]

Such stories were highly idealized and deflected attention from the poverty faced by many wounded veterans – even those in receipt of pensions – and the threat of disorder posed by demobilised ex-servicemen.[82] Nevertheless, stories of the heroic exploits of the dismembered were popular during the Napoleonic wars. Alongside famous examples such as Admiral Nelson were men like Colonel Calkin, who lost his leg at the Battle of Albuera in 1811. Despite languishing for some time, he drew upon the 'reserves of a mind never to be subdued' and 'gave an instance of heroism never paralleled' by returning to his brigade 'in this mutilated state' and was 'as active as any man in the peninsula with a cork leg and thigh'.[83]

Yet while the wooden-legged male amputee might possess a kind of heroism or romantic appeal, female amputees were regarded very differently. Lacking the honorific associations of prosthesis-wearing soldiers and sailors, amputation in women was anomalous and on the few occasions where it was discussed, usually invited moral judgement. Prosthetics were often associated with female duplicity.[84] Since artificial legs would have been hidden beneath skirts (in contrast to male prosthetics which were often in plain view), and therefore only discoverable on the wedding night, the revelation of a woman's prosthesis served as a metaphor for other kinds of artifice and matrimonial disappointment. A story printed in the *British Apollo* periodical in 1710 described how one man had been shocked to discover that his bride had a wooden leg, a glass eye, false teeth and 'stunk like a pole cat' when they undressed on their wedding night – revelations that presaged further disappointment when he found himself deceived in expectation of her reputed £20,000 fortune.[85] In a preliminary sketch for *A Harlot's Progress*, William Hogarth drew the bawd who inveigles the innocent Moll Hackabout into a life of prostitution as having a wooden leg – a feature that perhaps stood for the squalor of the life Moll was drawn into and the deception that got her there (Moll falsely believing that the bawd would provide her with work and refuge in a reputable establishment).[86]

While wooden legs in general were often seen as fit objects for derisive laughter in eighteenth-century humorous literature, wooden-legged women were the butts of particularly cruel jokes.[87] The only 'advantage' of marrying a woman with a wooden leg, consoled the friends of the man who discovered his wife's prosthetic limb on their wedding night, was that he could lock up her limb if he suspected her of 'gadding' abroad.[88] Similarly, in 1790 it was reported that one Mr W— of Cumberland was 'married to a beautiful woman, that has but one leg', but had turned his 'misfortune to his advantage', by locking up the prosthesis whenever he went from home 'to prevent his Lady from rambling' – a scheme that was reported to have caught the attention of the Marquis de Sade who had declared that 'a general system of amputation among the fair sex would be no bad thing'.[89] Prosthetics symbolised a woman's not being 'whole' or 'intact' and were indicative of loss of honour and chastity. Representations of prosthetics indicate the powerfully gendered meanings of disease and disability in eighteenth- and early nineteenth-century England.

## Conclusion

Physical disability took many different forms in eighteenth-century England, but arguably its most conspicuous manifestation was the wooden-legged amputee. Limb loss might proceed from various medical conditions in this period, from congenital defects to debilitating and life-threatening illnesses such as cancer, but unlike modern Britain and North America where disease causes (particularly vascular disease) predominate, most attention was paid to injury as a cause of dismemberment.[90] The focus on trauma in part reflected past surgical practices that found it easier to amputate rather than try to set compound fractures and the dangers of gangrenous infection from wounds in a period that lacked asepsis and modern standards of clinical hygiene. But the emphasis on trauma was also the product of cultural factors ranging from a taste for sensational reporting of 'casualties' in the period's newspapers, to a widespread interest in the military or naval amputee who, against the backdrop of eighteenth-century warfare, became a symbol of sentiment or patriotic pride.

The proliferation of new designs for artificial limbs, heralded in the press and in the advertisements of manufacturers, marked the late eighteenth and early nineteenth centuries as a period of fertile invention with regard to prosthetics. Alongside the cheaply fashioned peg legs doled out by hospitals or the Poor Law to amputees were new bespoke artificial limbs that were marketed not simply as medical appliances to restore lost function, but also as objects of 'admiration', indicative of consumers' taste and polite aspirations. Although all were developed as technological responses to a medical need, prostheses were not simply 'medical' devices and their users not simply medicalised 'patients'. At the high end of the market, as the advertising of products 'made in the genteelest manner' indicated, aesthetic considerations might be just as, or even more, important as assistive functionality in the marketing of products. At the dawn of the nineteenth century, prosthetics were becoming increasingly commercialised, although their manufacture remained dominated by particular artisans and innovators. But while suppliers boasted that their products would meet the consumer's needs in terms of both functionality and appearance, in practice individuals weighed up the value of particular products in terms of comfort and practicality, with simpler designs sometimes proving more popular than more fashionable and realistic-looking 'cork legs'. Users of prosthetics or assistive technologies were not passive recipients of 'medical' aid, but rather played an important role in evaluating products, modifying them and designing their own. George Webb Derenzy's invention of devices to give 'independence' to the mutilated based on his own experiences shows further the importance of regaining self-reliance for users over the aesthetic qualities of the 'life-like' prostheses that drew admiration in the press.

Prostheses were culturally symbolic as well as medically purposeful. Despite the difficulties faced by many who survived amputation, wearers of wooden legs were represented as enjoying a particular status among the impaired population of Georgian England. For those who conformed to a positive cultural stereotype of the patriotic, stoic war-wounded, the wooden leg was cast as a symbol of honourable sacrifice and certain virtuous qualities of character. Although amputation was potentially emasculating, artificial limbs presented a way of regaining control, and for some war heroes the wooden leg symbolised manly heroism. Yet cultural representations of prostheses indicate the ways in which technological responses to disease or injury were distinctly gendered. Despite there being 'many females' among the amputee population, in the misogynist popular culture of eighteenth-century England, the 'artifice' of the artificial limb came to stand for women's duplicity in general. By addressing their products to a masculine, war-wounded clientele, suppliers of artificial limbs who advertised in the late-eighteenth- and early-nineteenth-century press implicitly excluded women from the advances in prosthetic technology that they proudly proclaimed.

This case study of prosthetics in eighteenth- and nineteenth-century England raises a broader set of issues about the relationship between histories of disability and medicine, and about the relationship between 'disability' and 'disease' in the past. Although in recent years there has been a tendency for histories of disability and medicine to develop separately and with different agendas, the history of amputation and prostheses provides a means for reconciling them. The characterisation of amputees as the 'predictably impaired', as Linker puts it, whose experiences contrast with the 'unhealthy disabled' – those who suffer from debilitating medical symptoms and require more frequent medical intervention – is arguably more accurate for the modern era than it is for Britain in the eighteenth and early nineteenth centuries.[91] Accounts of prosthesis-wearing amputees in earlier periods emphasise the *unpredictability* of the dismembered body – its nervous sensitivity and awkward movement. Limb loss was often fatal, and survivors faced long periods of chronic pain and incapacity.

While historians of surgery have documented advances in amputation methods and prosthetic technology as evidence of scientific progress in the treatment of potentially life-threatening conditions and the restoration of ability to the 'victims' of limb loss, taking a disability approach 'transforms traditional disease histories by expanding them to include actors outside the clinic and larger political issues'.[92] Amputation in this period was the focal point of an early politics of disability in which the authority of practitioners was brought into conflict with the views and experiences of patients. The debate about whether amputation could be avoided by early and targeted intervention extended beyond medical circles to the popular press, where amputation became the target for criticisms of the arrogance of the surgical profession and the placing of expediency or even profit over the welfare of patients. Amputation, and the replacement of living tissue with artificial parts, represented the power of surgeons over the bodies of patients and, however humane the motives or necessary the operation, highlighted how medical intervention to treat disease or injury might lead to life-long functional incapacities.

However, the status of amputees as 'disabled' was not straightforward. As devices to restore physical ability or appearance by compensating for bodily loss, prostheses acknowledged both the wearer's impairment or physical difference and their quest for functional normality. Amputees were represented as both vulnerable and powerful. While amputee beggars might claim a 'disabled' state in order to press their case for support, accounts of the dismembered poor also presented amputees as intimidating figures, striking terror into citizens. Although medical texts often treated illness, mortification and trauma equally and neutrally, subsuming them into the broad category of 'disease', in popular culture the different causes of limb loss mattered greatly, with battlefield injury conveying a sense of heroic sacrifice that those who lost limbs as a result of progressive illness, ulceration or the diseases of old age were unable to claim. It is important, therefore, for historians of medicine to document not just the causes of surgical procedures such as amputation, but the different cultural meanings that attached to them and how these produced hierarchies of status among patients. Disability history, in drawing attention to the historical contingency of meanings of impairment, offers important ways forward in this regard.

The social and cultural visibility of amputees in late-eighteenth- and early-nineteenth-century England contributed to the fixing of the wooden leg as an archetypal symbol of impairment. It is impossible to count the number of amputees in this period, but the fact that commentators often drew attention to their social visibility in public spaces such as the streets of London shows the importance of display in creating clinical or social 'types'. Donning an artificial limb involved the wearer in a form of *performance* of impairment in which the feelings of others were engaged, whether in inviting sympathy, derision or patriotic passions. These performances were distinctly gendered. While dismemberment might be the fate of anybody, the relative *invisibility* of female amputees has implications for gender histories of health and disease. Observing and explaining the cultural *absence* of certain patient groups reveals the priorities of practitioners and the general public in documenting medical circumstances and investing them with wider cultural meanings.

## Notes

1 'An Account of the famous Sieur Rocquet, surgeon, just arrived from Paris', *The Scots Magazine*, 5 May 1749.

2 Jacques-Henri Meister, *Letters Written During a Residence in England. Translated from the French*, London: T. N. Longman and O. Rees, 1799, pp. 184–6.

3 Marquand Smith and Joanne Morea, 'Introduction', in Marquand Smith and Joanne Morea (eds), *The Prosthetic Impulse: From a Posthuman Present to a Biocultural Future*, Cambridge, MA and London: MIT Press, 2006, pp. 1–2.

4 Stanley Joel Reiser, *Medicine and the Reign of Technology*, Cambridge: Cambridge University Press, 1978.

5 Stephen Mihm, '"A Limb which shall be Presentable in Polite Society": Prosthetic Technologies in the Nineteenth Century', in Katherine Ott, David Serlin and Stephen Mihm (eds), *Artificial Parts, Practical Lives: Modern Histories of Prosthetics*, New York and London: New York University Press, 2002, p. 283.

6 Lisa Herschbach, 'Prosthetic Reconstructions: Making the Industry, Re-Making the Body, Modeling the Nation', *History Workshop Journal*, 44, 1997, 23–57.

7 Katherine Ott, 'The Sum of Its Parts: An Introduction to Modern Histories of Prosthetics', in Ott et al., *Artificial Parts*, pp. 1–42; Jaipreet Virdi-Dhesi, 'Curtis's Cephaloscope: Deafness and the Making of Surgical Authority in London, 1816–1845', *Bulletin of the History of Medicine*, 87(3), 2013, 347–77.

8 For recent approaches to these questions see: Claire Jones, *The Medical Trade Catalogue in Britain, 1870–1914*, London: Pickering and Chatto, 2013; Ben Curtis and Steven Thompson, '"A Plentiful Crop of Cripples Made by all this Progress": Disability, Artificial Limbs and Working Class Mutualism in the South Wales Coalfield 1890–1948', *Social History of Medicine*, 27(4), 2014, 708–27; and Seth Koven, 'Remembering and Dismemberment: Crippled Children, Wounded Soldiers, and the Great War in Great Britain', *American Historical Review*, 99(4), 1994, 1167–202.

9 Carsten Timmermann and Julie Anderson (eds), *Devices and Designs: Medical Technologies in Historical Perspective*, Basingstoke: Palgrave-Macmillan, 2006; Mara Mills, 'Hearing Aids and the History of Electronics Miniaturization', *IEEE Annals of the History of Computing*, 11, 2011, 24–44; Julie Anderson, Francis Neary and John V. Pickstone, *Surgeons, Manufacturers and Patients: A Transatlantic History of Total Hip Replacement*, Basingstoke: Palgrave-Macmillan, 2007.

10 Amboise Paré, *The Works of Ambrose Parey, Chyrurgeon to Henry II, Francis II, Charles IX and Henry III Kings of France*, London: J. Hindmarsh, 1691, pp. 524–7, 529–30.

11 David M. Turner and Alun Withey, 'Technologies of the Body: Polite Consumption and the Correction of Deformity in Eighteenth-Century Britain', *History*, 99, 2014, 775–96; Liliane Hilaire-Pérez and Christelle Rabier, 'Self-Machinery? Steel Trusses and the Management

of Ruptures in Eighteenth-Century Europe', *Technology and Culture*, 54, 2013, 460–502; Lynn Sorge-English, *Stays and Body Image in London: The Staymaking Trade, 1680–1810*, London: Pickering and Chatto, 2011; Alun Withey, *Technology, Self-Fashioning and Politeness in Eighteenth-Century Britain: Refined Bodies*, Basingstoke: Palgrave-Macmillan, 2016.

12  Reed Benhamou, 'The Artificial Limb in Preindustrial France', *Technology and Culture*, 35(4), 1994, 835–45; Timmermann and Anderson, 'Introduction: Devices, Designs and the History of Technology in Medicine', in Timmermann and Anderson, *Devices and Designs*, p. 8; Neil Handley, 'Artificial Eyes and the Artificialization of the Human Face', in Timmermann and Anderson, *Devices and Designs*, pp. 97–111.

13  For instance see Colin Barnes and Geof Mercer (eds), *Exploring the Divide: Disability and Illness*, Leeds: The Disability Press, 1996.

14  Beth Linker, 'On the Borderland of Medical and Disability History: A Survey of the Fields', *Bulletin of the History of Medicine*, 87(4), 2013, 499–535, especially p. 526.

15  G. Thomas Couser, 'Disease', in Susan Burch (ed.), *Encyclopedia of American Disability History*, 3 vols, New York: Facts on File, 2009, vol. 1, pp. 292–3.

16  William Buchan, *Domestic Medicine: Or, a Treatise on the Prevention and Cure of Diseases by Regimen and Simple Medicines*, London: W. Strahan, 1772 edition, p. 15.

17  See 'A table of the diseases with which the patients in the Royal Infirmary of Edinburgh in 1749 were afflicted', *The Scots Magazine*, 5 January 1750, which lists 10 amputations of limbs and 2 amputations of fingers, all 'cured'.

18  Anne Borsay, 'Returning Patients to the Community: Disability, Medicine and Economic Rationality before the Industrial Revolution', *Disability and Society*, 13(5), 1998, 645–63; David M. Turner, *Disability in Eighteenth-Century England: Imagining Physical Impairment*, New York: Routledge, 2012, pp. 42–4.

19  John Kirkup, *A History of Limb Amputation*, London: Springer, 2007.

20  Ott, 'The Sum of Its Parts', p. 14.

21  For a detailed discussion see Kirkup, *History of Limb Amputation*, pp. 13–52.

22  Johan Ulrich Bilguer, *A Dissertation on the Inutility of the Amputation of Limbs*, London: R. Baldwin, 1764; 'An Account of a Dissertation on the Inutility of the Amputation of Limbs: written originally in Latin by M. Bilguer', *The Gentleman's Magazine*, 34, September 1764, p. 403.

23  *The Derby Mercury*, 7 March 1727.

24  On reporting of accidents see Roy Porter, 'Accidents in the Eighteenth Century', in Roger Cooter and Bill Luckin (eds), *Accidents in History: Injuries, Fatalities and Social Relations*, Amsterdam and Atlanta GA: Rodopi, 1997, pp. 90–106.

25  *Derby Mercury*, 12 July 1754.

26  *Sussex Advertiser*, 3 April 1758.

27  Percivall Pott, 'Remarks on the Necessity and Propriety of the Operation of Amputation in certain cases and under certain circumstances' in Percivall Pott, *Remarks on that Kind of Palsy of the Lower Limbs, which is Frequently found to Accompany a Curvature of the Spine*, London: J. Johnson, 1779, p. 45.

28  Sylvester O'Halloran, *A Concise and Impartial Account of the Advantages Arising to the Public: From the General Use of a New Method of Amputation*, Dublin: S. Powell, 1763, p. 3 ('melancholy' nature of amputation); *Stamford Mercury*, 28 December 1738 (account of General Keith forced to 'submit to an amputation'). My emphasis.

29  Pott, 'Remarks on the Necessity', p. 56.

30  Ibid., p. 45.

31  Bilguer, *Dissertation on the Inutility*, p. 4.

32  *Oxford Journal*, 15 September 1753.

33  'Jackson's Tincture', *Derby Mercury*, 28 September 1750.

34  Meister, *Letters Written During a Residence in England*, pp. 184–6.

35  Anne Borsay, *Disability and Social Policy in Britain since 1750*, Basingstoke: Palgrave-Macmillan, 2005, p. 44.

36  Mary E. Fissell, 'The Disappearance of the Patient's Narrative and the Invention of Hospital Medicine', in Roger French and Andrew Wear (eds), *British Medicine in an Age of Reform*, London and New York: Routledge, 1991, pp. 92–109; Mary E. Fissell, *Patients, Power and*

*the Poor in Eighteenth-Century Bristol*, Cambridge: Cambridge University Press, 1991, ch. 7; Susan C. Lawrence, *Charitable Knowledge: Hospital Pupils and Practitioners in Eighteenth-Century London*, Cambridge: Cambridge University Press, 1996.

37  William Nolan, *An Essay on Humanity: Or a View of Abuses in Hospitals with a Plan for Correcting Them*, London: J. Murray, 1786, pp. 24–5.

38  'Review of William Nolan, *An Essay on Humanity*', *Gentleman's Magazine*, 57(3), March 1787, p. 254; Lawrence, *Charitable Knowledge*, pp. 71–2; 'Rules and Orders Governing the Government and Conduct of the House &c.', in Alured Clarke, *A Sermon Preached in the Cathedral Church of Winchester before the Governors of the County Hospital for Sick and Lame*, London: J. and J. Pemberton, 1737, p. 47.

39  O'Halloran, *Concise and Impartial Account*, p. 3; Kirkup, *History of Limb Amputation*, pp. 65–77.

40  Kirkup, *History of Limb Amputation*, p. 76.

41  *Salisbury and Winchester Journal*, 4 February 1751; see also *Newcastle Courant*, 10 August 1751 and 'A Short History of the Effects of the Agaric of the Oak in Stopping Bleedings, after some of the most capital Operations in Surgery. . . . Communicated to the Royal Society', *Newcastle Courant* 16 November 1754.

42  O'Halloran, *Concise and Impartial Account*, p. 7.

43  *The Tribune*, 15, 20 June 1795, p. 337.

44  Nolan, *Essay on Humanity*, p. 27.

45  Excerpts from 'Medical Facts and Observations, Volume the Second', *The Gentleman's Magazine*, 64(5), May 1794, p. 448.

46  Meister, *Letters Written During a Residence in England*, p. 187.

47  George Spearing, 'The Sufferings of Lieut. Geo. Spearing, in a Coal Pit, 1769', *The Gentleman's Magazine* 63(2), August 1793, pp. 697–700, quoting at p. 700.

48  Benhamou, 'Artificial Limb in Preindustrial France'.

49  Paré, *Works*, pp. 530, 532.

50  Benhamou, 'Artificial Leg in Preindustrial France', p. 844.

51  *London Evening Post*, 30 December 1760–1, January 1761.

52  *Public Advertiser*, 16 April 1772. This example is discussed further in Turner and Withey, 'Technologies of the Body', pp. 793–4.

53  *The Festival of Wit, Or the Small Talker*, London: M. Smith, 1788, p. 223.

54  Advertisements for Ranby Reid's patent artificial leg, *Sun*, 24 May 1794, 24 June 1794; *The Times*, 15 April 1800.

55  'CONSOLATION TO OUR BRAVE, BUT UNFORTUNATE DEFENDERS. THE PATENT ARTIFICIAL LEG', *Sun*, 24 June 1794; 'OF EVERY SPECIES OF INVENTION NONE ARE OF GREATER UTILITY THAN THOSE WHICH RELIEVE THE AFFLICTIONS OF HUMAN NATURE, *Observer*, 26 March 1797.

56  *The Morning Chronicle*, 16 September 1815.

57  Paul Youngquist, *Monstrosities: Bodies and British Romanticism*, Minneapolis: University of Minnesota Press, 2003, pp. 183–4.

58  For example, 'A DISCOVERY very interesting to MUTILATED PERSONS – THE PATENT ARTIFICIAL LEG', *Sun*, 24 May 1794.

59  'CONSOLATION TO OUR BRAVE, BUT UNFORTUNATE DEFENDERS. THE PATENT ARTIFICIAL LEG', *Sun*, 24 June 1794. For a fuller discussion of the points made in this paragraph see Turner and Withey, 'Technologies of the Body', pp. 790–4.

60  *Daily Courant*, 24 June 1731.

61  *Public Advertiser*, 17 July 1765.

62  *Public Ledger*, 26 September 1765.

63  Meister, *Letters Written During a Residence in England*, p. 184.

64  *Public Advertiser*, 1 September 1766.

65  Thomas Austin, *"Old Stick-Leg": Extracts from the Diaries of Major Thomas Austin*, arranged by Brigadier-General H. H. Austin, London: Geoffrey Ble, 1926, pp. 196–7. For a similar example see Thomas A. Foster, 'Recovering Washington's Body-Double: Disability and Manliness in the Life and Legacy of a Founding Father', *Disability Studies Quarterly*, 32(1), 2012. Online. http://dsq-sds.org/article/view/3028/3064 (accessed 28 May 2015).

66 George Webb Derenzy, *Enchiridion: Or, a Hand for the One-Handed*, London: T. and G. Underwood, 1822, p. 28.

67 Ibid., pp. 57, 11.

68 Ibid., p. 10.

69 Simon Dickie, *Cruelty and Laughter: Forgotten Comic Literature and the Unsentimental Eighteenth Century*, Chicago and London: University of Chicago Press, 2011, p. 93.

70 *London Chronicle or Universal Evening Post*, 9–12 February 1765.

71 'Reflections on publishing by subscription', *Gazetteer and New Daily Advertiser*, 20 May 1766.

72 'The Old Soldier', *The Lady's Monthly Museum*, December 1799, p. 441.

73 For examples, see Turner, *Disability in Eighteenth-Century England*, pp. 73–7, 125.

74 *Morning Herald and Daily Advertiser*, 23 December 1784; see also Turner, *Disability in Eighteenth-Century England*, pp. 90–4.

75 *London Evening Post*, 13–15 August 1765.

76 *Minutes of the Evidence Taken Before the Committee. . . . to Inquire into the State of Mendicity and Vagrancy in the Metropolis*, London: Sherwood, Neely, and Jones, 1816, p. 32.

77 Erin O'Connor, '"Fractions of Men:" Engendering Amputation in Victorian Culture', *Comparative Studies in Society and History*, 39(4), 1997, 742–77, at 744.

78 G. J. Barker-Benfield, *The Culture of Sensibility: Sex and Society in Eighteenth-Century Britain*, Chicago and London: University of Chicago Press, 1992.

79 Ott, 'The Sum of Its Parts', p. 16.

80 'A Lecture on Heads Concerning the English Sailor', *Morning Post and Daily Advertiser*, 30 June 1780.

81 *St James's Chronicle or the British Evening Post*, 10–12 June 1794.

82 Simon Parkes, 'Wooden Legs and Tales of Sorrow Done: The Literary Broken Soldier of the Late Eighteenth Century', *Journal for Eighteenth-Century Studies*, 36(2), 2013, 191–207.

83 *The European Magazine, and London Review*, March 1813, pp. 364–5.

84 Turner and Withey, 'Technologies of the Body', 739–40.

85 *British Apollo*, 17–19 April 1710.

86 The sketch dates from around 1731. Online. http://www.britishmuseum.org/research/collection_online/collection_object_details.aspx?assetId=95893001&objectId=753000&partId=1 (accessed 29 May 2015); Vic Gatrell, *The First Bohemians: Life and Art in London's Golden Age*, London: Allen Lane, 2013, pp. 280–1.

87 Dickie, *Cruelty and Laughter*, p. 51 and ch. 2 *passim*.

88 *British Apollo*, 17–19 April 1710.

89 *English Chronicle, or Universal Evening Post*, 2–5 January 1790.

90 See for example NHS Choices, 'Amputation'. Online. http://www.nhs.uk/conditions/amputation/Pages/Introduction.aspx (accessed 28 May 2015).

91 Linker, 'On the borderland', p. 526.

92 Beth Linker and Emily K. Abel, 'Integrating Disability, Transforming Disease History: Tuberculosis and Its Past', in Nancy J. Hirschmann and Beth Linker (eds), *Civil Disabilities: Citizenship, Membership and Belonging* (Philadelphia: University of Pennsylvania Press, 2015), p. 101; cf. Kirkup, *History of Limb Amputation*.

## Select bibliography

Barnes, Colin and Mercer, Geof (eds), *Exploring the Divide: Disability and Illness*, Leeds: The Disability Press, 1996.

Benhamou, Reed, 'The Artificial Limb in Preindustrial France', *Technology and Culture*, 35(4), 1994, 835–45.

Bilguer, Johan Ulrich, *A Dissertation on the Inutility of the Amputation of Limbs*, London: R. Baldwin, 1764.

Derenzy, George Webb, *Enchiridion: Or, a Hand for the One-Handed*, London: T. and G. Underwood, 1822.

Fissell, Mary E., *Patients, Power and the Poor in Eighteenth-Century Bristol*, Cambridge: Cambridge University Press, 1991.

Herschbach, Lisa, 'Prosthetic Reconstructions: Making the Industry, Re-Making the Body, Modeling the Nation', *History Workshop Journal*, 44, 1997, 23–57.

Kirkup, John, *A History of Limb Amputation*, London: Springer, 2007.

Lawrence, Susan C., *Charitable Knowledge: Hospital Pupils and Practitioners in Eighteenth-Century London*, Cambridge: Cambridge University Press, 1996.

Linker, Beth, 'On the Borderland of Medical and Disability History: A Survey of the Fields', *Bulletin of the History of Medicine*, 87(4), 2013, 499–535.

O'Connor, Erin, '"Fractions of Men": Engendering Amputation in Victorian Culture', *Comparative Studies in Society and History*, 39(4), 1997, 742–77.

Ott, Katherine, Serlin, David and Mihm, Stephen, *Artificial Parts, Practical Lives: Modern Histories of Prosthetics*, New York and London: New York University Press, 2002.

Parkes, Simon, 'Wooden Legs and Tales of Sorrow Done: The Literary Broken Soldier of the Late Eighteenth Century', *Journal for Eighteenth-Century Studies*, 36(2), 2013, 191–207.

Timmermann, Carsten and Anderson, Julie (eds), *Devices and Designs: Medical Technologies in Historical Perspective*, Basingstoke: Palgrave-Macmillan, 2006.

Turner, David M., *Disability in Eighteenth-Century England: Imagining Physical Impairment*, New York and London: Routledge, 2012.

Turner, David M. and Withey, Alun, 'Technologies of the Body: Polite Consumption and the Correction of Deformity in Eighteenth-Century Britain', *History*, 99, 2014, 775–96.

# 18

# DISEASE, REHABILITATION AND PAIN

*Julie Anderson*

In 1955, *The Lancet* reported on the case of a 'chronic' female patient who had been hospitalised for 20 years. During that time, the patient had been moved to several different hospitals until she arrived at the West Middlesex Hospital, where she was given 30 weeks of rehabilitative treatment, and from there sent to a welfare home.[1] The article notes the 'scanty records' that accompanied the patient, and criticised 'the negative therapeutic approach that cost the patient 20 years of freedom.'[2] The patient's hospital record did not outline her history or treatment regime, and did not offer her a way to articulate her experience, although she was interviewed during her 30 weeks of rehabilitation. For historians, details of the therapeutic experience provide vital information about patients, including their length of stay, the types of treatment they received, and their responses to it. Gerald Kutcher argues that the case study provides 'a counter story to the grand narratives of medical researchers and has much to tell us about the anecdotal character of clinical trials, the suffering and courage of patients as well as ethical behaviour'.[3] For many patients, their painful and alienating or positive and life-changing experiences of rehabilitation were articulated through the medical practitioner.

For this patient, her experience was evidence of the efficacy of rehabilitation. By 1955, when this patient was treated, rehabilitation was a well-developed system that sought to return patients to the fullest degree of health possible, given an individual's residual disability or underlying chronic condition. Owing to the rising costs of institutionalisation and partly because rehabilitation had been more fully developed as a medical speciality during the Second World War, patients who had previously been hospitalised were exposed to a range of physical and mental therapies, which meant that they could be released from hospitals, often into less costly institutions or back home. Special technologies and spaces were established for rehabilitation, including specific facilities in general hospitals as well as specialist hospitals. Rehabilitation methods were adopted by a number of branches of medical practice, treating those with a range of diseases, from cancer to tuberculosis. Its early association with physical medicine, physiotherapy and movement imbued rehabilitation with meanings and experiences. The uniqueness of rehabilitation was its physical and emotional processes, during which patients regained the use of their body or restoration of their minds following a range of treatment regimes. As the twentieth century progressed, rehabilitation became fundamental to a successful outcome, a means of completing other types of medical treatment, including surgery, radiation therapy and drug regimes.

This chapter focuses on pain and rehabilitation in disease. Rehabilitation was first used in a medical context in 1888, relating to children who had suffered from conditions due to parental neglect. The term was also used to describe the process that criminals underwent in efforts to reform them. During the Second World War, rehabilitation became an official organisational system that treated a range of physical and mental traumas, mainly of service personnel.[4] After the Second Word War, rehabilitation adopted a humanitarian perspective and referred to the process of settling children in different parts of war-torn Europe.

Concomitant with changed definitions of disease, a broad range of therapies became rehabilitative. Patients were rehabilitated from deviant types of behaviour, such as criminality or addiction to drugs and alcohol. Indeed, the term rehabilitation in English referred less to physical training and social intervention, and more to the treatment of drug addiction or criminality. For those with addiction, rehabilitation was carried out in a number of institutional settings, both public and private. Dedicated centres for rehabilitation from addiction were established in Britain, Europe and the United States. Rehabilitation was also subsumed into the medical mainstream, and became part of therapeutic regimes for diseases such as tuberculosis, polio and cancer. Within general medicine, rehabilitation referred to the process of taking people from illness to a condition of fullest function, although in some cases the outcome of rehabilitation was the creation of a chronic condition. Thus, regimes of rehabilitation changed significantly as a consequence of changed conceptions of disease, the range of patients, and its success as a therapeutic tool.

During the twentieth century, more diseases such as polio and tuberculosis were prevented through the use of vaccination, and increasing numbers of patients survived conditions that had previously been death sentences. As treatment regimes increased in complexity, rehabilitation at various stages was required for patients to recover fully from conditions such as cancer and tuberculosis. The complexity of diseases and treatments in the modern period meant that rehabilitation became fundamental to a patient's recovery. Rehabilitation was employed when diseases could not be fully cured, or when they left a lasting impact on the body, which needed to be managed. It was both an end-point and a process. Rehabilitation, and the time spent in specialist centres, as Seymour has noted, made the patient confront bodily frailty, and vulnerability.[5] It instilled discipline in the body and was important in the relief of permanent symptoms relating to disease.

One of the facets unique to rehabilitative treatment throughout the twentieth century was its duration. Rehabilitation took weeks, months, or years. Gloria Paris recounted her nine years of medical treatment for skeletal tuberculosis in the book *A Child of Sanatoriums*.[6] For those with bone cancer who had surgery, the process of rehabilitation took a considerable amount of time, no matter how many hours were spent in the operating theatre. The often painful physiotherapy for those affected with polio could take many months. For cancer patients, after surgical procedures, radiotherapy and chemotherapy, rehabilitation added many months of treatment. For others, such as those suffering from addiction, rehabilitation was a process lasting a lifetime.

## Rehabilitative spaces and technologies

In order for it to be successful, rehabilitation needed to be intensive and controlled. One of the ways that control was placed on patients undergoing rehabilitation for the recovery from the effects of disease was to regulate the space where rehabilitative therapy took place. These spaces formed a significant part of the patient's experience. The medical environment was a foreign country, and men, women and children forged their way through unfamiliar systems, rules and regulations. Recovery from disease moved a patient from their daily life into rehabilitation regimes. Rehabilitative spaces were associated with pain, struggle and failure, as well as success and recovery.

Specific spaces provided a locus for rehabilitation. As Dana Arnold has noted in her study of London's hospitals in the long eighteenth century, 'improvements in medical knowledge impacted on the design and layout of hospitals, with distinctive designs to create healthier environments for patients and staff'.[7] Throughout the twentieth century, mainly within hospitals, different spaces were created internally to accommodate spaces for rehabilitation. As hospitals were modernised, and new hospitals built, rehabilitative environments were established within them, as the regimes for rehabilitation became associated with standard medical practice. Particular spaces, such as gymnasiums, exercise rooms and swimming pools, were established in hospitals, thereby cementing the place of rehabilitation in a hospital or treatment centre.

In the late nineteenth and early twentieth centuries, specific spaces were established for rehabilitation from particular diseases. According to Helen Bynum, 'sanatoria and the sanatorium regime created a little world within a world in the first half of the twentieth century'.[8] Established from spaces as diverse as converted hospitals to private individual's homes, sanatoriums were places for rehabilitating individuals and limiting the effects of disease. Furthermore, Thomas Dormandy has argued that the English sanatorium was 'conceived as preventative rather than curative: it was to admit only early cases, heal them, make them fit for employment and keep them off the rates'.[9] Built to limit the effects of poverty mainly on children in the early twentieth century, the open-air regime of the sanatoriums provided a relatively inexpensive rehabilitative regime for significant numbers of patients with tuberculosis.

These spaces for rehabilitation were contested, where battles between patients and medical professionals were played out. The embodied experience of rehabilitation and pain was also located in these spaces as patients struggled to regain control of their bodies. In these spaces, bodies and minds were disciplined and an unruly body or mind, one that required rehabilitation, was trained in methods of management. Patients recovering from the effects of disease were also pitted against each other, competing to demonstrate to the medical professionals and indeed themselves, how well they had progressed in their therapy. Despite a significant number of patients undergoing rehabilitative therapy after they had recovered from the acute stages of their disease, the experience of rehabilitation was often a lonely one.

Feelings of isolation during the rehabilitative process were not just emotionally painful; they were also manifested physically. In particular, children affected by diseases such as polio and tuberculosis were often isolated from their families. The hospital was at the forefront of treatment, and isolation was one of the methods that hospitals employed in order to control the spread of disease. Even after infectious phases were over, patients were isolated while they completed their therapy. Accounts of children with polio remembered and told by adults often reveal the physical and mental pain; one child recalled being hugged by his mother in 1938 and was not touched by her again until 1941.[10]

Unfamiliarity with medical environments was alienating. Therefore, the familiarity of home occupied a central place in a patient's consciousness. 'Going home' was what many patients undergoing rehabilitation strove for, and was a milestone in the recovery process. Home was often seen as a place of refuge, and a target for the patient to aim for, a marker that the difficult and sometimes painful part of the rehabilitation process was complete. Rehabilitation often continued at home, but the familiar environment signalled a new phase of the process. As the twentieth century progressed, the home environment became more of a locus for rehabilitation and fewer rehabilitation units remained residential, although there were some that continued to be specifically geared towards rehabilitative therapy, particularly in Europe.

## Therapy and technology

Medical advances through the twentieth century altered the patient's experience of pain and rehabilitative regimes. Rehabilitation was associated with a wide range of therapeutic tools. From hydrotherapy to work therapy, different specialisms were included in medical regimes as patients were rehabilitated from the physical and emotional consequences of a range of diseases. The therapeutic process of rehabilitation was closely linked with developments in medical technology, including apparatus for exercising, physiology, support networks and aids and adaptations to living outside institutions after rehabilitation.

Therapeutic bathing in spas has a long history.[11] For rehabilitation, water provided an environment in which patients could regain bodily fitness without pain. In his study of polio, Gareth Williams has noted that 'the buoyancy provided by water cancels out the effects of gravity and gives weak muscles a better chance to start working again'.[12] Water therapy was often used as part of rehabilitation regimes, for a wide range of patients, and included, more recently, those with bone prosthetics as a result of cancer. Pools for hydrotherapy were built within hospitals, as rehabilitation became centrally located within the hospital environment. Indeed, Jane M. Adams has argued that in Britain the relocation of hydrotherapy to hospitals eventually closed spa buildings that had been so fundamental to the creation of health from the sixteenth century.[13] As technological advances ensured that water was consistently heated, an increasing number of swimming pools were built in hospitals throughout the 1950s and 1960s.[14]

Although swimming pools in hospitals were seen as preferable to spas in Britain, warm water therapy remained popular in Europe and the United States. In particular, water was used in rehabilitation regimes for polio. Warm Springs, Georgia, gained notoriety by the regular presence of the most famous person to contract polio, President Franklin Delano Roosevelt. Roosevelt was apparently inspired by a young engineer who had learned how to walk after swimming at Warm Springs.[15] Roosevelt's faith in hydrotherapy ensured that Warm Springs was heavily populated by patients who wanted to regain the ability to walk. The centrality of water to rehabilitation from the effects of polio was in some ways ironic. Fears surrounding infection from polio caused widespread panic and people were instructed to avoid crowds and especially water. These concerns forced the closure of swimming pools in the summer months until vaccination eradicated polio in industrialised nations.[16]

Rehabilitation and pain were embodied experiences. Disease was experienced through the body, both physically and emotionally. Occupational and work therapies were fundamental to the rehabilitative process. Occupational therapy, which was employed regularly in hospitals from the early twentieth century was often used as a diversion, and took attention away from the boredom that many patients experienced in hospital.[17] It often consisted of projects such as sewing and embroidery, or making rugs or baskets in bed. It required a small investment in materials and patients were kept under surveillance, especially in the early part of their treatment, when their bodies needed to be controlled in order for the rehabilitation process to run smoothly.

More active therapies were developed and became part of regimes of rehabilitation. Therapy that involved physical labour normalised the notion that people suffering from the effects of disease were curable, and that they would resume work once their rehabilitation was completed.[18] It also meant that state support, which was costly, was kept to a minimum. Work therapy was employed, particularly in diseases that required long-term treatment, such as tuberculosis.[19] Similarly to many other institutions, work therapy in sanatoriums varied according to a patient's social class. For those whose job had been mainly physical, graduated labour was used as therapy. This often consisted of digging, poultry keeping, and looking after the extensive grounds that surrounded the sanatorium. The outdoor environment was common for many patients undergoing rehabilitation. Not only was it inexpensive to treat patients with air, it was a fundamental part of the regime for treating patients with pulmonary tuberculosis.[20] Workshops and village settlements were established to treat patients out-of-doors. One of the best known of these was Papworth Village Settlement in Cambridge. Patients came for more than a short period of treatment and rehabilitation; they had access to financial support and paid work which facilitated longer-term treatment. Papworth started as an agricultural venture, but expanded to industrial production; Papworth Industries was established to provide employment for patients. The enterprise was financially successful and expanded to include furniture, travel goods, and even parts of the Green Goddess fire engines were produced by Papworth Industries.[21] Although the Settlement at Papworth was unique, graduated work therapy was used as part of rehabilitative treatment, for children as well as adults, throughout the twentieth century.

Aids and adaptations were also developed to help recovery. In the case of diseases such as polio and skeletal tuberculosis, massage and physiotherapy strengthened muscles to a point, but often bracing was needed so that patients could walk.[22] In many cases, the resumption of the ability to walk was a testament to the success of rehabilitation. Additional assistance from aids such as callipers served to support weakened leg muscles. Aids had varying levels of success, and patients experienced a range of experimental frames and supports while they were in rehabilitation. From Plaster of Paris casts to metal rods, external support came in myriad forms, some resembling instruments of torture.[23] As the twentieth century progressed, internal forms of support were also used to aid weakened muscles or replace bone. Cancerous bone was replaced by prostheses made from metals such as stainless steel and cobalt chrome.[24] Whether a prosthesis was internal or external, physical therapy and rehabilitative techniques in the exercise room, gymnasium or swimming pool were employed to ensure that the highest possible function was achieved.

Part of the purpose of rehabilitation was the achievement of a state of emotional as well as physical wellness. The healthy mind–body nexus was considered fundamental to full rehabilitation. Consequently, negative thoughts were an obstacle to rehabilitation. As rehabilitation became a more holistic practice, the mental health of patients became nearly as important as their physical health. In the case of patients with alcoholism or other forms of addiction, aids took the form of talking therapy, support groups such as Alcoholics Anonymous, and in more severe cases, institutionalisation.

## Pain

Much of rehabilitative therapy was associated with differing levels of pain. Whether a patient underwent amputation as a result of cancer, or regained the use of muscles with physiotherapy after polio, the physical nature of bodily rehabilitation meant that it was painful. Much of the rehabilitative process was about restoring weakened parts of the body. Those who narrated their experience of the rehabilitative aspects of polio wrote about the pain of learning to use muscles. Regaining function of the body meant that there was pain associated with new and unfamiliar forms of movement. Using body parts that were diseased, or that were painful from a lack of use, was often a challenging experience, and narrating this experience formed part of the therapeutic narrative. As Joanna Bourke has noted, 'pain-narratives can be productive'.[25] Arthur Frank goes further to suggest that in some cases they are necessary and assist the patient in the healing process.[26]

Nevertheless, pain caused by rehabilitative treatment was not the uncontrollable pain caused by the sudden onset of a disease or condition. The pain associated with rehabilitation regimes was controlled, medicalised and overseen by a practitioner, or by patients themselves. Indeed, pain caused by rehabilitation was measurable and could be stopped if the patient found it too unbearable. It could have a specific duration, lasting minutes or even seconds. Furthermore, treatment took place in controlled environments, many of which were especially reserved for rehabilitation and sometimes designed to alleviate the pain associated with movement. Locations associated with the rehabilitative process varied from swimming pools, treatment rooms and gymnasiums with weight-bearing apparatus to, even in the early stages, the patient's bed. While a patient often experienced a significant amount of pain during the process of rehabilitation, there was an end-point when the session of treatment was completed. Yet for some patients, the pain of rehabilitation was excruciating. The singer Ian Dury, who was infected with polio in 1949, detailed his rehabilitative therapy:

> Every Tuesday and Thursday they put you in this manipulatory situation. Obviously, they thought they were doing you good . . . But I mean this geezer used to get hold of my left ankle and my left thigh and . . . put my heel up to my arse . . . It was called the screaming ward and you could hear people screaming on the way there, and it was you when you was there and you could hear the others on the way back.[27]

Conversely, as rehabilitation often signalled a life-changing moment, patients experienced other types of pain, including emotional pain. Indeed, disease could have profound effects on a patient's mental health. According to Lisa Folmarson Kall, 'pain is the source of sorrow, suffering, hopelessness and frustration'.[28] In his edited volume on pain, Rob Boddice goes further, arguing that 'pain *is* sorrow, suffering, hopelessness and frustration'.[29] Although for those who were being rehabilitated from substance addiction there was often physical pain, the rehabilitation process was also significantly different emotionally from that endured by patients suffering from physical disease.

Rehabilitation and treatment methods were processes that lent themselves to documentation, such as patients' notes, as a record of the patient experience. Notes were made from communication between medical practitioners and patients, about levels of pain; and measurements such as the Oxford Hip Score assisted in charting a patient's experience of pain.[30] Medical practitioners assessed pain levels, as their familiarity with the painful processes of rehabilitation provided them with the means to note and understand a patient's expressions of pain. Pain was a shared experience, yet it was also a very individual and personal one.

## Patient experiences and narratives

The forms of patient narratives of rehabilitation and pain are diverse, but the written form is the most enduring. Whether details appeared in patient notes, in medical journals, or in books, patient narratives provided an additional way of understanding their experience of disease.[31] The method of dissemination of the narrative has altered over the twentieth century in particular. Indeed, public narratives of treatment methods and rehabilitation regimes became more common as the second half of the twentieth century progressed. The public 'illness narrative', the 'overcoming narrative' or the 'celebration narrative', became popular reading material, as people battled with treatment, pain and rehabilitation in a more public sphere.[32] There were ready audiences for this topic, as patients sought to learn, empower themselves or indeed identify with each other by differentiating themselves from the sick. Patient narratives of treatment regimes, pain and rehabilitation also informed other potential patients. The public enjoyed reading illness narratives and the redemptive power of the medical profession to bring a patient back from the brink of death. Instead of the private medium of letters or the ephemeral nature of conversation, books, films, and later blogs detailed the patient's experience as they negotiated their way through increasingly complex medical systems and rehabilitative treatments that at times seemed painful and brutal.

Identity was embodied and narratives of disease were associated with personal and indeed collective patient identity. In some stories of pain and rehabilitation, people readjusted to new forms of the self and narratives aided in this transformation.[33] In the past, narratives of disease existed, but were mediated through doctors. Patients responded to questions, were examined, and tests were undertaken. Indeed, the patient's role in the rehabilitative process was fundamental to its success, as noted by neurosurgeons such as Ludwig Guttmann.[34] As rehabilitation was a long and painful process, milestones and the documentation of success were important to the strategy of cure or alleviation of symptoms. Patients became discouraged should they feel they were not making progress, so improvement of any type was a morale-raising experience. Patient notes often provided witness to the success of rehabilitation for patient and practitioner. This meant that the medical practitioner was fundamental to the documentary process of rehabilitation.

It has been argued that this concentration on the medical profession's documentation of the rehabilitative process effectively missed out on an important component of illness – the patient's subjective experience. As extended hospital periods or periods of inactivity created boredom, writing was a type of therapy or passed the time between treatments. Writing about illness allowed both patients and doctors to gain perspective. Consequently, documenting a process could help to inform medical practice and refine treatment regimes. Although the readers of a patient narrative might not understand the process, or be unfamiliar with the medical spaces that a narrative described, most people had experienced pain and were able to relate to accounts of painful rehabilitation. Narratives collected by researchers endeavoured to understand the ways that patients understood and conceptualised their disease and their embodied experience of pain and rehabilitation.[35]

The following examples of cancer, tuberculosis, polio and addiction offer some insight into the ways that patient narratives were constructed, understood and presented. Specific diseases evoked different emotions and treatment regimes changed over time. Narratives of cancer and indeed its treatment have moved from being hushed conversations between medical practitioners and patients to a widely publicised, public fight against the disease. Conversely, the patient's experience of polio has all but disappeared, and become the subject of histories. Tuberculosis narratives were often used as the basis for fictionalised versions of narrative. Finally, narratives of addiction and alcohol dependency, which have proved highly popular, were affected by the reclassification of substance dependency and abuse as disease.

## Cancer

According to recent statistics, one out of two people born in the UK after 1960 will develop cancer.[36] As people live longer, their risk of getting cancer increases. As the twentieth century progressed, cancer became increasingly survivable and rehabilitative therapies were used more often to lessen the long-term impact of the disease. Prior to the 1970s, the subject of cancer was rarely discussed in the media, and an individual's experience of it was barely a topic for private, or indeed public consumption.[37]

Yet, by the late 1970s, experiences of treatment and recovery were becoming the subject of interest. Indeed, surviving cancer, the processes of cure, and the lengthening of life became common topics for biographies and autobiographies of disease. Some of them explored the experience of painful rehabilitative processes. Echoing the metaphors often used to describe cancer, such as 'fighting' and 'war', the theme of patient narratives is often that of overcoming, that the fight of this historically dread disease was one that could be won. Probably the best-known of these is Susan Sontag's *Illness as a Metaphor*, which paved the way for other authors to detail their own disease narrative generally, and their cancer story specifically. Cancer narratives remained popular, and well-known writers and journalists wrote about the treatment regimes and their reflections on the impact of the disease. Narratives of cancer and its treatment come in a number of forms and focus on different types of cancer. In her essay, *Welcome to Cancerland*, Barbara Ehrenreich wrote about her own experience of breast cancer. A number of well-known writers detailed their cancer treatment and rehabilitation and with it their feelings of alienation, hopelessness, confusion and pain. The British journalist John Diamond publicly detailed his fight against cancer in a regular column in the *Sunday Times* after he was diagnosed with throat cancer in 1997. His book, *C: Because Cowards Get Cancer Too* . . . which was published in 1999, was preceded by a documentary on the BBC's *Inside Story* where Diamond detailed his treatment and rehabilitation, including his difficulties with speech therapy after surgery.[38] This type of narrative explained treatment and rehabilitative regimes, and informed those with similar diseases.

## Polio

Poliomyelitis reached epidemic proportions in the 1950s in North America, Australia, South America and many other countries. Polio was a dreaded disease, and there was widespread panic surrounding modes of contagion until Salk's vaccine went into production after a large-scale clinical trial in 1955.[39] This was followed by Sabin's oral vaccine, and for the most part, polio was eradicated worldwide. During the epidemics, the effects of which were particularly severe in countries such as the United States and Australia, many thousands of people died, and for those who survived the disease, the rehabilitation regime was often long and painful.

Naomi Rogers has noted that many of those affected by polio before 1920 were less than five years old, so it is difficult to accurately present the patient's experience.[40] Yet, as Anne Borsay has argued, children were more often the recipients of rehabilitative therapy for diseases. The absence of what might be clear memories did not deter those affected by polio from telling their stories, and from these oral histories, historians have learned much about the rehabilitation therapies and pain experienced by these young patients. It has been argued that these retrospective stories lacked authenticity, as they were based on the memories of children, memories that, as John Tosh has noted, 'are filtered through subsequent experience'.[41] However, childhood memories provided an important context to painful and often alienating rehabilitation regimes.

Patient narratives also highlighted the different and individualised regimes that children experienced. These included the Australian memoir from Alan Marshall, *I Can Jump Puddles*. His rehabilitation took place outside the hospital because the medical facilities in rural Australia were not advanced.[42] Much of Marshall's rehabilitation, supported by his family, consisted of his undertaking physical activities in order to strengthen his weakened body. The regime of the hospital, with physiotherapy, exercise and aids such as callipers, experienced by children in an urban environment, were not part of Marshall's rehabilitative experience.

Different conceptions of the success of rehabilitation therapies for children affected by polio abounded. These notions informed treatment regimes, and a range of opinions divided the types of rehabilitative programmes available to patients. One of the most contested regimes surrounded the rehabilitation from the effects of polio, and was associated with one of the most controversial figures in the area of treatment and rehabilitation. Sister Kenny, an Australian nurse, used hot packs to reduce spasm in the acute stage of the disease, followed by movement of the affected limbs to treat children affected by polio. Although Kenny's methods were non-traditional and did not follow accepted medical theories, she toured the world with her revolutionary treatment. She established the Sister Kenny Foundation in 1943, which spread her message. Although there was criticism from a number of quarters, Kenny's methods were adopted by a number of practitioners, and she worked to rehabilitate polio patients at prestigious institutions including the Mayo Clinic. Kenny's book, which was reprinted four times in 1951, detailed her discovery of the treatment and her attempts to bring it to the world.[43] In the book, she represented herself as a caring individual, working against a sceptical and uninformed medical tradition. However, as Gareth Williams has pointed out, Kenny's young patients did not always enjoy her ministrations. He notes, 'there were no images of the children who began crying at the sound of the trolley carrying the hot packs being wheeled into the ward.'[44]

Although many children found the experience of hospital rehabilitation traumatic, some children had happy memories of their therapy and of hospital. One patient in a hospital in Ireland remembered:

> I can only say that I enjoyed my time there, as there were some great characters. I made some very good friends and remember them to this day. One of the lads, called Pat, used to play the mouth organ. One night he got on our nerves, so when he was asleep a few of us stole it and put a sausage in it. When he went to play it the next night he nearly choked himself.[45]

For the most part, however, the consistent theme running through many patient testimonies was that rehabilitation was often difficult and painful. As one patient recalled:

> What was extremely unpleasant was the physiotherapy. My physio used to walk in at nine o'clock, and I remember there was a central speaker . . . and the music for Housewives' Choice used to precede the physio by about three minutes; it was like the knell – or toll – of doom.[46]

Instead of focusing on medical rehabilitation in a hospital environment, adult patients saw a purpose in their therapy and accepted that pain, however unpleasant, was inevitable and perhaps necessary. Therefore, reports of their experiences often differed widely from those of children. An account of rehabilitative therapy was included in Paul Bates' *Horizontal Man*. Bates nearly died from polio when he contracted it in 1954 and documented his rehabilitation in the book. Originally in an iron lung, he was moved to a negative pressure chamber, and finally out of hospital. His book detailed his active life, which included running a radio station and representing polio survivors.[47] Another polio survivor, Harry Tildesley, wrote about his success as an artist, puppet maker, charity fundraiser and Group Scoutmaster.[48] These patient experiences focused more broadly on the painful regime of rehabilitation as part of the life-course, and were generally more positive than those affected by polio when they were children.

## Tuberculosis

Similarly to many other types of disease, tuberculosis relied on a therapeutic process to prolong life and limit the disease's effects, rather than on any cure. There were two types of tuberculosis, pulmonary and skeletal. Many patients died from tuberculosis of the lung, but the majority of patients survived skeletal tuberculosis. The impact of tuberculosis was alleviated when streptomycin was developed and successful experiments in Madras were undertaken in 1948.[49]

Disease, pain, and rehabilitation have all been the subject of novels. Yet tuberculosis has been one of the most enduring and best-remembered subjects of fictional disease narratives. The unfamiliar environment of the sanatorium provided a microcosm of human interaction, from the relationships between the patients and those who treated them, the landscape, the therapeutic regime, and their diseased bodies. Novels based on the experience of tuberculosis sufferers have documented patients' experiences. The most famous is Thomas Mann's *The Magic Mountain*, in which he details the lives and treatment regimes of a group of patients in Switzerland. In *The Magic Mountain* even though the patients have pulmonary tuberculosis, they do some active therapy including the prescribed daily walk followed by periods of rest. The British novel, *The Rack* also demonstrated the long and painful duration of treatment.

Many patients were ordered to rest in the early stages of the disease, so spent their time writing letters documenting their experiences which were sent to relatives and friends. Letters often detailed the treatment regime, the food, and other small incidents experienced by people suffering from disease. Letter writing provided a way for patients undergoing rehabilitation and treatment to pass the time as they endeavoured to recover. Although many sanatoriums of varying type and quality were established throughout the early part of the twentieth century, patients were distanced from those unaffected by disease, as institutions were often in rural environments, and there was a public fascination with the regime and the sanatorium as a place of rehabilitation.

Although novels, personal accounts and histories popularised pulmonary tuberculosis, there were few patient narratives that detail those whose bones were affected by it. Gloria Paris recounted her experience of skeletal tuberculosis in her book, *A Child of Sanatoriums*. Historically, those with tuberculosis of the bones were often able to function and therefore did not spend prolonged periods in sanatoriums. The only evidence of the infection was a weakened limb or a limp. Paris's account of her nine years undergoing treatment in the United States provides unique detail into the rehabilitation regimes for those with skeletal tuberculosis. She endured a range of treatment, from surgery to physical therapy. The book concentrates more heavily on her recovery and the success that she made of her life as a microbiologist and mother.

## Addiction

Throughout the twentieth century, a range of rehabilitative practices were developed that mirrored the increased range of conditions understood in medical terms. By the 1980s, addiction had been reconceptualised as a disease of genetic inheritance.[50] Other conditions were medicalised, and were classified as diseases, such as alcoholism and substance addiction, which resulted in the development of new methods for rehabilitating those affected. As rehabilitation began to centre more on changing modes of behaviour, as well as the reduction of impact on the physical body, it was also used to describe the process of recovering from addiction. Some treatment programmes were set up within the community such as Alcoholics Anonymous; others were established in residential clinical settings. Rehabilitation from addiction was an on-going process, which did not end with release from a hospital or clinic. Narratives of addiction were often situated outside of institutional settings. For addicts, going home was fraught with difficulty as diseases of addiction were often lifelong conditions that needed to be managed. Indeed, going home represented the potential for their disease to reassert itself, as surveillance and other forms of support were reduced or removed.

Recovery from addiction initiated different approaches to rehabilitation. Diseases of addiction presented a unique set of problems that resulted in certain types of therapeutic regimes. Rehabilitative treatment for addiction was physically painful, although not life-threatening, as patients suffered withdrawal symptoms. Removal from specific environments was also conducive to a patient's rehabilitation. Furthermore, discussion was important to a patient's recovery, especially in the talking therapies advocated by early practitioners such as Freud and Jung. Talking in groups or individually to a doctor or counsellor formed the basis of therapy for many patients, and their narratives were fundamental in establishing and maintaining treatment regimes. While some patient narratives were (and still are) published in order to inform, educate and entertain, the private narrative between medical practitioner and patient was also a significant part of the therapeutic process.

It has been argued by Angela Woods that there is a tendency in 'the field of narrative in medicine to treat mental and physical illnesses as resolutely distinct.'[51] The effective medicalisation of what had previously been regarded as behavioural 'weaknesses' served to broaden the scope of these newly constructed diseases. The stigma of rehabilitation from substance addiction, and its related issues, meant that the pain of rehabilitation for those struggling with addiction possessed an extra burden. Treatment regimes were physically and mentally painful, and patients often narrated these experiences in order to help themselves. The unfamiliarity of the residential institution, the pain and isolation of substance abuse as it affected one's friends and family, and the rehabilitative therapies were all part of the patient narrative. Many celebrities wrote books detailing their personal struggle with substance abuse and alcoholism, and the combination of interest in the lives of famous people, coupled with their perceived flaws, ensured that publishers maintained a market for these patient narratives. Betty Ford, wife of American president Gerald Ford, was one such celebrity. She co-authored a book about her substance abuse which was published in 1987, five years after she had opened the Betty Ford Clinic in 1982. Another book edited by Ford, *Healing and Hope*, detailed the experience of the recovery processes of several women at the renamed Betty Ford Center.[52] The 'addiction narrative' was a popular form of reading material for the public.

Narratives were communicated through a range of media. In particular, film, art and music have explored the life course of those who undergo rehabilitative treatment following cancer. Films such as *Champions* (1983), which starred the actor John Hurt as the jockey Bob Champion, followed his painful treatment and rehabilitation after his cancer diagnosis. Other film projects have followed, including *Opera Therapy* (2005), which narrates through music the experiences of the medical treatment of four Australians with cancer. Artists have also addressed treatment regimes and the pain and loneliness associated with cancer therapies. Canadian artist Robert Pope's 1991 book *Illness and Healing: Narratives of Cancer* detailed his treatment for Hodgkin's disease with a narrative essay accompanied by 92 illustrations.

During the twentieth century, more vehicles were used to explore the patient's experience of rehabilitation and pain. The Internet has opened further possibilities for patient narratives.[53] Patients have recounted their experiences through the use of blogs, self-publishing and other social media, which have also provided a space for documenting, campaigning and fundraising. Patients undergoing rehabilitation therapies found a ready audience through these outlets.

## Conclusion

During the twentieth century, the use of rehabilitation became firmly entrenched within medical therapies. Regimes of rehabilitation responded to the reconceptualisation of the boundaries of what was understood as disease. Whether it related to the addict, the cancer patient, or a child recovering from polio, rehabilitation was a fundamental feature of the process of ensuring that the effects of disease were mitigated. Different spaces, therapies, and technologies were developed to aid in rehabilitation. Nevertheless, despite pain management systems and new technologies, rehabilitation was often a long and painful process. Narratives were fundamental to the process of rehabilitation and pain and took many forms. For the medical profession, patient narratives facilitated understanding and were used to document medical practices and treatments. For some patients, writing about rehabilitation was part of the therapeutic process, serving either to help others understand or to facilitate their own recovery. New methods of publishing such as blogs, social media and self-publishing altered the nature of the presentation of patient narratives of rehabilitation, yet many of them followed similar tropes.

## Notes

1 J. R. D. Bayne and Marjory Warren, 'Disposal of the Chronic Case', *The Lancet*, 25 June 1955, 1317.
2 Ibid., 1317.
3 Gerald Kutcher, 'A Case Study in Human Experimentation: The Patient as Subject, Object and Victim', in Carsten Timmermann and Elizabeth Toon, *Cancer Patients, Cancer Pathways: Historical and Sociological Perspectives*, Basingstoke: Palgrave Macmillan, 2012, 57–77, 57.
4 Julie Anderson, *War Disability and Rehabilitation: 'Soul of a Nation'*, Manchester: Manchester University Press, 2011.
5 Wendy Seymour, 'Time and the Body: Re-embodying Time in Disability', *Journal of Occupational Science*, 9 (3), 2002, 135–42, 138.
6 See Gloria Paris, *A Child of Sanitariums: A Memoir of Tuberculosis Survival and Lifelong Disability*, Jefferson, NC: McFarland and Company, Inc., 2010.
7 Dana Arnold, *The Spaces of the Hospital: Spatiality and Urban Change in London 1680–1820*, Abingdon: Routledge, 2013, p. 7.
8 Helen Bynum, *Spitting Blood: A History of Tuberculosis*, Oxford: Oxford University Press, 2015, p. 128.
9 Thomas Dormandy, *The White Death: A History of Tuberculosis*, London: Hambledon, 1998, p. 166.
10 Nuala Harnett (ed.), *Polio and Us: Personal Stories of Polio Survivors in Ireland*, Dublin: Post-polio support group, 2007, p. 17.
11 See Roy Porter (ed.), 'The Medical History of Waters and Spas', *Medical History*, Supp. No. 10, London: Wellcome Institute for the History of Medicine, 1990.
12 Gareth Williams, *Paralysed With Fear: The Story of Polio*, Basingstoke: Palgrave Macmillan, 2013, p. 144.
13 Jane M. Adams, 'Healthy Places and Healthy Regimens: British Spas 1918–50', in Virginia Berridge and Martin Gorsky, *Environment, Health and History*, Basingstoke: Palgrave, 2012, 113–32, pp. 128–9.

14 For example, swimming pools were built at specialist rehabilitation units at Stoke Mandeville in Aylesbury, England, and orthopaedic hospitals across the UK, such as the Royal Orthopaedic Hospitals in Birmingham and Stanmore, the Robert Jones and Agnes Hunt Hospital in Shropshire from the early 1950s. Ludwig Guttmann, *Textbook of Sport for the Disabled*, Aylesbury, HM&M, 1976; Robert Jones and Agnes Hunt, *The Heritage of Oswestry, 1900–1961*, Oswestry: The Hospital, 1961; Maurice W. White, *Years of Caring: The Royal Orthopaedic Hospital*, Studley: Brewin Books, 1997.

15 Naomi Rogers, *Dirt and Disease: Polio before FDR*, New Brunswick, NJ: Rutgers University Press, 1996, p. 168.

16 Tony Gould, *A Summer Plague: Polio and Its Survivors*, New Haven: Yale University Press, 1997; Gareth Williams, *Paralysed with Fear: The Story of Polio*, Basingstoke: Palgrave Macmillan, 2013.

17 Judith Friedland, *Restoring the Sprit: The Beginnings of Occupational Therapy in Canada 1890–1930*, Montreal: McGill-Queens University Press, 2011; Catherine F. Paterson, *Opportunities Not Prescriptions: The Development of Occupational Therapy in Scotland 1900–1960*, Aberdeen: Aberdeen History of Medicine Publications, 2010.

18 This included therapies for people with mental illnesses. Vicky Long, *Destigmatising Mental Illness: Professional Politics and Public Education in Britain, 1870–1970*, Manchester: Manchester University Press, 2014.

19 German Sims Woodhead, Sir Clifford Allbutt, and P. C. Varrier-Jones, *Papworth: Administrative & Economic Problems in Tuberculosis*, Cambridge: Cambridge University Press, 1925.

20 Bynum, *Spitting Blood*, pp. 50–76.

21 Eleanor Birks, *Papworth Hospital and Village Settlement – Pendrill Varrier – Jones' Dream Realised*, Cambridge: Papworth Hospital, 1999, p. 25.

22 Ursula Keeble, *Aids and Adaptations*, London: The Social Administration Research Trust, 1979; Jean Barclay, *In Good Hands: The History of the Chartered Society of Physiotherapy 1894–1994*, Oxford: Butterworth-Heinnemann, 1994.

23 Williams, *Paralysed*, pp. 138–47.

24 Julie Anderson, Francis Neary and John V. Pickstone, *Surgeons, Manufacturers and Patients: A Transatlantic History of Total Hip Replacement*, Basingstoke: Palgrave Macmillan, 2007.

25 Joanna Bourke, *The Story of Pain: From Prayer to Painkillers*, Oxford: Oxford University Press, 2014, p. 28.

26 Arthur Frank, *The Wounded Storyteller*, Chicago: University of Chicago Press, 1995. See also the chapter by Arthur Frank in this volume.

27 Ian Dury quoted in Gould, *A Summer Plague*, p. 231.

28 Lisa Folksmarson Kall, *Dimensions of Pain: Humanities and Social Science Perspectives*, Basingstoke: Palgrave, 2013, p. 1.

29 R. Boddice (ed.), *Pain and Emotion in Modern History*, Basingstoke: Palgrave, 2014, p. 4.

30 Anderson, et al., *Surgeons, Manufacturers and Patients*, pp. 134–6.

31 A. H. Hawkins, *Reconstructing Illness: Studies in Pathography*, 2nd edn., West Lafayette, IN: Purdue University Press, 1999.

32 Margaret Somers, 'The Narrative Construction of Identity: A Relational and Network Approach', *Theory and Society*, 23, 1994, pp. 605–49.

33 H. Carel, 'Phenomenology as a Resource for Patients', *Journal of Medicine and Philosophy*, 37, 2, 2011, pp. 96–113.

34 Ludwig Guttmann said that the 'unit must be impregnated with enthusiasm'. Julie Anderson, '"Turned into Taxpayers": Paraplegia, Rehabilitation and Sport at Stoke Mandeville, 1944–56', *Journal of Contemporary History*, 38 (3), 461–75, 467.

35 Emm Barnes Johnstone, in her book on childhood leukaemia notes that although published narratives remain rare, they have been useful for researchers to study the lived experience of the disease. Emm Barnes Johnstone with Joanna Baines, *The Changing Faces of Childhood Cancer: Clinical and Cultural Visions since 1940*, Basingstoke: Palgrave Macmillan, 2015, p. 181.

36 A. S. Ahmad, N. Ormiston-Smith and P. D. Sasieni, 'Trends in the Lifetime Risk of Developing Cancer in Great Britain: Comparison of Risk for Those Born in 1930 to 1960', *British Journal of Cancer*, 112 (5), 2015, 943–7, p. 943.

37 Elizabeth Toon, 'The Machinery of Authoritarian Care: Dramatising Breast Cancer Treatment in 1970s Britain', *Social History of Medicine*, 27 (2014), 557–76.

38  'Tongue-Tied', *Inside Story*, BBC1 documentary, 1998.
39  Williams, *Paralysed*, p. 192.
40  Rogers, *Dirt and Disease*, p. 108.
41  John Tosh, *The Pursuit of History*, Abingdon: Routledge, 2015, p. 269.
42  Alan Marshall, *I Can Jump Puddles*, Melbourne: FW Cheshire, 1955.
43  Elizabeth Kenny, *And They Shall Walk: The Life Story of Sister Elizabeth Kenny*, London: Robert Hale, 1951.
44  Williams, *Paralysed*, p. 151.
45  Harnett, *Polio and Us*, p. 55.
46  Gould, *Summer Plague*, p. 234.
47  P. Bates and J. Pellow, *Horizontal Man: The Story of Paul Bates*, London: Longmans, Green and Co., 1964.
48  Harry Tildesley, *Polio and I: A Concise Biography*, Ipswich: WE Harrison and Sons, 1959.
49  Helen Valier, 'At Home in the Colonies: The WHO-MRC Trials at the Madras Chemotherapy Centre in the 1950s and 1960s', Flurin Condrau and Michael Worboys (eds), *Tuberculosis Then and Now: Perspectives on the History of an Infectious Disease*, Montreal: McGill and Queens University Press, 2010, 213–34.
50  Joanne Muzak '"They Say the Disease is Responsible": Social Identity and the Disease Concept of Drug Addiction', in Valerie Raoul et al., *Unfitting Stories: Narrative Approaches to Disease, Disability and Trauma*, Waterloo: Wilfred Laurier University Press, 2007, 255–64, 255.
51  Angela Woods, 'Beyond the Wounded Storyteller: Rethinking Narrativity, Illness and Embodied Self-Experience,' in Havi Carel and Rachel Cooper (eds), *Health, Illness and Disease: Philosophical Essays*, Durham: Acumen, 2013, 113–28, p. 120.
52  Betty Ford and Chris Chase, *Betty Ford: A Glad Awakening*, London: Doubleday, 1987; Betty Ford, *Healing and Hope*, New York: Putnam Adult, 2003.
53  See the chapter in this volume by Katherine Foxhall.

## Select bibliography

Anderson, Julie, *War, Disability and Rehabilitation: 'Soul of a Nation'*, Manchester, Manchester University Press, 2011.

Bates, Paul, and John Pellow, *Horizontal Man: The Story of Paul Bates*, London: Longmans, Green and Co., 1964

Boddice, Rob (ed.), *Pain and Emotion in Modern History*, Basingstoke: Palgrave, 2014.

Borsay, Anne, 'Disciplining Disabled Bodies: The Development of Orthopaedic Medicine in Britain, c1800–1939', in David Turner and Kevin Stagg (eds), *Social Histories of Disability and Deformity*, Abingdon: Routledge, 2006, 97–116.

Bourke, Joanna, *The Story of Pain: From Prayer to Painkillers*, Oxford: Oxford University Press, 2014.

Bynum, Helen, *Spitting Blood: A History of Tuberculosis*, Oxford: Oxford University Press, 2015.

Edwards, Laurie, *In the Kingdom of the Sick: A Social History of Chronic Illness in America*, New York: Walker and Company, 2013.

Harnett, Nuala (ed.), *Polio and Us: Personal Stories of Polio Survivors in Ireland*, Dublin: Post-polio Support Group, 2007.

Kutcher, Gerald, 'A Case Study in Human Experimentation: The Patient as Subject, Object and Victim', in Carsten Timmermann and Elizabeth Toon, *Cancer Patients, Cancer Pathways: Historical and Sociological Perspectives*, Basingstoke: Palgrave Macmillan, 2012, 57–77.

Marshall, Alan, *I Can Jump Puddles*, Melbourne: FW Cheshire, 1955.

Paris, Gloria, *A Child of Sanitariums: A Memoir of Tuberculosis Survival and Lifelong Disability*, Jefferson, NC: McFarland and Company, Inc., 2010.

Rogers, Naomi, *Dirt and Disease: Polio Before FDR*, New Brunswick, NJ: Rutgers University Press, 1992.

Williams, Gareth, *Paralysed With Fear: The Story of Polio*, Basingstoke: Palgrave Macmillan, 2013.

# 19

# FROM PARAFFIN TO PIP

## The surgical search for the perfect breast[1]

### *Fay Bound Alberti*

The novelist, poet and French man of letters Anatole France, winner of the Nobel Prize for Literature in 1921, is said to have observed that 'a woman without breasts is like a bed without pillows'. This is a provocative reduction of the female body to an object that is comfortable and useful while being (a) as domestic and controllable as a home's soft furnishings, and (b) incomplete if a woman is flat-chested, from birth, from breastfeeding or from disease. France's words deserve to be ridiculed, and yet they seem to strike at the heart of a wider, broadly acknowledged truth about attitudes towards women's breasts. Breasts symbolise womanhood in all its diverse manifestations: they are sexualised, nursed at, revered, mutilated, sniggered at and politicised, for how they look and feel, as well as for what they do.[2] Iris Marion Young has problematised the sexual *and* nutritive functions of breasts, which shatter the border between motherhood and sexuality in the West.[3]

This cultural obsession with breasts, with their size and shape, and intersections with femininity, has direct implications for women's health. The most obvious example is in the case of cancer, one of the most feared diseases in the Western world and a space where ideas about personal responsibility, disease, technological innovation and ethics coalesce around the female breast.[4] Consider the 2013 media storm surrounding the actor Angelina Jolie when she elected for a bilateral mastectomy after genetic testing revealed a high cancer risk.[5] The actor publicly exercised her right to fight a 'faulty gene' with an article about her experience in the *New York Times*.[6] Not only did the 'Jolie effect' inspire many thousands of women to ask their GPs for genetic testing, and open up the broader question of prophylactic surgery, it also raised the question of silicone breast implants: Should you? Could you? Would you look 'normal?' Is it 'safe'?[7] Unwittingly or otherwise, Jolie became an icon for women's right to choose medical treatment as well as those who bore their mastectomy scars with pride.[8] We are women, they say, with or without our breasts.[9]

Of course not all women have breast surgery and silicone implants as a result of preventative surgery or reconstruction. A recent study by the Institute of Medicine (IOM) found that of the 1.5 million American women with silicone breast implants, more than two-thirds chose implants because they were unhappy with the natural size and shape of their breasts.[10] In other words, they underwent a surgical increase in size or 'augmentation' as a consumer choice. In British and North American culture, big breasts are a ubiquitous symbol of sexual allure and individual attractiveness and more democratically available than ever before.[11] Elective surgery is so popular that the *Huffington Post* announced 2012 the 'year of the silicone breast implant', breast implants being far and away the most popular cosmetic procedure in the United States according to the American Society for Aesthetic Plastic Surgery (ASAPS).[12]

A similar trend is found in the UK where over 50,000 cosmetic procedures were performed in 2013. According to the British Association of Aesthetic Plastic Surgeons (BAAPS), cosmetic surgery is on the increase despite the recession, with a rise of 17 per cent in surgical procedures since the previous year. BAAPS puts the value of UK cosmetic procedures as a whole at £2.3 billion in 2010, predicting it will rise to £3.6 billion by 2015. Breast augmentation 'remains the top surgical procedure in the UK'.[13]

These trends raise important questions about the motivation of those who seek implants, as well as the ways in which cosmetic surgery as a specialism responds to, and arguably sustains, a culture of dissatisfaction with the human body. This chapter considers the history of silicone breast augmentation and the somatic ideals that it reflects, as well as the role of women as clients, patients and consumers. I recognise that the terms 'cosmetic' and 'plastic' are often used interchangeably. Here they are used to distinguish principally between reconstructive (the origin of 'plastic' surgery) and that which is purely aesthetic (cosmetic). I have chosen to focus on the latter because the pursuit of an objective beauty ideal opens up feminist and ethical issues around consumer choice and rights, medical safety, ethics and governmental regulation. These are typically peripheral concerns in the historiography of breast surgery, with its emphasis on surgical prowess, the 'cult of beauty', and the politics of the body.[14]

Between the 1950s and the present day and in North America and Britain, the concept of psychological 'need' emerged in relation to implants for perfectly healthy but perceptually deficient breasts through the medicalisation of breast size. That need was met by the evolution of plastic surgery techniques and materials, the development of specific standards of female beauty, and the growth of a mass market that viewed elective surgery as an opportunity for personal growth. Silicone implants also underwent several transformations: from pathological prostheses to coveted consumer objects to walking or ticking 'time bombs'.[15] An exploration of the evolution of silicone implants reveals a startling disjuncture between the narratives of suffering on the part of recipients and the evaluation of risk by a medico-scientific community invested in the continued use of silicone. The existence of these disputes, and the oft-competing interests of manufacturers, surgeons, media, patients and academics around the female breast, raises important questions about the gendering of health-care and about women's bodies as a site of conflict for broader medical, scientific and political concerns.

## The origins of cosmetic surgery

The history of cosmetic surgery for the purposes of reconstruction dates back to antiquity. Cosmetic surgery was then known as 'plastic surgery', which gives a clue to its restorative origins. Derived from the Greek *plastikos*, meaning 'to mould', plastic surgery has been identified in ancient Indian Sanskrit texts that describe procedures to repair noses and ears lost as punishment for crime or adultery. The Indian surgeon Suśruta, who compiled his knowledge as the *Suśruta Saṃhitā*, described in 600 BC a method of rhinoplasty that used skin from the cheek or the forehead of the patient. By the first century AD, the Romans were also undertaking a range of aesthetic surgical procedures, such as breast reduction and circumcision.[16]

Since the time of Galen, surgeons have attempted to correct eyes that drooped and noses that were considered misshapen. Yet the 'father of modern plastic surgery' is usually said to be Gasparo Tagliacozzi, who worked in Bologna, Italy, around 1590. His technique was to transfer a skin flap from the arm to the nose, an intervention necessitated by the number of street brawls and duels that took place during the period.[17] It is likely that his patients were men who had succumbed to some type of nose excision as a punishment; it was particularly emasculating for a man to have his nose sliced in two or removed entirely.[18] The history of aesthetics around the nose is a key theme of Sander Gilman's work on cosmetic surgery and the 'body of the Jew'.[19]

The predominantly reconstructive nature of plastic surgery was well established by the time the specialism came into regular practice during the two world wars. Dealing with the fall-out of trench combat enabled the practised expertise and skill of the surgeons to come into its own, as well as providing surgeons with a seemingly endless number of patients on whom to work.[20] New techniques were needed to deal with the carnage of the battlefields, and surgeons had to work together in teams quickly. The weapons of war – high-explosive shells, machine guns and poison gas – created huge numbers of wounded, many of whom required rehabilitative surgery to save and repair limbs and enable skin grafts. The trench warfare of the First World War encouraged the development of plastic surgery, as soldiers needed treatment for shattered jaws, skull wounds and demolished faces. Harold Gillies (later Sir Harold), a New Zealand otolaryngologist working in London developed many of the techniques of modern facial surgery in caring for soldiers with disfiguring facial injuries.[21] Having worked as a medical minder with the Royal Army Medical Corps and with the renowned French surgeon Hippolyte Morestin on skin grafts, Gillies and the army's chief surgeon, Sir William Arbuthnot-Lane, established a facial injury unit at the Cambridge Military Hospital in Aldershot and later a new hospital for facial injury reconstruction in Sidcup,[22] where they developed many plastic surgery techniques and performed thousands of operations on victims of explosives.

When the Second World War broke out, plastic surgery provision was divided between the different services of the armed forces. Gillies worked at Rooksdown House near Basingstoke, which became the principal army plastic surgery unit. Gillies' cousin Archibald McIndoe (also subsequently knighted for his services) worked with Gillies before founding a Centre for Plastic and Jaw Surgery in East Grinstead.[23] McIndoe developed new techniques for treating burned faces and hands, as well as skin grafts, and innovative methods to integrate ex-soldiers back into the community. By the end of the Second World War, plastic surgery covered the reconstruction of all areas of the body, and led to the development of a number of specialty and professional organisations. The British Association of Plastic Surgeons was founded in 1946, now named The British Association of Plastic Reconstructive and Aesthetic Surgeons (BAPRAS).

The activity of plastic surgeons and the institutionalisation of cosmetic surgery as a specialty gave it a rationale and an impetus that continued long after the Second World War. The British Society of Aesthetic Plastic Surgeons held its first meeting in 1979, reflecting a move away from reconstructive and towards explicitly cosmetic work by its members, becoming the British Association of Aesthetic Plastic Surgeons (BAAPS) in 1982. In North America, professional associations were formed earlier, though they followed a similar trajectory. The American Society of Plastic Surgeons (ASPS) was formed in 1931, and the American Board of Plastic Surgery was established in 1937. By contrast the American Society for Aesthetic Plastic Surgery (ASAPS) was not founded until 1967.

Plastic surgery on the breast was initially reconstructive. Implants as prostheses were used to enhance and increase the size of a woman's breast after reconstructive surgery and breast removal. Implants went through a wide variety of permutations with varying degrees of success: from lipoma auto-transplantation (the movement of fatty cells from one part of the body to another), paraffin injections, ivory and glass balls, ground rubber, ox cartilage, polyethylene chips, polyurethane sponge, silastic rubber, and liquid silicone.[24] In the late nineteenth century, the German-Bohemian surgeon Vincenz Czerny, head of the surgical departments of the universities of Freiburg and Heidelberg between 1871 and 1906, published his first account of a breast implant carried out to avoid asymmetry after he removed a breast tumour. The implant was created from the patient's own abdominal tissue.[25]

In reconstructive terms, either in the case of cancer patients or soldiers, the moral and ethical bases of plastic surgery are relatively straightforward. In each case, surgical reparation forms part of an individual's rehabilitation into the social world, explicitly engaging with physical appearance, body image and psychosocial functioning as an aspect of healing.[26] The situation is more problematic around purely aesthetic or cosmetic surgery, in which an individual perceives there to be something 'wrong' or lacking about his or usually her physical appearance, and seeks intervention through an elective surgical procedure – usually performed privately and as a response to a social and internalised ideal of bodily perfection.[27]

## Hypomastia and the pathologisation of small breasts

Medical interest in increasing the size of a woman's healthy breast was rare until the mid-1930s. This is perhaps unsurprising when one considers the changing fashions in women's body shape. In the 1920s the look was 'skinny – no hips, no breasts, just a straight figure'.[28] Cosmetic surgery was only considered in the case of overly large or 'hypertrophic' breasts, which could impact on the respiratory, circulatory and locomotor systems.[29] However some surgeons were already experimenting with fat transplants to alter the shape of women's bodies, which included moving fat from the buttocks or abdomen to the chest. Elizabeth Haiken has shown how this process of 'fat transfer' was used experimentally for a wide range of beautification purposes by the Chicago-based surgeon Robert C. Miller. Though the results were largely disappointing – transplanted fat normally being reabsorbed by the body, there was clearly a market for cosmetic surgery before World War Two.[30] There was already a professional distinction, too, between those who worked in 'reconstructive' surgery, and those who focused primarily on 'cosmetic' concerns, as well as a degree of rivalry.[31] As the American surgeon John Staige Davis put it, 'true plastic surgery' ('plastic' in the traditional sense of moulding and shaping) was absolutely distinct and separate from what is known as 'cosmetic or decorative surgery.'[32]

The ideal of a larger bust arose during the post-war years, arguably as a reaction to both the slenderness of 1920s fashion and the paucity imposed by rationing. The aspirational Hollywood bodies of the 1940s and 1950s were Marilyn Monroe, Jane Russell,

Lana Turner and Sophia Loren, all of whom were famed for their hourglass figures. The gamine look, as it would be popularised by Audrey Hepburn and Twiggy, was not yet as fashionable as breasts that defied gravity and pushed out college sweaters. Molly Haskell has written about the aggressive attitude towards women's bodies as epitomised through the Hollywood movie industry, as women's bodies were increasingly critiqued according to a largely unattainable set of norms and conventions.[33]

In the 1940s, women who failed to conform to the ideal were said to have bemoaned their fate: 'Doctor, why can I not have a breast which makes me look as good as any other woman, if not better?'[34] The 'need' was there, surgeons insisted, though the treatment was not yet surgical. In the 1930s, the cosmetic surgeon H. O. Bames described three kinds of 'anomalous' breasts: abnormally large and pendulous; normal in size but prolapsed, and abnormally small (the latter of which he viewed as an endocrinal rather than a surgical problem, requiring hormonal intervention).[35] By 1942 the possession of 'infantile' breasts was classified as 'hypomastia' – or underdevelopment – by the Hungarian surgeon Max Thorek in his 1942 textbook.[36] Thorek did not suggest any surgical corrections at this point; like Bames and other surgeons before him he suggested endocrinal solutions. Yet variations in size and shape were being medicalised and post-war plastic surgery journals showed a growing interest in breast augmentation. By 1950, Bames was writing not of endocrinal solutions, but of surgical ones, with 'a new approach to the problem of the small breast'.[37] He was already experimenting with fat transfer. No endocrine therapy had worked, he explained, 'hence the only recourse we have for their solution is Plastic Surgery'.[38]

Thus cosmetic surgery came to the rescue of women with an identified need for larger breasts, in a narrative of surgical advance by a profession that had grown in numbers, training and proficiency. In the youth-dominated culture of the West, there was a strong market among middle-aged, middle-class women with disposable income and all the 'pathological' body parts like wrinkly bellies and stretched skin that came with child-bearing, and breasts that were too small, shrunken or insufficiently perky. As the number of hair salons, manicurists and beauticians escalated, beauty 'contests' were established and mass consumerism became popularised; what Sander Gilman has called the 'cult of the body beautiful' dominated Britain and North America.[39] Modern day sociological concerns about the impact of the media on the self-esteem of women and girls, the existence of aspirational body ideals and the availability of tools by which individuals could achieve self-improvement and transformation, arguably originated in the post-war period.[40]

Indeed, it was the influence of the Austrian psychotherapists Alfred Adler and Sigmund Freud, with their respective work on the 'inferiority complex' and on concepts of the self in ego development, which justified the pursuit of physical perfection.[41] Looking good (or *right*), had been transformed into a psychological necessity, self-esteem and self-confidence irredeemably crushed if one possessed sticking-out ears or a flat chest.[42] As women's magazines and popular psychology stressed the attainability and obligation of pursuing one's 'best self', cosmetic surgery rose as a route to self-improvement.[43]

In this context 'hypomastia' became a psychiatric as well as a physical condition. It was no longer just a label used to describe breast size. The condition could produce intense inadequacy, shyness, neuroses, frigidity and depression. 'Literally thousands of women, in this country alone', cosmetic surgeons and psychiatrists from

Johns Hopkins reported in 1957, 'are seriously disturbed by feelings of inadequacy in regards to concepts of the body image . . . "augmentation" mammoplasty (for the small breast) is usually requested by patients with emotional problems'.[44] The positive impact of surgical alteration (and the removal of psychiatric problems) outweighed any potential physical risks of the surgery.[45]

Two of the most important considerations for cosmetic surgeons, as they weighed the potential for risk against the psychological needs of their patients, were aesthetic appearance and touch. What combination of technique and material might provide the most malleable, secure and 'life-like' alternative to human breast tissue? Without regulation the professional associations worked this out through trial and error, in collaboration with the women said to lead the market and the manufacturers who supplied the implants. During the 1950s a range of new techniques and tools became available, including such synthetic materials as polyvinyl alcohol and polyethylene. Cosmetic surgeons injected liquid substances straight into the breast tissue, just as they had done with fat, or moulded implants that could be inserted beneath the chest muscle. Robert Alan Franklyn, for instance, a Hollywood surgeon, implanted 'surgi-foam' in women's breasts (a kind of absorbable gelatin substance sheathed in teflon) as well as experimenting in the subcutaneous injection of liquid silicone through a technique he called 'Cleopatra's needle'. Other surgeons experimented with *Ivanol*, a polyvinyl substance related to plastic.[46] Synthetic implants gave a better overall result than injections, largely because the body could reabsorb injected substances, or even create ripples and lumps beneath the skin.

## The silicone age

It is against this backdrop of psychological need, experimentation, the availability of synthetic materials and a rising profession of cosmetic surgery that we need to situate the medical experience of Timmie Jean Lindsey, America's first silicone implant recipient. There is a perennial problem of women, and of poorer patients, being excluded from the historical record. It is telling, given the subsequent furore over the health and safety of silicone implants, that there is little about Lindsey in the academic or scientific literature. Her name appears as a footnote in a handful of papers, but she is otherwise anonymous, her experience not part of the official medical record.[47] Only with media interest in her case 50 years after her implants, as part of a broader historical retrospective, has her voice been heard.[48]

'When I had the implants put in,' Lindsey told a *Daily Mail* reporter in 2014, 'I would get wolf whistles when I walked down the street . . . [but] I started to get pain in the Eighties and sometimes it lasts for five to six weeks. It feels like I've broken a rib.'[49] Her story can be read in historical retrospect as an extraordinary tale of experimentation, profiteering and ethical bankruptcy.

The official narrative of the first breast implant is well known in the medical literature around breast augmentation.[50] Its language smacks of the locker-room and the boy's club, an example of the masculinist discourses that have historically defined both women and the language of scientific discovery.[51] In 1962 the American plastic surgeons Thomas Cronin and Frank Gerow were working with scientists from the Dow Corning Center for Aid to Medical Research on improving and refining breast implants. Dow Corning was an amalgamation of a glass works and a chemical company. By that time a number of different materials had been used as implants, mostly plastic, sponge-like materials such as Polystan (polyethylene tape and polyethylene) that produced a result 'pleasing to the eye', but unpleasantly 'rocklike' to the touch.[52] The inspiration for the first silicone gel implant is said to have been a blood transfusion bag. One of Cronin's employees, Thomas Biggs, recounted the events that would become famous. During the silicone gel implant's developmental stages, Cronin apparently visited a blood bank and, 'upon feeling the new and improved flexible plastic bag that contained the blood, he observed that it felt like a breast . . . The rest, as they say, is history.'[53]

Inspired by this haptic revelation, Cronin and Gerow used silicone to make a thin, flexible bag and a gel-like substance with which to fill it. Occurring in liquid, gel or solid forms, silicone is derived from silicon, a semi-metallic or metal-like element that in nature combines with oxygen to form silicon dioxide or silica.[54] Today silicone is widely used in the medical and cosmetic industries, in lubricants and oils, as well as suntan lotions, soaps, antiperspirants and chewing gum. As early as the 1950s, scientists debated the safety of silicone and warned against its widespread use without proper testing.[55] But Cronin and Gerow implanted one of the prototypes in the body of a dog. Jubilant that the new 'natural feel' silicone implant had arrived and 'the dog was fine', they implanted it in a person and she got along just great'.[56]

That person was Lindsey. Unlike the women discussed in surgical journals, she was apparently not clamouring for bigger breasts; nor mourning the loss of volume in her breasts after childbirth. Lindsey had married at 15 years old and was recently divorced. She met Gerow by happenstance: since she earned only £19 a week in an electronics factory, Lindsey qualified for free medical treatment. She arrived at the Jefferson Davis Hospital in Houston, Texas, with an altogether different problem. Impulsively, persuaded by a man she was dating, Lindsey had a tattoo of a rose inked onto her breast. She regretted it almost immediately, and wanted it removed. Gerow agreed to remove the tattoo with 'dermabrasion,' a procedure that takes away the top layers of the skin.

Was Lindsey also interested, Gerow asked Lindsey, in having her breasts fixed? Accompanied by a dozen medical students, he explained to Lindsey how his revolutionary new breast implants could improve the appearance of women's breasts that sagged like hers. It had apparently never occurred to Lindsey to feel self-conscious about her breasts. 'I told them I'd rather have my ears fixed than have new breasts', she replied, since she'd always been self-conscious about the way they 'stuck out'.[57] Lindsey agreed to have the implants inserted into her chest if Gerow pinned back her ears at the same time. Gerow agreed and the surgery was carried out in the spring of 1962. Gerow increased Lindsey's bust measurement from a B to a C cup with the use of two rubber envelope-sacs, shaped like teardrops and filled with viscous silicone-gel.

'When I came round from the anaesthetic, it felt like an elephant was sitting on my chest', Lindsey reported. When the bandages were removed, Gerow was pleased with the results. Although Lindsey was not interested in examining her breasts in the mirror, figuring the implants were 'out of sight and out of mind', Gerow and his team were intrigued. 'All the young doctors were standing around to look at "the masterpiece"', Lindsey said, while also remembering the years of 'pain and misery' that they had brought her.[58] The implants were viewed as a medical triumph. Their apparently smooth appearance and feel, and the widely reported success of the operation, were a tremendous boost to the Dow Corning Corporation, and to Cronin and Gerow, who sold their rights to the manufacturing company in exchange for royalties. By 1970, Dow Corning had sold 50,000 implants and the future of breast implants was secured.[59]

Yet silicone implants were to be plagued by controversy. For every woman happy with the outcome there was another complaining of deteriorating health, especially if the implant tore and leaked, spilling silicone into her chest.[60] But there was no proven link between silicone and illness. Lindsey had periodic consultations with Gerow where he maintained, even when she complained of severe pain and weakness, that silicone was a harmless material and could not make her sick. As discussed below, this remains the official response given to women who believe they have experienced silicone poisoning. After all, silicone is used in a range of medical devices including artificial valves, catheters, and as the lubricant in syringes.[61]

Debates over the safety and efficacy of silicone continued throughout the 1960s, 1970s and 1980s, reaching its peak in the 1990s with a flood of epidemiological studies on human populations.[62] Cosmetic surgery remained unregulated, and manufacturers like Dow Corning focused their attention on developing implants that were increasingly 'life-like', and able to capture more accurately the feel of a human breast. Polyurethane foam was used to hold the silicone gel, a process that helped prevent capsular contracture or the hardening of the collagen fibres in the breast as an immune response to the presence of a foreign object (the implant). The foam coating helped prevent capsular contracture, but an unfortunate side-effect was that the foam could disintegrate within the body and be difficult to remove.[63]

In North America, amidst calls for its involvement in the development of technological devices of all kinds from breast implants to pacemakers, the Food and Drug Administration (FDA) enacted the Medical Devices Amendment to the Federal Food, Drug and Cosmetic Act in 1976.[64] This meant that the FDA could approve the safety and effectiveness data of new medical devices.[65] Because breast implants were already in use, they were 'grandfathered in', which meant they could remain in use without any safety evidence being provided by manufacturers.[66]

Just over a decade later, however, the regulatory landscape shifted.[67] A number of important developments lay behind the change. Firstly, there was increased debate in the 1980s and 1990s over whether implants were safe and mounting evidence to suggest they were not. Medical articles citing anecdotal evidence from the USA and Japan began emphasising a link between the use of silicone in breast augmentation and the diagnosis of connective tissue disease.[68] A rising tide of multimillion-dollar lawsuits were issued against manufacturers by women who claimed their health suffered after their implants ruptured.[69] Sheila Jasanoff places the advent of breast implant litigation within the context of a flurry of mass torts in US federal courts, many of which involved 'large scale technological disasters' like breast implants.[70]

The national and international media responded loudly, investigating the plight of individual women and criticising the perceived negligence of surgeons and manufacturers.[71] The most famous and damaging example was an episode of the Connie Chung CBS television show *Face to Face with Connie Chung*, broadcast in 1990 and featuring the real-life experiences of women who claimed to have autoimmune diseases as a result of silicone implants.[72] As Marcia Angell has argued, the programme conveyed the message that unsuspecting women had had dangerous and untested devices foisted on them by an unthinking surgical profession. Furthermore, the FDA was blamed for permitting hazardous devices to be sold.[73]

In early 1991, in the face of mounting pressure from consumer and advocacy groups, the FDA shifted its position on implants, requiring manufacturers like Dow Corning to provide evidence of their safety. When the evidence proved inadequate, an FDA advisory panel met to discuss how to move forward. Around the same time Dow Corning was on the losing end of a lawsuit costing over $7 million from women who claimed the implants had caused connective tissue disease.[74] Most damagingly of all, Dow Corning was discovered to have withheld internal concerns about the safety of breast implants.[75] Bernard M. Patten, a neurologist and one-time colleague of Gerow and Cronin, published a series of articles in which he related breast implants to neurological and immunological problems.[76] Although his attempts to expose the dangers of silicone were rejected by aesthetic surgeons, it was revealed that as early as 1954 an in-house study by Dow Corning found that silicone had a high level of toxicity. It seems that not all the laboratory dogs were 'fine'. Silicone fluid injected into the dogs had migrated to major organs, ending in the death of one animal and the development of serious inflammatory disease in three others.[77] Patten was adamant that Dow Corning was aware that silicone caused both inflammation and autoimmune diseases. The company denied being aware of such studies, although it acknowledged the possibility that implants might rupture.[78] A full-scale reputational battle was played out between Dow Corning and the media as well as in the courts.[79]

According to his own emotive testimony, Patten was dismissed as a 'junk scientist' by the American Society of Plastic and Reconstructive Surgery (ASPRS) because he challenged the safety of silicone. He now offers support and advice to women seeking compensation and treatment for the side-effects of implantation.[80] Internal memos showed that not only Patten but also several other members of Dow Corning's staff questioned the safety of silicone implants, but that those concerns were kept under wraps.[81] When this evidence was exposed, the Commissioner of the FDA, American paediatrician David Kessler, issued an immediate moratorium on silicone implants. He explained the decision in an article for the *New England Journal of Medicine*, stating that too little was known about the dangers of silicone implant rupture:

> The link, if any, between these implants and immune-related disorders and other systemic diseases is also unknown. Serious questions remain about the ability of manufacturers to produce the device reliably and under strict quality controls. Until these questions are answered, the FDA cannot legally approve the general use of breast implants filled with silicone gel.[82]

Only women 'whose need for them is most urgent' were to be allowed silicone breast implants. In the case of cancer patients and others who needed implants for reconstruction purposes, the FDA judged that the risk-benefit ratio ruled in favour of the use of silicone implants.[83] 'Certainly as a society,' Kessler concluded, 'we are far from according cosmetic interventions the same importance as a matter of public health that we accord to cancer treatments.'[84] The moratorium on silicone implants was not because implants had been found to be dangerous, but because they had not been proved safe, a muddy logic that did little to pacify implant opponents. Litigation continued and, though breast implant manufacturers continued to deny any links between silicone implants and autoimmune diseases, the biggest class action settlement against manufacturers was finalised in 1994. The largest contributor to that settlement was Dow Corning. Amidst a further 20,000 pending lawsuits, the company subsequently filed for bankruptcy.[85]

However, less than a decade later silicone implants were back on the market. Since the ban numerous studies had been undertaken, directed by the FDA as well as individual interest groups, to establish whether there were scientific links between silicone implants and illnesses that included not only 'arthralgias, lymphadenopathy, myalgias, sicca symptoms, skin changes, and stiffness', but also 'cancer, definite or atypical connective tissue disease, adverse offspring effects, or neurologic disease'.[86] In 1997, the IOM carried out a comprehensive evaluation of the evidence for the association of breast implants – silicone and saline – with human health complaints.[87] It acknowledged that studies of the toxicology of silicone since the 1940s ought to have been more long-term and qualitative, but that the overall findings were not harmful in the usual quantities and exposure. The rupture and degeneration of breast implants was likely at some stage during their life-span. But there was no proof of a connection between ruptured implants and any kind of systemic condition. Reviewing seventeen epidemiological reports of connective tissue disease in women with breast implants, the authors found 'no elevated relative risk or odds ratio for an association of implants with disease'.[88] More specifically, the charge that there was a unique condition associated with breast implant rupture was entirely rejected:

> Evidence for this proposed disease rests on case reports and is insufficient or flawed. The disease definition includes, as a precondition, the presence of silicone breast implants, so it cannot be studied as an independent health problem. The committee finds that the diagnosis of this condition could depend on the presence of a number of symptoms that are nonspecific and common in the general population. Thus, there does not appear to be even suggestive evidence for the existence of a novel syndrome in women with breast implants. In fact, epidemiological evidence suggests that there is no novel syndrome.[89]

There was one caveat, relating to 'local problems' associated with silicone migration, and an acknowledgement that long-term quantitative studies were lacking. Overall the report found that the evidence silicone breast rupture could 'cause neurologic signs, symptoms or disease' was 'lacking or flawed'.[90] All that was certain were the usual local hazards linked with breast implant surgery: asymmetry, pain, atrophy of the skin, calcification and hardening of the implants, the appearance of chest wall deformity, delayed wound healing, extrusion of the implant through the skin, haematoma, damage to tissue or implant, toxic shock syndrome, inflammation, irritation and infection, and necrosis of skin around the implant.

Since there was no established link between silicone and disease, these hazards were not enough to keep silicone off the market. The complaints of many thousands of women were dismissed because they were not scientifically verifiable, or 'not consistently associated with objectively physical signs or laboratory abnormalities'. Subjective experiences listed by women were similar to chronic fatigue syndrome and fibromyalgia, both common to the wider female population (especially in 'young to middle aged women'), and also associated with 'concurrent psychiatric disorders'.[91] In short, it was now possible to argue that just as women sought cosmetic surgery because of psychiatric problems (principally feelings of inadequacy about having small or drooping breasts), those same psychiatric problems would lead them to be unhappy with the outcome and perhaps even to imagine a range of psychological and physical ailments.

Despite negative publicity and lawsuits there was a dramatic increase in the number of cosmetic breast implants when the moratorium was lifted, albeit under the auspices of the FDA and with a recommendation from the IOM of continued research and better tracking of recipients. According to ASPS, over 132,000 women received silicone breast implants in 1998 alone; an underestimate, one source has suggested, as increasing numbers of non-surgeon physicians began to perform cosmetic surgery.[92] Having reintroduced silicone implants, the FDA's website makes a significant disclaimer:

> Studies to date do not indicate that silicone gel-filled breast implants cause breast cancer, reproductive problems, or connective tissue disease, such as rheumatoid arthritis. However, no study has been large enough or long enough to completely rule out these and other rare complications.[93]

In other words, the FDA has pushed responsibility for implants back on the consumer. 'Breast implants are not lifetime devices', it warns. 'The longer a woman has them, the more likely she is to have complications and need to have the implants removed or replaced. Women with breast implants will need to monitor their breasts for the rest of their lives.'[94]

## PIP and déjà vu

Thus far I have dealt principally with the North American context, since that is where breast implants were first popularised. Yet in the 2000s, a remarkably similar debacle over the ethics and safety of silicone breast implants took place in the UK, Europe and parts of South America. This time the cause was a more explicit case of fraud and ethical negligence. There are parallels, too, with the North American case in the lack of regulatory framework in which cosmetic surgery has flourished. And an attitude towards cosmetic surgery that similarly promotes consumer choice as a positive way for women to overcome perceived physical defects, while subsequently rejecting many perceived physical and psychological problems. An added nuance in the UK has been the moralising judgement associated with aesthetic as opposed to reconstructive breast implants, thanks to the enforced involvement of the National Health Service (NHS) in rectifying the work of private clinics.[95]

Debbie Lewis, a hairdresser from Buckinghamshire, underwent breast augmentation surgery in 2004. Like Lindsey she had recently separated from her husband and she decided breast implants would increase her self-esteem. As in the case of Lindsey, we hear her story only as a result of media interest. Lewis's decision to have breast implants was psychological. She wanted to 'treat' herself after a difficult time in her life, seeking the 'pleasure' and 'self-confidence' that large breasts promised. The surgery cost Lewis £4,000. Within a year she had the implants changed twice, enduring the pain and discomfort of the surgery and its after-effects because of problems with the implant. First, Lewis had experienced capsular contraction. Then she discovered that one of her implants had ruptured and leaked its contents into a lymph node. After an operation lasting an hour and a half, in which her swollen lymph nodes were removed along with the implant, Lewis's surgeon reported that 'the implant shell looked like a thin beach ball. It was not a good quality product.'[96]

The implant removed from Lewis's body, at an additional cost of £6,000, was produced by Poly Implant Prothèse (PIP), the French company at the centre of the latest breast implant scandal. PIP's products have been banned throughout Europe from 2010, amidst complaints that rupture and degradation levels were double that of similar implants. Lewis was one of 40,000 British women who received PIP implants that were later discovered to be made not with medically approved silicone, but with industrial grade silicone – the kind used to stuff mattresses – as well as chemicals Baysilone, Silopren and Rhodorsil, most commonly used as fuel additives and in the manufacture of industrial rubber tubing.[97] In December 2013 the founder of the PIP distribution company, Jean-Claude Mas, was found guilty of aggravated fraud and sentenced to four years in prison by a court in Marseilles.[98]

The PIP incident prompted a UK review into whether the cosmetic surgery industry should be regulated, with the government eventually agreeing to remove PIP implants from affected women on the NHS. The case was of concern to more than 400,000 women throughout the world, not only in Britain but also in Brazil, Venezuela, Argentina, Italy, Germany and Spain. From 2009 an abnormally high level of ruptures began to be reported by other women, such as Debbie Davies from Liverpool, who described how one of her PIP implants had ruptured: 'so I have silicone going around my body. It is in my lymph nodes and I have lumps of it on my neck and in my chest. They are literally killing me.'[99] In this case, the lack of scientific 'proof' that silicone could cause illness was not reassuring – particularly since the implant did not contain only medical grade silicone. As a result of the scandal, recipients of PIP implants have been advised to have them removed in case of rupture. Though not on the US scale, class-action suits are being pursued by thousands of patients against clinics and surgeons that supplied the faulty implants.

How did this crisis happen? In the UK breast implants are regulated under a European Union medical device directive. Until the 1990s, each of the EU countries had its own approach to evaluating medical devices, though a Conformité Européenne (CE) mark given in any one country meant that the device could be sold throughout the EU. Device approval in each EU country is overseen by a governmental agency – in the UK that is the Medical and Healthcare Products Regulatory Authority (MHRA). The MHRA was formed in 2003 with the merger of the Medicines Control Agency (MCA) and the Medical Devices Agency (MDA): though it has acknowledged the limitations of the current system of regulation, the evidence on safety and efficacy of new devices and procedures at the time they are introduced into the UK is 'variable,' and the evidence base for most devices is 'poor'.[100] In the case of PIP implants, the content of the implants was changed *after* the original CE mark was awarded. This is only the tip of the regulatory iceberg: silicone implants used to enhance buttocks and calves, growth areas in cosmetic surgery, do not need to have a CE mark. They are not considered medical devices, and are therefore not covered under EU legislation.[101]

The MHRA did not release a Medical Device Alert concerning silicone gel implants manufactured by PIP until 2010, the year the implants were taken off the market. But France's medical safety watchdog, then the Agence Francaise de Sécurité Sanitaire des Produits de Santé (AFSSAPS), was aware of increasing risks in 2006, the same year in which UK surgeons had been reporting an overly high rupture rate. Yet British surgeons continued to use PIP despite their concerns. Had there been any self-regulation earlier, or any governmental intervention in the UK, thousands of women would have been prevented from receiving unsafe implants.[102] Like the American situation in the 1990s, the PIP scandal arguably occurred because the technologies involved were not appropriately monitored, and because the government in both instances was unwilling to get involved in the complex rights and responsibilities around cosmetic surgery.[103]

In 2010, the Department of Health (DH) estimated that 40,000 women in the UK had received PIP implants, though a more realistic estimate is 80,000.[104] Over 90 per cent were given for cosmetic reasons. Given that it was not only medical grade silicone that was in play, AFSSAPS, MHRA and other interested agencies carried out tests to assess the genotoxicity of the gel. So far, those tests have not revealed any hazards in humans.[105] The DH final expert report on PIP implants acknowledged that PIP implants were substandard and had an increased rate of rupture, but maintained that there was no evidence of any significant risk of clinical problems in that event.[106]

More than 95 per cent of PIP implants were fitted by private and unregulated clinics. Those clinics have since been criticised for cut-price deals, giving unrealistic expectations to patients, and aggressive selling practices that include time-limited deals, financial inducements, packages (like 'buy one get one free' or mother/daughter deals), and offering cosmetic procedures as competition prizes.[107] The DH report shows how a largely unregulated cosmetic surgery industry has emerged in the UK, where for a price a woman (or less frequently a man) is provided with aesthetic bodily improvements through a 'routine operation' that will change their lives for the better. The Harley Medical Group, one of the leading private clinics, advertises a range of 'popular' procedures to ensure that patients receive 'the body confidence and boost in self-esteem that [they] deserve'.[108]

In 2013, after a series of disputes about whether the NHS was morally obliged to remove implants inserted by private clinics, a judgement was made in favour of removal. The NHS in Wales will fund removal of PIP implants and their replacement with 'safe' alternatives. However, the NHS in England will only fund the explantation or removal of the implants; except in exceptional circumstances, any replacements are made at a patient's own expense.[109] This situation raises a series of further ethical dilemmas related to the moral obligation of the NHS and the problematic interrelation between private and national health-care sectors. Indeed, the PIP debacle raises several contentious questions. How did PIP sell thousands of implants worldwide for nearly a decade before its implants were found to be unsafe? There has been no consistent record kept of women who have received implants, nor of the impact of those implants on the psychological and physical health of those concerned – despite the fact that breast implants are increasingly associated with poor mental health. Meanwhile the subjective reports of thousands of women are dismissed out of hand.

In 2013 the Medical Director of the NHS, Sir Bruce Keogh, published a Review of the Regulation of Cosmetic Interventions (2013) that explored the condition of UK cosmetic surgical (and non-surgical) beauty treatments.[110] Though it was prompted by the PIP debacle, 'which exposed woeful lapses in product quality, after care and record keeping', the Review was also concerned with cosmetic interventions across the sector. It revealed 'widespread use of misleading advertising, inappropriate marketing and unsafe practices', not only in cosmetic interventions, but also in non-surgical interventions like dermal fillers and Botox. These can lead to scarring, tissue damage, blindness and facial disfigurement, but consumers have 'no more protection and redress' in that event 'than someone buying a ballpoint pen or a toothbrush'.[111]

Despite the findings of Keogh's Review, private clinics continue to be largely unregulated. Their advertising takes it for granted that women seeking breast augmentation suffer low self-esteem because of the way their bodies look or have problems with 'body confidence'. They might just want 'better shaped breasts that are in proportion with the rest of her body', as one cosmetic surgeon puts it on his website, or to 'restore the natural fullness to their breasts and feel confident in their bodies again', in the words of the Harley Medical Group – especially if they have 'aged' or have breast-fed children and endure drooping, sagging breasts.[112] For a 'no obligation' consultation with a specialist, those women could be well on the way to towards a younger and more attractive version of themselves. If a woman can 'fix' themselves this easily, this painlessly and conveniently, then why not turn to breast augmentation, just as she might try a new moisturiser or hair treatment?[113]

Cosmetic surgery adverts on bill-boards, at train and underground stations and in magazines reassure potential patients that these are everyday operations performed by highly skilled and experienced professionals. For example, surgeons at the Harley Medical Group 'perform hundreds of breast surgery operations every year'.[114] Surgical expertise is stressed above any social or psychological implications. It is extraordinary that serious medical procedures involving general anaesthetic and its accompanying risks have been framed as lifestyle consumer choices.[115] Perhaps unsurprisingly the ethics of the cosmetic surgery industry have come under attack as a public consultation showed strong support for a ban on cut-price deals and aggressive selling by sales staff that takes place before any consultation with a surgeon. Keogh's Review acknowledged that procedures should not be sold as a 'commodity' and also that that the current regulatory framework did not ensure patient safety or best practice.[116] The BAAPS has also argued that people have been misinformed about the risks and likely outcomes of cosmetic surgery by salespeople who work to targets rather than by qualified surgeons.[117] It was not until 2012 that medical revalidation was introduced to monitor surgeons' skills and abilities. Prior to that surgeons were able to certify themselves competent in a procedure, which most 'clients' seeking cosmetic surgery would take at face value. Keogh's Review found that over 91 per cent of people would presume the surgeon was 'fully qualified' in whatever procedure she – or usually he – was going to undertake.[118] As part of a regulatory overhaul Keogh recommended that the Royal College of Surgeons of England (RCS) should establish a Cosmetic Surgery Interspeciality Committee to regulate and monitor the UK cosmetic surgery industry, set standards for training and practice, ensure competence of surgeons, establish a clinical audit database, and work with the Parliamentary and Health Service Ombudsman on dispute resolution.

Despite Keogh's report, and the problems it highlights regarding the practice and monitoring of cosmetic surgery, there has been little impetus on the part of the British government to address its recommendations.[119] This is rather different from the American and even the French governmental response, where reactions to scandals have included a tightening up of surgical and non-surgical cosmetic treatments.[120] UK resistance to monitoring cosmetic surgery is not new. Melanie Latham has shown how breast implant risks were highlighted to the UK government over many years prior to the PIP scandal. A National Care Standards Commission (NCSC) Report, published in June 2003, catalogued the ways in which some cosmetic surgery clinics in London did not adhere to the most basic legal standards.[121] Better self-assessment and self-regulation was recommended rather than governmental intervention. Even earlier, fears about leaking silicone breast implants were rife in the UK in the 1990s just as they were in the US. In June 1994, Ann Clwyd MP introduced an unsuccessful cosmetic surgery bill to establish minimum standards of training and practice for cosmetic surgeons. As a response the government set up the National Breast Implant Register to collect data on those who had undergone breast implant surgery. This was closed in 2006 due to poor take-up and lack of government funding.[122] In 2012, Clwyd introduced another Private Members' Bill in pursuit of minimum standards in cosmetic surgery, but it got no further than its first reading.[123]

Whether things will be different in the wake of the PIP scandal remains to be seen. There has been an increase in demand for implants manufactured in the UK. The Glasgow-based company Nagor, for instance, experienced a 30 per cent increase in sales in 2012 alone.[124] Choose your implant as well as your surgeon carefully, women are now being told as 'good' consumers. Unfortunately the social and psychological reasons why women might be choosing implants are ignored and the emphasis placed on surgeons to self-regulate. The RCS has published its own statement of standards for cosmetic practice. The guidelines suggest individual surgeons should:

> Discuss relevant psychological issues (including any psychiatric history) with the patient to establish the nature of their body image concerns and their reasons for seeking treatment. They should not at any point imply that treatment would improve a patient's psychological wellbeing.[125]

Moreover, surgeons should consider whether a psychological assessment is necessary before operating, as well as ensuring that any advertising around cosmetic surgery is honest and not doctored in any way. These guidelines represent an apparent attempt by the RCS to protect its professional reputation by distancing itself from poor standards, false advertising and bad practice. Yet as Professor Norman Williams, former President of the RCS, acknowledged in 2013, the RCS is not a regulator, nor a legislator. Without governmental intervention it is unclear how far cosmetic surgery standards can or will be improved.[126]

## Conclusion: from patient to consumer – liberation or liability?

Timmie Jean Lindsey, the world's first silicone breast implant recipient, is now in her eighties and working nights at a care home. Many of the women that were subsequently enlisted in Cronin and Gerow's trial, including relatives and friends of Lindsey herself, apparently became ill after the procedure. Some died, allegedly because of the leaking silicone, though as this chapter has shown, such claims are unsubstantiated.[127] Lindsey has come to terms with her own implants, despite her pain and discomfort:

> You can feel the prostheses if you press, but I have enough breast tissue so it really isn't that noticeable. And they've aged, like I have. They haven't stayed straight up and perky; they've drooped, too. They feel like part of me now.[128]

Cosmetic surgery remains one of the fastest growing medical specialisms, with 'lunchtime' treatments for non-surgical interventions (Botox to stop lines from forming, fillers to 'fill them out') and surgical interventions being packaged as self-improvement in a way that only a century ago would have been unconvincing. Small or saggy breasts, bent penises, stretch-marked bellies and long labial lips have been redefined as clinical anomalies rather than facts of life. Living with such 'deformities' is linked to depression and a lack of confidence.

The PIP breast implant scandal, like others before it, has not prevented women and men from calculating that a desired outcome – a perkier, tighter, bigger, smoother or thinner appearance, which presumably is hoped to lead to a better relationship, job or quality of life – is worth the financial outlay and risks that include pain, dis-colouration, loss of nerve sensitivity, tissue necrosis and even death. What does it say about our discomfort and dis-ease as a nation that we take these risks? And how has it become a personal preference, or a matter of women 'choosing wisely' when under-taking breast implants in the private sector when their decisions are, according to the Keogh Review, often based on a lack of information or a misunderstanding about the risks and guarantees involved?[129]

This chapter has focused on the construction of need as identified by the medical profession – psychiatric and surgical – and there are important reasons for this. The story of why women might seek breast implants, now or in the 1950s, is a long and complex one. It is too simplistic to see women as passive recipients of cultural dictates – to be fatter, thinner, smaller or larger-breasted – and yet we know that the steady drip of negative images and self-hatred has devastating effects on a woman's psyche.[130] We also know that body image is a major factor in women's choices to have cosmetic surgery, and that this is a feminist issue. Around 91 per cent of all cosmetic surgery carried out in the UK takes place on women's bodies.[131]

Traditionally the story of silicone breast implants is told from a number of distinct, polarised perspectives; from the viewpoint of the surgeons, keen to deliver on wom-en's expectations and to develop their skills; from the manufacturers that develop and deliver products; or from women themselves, often left feeling damaged and debilitated by botched operations or mysterious illnesses.[132] We are used to think-ing of female patients as victims, with their bodies colonised, their minds conned into thinking that breast implants will lead to contentment.[133] But within feminist scholarship, the rise of breast implants is problematic, especially among women who do not self-identify as intellectually downtrodden or uneducated.[134] In *Reshaping the Female Body*, Kathy Davis has noted the phenomenon of rapid cosmetic surgery take-up among educated and avowedly feminist women.[135] Despite the fact that their surgeons are predominantly white, male and middle-class, some have seen this as evidence of women embracing the potential for personal transformation as a political act. Cosmetic surgery becomes a means for women to create new identities, as their notions of what is 'real' become transformed by surgically generated versions of themselves. Of course those new identities are inherently and indisputably gendered, as well as projecting a predominantly white and Western notion of attractiveness.[136]

It is hard not to see breast implants as symptomatic of the cultural and aesthetic demands of patriarchy, a mechanism that demands compliance by (literally) getting beneath the skin of women, and holding up impossible ideals of youth and beauty to which women feel subjected.[137] A disproportionately high number of implant patients report distress over their appearance. Studies have shown that most women who undergo breast augmentation have also undergone psychotherapy, suffered low self-esteem, depression and body dysmorphia.[138] Other studies report that women seeking breast implants were almost three times more likely to commit suicide than those who do not.[139] And breast implant patients are more frequently associated with depression and suicide than any other kinds of cosmetic surgery patients.[140] Susie Orbach has suggested that the self-hatred or criticism of an individual towards a specific body part reflects some latent and unresolved childhood conflict that is manifested or symbolised by that body part.[141] Like operations to alter 'unsightly' labia or replace the hymen, these kinds of procedures strike at the heart of femininity and suggest a degree of self-hatred towards the female body that is startling in its extremity.

As this chapter has shown, technological and surgical developments and decisions never take place within a vacuum but are inherently political. So, too, are ideas about psychological need and technological safety. The treatment of women as consumers and patients by the surgical profession needs to be addressed, especially since many breast implant recipients feel ignored or neglected by those professionals they have consulted. In this context, debates around scientific 'proof' are less important than the fact that women's physical and psychological suffering is hystericised in the twenty-first century just as it was in the nineteenth.[142] There are similarities in the gendering of power and authority around breast implants and other health concerns that affect women, including the use of oestrogen in cases of osteoporosis, the diagnosis and treatment of fibromyalgia as a disease entity and the treatment of pre-menstrual syndrome.[143] Few cosmetic surgeons today are women, and the profession remains male-dominated.[144]

The overwhelming evidence is that bigger breasts rarely, if ever, bring a better life, more luck in love or a more successful lifestyle. In fact they are more likely to bring additional pain and suffering, both psychological and physical. There is no 'quick fix' for most of the symptoms that take women to the cosmetic surgeon. And yet many of us still dream that a 'better', more youthful body will bring these things. We cannot dissociate breast implant surgery from the aspirations of the culture in which it takes place or the professional structures that perpetuate and maintain it. Nor can we ignore the profiteering that takes place on and around female bodies and the governmental and institutional frameworks that we expect to guide the ethical principles of a – still – unregulated industry.

Breasts are not unchanging and immutable signifiers of womanhood: they evolve from puberty, through sexual maturity, motherhood, childbirth and breastfeeding as well as during illness or fluctuations in weight and age. They change depending on what clothes are worn, what cosmetic enhancements are used, what body modifications or artwork might adorn them. In 2011 the photographer Laura Dodsworth interviewed 100 women between the ages of 19 and 101 to ask them how they felt about their breasts and to produce a photo gallery that was somewhat removed from ideal-ised surgery adverts or the airbrushed pages of *The Sun* newspaper.[145] In an interview with *The Guardian*, Dodsworth suggests that 'while breasts are interesting in them-selves, they are also catalysts for discussing relationships, body image and ageing'. Physically and psychically, breasts are marked with women's experiences as human beings. They are part of what Susan Bordo, in another context, calls the 'complex crystallizations of culture'; lenses through which we can see both the expectations and pathologies of cultural ideals and the ways in which individual women respond to those ideals as evolving and embodied human beings.[146] Breasts are far more than pillows, then. They are stamped passports of women's interactions with the world.

## Notes

1   With thanks to Sam Alberti, Brian Morgan and Mark Jackson. I have rehearsed some of these ideas in *This Mortal Coil: The Human Body in History and Culture* (Oxford University Press, forthcoming 2016).

2   M. Yalom, *A History of the Breast*, London: Pandora, 1998.

3   I. M. Young, 'Breasted experience: The look and the feeling', in D. Leder (ed.) *The Body in Medical Thought and Practice*, Dordrecht: Kluwer, 1992, pp. 215–30.

4   See D. Lupton, 'Femininity, responsibility, and the technological imperative: Discourses on breast cancer in the Australian press', *International Journal of Health Services*, 1994, vol. 24, 73–90.

5   See for example: http://www.independent.co.uk/life-style/health-and-families/health-news/the-angelina-jolie-effect-her-mastectomy-revelation-doubled-nhs-breast-cancer-test ing-referrals-9742074.html, accessed 1 December 2014.

6   A. Jolie, 'My medical choice', *The New York Times*, 14 May 2013, Opinion Pages.

7   D. G. Evans, *et al.*, 'The Angelina Jolie effect: How high celebrity profile can have a major impact on provision of cancer related services', *Breast Cancer Research*, 2014, vol. 16, 442–7.

8   See for example the award-winning poster for Breast Cancer Care: http://www.wale sonline.co.uk/news/wales-news/cancer-survivor-jill-hindley-proudly-7987997, accessed 12 December 2014.

9   L. R. Rubin and M. Tanenbaum, '"Does that make me a woman?" Breast cancer, mastec-tomy, and breast reconstruction decisions among sexual minority women', *Psychology of Women Quarterly*, 2011, vol. 35, 401–14.

10  M. Grigg *et al.*, 'Information for women about the safety of silicone breast implants', *NIH Consensus Statement*, 1997, vol. 15, 1–35.

11  H. Wijsbek, 'The pursuit of beauty: The enforcement of aesthetics or a freely adopted life-style?' *Journal of Medical Ethics*, 2000, vol. 26, 454–8.

12  http://www.huffingtonpost.com/anthony-youn-md-facs/the-prespent-and-future-of_1_b_2864541.html, accessed 21 July 2014.

13  http://baaps.org.uk/about-us/press-releases/1833-britain-sucks, accessed 1 November 2013.

14  On reconstructive surgery, see: R. A. Aronowitz, *Unnatural History: Breast Cancer and American Society*, Cambridge: Cambridge University Press, 2007. See also S. L. Gilman, *Making the Body Beautiful: A Cultural History of Aesthetic Surgery*, Princeton: Princeton University Press, 1999; E. Haiken, *Venus Envy: A History of Cosmetic Surgery*, Baltimore: Johns Hopkins University Press, 1997. See also K. Davis, *Reshaping the Female Body: The Dilemma of Cosmetic Surgery*, Abingdon: Routledge, 2013.

15  The image of silicone as a time bomb is discussed in M. Rapaport, 'Silicone injections revis-ited', *Dermatologic Surgery*, 2002, vol. 28, 594–5.

16  See P. Santoni-Rugiu and P. J. Sykes, *A History of Plastic Surgery*, Berlin: Springer, 2007.

17  Haiken, *Venus Envy*, p. 5.

18  V. Groebner, 'Losing face, saving face: noses and honour in the late medieval town', trans. Pamela Selwyn, *History Workshop Journal*, 1995, vol. 40, 1–15.

19 S. L. Gilman, *The Jew's Body*, New York: Routledge, 1991.
20 D. J. Hauben and G. J. Sonneveld, 'The influence of war on the development of plastic surgery', *Annals of Plastic Surgery*, 1983, vol. 10, 65–9; J. A. Chambers *et al.*, 'A band of surgeons, a long healing line: Development of craniofacial surgery in response to armed conflict', *Journal of Craniofacial Surgery*, 2010, vol. 21, 991–7.
21 See S. J. M. M. Alberti (ed.), *War, Art and Surgery: The Work of Henry Tonks & Julia Midgley*, London: Royal College of Surgeons, 2014; H. D. Gillies and D. R. Millard, *The Principles and Art of Plastic Surgery*, vol. 2, Boston, MA: Little, Brown, 1957.
22 H. D. Gillies, *Plastic Surgery of the Face, Based on Selected Cases of War Injuries of the Face Including Burns*, Oxford: Oxford University Press, 1920.
23 See M. C. Meikle, 'The evolution of plastic and maxillofacial surgery in the twentieth century: The Dunedin connection', *The Surgeon*, 2006, vol. 4, 325–34.
24 M. G. Berry and D. M. Davies, 'Breast augmentation: Part I – A review of the silicone prosthesis', *Journal of Plastic, Reconstructive & Aesthetic Surgery*, 2010, vol. 63, 1761–8, 1761.
25 A. Losken and M. J. Jurkiewicz, 'History of breast reconstruction', *Breast Disease*, 2002, vol. 16, 3–9.
26 T. Cash *et al.*, *Psychological Aspects of Reconstructive and Cosmetic Plastic Surgery: Clinical, Empirical, and Ethical Perspectives*, Philadelphia: Lippincott Williams & Wilkins, 2006.
27 S. Askegaard *et al.*, 'The body consumed: Reflexivity and cosmetic surgery', *Psychology & Marketing*, 2002, vol. 19, 793–812.
28 S. T. Phelan, 'Fads and fashions: The price women pay', *Primary Care Update for Ob/Gyns*, 2002, vol. 9, 138–43.
29 N. Jacobson, 'The socially constructed breast: Breast implants and the medical construction of need', *American Journal of Public Health*, 1998, vol. 88, 1254–61; I. Pitanguy, 'Surgical treatment of breast hypertrophy', *British Journal of Plastic Surgery*, 1967, vol. 20, 78–85.
30 E. Haiken, 'Modern miracles', in K. Ott *et al.* (eds), *Artificial Parts, Practical Lives: Modern Histories of Prosthetics*, New York: New York University Press, 2002, pp. 171–98.
31 V. L. Blum, *Flesh Wounds: The Culture of Cosmetic Surgery*, Berkeley and London: University of California Press, 2003, p. 14.
32 Cited in Haiken, 'Modern miracles', p. 177.
33 Molly Haskell, *From Reverence to Rape: The Treatment of Women in the Movies*, 2nd edn, Chicago: University of Chicago Press, 1987, pp. 235–52.
34 H. O. Bames, 'Correction of abnormally large or small breasts', *Southwestern Medicine*, January 1941, 11.
35 H. O. Bames, 'Plastic reconstructive of the anomalous breast', *Revue de Chirurgie Structive*, July 1936, 294; and Jacobson, 'The socially constructed breast', 1255.
36 Max Thorek, *Plastic Surgery of the Breast and Abdominal Wall*, Springfield, IL: Charles C. Thomas, 1942.
37 H. O. Bames, 'Breast malformations and a new approach to the problem of the small breast', *Plastic and Reconstructive Surgery*, 1950, vol. 5, 499–506.
38 Bames, 'Breast malformations', p. 499.
39 Gilman, *Making the Body Beautiful*, p. 242.
40 C. E. Martin, *Perfect Girls, Starving Daughters: The Frightening New Normalcy of Hating Your Body*, New York: Simon and Schuster, 2007.
41 Alfred Adler, *The Education of Children*, trans. by Eleanore and Friedrich Jensen, London: George Allen and Unwin, 1935; A. C. Traub and J. Orbach, 'Psychophysical studies of body-image: I. The adjustable body-distorting mirror', *Archives of General Psychiatry*, 1964, vol. 11, 53–66.
42 Alfred Adler identified the 'inferiority complex' in *The Education of Children*, trans. by E. Jensen and F. Jensen, London: George Allen and Unwin, 1935.
43 See J. Leman, '"The advice of a real friend". Codes of intimacy and oppression in women's magazines 1937–1955', *Women's Studies International Quarterly*, 1980, vol. 3, 63–78; H. Dittmar *et al.*, 'Understanding the impact of thin media models on women's body-focused affect: The roles of thin-ideal internalization and weight-related self-discrepancy activation in experimental exposure effects', *Journal of Social and Clinical Psychology*, 2009, vol. 28, 43–72.
44 Subsequently published as M. T. Edgerton and A. R. McClary, 'Augmentation Mammaplasty: Psychiatric implications and surgical indications', *Plastic and Reconstructive Surgery*, 1958, vol. 21, 279–305.

45 M. T. Edgerton, E. Meyer and W. E. Jacobson, 'Augmentation mammaplasty II: Further surgical and psychiatric evaluation', *Plastic and Reconstructive Surgery*, 1961, vol. 27, 279–302.

46 P. Conrad and H. T. Jacobson, 'Enhancing biology? Cosmetic surgery and breast augmentation' in S. J. Williams *et al.* (eds), *Debating Biology: Sociological Reflections on Health, Medicine and Society*, London: Routledge, 2003, pp. 223–35.

47 M. G. Berry and D. M. Davies. 'Breast augmentation: Part I–A review of the silicone prosthesis', *Journal of Plastic, Reconstructive & Aesthetic Surgery*, 2010, vol. 63, 1761–8; L. Miller, 'Mammary mania in Japan', *Positions: East Asia Cultures Critique*, 2003, vol. 11, 271–300.

48 http://www.theguardian.com/lifeandstyle/2008/may/03/healthandwellbeing.health, accessed 1 November 2014.

49 Timmie Jean Lindsey, quoted in http://www.dailymail.co.uk/femail/article-484674/I-worlds-breast-job—endured-years-misery-says-Texan-great-grandmother.html, accessed 1 July 2014.

50 T. D. Cronin and F. J. Gerow, 'Augmentation mammoplasty: A new "natural feel" prosthesis', in *Transactions of the Third International Congress of Plastic Surgery*, 1964, 41–9; M. L. Vanderford and D. H. Smith, *The Silicone Breast Implant Story: Communication and Uncertainty*, Abingdon: Routledge, 2013, p. 11.

51 M. Jacobus *et al.* (eds), *Body/Politics: Women and the Discourses of Science*, New York and London: Routledge, 1989; and L. Schiebinger, *Nature's Body: Gender in the Making of Modern Science*, New Brunswick, NJ: Rutgers University Press, 1993.

52 T. Cronin and R. Greenberg, 'Our experiences with the silastic gel breast prosthesis', *Plastic and Reconstructive Surgery*, 1970, vol. 46, 1–7, 1.

53 Interview with Dr Biggs in http://cosmeticsurgerytimes.modernmedicine.com/cosmetic-surgery-times/news/modernmedicine/modern-medicine-feature-articles/modern-breast-implants-e, accessed 1 December 2014.

54 Grigg *et al.* 'Information for women about the safety of silicone breast implants'.

55 R. Blocksma and S. Braley, 'The silicones in plastic surgery', *Plastic and Reconstructive Surgery*, 1965, vol. 35, 366–70.

56 Ibid., and Cronin and Gerow, 'Augmentation mammoplasty'.

57 http://www.ocregister.com/articles/implants-354332-breast-silicone.html, accessed 1 November 2014 and http://www.dailymail.co.uk/femail/article-484674/I-worlds-breast-job—endured-years-misery-says-Texan-great-grandmother.html, accessed 1 July 2014.

58 http://www.dailymail.co.uk/femail/article-484674/I-worlds-breast-job—endured-years-misery-says-Texan-great-grandmother.html, accessed 1 July 2014.

59 Jacobson, 'The socially constructed breast', p. 1256.

60 M. Angell, 'Evaluating the health risks of breast implants: The interplay of medical science, the law, and public opinion', *New England Journal of Medicine*, 1996, vol. 334, 1513–18, on 1513–14.

61 Vanderford and Smith, *The Silicone Breast Implant Story*, p. 11.

62 A. R. Muzaffar and R. J. Rohrich, 'The silicone gel-filled breast implant controversy: An update', *Plastic and Reconstructive Surgery*, 2002, vol. 109, 742–8, on 743; J. M. Spanbauer, 'Breast implants as beauty ritual: Woman's sceptre and prison', *Yale Journal of Law & Feminism*, 1997, vol. 9, 157.

63 T. M. Sinclair *et al.*, 'Biodegradation of the polyurethane foam covering of breast implants', *Plastic and Reconstructive Surgery*, 1993, vol. 2, 1003–13.

64 S. B. Foote, 'Loop and loopholes: Hazardous device regulation under the 1976 medical device amendments to the food, drug and cosmetic act', *Ecology Law Quarterly*, 1978, vol. 7, 101.

65 Medical Device Amendments of 1976, P. L. 94-295 coded at 21 U.S.C. Section 360.

66 S. B. Foote *et al.*, 'The impact of public policy on medical device innovation: A case of polyintervention', in A. C. Gelijins and E. A. Halm (eds), *The Changing Economics of Medical Innovation in Medicine*, 1991, vol. 2, 69–88.

67 R. E. Stombler, 'Breast implants and the FDA: Past, present, and future', *Plastic Surgical Nursing*, 1993, vol. 13, 185–7.

68 For examples Y. Kumagai *et al.*, Scleroderma after cosmetic surgery: Four cases of human adjuvant disease, *Arthritis & Rheumatology*, 1979, vol. 22, 532–7.

69 Marcia Angell, 'Evaluating the health risks of breast implants: The interplay of medical science, the law, and public opinion', *New England Journal of Medicine*, 1996, vol. 334, 1513–18.

70 S. Jasanoff, 'Science and the statistical victim modernizing knowledge in breast implant litigation', *Social Studies of Science*, 2002, vol. 32, 37–69.

71 A. Powers and J. L. Andsager, 'How newspapers framed breast implants in the 1990s', *Journalism & Mass Communication Quarterly*, 1999, vol. 76, 551–64.

72 *Face to Face with Connie Chung*, CBS Broadcasting, 10 December 1990.

73 Angell, 'Evaluating the health risks', 1513–14; C. Allen, 'Jurisprudence of breasts', *Stanford Law and Policy Review*, 1993, vol. 5, 83.

74 Angell, 'Evaluating the health risks', 1514.

75 Vanderford and Smith, *The Silicone Breast Implant Story*, pp. 12–13.

76 For example B. O. Schoaib and B. M. Patten. 'A motor neuron disease syndrome in silicone breast implant recipients', *Journal of Occupational Medicine and Toxicology*, 1995, vol. 4, 155–63.

77 T. E. Olive, 'Vilification stories: The fall of Dow Corning', in Vanderford and Smith, *The Silicone Breast Implant Story*, pp. 148–76, on p. 150.

78 R. Weisman, 'Reforms in medical device regulation: An examination of the silicone gel breast implant debacle', *Golden Gate University Law Review*, 1993, vol. 23, 973.

79 R. R. Cook *et al.*, 'The breast implant controversy', *Arthritis & Rheumatism*, 1994, vol. 37, 153–7.

80 http://www.humanticsfoundation.com/bernard_patten.htm, accessed 1 December 2014.

81 Olive, 'Vilification stories', 158.

82 D. A. Kessler, 'The basis of the FDA's decision on breast implants', *New England Journal of Medicine*, 1992, vol. 326, 1713–15.

83 Ibid., pp. 1713–14.

84 Ibid., p. 1714.

85 M. Angell, *Science on Trial: The Clash of Medical Evidence and the Law in the Breast Implant Case*, London: Norton, 1997.

86 J. K. McLaughlin, *et al.*, 'The safety of silicone gel-filled breast implants: A review of the epidemiologic evidence', *Annals of Plastic Surgery*, 2007, vol. 59, 569–80, abstract; P. Tugwell *et al.*, 'Do silicone breast implants cause rheumatologic disorders? A systematic review for a Court-Appointed National Science Panel', *Arthritis & Rheumatism*, 2001, vol. 44, 2477–84.

87 S. Bondurant, *et al.*, *Safety of Silicone Breast Implants: Report of the Committee on the Safety of Silicone Breast Implants*, Washington, DC: Institute of Medicine, National Academy Press: 1999.

88 Ibid., p. 7.

89 Ibid., p. 7.

90 Ibid., p. 8.

91 Ibid., p. 229.

92 D. B. Sarwer *et al.*, 'Cosmetic breast augmentation surgery: A critical overview', *Journal of Women's Health & Gender-Based Medicine*, 2000, vol. 9, 843–56.

93 http://www.fda.gov/ForConsumers/ConsumerUpdates/ucm259825.htm, accessed 17 November 2014.

94 Ibid.

95 A. O'Dowd, 'Around 1000 women with private sector PIP implants seek NHS help', *British Medical Journal*, 2012, vol. 344, e972; I. Torjesen, 'Hundreds of thousands of pounds of NHS funds have been spent on care of private patients with PIP implants', *British Medical Journal*, 2012, vol. 344, e1259.

96 See http://www.bbc.co.uk/news/health-16749773, accessed 7 September 2014.

97 M. G. Berry and J. J. Stanek, 'The PIP mammary prosthesis: A product recall study', *Journal of Plastic, Reconstructive & Aesthetic Surgery*, 2012, vol. 65, 697–704.

98 P. Benkimoun, 'Founder of PIP breast implant company gets four year prison sentence', *British Medical Journal*, 2013, vol. 347, f7528.

99 http://www.liverpoolecho.co.uk/news/liverpool-news/pip-breast-implants-killing-me-3354114, accessed 21 July 2014.

100 D. Cohen and M. Billingsley, 'Europeans are left to their own devices', *British Medical Journal*, 2011, vol. 342, d2748.

101 *Review of the Regulation of Cosmetic Interventions*, 2013, available at: https://www.gov.uk/government/uploads/system/uploads/attachment_data/file/192028/Review_of_the_Regulation_of_Cosmetic_Interventions.pdf, accessed 1 October 2014.

102 http://www.theguardian.com/world/2013/oct/02/french-breast-implant-scandal-france, accessed 1 February 2015.

103 R. Horton, 'Offline: A serious regulatory failure, with urgent implications', *The Lancet*, 2012, vol. 379, 106.

104 S. Chummun and N. R. McLean, 'Poly implant prothèse (PIP) breast implants: Our experience', *The Surgeon*, 2013, vol. 11, 241–5.

105 Ibid.

106 *Poly Implant Prothese (PIP) Breast Implants: Final Report of the Working Group*, 18 June 2012, Department of Health, NHS Medical Directorate.

107 Ibid., recommendation 31.

108 http://www.harleymedical.co.uk, accessed 2 October 2013.

109 B. Schofield, 'The role of consent and individual autonomy in the PIP breast implant scandal', *Public Health Ethics*, 2013, vol. 6, 220–3.

110 Bruce Keogh, *Review of the Regulation of Cosmetic Interventions*, London: Department of Health, 2013.

111 Ibid., foreword.

112 http://www.cc4plasticsurgery.com/cosmetic/breast/breast-reduction/, accessed 14 July 2014 and http://www.harleymedical.co.uk/cosmetic-surgery-for-women/breast-surgery/breast-surgery-overview, accessed 14 July 2014.

113 M. L. Richins, 'Social comparison and the idealized images of advertising', *Journal of Consumer Research*, 1991, vol. 18, 71–83.

114 http://www.harleymedical.co.uk/cosmetic-surgery-for-women/breast-surgery/breast-surgery-overview, accessed 14 July 2014.

115 F. G. Miller *et al.*, 'Cosmetic surgery and the internal morality of medicine', *Cambridge Quarterly of Healthcare Ethics*, 2000, vol. 9, 353–64.

116 https://www.gov.uk/government/publications/review-of-the-regulation-of-cosmetic-interventions, accessed 14 July 2014.

117 http://baaps.org.uk/about-us/press-releases/1624-one-in-five-unsuitable-for-cosmetic-surgery-patients-dangerously-misinformed-by-salespeople, accessed 14 July 2014.

118 Keogh, 'Review', 3.3.

119 See the government's response: https://www.gov.uk/government/uploads/system/uploads/attachment_data/file/279431/Government_response_to_the_review_of_the_regulation_of_cosmetic_interventions.pdf, accessed 1 December 2014.

120 M. Latham, '"If it ain't broke, don't fix it?" Scandals, "risk", and cosmetic surgery regulation in the UK and France', *Medical Law Review*, 2014, vol. 22, 384–408.

121 National Care Standards Commission, Report to the Chief Medical Officer for England on the Findings of Inspectors of Private and Cosmetic Surgery Establishments in Central London during March/April 2003, June 2003, discussed in ibid.

122 Latham, 'If it ain't broke'.

123 http://services.parliament.uk/bills/2012-13/cosmeticsurgeryminimum standards.html.

124 http://www.express.co.uk/finance/city/412963/Silicon-scare-helps-plump-up-sales-at-UK-s-Nagor, accessed 2 October 2013.

125 https://www.rcseng.ac.uk/news/surgeons-publish-landmark-standards-for-cosmetic-practice#.VL5DTVrz1Ec, accessed 20 January 2013.

126 Ibid.

127 http://www.theguardian.com/lifeandstyle/2008/may/03/healthandwellbeing.health, accessed 1 November 2014.

128 Ibid.

129 Keogh, 'Review', 3.3.

130 D. B. Sarwer, *et al.*, 'An investigation of changes in body image following cosmetic surgery', *Plastic and Reconstructive Surgery*, 2002, vol. 109, 363–9.

131  M. Latham, 'The shape of things to come', *Medical Law Review*, 2008, vol. 16, 437–57.
132  Vanderford and Smith, *The Silicone Breast Implant Story*, introduction.
133  Morgan, 'Women and the knife'.
134  For a timely intervention, see C. Hayes and M. Jones (eds), *Cosmetic Surgery: A Feminist Primer*, Farnham: Ashgate, 2009.
135  Davis, *Reshaping the Female Body*.
136  A. Balsamo, *Technologies of the Gendered Body: Reading Cyborg Women*, Durham, NC: Duke University Press, 1996, p. 78.
137  C. Chambers 'Are breast implants better than female genital mutilation? Autonomy, gender equality and Nussbaum's political liberalism', *Critical Review of International Social and Political Philosophy*, 2004, vol. 7, 1–33.
138  D. B. Sarwer *et al.*, 'Body image concerns of breast augmentation patients', *Plastic and Reconstructive Surgery*, 2003, vol. 112, 83–90.
139  L. Lipworth *et al.*, 'Excess mortality from suicide and other external causes of death among women with cosmetic breast implants', *Annals of Plastic Surgery*, 2007, vol. 59, 119–23.
140  See Cosmetic surgery special, 'When looks can kill', *New Scientist*, 2006, vol. 2574, 18; V. C. M. Koot *et al.*, 'Total and cause specific mortality among Swedish women with cosmetic breast implants', *British Medical Journal*, 2003, vol. 326, 527.
141  S. Orbach, *Bodies*, London: Profile Books, 2009, 22–3.
142  Vanderford and Smith, *The Silicone Breast Implant Story*, pp. 8–9: E. Showalter, *The Female Malady: Women, Madness and Culture in England, 1830–1980*, New York: Pantheon, 1986.
143  Vanderford and Smith, *The Silicone Breast Implant Story*, p. 9; and N. Cunningham, 'Power, the meaning of subjectivity and public education about fibromyalgia', *Journal of Philosophy and History of Education*, 59, 2009, 186; K. Kendall, 'Masking violence against women: The case of PMS', *Canadian Woman Studies*, 1991, vol. 12, 17–20.
144  http://surgicalcareers.rcseng.ac.uk/wins/statistics accessed 22 January 2015.
145  See http://www.barereality.net/about, accessed 9 September 2014.
146  S. Bordo, *Unbearable Weight: Feminism, Western Culture, and the Body: Tenth Anniversary Edition*, Berkeley: University of California Press, 2003.

## Select bibliography

M. Angell, 'Evaluating the health risks of breast implants: The interplay of medical science, the law, and public opinion', *New England Journal of Medicine*, 1996, vol. 334, 1513–18.

S. Bondurant, V. Ernster and R. Herdman, *Safety of Silicone Breast Implants: Report of the Committee on the Safety of Silicone Breast Implants*, Washington, DC: Institute of Medicine, National Academy Press: 1999.

K. Davis, *Reshaping the Female Body: The Dilemma of Cosmetic Surgery*, Abingdon: Routledge, 2013.

E. Haiken, *Venus Envy: A History of Cosmetic Surgery*, Baltimore: Johns Hopkins University Press, 1997.

N. Jacobson, 'The socially constructed breast: breast implants and the medical construction of need', *American Journal of Public Health*, 1998, vol. 88, 1254–61.

S. Jasanoff, 'Science and the statistical victim modernizing knowledge in breast implant litigation', *Social Studies of Science*, 2002, vol. 32, 37–69.

B. Keogh, *Review of the Regulation of Cosmetic Interventions*, London: Department of Health, 2013.

M. Latham, '"If it ain't broke, don't fix it?" Scandals, "risk", and cosmetic surgery regulation in the UK and France', *Medical Law Review*, 2014, vol. 22, 384–408.

B. Schofield, 'The role of consent and individual autonomy in the PIP breast implant scandal', *Public Health Ethics*, 2013, vol. 6, 220–3.

M. L. Vanderford and D. H. Smith, *The Silicone Breast Implant Story: Communication and Uncertainty*, Abingdon: Routledge, 2013.

# 20

# CANCER SCREENING

*David Cantor*

When in 1951 the United States Commission on Chronic Illness defined screening as 'the presumptive identification of unrecognized disease or defect by the application of tests, examinations, or other procedures which can be applied rapidly', it sought to promote a relatively new way of detecting disease.[1] Screening, the Commission explained, aimed to sort out apparently well persons who probably had a disease from those who probably did not. It was not a diagnostic test since persons with suspicious findings had to be referred to their physicians for diagnosis and necessary treatment, nor was it yet available for many diseases. The Commission noted that the most commonly used screening tests were blood glucose determination for diabetes, a serological test for syphilis, radiography for chest pathology, and cytology for cancer detection. The last of these referred to the Pap test used to screen for cervical cancer.

The Commission on Chronic Illness constituted an unprecedented post-war effort to establish chronic illness, including cancer, at the heart of American health policy.[2] As the historian George Weisz argues, it pressed for massive investment in research into chronic disease, regular physical examinations, support for more rehabilitation facilities, and a new emphasis on prevention rather than cure. It also pushed for keeping the chronically ill at home instead of institutionalizing them; for providing care for them in general, rather than chronic-care, hospitals; for building nursing homes for those unable to live at home but not requiring medical care; and – most important for this chapter – for developing mass multiphasic screening technologies to detect diseases cheaply in their early and treatable stages. The history of cancer screening was part of this broader shift in interest towards chronic disease in the mid-twentieth century, and was to be supported by an extraordinary growth of government and private funding.

The focus of this chapter will be the development of screening programmes in the United States, and especially those for breast and cervical cancers. But screening was not simply an American development, nor was it to be limited to these two cancers. Advocates of the practice in North America and Europe argued that there was a vast reservoir of undetected cancers or pre-cancers, which posed a real threat to the success of efforts to reduce mortality and morbidity. It was not sufficient, in their view, to interpret the absence of clinical cases as the absence of disease or a precondition, nor was it sufficient to view rises in clinical cases over time as evidence of the effectiveness of cancer screening programmes. The presence of cancer or precursors in a population was like an 'iceberg' in that most cases were invisible, hidden below the surface.

Screening seemingly healthy populations would, the argument went, help to detect more of these hidden cases at an earlier phase in their natural history than could be achieved by other methods of early detection. The promise seemed to be borne out when mammography joined cancer screening in the 1960s, followed by colonoscopy for colon cancer, the PSA test for prostate cancer, and genetic screening for some cancers. Yet by the beginning of the twenty-first century this promise had not been realized; only a few cancers were subject to screening programmes.

## Before screening

Screening for cancer had roots in early-twentieth-century programmes of early detection and treatment, which aimed to catch a growth as early in its life as possible, ideally before it turned malignant, and to treat it the moment cancer or something that might turn into cancer was discovered.[3] Physicians had long urged people to see their doctors as soon as they spotted something wrong, but until the twentieth century the message was not promoted consistently, nor was there agreement as to the best therapeutic interventions. Some advocated surgery. Others saw it as risky and to be avoided, especially in advanced and internal cancers. Physicians thus used a mix of other interventions – including herbal remedies, caustics (which ate away surface tumours and surrounding tissue), palliation and advice on prevention, including recommendations on diet, lifestyle, and exposure to environmental and occupational carcinogens. In the early twentieth century, however, surgery came to take centre stage, joined by x-rays (discovered in 1895) and radium (1898). Other therapeutic and preventive interventions began to be marginalized.

The dominance of surgical approaches to cancer was to be a major factor in the emergence of programmes of early detection and treatment. A key moment was the late-nineteenth-century development of the radical mastectomy by the Johns Hopkins surgeon William Stewart Halsted.[4] Halsted based his operation on the belief that breast cancer spread in a centrifugal manner from the primary tumour to nearby structures. In his view, most surgeons did not remove sufficient mass, and so left some cancer tissue within the body to become the source of recurrences. To catch these remnants, surgeons had to cut more aggressively, hc claimed. They had to remove not only the breast but also surrounding body tissues including the skin, neighbouring lymph nodes, muscles, and parts of the rib cage or shoulder.

Halsted's methods rapidly became central to breast cancer therapy in the United States. His students colonized leading American hospitals and medical schools, trained their own students in his approach, and applied it to other cancers. Halsted's students were among the founders in 1913 of the first US national campaign against cancer, the American Society for the Control of Cancer (ASCC). Adapting the local, centrifugal model of breast cancer to broader anti-cancer programmes, the ASCC began to promote early detection and treatment of the disease or its possibility. The window of opportunity for successful treatment was short, the ASCC claimed; patients had to go to their physicians as soon as the disease or its possibility was identified, before it spread out from the local site, and physicians had to act with urgency.[5]

The problem, the ASCC argued, was that many patients got to their physicians too late: too many were ignorant of the early signs of cancer, it claimed, which could be subtle, and without sufficient pain or debility to prompt them to consult their physicians. But even if cancer was suspected, the ASCC worried that patients delayed out of fear of disease or its treatments, from fatalism, or because they were swayed by quacks, purveyors of patent medicines, or friends and family to seek inappropriate help. Even regular physicians were a problem. Too often, the ASCC complained, they were ignorant of the disease and its treatment, pessimistic about the possibility of successful treatment, or hesitated too long when confronted with cancer.[6]

The ASCC began a nationwide education programme to transform public attitudes and behaviours towards cancer. It taught people about the 'early warning signs' of the disease, to go to a regular physician at the first suggestion of cancer and, from the late 1910s, to go for regular medical check-ups even if they felt well. They also warned the public of the dangers of quackery, of ineffective and hazardous home remedies, and of the bad advice of friends, family, the media, and ignorant physicians. Yet, as its educational programme took off in the 1920s and 1930s, the ASCC became alarmed that its own efforts were actually exacerbating fears of the disease and its treatments and consequently encouraging delay. It also came to worry about the rare voices from within the organization that questioned the focus on early detection.[7]

One such voice in the 1920s was James Ewing, the director of the Memorial Hospital in New York, who argued that early diagnosis alone was not sufficient to reduce the death rate from cancer. Surgeons, he claimed, knew that patients who came with an early diagnosis all too often did not undergo a complete cure. The cancer returned, or failed to go away, even after therapy. He further urged that public education campaigns should acknowledge that early diagnosis and the best treatment would still leave cancer a significant menace. Yet ASCC public education materials rarely, if at all, acknowledged such uncertainties. Diluting the message of early detection and treatment, the ASCC feared, would mean that many would to fail to go to their physicians.[8]

So the ASCC's education programmes continued to promote early detection and treatment. In addition, together with public and private agencies, it sought to establish networks of cancer clinics and hospitals into which patients might be channelled. The economic depression of the 1930s frustrated these efforts, and in 1937 the Federal government established the National Cancer Institute (NCI) in part to help struggling cancer programmes by providing free loans of highly rare and expensive radium. The radium loan programme fell victim to medical anxieties about socialized medicine, but the NCI vigorously promoted the ASCC's message of early detection and treatment in its public educational and medical training programmes in the 1940s and provided other support to help state and private organizations develop cancer control programmes.[9] 'The whole concept of "control" is different in cancer from the concept in most of the communicable diseases', explained Austin Diebert, head of cancer control at the NCI in 1950:

> We have no reliable mass screening technique for cancer, and its diagnosis and management ordinarily call for the services not of a single physician, but for a highly qualified team. Moreover, the cancer case is usually seen first by a general practitioner – and it is his diagnostic training and experience that often determine the outcome.[10]

Unlike tuberculosis, common practice in cancer was for referrals to go not to the specialist, but to the general practitioner. All too often, Diebert suggested, the general practitioner's training and experience was insufficient. Something had to change.

## Mass screening begins

In the 1940s, mass screening joined the armamentarium of early detection and treatment, first in cervical cancer and then breast and other cancers. Its focus, to begin with, was almost entirely on women, and on what came to be known as pre-cancers – lesions that might (or might not) turn into cancer proper. Efforts to tackle cancer in men looked much more like the pre-war early detection and treatment campaigns than the screening campaigns that targeted female cervical and breast cancers. Some proposals for screening for lung cancer in the 1960s targeted cancers that then predominantly affected men. But, it was only in the 1990s, with the development of screening programmes against prostate cancer, that men became the object of screening. But even then there were differences with the screening programmes that developed in the 1940s for female cancers. Prostate cancer screening programmes tended to focus on detecting the presence of cancers rather than pre-cancers.[11]

Screening aimed to extend the reach of early detection and treatment programmes from the individual patient to the population. Much as general practitioners and family physicians had earlier been encouraged to identify cancer (or its possibility) in an individual patient, who would then be referred on for more specialist diagnostic tests, so screening programmes aimed to identify cancer (or its possibility) among individuals within a given population, who would then undergo further diagnostic testing. Screening often happened in specialist centres rather than the general practitioner's office, and the patient was sent on for further investigation only if cancer or a risk of it was identified.

As before the war, surgery remained a major therapy, which became increasingly radical in the 1950s and 1960s, along with ionizing radiation which was often produced by new technologies such as cobalt, artificial radioisotopes, betatrons, linear accelerators and later cyclotrons.[12] From the 1960s, these were joined by chemotherapy (initially for leukaemia and lymphomas, and later other cancers), immunotherapy and, from the 1990s, targeted therapy that was aimed at certain molecular 'lesions'. Advocates argued that the effectiveness of screening programmes and of therapeutic interventions tended to be co-dependent: screening programmes would only reduce cancer mortality if there were effective therapeutic interventions and, conversely, therapeutic interventions were most effective if screening programmes identified cancers or pre-cancers at an early stage.

Often screening was presented as a form of emotional management, reassuring patients between their tests that they did not have cancer or conditions that predisposed to it, or more problematically that they had cancers that were unlikely to do harm in their lifetime. This was not entirely a new idea. Since the 1910s, regular medical check-ups had been used to counter such fears, by reassuring patients that they did not have cancer in intervals between examinations or that a bodily anomaly was not cancerous. Ideally, screening offered similar reassurance.[13] However, from the 1970s public debates about the hazards of screening technologies (such as x-rays), the dangers of false positives (people identified by the screening test with the disease, but who did not have it) and negatives (people who had the disease, but were not identified by the screening test), and concerns about over-diagnosis tended to complicate such efforts at emotional management.

Screening built upon older programmes of early diagnosis and treatment in another way in that it was often presented as a way of targeting conditions that might one day turn malignant or promote malignancy. Since the late nineteenth century, physicians had undertaken surgical interventions to remove sources of chronic irritation (such as warts, moles, or bad teeth, often thought to cause cancer) or more radical procedures such as the removal of a woman's breasts upon the identification of pre-cancerous conditions such as cystic mastitis, proliferative lesions of the breast, and carcinoma in situ, all conditions that might (or might not) turn malignant.[14] Mass screening would intensify this focus on pre-cancers in women, and coincided with a growing radicalization of cancer surgery in the United States and the promotion of the one-step method for breast cancer whereby surgery happened immediately the biopsy indicated cancer or pre-cancer, while the patient was still under anaesthetic. Tens of thousands of women went into surgery for a biopsy and came out without a breast.

None of this should be read to suggest that the transition from early detection and screening was unproblematic.[15] Although screening programmes often built upon the institutional structures created to promote early diagnosis and treatment, there were a host of organizational and technical questions involved in the transition to mass screening. Who should do the screening? Should resources be devoted to training a cadre of specialists for this purpose or should they rely on physicians to screen healthy people given fears that doctors were not interested in this work, would become bored, and do an inadequate job, if they did it at all? Could they develop mechanical methods of screening out individuals for specialist referral along the lines of mass radiography for tuberculosis? Finally, could cancer screening be combined with other screening programmes? In San Jose (California), and Brookline (Massachusetts) efforts were underway in the late 1940s to determine whether it made sense to combine screening for several diseases in one centre, since it seemed that some people who appeared for screening for one disease later turned up in screening clinics for others. Yet it was often unclear how cancer might be combined with such screening programmes.

## The Pap test

The history of the Pap test – the first mass screening technology for cancer – illustrates the complexities of initiating screening programmes for cancer and the porous boundary between diagnosis and screening.[16] The Pap test or Papanicolaou test, named for its inventor, the Greek physician George Papanicolaou, began life as a biological research tool, before becoming a diagnostic one, and then a tool for screening. The story begins in 1917 when Papanicolaou published an article on the vaginal smear as an indicator of the stages of the oestrous cycle in guinea pigs. Papanicolaou wanted ova at particular stages of development for his research, and he turned to guinea pigs in the hope of obtaining clear indicators of cyclicity. He took smears of cells from the vaginas of guinea pigs over time, and found it was possible to establish the precise stage of oestrus. The Pap test thus began life as an indicator of biological activity.

Papanicolaou's cytological smear was quickly taken up by scientists interested in questions of reproduction, and emerged as a key tool in reproductive endocrinology in the 1920s and 1930s. Then in the 1920s he identified what he called exfoliated cancer cells in human vaginal smears. Papanicolaou came to hope that this discovery would provide an opportunity to diagnose cervical cancer at an early stage in its development. Yet, the new technique did not get the same enthusiastic reception as his earlier biological test. His 1928 presentation to the Third Race Betterment Conference was poorly received. Papanicolaou blamed technical problems such as a lack of clarity in his staining method; the fact that pathologists preferred specifically localized tissue biopsy materials, with which they were more familiar than the exfoliated (free floating) cells of a Pap test; and because gynaecologists were more interested in cyclicity itself than cancer. It cannot have helped that the published version of the paper was full of typographical mistakes and badly reproduced photographs.[17]

Papanicolaou largely abandoned his interest in cancer detection and returned to reproductive endocrinological research that used the vaginal smear as a biological indicator. Then, in 1939, Joseph C. Hinsey, the new Chair of the Department of Anatomy at Cornell, encouraged him to turn once again to the problem of using the vaginal smear as a cancer detection tool, and he began a new collaboration with the gynaecological pathologist Herbert F. Traut. For much of the next decade Papanicolaou and Traut continued their research, supported by the Commonwealth Fund as part of a broad programme of research funding to reduce maternal mortality, and helped by a policy that required all women admitted to the gynaecological department of the New York Hospital to have a routine vaginal smear. Papanicolaou came to argue that the vaginal smear permitted diagnosis at a much earlier date than would have been possible with the biopsy technique that pathologists still preferred, and that many cases of cervical cancer at New York Hospital would have been missed without the practice of undertaking a routine vaginal smear of all patients.[18]

At this point, Papanicolaou framed his work as a tool for diagnosing malignancies at an early stage, when physicians claimed they could best treat them by means of surgery and radiotherapy. Nevertheless, pathologists continued to doubt the value of cytology in cancer detection before invasion occurred, arguing that, even if the smear technique was capable of such detection, they would not have time to review the large number of slides necessary to find a positive case. At a time when demand for all kinds of testing had expanded dramatically, adding cytological analyses to the workload of existing departments of pathology (which were generally located in hospitals and large out-patient facilities, where work centred on the study of suspicious tissue samples surgically removed for biopsy) seemed particularly formidable given that about 75 per cent of slides were expected to be 'normal'.[19]

The turning point came in the mid-to-late 1940s, when the American Cancer Society (ACS – successor to the ASCC) and the NCI turned to the Pap test as a simple and cheap technique that could expand early detection and treatment programmes. Both organizations sponsored training programmes to teach how to read slides, and encouraged the use of the smear as a screening technology through conferences and disease incidence discovery studies. In the 1950s, the ACS and the NCI initiated large-scale Pap smear screening studies including, for example, one study of 95,000 women in Memphis.[20]

With ACS and NCI support, the Pap test began to mutate from a diagnostic test into a mass screening technology. The test came to involve fixing a cellular specimen from the cervix, which would be stained on a slide for visual interpretation, followed by efforts to identify morphologic changes of pre-cancerous cells. Patients with abnormal findings on Pap smears would then be referred for colposcopy (a procedure involving the use of a colposcope to examine a woman's vulva, vagina, and cervix), and if necessary, cervical biopsy. Women would then be recalled for results and treatment. Earlier hopes of using pathologists and physicians to undertake screening gave way to a gendered division of labour between pathologists (often men) and lower-status cytotechnicians (often women) who did the screening. In addition, efforts to create universally valid classifications gave way to locally negotiated orders, since physicians found it difficult to agree on a standard diagnosis.[21]

The routine use of colposcopy had other consequences. It reduced the need to make the Pap test more accurate, countered criticism that exfoliated cells were not representative of the tumour, and formed the focus of a new specialty of gynaecologists who specialized in colposcopic diagnosis of cervical lesions. In some clinics a 'see and treat' approach emerged, in which patients with an abnormal Pap smear would be evaluated and treatment determined by colposcopy in the same visit – often promoted in low-resource areas or when patient non-compliance was anticipated. Such an approach provided not only a means of monitoring and detecting cancers and pre-cancers in women, but also of managing scarce resources since, the argument went, 'see and treat' meant that the overall cost of treatment should be less.[22]

Today, it is claimed that the Pap test has decreased the mortality of cervical cancer: the decline in cervical cancer mortality has closely paralleled the implementation of the Pap smear during the past 50 years.[23] However, controversy exists as to the role of the Pap test in this decline, the frequency with which the test should be performed, how to incorporate human papillomavirus (HPV) DNA testing into screening procedures (HPV being recognized as a cause of some forms of cervical cancer in the 1980s), and the lack of applicability of programmes developed in the West for Third World countries or for low-resources areas, whose populations were often 'under-screened'. Critics also came to worry about a tendency to over-treat given the relative ease with which suspicious cervical lesions could be eliminated and the popularity of 'see and treat'. Cytological diagnosis cannot predict the course of a given lesion, and critics complained that a consensus had emerged that all suspected lesions should be destroyed if screening revealed abnormal readings.

## Mammography

As with the Pap smear, mammography has roots in older programmes of early detection and treatment, in particular, efforts from the 1930s to promote x-ray examination as an accessory to physical examination for the diagnosis of breast cancer.[24] Yet at first x-ray examinations were not widely taken up. Despite technical innovations, such as double-emulsion film and breast compression to produce higher-quality images, mammographic films often remained dark and hazy, and the new techniques were not easily reproduced by others. In the late 1950s, the Houston radiologist Robert L. Egan used a high milliamperage-low voltage technique, a fine-grain intensifying screen, and industrial film to produce much clearer images that were easier to interpret and reproduce; Egan confirmed their value as a diagnostic tool in prospective clinical trials. Between 1956 and 1959, he and his colleagues at the M. D. Anderson Cancer Hospital took films of 1,000 women evaluated in the breast clinic who did not have obvious cancer on physical examination. Of the 245 breast cancers ultimately confirmed by biopsy, Egan had identified 238 by mammography. Nineteen of these cancers were in women whose physical examinations had revealed no breast pathology. With the backing of the ACS and the NCI, Egan's work provided a basis for new efforts to routinely screen for breast cancer in otherwise healthy women.[25]

Mammography started with high expectations of a new beginning in detecting early breast cancer or pre-cancers, but in the following decades it became the focus of highly politicized and public controversies over the value of screening to women in their forties, the dangers of x-rays, and the possibilities of over-treatment. What made this controversy all the more difficult was the multiplicity of US organizations making different recommendations, the involvement of Congress and the media, and the emergence in the 1960s of mammography as a sub-specialty within radiology, with a vested interest in the procedure.

The difficulties can be traced back to 1963, when the National Cancer Institute began a trial of mammography screening. Conducted by the Health Insurance Plan (HIP) of New York, the 1963 trial randomly allocated 61,000 women aged 40–64 either to annual screening for five years or to a control group that received usual care. The first results were published in the early 1970s, and suggested that screening reduced mortality from breast cancer in women over 50, but that there was no benefit in screening women in their forties.[26] In 1973, the Breast Cancer Detection Demonstration Project (BCDDP) – co-sponsored by the NCI and the ACS – established 29 centres across the country which screened 280,000 women over five years, without a control group, in an attempt to show the usefulness of mammography.

By the mid-1970s, however, critics, such as the physician and statistician, John C. Bailar III, questioned the value of the HIP and BCDDP studies, especially in women under 50. Biases in the HIP study, Bailar claimed, led physicians to exaggerate the value of screening tests: routine mammography was likely to detect many slow growing lesions that were unlikely ever to become clinically significant, and there was the possibility the x-rays used in mammography might cause breast cancer. He concluded that the routine use of mammography in screening asymptomatic women might eventually take almost as many lives as it saved.[27]

With Bailar's criticism the focus of media reporting, the NCI appointed experts to review his criticism of the HIP trial.[28] They reported to a public meeting of the President's Cancer Panel on 19 July 1976 which failed to achieve agreement.[29] Then the NIH convened a Consensus Development Conference in 1977 on screening mammography.[30] The consensus panel recommended that women aged 35–39 should be screened only if they had a personal history of breast cancer; women aged 40–49 should be screened if they had a family history of breast cancer; and women aged 50 and older should be screened annually. The NCI ordered its contractors to follow the panel's findings in screening women enrolled in the BCDDP, but did not issue a public health recommendation or guidelines about screening mammography for women outside of the BCDDP. The ACS endorsed the panel's recommendations.

Public consensus over screening for women under 50, however, broke down in the 1980s. The ACS diverged from the NCI and other organizations and recommended that women aged 35–39 receive a baseline mammogram and that women aged 40–49 have a screening mammogram every one to two years. In 1987, the NCI endorsed the new ACS recommendations, as in 1989 did the American College of Radiology and 11 other medical organizations.[31]

But then, in 1993 the NCI withdrew its support for this policy following a meta-analysis of eight randomized controlled trials of mammography that included women in their forties which concluded that mammography had no demonstrable benefit. The decision led to a public rift between the ACS and the NCI.[32]

A second consensus conference was appointed in 1997 to assess new data from randomized trials in Sweden.[33] The Swedish data suggested a statistically significant reduction in mortality for women who underwent mammography in their forties, and had prompted suggestions that the NCI's 1993 decision should be reversed. However, the consensus panel reported in 1997 that the available data did not warrant a single recommendation for mammography for all women in their forties: each woman should decide for herself whether to undergo mammography in consultation with her physician.[34]

The result was a political firestorm.[35] As the *Cancer Letter* put it: '"Consensus" would be the wrong word to describe the aftermath' of the conference.[36] Amid a blizzard of media reporting on the subject, the US Senate unanimously passed a non-binding resolution on 4 February 1997 urging the NCI to recommend regular mammography for women aged 40–49.[37] The following day its director was summoned before a hearing of the Senate subcommittee on Labor, Health, and Human Services where he was asked to reject the panel's conclusions.[38]

The consensus panel's role was only advisory: the next step was for the NCI to decide whether to implement the conclusions. The National Cancer Advisory Board (NCAB), which oversaw the NCI, formed a subcommittee to make a recommendation for endorsement by the full board against a background of continued Congressional pressure.[39] The House Labor, Health and Human Services and Education Appropriations Subcommittee, which had oversight of NCI funding, held hearings at which Arlen Specter, the subcommittee chair, made it clear that decisions on the budget would be postponed until the NCI's final recommendation about screening for women in their forties was issued. The political message was clear that the NCI's budget depended on it, any further delay was unacceptable, and the NCI should reverse the decision of the panel which threatened insurers' willingness to cover mammograms.[40] Thirty-nine women members of Congress wrote to the NCAB claiming that without definitive guidelines women's lives would be at risk.[41]

There was also pressure from professional associations, advocacy groups, and continuing media interest.[42] The ACS expressed disappointment in the panel's findings, and others, especially radiologists, strenuously criticized them.[43] Hitherto the ACS had recommended a screening mammogram every one to two years for women in their forties. On 21 March 1997 it increased the recommended frequency to an annual mammogram.[44] A few days later on 27 March the NCAB's subcommittee issued a statement (agreed by all members of the board but one) recommending that women aged 40–49 who were considered to be at normal risk should be screened every one to two years, that women at higher risk should seek expert advice on the frequency of screening and whether they should begin regular screening before age 40, and that women over 50 should continue to be screened every one to two years.[45] This recommendation was adopted by the National Cancer Institute the same day, and a joint statement was issued by the Institute and the American Cancer Society agreeing that 'mammography screening of women in their forties is beneficial and supportable with the current scientific evidence'.[46] Statements of approval came quickly from President Bill Clinton and Department of Health and Human Services (DHHS) secretary Donna Shalala.[47] The political crisis was diffused.[48]

For US women – as Figure 20.1 suggests – such machinations could be confusing, given the plethora of recommendations and of organizations offering recommendations and accusations of political and vested interests behind them, and the fact that European recommendations tended to favour starting screening at 50, and with longer intervals between screenings than in the United States. Whom should these women trust? Indeed, their confusion was not helped by continuing questions about the procedure.[49] Thus, in 2007, the American College of Physicians issued guidelines that noted that while regular mammograms for women in their forties could reduce the risk of dying from breast cancer by a modest amount, a very high percentage of women could get false positive results that resulted in unnecessary biopsies, increased costs, and risks of injury. The College recommended that women in their forties and their doctors periodically evaluate their risk to guide screening decisions.[50] Other studies also raised alarm bells. A 2008 Norwegian study in the *Archives of Internal Medicine* suggested that some invasive breast cancers might disappear without treatment, raising the possibility that some cancers detected by mammograms spontaneously regressed.[51]

The following year, a report in the *Journal of the American Medical Association* argued that after 20 years of widespread breast screening the promised health benefits had not materialized.[52] Population-based screening had resulted in a substantial increase in the incidence of early disease with only a very slight decrease in late-stage disease.

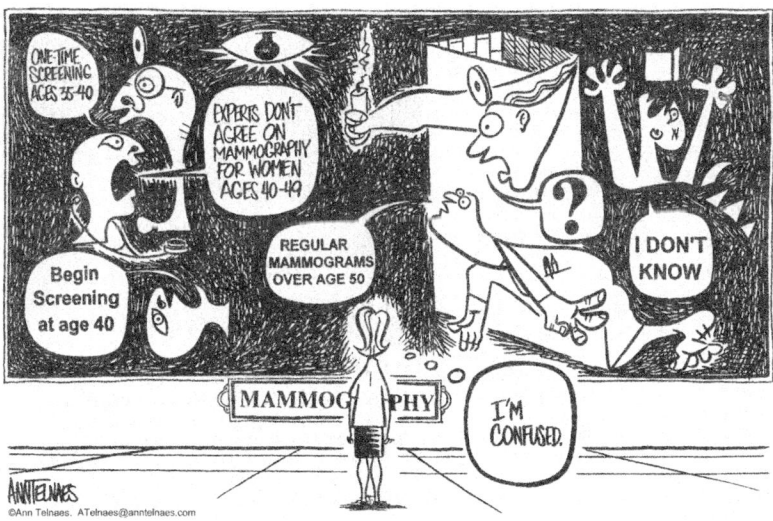

*Figure 20.1* 'I'm Confused', *Times Herald Tribune* (Late Final South Edition), Middletown, NY, 13 February 1997, clipping from NIH, Office of the Director, Director of Disease Prevention, Office of Medical Applications of Research (OMAR). Image enhanced by Hank Grasso and the Cartoonist Group. Ann Telnaes Editorial Cartoon used with the permission of Ann Telnaes and the Cartoonist Group. All rights reserved.

The latter finding was significant, because a crucial justification for screening was that identifying and treating early disease would lead to a commensurate fall in the number of late-stage cancers. But this had not happened, suggesting that screening detected many non-aggressive cancers that would not have progressed if not detected: there had thus been a dramatic increase in the numbers of individuals treated for cancer without necessarily improving the health of the population. Mammography had resulted in over-diagnosis and over-treatment.

In November that year, an influential panel of medical experts, the US Preventive Services Task Force (USPSTF), kept the controversy alive when they published recommendations that most women should not begin routine mammograms to screen for breast cancer until the age of 50, reversing guidelines they had issued just seven years before when they recommended 40 as the optimal age to start getting mammograms.[53] Women aged 50 to 74, the report noted, should have mammograms less frequently – every two years, rather than every year. Doctors should also stop teaching women to examine their breasts on a regular basis, according to the guidelines issued by an independent panel of experts in prevention and primary care appointed by the DHHS. While some praised the new recommendation as sensible given the smaller benefit women under 50 derived from mammography, Health and Human Services Secretary, Kathleen Sebelius, urged women to make no changes in their screening strategies, the Senate approved an amendment to nullify the breast cancer screening recommendation, and the American College of Radiology asked that the recommendations on mammography be specifically excluded from health-care reform legislation – a reference to what became the Affordable Care Act of 2010.[54] Some women's groups, health-care advocates, and individual women saw the guidelines as privileging financial considerations over women's health and a setback to decades-long efforts to reduce the mortality rate of breast cancer.[55] Ironically, these recommendations brought the USA closer to many European countries, where support, including among women's groups, for mammography for women in their forties had never been as strong as in the USA.

## Expert evaluation and public responses

As screening programmes for cervical and breast cancer were established, the hope emerged that screening would be generalized across many cancers, eventually replete with algorithms for screening, follow-up, recall of patients for regular routine screening, and quality control measures.[56] Programmatic screening of this sort came to be seen as more effective than opportunistic screening in which a patient sees a physician who chooses to screen or not to screen. But, by the 2010s, only a few other cancers had become the focus of mass screening programmes. There are no widely accepted screening guidelines for cancers of the oesophagus, stomach, pancreas, and liver. In ovarian, endometrial and lung cancers the focus is on people in high-risk groups, such as women with a known hereditary ovarian cancer syndrome (ovarian cancer), a suspected autosomal-dominant predisposition to colon cancer (endometrial), and long-term heavy smokers (lung).

Of those cancers that are the subject of screening, some were little more than extensions of the pre-war programmes that recommended routine physical examinations by a physician. Skin cancer screening, for example, involved routine examination by a physician, generally a dermatologist, and the removal of cancers and precancerous growths, such as actinic keratosis and leucoplakia, through topical creams or minor surgery.[57] Others involved specialist interventions, such as for colon cancer, which involved routine colonoscopy every 10 years for people over 50, or shorter intervals for those with a risk factor for the disease. Colonoscopy began as a means of detecting malignancies and was later adapted to allow removal of suspicious polyps for microscopic examination.[58]

The reasons why only a few cancers emerged for screening were complex and evolved as more cancers became candidates for screening. Some cancers seemed too rare to justify screening an asymptomatic population, while others (like prostate and some lung cancers) might do little harm to an individual with the disease during the reminder of their lifetime, making them controversial candidates for screening.[59] In the case of prostate cancer, the identification of a link between prostate cancer and rising levels of prostate-associated antigen (PSA) in the blood seemed to promise a cheap, relatively reliable test for the disease, even allowing for a needle biopsy follow-up for those men who screened positive.[60] But, it was not to be. Few efforts to extend screening to other cancers were to be as contentious as those for prostate cancer. Because of the high prevalence of latent prostatic cancer in older men, PSA tests detect some cancers that, in the absence of screening, would have remained asymptomatic during the remainder of the person's lifetime. The problem was (and is) that there is often no way to know this from the PSA test alone, and as a result of the testing such men would be identified and subsequently treated in the same manner (and perhaps with the same urgency) as all other prostate cancers, sometimes with serious consequences – short-term disability, unnecessary financial costs, and long-term treatment-related side-effects.

Debates also focused on the window of opportunity for early detection. Cancer experts had initially hoped that screening would extend efforts to identify cancers and pre-cancers early in their life, ideally before they turned malignant, what eventually came to be called a detectable preclinical phase (DPCP).[61] Part of the problem was to determine what constituted this pre-clinical phase, and the risks that a pre-malignant condition might turn malignant. If a screening programme was to work, the DPCP had to be longer than the recommended interval between screening tests. A shorter DPCP meant that screening tests could fail to detect pre-cancers before they turned cancerous. Other questions focused on the available therapeutic interventions – mainly surgery, radiotherapy, chemotherapy, immunotherapy and targeted therapy. Did they justify the costs and risks of screening asymptomatic populations? Experts struggled over whether available therapeutic interventions favourably changed the natural history of the disease, for example by reducing cause-specific mortality; and whether treatments were more effective if begun at a pre-symptomatic stage than at a symptomatic stage. What about those cases where treating at an early (pre-symptomatic) stage had no advantage over treating late (symptomatic) stage? Was screening (or treatment) justified? And who was to determine advantage, risk and cost? Some commentators transcended such equations to argue that screening was worth it if it only saved one life, or because it meant that patients were more likely to see the physician.

Other questions focused on the availability of a suitable screening test, one that was accurate, acceptable to the population, fairly easy to administer, safe, and relatively inexpensive; an appropriate screening strategy for the target population (that is, the age to begin screening and a screening interval, not always an easy task to solve as the breast cancer controversies suggest); the scientific grounding of screening guidelines, ideally randomized controlled clinical trials, and whether they were economically feasible.[62] For organizers of screening programmes there were practical questions of how to ensure high rates of participation from the target populations so that the programme had a chance of success; how to match national recommendations with local circumstance; how to promote programmes that were sensitive to individual patient and provider concerns while focused on issues at a population level; how to ensure prompt follow-up of positive tests with a diagnostic examination and prompt treatment of cases; how to ensure cost-effectiveness of the programme; and how to monitor and regularly evaluate screening programmes.[63]

The evaluation of screening was (and is) technically difficult, in part because of the problem of false positives and negatives.[64] It was often easier to measure the effectiveness of efforts to identify people without the disease: individuals with positive tests tended to be subjected to further testing which meant that false-positive outcomes could be identified relatively shortly after the initial positive test results. Measuring efforts to identify people with the disease was more of a problem: although true positives, like false-positives, would be subject to further diagnostic evaluation and likely identified at an earlier stage, those with cancer but who tested negative would often be overlooked, because they generally would not be subjected to a diagnostic evaluation, which made it extremely difficult to calculate the false-negative rate generally. In practice, estimates of the ability of a test to identify those with the disease relied on long-term follow-up through cancer registries to determine which individuals were diagnosed with cancer within a fixed interval after a negative screening test, using cases diagnosed between screenings (interval cancers) as the criterion for a false-negative test result.

Other questions focused on statistical biases that were likely to confound evaluation.[65] One of the goals of a screening programme was to ensure that diagnosis occurred at an earlier stage of the disease or a precondition, and so reduce the diagnosis of advanced cancer and death from the disease. Statisticians, however, pointed out that studies that showed higher *survival* in screening-detected cancers compared with symptom-detected cancers did not provide sufficient evidence to support efforts to provide screening to the population. Survival might have been extended because the disease was identified earlier in its lifetime, and because survival was measured from the point of detection onwards. Put another way, if screening resulted in earlier detection of disease, but death occurred at the same time as it would have in the absence of screening, then there would appear to be an increase in mean survival associated with screening. But, this increase would be an artefact, and no increase had actually happened. This was (and is) referred to as *lead-time bias*.

Other possible biases also emerged from the nature and goals of screening.[66] One related to the tendency for screening to be more successful at detecting slow-growing, less aggressive types of cancer, and to be less successful at detecting more aggressive, faster-growing types: *length bias* referred (and refers) to the greater likelihood of screening-detected cancers having a longer DPCP, and hence a greater likelihood of being detected. A second bias related to the possibility that screening might result in the detection of a cancer that would not have progressed to become symptomatic in the person's lifetime (over-diagnosis). Such lesions, when detected, were indistinguishable from lesions that are, or evolve, to become clinically significant. A final bias – *selection bias* – occurred because individuals who participate in cancer screening are usually different from those who do not participate, and these differences affect disease outcomes. For example, it is claimed that compared with the population who does not undergo screening, those who do will be generally more health-conscious and healthier, more aware of the signs and symptoms of disease, have access to better health-care, and be more adherent to treatment.

The development of such efforts to evaluate programmes, and the increasingly complex statistical and methodological efforts to count various biases, as well as to distinguish between other criteria, such as between the ability of a specific test to identify people with the disease and the ability of the programme in which that test was embedded to identify them, were further complicated by concerns about the impact of screening on patients. For example, over-diagnosis and false-positives raised the possibility or probability that patients might be subjected to further unnecessary diagnostic and therapeutic interventions. The problem was that such interventions were often necessary for the system to work: screening was about populations of patients, as much as it was about detecting cancers in individuals, and false-positives and over-diagnosis were almost inevitable.[67]

It should be clear by now that since the early twentieth century cancer authorities had attempted a broad social experiment which aimed to transform people's attitudes and behaviours towards cancer by encouraging them to abandon earlier ideas and practices and to seek early detection and treatment for the disease from recognized experts, even in the absence of overt symptoms. From the start, the ACS and other cancer organizations had worried that such efforts were undermined by public ignorance and fear and by bad advice proffered by the media, family and friends, 'quacks' and ignorant physicians. Screening programmes developed after World War II in part in an effort to catch even more cancers at an early stage, but also to counter various groups and attitudes that undermined public faith in early detection and treatment programmes. Yet cancer organizations found themselves struggling at times to maintain public faith in screening. Concerns about the dangers of x-rays, over-diagnosis, unnecessary treatments, and the confusion of advice and recommendations from the many cancer organizations raised public anxieties about screening. Against such a backdrop, cancer agencies came to worry that they were failing to maintain public faith in screening programmes which threatened to undermine such programmes since they depended for their success on high rates of public participation.

Paradoxically, at the same time that cancer agencies worried about maintaining public faith and participation in screening programmes, they also worried about impractical expectations and demands for services that agencies felt were unwarranted. The problem of managing excessive demands was not new, but can be traced back to the 1910s and concerns that unrealistic hopes of a cure would lead to disappointment and so undermine faith in the ASCC's programmes of early detection and treatment and drive people to alternative treatments. Such concerns persisted into the era of mass screening, the anxiety being that screening programmes would be undermined by excessive hopes, or that people would undergo unnecessary or potentially harmful interventions.[68] For example, although the USPSTF recommended in 2009 that women under 50 years of age not undergo routine mammography screening, and that those between 50 and 75 years of age be screened less frequently, commentators lamented that by 2014 screening rates had apparently held steady or perhaps even increased, a trend blamed variously on conflicting guidelines regarding mammography, the enduring habits of physicians who continued to recommend the procedure, radiologists' preference for the status quo, fears of litigation should breast cancer be detected in the absence of screening, the mandating of screening coverage for women of all ages under the 2010 Affordable Care Act, and women's emotional response to mammography.[69] From this perspective the hopes vested in mammography led many to seek unnecessary, and sometimes dangerous, medical interventions.

The problem was heightened by growing suspicions that screening had been oversold: the method had gained widespread acceptance despite a general lack of evidence for screening efficacy.[70] While mortality from cervical cancer had fallen 75 per cent in the first three decades after the introduction of the test, the results in other cancers were not so clear, and controversies over prostate and lung cancer and backtracking over breast cancer continued to undermine confidence in the technology. Experts responded, at times, that the problem was that the public was not able to deal with the complicated sorts of decisions that screening interventions now demanded, and that emotions overwhelmed the public. This was (and is) a technology that might or might not help; it might lead to a cure of disease, or to an unnecessary surgical procedure, or it might miss a dangerous pre-cancer. And there were disagreements as to which populations might benefit from screening, and whether the price for individuals subjected to unnecessary procedures was worth it.

To complicate matters further in the United States, as the discussion of mammography illustrates, there were multiple agencies offering recommendations and advice, and at times these recommendations did not concur about what counted as orthodoxy. Some agencies proposed screening for certain populations, others were more cautious, and the worry was that in the absence of agreement the public would be confused. In such circumstances the fine distinctions that experts developed in their efforts to evaluate screening tests and programmes could easily be swept aside. Experts feared that the increasingly public and politicized nature of the debate was undermining faith in any scientific evaluation of screening tests and programmes, at a time when mistrust of science and government was said to be growing. Yet experts were not immune from promoting such pressures when their own scientific evaluations were over-ruled, and commercial interests appeared to be fuelling public demands for screening tests that some experts felt were unwarranted or, as in the case of insurance companies, dampening demand that other experts felt were warranted.[71]

## Conclusion

Issues of uncertainty have been at the heart of disputes over cancer screening. Experts often disagreed over the effectiveness of screening, its safety, who should be screened and when, and over what to do if cancer or something that might turn cancerous was discovered. Confronted with a lack of expert consensus, how were policymakers or patients to decide which experts to trust? And how were they to deal with the chronic uncertainty that dogged the procedures?

The question of what to do about growths and abnormalities that might or might not turn cancerous illustrates the point. Since the early twentieth century, pathologists had generally been able to identify the cytological signs of cancer in malignant growths. But small lumps composed of slightly unusual cells, abnormal cervical smears or polyps in the colon were more of a problem. Would they develop into cancer? How could anyone know? Should they lead to prophylactic surgery or other interventions? The general response in the USA was to act aggressively, but elsewhere and at other times responses could be very different, from conservative clinical observation to radical surgery, radiotherapy and chemotherapy. Some suggested that science would resolve such differences in the course of time. Others argued that uncertainty was inherent in any decision-making, that science would never provide an unerring guide to prophylactic therapy, that patients often did not know how to respond to such ambiguity, and that physicians were bad at explaining it.[72]

The introduction of screening for genetic markers of cancer risk did little to solve the issue.[73] Uncertainty and a lack of consensus continued to plague the procedures, since it was often unclear whether having genes associated with a particular cancer, for example, meant a person would develop cancers or what he or she should do about it. In the USA this decision is often given to the patient who has to choose whether to do nothing but live with the doubt that the cancer may occur, or to have a prophylactic operation and live with the doubt that the operation may have prevented something that would never have happened. Where earlier generations of physicians used the illusion of certainty to make a virtue of their own therapeutic decisions, now physicians use their uncertainties to make a virtue of patient choice and shared decision-making. In this new world of constant uncertainty, the conceit is that an informed patient, given all the facts, can, in collaboration with a physician, be equipped to decide whether to undergo prophylactic therapy and/or to submit to regular tests.

## Notes

1 Commission on Chronic Illness, *Chronic Illness in the United States.* Vol. I. *Prevention of Chronic Illness*, Cambridge, MA: Harvard University Press, 1957, p. 45.
2 G. Weisz, *Chronic Disease in the Twentieth Century: A History*, Baltimore: Johns Hopkins University Press, 2014, Chapter 5.
3 D. Cantor, 'Cancer Control and Prevention in the Twentieth Century', in D. Cantor (ed.), *Cancer in the Twentieth Century*, Baltimore and London: Johns Hopkins University Press, 2008, pp. 1–38.
4 B. Lerner, *The Breast Cancer Wars: Hope, Fear, and the Pursuit of a Cure in Twentieth-Century America*, New York: Oxford University Press, 2003. R. A. Aronowitz, *Unnatural History: Breast Cancer and American Society*, Cambridge and New York: Cambridge University Press, 2007. I. Löwy, *Preventive Strikes: Women, Precancer, and Prophylactic Surgery*, Baltimore: Johns Hopkins University Press, 2010.
5 Aronowitz, *Unnatural History.*
6 Cantor, 'Cancer Control and Prevention'.
7 Ibid.
8 Löwy, *Preventive Strikes*, p. 118.
9 D. Cantor, 'Radium and the Origins of the National Cancer Institute', in C. Hannaway (ed.), *Biomedicine in the Twentieth Century: Practices, Policies, and Politics*, Amsterdam: IOS Press, 2008, pp. 95–146.
10 A. V. Deibert, 'Control Features of the Cancer Problem', *Journal of the National Medical Association*, 1950, vol. 42, 279–86, at 279.
11 Peter C. Albertsen, 'Screening for Prostate Cancer', in V. T. DeVita, Jr., T. S. Lawrence, and S. A. Rosenberg (eds), *DeVita, Hellman, and Rosenberg's Cancer: Principles & Practice of*

*Oncology*, 9th edition, Philadelphia: Wolters Kluwer Health/Lippincott Williams & Wilkins, 2011, pp. 617–24.

12  Cantor, 'Cancer Control and Prevention'.

13  Cantor, Ibid. L. J. Reagan, 'Projecting Breast Cancer: Self-Examination Films and the Making of a New Cultural Practice', in L. J. Reagan, N. Tomes, and P. A. Treichler (eds), *Medicine's Moving Pictures: Medicine, Health, and Bodies in American Film and Television*, Rochester, NY: University of Rochester Press, 2007, pp.163–95.

14  Cantor, Ibid.

15  Proceedings: First Conference on Cancer Teaching, Billings Memorial Hospital, Chicago, Oct 8–19 1948, pp. 143–4, NCI Lion Database: AR-4810-002652. On the Lion database see D. Cantor, 'Finding Historical Records at the National Institutes of Health', *Social History of Medicine*, 2015, vol. 28, 617–37, pp. 627–8.

16  E. Vayena, 'Cancer Detectors: An International History of the Pap test and Cervical Cancer Screening, 1928–1970', University of Minnesota PhD, 1999; M. J. Casper and A. E. Clarke, 'Making the Pap Smear into the "Right Tool" for the Job: Cervical Cancer Screening in the USA, circa 1940–95', *Social Studies of Science*, 1998, vol. 28, 255–90; A. E. Clarke and M. J. Casper, 'From Simple Technology to Complex Arena: Classification of Pap Smears, 1917–90', *Medical Anthropology Quarterly*, 1996, vol. 10, 601–23; M. Murphy, *Seizing The Means of Reproduction: Entanglements of Feminism, Health, and Technoscience*, Durham, NC: Duke University Press, 2012, pp. 102–49; I. Löwy, *Preventive Strikes*; I. Löwy, *A Woman's Disease: The History of Cervical Cancer*, Oxford and New York: Oxford University Press, 2011; K. E. Gardner, *Early Detection: Women, Cancer, and Awareness Campaigns in the Twentieth-Century United States*, Chapel Hill: University of North Carolina Press, 2006.

17  Vayena, Ibid.; Casper and Clarke, Ibid.; Clarke and Casper, Ibid.

18  Vayena, Ibid.; Casper and Clarke, Ibid.; Clarke and Casper, Ibid.

19  Vayena, Ibid.; Casper and Clarke, Ibid.; Clarke and Casper, Ibid.

20  Vayena, Ibid. Casper and Clarke, Ibid.; Clarke and Casper, Ibid.

21  Vayena, Ibid.; Casper and Clarke, Ibid.; Clarke and Casper, Ibid.; Löwy, *Preventive Strikes*, p. 130;. Löwy, *A Woman's Disease*, pp. 79–81, 95–8.

22  Vayena, Ibid.; Casper and Clarke, Ibid.; Clarke and Casper, Ibid.; Löwy, *Preventive Strikes*, p. 130. Löwy, *A Woman's Disease*, pp. 79–81, 95–8. M. B. Daly and J. S. Rader, 'Screening for Gynecologic Cancers', in *DeVita, Hellman, and Rosenberg's Cancer*, 9th edition, pp. 603–9.

23  Daly and Rader, 'Screening for Gynecologic Cancers'.

24  B. H. Lerner, '"To See Today with the Eyes of Tomorrow": A History of Screening Mammography', *Canadian Bulletin of Medical History*, 2003, vol. 20, 299–321; Lerner, *Breast Cancer Wars*, pp. 196–222, 242–50; Aronowitz, *Unnatural History*, Chapters 9 and 10. Gardner, *Early Detection*, pp. 179–86; J. Wells, 'Mammography and the Politics of Randomised Controlled Trials', *British Medical Journal*, 31 October 1998, vol. 317, 1224–9; V. L. Ernster, 'Mammography Screening for Women Aged 40 through 49: A Guidelines Saga and a Clarion Call for Informed Decision Making', *American Journal of Public Health*, 1997, vol. 87, 1103–6; K. B. Goldberg, 'Clarity Was the First Casualty in 30-Year War over Mammography for Younger Women', *Cancer Letter*, 20 November 2009, vol. 35, no. 43, 1 and 8–11.

25  Lerner, 'To See Today with the Eyes of Tomorrow'; Aronowitz, *Unnatural History*, pp. 223–32.

26  S. Shapiro, P. Strax, and L. Venet, 'Periodic Breast Cancer Screening in Reducing Mortality from Breast Cancer', *Journal of the American Medical Association*, 1971, vol. 215, 1777–85; S. Shapiro, W, Venet, P. Strax, and L. Venet, *Periodic Screening for Breast Cancer: The Health Insurance Plan Project and its Sequelae, 1963–1986*, Baltimore and London: Johns Hopkins University Press, 1988; Aronowitz, *Unnatural History*, pp. 232–4.

27  J. C. Bailar III, 'Mammography: A Contrary View', *Annals of Internal Medicine*, 1976, vol. 84, 77–84. Aronowitz, *Unnatural History*, pp. 235–42.

28  This account of events is based on *Background Statement: NIH/NCI Consensus Development Meeting On Breast Cancer Screening, September 14–16, 1977*, NIH/NCI Consensus Development Meeting Packet, 1977, NCI Lion Database: PB025423. Aronowitz, *Unnatural History*, pp. 242–51.

29  President's Cancer Panel/NIH, *Breast Cancer Detection and Mammography (Transcript of Proceedings, President's Cancer Panel held July 19, 1976, in Bethesda, Maryland)*, NCI Lion Database: PB030564.

30 *Breast Cancer Screening.* NIH Consensus Statement Online 1977 September 14–16; 1(1): 5–8, http://consensus.nih.gov/1977/1977BreastCancer001html.htm, accessed 15 August 2015. For verbatim reports of this meeting see National Institutes of Health/National Cancer Institute, *Consensus Development Meeting on Breast Cancer Screening September 14–16, 1977,* Breast Cancer Screening, NIH Consensus Development Conference Meeting Packet, DHEW Pub No (NIH) 78-1257, September 14–16, 1977, NCI Lion Database: PB025420.

31 A. M. Leitch, 'Breast Cancer Screening: Success amid Conflict', *Surgical Oncology Clinics of North America*, 1999, vol. 8, 657–72.

32 'NCI Drops Breast Screening Guidelines, Issues "Summary of Scientific Fact"', *Cancer Letter*, 10 December 1993, vol. 19, no. 48, 1–4. S. W. Fletcher, W. Black, R. Harris, B. K. Rimer, S. Shapiro, 'Report of the International Workshop on Screening for Breast Cancer', *Journal of the National Cancer Institute*, 1993, vol. 85, 1644–53.

33 These studies were presented to the 1997 Consensus Development Conference; see Office of the Director, National Institutes of Health, *NIH Consensus Development Conference. Breast Cancer Screening for Women Ages 40–49. Program and Abstracts*, Bethesda: National Institutes of Health 1997, pp. 55–69. 'NCI Plans Conference to Reconsider Breast Cancer Screening in Women 40–49', *Cancer Letter*, 19 April 1996, vol. 22, no. 16, 1–3. On the 1997 controversy see Wells, 'Mammography'. Ernster, 'Mammography'.

34 'Mammography Screening for Ages 40–49 Not Supported by Data, NIH Panel Says', *Cancer Letter*, 31 January 1997, vol. 23, no. 4, 1–5.

35 '"Tone" of Panel's Statement Lacks Balance, Klausner Says', *Cancer Letter*, 31 January 1997, vol. 23, no. 4, 6–8. For a typescript of Klausner's response see 'Press Conference: Consensus Development Conference on Breast Cancer Screening for Women Ages 40–49', 23 January 1997. (Response to a Question on His View of the Consensus Panel's Conclusion by Richard Klausner, the NCI Director), NCI Lion Database: DC-9701-012352.

36 'NIH Statement Doesn't Resolve Mammography Controversy,' *Cancer Letter*, 31 January 1997, vol. 23, no. 4, 8–11, p. 8.

37 G. Kolata, 'Stand on Mammograms Greeted by Outrage', *New York Times*, 28 January 1997, p. A20. M. Cimons, 'Panel Offers Mixed Mammogram Message', *Los Angeles Times*, 24 January 1997, A:1. R. Boyd, 'New Mammogram Report Leaves Women Adrift', *Chicago Tribune*, 24 January 1997, 1. S. Okie, 'Mammograms for 40-somethings?', *Washington Post*, 18 February 1997, sect Z:12. Senate resolution 47: relative to accurate guidelines for breast cancer screening. *Congressional Record*, 4 February 1997, 143:S970.

38 Ernster, 'Mammography'.

39 101st National Cancer Advisory Board Meeting Summary Held February 25–26 1997, pp. 8–13, NCI Lion Database: DC-9702-012938.

40 Arlen Specter letter to Donna Shalala, c.30 January 1997, NIH, Office of the Director (OD), records provided by the OD Records Officer.

41 Ernster, 'Mammography'. For other Congressional criticism see the article by Senator Olympia Snowe in the *Washington Post* which argued that scientific evidence unequivocally supported routine screening for younger women O. Snowe, 'Mammograms Save Lives,' *Washington Post*, 11 February 1997, A:21.

42 For a sample of such reports collected by Bill Hall (Director of Communications, OMAR) and sent to the Panel, 24 February 1997 see records provided by NIH, Office of the Director, Director of Disease Prevention, Office of Medical Applications of Research (OMAR).

43 'Mammography Screening for Ages 40–49 Not Supported by Data', *Cancer Letter*. 'NIH Statement Doesn't Resolve Mammography Controversy', *Cancer Letter*.

44 Ernster, 'Mammography'. For the NCI's immediate response to the ACS report see NCI Press Office Statement, 'NCI Responds to ACS Announcement', 23 March 1997, NCI Lion Database, (no accession number).

45 NCI Press Office, 'Statement from the National Cancer Institute on the National Cancer Advisory Board Recommendations on Mammography', 27 March 1997, NCI Lion Database: DC012399.

46 The National Cancer Institute and the American Cancer Society, 'Joint Statement on Breast Cancer Screening for Women in Their 40s', 27 March 1997, NCI Lion Database: DC012401.

47 The White House, Office of the Press Secretary, 'Remarks by the President in Mammogram Announcement', 27 March 1997, 12.17 PM EST, NCI Lion Database: DC-9703-012419.

48 A final (although not unanimous) statement was also issued by the consensus panel in July. This reinforced its support for shared decision-making rather than universal screening and supported reimbursement for screening for women in their forties who chose to have mammography. This statement went largely unnoticed. National Institutes of Health Consensus Development Panel, 'National Institutes of Health Consensus Development Conference Statement: Breast Cancer Screening for Women Ages 40–49, January 21–23, 1997', *Journal of the National Cancer Institute*, 1997, vol. 89, 1015–20.

49 American Academy of Family Physicians, *Summary of Policy Recommendations for Periodic Health Examinations*, Rev 5.4, Leawood, KS: American Academy of Family Physicians, 2003. P. C. Gøtzsche and M. Nielsen, 'Screening for Breast Cancer with Mammography', *Cochrane Database of Systematic Reviews*, 2006, Issue 4, Art. No.: CD001877.

50 A. Qaseem, V. Snow, K. Sherif, M. Aronson, K. B. Weiss and D. K. Owens, 'Screening Mammography for Women 40 to 49 Years of Age: A Clinical Practice Guideline from the American College of Physicians', *Annals of Internal Medicine*, 2007, vol. 146, 511–15.

51 P. Zahl, J. Mæhlen, H. Welch, 'The Natural History of Invasive Breast Cancers Detected by Screening Mammography', *Archives of Internal Medicine*, 2008, vol. 168, 2311–16.

52 L. Esserman, Y. Shieh and I. Thompson, 'Rethinking Screening for Breast Cancer and Prostate Cancer', *JAMA*, 21 October 2009, vol. 302, 1685–92.

53 US Preventive Services Task Force, 'Screening for Breast Cancer: US Preventive Services Task Force Recommendation Statement', *Annals of Internal Medicine*, 2009, vol. 151, 716–26. H. D. Nelson, K. Tyne, A. Naik, C. Bougatos, B. Chan, P. Nygren and L. Humphrey, *Screening for Breast Cancer: Systematic Evidence Review Update for the U.S. Preventive Services Task Force*, Agency for Healthcare Research and Quality, AHRQ Publication No 10-05142-EF-1, Evidence Review Update No 74, November 2009.

54 P. Goldberg, 'HHS Secretary Rebukes Task Force Guidelines on Breast Cancer Screening', *Cancer Letter*, 20 November 2009, vol. 35, no. 43, 1–5. P. Goldberg, 'Task Force Alters Web Site to Clarify Recommendation on Mammography', *Cancer Letter*, 4 December 2009, vol. 35. No. 44, 1–4. P. Goldberg, 'Task Force Brings *Mea Culpa* to Congressional Hearing', *Cancer Letter*, 4 December 2009, vol. 35. No. 44, 4–6.

55 For background to the 2009 controversy see The Editors, 'When Evidence Collides with Anecdote, Politics, and Emotion: Breast Cancer Screening,' *Annals of Internal Medicine*, 2010, vol. 152, 531–2. F. M. Visco, 'Translation and Communication Needs for Care in the Face of Uncertain Evidence', in L. A. Olsen, R. S. Saunders and J. M. McGinnis, (eds), *Patients Charting the Course: Citizen Engagement and the Learning Health System: Workshop Summary*, Washington, DC: National Academies Press, 2011, pp. 177–86. 'Reactions: Outrage from Believers Drowns Out Skeptics', *Cancer Letter*, 20 November 2009, vol. 35, no. 43, 5–8.

56 For a survey of the current state of cancer screening, see O. W. Brawley and H. L. Parnes, 'Cancer Screening', in V. T. DeVita, Jr., T. S. Lawrence and S. A. Rosenberg (eds), *DeVita, Hellman, and Rosenberg's Cancer: Principles & Practice of Oncology*, 10th edition, Philadelphia: Wolters Kluwer, 2015, pp. 370–88.

57 Ibid., p.385.

58 I. Löwy, *Preventive Strikes*, pp. 159–62. T. R. Church and J. S. Mandel, 'Screening for Gastrointestinal Cancers', in *DeVita, Hellman, and Rosenberg's Cancer*, 9th edition, pp. 596–602.

59 On lung cancer see Ibid., pp. 155–7. C. Timmermann, *A History of Lung Cancer. The Recalcitrant Disease*, Houndsmills, Basingstoke: Palgrave MacMillan, 2014, pp. 112–13, 126–32 and 167–8. D. E. Midthun and J. R. Jett, 'Screening for Lung Cancer', in *DeVita, Hellman, and Rosenberg's Cancer*, 9th edition, pp. 625–30.

60 I. Löwy, ibid., pp. 157–9. Albertsen, 'Screening for Prostate Cancer'.

61 J. S. Mandel and R. Smith, 'Principles of Cancer Screening', in *DeVita, Hellman, and Rosenberg's Cancer*, 9th edition, pp. 582–6, p. 582.

62 J. S. Mandel and R. Smith, 'Principles of Cancer Screening'. J. M. Croswell, D. F. Ransohoff and B. S. Kramer, 'Principles of Cancer Screening: Lessons from History and Study Design Issues', *Seminars in Oncology*, 2010, vol. 37, 202–15.

63 Croswell, Ransohoff and Kramer, ibid.

64  Ibid.; Mandel and Smith, 'Principles of Cancer Screening'.
65  Croswell, Ransohoff, and Kramer, 'Principles of Cancer Screening'.
66  Ibid.
67  Ibid. and Mandel and Smith, 'Principles of Cancer Screening'.
68  D. Cantor, 'Cancer Control and Prevention in the Twentieth Century'.
69  L. Rosenbaum, 'Invisible Risks, Emotional Choices: Mammography and Medical Decision Making', *New England Journal of Medicine*, 2014, vol. 371, 1549–52.
70  D. F. Ransohoff, 'Have We Oversold Colonoscopy?' *Gastroenterology*, 2005, 129, 1815. 'Save Your Life: Cancer Screening is Oversold. Know the Tests to Get – And Those to Skip', *Consumer Reports*, March 2013, vol. 78, No. 3, 28–33. See also M. J. Yaffe and K. I. Pritchard, 'Overdiagnosing Overdiagnosis', *The Oncologist* 2014, vol. 19, 103–6. D. B. Kopans, 'Arguments Against Mammography Screening Continue to be Based on Faulty Science', *The Oncologist*, 2014, vol. 19, 107–12. A. Bleyer, 'Were Our Estimates of Overdiagnosis With Mammography Screening in the United States "Based on Faulty Science"?' *The Oncologist*, 2014, vol. 19, 113–26.
71  Visco, 'Translation and Communication'.
72  Löwy, *Preventive Strikes*.
73  Ibid.

## Select bibliography

Aronowitz, R. A., *Unnatural History: Breast Cancer and American Society*, Cambridge and New York: Cambridge University Press, 2007

Cantor, D. (ed.), *Cancer in the Twentieth Century*, Baltimore and London: Johns Hopkins University Press, 2008.

Casper, M. J. and Clarke, A. E., 'Making the Pap Smear into the "Right Tool" for the Job: Cervical Cancer Screening in the USA, Circa 1940–95', *Social Studies of Science*, 1998, vol. 28, 255–90.

Clarke, A. E. and Casper, M. J., 'From Simple Technology to Complex Arena: Classification of Pap Smears, 1917–90', *Medical Anthropology Quarterly*, 1996, vol. 10, 601–23.

Gardner, K. E., *Early Detection. Women, Cancer, and Awareness Campaigns in the Twentieth-Century United States*, Chapel Hill: University of North Carolina Press, 2006.

Lerner, B. H., *The Breast Cancer Wars: Fear, Hope and the Pursuit of a Cure in Twentieth-Century America*, New York: Oxford University Press, 2003.

Löwy, I., *Preventive Strikes: Women, Precancer, and Prophylactic Surgery*, Baltimore: Johns Hopkins University Press, 2009.

Löwy, I., *A Woman's Disease: The History of Cervical Cancer*, Oxford: Oxford University Press, 2011.

Murphy, M., *Seizing The Means of Reproduction: Entanglements of Feminism, Health, and Technoscience*, Durham, NC: Duke University Press, 2012.

Patterson, J. T. *The Dread Disease: Cancer and Modern American Culture*, Cambridge, MA: Harvard University Press, 1987.

Vayena, E., 'Cancer Detectors: An International History of the Pap Test and Cervical Cancer Screening, 1928–1970', University of Minnesota PhD, 1999.

Wailoo, K., Livingston, J., Epstein, S., and Aronowitz, R. (eds), *Three Shots at Prevention: The HPV Vaccine and the Politics of Medicine's Simple Solutions*, Baltimore: Johns Hopkins University Press, 2010.

# 21

# MEDICAL BACTERIOLOGY

## Microbes and disease, 1870–2000

*Christoph Gradmann*

It is considered naïve historiography to judge a story by its ending. Any attempt, by contrast, to write the history of modern medical bacteriology, suffers from a reverse problem in which the outset of a story has come to dominate its further development. We know a lot about the golden days of medical bacteriology between 1860 and 1900,[1] but we struggle when it comes to later developments. Attempting to discuss the history of medical thinking on infectious disease in relation to the development of medical microbiology, this chapter first sketches out the main lines in the history of that field, and then places more accent on developments after 1900.

If we accept the notion that it was in the decades before 1900 that medical microbiology came into being as a corpus of technologies and methods to study and control infectious disease, it will help to frame our story if we contrast important features of this emergent science with what has become familiar around 2000.[2] Three important changes are evident. First, there have been drastic changes in the object of investigation. Nineteenth-century epidemiology was dominated by infectious disease. The emergent discipline investigated a set of high-prevalence, high-mortality, common bacterial infections, such as typhoid, cholera and tuberculosis.[3] Its prestige rested on the promise of their control. Much of that has changed as a result of what is commonly known as the epidemiologic transition.[4] It has long been claimed that the declining prevalence of infectious disease was brought about by medical bacteriology. Yet, this claim has been debated ever since Thomas McKeown's *The Role of Medicine: Dream, Mirage or Nemesis* was published in 1979.[5] The outcome of the epidemiologic transition was also far from what had been projected in several ways. Chronic conditions and cancers came to dominate epidemiology and public health.[6] The declining prevalence of common infections is a phenomenon that – globally speaking – remained localized to some high-income countries.[7] Finally, there are now novel infectious diseases that have resulted from the interaction of their biology with technologies invented for their control. Antibiotics have bred antibiotic resistance and modern hospitals have bred health-care-related infections.[8]

The second feature relates to developments in understanding the biology of infections. At first sight it appears that there has been a lot of change. Around 1900 attention almost exclusively focused on bacteria while the study of other microbial pathogens such as unicellular parasites or viruses was only nascent. During the twentieth century, however, new classes of infectious agents from viruses to prions were defined, often by employing novel investigative technology: viruses appeared to be linked to electron microscopy and the ultracentrifuge; sub-dividing bacteria into strains relied on typing with phages or serology; and prions could hardly be imagined without molecular biotechnology.[9] Yet, the development of investigative technology can easily divert attention from the fact that concepts of pathogenesis seem to have changed more in detail than in substance. There are microbiological research networks that cover almost the whole century.[10] In 1992 Andrew Cunningham argued that late-nineteenth-century medical bacteriology created a worldview in which the pathogenic microbe has ruled ever since.[11] With the notion of the pathogenic, invading microbe, Robert Koch, Louis Pasteur and others provided an ontological and highly graphic framing of infectious disease that has – despite many malcontents – proved to be of remarkable tenacity. The continued popularity of Koch's postulates, a set of criteria serving to identify microbes as pathogens, points to the tenacity of traditions in shaping the field. Given the tendency to refer to methodologies that date back to the heroic days of medical bacteriology, we may assume that the notion of the invading microbe is a tradition still alive in medical microbiology.[12]

Thirdly, historians need to reflect on what can be called the therapeutic promise that medical microbiology has inspired over our period of study, but which it has only occasionally lived up to. One of the core disciplines of the late-nineteenth-century laboratory revolution in medicine, the prestige of microbiology rested on its ability to bring laboratory knowledge to work in public health and at the bedside. Pathogens became popular because their identification seemed to entail a promise of cure. Antibacterial therapy was thus anticipated in the medical bacteriology of the heroic days, yet by and large failed to materialize until the 1930s. Likewise, the dynamic development of anti-infectious therapies that we associate with sulphas and antibiotics from the mid-1930s did not last very long: the respective industries have been in a crisis of innovation since at least the 1980s. The expectation that identifying a pathogen opens an alley of control was, however, not eliminated by such a crisis. Instead it has taken a gloomy, dystopian form that we are all used to: that of an empty pipeline in a post-antibiotic age where the riverbed is still there but the source has run dry.

With these three considerations in mind we can structure the history of medical microbiology since the mid-nineteenth century into four partially overlapping periods. An early phase, in which cognitive and technological foundations were laid, rapid institutional growth occurred and the prestige of the field was created. Secondly, a phase from World War One to the mid-1930s, in which this corpus of knowledge and practices was challenged and expanded at the same time. While certain limitations of medical bacteriology where commonly addressed in the inter-war years, this was rarely the case in the decades that followed. The arrival of anti-infective therapies shortly before and during the Second World War indicates the outset of the third phase. Such therapies seemed a fulfilment of earlier promises and marked the beginning of several waves of innovative drug research that continued into the 1970s. Understanding the consequences of an on-going therapeutic revolution in infection medicine became important for the development of medical bacteriology in the period. During the fourth phase, from the 1980s, the eclipse of the therapeutic revolution was making itself felt.[13] This coincided with developments in the basic biological sciences and a transformed epidemiological landscape, in which the management of chronic infection and adapting to novel pathologies seems more important than dealing with the control of an established set of common infectious diseases.

## Making microbiology

The late nineteenth century did not invent the microbiology of infectious disease. Yet, it was from about 1860 that botanists and physicians developed a broadly shared conviction that in both reality and imagination transformed medicine and public health. Spontaneous generation of microbial life had been disputed (and remained so) after Louis Pasteur proved its non-existence in 1860–61.[14] Robert Koch was not the first to cultivate microbes or employ animal experiments to prove the transmissible character of anthrax or tuberculosis around 1880. Neither was Joseph Lister the first who attempted to sanitize surgical theatres when he brought carbolic acid to those places.[15] However, it was in the final third of the nineteenth century that such observations and practices began to transform medicine: Pasteur's disproof of spontaneous generation was connected to his ability to control fermentative processes in the production of beer or vaccines; Koch's pathogenic bacteria became influential when, after being sanctioned by the German Imperial Health Office,[16] they became the dominant scientific basis of German public health; Lister's antisepsis gained importance by its connection to the expansion of surgery that characterized the medical world of the late nineteenth century.[17]

The emergence of medical microbiology has often been attributed to the rise of a germ theory of infectious disease.[18] However, while such a phenomenon is still discernible as a whole, it has turned out to be elusive in detail. Michael Worboys has demonstrated that there were multiple germ theories which often mixed easily with traditional thought.[19] To name the most obvious example, Pasteur's rabies vaccine of 1885 did not sweep away an ocean of presumably ineffective treatments. Instead Pasteur's understanding of his vaccine competed and mingled with existing knowledge. Even in the arguably more monolithic world of Imperial Germany bacteriology's prestige rested on technological arrangements, such as microphotography that implied rather than demonstrated infectious causality.[20] Alliances and compromises with the older hygienic movement also needed to be made.[21] Koch's famous demonstration of the tubercle bacillus on 24 March 1882 not only stood out as work on the presumably epidemiologically dominant infectious disease of the age, but also presented a menagerie of laboratory technologies, from Petri dishes to guinea pigs, that came to be influential in medicine and public health in the decades to come.[22] Something comparable can be said of Pasteur's rabies vaccine. It was not the first treatment for rabies but, as Gerald Geison has shown, it advertised in grandiose fashion the skills of Pasteur's laboratory, namely the production of vaccines from attenuated microbial cultures.[23]

Beyond laboratory science it was industrial technology that added scale to practices employed in laboratories and clinics and made them pervasive. Koch's pathogenic bacteria thus combined swiftly with the timely medium of photography that lent both objectivity and popularity to them.[24] Industrially mass-produced microscopes facilitated the proliferation of work that could be done with them.[25] Pasteur's vaccines became the pillars of a worldwide network of institutes that usually started from their production.[26] The diphtheria antiserum, first marketed in 1892, became a success because it translated laboratory phenomena – immunological reactions including toxins and antitoxins – into a standardized medicine, produced on industrial scale.[27]

However, like other technologies that entered clinical practice in diagnosis and treatment – think of the thermometer – the blessings of medical microbiology were not immediately obvious to contemporaries and instead needed to be promoted, often in direct opposition to existing skills. A microbiological diagnosis of tuberculosis, claimed to be simple and secure by Robert Koch in 1883,[28] was not widely practised before the twentieth century.[29] As an antidote to older Whiggish accounts, scepticism against such 'incommunicable knowledge' has long received attention in the historiography,[30] yet more recent work on British, French and German clinical medicine has also provided examples where such tools were embraced rather quickly – such as when *fin-de-siècle* clinicians in Paris were neither antimodernists nor in fact Pasteurized, but happily embraced German-style medical bacteriology in their wards.[31] Public imagination, at any rate, seems to have responded swiftly and enthusiastically to bacteriology. Bacteria and the horrors they seemed to embody have been part of our imagination ever since.[32]

From the 1880s the rising prestige of medical microbiology facilitated the foundation of a discipline with all the regalia that go with it: textbooks, journals and most visibly the foundation of prestigious institutes like the Institut Pasteur in Paris in 1887 and the Robert Koch Institute in Berlin in 1891.[33] Others like the School of Hygiene and Public Health at Johns Hopkins University from 1916 or the German Imperial Health Office from 1876 symbolize the standing that bacteriological hygiene enjoyed in public health. The dominance of bacteriological hygiene in public health, however, was a phenomenon that often mingled with rather than pushed aside older sanitarian traditions in hygiene. Moreover, results differed enormously from country to country, being rather sweeping in France and Germany and more uneven in Great Britain.[34]

Medical historians have struggled to come to a conclusion as to what such changes amounted to in their entirety. While in older accounts they would have been heralded as a revolutionary breakthrough powered by the germ theory of infectious disease,[35] more recently this bacteriological revolution seems to have evaporated into a phenomenon that had 'no transcendent reality, but rather had many meanings'.[36] Few authors would deny that fundamental change happened in the late nineteenth century. Yet change took a long time, was far from complete at any point in time, and results differed with locality; thus the term 'revolution' seems inappropriate.[37] The most interesting feature of this revision was less the demise of a heroic historiography, and more the fact that a phenomenon that used to be seen in terms of dissemination is now framed in terms of appropriation, the result of which was shaped by clinicians, public health officers, engineers or lay people rather than heroic laboratory experts.

Although change was uneven and slow, it was fundamental nonetheless. As Andrew Mendelsohn has argued, in the late nineteenth century infectious diseases could be understood as microbial invasion of human bodies.[38] Notions of pathogenesis – circling around invasion and consumption – bore a striking analogy to the process of cultivating bacterial species. While originally being identified as causes, microorganisms could thus easily be conceived as an embodiment of infectious disease as such. Interestingly, this is a point made by older intellectual history: namely that medical bacteriology was a re-affirmation of an ontological conception of infectious disease – embodied in the notion of a pathogenic microbe.[39] As Andrew Cunningham has reminded us, such thinking has proven to be pervasive and persistent. Explaining infection without referring to the involvement of microorganisms became almost unimaginable: 'To oppose the claims of bacteriology is now not a rival view, nor an alternative view, nor even a dissident view. It is now a lunatic view.'[40] However, it is important to recognise that the notion of infectious causality based on microbiology is something that cannot be understood from the angle of its intellectual history alone.[41] Instead, the concept of pathogenesis arrived in conjunction with investigative technology, a technomorphic image. In Koch's writings, for example, it is hard to find arguments that go beyond discussion of proper investigative technologies for studying disease. Yet, what counted as the traits of a microbe in such studies implies, rather than argues, far-reaching ideas about the nature of infectious disease. As Georges Canguilhem has suggested, medical bacteriology was more a technological than an intellectual innovation and, following him, we have every reason to assume that the connections between industrial culture and one of its most prominent pieces of medical science were of an essential nature.[42]

What followed the golden age of the 1880s was a period of transformation and growth. The first step in that direction was the internationalization of national, and occasionally even nationalist, disciplines from the 1890s. Differences between Koch's and Pasteur's schools had initially been marked: the phenomenon upon which Parisian vaccines were based – attenuation – would be regarded as a laboratory artefact in Berlin in the early 1880s.[43] Gaining insight into what happened on either side of the Rhine was no easy matter either. In 1886 Alexandre Yersin, a faithful Pasteurian with the added advantage of being fluent in German, enrolled into Koch's training courses in 1886 to deliver under-cover information about Koch's bacteriology.[44] Just a few years later Roux and Behring, working on the diphtheria antitoxin, enjoyed rather cordial relations and Koch silently buried his critique of attenuation.[45] What had started in a few sometimes rather private laboratories became public with the existence of textbooks, journals and large-scale research and training institutions. That being said, homogeneity was never achieved. Instead expansion and proliferation became driving forces in transformations.

In the years before World War One, three lines of development were of particular significance. Firstly, medical microbiology became a technological and ideological resource of colonialism. Tropical hygiene would serve as a technology in the colonization of lands, while at the same time providing even the most brutal campaigns such as the sleeping sickness camps in the Belgian Congo with the gloss of civilization.[46] Secondly, it was an expansion into immunology that resulted in a host of therapeutic and diagnostic applications.[47] The description of toxin–antitoxin interactions resulted in what arguably became the first therapeutic intervention that late-nineteenth-century medical microbiology could claim as its own, the diphtheria antiserum that was developed in Germany and France in parallel and became applied from 1892. This and other therapeutic vaccines became the starting point of the industrialization of biological medicines that followed.[48] Serological tests became notoriously hard to calibrate but highly graphical applications of the technological image of infectious processes that the laboratory had created. Its most famous example is the Wassermann test for syphilis. Applied from 1905 it became popular in clinical medicine and eventually also in philosophy of science.[49] Finally, medical bacteriology became more engaged in epidemiology in the decades before 1914.[50] Having initially pictured epidemics as a simple chain of infections and having obliterated much of the older epidemiology of places and spaces, observations of subclinical or symptomless infection and the notion of healthy disease carriers, spelled out in 1902, turned out to be a powerful albeit problematic tool in epidemiology and public health. Where the spreading of epidemics relied on infected, but healthy, individuals, hunting for microbes became supplemented by hunting for their carriers.[51] The sad fate of the New York cook Mary Mallon (or Typhoid Mary), who was diagnosed as a healthy carrier in 1907 and eventually imprisoned because of her refusal to comply with public health authorities' demands, serves as an example of how the older conflict of individual and public health took a new form in the bacteriological age.[52]

## Laboratories of complexity

The charms of medical bacteriology as it had developed from the 1870s lay in simplicity and utility. However, both of these features came to be contested during World War One and the inter-war years. As a result, both the biology and the epidemiology of infectious disease became more complex in these years. Likewise the therapeutic interventions that seemed to flow from medical bacteriology received some scrutiny. The First World War acted as a catalyst in developments, bringing about large-scale application of much that had been invented before.[53] For instance, the tetanus antitoxin that would be considered a fairly experimental prophylaxis indicated in a limited range of situations became commonly used during the war.[54] For many the war was indeed the first hygienic war, in which casualties resulting from battle for the first time superseded those from war epidemics, even if some of the hygienic measures, especially in the Eastern war theatre – subjecting Jews as supposed carriers of war diseases to humiliating hygienic measures – look like prototypes of racially argued persecution and ultimately genocide in retrospect.[55]

The biggest challenges to bacteriological hygiene, however, originated not so much in the strains on civilian public health in many European countries during the war, but from the influenza pandemic of 1918–19.[56] It is not possible to give precise numbers of how many were affected, but we can assume that globally the number was around half a billion and that at least 30 million died, 2 million of whom in Europe.[57] Yet, it was not so much the sheer number of deaths, which easily superseded those of the war itself, nor the unusually high death rate of 3 per cent in the second wave of the epidemic in autumn 1918, but the futility of all attempts at control that threw a shadow over bacteriological hygiene. 'I know of no public health measures which can resist the progress of pandemic influenza',[58] wrote Britain's Chief Medical Officer Arthur Newsholme in November 1918. Perplexity beset medical scientists: in the middle of a visible recession of common infectious diseases there arrived an epidemic that was completely beyond control. In fact, influenza had been one of those diseases for which a bacterial pathogen, *Haemophilus influenzae*, had been identified.[59] Such knowledge, however, was swept away by observations in wartime hospitals when it became clear that the bacterium was linked to a common complication of influenza, pneumonia, rather than the disease itself. It seemed now that the pathogen was something much smaller than a bacterium, a filterable virus as it would have been called in those days, being defined by its capacity to pass through bacteria-tight, clay filters. Such microorganisms had been observed before but the influenza pandemic transformed them from biological curiosities into important objects of medical research.[60] Medical microbiology, it was dawning on many, had focused too much on bacteria and had had a too static concept of its objects: swift change on the side of the pathogen, as had seemingly occurred during successive waves of the influenza pandemics, had not been highlighted in medical bacteriology prior to the war.[61]

The 1920s became a decade of malcontent with classical bacteriology. Some discontent was of a more scientific nature, but it is easiest to start with institutional changes. German medical bacteriology, which had grown to dominance in an alliance with state-level public health institutions and military medicine, faltered. This was partly because it was excluded from international collaborations, as in the case of serology where the functions of Paul Ehrlich's Institute in Frankfurt were taken over by Denmark's State Serum Institute, and partly because of the economic collapse of post-war Germany, which could not sustain the growth of institutions like the Imperial Health Office, the Robert Koch Institute, or the Academy of Military Medicine, the Kaiser-Wilhelm-Akademie.[62] While such places now became stagnant, competing disciplines like social hygiene or, more relevant for us, other branches of microbiology that had enjoyed less support in the pre-war years, flourished. While other institutions like the Institut Pasteur and its network saw less change as an outcome of war they would still be affected by the arrival of new actors: the Rockefeller Foundation came to play a dominant role in inter-war public health and a formerly provincial institution like the Danish State Serum Institute came to replace Germany's Paul Ehrlich Institute as a hub of international serology.[63] British medical microbiology that had so far, with the exception of tropical medicine, not been in the forefront, began to expand through the National Institute for Medical Research and the Medical Research Council (MRC).[64]

A corresponding description of scientific developments can be made. Earlier on, epidemiological models and hygienic practices, such as the chain of bacteriological laboratories guarding the German empire's Eastern borders, had been based on the notion that a hygienic space had to be protected against microbial intruders.[65] As Silvia Berger has shown, despite its earlier successes this system collapsed at the end of the war.[66] It required, for instance, a whole network of delousing stations to protect Germany from importing typhus from a supposedly infested East. The winter of 1918–19, when millions of German soldiers were returning home, obliterated that system. Under dominant hygienic thinking, war epidemics were now bound to hit the defeated empire. Yet, this did not happen: 'What was most impressive was the failure of after-war epidemics to materialize.'[67] It was this non-event in conjunction with the unexpected flu pandemic and epidemics such as encephalitis lethargica that disproved medical bacteriology's obsession with hunting bacterial invaders and cleansing spaces. Explaining epidemics required more attention to such processes – be it on the individual level as immunology or on that of populations as epidemiology.

In a notable parallel to the cultural history of the age where the outsiders of the imperial days found themselves centre stage,[68] the 1920s became a period where dissident science held sway, some of it from pre-war days, some of it novel. Andrew Mendelsohn has described this masterfully for epidemiology where the crisis of microbe hunting helped ecological explanations to gain ground.[69] Allowing for host interaction paved the way for a wide array of medical knowledge – from constitutional pathology to immunology – to become meaningful in the explanation of infection. Another element of innovation was the appropriation of statistical tools by epidemiology, an innovation in which British scholars took the lead. Epidemic outbreaks could now be explained and forecasted by observing the density of host and pathogen populations in a given space rather than by focusing on invading microbes.[70] Such knowledge could have practical implications, such as when Fred Neufeld, director of Germany's Robert Koch Institute, remarked that large-scale chemical delousing could be dropped if the prospective hosts of a typhus infection would change their shirts once in a while. However, microbe hunting was not gone for good, but drew inspiration from other fields, such as tropical medicine. Vector-borne infections provided a multitude of non-bacterial pathogens and intricate epidemiologies to be studied. As Helen Tilley has shown, such research went back to pre-war traditions but it flourished between the wars.[71] Another matter was the microbes themselves or what counted as their traits. Famously, Ludwik Fleck characterized pre-war bacteriology by its insistence on bacterial fixity in all its traits:

> We see the same thing happening in the classical theory of infectious diseases. Every infectious disease was supposed to be caused by very small living 'agents'. Nobody could see that these 'agents' were also present in healthy persons . . . At the time when Koch's theory of specificity held complete sway, any variability was unthinkable.[72]

Rather than adding to this, inter-war microbiology attempted to make microbes more variable. Some of the research, such as on the physiological variation of bacteria in relation to environmental stimuli, remained inconclusive.[73] Based in pre-war traditions, serology facilitated the differentiation of bacteria of the same species according to their serotypes and resulted in impressive therapeutic advances. Fine-tuning the sero-therapy of pneumonia to a serotype diagnosed in a patient would result in a markedly more effective therapy.[74] Other research, such as that on phages, an inter-war discovery, was also suited to establish intra-species differentiation. The application of phages in therapy was envisioned but remained without tangible results in the early 1930s.[75]

Other inventive work related to bacterial genetics and evolutionary biology. Despite the fact that modern German-style medical bacteriology had been based on the work of a Darwinist botanist, Ferdinand Julius Cohn,[76] and that Louis Pasteur had held elaborate views on the natural history of infectious diseases,[77] evolutionary and medical biology had surprisingly little contact initially. 'The evolution of serological types in nature is an interesting subject for speculation', wrote Fred Griffith, a bacteriologist, in 1934.[78] The natural history of diseases did not seem to matter clinically; it was a case for biologists. The 'law of declining virulence' had been proposed by the American biologist Theobald Smith before the war.[79] It envisioned the evolution of pathogens as a successive adaption to the host. Manifest disease or even death of the host stood for early stages and incomplete adaption respectively, and symbiosis replacing pathogenesis was the outcome of such a process. Smith's ecological vision of a co-evolution of host and microbe was discussed by specialized researchers between the wars,[80] but came to enjoy popularity in medicine after World War Two much like the work of Karl Friedrich Meyer, another pioneer of disease ecology.[81] In the 1920s bacterial genetics was also beginning to grow as a field, when researchers were beginning to employ bacteria's short generational cycles to develop laboratory models of heredity.[82] A tool to be employed in such work was induced bacterial resistance with microbes such as E. coli or trypanosomes. Interestingly, for inter-war researchers, resistance was a phenomenon almost devoid of clinical relevance. It would instead be considered a laboratory tool that could be employed in bacterial genetics or experimental pharmacology.[83]

Nineteenth-century medical bacteriology had entailed a promise that identifying pathogens would offer opportunities for control. Paul Ehrlich's chemotherapy seemed to provide a master plan. Starting from specific affinities of certain synthetic dyes to prokaryotic cells, he explored the pharmacological dimension of what had originally been a technique for staining bacteria in tissues. Therapies based on that mechanism 'would then be able to exert their full action exclusively on the parasite harboured within the organism and would represent, so to speak, magic bullets, which seek their target of their own accord'.[84] Salvarsan, a chemotherapy for syphilis, when it was marketed from 1906 could thus be seen as the first drug in a long line of chemotherapies. Instead, however, the inter-war years became what a medical historian called 'the doldrum years',[85] a period during which the search for antibacterial medicines remained mostly futile. There were only a few anti-parasiticals for tropical diseases, such as Bayer 205 for sleeping sickness, which was found during the war and subsequently became marketed as Germanin.[86]

## The age of control

The years down to 1935 can in many ways be characterized by the absence of working antibacterial therapies of industrial origin. The decades to follow saw, as we all know, the reversal of that trend. Starting with the sulfas in 1935, several waves of antibacterial medicines arrived: injectable penicillin during the war, its preparation in oral form and broad spectrum antibiotics in the 1950s, and finally from the 1960s preparations specifically designed to tackle problems created by widespread use of earlier preparations, that is, designed to control antibiotic resistance. The following discussion will explore their impact from the perspective of changing notions of infectious disease that resulted from the widespread availability of effective anti-infective medicines.

One dynamic factor in the period in question was the changing relations between basic biological sciences and medicine. That distinction had still been difficult to make in the inter-war years.[87] One of the most influential research institutes, the Rockefeller Institute, had indeed been based on their close proximity. The same cannot be said of the after-war years. It was the emergence of bacterial genetics in the inter-war period that indicated the change that was underway. Where the study of bacteria had been dominated by medical perspectives for decades,[88] such studies began to follow their own trajectories, leading from bacterial genetics to molecular biology.[89] This in turn favoured the development of a specialized, medical bacteriology that we will meet on the subsequent pages.

After they arrived in 1935, sulfa drugs, 'the first miracle drugs',[90] became a peculiarly rapidly proliferating class of drugs. Soon after Prontosil had been marketed as treatment for streptococcal infections, its active molecule, sulfanilamide, was isolated within a year. The instance that sulfas were based on synthetic dyes that had been handled by a mature industry for decades made modifications of the original molecule and mass-production comparatively easy. As a result, thousands of derivatives were synthesized, resulting in a dozen drugs, active against most major infective diseases, to be invented by an international industry in the short period before the arrival of penicillin. With fungal antibiotics arriving from about 1942, the situation was in many ways reversed. The antibiotic effects of certain fungal extracts had been known for decades, yet they had not appeared as attractive candidates for drug development – not even to Alexander Fleming in 1929 after he repeated an observation that had been made before, namely that fungal and bacterial cultures could hold each other in check.[91] It was purification in Oxford laboratories and subsequent mass production, employing American agrarian mass fermentation technology, which turned an odd phenomenon that could not be reproduced safely in clinical practice into an applicable medicine from 1942.[92]

The output of pharmaceutical industries grew about tenfold during World War Two and anti-infective medicines, in particular those produced by American companies, skyrocketed during this period, leading to new conceptualizations and control of infectious disease. We can summarize the effects under the headings of normalization and standardization. While there is little indication that anti-infective therapies had a noticeable effect on overall mortalities – vaccines were more important here – they did influence patient experience and physicians' practice. From the patient's perspective, conditions that had held existential threats or the prospect of long treatments, like pneumonia or syphilis, were normalized into manageable conditions which, given proper patient behaviour and medication, could be controlled.[93] 'In case of syphilis the answer is penicillin', was the promise of a 1950s advertisement. Effective therapies based on industrially produced medicines also heralded efficient, shorter and more standardized treatments. To date we know little about how this happened in practice, but it was certainly how such medicines were advertised.[94]

The application of anti-infective therapies may, at first sight, be seen as a historical event that falls into the histories of pharmacology and clinical medicine. Yet, the consequences for medical bacteriology were massive. Antibiotic resistance that had been known as a fascinating laboratory phenomenon for decades now became a clinical phenomenon to reckon with. 'Sulphonamide resistance, at first a test tube curiosity, is rapidly becoming of considerable importance in clinical medicine',[95] noted William Kirby in 1942, and indeed the mass application of sulfas and antibiotics resulted in an induced and speedy evolution of infectious disease. As Alexander Fleming prophesized when he received the Nobel Prize in 1945, resistance was to evolve quickly from a laboratory phenomenon into a feature of infectious disease.[96] As we know from contemporary studies, resistant *Staphylococcus* was already becoming widespread in clinical wards during the 1940s.[97] The consequences of this were manifold. First, we see an adaption of basic science laboratory equipment into diagnostic tools of hospital laboratories. Important examples are the typing of resistant strains employing phages, which became common during the 1950s,[98] and the refining and standardizing of the disc diffusion test for antibiotic sensitivity from the late 1950s.[99] Both instances reflected something that had been all but nascent in inter-war medical bacteriology: namely the need to define bacteria deeper than on species level. Of course, serum therapy of the inter-war years had also been tuned to bacterial strains – in this case mostly using serology rather than phages.[100] What was novel in the 1950s was not just the more widespread character of the phenomena studied, but that these were evolving under the eyes of medical researchers. With soaring rates of resistance, the biology of infectious disease was changing rapidly. In Robert Koch's days it had been sufficient to determine bacterial species in microbiological diagnostics. Now diagnostic tools were aimed at sub-species levels and they described a biology of disease that was changing while being under study. By the late 1950s, a first global epidemic of resistant staphylococcus heralded the arrival of a novel medical discipline, clinical microbiology.[101]

The effects of mass application of antibiotics were not limited to issues related to drug resistance. 'The problem of hospital infection has been complicated rather than solved by chemotherapy'.[102] Hospitals had been harbingers of therapeutic progress for the entire period we have covered so far and the prestige of medical bacteriology rested on its assumed utility to clinical medicine. Now widespread antibiotic resistance combined with the arrival of patients with reduced immunity – the elderly and those who had received transplants or chemotherapies for cancer – seemed to reverse a century of therapeutic progress.[103]

How did medicine and the pharmaceutical industry react to these challenges? In industry, the existing dynamics of innovation intensified. The spread of hospital infections and the rise of antibiotic resistance triggered several waves of new antibiotics.[104] In the 1950s, broad spectrum antibiotics like tetracycline became available and penicillin was produced in oral form. In the context of respiratory tract infections, increased consumption was driven by promises of increased clinical efficiency and of expanding indications: rather than curing pneumonia, antibiotics should now be given in cases of bronchitis to prevent its progression to a severe condition: 'Today the general practitioner does not wait for the development of the classical signs of pneumonia, or seek in his treatment to discriminate between acute bronchitis and broncho-pneumonia . . . his aim must be . . . to hit the invading germs with his most potent weapons.'[105] In the 1960s, pharmaceutical innovation began to target the evolving biology of infectious disease as a distinct market. Since a bacterial enzyme, penicillinase, could block antibiotics' action, so-called penicillinase-resistant, second-generation antibiotics were in demand. The quintessential drug of that type became methicillin. According to an editorial in *The Lancet* in 1960, it was 'a means of controlling staphylococcal infections which have plagued hospitals throughout the world during the last ten years'.[106]

However, there was also a growing critical chorus of an approach based on expansive technology, of treating problems created by antibiotics with more and novel pills alone. 'Therapeutic rationalism', actually an older movement, now focused on the consequences of antibiotic consumption.[107] Proposals included much of what we today associate with modern principles in the application of such medicines: use based on proper diagnosis, restraint in application, diagnostic standards that could serve clinical work and epidemiology, and a pursuit of hospital hygiene that avoided rather than encouraged the use of antibiotics.

## Ecology, emergency and evolution

From its beginnings in the 1870s down to the antibiotic age, the story of medical microbiology as it has been sketched out here was intrinsically linked to the technological optimism of its age. Where railways and telegraphs promised control of space and time, medicines implied in the bacteriological concept of infectious disease promised control of the biology of such conditions. Evidence for the technological character of medical bacteriology is not just to be found in the interventions that it brought about, such as anti-sera or antibiotics. Technological optimism is evident in contemporary thinking on disease. Koch's bacteriology was descendant from the microbiology of Ferdinand Julius Cohn, whose system of bacteria was deeply Darwinist and contained a taxonomy of bacteria that was assumed to have resulted from evolution. Yet, once such thinking became medical, evolutionary biology was almost totally absent; it did not seem to matter in medical work of the generation of Koch and Pasteur where the history of disease was expected to be one of extermination brought about by human technology.[108] In subsequent developments, attempts to align medical bacteriology and evolutionary biology, such as in Theobald Smith's 'law of declining virulence', remained without noticeable impact in the medicine of their times. Given that, it is hardly surprising that the mass arrival of antibiotics fuelled dreams of disease extermination. Yet, this has changed in recent decades and the following pages will describe how towards the end of the twentieth century several interconnected events fundamentally changed medical microbiology. The notion of a mostly static microbial world gave way to one where the natural history of microbes mattered in clinical practice,[109] and the history of infectious diseases appeared no longer to be what Joshua Lederberg despairingly called 'the last refuges of the concept of special creationism'.[110]

A glimpse of what was to change can be inferred from a 1959 paper by David Rogers, an American specialist in internal medicine. Studying the changing epidemiology of infectious diseases, he described the hospital as a place where the history of infectious disease took a new turn. Comparing statistically deaths from infections in a hospital in 1938 and 1958, that is before and after the antibiotic revolution, he observed that while previously people had mostly died from infections they had contracted outside the hospital, now infection tended to occur inside hospital walls. Fatal infections in 1958 shared two common features:

> disease was produced by microbes, ordinarily of low invasiveness, that commonly form part of the normal human flora. Infections arose in patients already compromised by serious disease who were often receiving therapies known to affect host resistance to infection.[111]

Attempts to combat infectious disease in hospitals had fostered disease evolution. Rogers employed an ecological understanding of infectious disease causation and emphasized that eradicating one bacterial species would open an ecological niche for others to colonize – even more so if that niche was a hospital full of patients whose immune defences had been weakened by antibiotic therapies. What the antibiotic revolution had resulted in were 'dramatic shifts in the nature of life threatening microbial infection'.[112] Rogers did not focus on antibiotic resistance in isolation but viewed that challenge in close proximity to the spread of hospital infection that began to attract more attention around 1960. Research on that history has so far been sketchy but from the available information we can conclude that there were a variety of causes for the rise of infections of hospital patients with so-called non-classical pathogens like pseudomonas which would not give healthy humans disease symptoms.[113] Be it the arrival of the elderly or of those who had received transplants or chemotherapy, the result was the presence of a patient population in hospitals that was prone to suffer from hospital infections.

An important backdrop to the rising awareness of disease evolution in the 1960s was scepticism in relation to disease extermination programmes and the nascent environmentalism of those days. A popular read was René Dubos' book *The Mirage of Health* of 1959,[114] which put forward a radical critique of the technological optimism that lay behind dreams of disease eradication: 'The belief that disease can be conquered through the use of drugs is comparable to the naïve cowboy philosophy that permeates the Wild West thriller'.[115] Dubos encouraged doctors to learn from history, which contained many examples of failed disease control via eradication. For hospitals this meant that while they had been intended as healthy environments, mass antibiotic therapy had turned them into hothouses of disease evolution. Rogers, in fact, concluded with an explicit reference to Dubos:

> The present observations reinforce the thesis recently elaborated by Dubos . . . The nature of underlying disease appears in part to determine the risk of acquiring an intra-hospital infection. Antimicrobial drugs appear to determine the particular microbes that can establish fatal disease under such circumstances.[116]

To understand the scope of developments that were underway from the 1960s we need to look more closely at how the scientific basis for studying infections and their evolution was changing in those years. Genetic analysis of bacteria, adapted to medical purposes from nascent gene technology, pointed to phenomena that could account for rapid disease evolution. The most important example is the study of what were called episomes or plasmids in molecular biology around 1960.[117] These denoted snippets of genetic information that could be traded between different species and could thus account for the acquisition of genetic traits that exceeded the slow path of mutation, selection and hereditary transmission between generations. If such snippets coded for antibiotic resistance, this property could now be propagated swiftly between bacterial species. The phenomenon explained the existence of so-called resistance factors in bacteria that had not been addressed with antibiotic therapies. Starting from E. S. Anderson's work on resistance transfer in E. coli – a widespread phenomenon in animal husbandry – the threat that antibacterial resistance posed now seemed to increase massively. It was this context that brought the *New England Journal of Medicine* in 1967 to introduce a new notion into debates, that of a future without antibiotics: 'It appears that unless drastic measures are taken very soon, physicians may find themselves back in the preantibiotic Middle Ages in the treatment of infectious diseases.'[118]

One popular textbook summarizing such work was Stanley Falkow's *Infectious Multiple Drug Resistance* of 1975.[119] Those who consider this exotic knowledge should be reminded of the 2011 E. coli crisis, in which European consumers were threatened by a strain of E. coli that had two interesting traits: it had rather suddenly acquired higher pathogenicity and – despite never being routinely addressed in antibiotic therapy – turned out to be resistant to 12 different antibiotics.[120] Horizontal transfer of genetic information, first elucidated for antimicrobial resistance, proved to be relevant for other phenomena such as increases in virulence or the emergence of pathogenicity in previously benign microbes.[121] 'New mechanisms of genetic plasticity of one microbe species to another are uncovered almost daily', summarized Joshua Lederberg in 1997. He continued: 'Lateral transfer is very important in the evolution of microorganisms.

Their pathogenicity, their toxicity, their antibiotic resistance do not rely exclusively on evolution within a single clonal proliferation.'[122] One important consequence of the notion of horizontal gene transfer lay in its impact on the framing of disease evolution. Previously believed to proceed along the comparatively slow lines of mutation and selection, the notion of horizontal transfer of genetic information allowed the swift and discontinuous development of the biology of infectious disease. As Stanley Falkow put it in 2006, 'pathogenicity in all microbes . . . can occur in big genetic jumps, rather than through slow, adaptive evolution; HIV is the case in point'.[123]

The history of medical bacteriology in recent decades has not been written, but there is evidence that its framing of human–microbial relations differs markedly from the dreams of control or extermination of a more or less stagnant microcosm that dominated earlier on. In what one author called 'the golden age of genetics and the dark age of infectious disease',[124] accent shifted onto evolution and emergence. From the mid-1980s, the popular term 'emerging infections' directed attention to rather different phenomena.[125] 'Emergence' could be understood as evolutionary-biological emergence or, in a more classical meaning, as the (re)emergence of epidemic infectious disease. What connected both cases was the (re)creation of a world on alert in relation to infectious threats. Triggered in part by the AIDS pandemic, this notion can be identified in events such as the SARS epidemic of late 2002, fears of bioterrorism or the recent career of influenza as a common infectious disease.[126] Yet, the scientific perspective was not limited to despair and gloom, but included scenarios of a co-evolution and occasionally even symbiosis of men and microbes. For instance, genetic analysis showed that a substantial part of human physiology rested on the incorporation of bacterial DNA, as so-called mitochondria, into human cells.[127] In other instances the natural and the human histories of diseases were shown to have evolved in close connection. The best-researched example of that type is influenza. From the 1970s, the existence of a large reservoir of flu viruses in aquatic birds was revealed. Through mutation or recombination these could switch host species rather suddenly and would then result in pandemics.[128]

Another trend in studying infections was a pervasive interest in chronic infection. In part this can be seen to reflect rising interest in viral pathologies from warts to certain cancers after World War Two.[129] The wider historical significance can be explained in relation to a chronic condition that was identified as a bacterial infection in the 1980s, the common stomach ulcer.[130] The world of classic bacteriology had focused on infectious diseases that mostly shared some important traits: high prevalence, obvious symptoms, swift course and a high death rate. From that vantage point, infections could be easily distinguished from chronic conditions and cancers that would only rarely be considered infectious.[131] Microbes in intestinal organs had in fact been observed for most of the twentieth century, but the common ulcer remained a chronic disease of civilization for all that time, usually thought to be caused by stress. It was treated with a colourful set of therapies, ranging from surgery through pills to psychotherapy. In an intellectual climate concerned about chronic conditions from sarcoma to sickle cell anaemia,[132] chronic infection that occasionally erupted in manifest disease now seemed more plausible. The 1981 discovery of *Helicobacter pylori* as the pathogen in gastric ulcers is thus indicative of a different way of framing infection: the outside intruder of classical bacteriology had become supplemented with the occasionally nasty tenant.

Change was bound up with innovations in diagnostic technology. Where classical bacteriology had usually identified single bacterial species or their strains through cultivation, serology or phages, the use of polymerase chain technology from the late 1980s quickly populated the human body with an unprecedented number of microbial inhabitants, few of which were ever pathogenic.[133] Relations between host and microbe, as a recent attempt to theorize the field has framed it, are less a matter of cause than of circumstance. In Casadevall's and Pirowski's *damage-response-framework* contacts between humans and microbes range all the way from being deadly to being beneficial and the outcome is dependent on the host, the microbe and the conditions under which they meet.[134] Gone were the days when, as Stanley Falkow remembered, 'medical bacteriology . . . focused on differentiating the "good guys" from the "bad guys"'.[135] The term 'microbiome' describes what now appears as a vast multitude of microbes inhabiting a human body.[136] Casadevall and Pirowski's approach also explicitly reflects the sense that challenges to infection medicine in high-income countries often lie with patients suffering from opportunistic infections, where ubiquitous microbes produce diseases when the host's immune system is weak.

Traditional phenomena, such as the conflict of public and individual health that related to screening for healthy carriers of infectious diseases, took a new shape under such conditions. Where Typhoid Mary was identified by screening for healthy carriers of bacterial infections in 1907,[137] her modern counterparts can be identified with increased refinement. As David Barnes has shown, molecular diagnostics of drug-resistant tuberculosis facilitated the identification of Patient Zeros of the late twentieth century.[138]

In the broader picture it is interesting to see that developments took a distinct turn in the early 1990s. What had so far mostly been expert debate reached the public in a dystopian awareness of a post-antibiotic age and the return of infectious disease.[139] In Joshua Lederberg's report on emerging infections of 1992, the perspective on disease history was quite different from Dubos' earlier framing.[140] Where the latter had envisioned a co-existence of men and microbes along ecological principles, Lederberg argued that mankind was threatened with extermination and that the relation between men and microbes was in fact an arms race: 'If we were to rely strictly on biologic selection to respond to the selective factors of infectious disease, the population would fluctuate from billions down to perhaps millions before slowly rising again.'[141] This alarmist tone combined attention to 'a host of apparently "new" infectious diseases . . . that affect more and more people every year',[142] with the notion of an antibiotic development pipeline that was running dry. In *The Antibiotic Paradox* of 1992, Stuart Levy warned his readers that 'antibiotic usage has stimulated evolutionary changes that are unparalleled in biologic history. This situation raises the staggering possibility that a time will come when antibiotics as a mode of therapy will only be a fact of historic interest.'[143] The notion of a post-antibiotic age became popular and it has remained with us ever since, recently being compared to bioterrorism by Britain's Chief Medical Officer, Sally Davies.[144]

## Conclusion

It seems attractive to sum up developments of more than a century under the heading of shifting relations of natural and human histories of infectious diseases, their understanding and control. Late-nineteenth-century medical bacteriology had focused on conditions such as syphilis, tuberculosis or typhoid, which had been framed as traditional challenges: they could, as in the case of cholera, travel and, as syphilis had done in the fifteenth century, they could (re)appear, but mostly they had been around for a considerable period of time. That is why knowing the history of epidemics provided insight into the panorama of infectious disease that late-nineteenth-century medical bacteriology acted upon. The notion of an evolving biology of infectious disease in those days was a matter for specialized biology and hardly impinged on daily medical practice. We can assume that most physicians sorted microbes into much the same camps as the great historian of infectious disease, Hans Zinsser, still did in 1935 – into saprophytes and parasites, good guys and bad guys.[145] Against the backdrop of this slowly evolving natural history of infections it was man-made technology that brought about historical change. Medical bacteriology was part of the respective technologies, from disinfection to drugs.

By comparison, medical microbiology at the outset of the twenty-first century is attempting to develop tools to identify and monitor on-going change in the biology of infectious disease. In other words, the natural history of infectious diseases, which for a long time had remained a specialty for biologists and epidemiologists, has come to matter in clinical medicine. Clinical microbiology, as an expert on hospital infections put it, has become a matter of 'stalking microbes',[146] of knowing the present and anticipating future developments, rather than answering long-established challenges.

The history of medical microbiology sketched out in this chapter is technological on two important levels. First, from its creation in the nineteenth century, the field was deeply influenced by industrial culture. Therapeutic interventions that had been entailed in the science of medical microbiology materialized as industrially mass-produced medicines. This dynamic continued into the twentieth century, yet there has been a crisis of innovation when it comes to therapeutic interventions since about 1980. Medical microbiology continued to appropriate cognitive and technological tools from neighbouring disciplines such as epidemiology or biology. Second – and this is a unique feature of the history of infectious disease in modernity – the biology of the phenomena under study evolved in response to medical practices. While this is quite graphic in the case of antimicrobial resistance, the most radical development is to be noted in relation to infection control in hospitals, where an epidemiological landscape has evolved that was perhaps beyond the imagination of nineteenth-century medicine.

Medical microbiology set out to control the natural history of infectious diseases. Yet ironically, its history in the twentieth century came to be dominated by the natural history of disease. As Susan Jones has observed in her history of anthrax, the picture of infectious disease that gene-technology draws provides rich information on a lively and speedy natural history of such conditions. For instance, the attempt to elucidate the origin of the anthrax spores used in the 2001 attacks in the USA by genetic analysis produced a document of recent history. While researchers were trying to identify a single source of certain strains, they eventually discovered that any contact with the human species had left multiple traces on the bacteria's genome: 'The level of genomic resolution has now increased to the point to where we can read the history of the human interactions with the organism and the genome . . . has become a complex historical text.'[147]

## Acknowledgements

Sylvia Berger (Zürich), Mark Honigsbaum (London) and Pierre-Olivier Méthot (Québec City) delivered helpful critical readings of an early version of this text. Valuable criticism was also received from colleagues in the Section for Medical Anthropology and Medical History at the University of Oslo during a staff seminar in February 2015.

## Notes

1 A. M. Brandt and M. Gardner, 'The Golden Age of Medicine?', in *Companion to Medicine in the Twentieth Century*, R. Cooter and J. Pickstone (eds), London and New York: Routledge, 2000, pp. 21–37; T. D. Brock (ed.), 'Milestones in Microbiology', Washington, DC: American Society for Microbiology, 1975; W. Bulloch, *The History of Bacteriology*, London: Oxford University Press, 1960 (1938).

2 This approach is inspired by W. F. Bynum, *Science and the Practice of Medicine in the Nineteenth Century*, Cambridge: Cambridge University Press, 1994.

3 S. Berger, *Bakterien in Krieg und Frieden: Eine Geschichte der medizinischen Bakteriologie in Deutschland 1890–1933*, Göttingen: Wallstein, 2009.

4 Summarized in A. Mercer, *Infections, Chronic Disease, and the Epidemiological Transition*, Rochester, NY: University of Rochester Press, 2014.

5 T. McKeown, *The Role of Medicine*, Oxford: Blackwell, 1979. For an introduction to the debate, see E. Grundy, 'Commentary: The McKeown Debate: Time for Burial', *International Journal of Epidemiology* 34, 2005, 529–33.

6 It is worth noting that the theory of the epidemiologic transition that originally had three phases (age of pestilence, decline of infectious diseases, modernity) has been expanded by a fourth phase, covering the rise of non-communicable diseases on a global scale.

7 The WHO website on the global presence of infectious disease is instructive in that respective: http://www.who.int/topics/infectious_diseases/en/.

8 G. A. J. Ayliffe and M. P. English, *Hospital Infection From Miasmas to MRSA*, Cambridge: Cambridge University Press, 2003; S. H. Podolsky, *The Antibiotic Era: Reform, Resistance, and the Pursuit of Rational Therapeutics*, Baltimore: Johns Hopkins University Press, 2015.

9 A splendid study of the impact of investigate technology is N. Rasmussen, *Picture Control: The Electron Microscope and the Transformation of Biology in America, 1940–1960*, Stanford, 1997.

10 A. Hardy, *Salmonella Infections, Networks of Knowledge, and Public Health in Britain 1880–1975*, Oxford: Oxford University Press, 2015.

11 A. Cunningham, 'Transforming Plague. The Laboratory and the Identity of Infectious Disease', in *The Laboratory Revolution in Medicine*, A. Cunningham and P. Williams (eds), Cambridge: Cambridge University Press, 1992, pp. 209–24.

12 C. Gradmann, '"A spirit of scientific rigour": Koch's Postulates and 20th Century Medicine', *Microbes and Infection* 16, 2014, pp. 885–92.

13 V. Quirke and J-P. Gaudillière, 'The Era of Biomedicine: Science, Medicine, and Public Health in Britain and France after the Second World War', *Medical History* 52, 2008, 441–52. Historical studies on recent anti-infective drug development do not exist, but see: H. P. Rang, 'The Development of the Pharmaceutical Industry', in *Drug Discovery and Development: Technology in Transition*, H. P. Rang (ed.), London: Churchill Livingstone, 2005, pp. 3–18.

14 J. Farley, *The Spontaneous Generation Controversy from Descartes to Opain*, Baltimore and London: Johns Hopkins University Press, 1977; G. Geison, *The Private Science of Louis Pasteur*, Princeton, NJ: Princeton University Press, 1995, pp. 110–42.

15 M. Worboys, *Spreading Germs: Disease Theories and Medical Practice in Britain, 1985–1900*, Cambridge: Cambridge University Press, 2000, pp. 73–107.

16 A. Hüntelmann, *Hygiene im Namen des Staates. Das Reichsgesundheitsamt 1876–1933*, Göttingen: Wallstein, 2008.

17 On surgery and bacteriology, see T. Schlich, 'Asepsis and Bacteriology: A Realignment of Surgery and Laboratory Science', *Medical History* 56, 3, 2012, 308–34.

18  K. C. Carter, *The Rise of Causal Concepts of Disease: Case Histories*, Aldershot: Ashgate, 2003.
19  Worboys, *Spreading Germs*.
20  T. Schlich, 'Linking Cause and Disease in the Laboratory: Robert Koch's Method of Superimposing Visual and "Functional" Representations of Bacteria', *History and Philosophy of the Life Sciences* 22, 2000, 43–58.
21  A. I. Hardy, *Ärzte, Ingenieure und städtische Gesundheit: Medizinische Theorien in der Hygienebewegung des 19. Jahrhunderts*, Frankfurt: Campus, 2005.
22  C. Gradmann, *Laboratory Disease: Robert Koch's Medical Bacteriology*, Baltimore: Johns Hopkins University Press, 2009, pp. 69–90.
23  Geison, *Private Science*, pp. 145–76.
24  Schlich, 'Linking Cause and Disease'; C. Gradmann, 'Invisible Enemies: Bacteriology and the Language of Politics in Imperial Germany', *Science in Context* 13, 2000, 9–30.
25  B. Bracegirdle, *A History of Microtechnique*, Ithaca, NY: Cornell University Press, 1978.
26  I. Löwy, 'On Hybridizations, Networks and New Disciplines: The Pasteur-Institute and the Development of Microbiology in France', *Studies in the History and Philosophy of Science* 25, 1994, 655–88. On Pasteur Institutes outside France, see: P. Chakrabarti, *Bacteriology in British India: Laboratory Medicine and the Tropics*, Rochester, NY: University of Rochester Press, 2012; K. Pelis, *Charles Nicolle, Pasteur's Imperial Missionary: Typhus and Tunisia*, Rochester, NY: University of Rochester Press, 2006.
27  J. Simon, 'Monitoring the Stable at the Pasteur Institute', *Science in Context* 21, 2008, 181–200; C. Gradmann, 'Locating Therapeutic Vaccines in 19th Century History', *Science in Context* 21, 2008, 145–60; D. Linton, *Emil von Behring. Infectious Disease, Immunology, Serum Therapy*, Philadelphia: American Philosophical Society, 2005.
28  R. Koch, 'Kritische Besprechung der gegen die Bedeutung der Tuberkelbazillen gerichteten Publikationen', in *Gesammelte Werke von Robert Koch*, J. Schwalbe (ed.), Leipzig: Verlag von Georg Thieme, 1912 (1883), vol. 1, pp. 454–66.
29  F. Condrau, *Lungenheilanstalt und Patientenschicksal. Sozialgeschichte der Tuberkulose in Deutschland und England im späten 19. und frühen 20. Jahrhundert*, Göttingen: Vandenhoeck und Ruprecht, 2000, pp. 119–63. In typhoid it could be rather quick (R. Wall, *Bacteria in Britain, 1880–1939*, London: Pickering & Chatto, 2013).
30  C. Lawrence, 'Incommunicable Knowledge: Science, Technology and the Clinical Art in Britain 1850–1914', *Journal of Contemporary History* 20, 1985, 503–20; S. Sturdy, 'Looking for Trouble: Medical Science and Clinical Practice in the Historiography of Modern Medicine', *Social History of Medicine* 24, 2011, 739–57.
31  A. Contrepois, 'The Clinician, Germs an Infectious Diseases: The Example of Charles Bouchard in Paris', *Medical History* 46, 2002, 197–220.
32  Gradmann, 'Invisible Enemies'; B. Hansen, *Picturing Medical Progress from Pasteur to Polio: A History of Mass Media Images and Popular Attitudes in America*, New Brunswick, NJ: Rutgers University Press, 2009; N. Tomes, *The Gospel of Germs: Men, Women, and the Microbe in American Life*, Cambridge, MA: Harvard University Press, 1998.
33  P. Weindling, 'Scientific Elites and Laboratory Organisation in Fin de Siècle Paris and Berlin: The Pasteur Institute and Robert Koch's Institute for Infectious Diseases Compared', in *The Laboratory Revolution in Medicine*, Cunningham and Williams (eds), pp. 170–88; I. Löwy, '"A River That Is Cutting Its Own Bed": The Serology of Syphilis between Laboratory, Society and the Law', *Studies in History and Philosophy of Science Part C: Studies in History and Philosophy of Biological and Biomedical Sciences* 35, 2004, 509–24; A. Hinz-Wessels, *Das Robert Koch-Institut im Nationalsozialismus*, Berlin: Kadmos, 2008, pp. 9–20.
34  C. Hamlin, *A Science of Impurity: Water Analysis in Nineteenth Century Britain*, Bristol: Adam Hilger, 1990; A. Hardy, *The Epidemic Streets: Infectious Disease and the Rise of Preventive Medicine, 1856–1900*, Oxford: Clarendon Press, 1993. On the US-American case see P. P. Gossel, 'Pasteur, Koch and American Bacteriology', *History and Philosophy of the Life Sciences* 22, 2000, 81–100.
35  For example E. H. Ackerknecht, *A Short History of Medicine*, Baltimore: Johns Hopkins University Press, 1982 (1955), pp. 175–85.
36  N. J. Tomes and J. H. Warner, 'Introduction to the Special Issue on Rethinking the Reception of the Germ Theory of Disease: Comparative Perspectives', *Journal of the History of Medicine and Allied Sciences* 52, 1997, 7–16, at 12.

37  M. Worboys, 'Was There a Bacteriological Revolution in Late Nineteenth-century Medicine?', *Studies in History and Philosophy of Biological and Biomedical Sciences* 38, 2007, 20–42.

38  A. Mendelsohn, '"Like All That Lives": Biology, Medicine and Bacteria in the Age of Pasteur and Koch', *History and Philosophy of the Life Sciences* 24, 2002, 3–36.

39  K. Faber, *Nosography: The Evolution of Clinical Medicine in Modern Times*, New York: Paul B. Hoeber, 1930.

40  Cunningham, 'Transforming Plague', p. 239.

41  Carter argues from a Lakatosian theoretical background for a theory driven development: Carter, *Causal Concepts.*

42  G. Canguilhem, 'Der Beitrag der Bakteriologie zum Untergang der "medizinischen Theorien" im 19. Jahrhundert', in *Wissenschaftsgeschichte und Epistemologie*, G. Canguillhem, Frankfurt: Suhrkamp, 1979, pp. 110–32.

43  J. A. Mendelsohn, 'Cultures of Bacteriology: Formation and Transformation of a Science in France and Germany, 1870–1914', PhD, Princeton University, 1996, pp. 255–63.

44  Ibid., pp. 280–5; A. Métraux, 'Reaching the Invisible: A Case Study of Experimental Work in Microbiology (1880–1900)', in *Social Organisation and Social Process: Essays in Honor of Anselm Strauss*, D. R. Maines (ed.), New York: Aldine de Gruyter, 1991, pp. 249–60.

45  Koch produced vaccines for veterinary infections in the early 1900s: C. Gradmann, 'Robert Koch and the Invention of the Carrier State: Tropical Medicine, Veterinary Infections and Epidemiology around 1900', *Studies in History and Philosophy of Biological and Biomedical Sciences* 41, 2010, 232–40. On Behring's work J. Simon, 'Emil Behring's Medical Culture: From Disinfection to Serotherapy', *Medical History* 51, 2007, 201–18.

46  W. U. Eckart, 'The Colony as a Laboratory: German Sleeping Sickness Campaigns in German East-Africa and in Togo, 1900–1914', *History and Philosophy of the Life Sciences* 24, 2002, 69–89; M. Lyons, *The Colonial Disease: A Social History of Sleeping Sickness in Northern Zaire, 1900–1940*, Cambridge: Cambridge University Press, 1992; M. Worboys, 'The Comparative History of Sleeping Sickness in East and Central Africa, 1900–1914', *History of Science* 32, 1994, 89–102. On tropical medicine: D. Arnold (ed.), 'Warm Climates and Western Medicine: The Emergence of Tropical Medicine, 1500–1900', Amsterdam: Rodopi, 1996; W. U. Eckart, *Medizin und Kolonialimperialismus in Deutschland 1884–1945*, Paderborn: Schöningh, 1997; D. Neill, *Networks in Tropical Medicine: Internationalism, Colonialism and the Rise of the Medical Specialty*, Palo Alto: Stanford University Press, 2012; M. A. Osborne, *The Emergence of Tropical Medicine in France*, Chicago: Chicago University Press, 2014.

47  P. M. H. Mazumdar, *Species and Specificity: An Interpretation of the History of Immunology*, Cambridge: Cambridge University Press, 1995; A. Silverstein, *A History of Immunology*, San Diego, CA: Academic Press, 1989.

48  Gradmann, 'Therapeutic Vaccines'. By contrast, small-pox vaccine was not standardized in industrial production but calibrated continuously in practice.

49  Löwy, 'Serology of Syphilis'; L. Fleck, *Genesis and Development of a Scientific Fact*, Chicago: University of Chicago Press, 1988 (1935).

50  A. Hardy, 'Methods of Outbreak Investigation in the "Era of Bacteriology" 1880–1920', in *A History of Epidemiologic Methods and Concepts*, A. Morabia (ed.), Basel: Birkhäuser Verlag, 2004, pp. 199–206.

51  Gradmann, 'Carrier State'; Mendelsohn, *Cultures*, pp. 556–774.

52  P. Wald, *Contagious: Cultures, Carriers, and the Outbreak Narrative*, Durham, NC: Duke University Press, 2008, pp. 68–113.

53  M. Harrison, *The Medical War: British Military Medicine in the First World War*, Oxford: Oxford University Press, 2010; W. U. Eckart, *Medizin und Krieg. Deutschland 1914–1924*, Paderborn: Schöningh, 2014.

54  Linton, *Behring*, pp. 357–61. On typhoid vaccination, see: Harrison, *The Medical War*, pp. 142–52; D. S. Linton, 'Was Typhoid Inoculation Safe and Effective during World War I? Debates within German Military Medicine', *Journal for the History of Medicine and Allied Sciences* 55, 2000, 101–3; Eckart, *Medizin und Krieg*, pp. 173–8.

55  P. Weindling, *Epidemics and Genocide in Eastern Europe, 1890–1945*, Oxford: Oxford University Press, 2000, pp. 73–108.

56 M. Bresalier, *Short History of Flu*, London: Continuum, 2010; M. Honigsbaum, *A History of the Great Influenza Pandemics: Death, Panic and Hysteria, 1830–1920*, London: I. B. Tauris, 2014.

57 N. P. A. S. Johnson and Mueller, J., 'Updating the Accounts: Global Mortality of the 1918–1920 "Spanish" Influenza Pandemic', *Bulletin of the History of Medicine* 76, 2002, 105–15.

58 Quoted in J. A. Mendelsohn, 'From Eradication to Equilibrium: How Epidemics Became Complex after World War I', in *Greater than the Parts: Holism in Biomedicine, 1920–1950*, C. Lawrence and G. Weisz (eds), New York/Oxford: Oxford University Press, 1998, pp. 303–31, p. 311.

59 M. Bresalier, '"A Most Protean Disease": Aligning Medical Knowledge of Modern Influenza, 1890–1914', *Medical History* 56, 4, 2012, 481–510; E. Tognotti, 'Scientific Triumphalism and Learning from Facts', *Social History of Medicine* 16, 2003, 97–110.

60 M. Bresalier, 'Uses of a Pandemic: Forging the Identities of Influenza and Virus Research in Interwar Britain', *Social History of Medicine* 25, 2012, 400–24.

61 O. Amsterdamska, 'Achieving Disbelief: Thought Styles, Microbial Variation, and American and British Epidemiology, 1900–1940', *Studies in History and Philosophy of Science Part C: Studies in History and Philosophy of Biological and Biomedical Sciences* 35, 2004, 483–507; T. V. Helvoort, 'A Bacteriological Paradigm in Influenza Research in the First Half of the Twentieth Century', *History and Philosophy of the Life Sciences* 15, 1993, 3–21.

62 A. Hardy, 'Questions of Quality: The Danish State Serum Institute, Thorvald Madsen and Biological Standardization', in *Evaluating and Standardizing Therapeutic Agents 1890–1950*, C. Gradmann and J. Simon (eds), Basingstoke: Palgrave Macmillan, 2010, pp. 139–52; Hinz-Wessels, *Robert Koch-Institut*, pp. 9–20; Hüntelmann, *Hygiene*, pp. 130–43.

63 J. Farley, *To Cast Out Disease: A History of the International Health Division of the Rockefeller Foundation (1913–1951)*, Oxford University Press, 2004; Hardy, 'Questions of Quality'.

64 J. Austoker and Bryder, L. (eds), *Historical Perspectives on the Role of the MRC*, Oxford: Oxford University Press, 1989.

65 Weindling, *Epidemics*, pp. 49–72.

66 Berger, *Bakterien*, pp. 267–90.

67 Quoted in ibid., p. 312.

68 P. Gay, *Weimar Culture: The Outsider as Insider*, New York: Harper & Row, 1969.

69 Mendelsohn, 'Eradication'.

70 O. Amsterdamska, 'Standardizing Epidemics: Infection, Inheritance, and Environment in Prewar Experimental Epidemiology', in *Heredity and Infection: The History of Disease Transmission*, J. P. Gaudilliere and I. Löwy (eds), London and New York: Routledge, 2003, pp. 135–79; A. Hardy and Magnello, M. E., 'Statistical Methods in Epidemiology: Karl Pearson, Ronald Ross, Major Greenwood and Austin Bradford Hill, 1900–1945', in Morabia (ed.), *A History of Epidemiologic Methods*, pp. 207–21.

71 H. Tilley, *Africa as a Living Laboratory: Empire, Development, and the Problem of Scientific Knowledge, 1870–1950*, Chicago: University of Chicago Press, 2011.

72 Fleck, *Scientific Fact*, pp. 29–30.

73 Amsterdamska, 'Stabilizing Instability: The Controversy over Cyclogenic Theories of Bacterial Variation during the Interwar Period', *Journal of the History of Biology* 24, 1991, 191–222.

74 S. H. Podolsky, *Pneumonia before Antibiotics: Therapeutic Evolution and Evaluation in Twentieth-century America*, Baltimore: Johns Hopkins University Press, 2006, pp. 9–50.

75 W. C. Summers, 'The Strange History of Phage Therapy', *Bacteriophage* 2, 2012, 130–3.

76 On Cohn, see Mazumdar, *Species*.

77 Mendelsohn, *Cultures*, pp. 99–110.

78 Cited in P.-O. Méthot, 'Bacterial Transformation and the Origins of Epidemics: The Epidemiological Significance of Fred Griffith's "Transforming Experiment" in the Interwar Period', *Journal of the History of Biology* 49, 2016, p. 46.

79 W. Anderson, 'Natural Ecologies of Infectious Disease: Ecological Vision in Twentieth-Century Biomedical Science', *Osiris* 19, 2004, 39–61; C. E. Dolman and Wolfe, R. J., *Suppressing the Diseases of Animals and Man: Theobald Smith, Epidemiologist*, Boston: Boston Medical Library, 2003; P.-O. Méthot, 'Why Do Parasites Harm Their Host? On the Origin and Legacy of Theobald Smith's "Law of Declining Virulence" – 1900–1980', *History and Philosophy of the Life Sciences* 34, 4, 2012, 561–601.

80 Berger, *Bakterien*, pp. 346–54.

81 M. Honigsbaum, '"Tipping the Balance": Karl Friedrich Meyer, Latent Infections, and the Birth of Modern Ideas of Disease Ecology', *Journal of the History of Biology* 49, 2016, 261–309.

82 T. D. Brock, *The Emergence of Bacterial Genetics*, Cold Spring Harbour, NY: Cold Spring Harbour Laboratory Press, 1990.

83 C. Gradmann, 'Magic Bullets and Moving Targets: Antibiotic Resistance and Experimental Chemotherapy 1900–1940', *Dynamis* 31, 2011, 305–21.

84 P. Ehrlich, 'Address Delivered at the Dedication of the Georg-Speyer-Haus', in *The Collected Papers of Paul Ehrlich*, F. Himmelweit (ed.), London: Pergamon Press, 1960 (1906), pp. 53–63, at 59; see C.-R. Prüll, A.-H. Maehle and R. F. Halliwell, *A Short History of the Drug Receptor Concept*, Basingstoke: Palgrave, 2009.

85 I. Galdston, *Behind the Sulfa Drugs: A Short History of Chemotherapy*, New York; London: D. Appleton-Century, 1943, p. 137.

86 Eckart, *Kolonialimperialismus*, pp. 509–14.

87 Amsterdamska, 'Inventing Utility'; R. E. Kohler, 'Bacterial Physiology: The Medical Context', *Bulletin for the History of Medicine* 59, 1985, 54–74; T. V. Helvoort, 'Bacteriological and Physiological Research Styles in the Early Controversy on the Nature of the Bacteriophage Phenomenon', *Medical History* 26, 1992, 243–70.

88 Mendelsohn, 'Like All That Lives'.

89 Méthot, 'Bacterial Transformation', p. 39. See S. D. Chadarevian, *Designs for Life. Molecular Biology after World War II*, Cambridge: Cambridge University Press, 2002; M. Morange, *A History of Molecular Biology*, Cambridge, MA and London: Harvard University Press, 2000 (1994).

90 J. E. Lesch, *The First Miracle Drugs: How the Sulfa Drugs Transformed Medicine*, Oxford: Oxford University Press, 2007.

91 K. Brown, *Penicillin Man: Alexander Fleming and the Antibiotic Revolution*, Gloucestershire: Sutton, 2004; R. Bud, *Penicillin: Triumph and Tragedy*, Oxford: Oxford University Press, 2007. On Lister's use of mould extracts, see M. Wainwright, *Miracle Cure. The Story of Penicillin and the Golden Age of Antibiotics*, Cambridge, MA: Basil Blackwell, 1990, pp. 36–7.

92 Bud, *Penicillin*, pp. 23–53.

93 S. H. Podolsky, 'Antibiotics and the Social History of the Controlled Clinical Trial', *Journal for the History of Medicine and Allied Sciences* 65, 2010, 327–67; C. Connolly, J. Golden, and B. Schneider, '"A Startling New Chemotherapeutic Agent": Pediatric Infectious Disease and the Introduction of Sulfonamides at Baltimore's Sydenham Hospital', *Bulletin of the History of Medicine* 86, 2012, 66–93.

94 Podolsky, *Antibiotic Era*, pp. 43–72; J. T. MacFarlane and M. Worboys, 'The Changing Management of Acute Bronchitis in Britain, 1940–1970: The Impact of Antibiotics', *Medical History* 52, 2008, 47–72.

95 W. M. M. Kirby, 'Sulfonamide Resistance', *California and Western Medicine* 57, 1942, 174–5, p. 174.

96 Brown, *Penicillin Man*, p. 196.

97 For an introduction to Mary Barber's work, see M. Tansey (ed.) 'Post Penicillin Antibiotics: From Acceptance to Resistance: A Witness Seminar Held at the Wellcome Institute for the History of Medicine, London, 12 May 1998', *Wellcome Witnesses to Twentieth Century Medicine*, London, 2000.

98 K. Hillier, 'Babies and Bacteria: Phage Typing, Bacteriologists, and the Birth of Infection Control', *Bulletin of the History of Medicine* 80, 2006, 733–61.

99 C. Gradmann, 'Sensitive Matters: The World Health Organisation and Antibiotics Resistance Testing, 1945–1975', *Social History of Medicine* 26, 2013, 555–74.

100 For an overview, see Hardy, *Salmonella Infections*.

101 Hillier, 'Babies'; A. K. Lie, 'Producing Standards, Producing the Nordic Region: Antibiotic Susceptibility Testing, from 1950–1970', *Science in Context* 27, 2, 2014, 215–48.

102 E. J. Lowbury *British Medical Journal*, 1955 April 23, 1(4920), 985–90, quoted in L. Colebrook, 'Infection Acquired in Hospitals', *The Lancet* 266, 1955, 885–91, at 885.

103 Ayliffe and English, *Hospital Infections*; F. Condrau and R. Kirk, 'Negotiating Hospital Infections: The Debate between Ecological Balance and Eradication Strategies in British Hospitals, 1947–1969', *Dynamis* 31, 2011, 385–405.

104 D. Greenwood, *Antimicrobial Drugs. Chronicle of a Twentieth Century Triumph*, Oxford: Oxford University Press, 2008.

105 A. U. Mackinnon, 1960, quoted in MacFarlane and Worboys, 'Bronchitis', p. 59.

106 Anon, 'Penicillinase-Resistant Penicillin', *The Lancet* 276, 7150, 1960, 585–6. On MRSA: Bud, *Penicillin*, pp. 116–39; M. McKenna, *Superbug: The Fatal Menace of MRSA*, New York: Free Press, 2010; L. A. Reynolds and Tansey, E. M. (eds), 'Superbugs and Superdrugs: A History of MRSA', *Wellcome Witnesses to Twentieth Century Medicine*, London: Wellcome Trust Centre for the History of Medicine at UCL, 2008.

107 Podolsky, *Antibiotic Era*, focuses on the US. Many of the critical voices from Europe can be gathered from Ayliffe and English, *Hospital Infections*, pp. 153–62.

108 On Cohn, see Mazumdar, *Species*.

109 F. M. Snowden, 'Emerging and Reemerging Diseases: A Historical Perspective', *Immunological Reviews* 225, 1, 2008, 9–26.

110 Quoted in the instructive paper P.-O. Méthot and Alizon, S., 'What is a pathogen? Towards a process view of host-parasite interaction', *Virulence* 5, 8, 2014, 775–85, p. 781.

111 Ibid., 682.

112 Ibid., 683.

113 Ayliffe and English, *Hospital Infections*; P. Suter and J. L. Vincent (eds), *Milestones in Hospital Infections*, Egham: Medical Action Communication Ltd., 1993.

114 R. Dubos, *Mirage of Health: Utopias, Progress, and Biological Change*, London: Allen and Unwin, 1959; C. L. Moberg, 'René Dubos: A Harbinger of Microbial Resistance to Antibiotics', *Perspectives in Biology and Medicine* 42, 1999, 559–80.

115 Dubos, *Mirage*, p. 132.

116 Rogers, 'Changing Patterns', 682–3.

117 Brock, *Bacterial Genetics*, pp. 106–8; M. Grote, 'Hybridizing Bacteria, Crossing Methods, Cross-Checking Arguments: The Transition from Episomes to Plasmids (1961–1969)', *History and Philosophy of the Life Sciences* 30, 2008, 407–30.

118 'Infectious Drug Resistance', *New England Journal of Medicine* 275, 1966, 277.

119 S. Falkow, *Infectious Multiple Drug Resistance*, London: Pion, 1975.

120 'The Reason Why This Deadly E Coli Makes Doctors Shudder', *The Guardian*, 5 June 2011.

121 Méthot and Alizon, 'Pathogen', 779–80.

122 J. Lederberg, 'Infectious Disease As an Evolutionary Paradigm', *Emerging Infectious Diseases* 3, 4, 1997, 417–23, p. 419.

123 'The Ecology of Pathogenesis', in *Ending the War Metaphor: The Changing Agenda for Unraveling the Host-Microbe Relationship: Workshop Summary*, Forum on Microbial Threats and Board on Global Health (eds), Washington, DC: National Academies Press, 2006, pp. 102–15, at 113.

124 M. Tibayrenc, 'The Golden Age of Genetics and the Dark Age of Infectious Diseases', *Infection, Genetics and Evolution* 1, 2001, 1–2.

125 N. B. King, 'The Scale Politics of Emerging Diseases', *Osiris* 19, 2004, 62–76; Snowden, 'Emerging and Reemerging Diseases'.

126 P.-O. Méthot and Fantini, B., 'Medicine and Ecology: Historical and Critical Perspectives on the Concept of "Emerging Disease"', *International Archive of the History of Science* 64, 2014, 213–30; P. Sarasin, *"Anthrax" Bioterror als Phantasma*, Frankfurt: Suhrkamp, 2004.

127 M. Bernt, Machné, R., Sahyoun, A. H., Middendorf, M. and Stadler, P. F., 'Mitochondrial Genome Evolution', in *Encyclopedia of the Life Sciences*: John Wiley & Sons, Ltd, 2001, http://dx.doi.org/10.1002/9780470015902.a0025142; J. Sapp, *Evolution by Association: A History of Symbiosis*, New York: Oxford University Press, 1994.

128 W. G. Laver, N. Bischofberger and R. G. Webster, 'The Origin and Control of Pandemic Influenza', *Perspectives in Biology and Medicine* 43, 2000, 173–92.

129 A. Creager and Gaudillière, J.-P., 'Experimental Platforms and Technologies of Visualisation: Cancer as a Viral Epidemic, 1030–1960', in *Heredity and Infection: The History of Disease Transmission*, I. Löwy and J.-P. Gaudillière (eds), London: Routledge, 2001,

pp. 203–41; M. W. Taylor, *Viruses and Man: A History of Interactions*, Champaign: Springer International Publishing, 2014, pp. 121–41, 267–307.

130 P. Thagard, *How Scientists Explain Disease*, Princeton: Princeton University Press, 1999, C. Gradmann, '*Helicobacter pylori*', in *Eine Naturgeschichte für das 21. Jahrhundert*, S. Azzouni, C. Brandt, B. Gausemeier, J. Kursell, H. Schmidgen and B. Wittmann (eds), Berlin, 2011, pp. 241–3.

131 E. Becsei-Kilborn, 'Scientific Discovery and Scientific Reputation: The Reception of Peyton Rous' Discovery of the Chicken Sarcoma Virus', *Journal of the History of Biology* 43, 2010, 111–57.

132 P. W. Ewald, *Evolution of Infectious Disease*, Oxford: Oxford University Press, 1994.

133 Méthot and Alizon, 'Pathogen'.

134 A. Casadevall and Pirofski, L.-A., 'The Damage-Response Framework of Microbial Pathogenesis', *Nature Reviews / Microbiology* 1, 2003, 17–24.

135 S. Falkow, 'I Never Met a Microbe I Didn't Like', *Nature Medicine* 14, 10, 2008, 1053–7, at 1053.

136 Méthot and Alizon, 'Pathogen', 778.

137 Leavitt, *Typhoid Mary*.

138 D. S. Barnes, 'Targeting Patient Zero', in *Tuberculosis Then and Now: Perspectives on the History of an Infectious Disease*, F. Condrau and M. Worboys (eds), Montreal: McGill-Queens University Press, 2010, pp. 49–71. The quintessential patient 0 of our times was an airline steward presented as a super-spreader of HIV R. A. McKay, '"Patient Zero": The Absence of a Patient's View of the Early North American AIDS Epidemic', *Bulletin for the History of Medicine* 88, 2014, 161–94.

139 A. H. K. Lie and S. H. Podolsky, 'Futures and Their Uses: Antibiotics and Therapeutic Revolution', in *The Eclipse of the Therapeutic Revolution*, F. Condrau (ed.), forthcoming.

140 J. Lederberg, R. E. Shope, and S. C. Oaks (eds), 'Emerging Infections. Microbial Threats to Health in the United States', Washington, DC: National Academic Press, 1992.

141 Lederberg, 'Infectious Disease'.

142 Lederberg, Shope and Oaks, *Emerging Infections*, p. 26. Emphasis in the original.

143 S. B. Levy, *The Antibiotic Paradox. How Miracle Drugs Are Destroying the Miracle*, New York; London: Plenum Press, 1992, p. 183.

144 I. Sample, 'Antibiotic-Resistant Diseases Pose "Apocalyptic" Threat, Top Expert Says', *The Guardian*, 23 January 2013.

145 H. Zinsser, *Rats, Lice and History*, London: Routledge, 1935.

146 R. P. Wenzel, *Stalking Microbes: A Relentless Pursuit of Infection Control*, Bloomington, IN: AuthorHouse, 2005.

147 S. D. Jones, *Death in a Small Package: A Short History of Anthrax*, Baltimore: Johns Hopkins University Press, 2010, p. 262.

## Select bibliography

O. Amsterdamska, 'Achieving Disbelief: Thought Styles, Microbial Variation, and American and British Epidemiology, 1900–1940', *Studies in History and Philosophy of Science Part C: Studies in History and Philosophy of Biological and Biomedical Sciences* 35, 2004, 483–507.

S. Berger, *Bakterien in Krieg und Frieden: Eine Geschichte der medizinischen Bakteriologie in Deutschland 1890–1933*, Göttingen: Wallstein, 2009.

P. Chakrabarti, *Bacteriology in British India: Laboratory Medicine and the Tropics*, Rochester, NY: University of Rochester Press, 2012.

G. Geison, *The Private Science of Louis Pasteur*, Princeton, NJ: Princeton University Press, 1995.

C. Gradmann, *Laboratory Disease: Robert Koch's Medical Bacteriology*, Baltimore: Johns Hopkins University Press, 2009.

A. Hardy, 'Methods of outbreak investigation in the "Era of Bacteriology" 1880–1920', in A. Morabia (ed.), *A History of Epidemiologic Methods and Concepts*, Basel: Birkhäuser Verlag, 2004, pp. 199–206.

A. Hardy, *Salmonella Infections, Networks of Knowledge, and Public Health in Britain 1880–1975*, Oxford: Oxford University Press, 2015.

K. Hillier, 'Babies and Bacteria: Phage Typing, Bacteriologists, and the Birth of Infection Control', *Bulletin of the History of Medicine* 80, 2006, 733–61.

S. D. Jones, *Death in a Small Package: A Short History of Anthrax*, Baltimore: Johns Hopkins University Press, 2010.

J. A. Mendelsohn, 'From Eradication to Equilibrium: How Epidemics Became Complex after World War I', in C. Lawrence and G. Weisz (eds), *Greater than the Parts: Holism in Biomedicine, 1920–1950*, New York and Oxford: Oxford University Press, 1998, pp. 303–31.

P.-O. Méthot and Alizon, S., 'What Is a Pathogen? Towards a Process View of Host-Parasite Interaction', *Virulence* 5, 8, 2014, 775–85.

S. H. Podolsky, *The Antibiotic Era: Reform, Resistance, and the Pursuit of Rational Therapeutics*, Baltimore: Johns Hopkins University Press, 2015.

R. Wall, *Bacteria in Britain, 1880–1939*, Pickering & Chatto, 2013.

P. Weindling, *Epidemics and Genocide in Eastern Europe, 1890–1945*, Oxford: Oxford University Press, 2000.

M. Worboys, *Spreading Germs: Disease Theories and Medical Practice in Britain, 1985–1900*, Cambridge: Cambridge University Press, 2000.

# 22

# TECHNOLOGY AND THE 'SOCIAL DISEASE'

*Helen Bynum*

Tuberculosis has served historians well. Histories of disease are a means of under-standing the social milieu and political order in which they occur.[1] The association of tuberculosis with creativity and the arts has been a fertile field of investigation.[2] Reviewing past attempts to heal meshes with histories of travel, spas, architecture, quackery and drug discovery.[3] The history of preventing tuberculosis, from vaccines to anti-spitting legislation, incorporates a wide range of approaches to public health.[4] The involvement of the closely related disease of cattle engages with the study of zoonoses and the growing scrutiny of the animal/human health nexus.[5] Unravelling the causative micro-organism's past is part of the increasing use of molecular technologies informing history.[6]

Writing the history of tuberculosis presents the usual problems of retrospective diagnosis, but if we accept that it is possible to understand the disease as an entity over the *longue durée*, we have a tool to study the impact of one of the major causes of death throughout human history.[7] Used in this way, the disease featured in a par-ticular dialogue that appeared in the 1970s among those interested in histories of health and illness, demographic change, and the relationship between technology and medical progress: the McKeown debate. While there are calls among historians that this flawed work has run its course, served its purpose and should be put to rest, it supports those who continue to define medicine as a technology and to measure its progress accordingly.[8] It still features in policy history and is invoked by those working in the tuberculosis community.[9]

McKeown's weighing of the relative merits of social and technological disease par-adigms can also be read as part of today's much subtler and more comprehensive social determinants of health movement. A key social determinant is now considered to be the ability or inability to access health services to make use of technologies that do exist. Access can be possible for those classed as living in poverty, but it remains a challenge to the international community and national programmes. This is particu-larly true in areas of the world where tuberculosis rates are high – the 22 high-burden countries – whose health infrastructures tend to be unequally distributed and weak.[10] A history of tuberculosis diagnostics offers a way to explore the social and technolog-ical pasts of this disease and their interactions.

By way of introduction this chapter briefly reviews the McKeown moment, the social paradigm in medicine associated with it, and its place in the narrative of tuberculosis control in the twentieth century. The focus shifts to the elaboration of a series of labo-ratory tests for tuberculosis, the forging of a diagnostic tool kit as part of the unfolding bacterial aetiology, the use of x-rays and molecular diagnostics. Unsurprisingly, there is a profound difference between the experience of wealthier countries and the poorest countries. From the mid-twentieth century, wealthier countries typically had very low levels of tuberculosis (despite hot spots) and little transmission. Their strategies for diagnosis and screening could be tailored to this epidemiological profile. These countries were generally able to employ the benefits of developments in diagnostics incrementally, as they became available. This was most successful where the healthcare system was comprehensive and integrated.[11] For those countries at the other end of the spectrum – weak health infrastructure, high incidence, high transmission and high co-morbidities – cost, efficiency, validity of the results and deliverability in sub-optimum circumstances are recurrent themes.

What also emerges from the revitalised attempts at tuberculosis control since the 1990s are a series of further tensions, in managing technology transfer. The World Health Organization (WHO) and its partners in tuberculosis control devised a global plan, with an approved diagnostic algorithm that could be marketed to donors. The local experiences of countries implementing such a plan at the national level were problematic. What appears to be new, more recently, is a franker recognition of this problem and attempts to properly trial the diagnostics in operational situations.

## The social paradigm?

Thomas McKeown (1912–1988) wanted to know whether doctor-led scientific medicine, public health interventions focusing on the transmission of infectious diseases, or other factors were responsible for the sustained growth of Britain's population from the 1770s.[12] Inquiring rather clumsily about the relative merits of the role of technology, environmental legislation and economic development in determining health, McKeown argued that it was declining death rates from infectious diseases that had reshaped population history. He emphasised the role of air-borne diseases and identified tuberculosis as the leading cause of death. He argued that long before immunisation, effective drugs and dedicated institutional facilities were available, this disease was losing its severe grip on the population. Water and food-borne diseases were the only classes of disease to benefit from an emerging sanitary infrastructure but their role in the mortality transition was less significant, he argued. It was primarily neither what doctors could prescribe nor what public health mandarins could legislate for that mattered.

By deduction McKeown was left with what he termed the 'invisible hand': the positive effects of widespread economic progress. Of the possible socio-economic factors, he championed the rising wages of the working population, which he believed were spent on an improved diet. He saw these changes as predating effective measures in public health, just as he saw declining deaths from tuberculosis as leading the downward fall in mortality. Even though it was later shown to be flawed, it was an attractive thesis for McKeown.[13] As professor of social medicine in Birmingham University, he had a clear 'professional and political' agenda. He was working within a system he thought was committed to the principles of income redistribution and improved social infrastructure via the welfare state. With little time for what he saw as old-fashioned public health, his concern was to sway the resources of the National Health Service (NHS) away from 'curative technical medicine – invasive surgery and biochemical "treatments" – towards preventative, humanist medicine and efforts to understand and modify the health implications of the environment in its widest sense, including lifestyle, behaviour, and diet'.[14]

It perhaps helped McKeown's case that in Britain, in the 1970s, tuberculosis did not seem particularly germane any more. He did not see the need to untangle the more recent history of the disease. He largely ignored the early role of segregation in poor law infirmaries. Nor did he pay much attention to the concerted efforts that had been made to provide assistance to those with tuberculosis following the 1911 National Insurance Act and its subsequent extension via sanatorium benefits. While bed rest and other sanatoria treatments were not cures, debate would continue about the possible role of closeting infectious patients and preventing transmission.[15] All the while technology had played an important role in diagnosis and prognosis.

After the advent of effective chemotherapy, in the 1940s and 1950s, the rapid detection and cure of infectious patients was considered a successful strategy to control tuberculosis. In this, tuberculosis is similar to the other significant mycobacterial disease, leprosy, but unlike vaccine-preventable infectious diseases and those spread by water or food. Similarly, while barrier protection can prevent the transmission of sexually transmitted diseases such as syphilis and gonorrhoea, the only other option to interrupt their spread is treatment. HIV/AIDS, which has such a powerful co-morbidity with tuberculosis, is perhaps the most important disease currently sharing this profile. As a public health strategy, treatment-as-control for tuberculosis was and remains reliant upon efficient diagnosis and care in delivering the therapy to ensure that it is taken properly until cure is realised. Given the duration of treatment, which had to be continued long after the patient had ceased to feel the symptoms of tuberculosis but had to endure the side-effects of the drugs, help and support at this stage were critical. This aspect of the social dimension of tuberculosis reinforced the need for a functional, integrated health-care infrastructure.

McKeown was fortunate to be writing when tuberculosis in Britain had lost what remained of its bite. Post-war Britain had hosted some of the worst tuberculosis black spots in Europe and cities such as Liverpool and Glasgow were characterised by high levels of both poverty and disease. But the combination of medical advances that were being made available in the UK, as and where needed, were matched by a welfare state that provided a 'visible' hand. It was helping many to gradually improve their socio-economic status amid the continuing post-war reconstruction. Poverty remained, but there were improvements in its absolute level and continued pressure for further progress.[16]

Without trivialising local problems, the tools and practices of 'scientific medicine', delivered through a variety of targeted measures and agencies, had been able to manage the end of a long history of tuberculosis in Britain. It did not go away but with far fewer cases of transmitted disease rather than imported or reactivation cases, it ceased to be classed as a public health priority and then a problem at the national level. Useful initiatives such as the Trades Union Congress's long campaign to have tuberculosis classified as an occupational disease (1951) were important concessions, but came at the eleventh hour.[17] By the 1970s active case finding (ACF) had given way to referrals to more generalised chest clinics, sanatoria were closing or serving other purposes, and short-course chemotherapy could be offered as a relatively established protocol on a supervised outpatient basis.[18] Milk and meat inspection had essentially stopped infection with *M. bovis* at source.[19] BCG vaccination had been rolled out for schoolchildren.[20] Among immigrants tuberculosis posed problems.[21] They were not dealt with particularly effectively or sensitively but the numbers remained small, particularly when compared with other national morbidity statistics. Emphasis was placed on alternative areas of ill-health. Interest and resources would be increasingly focused on the rising morbidity and mortality of non-communicable diseases – cancers, cardiovascular and neurodegenerative diseases, the effects of alcohol, tobacco and belatedly mental health – with some consideration of their social determinants.

McKeown's (flawed) tale of tuberculosis became prescient because his optimism for a possible future was shared with others who wanted to see health as a positive state of existence, not mere freedom from disease.[22] The overlap in thinking moved McKeown beyond parochialism. It seemed apposite, in the context of the Canadian Lalonde report (1974) and the subsequent WHO redefinition of health (1978). The WHO and its staff in the field were fatigued by poor returns from its costly vertical disease programmes. They sought a new direction via 'health for all', health services strengthening, and a greater focus on primary health care (PHC).

Despite this optimism, McKeown's ideas also fed into interpretations and consequences he had not foreseen and would certainly not have applauded. The later 1970s witnessed frequent, often politically motivated, criticism of all the professions, including medicine. Those on the radical left welcomed the blow he had struck against scientific medicine in the past, but were often stronger on rhetoric than historical accuracy. There were also increasingly intense discussions about how to spend health-related budgets emerging from the right. Issues included deciding on the appropriate balance between funding medical care and funding wider improvements in the social infrastructure. If economic growth had led to improved health, a free-market approach and neoliberal's 'trickle-down effect' was the best strategy. These strategies lacked rigour, but the overriding importance of economic growth struck a chord with those concerned with decreasing the amount that governments spent on health-care at home and on international health and development abroad. In the 1970s and 1980s, the aims of reducing poverty and increasing welfare by equitable distribution of wealth and investment in health-care were replaced by loans and programmes of structural adjustment intended to boost nascent economies.[23]

There were many negative effects of these changes. Even the well-intentioned emphasis on PHC would prove precipitous for tuberculosis. It contributed to a disengagement with the disease in places where active specialist involvement was most needed and there was little in the way of funding for the infrastructure that good PHC required.[24] These were difficult times after the OPEC countries increased the price of oil and the world economy stalled. Tuberculosis resurfaced again in the developed world, most famously among the urban underclass of major cities such as New York, San Francisco and Seattle in the 1980s and disease rates climbed into the 1990s. In Europe, the incidence rose dramatically in those countries subjected to the tumultuous conditions following the collapse of the USSR, but nowhere was immune.

In many of the developing countries of Asia, sub-Saharan Africa and South America much higher initial rates would rise precipitously. Where tuberculosis thrived it was grounded in poverty, deprivation and dislocation. The disease had also changed with the rise of drug resistance to the frontline drugs and co-infection with HIV/AIDS. Those who needed it most were often unable to access health-care. Surveillance and case detection tended to be poorly managed and the supply of drugs was uncertain. Anything beyond clinical diagnosis relied upon old technologies, some of which had been generated at the end of the nineteenth century.

## The laboratory disease

Robert Koch's discovery of the germ of tuberculosis in 1882 was not a moment of transcendence but it did open a new chapter in the disease's long history.[25] Case reports in the medical press show how clinicians educated in the pre-bacteriology era incorporated their numerous understandings of various consumptions and types of phthisis into the new single disease of tuberculosis.[26] One of the ways in which this was negotiated was to embrace the diagnostic potential that Koch's work offered. As Andrew Cunningham has explained for plague, the reshaping process that took place in the laboratory transformed 'a disease whose identity was symptom-based' into one 'whose identity was cause-based' – the germ became the disease.[27] At its most simplistic, sickness was marked by the causative germ's presence and health by its absence. For tuberculosis, the complexities of microbial virulence, exposure without disease, seroconversion and latency would come to challenge such an elegant and appealing simplicity, but its explanatory power was effectively exploited by its early evangelists and turned to practical use. Christoph Gradmann has expertly applied this thinking to Robert Koch and his work on tuberculosis, analysing Koch's laboratory methods as technological processes.[28] Koch had honed his bacteriological skills on anthrax and wound infections. In his tuberculosis work, he was obliged by the biological characteristics of mycobacteria to adapt the techniques, equipment and processes he had adapted from others and established for himself.

Koch's postulates were the cornerstone of his proof of the role of the tubercle bacillus. He identified, visualised and cultured, as a pure culture, the organism he believed to be the causative germ.[29] He then reintroduced the pure culture into an experimental animal and followed the development of the expected pathologic state, before recovery of the same organism completed the circle of proof. Broken down into its constituent parts, the stages of this experimental protocol provided various techniques of laboratory diagnosis to augment traditional clinical appraisals. The thermometer already offered precision for assessing fever and scales for weight-loss, but these were now likely to become supplementary measures of symptoms. The stethoscope also straddled the pre- and post-germ aetiologies of tuberculosis. Born of the clinic and the morgue, it was a widely applied listening device, opening up to the ear what lay within the chest cavity.[30] In comparison, the tools and techniques of the bacteriological laboratory were concerned exclusively with infectious diseases and employed to provide a previously unachievable degree of certainty in diagnosis. Such tests could be transferred from the experimental laboratory to laboratories in hospitals, or attached to the public health bureaucracies and act as diagnostic tests. A good example is the free diagnosis in public health laboratories offered as an incentive to registration of cases in late nineteenth-century New York City.[31] In the community the same tests could be used for screening either by enhanced or active case-finding programmes.

## Identifying and seeing: the sputum smear

Koch developed Rudolph Virchow's injunction that life scientists must 'learn to see microscopically', extending this idea in new directions via his diagnostic tests (although there is an irony here since Virchow resisted the germ theories at the heart of these innovations). Koch combined the latest developments in microscopic hardware (the oil-immersion lens for improved resolution combined with a light condenser to better illuminate the microscope's field) with innovative, if quickly superseded, staining methods. Koch began with a solution of methylene blue and a mordant (or fixative) to develop the colouration of his sample of tuberculous tissue. After applying heat via a water bath, the blue over-stained preparation was then treated with vesuvin (Bismarck brown) and washed. Everything now appeared a faint brown except the bacilli, which remained a conspicuous blue and could be much better differentiated.[32] Koch believed it was only the tubercle bacilli (and the closely related leprosy bacilli) that took on this unique coloured identity, one of its various laboratory identities. The lipids in the cell walls of *Mycobacteria* make this a difficult organism to stain, but once the colour is taken up, it cannot be stripped out with an acid wash. Subsequently labelled 'acid-fast', the technology of the staining protocol shaped the object under scrutiny, reducing it to a property of the bacterial cell wall.

In trying to see what Koch had seen, others quickly improved on the way he had made his mycobacterium visible. A flurry of papers appeared during 1882, 1883 and 1884. Some were testy exchanges on methodology indicating the difficulties experienced in working out these methods. They came from a variety of sources – research laboratories, sanatoria and public health agencies – and from various countries, as the bacteriological aetiology of tuberculosis was subject to widespread independent investigation. In Berlin, Paul Ehrlich (1854–1915) experimented with different stains and acid decolourants. Rather than taking tuberculous matter from a tissue sample, Ehrlich sought his mycobacteria in samples of sputum. He used a dissecting needle to transfer a tiny amount of this material, placed it between two glass cover slips and squeezed. Prising the two apart he had created two thin layers ready to be dried either by heating or more quickly by passing through the flame of a Bunsen burner – a direct smear.[33] Sputum coughed up from the lungs or larynx, as if one were spitting into a spittoon or bottle containing disinfectant, was the most easily accessible source of mycobacteria from the commonest form of the disease. It offered great potential as a diagnostic material in cases of tuberculosis. In the community, this body fluid became the target of public health campaigns to prevent spitting.[34] In the laboratory, it was transformed into the customary diagnostic biological substrate in tuberculosis, providing material for smears, culture and inoculation.

Interest in the diagnostic and prognostic implications of visualising mycobacteria in the sputum accompanied the technical innovations. These were driven by necessity and pragmatism as much as invention. Franz Ziehl (1857–1926), a Lübeck neurologist, was unhappy with the quality of aniline oil available to him and he used carbolic acid instead. Friedrich Neelsen (1854–94), pathologist and prosector at Dresden, used fuchsin in carbolic acid and sulphuric acid as decolourant and reported that 'This method, which I have used almost exclusively for a long time, gives in my hands better results than the other methods (perhaps only because I am most accustomed to it)'.[35] Although others were involved in modifying Koch's original methods, the Ziehl-Neelsen (ZN) stain, first described eponymously in the 1890s, became the definitive protocol to see mycobacteria and, by seeing, to diagnose tuberculosis from direct sputum smears.[36]

ZN staining remained for some time the mainstay of first-line tuberculosis diagnosis. It was the test most easily delivered at or near a point of care such as a primary health-care setting. Performed correctly it proved to be highly specific (there are few false positives) but carried a lower sensitivity (the false negative rate). If it began life as cutting-edge bacteriological science, it would owe its longevity to factors that meant it could be performed in resource-poor settings where equipment, skill sets and overburdening of weak laboratory systems were limiting factors. Later used as part of a standardised protocol of three deliberately spaced smears ('smear-positive' was defined as two of the three being positive), it became the baseline against which innovations in sputum smear diagnosis were measured in terms of cost and efficiency for high-prevalence low-income countries.[37]

## Pure culture: tubes, dishes and media

The development of the solid media 'plate technique' was significant for the laboratory production of pure cultures of disease-causing bacteria, and is seen as one of the key technological innovations of the early years of bacteriological research.[38] Pasteur had relied upon liquid nutrient to culture his pathogenic microorganisms, but a solid medium avoided the uncontrollable mixing of different species of microorganisms, which were difficult to separate. There were various solid media precedents but the addition of gelatine to a sterile nutritive liquid, a broth or bouillon, which could then be poured onto a plate to set created the basis for a string of technical improvements in equipment and the composition of the growth medium. To culture tubercle bacilli, Koch settled on a different medium of coagulated blood serum and tightly stopped test tubes, tilted to one side to produce a nutrient slope. With appropriate care this could overcome the difficulties imposed by the fastidious growth requirements of *M. tuberculosis.*

Early examples of Koch's cultures are historic objects, icons of germ theory.[39] As with developments in staining technology, the techniques Koch had devised were rapidly improved upon. Better media increased the reliability of diagnostic cultures and the kinds of information that could be gained, including ultimately the ability to determine the strain infecting the patient and sensitivity to various anti-tuberculosis drugs. Perhaps the most important innovation in terms of its sustained use would prove to be the Löwenstein-Jensen solid egg-based growth media, introduced in the 1930s. This used dyes to prevent the growth of contaminating bacteria and targeted nutrients to improve growth of the target mycobacteria.

In the 1940s, René Dubos and others initiated a successful series of liquid media that allowed initial rapid bacterial multiplication from a smaller amount of source material, but problems of contamination remained. In the 1970s further reduction in 'time to detection' was achieved by a liquid media method involving radioactive carbon 14. This automated process reduced the waiting time from three to four weeks to 10 to 14 days on solid medium. The radiometric assay was expensive and required more sophisticated laboratory equipment and highly trained staff. It was only practical as part of a centralised laboratory system processing large numbers of samples. Even the non-radioactive carbon assays that followed were beyond the resources of many. In low-income settings these and other advances in culture tests were compared to the Lowenstein-Jensen method, which if slower was much cheaper and easier to store and use. After ZN staining the 'gold standard' of tuberculosis diagnosis was pure culture. As with the staining protocol, cost and efficiency were recurrent problems in laboratory provision.

## Living proof: animals and tuberculin testing

In addition to the chemical technologies of staining, physical technologies of glassware and solid media, microscope and microphotograph, Koch utilised experimental animals as a living technology. Suitably prepared animals, that is those free from previous infection, acted as *in vivo* culture experiments. In favoured species – Koch preferred the guinea pig, but rabbits were also used – the pathological organism produced a recognisable disease state from which the infectious agent could be recovered again. Experimental animals provided Koch with what Schlich terms a 'functional representation' of the pathogenic microorganisms.[40] In the diagnostic laboratory, such a process could be used as a confirmatory diagnostic procedure after direct sputum smear and concurrently with culturing, but its longevity was in the research and reference laboratory.[41]

The principle of bacterial agglutination (1896), using predetermined dilutions of sera taken from immunised animals, was successfully developed into a diagnostic test for various diseases but early efforts for tuberculosis were disappointing. More success came from *in vivo* methods and tuberculin testing would prove another form of living assay. Tuberculin testing was an offshoot of Koch's abortive tuberculin therapy. Gradmann has stressed that Koch was well aware of the diagnostic potential of the 'tuberculin reaction' and it was actively discussed by those testing tuberculin as a cure immediately following its release.[42] The diagnostic test involved injecting a very small amount of tuberculin and measuring the febrile reaction. It was the first of a series of immunological hypersensitivity reactions promoted as a means of diagnosing early, pre-clinical tuberculosis. Among prominent early versions were those developed by Clemens von Pirquet (1874–1929), Charles Mantoux (1877–1947), Ernst Moro (1874–1951) and Albert Calmette (1866–1933). Each used the body's immune reaction as the diagnostic assay, reading the result directly on the patient's skin.

Moro (1908) combined tuberculin with lanolin to form an ointment that could be rubbed into a small area of the skin.[43] After 24 hours the appearance of raised papules where the ointment had been applied was read as a positive result. Moro's technique was subsequently refined into a more routine patch test. Calmette (1906) instilled tuberculin solution into the least sensitive part of the eye, the conjunctival sac. Von Pirquet (1907) and Mantoux (1908) introduced tuberculin under the skin, developing the tuberculin skin tests (TST). Von Pirquet placed a drop on the skin and then lightly scratched the skin through the drop and left the liquid to dry. Mantoux injected a measured amount of a known dilution of tuberculin intracutaneously. In both a positive reaction – a raised pink area – was visible after 48–72 hours.

The size and hardness of the reaction indicated exposure to the bacilli, but appreciating its meaning was a learned clinical skill not an absolute measure. Debate ensued about whether the test indicated exposure, naturally acquired immunity, or a latent infection that could in time, and in the right circumstances, progress to active disease.[44]

Beginning in the 1900s, it was among the young that TST was initially used as a survey tool. The aim of these early surveys was to better gauge the exposure in a given population and identify more precisely a new category of vulnerability, termed the 'pre-tuberculous child'. The value of tuberculin testing was most appreciated by those concerned with the health of the next generation as fears continued about national degeneration.[45] The possibility of latent cases of respiratory tuberculosis fuelled the establishment of preventoria, open-air schools and other specialist institutions.[46] TST was not necessarily preferred to clinical and social assessments when screening for 'at risk' children, because it could be cumbersome. Up to five injections of increasing strength were recommended until the early 1930s; thereafter two were still required and there were a number of reasons why false positives were produced.[47] Teemu Ryymin discusses why school inspectors in Norway in the 1920s and 1930s did not favour the TST, as well as the shift to its mandatory use during WWII.[48]

TST was also employed as a preparatory tool for vaccination. After the war, Norway and its Scandinavian neighbours, keen advocates of BCG vaccination, promoted TST and BCG as part of their efforts to rebuild Europe. Success in this arena led to expansive programmes involving children and adults beyond Europe.[49] The International Tuberculosis Campaign (ITC) began its work in India in 1948 using the newer Purified Protein Derivative (PPD) tuberculin (developed just before war), and the Moro patch test for children under 12. The aim was to differentiate between reactors, who would not be vaccinated, and non-reactors who would receive BCG. The ITC wanted to develop a standardised procedure to match their standardised tuberculin, a good example of the importance of understanding the 'practice turn' in medicine.[50] India provided a valuable testing ground for large-scale programmes in low-income countries, but the 'huge population and a weak infrastructure' was one of a series of problems that led to a pragmatic decision to remodel the programme and adopt a single TST injection before vaccination of non-reactors.[51] As Niels Brimnes has argued, the difficulties surrounding transfer of TST technology to low-income countries shared many similarities with the final diagnostic tool in the kit: x-rays.[52]

## Seeing in real time: x-rays

While both bacteriological and immunological methods of diagnosis arose as a direct result of germ theory, the advent of x-rays was unrelated to medicine in general and tuberculosis in particular. This horizontal transfer was initially expensive and of only limited value, and the first use of x-rays tended to be an additional confirmation of disease in high-end sanatoria. Better results were reported with a fluorescent screen rather than a photographic plate but this involved large doses of radioactivity and left no record that could be referred to again. Where the aim was to find early, pre-symptomatic or latent cases this was a serious disadvantage. Although costly, a few early detection surveys explored the potential of x-rays but it was Manuel Dias de Abreu's (1884–1962) development of a reliable means of mass radiography in 1936 that proved crucial. He put together a package of recent advances fine-focus, rotating anode tubes; a highly luminescent screen; wide aperture lenses and fine-grain, fast films – and used these to conduct thoracic surveys in his native Brazil. Refinements, such as a reduction in size and weight of the equipment and allowing people to remain clothed, increased the appeal of taking mass x-ray to the target group. While there were heroic population surveys, it quickly became clear that more focused surveys of 'at risk' groups provided the best results. When incidence was very high, for instance among Alaskan Native Americans (1945–57), or a government determined enough, such as in Reykjavik, Iceland (1945) and Denmark (1950–52), this could still involve most of the population in a given area. Measured as the percentage of the target population x-rayed, success depended greatly on the pre-survey campaign alerting people to come forward and following up those who did not in order to persuade them that they should.[53]

Mass x-ray, it was hoped, could detect active disease at an earlier stage and bring previously undetected cases to the attention of prevention programmes. Where falling incidence made mobile x-ray programmes uneconomic, referral to medical centres with standard x-ray equipment was considered to be most cost-effective and efficient. While this refinement in policy was happening in high-income countries, efforts to roll out mass x-ray in low-income countries began to founder. India again provided a salutary lesson. Targeted surveys among army recruits (1945), Delhi civil servants (1953–54, 1960–62) and 200 rural villages in South India (1950–51) appeared to give good results. Analysis of these and subsequent studies indicated that 80 per cent of those identified by mobile x-ray units as suffering from tuberculosis were already aware of their symptoms (most often this was a persistent cough) and 50 per cent had sought help.[54]

That this latter group had unmet needs sent out a disturbing message about resource allocation and vertical health programmes versus health-systems strengthening. Once (re)diagnosed, patients often found it hard to access regular treatment. They joined the pejoratively labelled group of 'defaulters': those who were unable to complete the long course of treatment recommended for a variety of reasons. Rather than expanded investment in expensive mobile technology as part of a vertical programme, alternative resourcing improved passive case detection at health centres. Once diagnosed through this route, those who came forward because they felt ill and wanted help might receive better support throughout their treatment regime, to ensure it was completed.[55] Mass x-ray can be seen as an example of an inappropriate technology transfer.[56] In this it echoed similar transfers during the colonial era. The flaw was not in the machinery per se but in its imposition on a system not able to cope – what the WHO's director general Marcolino Candau termed the 'economic "have" and the "have not" countries'.[57] Patients who voiced their dissatisfaction were speaking out against both local limitations and the external interventions of the international community with their 'purposeful' or 'planned modernization'.[58] Implicit in its 1964 report, the 1974 WHO Expert Committee on Tuberculosis positively discouraged this technology where the health infrastructure still struggled to help those who came forward. This fed into the rationale for strengthening health systems and integrating the detection of tuberculosis within public health programmes, accessed at a point of care, rather than continuing vertical disease programmes. Chest x-ray screening continued to be 'officially discouraged' when the WHO launched its DOTS

411

(directly observed treatment, short-course) strategy in 1993 in response to the 'global emergency' that tuberculosis represented.[59]

## A 'global emergency' in diagnostics

DOTS was introduced both as a programme and a brand. A unified strategy was considered necessary to streamline care based on best-practice guidelines. At the same time it was also recognised that a concept that could attract both funding and partner organisations was essential in the increasingly competitive international health arena. Funds were sought to run the programme and innovate new tools. In the same year the World Bank's *World Development Report* explained how treating tuberculosis could actually lead to economic development since tuberculosis affected young wage-earners – those in their 'economic prime'. Using the disability-adjusted life year (DALY), their calculations made tuberculosis treatment the most cost-effective of its low-cost public health interventions.[60] The immediate response sought to intensify diagnosis to utilise this technology. At its launch, the DOTS strategy included clear diagnostic and clinical algorithms to help countries take their national tuberculosis control plans forward.[61]

Cognisant of the scale of the problem and the technical limitations in many of the high-burden countries, preliminary testing, preferably at the point of care, relied upon direct ZN sputum smear. In high-prevalence, low-income settings this was the best primary test available. It was most effective where bacterial load was high, successfully catching patients with the severest infections who were considered to be high transmitters. Patients could begin treatment immediately once a positive smear was detected. Under guidelines for national tuberculosis programmes using DOTS, the aim was to detect 70 per cent of sputum smear-positive cases in a country and treat 85 per cent of those detected with a minimum six-month course of drugs, to be supervised for at least some of the time. Diagnosis would rely upon passive case detection with some enhanced case-finding at well-equipped locations. There were continuing concerns about overburdening fragile delivery systems with active case-finding. Variations in the existing diagnostic tests had histories of use but unlike the drug regimes had not been subject to the same rigorous testing.[62] In addition the added burdens of co-infection with HIV, drug resistance and the growing importance of tuberculosis in children subsequently added to the complexities of diagnosis.[63]

Countries with the highest burden of tuberculosis have tended to be those with the poorest disease surveillance and vital registration systems. Their laboratory facilities and levels of trained staff are also weaker. This has historically kept diagnostic technology in these countries at the lowest level. In the aftermath of the launch of DOTS, it became a recurrent theme in the tuberculosis advocacy and policy literature that the diagnostics then in routine use were outmoded and inadequate: 'for most TB patients worldwide there is a simple answer to the question "What is new in TB diagnostics?" Not much.'[64]

An assessment of the success of the DOTS programme over the period 1995 to 2000 revealed that although treatment rates were promising (the cure rate was cited as a global average of 80 per cent of the 1999 cohort), case detection was less effective. Over the same period only 40 to 50 per cent of the hoped-for smear-positive cases were detected. Projections indicated that without enhanced ways of reaching those who needed diagnosis, case detection would remain static and leave an estimated 3 million undetected cases worldwide. The authors appealed for a comprehensive approach: 'The methods of extending DOTS must be as diverse as the reasons why case detection is low: these methods will include building links between public and private practitioners, targeting populations at risk, ensuring best use of current diagnostic methods as well as introducing new ones, and providing health facilities where none have previously existed'.[65]

During the 2000s, the urgency to improve diagnosis to try and meet the revised Millennium Development Goals garnered increasing attention. In addition to realising that large numbers of smear-positive cases of disease were not being reached, the realisation that smear-negative cases could transmit the disease posed a new challenge. According to Dr Rowan Gillies, president of Médecins Sans Frontières (MSF) in 2004: 'We rely on a 19th century tool – it doesn't detect paediatric, extra pulmonary, or smear negative tuberculosis. People who get into a well managed DOTS programme do well; the problem is that DOTS excludes many people [through poor diagnosis] with active tuberculosis.'[66]

A survey paper on behalf of the World Health Organization Tuberculosis Diagnostic Initiative (WHO/TDR) referred to 50 academic and industrial groups actively working towards improving current tests and exploring new technologies. Given the socio-economic variations in countries with high disease burdens, the survey's author, Dr Mark Perkins, concluded that it was unlikely to be a 'one-size-fits-all diagnostic solution'. His suggestion that tuberculosis laboratories needed to correctly perform ZN smear microscopy, 'which includes microscope maintenance, training for technologists, and laboratory quality control', hinted at problems facing the most basic diagnostic test for tuberculosis and the one that formed the primary DOTS diagnostic.[67] In a more transparent era, there was a tricky pathway to be negotiated between overt criticism of the laboratory infrastructure of individual countries and highlighting weaknesses to stimulate improvement.

The Global Laboratory Initiative (2008) was a structural response. Meanwhile, there were ongoing technical and operational innovations in sputum smear technology and trials to determine their suitability in the field. Technical means included centrifuging the sample and chemical pre-treatments, and the introduction of low-cost fluorescent LED microscopes.[68] Operational assessments evaluated clinical predicators and scores used in advance of sputum smear. The rationale for a patient to remain at a testing centre to produce repeated sputum samples over as long as three days was questioned. Such protracted diagnoses had led to 'diagnostic default', a familiarly pejorative term. Research indicated that it was possible to reduce the number of samples to two and change the definition of smear-positive accordingly. The design of the trials and published debates by expert consultations to the TDR programme showed increasing sensitivity to both the needs of patients and those working in the field.[69]

Perkins (WHO/TDR) also looked forward to potential diagnostics including molecular assays that relied upon the burgeoning technological applications of the polymerase chain reaction (PCR). Just as x-rays had not been developed with tuberculosis diagnostics in mind, PCR was a ubiquitous technology used for genotyping, cloning, mutation detection, sequencing, microarrays, forensics and diagnostics.[70] Although this was a technical advance for tuberculosis laboratories in 'industrialised countries', the outlook in developing country settings was poor because the required degree of 'technical support and quality control' could not be achieved. The role of the new technology was not 'in doubt' but it must be subject to innovation and thorough field testing to allow the drawing up of appropriate diagnostic algorithms. In low- and middle-income, high-burden countries, 'there are no funds to be wasted on inefficient or misapplied technologies'.[71]

Developments in molecular diagnostics continued. The Gates Foundation provided extensive research funding as part of its commitment to tuberculosis control. Through a new public/private/philanthropic partnership, characteristic of international aid in the twenty-first century, FIND (Foundation for Innovative Diagnostics) and Cepheid were the major partners in the development of an automatic cartridge-based molecular diagnostic system. This adapted technologies developed for the Department of Homeland Security in the USA in the wake of the anthrax threat.[72] The GeneXpert platform was designed to take a single sample of sputum and in a fully automated process extract the bacteria from the sample, prepare it for DNA amplification and detection, and provide a result in less than two hours. The commercial Xpert MTB/RIF is designed to detect the presence of *M. tuberculosis* and rifampicin-conferring resistant mutations in the sample bacilli, providing immediate information on drug resistance.

In December 2010 the WHO felt sufficiently confident to endorse the Xpert MTB/RIF test and the FIND consortium subsequently announced a subsidised price for the cartridges ($9.98). The cost of the automated instrument was $17,000 and in addition there were warranty and calibration add-ons. For comparison, the same website costed an LED microscope, with battery supply unit and carry case, at €1,637 (c. $2,136).[73] Continued reports of trials to measure the impact of the technology against existing microscopy and culture tests followed these policy announcements. Among the positives, Xpert was reported to be more sensitive than microscopy, although less so than culture. Where cases typically have few bacilli and microscopy would generally give a sputum-negative result this was considered significant. Assessments of the operational impact proved to be mixed. Improvements in the case detection rate were inconclusive. The equipment, which must also be purchased and maintained, placed extra demands on laboratory facilities. These ranged from ensuring a stable electricity supply, the provision of air-conditioning to maintain the equipment in working order and the training of computer-literate technical staff. Despite increases in the number of cases beginning treatment on the day of their test, this had no effect on lowering tuberculosis morbidity. It is suggested that false positives for rifampicin resistance could result in patients being sent away from home for inappropriate treatment and discourage patient participation.[74] Since a diagnosis of tuberculosis remains a stigmatising experience anything that serves to further de-incentivise patients to come forward for treatment poses a serious problem.[75]

## Conclusion

The Stop TB Partnership slogan for the 2015 World TB day (24 March), 'Reach the 3 Million: Reach, Treat, Cure Everyone', referred to the 3 million active cases of disease that remain undiagnosed each year. Its promoters called for an intensification of efforts by those running national programmes and for donors to come forward to meet a current funding deficit.[76] Other advocates had already suggested that calling tuberculosis a social disease was an 'excuse for complacency', and that 'the most important ingredient of a new paradigm [for tuberculosis control] is not biomedical or social, it's urgency'.[77] While this might be a much needed boost, urgency has been a familiar refrain since the development of effective drug protocols made control appear feasible.

The direct-stained sputum smear and standardised methods for growth of pure cultures of *M. tuberculosis* were developed as part of the germ theory of tuberculosis. There were routinised and commercialised improvements, but these tests remained essentially the same. Tuberculosis screening benefited from TST, an early development from the nascent understanding of tuberculosis immunology. The adjunct technology of x-rays provided a second string for screening and diagnosis. These formed a diagnostic tool kit to augment clinical assessments and were looked upon as a prerequisite to effective tuberculosis control, although there was little rigorous investigation of their relative merits or applicability in more challenging situations. There were considerable successes in their targeted use where health systems were well developed or where resources were concentrated on a small number of patients. By contrast their initial use as part of the ambitious plans of the 1950s and 1960s to deal expeditiously with tuberculosis in the developing world, as 'a public health and medical administrative problem' rather than a 'social and economic' one, yielded disappointing results.[78] Structural weaknesses in the provision of health-care impeded the deployment of isolated technologies introduced as part of top-down vertical programmes in resource-poor settings. The poverty in which a patient lived was not the only impoverishment that needed to be factored into attempts to control a social disease such as tuberculosis.[79]

In the wake of the declaration of tuberculosis's global emergency in 1993, systematic reviews of the DOTS programme revealed the limitations of the recommended diagnostic tests. The aim of the international partnership behind DOTS was to deliver quality diagnosis and drug treatments in high-incidence countries whatever the socio-economic status of the patient. Despite increases in the numbers of cases reached, weak laboratory infrastructure remained a stumbling block. Poor laboratories tended to force reliance upon the simplest test at a programme level and impeded the deployment of newer technologies. The realisation of these difficulties has been part of the drive to develop diagnostics that offer better results without making impossible demands on infrastructure.

The emergence of new diagnostic technologies in the twenty-first century has been greeted with enthusiasm. Efforts have been made to make these affordable to national programmes to ensure their use, but concerns about introducing the favoured Xpert TB/FIR have a familiar ring – that the costs may not bring the intended benefits and that such technology if inappropriately applied may cause problems as well as solve them. In the scientific and policy literature there has been a greater commitment to assessing these issues. An appreciation of the suitability of diagnostic technology is far more prominent than in the past. There is a growing awareness of cross-disciplinary research to better understand why deployment might fail. Until a new vaccine is developed and introduced, rapid diagnosis and apposite treatment will continue to remain the technical key to tuberculosis control and their deployment in turn will present pressing technical, social and political problems.

# Notes

1 S. Sontag, *Illness as Metaphor*, New York: Farrar, Straus & Giroux, 1978; W. Johnston, *The Modern Epidemic: A History of Tuberculosis in Japan*, Cambridge, MA: Council on East Asian Studies, Harvard University, distributed by Harvard University Press, 1995; R. Packard, *White Plague, Black Labor: Tuberculosis and the Political Economy of Health and Disease in South Africa*, Pietermaritzburg: University of Natal Press, 1990.

2 C. Lawlor, *Consumption and Literature: The Making of the Romantic Disease*, Basingstoke and New York: Palgrave Macmillan, 2006.

3 P. Pringle, *Experiment Eleven: Deceit and Betrayal in the Discovery of the Cure for Tuberculosis*, London: Walker and Company, 2012.

4 L. Bryder, '"We shall not find salvation in inoculation": BCG vaccination in Scandinavia, Britain and the USA, 1921–1960', *Social Science & Medicine* 49, 1999, 1157–67; D. Barnes, *The Making of a Social Disease: Tuberculosis in Nineteenth-Century France*, Berkeley: University of California Press, 1995.

5 K. Waddington, *The Bovine Scourge: Meat, Tuberculosis and Public Health, 1850–1914*, Woodbridge and Rochester, NY: Boydell Press, 2006.

6 C. Roberts and J. Buikstar, *Bioarcheology of Tuberculosis: A Global View on a Re-emerging Disease*, Gainseville: University Press of Florida, 2003.

7 C. Timmermann, 'Chronic illness and disease history', in M. Jackson (ed.), *The Oxford Handbook of The History of Medicine*, Oxford: Oxford University Press, 2011, pp. 393–410.

8 M. Worboys, 'Before McKeown: explaining the decline of tuberculosis in Britain, 1880–1930' in F. Condrau and M. Worboys (eds), *Tuberculosis Then and Now: Perspectives on the History of an Infectious Disease*, Montreal and London: McGill-Queen's University Press, 2010; D. Wootton, *Bad Medicine*, Oxford: Oxford University Press, 2006; R. Woods, 'Medical and demographic history: inseparable?' *Social History of Medicine* 20(3), 2007, 483–503.

9 A. Fairchild and G. Oppenheimer, 'Public health nihilism vs pragmatism: history, politics, and the control of tuberculosis', *American Journal of Public Health* 88(7), 1998, 1105–17; J. Colgrove, 'The McKeown Thesis: a historical controversy and its enduring influence', *American Journal of Public Health* 92(5), 2002, 725–9; M. Selgelid, 'Ethics, tuberculosis and globalization', *Public Health Ethics*, 1(1), 2008, 10–20.

10 'Tuberculosis Profiles by Country', http://www.stoptb.org/countries/tbdata.asp, accessed 26 May 2015.

11 R. Coker, *From Chaos to Coercion: Detention and the Control of Tuberculosis*, New York: St Martin's Press, 2000.

12 McKeown's papers from 1955 onwards (co-authored variously with R. G. Brown, R. G. Record and R. D. Tuner) were followed by his books, *The Modern Rise of Population*, New York: Academic Press, 1976 and *The Role of Medicine: Dream, Mirage, or Nemesis?*, London: Nuffield Provincial Hospitals Trust, 1976.

13 S. Szreter, 'The importance of social intervention in Britain's mortality decline c. 1850–1914: a reinterpretation of the role of public health', *Social History of Medicine* 1(1), 1988, 1–38; G. Mooney, 'Historical demography and epidemiology: the meta narrative, in Mark Jackson (ed.), *The Oxford Handbook of the History of Medicine*, pp. 373–92.

14 Szreter, 'The importance of social intervention', p. 33.

15 L. Wilson, 'The historical decline of tuberculosis in Europe and America: its causes and significance', *Journal of History of Medicine and Allied Sciences* 45, 1990, 366–96.

16 S. Sheard, *The Passionate Economist: How Brian Abel-Smith Shaped Global Health and Social Welfare*, Bristol: Policy Press, 2014.

17 A. McIvor, 'Germs at work: establishing tuberculosis as an occupational disease in Britain, c. 1900–1951', *Social History of Medicine* 25(4), 2012, 812–29.

18 Claire Latham-Leeming 'Unravelling the "tangled web": chemotherapy for tuberculosis in Britain, 1940–1970', *Medical History* 59(2), 2015, 156–76.

19 Waddington, *The Bovine Scourge*.

20 Bryder '"We shall not find salvation in inoculation"'.

21 J. Welshman, 'Importation, deprivation, and susceptibility: tuberculosis narratives in post-war Britain', in F. Condrau and M. Worboys (eds), *Tuberculosis Then and Now*, pp. 123–47.

22  S. Szreter, 'The population health approach in historical perspective', *American Journal of Public Health* 93(3), 2003, 421–31, at 428.

23  S. Keshavjee, *Blind Spot: How Neoliberalism Infiltrated Global Health*, Berkeley: University of California Press, 2014.

24  S. Amrith, *Plague of Poverty? The World Health Organization, Tuberculosis and International Development 1945–1980*, Cambridge: King's College, University of Cambridge, 2002.

25  H. Bynum, *Spitting Blood: The History of Tuberculosis*, Oxford, Oxford University Press, 2012.

26  M. Worboys, *Spreading Germs: Disease Theories and Medical Practice in Britain, 1865–1900*, Cambridge: Cambridge University Press, 2000.

27  A. Cunningham, 'Transforming plague: the laboratory and the identity of infectious disease' in A. Cunningham and P. Williams (eds), *The Laboratory Revolution in Medicine*, Cambridge: Cambridge University Press, 1992, pp. 209–44 at 224.

28  C. Gradmann, *Laboratory Disease: Robert Koch's Medical Bacteriology* (trans. E. Forster), Baltimore: Johns Hopkins University Press, 2009; T. Brock, *Robert Koch: A Life in Medicine and Bacteriology*, Madison, WI: Science Tech, 1988.

29  See: the chapter in this by Christoph Gradmann; and T. Schlich, 'Linking cause and disease in the laboratory: Robert Koch's method of superimposing visual and "functional" representations of bacteria', *History and Philosophy of the Life Sciences* 22, 2000, 43–58.

30  S. Reiser, *Technological Medicine: The Changing World of Doctors and Patients*, New York: Cambridge University Press, 2009; J. Duffin, *To See with a Better Eye: A Life of R. T. H. Laennec*, Princeton: Princeton University Press, 1998.

31  D. Fox, 'Social policy and city politics: tuberculosis reporting New York, 1889–1900', *Bulletin of the History of Medicine* 49(2), 1975, 169–95.

32  Koch reported his method in 'Die Aetiolozie der Tuberkulose', *Berliner Klinische Wochenschrift* 19, 1882, 221; it is discussed in detail in Gradmann *Laboratory Disease*, pp. 75–6 and Brock, *Robert Koch*, pp. 119–20.

33  P. Ehrlich, 'Aus dem Verein für innere Medizin zu Berlin. Sitzung vom 1. Mai 1882', *Deutsche Medizinische Wochenschrift*, 8, 1882, 269.

34  Barnes, *The Making of a Social Disease*, pp. 83–6, discusses the renewed abhorrence of sputum.

35  F. Neelsen, 'Ein casuistischer Beitrag zur Lehre von der Tuberkulose', *Centralblatt für die medicinischen Wissenschaften* 28, 1883, 497–501, at 500, cited in P. Bishop and G. Neumann, 'The history of the Ziehl-Neelsen stain', *Tubercle* 51, 1970, 196–206, at 202.

36  Bacilli from other samples – gastric secretions, urine, faeces and biopsy or pathology samples – could also be treated in this way.

37  K. R. Steingart *et al.*, 'Sputum processing methods to improve the sensitivity of smear microscopy for tuberculosis: a systematic review', *Lancet Infectious Diseases* 6(10), 2006, 664–74.

38  Brock, *Robert Koch*, pp. 96–104.

39  G. Taylor *et al.*, 'Koch's Bacillus: a look at the first isolate of *Mycobacterium tuberculosis* from a modern perspective', *Microbiology* 149(11), 2003, 3213–20.

40  Schlich, 'Linking cause and disease', p. 55.

41  J. Marks, 'Ending the routine guinea-pig test', *Tubercle* 53, 1972, 31–4.

42  Gradmann, *Laboratory Disease*, p. 101; 'Official report on the report of Koch's treatment in Prussia', *British Medical Journal* March 14, 1891, 598–600 summarised parts of the 900-page report which appeared as a supplement to the *Klinische Jahrbuch* of 1891.

43  E. Moro, 'Ueber eine diagnostisch verwertbare Reaktion der Haut auf Einreibung mit Tuberkulinsalbe', *Münchner medizinische Wochenschrift* 55, 1908, s. 216–18.

44  T. Ryymin, '"Tuberculosis-threatened children": the rise and fall of a medical concept in Norway, c. 1900–1960', *Medical History* 52(3), 347–64.

45  TST's other vital role was in determining the status of cattle – see Waddington, *The Bovine Scourge*.

46  C. Connolly, *Saving Sickly Children: The Tuberculosis Preventorium in American Life, 1900–1970*, New Brunswick, NJ and London: Rutgers University Press, 2008; L. Bryder, '"Wonderlands of buttercup, clover and daises": Tuberculosis and the open-air school movement in Britain, 1907–1939' in R. Cooter (ed.), *In the Name of the Child: Health and Welfare, 1880–1940*, London: Routledge, 1992, pp. 72–95.

47  E. Lee and R. Holzman, 'Evolution and current use of the tuberculin test', *Clinical Infectious Diseases* 34, 2002, 365–70.

48  Ryymin, '"Tuberculosis-threatened children"', p. 353. TST had been used as a survey tool to understand tuberculosis among discreet populations such as Native Americans before WWII; see C. McMillen, *Discovering Tuberculosis: A Global History, 1900 to the Present*, New Haven and London: Yale University Press, 2015, pp. 71–82.

49  C. W. McMillen '"The red man and the white plague": rethinking race, tuberculosis, and American Indians, ca. 1890–1950', *Bulletin of the History of Medicine* 82, 2008, 608–45.

50  M. Worboys, 'Practice and the science of medicine in the nineteenth century', *Isis* 102(1), 2011, 109–15; I. Löwy, 'Historiography of biomedicine: "bio", "medicine", and in between', *Isis* 102(1), 2011, 116–22.

51  N. Brimnes, '"Vikings against tuberculosis": the international tuberculosis campaign in India, 1948–1951', *Bulletin of the History of Medicine* 81, 2007, 407–30.

52  N. Brimnes, 'Another vaccine, another story: BCG vaccination against tuberculosis in India, 1948 to 1960', *Ciência & Saúde Coletiva* 16(2), 397–407.

53  A. Cochrane, 'The detection of pulmonary tuberculosis in a community', *British Medical Bulletin* 10(2), 1954, 91–5.

54  J. Golub *et al.*, 'Active case finding of tuberculosis: historical perspective and future prospects', *International Journal of Tuberculosis and Lung Disease* 9(11), 2005, 1183–1203.

55  H. Mahler, 'The tuberculosis programme in the developing countries', *Bulletin of the International Union against Tuberculosis* 37, 1966, 77–82.

56  S. Irfan Habib and D. Raina (eds), *Social History of Science in Colonial India*, New Delhi, New York: Oxford University Press, 2007.

57  WHO Expert Committee on Tuberculosis: Eighth Report, World Health Organization Technical Report Series No. 290. Geneva: WHO, 1964, p. 3; see also the discussion in M. Jones, 'Policy innovations and policy pathways: the vagaries of tuberculosis control in Sri Lanka, 1948–2010', paper presented at 'TB and its challenges: perspectives on the resurgence and elimination of a deadly disease', Maulana Azad Medical College, New Delhi, India 27/28 February 2015.

58  C. W. McMillan and N. Brimnes, 'Medical modernization and medical nationalism: resistance to mass tuberculosis vaccination in postcolonial India, 1948–1955', *Comparative Studies in Society and History* 52(1), 2010, 180–209, at 184.

59  M. van Cleef *et al.*, 'The role and performance of chest X-ray for the diagnosis of tuberculosis: a cost-effectiveness analysis in Nairobi, Kenya', *BioMed Central Infectious Diseases* 5, 2005, 111.

60  *Ancient Enemy, Modern Imperative: A Time for Greater Action against Tuberculosis*, http://www.economistinsights.com/sites/default/files/Ancient%20enemy%20modern%20imperative.pdf, p. 6, accessed 3 April 2015.

61  World Health Organization, 'Framework for effective tuberculosis control WHO/TB/94', Geneva: WHO, 1994.

62  Golub *et al.*, 'Active case finding of tuberculosis'.

63  S. Swaminathan *et al.*, 'HIV-associated tuberculosis: clinical update', *Clinical Infectious Diseases* 50(10), 2010, 1377–86.

64  M. Perkins, 'New diagnostic tools for tuberculosis', *International Journal Tuberculosis and Lung Disease* 2000, 4(12), S182–S188, at S182.

65  C. Dye *et al.*, 'What is the limit to case detection under the DOTS strategy for tuberculosis control?', *Tuberculosis* 83, 2003, 35–43, at 42.

66  G. Mudur, 'Medical charity criticises shortcomings of DOTS in management of tuberculosis', *British Medical Journal* 328(7443), 2004, 784.

67  Perkins, 'New diagnostic tools for tuberculosis', SS182–4.

68  World Health Organization, *Fluorescent light-emitting diode (LED) microscopy for diagnosis of tuberculosis policy*, 2011, http://www.who.int/tb/publications/2011/led_microscopy_diagnosis_9789241501613/en/, accessed 2 April 2015.

69  World Health Organization, Tropical Diseases Research, Approaches to improve sputum smear microscopy for tuberculosis diagnosis, Expert Group Meeting Report, Geneva 31 October 2009, http://www.who.int/tb/laboratory/egmreport_microscopymethods_nov09.

pdf, accessed 1 April 2015; D. Walker *et al.*, 'An incremental cost-effectiveness analysis of the first, second and third sputum examination in the diagnosis of pulmonary tuberculosis', *International Journal of Tuberculosis and Lung Disease* 4(3), 2000, 246–51; B. Marais, M. Raviglione, P. Donald *et al.*, 'Scale-up of services and research priorities for diagnosis, management, and control of tuberculosis: a call to action', *The Lancet* 375, 2010, 2179–91.

70  P. Rabinow, *Making PCR: A Story of Biotechnology*, Chicago: University of Chicago Press, 1996.

71  Perkins, 'New diagnostic tools for tuberculosis', SS183–6.

72  M. Ulrich *et al.*, 'Evaluation of the Cepheid GeneXpert system for detecting *Bacillus anthracis*', *Journal of Applied Microbiology* 100, 2006, 1011–16.

73  'Product Prices', http://www.finddiagnostics.org/about/what_we_do/successes/find-negotiated-prices/, accessed 1 July 2015.

74  R. McNerney *et al.*, 'New tuberculosis diagnostics and rollout', *International Journal of Infectious Diseases* 32, 2015, 81–6.

75  P. Mason *et al.*, 'Social, historical and cultural dimensions of tuberculosis', *Journal of Biosocial Science* 22, 2015, 1–27.

76  *World TB Day*, http://www.stoptb.org/events/world_tb_day/2015/, accessed 1 June 2015.

77  P. Isaakidis *et al.*, 'Calling tuberculosis a social disease: an excuse for complacency?', *The Lancet* 384, 2014, 1095.

78  F. L. Soper, 'Problems to be solved if the eradication of tuberculosis is to be realized', *American Journal of Public Health* 52, 1962, 734–45, at 735.

79  K. Lönnroth *et al.*, 'Drivers of tuberculosis epidemics: the role of risk factors and social determinants', *Social Science & Medicine* 38, 2009, 2240–6.

## Select bibliography

Amrith, S., *Plague of Poverty? The World Health Organization, Tuberculosis and International Development 1945–1980*, Cambridge: King's College, University of Cambridge, 2002.

Bynum, H., *Spitting Blood: The History of Tuberculosis*, Oxford: Oxford University Press, 2012.

Coker, R., *From Chaos to Coercion: Detention and the Control of Tuberculosis*, New York: St Martin's Press, 2000.

Condrau, F. and Worboys, M. (eds), *Tuberculosis Then and Now: Perspectives on the History of an Infectious Disease*, Montreal and London: McGill-Queen's University Press, 2010.

Farmer, P., *Infections and Inequalities: The Modern Plagues*, Berkeley: University of California Press, 1999.

Gandy, M. and Zumla, A. (eds), *The Return of the White Plague: Global Poverty and the 'New' Tuberculosis*, London: Verso, 2003.

Gradmann, C., *Laboratory Disease: Robert Koch's Medical Bacteriology* (trans. E. Forster), Baltimore: Johns Hopkins University Press, 2009.

Keshavjee, S., *Blind Spot: How Neoliberalism Infiltrated Global Health*, Berkeley: University of California Press, 2014.

McMillen, C. W., *Discovering Tuberculosis: A Global History, 1900 to the Present*, New Haven and London: Yale University Press, 2015.

Mason, P. H., Roy, A., Spillane, J. and Singh, P. 'Social, historical and cultural dimensions of tuberculosis', *Journal of Biosocial Science* 22, 2015, 1–27.

# 23

# REORGANISING CHRONIC DISEASE MANAGEMENT

## Diabetes and bureaucratic technologies in post-war British general practice

*Martin D. Moore*

On 1 April 1990, the third Thatcher administration imposed a new contract upon British general practitioners (GPs). As part of its plans to remake both the National Health Service (NHS) and UK public health, the Conservative Government made available to GPs for the first time specific remuneration for undertaking special health promotion clinics – a service which covered a range of chronic conditions and their risk factors.[1] By 1993, subsequent Conservative governments had introduced distinct incentives for 'chronic disease management' clinics, issued protocol to outline process measures required for chronic disease payments, and incorporated a number of chronic diseases in new target-based public health programmes.[2]

Taken together, these initiatives marked the first major and sustained policy concern with 'chronic disease' at the central government level in Britain. As recent work has shown, of course, this was not the first time that policy-makers had demonstrated an interest in long-term sickness or in specific chronic conditions.[3] The care of the 'chronic sick', populations of largely elderly and infirm patients institutionalised in Britain's old Poor Law Hospitals, had been a service concern since at least the 1940s.[4] The GP policies were, nonetheless, the first time that the Department of Health had based the reform of services around chronic disease as a distinct – if elastic – category and object of policy. It was a trend, George Weisz makes clear, which gained a contested momentum over the 2000s, in line with various international developments.[5]

Below the level of central government policy, however, concepts of chronic disease – and the organisational challenges of long-term care – had been influential in encouraging reform of local services for decades prior to these changes. New models for chronic disease care based in general practice, though often closely integrated with hospital clinics, had first emerged in the early 1970s, and began to spread widely during the 1980s. In diabetes, the new systems of care were predicated upon proactive disease surveillance of 'routine' patients in community settings, with specialist hospital outpatient clinics reserved for consultation or more 'complicated' cases. GPs slowly assumed the position of the head of the 'primary care health team', having direct access to nursing, ancillary and clerical staff. Finally, the whole system was facilitated by the deployment of numerous forms of managerial bureaucratic instruments, ranging from patient register and appointment systems (allowing the management of time and attendance), to codified treatment protocol and mobile clinical records (directing labour and facilitating shared data storage and retrieval).[6] Similar developments not only occurred in relation to hypertension and asthma during the same decades, but clinicians also consciously drew upon experiences across conditions.[7] Shared 'natural histories', intellectual networks, and sites of care facilitated the exchange of organisational and bureaucratic technologies.[8]

Using a case study of non-insulin-dependent diabetes mellitus, this chapter examines the emergence of this disease management model in post-war Britain. While recognising the influence of shifting epidemiological patterns, it nonetheless situates developments in relation to resource constraints, professional politics, and innovations in medical knowledge and technology. In the face of stagnant resources, consultants felt that diabetes care was no longer sustainable as a solely hospital specialty, especially as changing diagnostic and therapeutic patterns transformed the character of both disease and specialty clinic. The proposed solution of moving greater patient care to – and systematising patient management in – general practice was only accepted nationally because it fitted political and ideological projects of general practice reform. By the 1980s, discussions of diabetes care – and its proactive and bureaucratised surveillance of risk – increasingly mapped onto broader considerations of modern general practice. And by 1990, new models of GP care received formal political backing.

In telling this story, this chapter seeks to provide insight into the adaptation of British health services – and general practice in particular – to the challenges of chronic disease after the Second World War. Whilst historians have produced excellent studies of how public health ideology and central government policy engaged with chronic disease in the post-war period, little has been written in this regard about service innovation.[9] Similarly, although there are numerous accounts detailing the political and institutional histories of the NHS, its scholars have largely neglected the study of post-war general practice and the emergence of new models of care outside of key policy years.[10] Diabetes mellitus provides an effective case study for many of the key developments in post-war service changes in relation to chronic disease. This is not because of any essential features of the disease itself, but rather because doctors discussed approaches to diabetes management as models for other conditions.[11] Moreover, as diabetes care became embedded within visions for reforming general practice more broadly, its study also opens a vista onto the broader transformation of general practice as a distinct form of medicine.

## Diabetes and hospital management

By the early 1950s, diabetes was considered a metabolic disorder or syndrome, primarily caused by a deficiency of, or insensitivity to, insulin.[12] The cardinal markers for the disturbance were taken to be high levels of glucose in the blood (hyperglycaemia) and urine (glycosuria), while clinicians had described a range of prominent symptoms in relation to the condition (including excessive urination, thirst and hunger) across a number of centuries.[13] Since the 1930s, doctors had also noted patients with diabetes developing a plethora of renal, micro- and macro-vascular 'complications' over time, so that by the 1950s major concerns existed about neuropathy, nephropathy and retinopathy in patients as much as their liability to various types of coma.[14]

It was in relation to coma, along with age of onset and weight, that clinicians roughly 'typed' patients from the 1920s onwards. That is, while researchers had posited various causes for (and exceptions to) such a division, doctors tended to discuss patients as either 'severe' – typically young and thin at onset, and needing insulin to prevent rapid hyperglycaemia, wasting and ketosis – or 'mild' – individuals who were largely overweight and above 45 years old at onset, but who were able to maintain normal metabolic function through various forms of diet alone.[15] Alternative terms for dividing patients existed over the post-war period – for instance, 'juvenile' and 'maturity-onset' diabetes, and later 'insulin-dependent diabetes mellitus' and 'non-insulin-dependent diabetes mellitus' – but in essence it was a patient's symptoms, ketones, and response to treatment that determined clinical course and categorisation.[16] The development of oral hypoglycaemic drugs during the 1950s further consolidated therapeutic and clinical divisions, proving efficacious in only certain classes of supposedly 'mild' patients.[17]

It was in relation to the clinically more challenging severe patients that specialist clinics had been developed in Britain after the introduction of insulin therapy in 1922.[18] On the one hand, clinics were praised for their research and teaching potential, and the invention of insulin therapy helped generate the financial and cultural capital necessary for institutionalising specialist research and practice.[19] On the other, interested doctors justified these clinics into the 1940s and 1950s by suggesting that their concentration of expertise and laboratory facilities were central to maintaining patient health.[20] Such claims recognised that management of diabetes had come to centre on long-term monitoring of various metabolic and clinical markers of disease progression, despite the growth of significant doubts about the long-term benefits of 'normoglycaemic' control over previous decades.[21] And clinics, at least initially, were designed to facilitate more organised access to the required laboratory facilities for surveillance, as well as to offer ongoing consultation and education for patients in their self-management.[22]

Although GPs continued to offer varying levels of care, by the 1950s hospital clinics had been widely – if reluctantly – accepted as the ideal place for long-term and specialist surveillance of patients with diabetes.[23] GPs and hospital clinicians alike noted how 'diabetes mellitus . . . and its management is gradually being taken out of the hands of the general practitioner by the establishment of diabetic clinics', a trend that even resentful GPs described as a 'relief' in terms of workload.[24] Such observations find support in the rapidly increasing numbers of clinics through the 1940s and 1950s, rising from 40 clinics in 1940 to 194 in 1955.[25] The importance of clinics and specialist supervision received recognition from the Ministry of Health, which issued guidance on the regional organisation of inpatient and outpatient services in 1953.[26]

The 1950s, however, in many ways marked the zenith of the clinic's reputation. Over the following decade and a half, doctors began to note a number of problems with clinic practice. Whereas doctors had previously lauded the clinic's concentration of patients, many clinicians now voiced concerns about the strains that such centralisation placed on care as patient numbers grew. At one major centre in the Leicester Royal Infirmary (LRI), for instance, registered attendances rose from 6,379 in 1949 to 9,854 in 1965.[27] It was in response to similar growth that John Malins, an eminent mid-century authority in the famous Birmingham clinic, suggested that: 'the size of the diabetic problem . . . is apt to oust all other work unless the intake of patients is strictly controlled – no easy matter', and 'as a result the clinic is apt to become large and unwieldy'.[28] In part, such increases may have resulted from patients living longer in line with new therapies and improved facilities.[29] However, greater detection and referral from general practice also played a role, with the average number of *new* patient attendances yearly at the LRI increasing by over 60 per cent from 286 for 1949–52 to 483 for 1959–62.[30]

During the 1950s and 1960s, clinical and public health doctors identified a range of factors supposedly responsible for the increase in the incidence and prevalence of diabetes. They implicated more sedentary lifestyles, rising affluence, rapidly altered diets, the 'conquest' of infectious disease, and a general ageing of the population as potential causes.[31] Such explanatory frameworks formed part of wider concerns with reframing public health in terms of both chronic disease and lifestyle change, a development itself intimately tied to the creation of the new tools of risk factor epidemiology.[32] Increases in clinic patient load also had some relation to the development of new diagnostic technologies, with two innovations in particular worthy of mention. The first was a new heat-producing tablet, available from 1944, which removed the need for doctors to measure solutions and heat urine for testing.[33] The second, a variety of enzyme-loaded paper, was developed in the 1950s and allowed practitioners to identify initial glycosuria by the strip changing colour when dipped in a urine sample.[34] Although each took time to become widely available to GPs, both nonetheless made initial detection of diabetes considerably easier, with one trial even suggesting that tablets more than halved testing times.[35]

The enzyme-loaded strips in particular were central to a wave of major diabetes population surveys taking place in Britain between the early 1950s and mid-1960s, with substantial effects on awareness and understandings of diabetes. The surveys had an international impetus. Many were inspired by equivalent American undertakings in the 1940s, and they later formed part of efforts to assess global prevalence through research networks established as part of colonial expansion.[36] They reflected, nonetheless, a growing interest in epidemiological surveys of chronic disease in Britain, and were underpinned by the significant growth of post-war finance for biomedical research.[37] In the short term these surveys were designed to assess the extent of unmet need for health-care services, and to hint at potential physiological precursors or socio-cultural causes to be longitudinally assessed.[38] With regards to diabetes, they uncovered significant numbers of undiagnosed cases to add to local clinics, roughly doubling the number of patients in any given community.[39] In the longer term, increased numbers were likely compounded by raised professional and public awareness. In the national press, specialists increased estimates of national prevalence from between 3 and 6 cases per thousand population in 1950, to around 12 per thousand by 1960, and the potential existence of significant amounts of hidden disease was widely reported.[40] National figures, along with the novel notion of submerged disease, concerned the profession and were repeated in the press. The subject of diabetes detection also featured more prominently at national medical conferences and in major medical journals, including one exchange in *The Lancet* during 1963 taking in seven letters over six editions and three months.[41]

Unfortunately for hospital doctors, the rising numbers of patients under their care were not matched by increasing finances. Resources under the new National Health Service were scarce and under intense political scrutiny.[42] Initial projections for resource requirements had been wildly inaccurate, and the resulting disconnect with actual spending frightened politicians into introducing new rationing measures and subjecting the Service to consistent financial review.[43] While the most politically significant investigation into NHS expenditure during this decade – the Guillebaud Report (1956) – defended the NHS as excellent value for money, searches for savings in resource use intensified over subsequent decades.[44] By the 1960s, this scrutiny had turned to a search for improved management of expensive hospital facilities, with voluntary health organisations and central government alike examining means for better organisation of resources and outpatient clinics.[45] Under such circumstances, the retention of large numbers of patients under specialist care became both financially challenging and politically problematic. Concern with overcrowding in outpatient departments peaked in the early 1970s, and led *The Lancet* to blame consultants for confusing quantity with quality in assessing their worth, and for undertaking work which was 'quite unnecessary'.[46]

In the face of this combination of rising patient numbers and strained resources, clinicians became increasingly dissatisfied with both standards offered to patients and the conditions of their work. Initial attempts to compensate for changed conditions involved drafting in more junior staff to assist. However, their characteristically short tenures broke the continuity of care that long underpinned the logic of clinics.[47] Equally, the drive for efficiency in terms of patient turnover left practitioners disgruntled, and they claimed that 'the aims of treatment were becoming increasingly frustrated' in light of resultant 'overcrowding'.[48]

Hospital clinicians also grew tired of the more 'routine' patient being seen. These tended to be patients on diet alone or oral drugs, with physicians complaining that 'management is not difficult and they would have been discharged to their general practitioner if it was known they would have regular supervision of their diabetes in the practice'.[49] This tilting towards the predominance of 'mild' patients had been noted in the early 1950s, but was possibly accelerated by the application of the survey and its mutation of diabetes into a less symptomatic and more quantitative disease entity.[50] Physiological research, that is, had long shown that there was no single norm for average blood glucose levels in humans, and even during the 1920s and 1930s, models of diabetes had suggested that clinical symptoms appeared often only after a long asymptomatic onset.[51] However, along with uncovering 'hidden' cases of clear diabetes, new surveys suggested that far more people experienced blood glucose levels outside a clearly bounded 'normal range' than had been expected.[52] The meaning of these quantitative deviations was unclear, correlating neither with symptoms, clear lesions, nor a definite increased likelihood of pathological change in the future.[53] Indeed, while some patients with 'abnormal' results might go on to develop symptomatic diabetes, when retested a number might revert to 'normal' tolerance.[54]

When found in individuals, these 'borderline' results left clinicians unclear as to what level of hyperglycaemia could be considered pathological, especially in older age groups where glucose tolerance appeared to decline.[55] Conceptually borrowing from hypertension, some clinicians and researchers in the mid-1960s even wondered whether 'benign and malignant hyperglycaemia would be preferable terms' to diabetes.[56] By 1980, follow-up studies of the relationship between persistent hyperglycaemia and complications in populations would fix diagnostic criteria for individuals in relation to quantitative risk for diabetic retinopathy.[57] In the meantime, however, it is possible that this changed character of disease and diagnostic uncertainty led to even more 'mild' hyperglycaemic patients appearing in clinics, even if symptoms continued to form the major prompt for diagnosis into the 1970s.[58] With such patients more common than before, specialists complained that they were unable to focus on the 'difficult problems' suited to their training.[59] Experience, efficiency drives and an expanded disease concept had turned diabetes into a routine disease; such a change frustrated hospital doctors who felt such patients less requiring of their skills than others.

## New models of care and bureaucratic tools in general practice

By the late 1960s and early 1970s, a number of hospital consultants and diabetes specialists had decided that easing the pressures on clinics would require more radical solutions than tried previously, primarily involving moving patients out of the clinics. At the same time, there were GPs who felt that standards in general practice could also be improved through better organisation and co-operation, particularly where GP responsibility for patients had been retained.[60] Through such drives for reform, diabetes care underwent significant change during the following decades.

In terms of clinics, during the 1970s and 1980s physicians tried a number of systems to relieve patient-load, and to extend advice and educational facilities outside of clinic hours.[61] The most widely pursued innovation in care during this period was that of GPs assuming greater responsibility for the care of routine patients, and those deemed well-managed on insulin. In most instances, the initial step in establishing new forms of care came from consultants reaching out to GPs. Programmes for GP education and care protocol thus developed out of friendly collaboration with GPs.[62] Where such cordiality did not exist, however, hospital-led schemes might meet resistance from local practitioners and it was here that specialist nurses were essential, building relationships and refining schemes in situations where tensions existed.[63] As with other forms of service innovation, local politics and culture were clearly important in influencing outcomes and trajectories at a micro-level.[64]

The form of scheme that GPs and consultants entered into varied according to these local circumstances, and could involve multiple practices joining with a clinic in a large scheme, or establishing individual relations between consultant and GP in smaller ones.[65] Likewise, the organisation of GP-care itself also differed across practices. Some practitioners sought to run special 'mini-clinics' in protected surgery time, headed by one or two interested GPs if operating within a group surgery. Here, patients were registered and recalled for regular follow-up consultations and tested at their local practice in a single sitting. Nursing staff would undertake patient education and take samples, whilst GPs performed more complex screening procedures, analysed results and offered advice. Patients were generally requested to attend at intervals of three to six months for at least glucose, ketone, weight, and visual acuity tests, with a special check-up – including screening for complications – performed annually.[66] A mixture of automated testing equipment and contractual arrangements for direct access to off-site diagnostic facilities fostered independence, though some mini-clinics directly involved hospital staff.[67]

Other schemes, by contrast, encouraged all GPs to look after their own listed patients. Here, the routine surveillance of patients would be undertaken during regular surgery time and the link between patient and single practitioner would not be broken. In perhaps the most comprehensive of these shared care schemes, established in Poole during the early 1970s, some patients had no direct contact with the hospital, except for visits to the laboratory for glucose estimations. The rest of the surveillance tests and consultation were performed in surgery time, albeit with connections maintained to community opticians, while community nurses and health visitors were also attached to surgeries for domestic visiting and dietetic advice. Technological innovation made possible new forms of team work – with communication between care sites achieved via computer, as well as letters and patient-held records – while all practitioners in the scheme signed up to shared care protocol that detailed responsibilities, referral criteria, and the aims and metabolic targets of therapy.[68] Similar models were tried elsewhere, but GPs were able to organise their practice idiosyncratically so long as minimum standards of care were agreed and shared records were completed.[69]

As time passed a number of other alternatives were created, such as travelling clinics for more remote communities, and diabetic days or hours in certain London surgeries.[70] Learning about these innovations through publication, education, or participation in professional conferences, some GPs sought to raise their own standards of care for diabetes patients, and had adopted some of these organisational forms independently of clinics.[71] Although more isolated from clinic teams, some GPs might still maintain access to hospital or health authority expertise, such as dietetic advice or chiropody care.[72] Along with practice nursing and secretarial staff, that is, some GPs operated as the head of a health-care team with referral as an option for more specialist services.

Central to new forms of GP-based care were a series of tools designed to make surveillance of the patient population more effective, and to monitor and reform professional activity. Common to almost all shared and structured general practice care programmes, for instance, were patient registers to help track the size of the observed population, and recall systems – buttressed by home visits to non-attenders – to ensure regularity and frequency of patient attendance.[73] Similarly, advocates of these schemes recommended the use of highly structured patient records, generally comprising a checklist of tests to be undertaken, a follow-up section for results to be recorded, and additional spaces to record treatment notes and patient information.[74] These specially designed cards would allow for 'quick accurate recording and recall of information', thus making it easier to account for a patient's progress and facilitate communication between staff members and over time.[75] Moreover, along with the agreed practice protocol, designers of these records believed that they would improve the standard of care, directing professional action and providing 'built in reminders so that both patient and professional will remember to carry out all the routine and sometimes tedious checks which are part and parcel of good diabetic care'.[76] As well as helping to review the care of individual patients, these records therefore facilitated the shift of surveillance onto practitioners. Practice protocol set team responsibilities and 'the basic standard for general practitioner and hospital care', while records were considered key to providing the raw material for regular and research-based audits.[77] This was so much so that the quality of record completion was a prominent audit measure or discussion point in research assessments of GP care.[78] Although such highly structured, reviewed and bureaucratised care might be thought of as anathema to professionals who so frequently spoke of clinical autonomy, as discussed below, this organisation of care was constructed as one of the major benefits to systematic general practice involvement in diabetes management.

The extent of GP engagement in structured independent, shared or community diabetes management during subsequent years is hard to uncover. From archival material and publications it appears that GP-based care had spread right across Britain by 1990, with references to practice in: Kirkcaldy and Stirling in Scotland;[79] Newcastle,[80] Manchester,[81] and Sheffield in the north of England;[82] Wolverhampton,[83] parts of Staffordshire,[84] Birmingham and Warwickshire,[85] Ilkeston,[86] Leicester,[87] Nuneaton,[88] and parts of Oxfordshire in the English Midlands;[89] Norwich,[90] Kings Lynn,[91] Newmarket,[92] and Ipswich, in the east of England;[93] London,[94] Southampton,[95] and parts of Surrey and Hampshire in the south of England;[96] Poole,[97] Bristol,[98] and Exeter in the South West;[99] and Cardiff,[100] the Upper Afan Valley,[101] Powys and Gwent in Wales.[102] Thus, while initial experiments appeared in the English Midlands in relation to expertise in the Birmingham area, by the early 1990s shared and structured care in general practice had spread widely – if not necessarily deeply – beyond this base.

## Diabetes and general practice management

Why did GPs want to take on such care? And what does the spread of GP-based diabetes management tell us about the changing basis of post-war general practice? Of course, like hospital clinicians, the authors of early texts about general practice diabetes care believed that GP involvement would raise standards, and certainly make care more convenient for patients. These doctors pointed to the inefficiency and depersonalisation of 'the diabetic clinic scrummage', highlighting its fleeting consultations, costs to patients, and broken continuity of care.[103] They deemed general practice care, by contrast, to be far more amenable to patients, who would have shorter waiting times and greater 'comfort'.[104] Familiarity with medical teams, proponents suggested, would also make oversight more regular, and thus more effective.[105] And indeed, the willingness of patients to accept greater GP involvement was itself important in helping new schemes gain momentum.

Yet, these same texts frequently included references to a number of advantages for GPs themselves, highlighting a clear political and professional interest in clinical change. For instance, initially – and somewhat ironically given references to 'routine' patients – hospital doctors and GPs pointed to the intellectual satisfaction gained from looking after patients with diabetes. Diabetes was portrayed as a complex condition, affecting almost every bodily system, which would provide a 'wide spectrum of experience in symptomatology, pathology and treatment'.[106] It was, according to its proponents, 'the ideal disease for the general practitioner to diagnose, observe and treat with interest'.[107] Diabetes thus sat in opposition to the supposed wave of 'trivial' cases that some felt characterised general practice, and which even resulted in feelings of 'professional humiliation' over 'wasted' training.[108] Although such feelings were not universal, strong references to work satisfaction and clinical complexity clearly sought to make capital from anxieties of inferiority and dissatisfaction, even while other doctors were trying to rehabilitate such trivia as central to primary care.[109]

Discursive appeals to GP interests were multivalent and it was perhaps their ambiguity and eclecticism that proved persuasive. While at once emphasising skill, early proponents of GP care also pointed to the need to care for the whole patient in such an 'all embracing condition'.[110] GPs knew about the intimate aspects of their patients' lives, and diabetes' status as a 'lifelong disease' combining 'symptoms and physical and biochemical findings with social and emotional problems' was therefore considered to strongly 'interest the general practitioner'.[111] Although managing social and emotional problems was clearly seen as a 'skill' – and by the 1950s a skill seen as an integral part of diabetic clinic care – it was not one necessarily associated with the hospital form of clinical medicine.[112] Such references were, therefore, likely alluding to the 'whole person' rhetoric associated with views that GPs might become specialists in the psycho-social medicine of individuals.[113]

As time passed, the basis of these claims transformed in line with shifting politics of, and visions for, general practice. Alongside references to patient satisfaction and improving clinical skill, some advocates for general practice care in the 1970s and 1980s began to cast diabetes as a model of proactive, preventive medicine. For example, one GP suggested that, though 'extra time is needed to run a clinic', their team felt 'that diabetics are such a high-risk group tha[t] an average of five minutes per day for prevention and treatment is an efficient use of a doctor's time'.[114] Another retrospectively agreed, suggesting that the programme for diabetes in his practice emerged due not only to 'an impression that we could do a lot better with diabetes', but also to a desire amongst his partners to 'do more about preventing people becoming ill, rather than just reacting to the crises'.[115] This began with opportunistic screening and follow-up care for patients with hypertension and non-insulin-requiring diabetes during the 1970s and 1980s, but later took in individuals with conditions requiring similar long-term maintenance therapy, like hypothyroidism and epilepsy.

The notion that GPs should engage in preventive medicine was not new. Armstrong and Lewis, for instance, have pointed to similar claims on prevention during the early twentieth century, with Armstrong suggesting that engagement with epidemiology and social medicine resulted in an expanded surveillance ethos in general practice.[116] However, in terms of proactive screening for disease, it seems that resource and knowledge limitations dissuaded many GPs from engaging in anything beyond opportunistic work during the early post-war decades.[117] Furthermore, across the 1960s and early 1970s, some influential epidemiologists and public health doctors sought to link the detection and prevention of chronic diseases to the political future of the Medical Officer of Health (MOH).[118] Politically, that is, GPs faced competition for space in preventive medicine, albeit within a context where collaboration was being encouraged from both academic and practitioner perspectives.[119] After 1974, however, the role of the MOH had been abolished and public health doctors were incorporated as managers and service planners into the NHS as Community Physicians.[120] GPs were thus afforded space to move into expanded preventive health work, including chronic disease management.[121] In fact, work to improve service delivery and co-ordination in conditions like diabetes was even performed out of academic departments of community medicine and general practice, marking an institutional collaboration on public health from the old and new public health workers.[122]

That managing non-insulin-dependent diabetes in general practice was included in considerations of preventive medicine is revealing. According to the GPs above, detecting disease early was no longer itself at the centre of discussions about disease prevention. Rather, in the context of patients being considered 'at exceptional risk' of arterial disease, managing diabetes patients more effectively – meaning proactive, organised surveillance and therapeutic titration – had in itself become a core component of preventive medicine.[123] The increasing centrality of epidemiological methods as clinical and public health tools was central to such a transformation. Uncovering large amounts of undetected and 'subclinical' disease, along with calculations of individual risk, cast prevention into three stages. Academic GPs and epidemiologists began to talk of primary prevention (preventing occurrence by treating a precursor state, targeting those at risk, or promoting health in the population), secondary prevention (preventing the progression of a subclinical disease into a clinical state by detecting asymptomatic cases and managing patients early), and tertiary prevention (preventing the worsening of a clinical condition into severe disability). Each stage was linked via a series of interconnected interventions.[124] In the light of such discussions, interlocutors claimed that 'poles of curative medicine on the one side and prevention on the other no longer apply', broken down in suggested webs of risk and surveillance.[125] As noted, given its serious complications and its status as a risk factor for other conditions, the detection and treatment of diabetes could be considered as secondary or tertiary prevention. Equally, targeting overweight or elderly individuals with preventive lifestyle advice as a means to avoid the condition could be classified as primary prevention.[126] But it was this division between complete prevention, and the prevention of complications that turned diabetes management itself into a preventive health activity. It was a distinction that became sharper as acceptable – meaning more large-scale clinical and epidemiological – evidence emerged to buttress traditional faith in metabolic control as a deterrent to the emergence of diabetic complications.[127]

## Conclusion

Historians have long used the construction of, and responses to, specific diseases as lenses for investigating medicine and society.[128] In this chapter, I have sought to connect changing understandings of a particular disease – diabetes mellitus – with shifting strategies and models of care as a means to draw out the drivers for service innovation in a period of broad epidemiological change. By focusing our gaze in this manner, it has been possible to demonstrate that clinical and organisational change in British chronic disease care was driven by more than shifting disease profiles and demographic patterns. Rather, a combination of financial constraints within a new health and welfare system, the deployment of novel diagnostic and surveillance technologies, changing meanings of disease and work, and fluctuating professional politics all shaped and helped generate a new model of care for long-term illness.

Beyond this, however, a case study approach to disease management has also highlighted how responses to chronic disease care during the late twentieth century revolved around a process of bureaucratisation. In many respects, doctors and health care providers were building on a late-nineteenth and early-twentieth-century heritage. As Steve Sturdy and Roger Cooter amongst others have noted, the development of mass health care provision – alongside concerns over population quality – generated a significant momentum for the bureaucratisation of medical care and knowledge at the turn of the century.[129] Doctors, clinical researchers, insurance bodies and states alike converged over efforts to standardise diagnostic labels and processes, as well as to divide and reintegrate bodies (of knowledge, individuals, populations and medical labour) in pursuit of maximum efficiency.[130] In post-war Britain, the need for efficiency grew ever more important under the resource-strapped NHS, while the integration of clinical and epidemiological research cultures in elite institutions fostered understandings of the care process as standardisable and reviewable.[131] In terms of diabetes, these pressures extended beyond the initial hospital cocoon, following chronic disease care into general practice. Now spatially redistributed, medical teams drew on well-developed bureaucratic cultures and practices in pursuit of more efficient and effective care: more tightly prescribing roles, documenting work, and reviewing activities than ever before within new hierarchies.[132] This process was facilitated by an expanding range of tools for prescription and surveillance, as well as changing political projects in British general practice.

These changes in diabetes care have had long-lasting legacies, despite research over the 1980s and early 1990s offering mixed results about their efficacy.[133] In part, this is because models of diabetes management formed a symbolic part of broader arguments about raising standards in general practice into the 1980s and 1990s. By 1985, for instance, diabetes had formed an early target for the RCGP's 'Quality Initiative', and the College had begun discussing new models of care in terms of good general practice more broadly:[134]

> Teamwork and practice management are gaining recognition as essential rather than desirable for effective patient care in all types of practice, especially in the management of chronic diseases such as diabetes mellitus, hypertension and asthma, and in anticipatory care. Agreed protocols and standards, the registered list of patients, the continuous clinical record, the microcomputer, practice leaflets and the practice annual report are seen as some of the tools [central to such care].[135]

A discussion of the politics of this 'quality' agenda is beyond the scope of this chapter.[136] Nonetheless, the equation of managed chronic disease care with good general practice underlined both the extent to which general practice was reforming around new sets of disease, and how far end-of-the-century visions of general practice had shifted away from earlier interest in facilities, practitioner intelligence and time management.[137] Such has been the strength of this integration that the links between chronic disease care, bureaucratised managerialism and good general practice continue to the present day in contemporary discourse and policy. In 2004, for instance, the General Medical Services contract introduced the Quality and Outcomes Framework (QOF) for general practice, a voluntary pay-for-performance scheme to encourage improved organisation, expanded preventive health service delivery, and developed GP-management of common chronic diseases. The QOF is vastly more complex than the earlier 1990s contracts, awarding gradated points for various levels of compliance with selected indicators. The premise of the scheme nonetheless shares much in common with the earlier contracts, and in particular with the emphases on bureaucracy, review, and GP chronic disease management. Recent discussions of the QOF, while equivocal – and even critical – about its impact, have reinforced ideas that incentives are effective, and that organised GPs should provide vital preventive health care to local populations.[138]

There is significant scope for further accounts of evolving architectures of care, as well as for more expansive discussions about how models of care and management moved between sites, specialties and conditions.[139] Currently, for example, historians have paid very little attention to the histories of surveillance tools, and while the early-twentieth-century medical record has received significant attention, scholars have generally restricted their interest to this instrument.[140] The histories of recall systems, of clinical computers, of care protocol, and clinical audit and guidelines – and, crucially, their connections to shifting understandings of disease, treatment and labour – are greatly under examined.[141] Such studies are likely to take in significant changes in post-war medicine, politics and society in specific (and interlinked) locations, and in the British case they will require closer attention to the interactions between private companies, public health care architectures, and large-scale research infrastructure than hitherto attempted.[142] This work will undoubtedly prove challenging to historians, and require flexible, interdisciplinary frameworks for analysing disease and formations of medical labour. It will nonetheless likely also prove very valuable, and elucidate some of the defining features of post-war clinical practice.

## Notes

1 T. Scott and A. Maynard, 'Will the new GP contract lead to cost effective medical practice?', Discussion Paper 82, University of York, 1991, p.ii, pp. 14-34; R. Klein, *The New Politics of the NHS: From Creation to Reinvention*, 5th edn, Oxford: Radcliffe, 2006, pp. 140–86.

2 P. Selby, 'W(h)ither Diabetes Care?', *Diabetic Medicine*, 10, 1993, 791–2; Department of Health, *The Health of the Nation: A Strategy For Health in England*, Cm. 1986, London: HMSO, 1992.

3 M. Bufton and V. Berridge, 'Post-war nutrition science and policy making in Britain c. 1945–1994: The Case of Diet and Heart Disease', in D. Smith and J. Phillips, (eds), *Food, Science, Policy and Regulation in the Twentieth Century: International and Comparative Perspectives*, London: Routledge, 2000, pp. 207–22.

4 M. Denham, 'The surveys of the Birmingham chronic sick hospitals, 1948–1960s', *Social History of Medicine*, 19, 2006, 279–93.

5 G. Weisz, *Chronic Disease in the Twentieth Century: A History*, Baltimore: Johns Hopkins University Press, 2014, pp. 233–45.

6 M. Berg, 'Practices of reading and writing: the constitutive role of the patient record in medical work', *Sociology of Health and Illness*, 18, 1996, 499–524.

7 S. Ezedum and D. Kerr, 'Collaborative care of hypertensives, using a shared record', *British Medical Journal*, 2, 1977, 1402–3.

8 J. D. Howell, *Technology in the Hospital: Transforming Patient Care in the Early Twentieth Century*, Baltimore: Johns Hopkins University Press, 1995, pp. 30–60.

9 V. Berridge, *Marketing Health: Smoking and the Discourse of Public Health in Britain, 1945–2000*, Oxford: Oxford University Press, 2007; G. Weisz, *Chronic Disease in the Twentieth Century*, pp. 176–203. Though note: H. Valier and R. Bivins, 'Organization, ethnicity and the British National Health Service', in J. Stanton, (ed.), *Innovations in Health and Medicine: Diffusion and Resistance in the Twentieth Century*, London: Routledge, 2002, pp. 37–64.

10 For a thorough review of NHS historiography: M. Gorsky, 'The British National Health Service 1948–2008: a review of the historiography', *Social History of Medicine*, 21, 2008, 437–60. On post-war general practice, see: D. Armstrong, 'Space and time in British general practice', *Social Science and Medicine*, 20, 1985, 659–66; I. Loudon, J. Horder and C. Webster (eds), *General Practice Under the National Health Service, 1948–1997*, Oxford: Oxford University Press, 1998; G. Smith and M. Nicolson, 'Re-expressing the division of British medicine under the NHS: the importance of locality in general practitioners' oral histories', *Social Science and Medicine*, 64, 2007, 938–48. For care models and variations: J. Stanton, 'Intensive care: measurement and audit in an expansive growth area of medicine', in

V. Berridge, (ed.), *Making Health Policy, Networks in Research and Policy after 1945*, Amsterdam: Rodopi, 2005, pp. 243–75.

11 J. Hasler, 'The size and nature of the problem', in J. Hasler and T. Schofield, (eds), *Continuing Care: The Management of Chronic Disease*, 2nd edn, Oxford: Oxford University Press, 1984, 3–15, p. 10.

12 H. P. Himsworth, 'The syndrome of diabetes mellitus and its causes', *The Lancet*, 253, 1949, 465–73. Insulin was acknowledged to be the hormone catalyzing carbohydrate use.

13 W. Oakley, 'The management of diabetes mellitus', *British Medical Journal*, 2, 1949, p. 1345; R. B. Tattersall, *Diabetes: The Biography*, Oxford: Oxford University Press, pp. 10–30.

14 British Medical Journal, 'Complications of diabetes', *British Medical Journal*, 1, 1953, 1438–9.

15 Oakley, 'Management of diabetes mellitus', 1345. For review of theories: J. Lister, *The Clinical Syndrome of Diabetes Mellitus*, London: H. K. Lewis, 1959, pp. 4–9.

16 W. G. Oakley, D. A. Pyke, and K. W. Taylor, *Diabetes and Its Management*, 3rd edn, Oxford: Blackwell Scientific Publications, 1978, pp. 61–6, pp. 76–102. Also: R. D. Lawrence, *The Diabetic Life: Its Control by Diet and Insulin, A Concise Practical Manual*, 15th edn, London: J. A. Churchill, 1955, pp. 38–41, 48–54. Classification debates were more intense in research circles in later decades: A. M. Hedgecoe, 'Reinventing diabetes: classification, division and geneticization of disease', *New Genetics and Society*, 21, 2002, 7–27.

17 J. Greene, *Prescribing by Numbers: Drugs and the Definition of Disease*, Baltimore: Johns Hopkins University Press, 2007, esp. 83–114.

18 M. Bliss, *The Discovery of Insulin*, 25th Anniversary Edition, Chicago: University of Chicago Press, 2007; J. Liebenau, 'The MRC and the pharmaceutical industry: the model of insulin', in J. Austoker and L. Bryder, (eds), *Historical Perspectives on the Roles of the MRC: Essays in the History of the Medical Research Council and Its Predecessor the Medical Research Committee, 1913–1953*, Oxford: Oxford University Press, 1989, pp. 163–80.

19 'Scotland: from our own correspondent', *The Lancet*, 140, 1924, 461. On the history of specialist clinics: R. Cooter, 'The politics of spatial innovation: fracture clinics in inter-war Britain', in J. Pickstone (ed.), *Medical Innovations in Historical Perspective*, Basingstoke: Macmillan, 1992, pp. 146–64. Specialists here were not certified full-time practitioners so much as researchers and clinical generalists receiving referrals on the basis of a special interest and acknowledged expertise. This model continued into the NHS: G. Weisz, *Divide and Conquer: A Comparative History of Medical Specialization*, Oxford: Oxford University Press, 2006, pp. 164–87.

20 R. D. Lawrence, 'Special clinics for diabetics', *British Medical Journal*, 2, 1942, 322.

21 G. F. Walker, 'Reflections on diabetes mellitus: answers to a questionary', *The Lancet*, 262, 1953, 1329–32. Even those who doubted control still observed patients for clinical symptoms, weight gain, and ketosis: D. M. Dunlop, 'Are diabetic degenerative complications preventable?', *British Medical Journal*, 2, 1954, p. 383. Many clinicians were agnostic, but continued close control on the principles of physiology and harm reduction: 'Detection of diabetes', *British Medical Journal*, 1, 1962, p. 1537. Later evidence from observational studies was not considered proof but strong enough to be convincing: M. T. Wojciechowski, 'Systematic care of diabetic patients in a general practice', *Journal of the Royal College of General Practitioners*, 32, 1982, p. 531.

22 C. J. C. Earl, 'The treatment of diabetics as hospital out-patients', *British Medical Journal*, 1, 1927, 831–3; Walker, 'Reflections on diabetes mellitus', 1329–32.

23 Though compare practice and beliefs in: I. H. Redhead and J. J. A. Reid, 'Diabetic clinics and the general practitioner', *The Lancet*, 281, 1963, 159–60.

24 D. G. French, 'Advances in general practice', *The Practitioner*, 183, 1959, p. 514; The National Archives [TNA], BD 18/793, F. L. Dyson to Dr. A. Trevor Jones, 'Regional planning of diabetic services', 12 August 1953, 1.

25 The Diabetic Association, 'London diabetic clinics, provincial diabetic clinics', *The Diabetic Journal*, 3, 1940, 32–4; TNA, BD 18/793, 'Diabetic clinics', 1955.

26 'Regional diabetic services', *British Medical Journal*, 2, 1953, 160.

27 J. Walker, *Chronicle of a Diabetic Service*, London: British Diabetic Association, 1988, Appendix B, p. 91.

28 J. Malins, *Clinical Diabetes Mellitus*, London: Eyre and Spottiswood, 1968, p. 456.

29 See shifting mortality patterns in the Birmingham clinic: ibid., 466–8. However, discussions also reflexively highlighted problems with data, especially when considering national patterns: J. M. Malins, 'Food and death-rates from diabetes', *The Lancet*, 304, 1974, 1201.

30 Walker, *Chronicle of a Diabetic Service*, Appendix B, p. 91.

31 H. P. Himsworth, 'Diet in the aetiology of human diabetes', *Proceedings of the Royal Society of Medicine*, 63, 1949, pp. 323–6; J. J. A. Reid, 'A new public health: the problems and the challenge', *Public Health*, 79, 1965, pp. 183–4; T. L Cleave and G. D. Campbell, *Diabetes, Coronary Thrombosis and the Saccharine Disease*, Bristol: John Wright & Sons, 1966; Malins, *Clinical Diabetes Mellitus*, pp. 456–7.

32 D. Porter, *Health Citizenship: Essays in Social Medicine and Biomedical Politics*, Berkeley: University of California Press, 2011, pp. 154–81; L. Berlivet, '"Association or causation?" The debate on the scientific status of risk factor epidemiology, 1947–c.1965', in V. Berridge (ed.), *Making Health Policy*, pp. 39–74; G. M. Oppenheimer, 'Profiling risk: the emergence of coronary heart disease epidemiology in the United States (1947–70)', *International Journal of Epidemiology*, 35, 2006, 720–30.

33 Tattersall, *Diabetes*, pp. 89–90.

34 J. A. Hunt, C. H, Gray, and D. E. Thorogood, 'Enzyme tests for the detection of glucose', *British Medical Journal*, 2, 1956, 586–8.

35 C. H. Gray and H. R. Millar, 'Tests for glycosuria: a comparison of Benedict's Test, Clinitest and Glucotest', *British Medical Journal*, 1, 1953, p. 1363.

36 Working Party of the College of General Practitioners, 'A diabetes survey', *British Medical Journal*, 1, 1962, p. 1497. For colonial research connections and international interests: J. A. Tulloch, *Diabetes Mellitus in the Tropics*, Edinburgh: E&S Livingstone, 1962.

37 V. Quirke and J-P. Gaudillière, 'The era of biomedicine: science, medicine and public health in Britain and France after the Second World War', *Medical History*, 52, 2008, pp. 441–4. The presence of commercial funding for research in Britain has been vastly underplayed, but for the US: Greene, *Prescribing by Numbers*. For the emergence of 'the survey' as a tool of community surveillance in Britain: D. Armstrong, *Political Anatomy of the Body: Medical Knowledge in Britain in the Twentieth Century*, Cambridge: Cambridge University Press, 1983.

38 C. Timmermann, 'A matter of degree: the normalisation of hypertension, c.1940–2000', in W. Ernst (ed.), *Histories of the Normal and the Abnormal: Social and Cultural Histories of Norms and Normativity*, London: Routledge, 2006, pp. 252–5.

39 Working Party, 'A diabetes survey', 1497–1503.

40 R. D. Lawrence, 'Regional centres for the treatment of diabetes', *The Lancet*, 257, 1951, p. 1318; 'Detection of diabetes', *British Medical Journal*, 2, 1959, 555–6; '350,000 may have diabetes', *The Times*, London, 17 July 1962, p. 6, http://find.galegroup.com/ttda/basic-Search.do;jsessionid=7C31A3032D21FA2463799CE089BDCB10, accessed 11 December 2014.

41 For instance: 'BMA Annual Meeting: panel discussion on diabetes', *British Medical Journal*, 2, 1964, 302–3; J. W. Laurie, 'Detection of diabetes', *The Lancet*, 282, 1963, 390; H. Cairns, 'Detection of diabetes', *The Lancet*, 282, 1963, 412; H. Creditor, 'Detection of diabetes', *The Lancet*, 282, 1963, 580; W. L. Ashton, 'Detection of diabetes', *The Lancet*, 282, 1963, 684; A. V. Neale, 'Detection of diabetes', *The Lancet*, 282, 1963, 786–7; W. Oakley, 'Detection of diabetes', *The Lancet*, 282, 1963, 787; J. W. Laurie, 'Detection of diabetes', *The Lancet*, 282, 1963, 838.

42 T. Cutler, 'A double irony? The politics of National Health Service expenditure in the 1950s', in M. Gorsky and S. Sheard (eds), *Financing Medicine: The British Experience since 1750*, London: Routledge, 2006, pp. 200–20.

43 T. Cutler, 'Dangerous yardstick? Early cost estimates and the politics of financial management in the first decade of the National Health Service', *Medical History*, 47, 2003, 217–38.

44 C. Webster, *The Health Services Since the War, Volume I: Problems of Health Care. The National Health Service Before 1957*, London: HMSO, 1988, pp. 133–226. The topic of NHS spending is, of course, contentious. Deflated by a general price index, NHS spending did double between 1957 and 1979. However, adjusting in light of an NHS-specific price index, Webster has suggested that this growth generally fell below the 2.5 per cent annual growth

that ministers suggested was required to provide a constant level of service. C. Webster, *The Health Services Since The War, Volume II: Government and Health Care, The British National Health Service 1958–1979*, London: HMSO, 1996, pp. 760–61, Appendices 3.3.4–3.5.

45 Ministry of Health, *First Report of the Joint Working Party on the Organisation of Medical Work in Hospitals*, London: HMSO, 1967; G. Forsyth and R. Logan, *Gateway or Dividing Line? A Study of Hospital Outpatients in the 1960s*, Oxford: Oxford University Press for the Nuffield Provincial Hospitals Trust, 1968.

46 'To come again, 3 months', *The Lancet*, 307, 1976, 1168–9.

47 P. A. Thorn and R. G. Russell, 'Diabetic clinics today and tomorrow: mini-clinics in general practice', *British Medical Journal*, 2, 1973, pp. 535–6.

48 R. D. Hill, 'Community care service for diabetics in the Poole area', *British Medical Journal*, 1, 1976, p. 1137.

49 Thorn and Russell, 'Diabetic clinics today and tomorrow', 536.

50 R. D. Lawrence, 'Types of human diabetes', *British Medical Journal*, 1, 1951, 374.

51 R. D. Lawrence, *The Diabetic Life: Its Control by Diet and Insulin, A Concise Practical Manual For Practitioners and Patients*, 6th edn, London: J. A. Churchill, 1931, pp. 14, pp. 18–19.

52 'Detection of diabetes', 1537.

53 In certain cancers, doctors had begun to debate deviations of normality earlier in the century. Differences in disease politics, materiality of lesions, and diagnostic and therapeutic modalities provided a different trajectory across and between cases: Ilana Löwy, *Preventive Strikes: Women, Precancer, and Prophylactic Surgery*, Baltimore: Johns Hopkins University Press, 2010.

54 Malins, *Clinical Diabetes*, pp. 68–76.

55 On diabetes and concepts of pathology and normality in quantitative terms: Georges Canguillhem [trans. Caroline Fawcett et al.], *The Normal and the Pathological*, New York: Zone Books, 1989.

56 'British Diabetic Association', *British Medical Journal*, 2, 1962, p. 1252.

57 R. J. Jarrett, *Diabetes Mellitus*, London: Croom Helm, pp. 1–3.

58 Royal College of General Practitioners Archives, John Fry, 'The management of maturity onset diabetes in a general practice (1951–1971)', MSS, B Fry C6-1 (i), 8–9. Undated.

59 Hill, 'Community care service', p. 1139.

60 Adrian P. Kratky, 'An audit of the care of diabetics in one general practice', *Journal of the Royal College of General Practitioners*, 27, 1977, 536–42.

61 Though they often co-existed with extended GP care: Valier and Bivins, 'Organization, ethnicity and the British National Health Service', pp. 37–64.

62 C. E. Upton, 'Diabetic community care', *The Practitioner*, 215, 1975, 83–5.

63 Oral History Interview with Mary Mackinnon performed by the University of Oxford, *Diabetes Stories*, 23 April 2007, http://diabetes-stories.com/interview.asp?UID=62, accessed 8 February 2015.

64 M. Gorsky, 'To regulate and confirm inequality? A regional history of geriatric hospitals under the English National Health Service, c.1948–1975', *Ageing & Society*, 33, 2013, 598–625.

65 J. M. Malins and J. M. Stuart, 'Diabetic clinic in a general practice', *British Medical Journal*, 4, 1971, 161; D. R. R. Williams, C. Munroe, C. J. Hospedales and R. H. Greenwood, 'A three-year evaluation of the quality of diabetes care in the community care scheme', *Diabetic Medicine*, 7, 1990, p. 74.

66 Wojciechowski, 'Systematic care of diabetic patients', 531–3.

67 I. Benett, 'Diabetes mini-clinic', *Journal of the Royal College of General Practitioners*, 38, 1988, 76–7.

68 Hill, 'Community care service', 1137–9.

69 J. L. Day, H. Humphries and H. Alban-Davies, 'Problems of comprehensive shared diabetic care', *British Medical Journal*, 294, 1987, 1590–2.

70 B. Hurwitz and J. Yudkin, 'Diabetes care: whose responsibility?', *British Medical Journal*, 289, 1984, 1000–1; A. K. Baksi, J. Brand, M. Nicholas, A. Tavabie, B. J. Cartwright, M. R. Waterfield, 'Non-consultant peripheral clinics: a new approach to diabetic care', *Health Trends*, 16, 1984, 38–40.

71  P. R. W. Tasker, 'General practice diabetic care', *Journal of the Royal College of General Practitioners* 1983, 33, 828.
72  D. Chesover, P. T. Miles, and S. Hilton, 'Survey and audit of diabetes care in general practice in South London', *British Journal of General Practice*, 41, 1991, 282–5.
73  J. M. Wilks, 'Diabetes: a disease for general practice', *Journal of the Royal College of General Practitioners*, 23, 1973, pp. 51–2; Wojciechowski, 'Systematic care of diabetic patients', p. 532.
74  R. D. Hill, *Diabetes Health Care: A Guide to the Provision of Health Care Services*, London: Chapman and Hall, 1987, pp. 95–7.
75  B. R. G. Fletcher, 'Looking after diabetics in general practice: a trainee project', *Journal of the Royal College of General Practice*, 27, 1977, p. 87.
76  M. Hall, 'Diabetic care in general practice: the Exeter project', in Royal College of General Practitioners (ed.), *Diabetes Information Folder*, London: Royal College of General Practitioners, 1988, Appendix 8, p. 5.
77  R. L. Gibbins and J. Saunders, 'Develop diabetic care in general practice', *British Medical Journal*, 297, 1988, p. 188.
78  Williams et al., 'A three-year evaluation', pp. 76–8.
79  Scottish Home and Health Department, *Report of the Working Group on the Management of Diabetes*, Edinburgh: HMSO, pp. 77–8.
80  F. K. E Tunbridge, J. P. Millar, P. J. Schofield, J. A. Spencer, G. Young, and P. D. Home, 'Diabetes care in general practice: an approach to audit of process and outcome', *British Journal of General Practice*, 43, 1993, 291–5.
81  Benett, 'Diabetes mini-clinic', 76–8.
82  Oral History Interview with Mary Mackinnon performed by the University of Oxford.
83  B. Singh, M. Holland and P. Thorn, 'Metabolic control of diabetes in general practice clinics: comparison with a hospital clinic', *British Medical Journal*, 289, 1984, 726–8.
84  D. G. Garvie, 'The diabetic, the hospital, and primary care', *Journal of the Royal College of General Practitioners*, 30, 1980, 440.
85  Malins and Stuart, 'Diabetic clinic in a general practice', 161.
86  TNA, FD 7/2303, P. R. W. Tasker to D. A. Pyke, 'RE: the provision of medical care for diabetic patients in the UK', 1985, 1–2.
87  J. G. Mellor, A. Semanta, R. L. Blandford, A. C. Burden, 'Questionnaire survey of diabetic care in general practice in Leicestershire', *Health Trends*, 17, 1985, 61–3.
88  TNA, Tasker to Pyke', p. 2.
89  C. Dornan, G. Fowler, J. I. Mann, A. Markus, and M. Thorogood, 'A community study of diabetes in Oxfordshire', *Journal of the Royal College of General Practitioners*, 33, 1983, 151–5.
90  Williams et al., 'A three-year evaluation', 74–9.
91  TNA, Tasker to Pyke, p. 2.
92  Ibid.
93  Day et al., 'Problems of comprehensive shared care', 1590–2.
94  Chesover et al., 'Survey and Audit', 282–5.
95  A. Foulkes, A. L. Kinmonth, S. Frost and D. MacDonald, 'Organised personal care: an effective choice for managing diabetes in general practice', *Journal of the Royal College of General Practitioners*, 39, 1989, 444–7.
96  TNA, Tasker to Pyke, p. 2.
97  Hill, 'Community care service', 1137–9.
98  T. J. Kemple and S. R. Hayter, 'Audit of diabetes in general practice', *British Medical Journal*, 302, 1991, 451–3.
99  Hall, 'Diabetic care in general practice', in Royal College of General Practitioners, *Diabetes Clinical Information Folder*, Appendix 8.
100  At least on a trial basis: G. F. Morgan, D. A. Cadman, P. H. Edwards, T. C. O'Dowd, and R. H. Davis, 'Diabetes care: whose responsibility?', *British Medical Journal*, 289, 1984, 1309–10.
101  Wojciechowski, 'Systematic care of diabetic patients', 531–3.
102  R. L. Gibbins and J. Saunders, 'Characteristics and pattern of care of a diabetic population in Mid-Wales', *Journal of the Royal College of General Practitioners*, 39, 1989, 206–8.
103  Wilks, 'Diabetes: a disease for general practice', p. 46.
104  Thorn and Russell, 'Diabetic clinics today and tomorrow', p. 536.

105 Singh et al., 'Metabolic control of diabetes in general practice clinics', p. 728.
106 C. H. Stewart-Hess, 'The management of maturity onset diabetes in general practice', *Journal of the Royal College of General Practitioners*, 23, 1973, p. 859.
107 Wilks, 'Diabetes: a disease for general practice', p. 46.
108 Quote from an interview published in Ann Cartwright, *Patients and Their Doctors*, London: Routledge and Kegan Paul, 1967, p. 58; see also pp. 39–62 for an overview of doctors' frustrations.
109 Armstrong, *Political Anatomy of the Body*, pp. 79–82.
110 Thorn and Russell, 'Diabetic clinics today and tomorrow', p. 534.
111 Kratky, 'An audit of the care of diabetics in one general practice', p. 536.
112 R. E. Tunbridge, 'Sociomedical aspects of diabetes mellitus', *The Lancet*, 262, 1953, 893–9.
113 M. Marinker, '"What is wrong" and "How we know it": changing concepts of illness in general practice', Loudon, Horder and Webster, *General Practice Under the National Health Service*, pp. 71–8. Though cf: M. Perry, 'Academic general practice in Manchester under the early National Health Service: a failed experiment in social medicine', *Social History of Medicine*, 13, 2000, 111–29.
114 Wojciechowski, 'Systematic care of diabetic patients', p. 533.
115 Oral History Interview with Richard Gee performed by the University of Oxford, *Diabetes Stories*, 22 April 2008, http://diabetes-stories.com/interview.asp?UID=98, accessed 8 February 2015.
116 J. Lewis, *What Price Community Medicine? The Philosophy, Practice and Politics of Public Health since 1919*, Brighton: Wheatsheaf, 1986, pp. 18–26; Armstrong, *Political Anatomy of the Body*, pp. 48–53, pp. 73–8, pp. 93–100.
117 D. L. Crombie, 'Testing in family practice for diabetes', *Journal of the Royal College of General Practitioners*, 7, 1964, 379–85.
118 Reid, 'A new public health', 183–95.
119 D. E. Cullington, 'The orientation of the general practitioner', *Public Health*, 79, 1965, 215–18.
120 Lewis, *What Price Community Medicine?*, pp. 100–59.
121 Royal College of General Practitioners, *Report From General Practice 19. Prevention of Arterial Disease in General Practice: A Report of a Sub-Committee of the Royal College of General Practitioners' Working Party on Prevention*, London: Royal College of General Practitioners, 1981.
122 Dornan et al., 'A community study of diabetes in Oxfordshire', 151–5.
123 Royal College of General Practitioners, *Prevention of Arterial Disease in General Practice*, p.7.
124 J. N. Morris, 'The prevention of disease in middle age', *Public Health*, 77, 1963, p. 238.
125 R. F. L. Logan, 'Control of chronic disease in general practice and industry', *Journal of the College of General Practitioners*, 11, Supplement 1, 1966, p. 95. On risk and surveillance: D. Armstrong, 'The rise of surveillance medicine', *Sociology of Health and Illness*, 17, 1995, 393–404.
126 Stewart-Hess, 'The management of maturity onset diabetes in general practice', pp. 843–4.
127 For instance: H. Al Sayegh and R. J. Jarrett, 'Oral glucose-tolerance test and the diagnosis of diabetes: results of a prospective study based on the Whitehall Survey', *The Lancet*, 314, 1979, 431–3.
128 Or, in the words of one doyen of the field, 'as sampling technique as well as subject' – cited in C. Rosenberg, *The Cholera Years: The United States in 1832, 1849, and 1866*, Chicago: University of Chicago Press, 1962, p. 4.
129 S. Sturdy and R. Cooter, 'Science, scientific management and the transformation of medicine in Britain, c.1870–1950', *History of Science*, 36, 1998, 421–66.
130 Weisz, *Divide and Conquer*, pp. xix–xxi, pp. xxvii–xxviii.
131 H. Valier and C. Timmermann, 'Clinical trials and the reorganization of medical research in post-Second World-War Britain', *Medical History*, 2008, 52, 493–510; G. Weisz, A. Cambrosio, P. Keating, L. Knaapen, T. Schlich and V. J. Tournay, 'The emergence of clinical practice guidelines', *Milbank Quarterly*, 85, 2007, 691–727.

132 This generalised view of the principles of bureaucratisation is adapted from a Weberian model, though abstracted from Weber's universalist theories of historical development. By using the term bureaucratisation, I am here referencing an ongoing process – or, rather, a set of processes – and acknowledging that British medicine is clearly not, nor ever has been, a model of bureaucracy. This is in line with Weber's own work, in which his methodology often centred on exploring phenomena and what we might term 'big pictures' by drawing out 'ideal types', rather than producing complex case studies rich in minute details. For Weber's translated works on bureaucracy: M. Weber [edited by G. Roth and C. Wittich (eds), translated by E. Fischoff et al.], *Economy and Society: An Outline of Interpretive Sociology*, Vols. 1 and 2, Berkeley: University of California Press, 1978, pp. 217–26, pp. 956–1005, esp. pp. 956–8, pp. 973–5.

133 Hurwitz and Yudkin, 'Diabetes care', 1000–1001; Day et al., 'Problems of comprehensive shared care', 1590–2.

134 Colin Waine, *Why Not Care for Your Diabetic Patients*, 2nd edn, London: Royal College of General Practitioners, 1988, p. 2.

135 Royal College of General Practitioners, *Quality in General Practice: Policy Statement 2*, London: Royal College of General Practitioners, 1985, p. 1.

136 I. Kirkpatrick, and M. Martinez Lucio (eds), *The Politics of Quality in the Public Sector: The Management of Change*, London: Routledge, 1995.

137 Of course, even during the 1950s, emphases on quality varied between interlocutors in the issue and organisation featured in some discussions. Nonetheless, the links with chronic disease and management were significant novelties of later decades. Cf above with: S. Taylor, *Good General Practice: A Report on a Survey*, London: Oxford University Press, 1954.

138 A. Dixon, A. Khachatryan, A. Wallace, S. Peckham, T. Boyce, S. Gillam, *Impact of Quality and Outcomes Framework on Health Inequalities*, King's Fund: London, 2011.

139 Work in the history of the laboratory, scientific medicine and in Science and Technology Studies made important strides here, but this has not been followed up on. For a review of work in STS on local knowledge and making practices universal: J. Golinski, *Making Natural Knowledge: Constructivism and the History of Science*, Chicago: University of Chicago Press, 2005, esp. pp. 162–85. For histories of laboratories and scientific medicine: A. Cunningham and P. Williams (eds), *The Laboratory Revolution in Medicine*, Cambridge: Cambridge University Press, 1992; S. Sturdy, 'The political economy of scientific medicine: science, education and the transformation of medical practice in Sheffield, 1890–1922', *Medical History*, 36, 1992, 125–59; A. Hull, 'Hector's house: Sir Hector Hetherington and the academicization of Glasgow hospital medicine before the NHS', *Medical History*, 45, 2001, 207–42.

140 J. H. Warner, *The Therapeutic Perspective: Medical Practice, Knowledge and Identity in America, 1820–1885*, Cambridge, MA: Harvard University Press, 1986, pp. 83–162; B. L. Craig, 'The role of records and of record-keeping in the development of the modern hospital in London, England and Ontario, Canada', *Bulletin of the History of Medicine*, 65, 1991, pp. 383–91; Howell, *Technology and the Hospital*, pp. 42–56; M. Berg and P. Harterink, 'Embodying the patient: records and bodies in early 20th-century US medical practice', *Body and Society*, 10, 2004, 13–41.

141 Some work has begun here, but there remains much to scope out in terms of transfer between sites, and in terms of different contexts: Weisz et al., 'The emergence of clinical practice cuidelines', 691–727; P. Day, R. Klein and F. Miller, *A Comparative US-UK Study of Guidelines*, London: Nuffield Trust, 1998; S. Mars, 'Peer pressure and imposed consensus: the making of the 1984 *Guidelines of Good Clinical Practice in the Treatment of Drug Misuse*', in Berridge, *Making Health Policy*, pp. 149–84. On audit, see Stanton, 'Intensive care', 243–75.

142 This is far more common in histories of US medicine, even where British companies and international connections are involved: Greene, *Prescribing by Numbers*; V. Quirke, 'Targetting the American market for medicines, ca. 1950s–1970s: ICI and Rhône-Poulenc compared', *Bulletin for the History of Medicine*, 88, 2014, 654–96.

## Select bibliography

Armstrong, D., 'The rise of surveillance medicine', *Sociology of Health and Illness*, 17, 1995, 393–404.

Armstrong, D., 'Chronic illness: a revisionist account', *Sociology of Health and Illness*, 36, 2014, 15–27.

Berg, M., 'Practices of reading and writing: the constitutive role of the patient record in medical work', *Sociology of Health and Illness*, 18, 1996, 499–524.

Berridge, V., *Marketing Health: Smoking and the Discourse of Public Health in Britain, 1945–2000*, Oxford: Oxford University Press, 2007.

Porter, D., *Health Citizenship: Essays in Social Medicine and Biomedical Politics*, Berkeley: University of California Press, 2011.

Greene, J. *Prescribing by Numbers: Drugs and the Definition of Disease*, Baltimore: Johns Hopkins University Press, 2007.

Howell, J. D., *Technology in the Hospital: Transforming Patient Care in the Early Twentieth Century*, Baltimore: Johns Hopkins University Press, 1995.

Stanton, J., 'Intensive care: measurement and audit in an expansive growth area of medicine', in V. Berridge (ed.), *Making Health Policy, Networks in Research and Policy after 1945*, Amsterdam: Rodopi, 2005, pp. 243–75.

Sturdy, S., and Cooter, R., 'Science, scientific management and the transformation of medicine in Britain, c.1870–1950', *History of Science*, 36, 1998, 421–66.

Weisz, G., *Chronic Disease in the Twentieth Century: A History*, Baltimore: Johns Hopkins University Press, 2014.

Weisz, G., *Divide and Conquer: A Comparative History of Medical Specialization*, Oxford: Oxford University Press, 2006.

Weisz, G., Cambrosio, A., Keating, P., Knaapen, L., Schlich, T., and Tournay, V. J., 'The emergence of clinical practice guidelines', *Milbank Quarterly*, 85, 2007, 691–727.

Valier, H., and Bivins, R., 'Organization, ethnicity and the British National Health Service', in J. Stanton, (ed.), *Innovations in Health and Medicine: Diffusion and Resistance in the Twentieth Century*, London: Routledge, 2002, pp. 37–64.

Valier, H., and Timmermann, C., 'Clinical trials and the reorganization of medical research in post-Second World-War Britain', *Medical History*, 52, 2008, 493–510.

# 24

# BEFORE HIV

## Venereal disease among homosexually active men in England and North America

*Richard A. McKay*

The transmission of sexually transmitted infections among homosexually active men would not regularly command headlines in England and North America until the emergence of the Acquired Immune Deficiency Syndrome (AIDS) epidemic in the early 1980s. Yet this topic had, by then, been a developing social and public health concern for over 30 years; it remains a pressing issue today, over 30 years later. This chapter aims to sketch the contours of this pre-AIDS history, one which has languished in the shadow of HIV and has received relatively little attention from historians.[1]

This overview presents interim findings from an on-going historical study of venereal disease (VD) – the term most often employed before 1980 for infections spread through sexual contact – and sexual health among men who had sex with men. In describing these men, I follow the work of other historians of sexuality by employing the word 'gay' for those individuals who organized their personal identities to a significant degree around their sexual attraction to other men. I use 'queer' as a more broad descriptor, a category which included gay men as well as those who might not identify as gay but who were attracted to and had sex with other men. The project concentrates on three predominantly English-speaking countries – England, the United States, and Canada – in the twentieth century's middle decades, and the current chapter reflects these geographic and temporal foci.[2] Though these countries present obvious differences in terms of their populations, health-care delivery systems, and immigration patterns, among other variations, they share a number of important similarities. Common religious and legal traditions generated similar social and legal prohibitions on same-sex relationships. In response, first homophile and later gay-rights activists drew encouragement and assistance from supportive transnational networks.[3] Furthermore, each country also had an established corps of health-care professionals dealing with a growing venereal disease problem, many of whom were attentive to international developments. A working assumption, therefore, is that sufficient similarities exist to justify drawing together examples from these three countries to illuminate shared phenomena and chronologies. My hope is that interested readers will pursue the cited sources for more detailed study, and to determine the extent to which local conditions might complicate the general picture I present here.

## Earlier awareness and tacit knowledge

An awareness that same-sex sexual practices could communicate infection circulated within European medical and forensic communities – and, of course, among those men unfortunate enough to become so infected – as early as the 1490s.[4] As cities across Europe prosecuted individuals for sodomitical practice and assault, and confessions by torture gradually gave way to the use of forensic evidence in trials, medical professionals' testimony assumed greater importance in the courtroom. Paolo Zacchia, a prominent seventeenth-century medico-legal authority, consolidated and disseminated a body of logical knowledge about anal intercourse which medical experts could then read onto the bodies of those accused of such crimes. Over time, venereal maladies like condylomous bumps, chancres, and foul discharges joined penile traumas, traces of anal inflammation, tears, and excrudences – or a smoothened rectal passage in the case of habitual sodomites – as suggestive signs of anal intercourse.[5] In London, a Turkish man was convicted of sodomizing a Dutch youth in 1694; the surgeon testifying in this assault case pointed to the victim's anal venereal ulcers and corresponding chancres on the accused's penis as evidence of the crime.[6] Just over a decade later, John Marten, another London surgeon, wrote that, 'in this dissolute Age', sex between men transmitted venereal infections 'very frequently'. Marten noted disapprovingly that one patient he treated had become sick with the clap and the pox after another man with mouth ulcers sucked his penis – an act which, before the onset of a venereal distemper, brought both men 'great Pleasure'.[7] The surgeon later highlighted, in an expanded version of his treatise, the tensions experienced by medical men treating such individuals, their loyalties divided between a professional duty to aid these patients and a legal responsibility to report them to a magistrate.[8] Marten's contemporaries viewed him as unusually scandalous, however, for his explicit mention of sex between men in print; most practitioners maintained a well-practised silence on the topic.[9]

Perhaps the dissolute relations that Marten's contemporaries avoided discussing, and which dismayed urban reformers in the eighteenth and nineteenth centuries – in London and other large cities of Western Europe, and later in North America – stemmed from the disruptive agricultural improvements linked with the industrial revolution's onset. Thousands of young workers, displaced from their traditional rural employment, streamed to urban areas. Young people in their teens and twenties dominated London's population in the late seventeenth and early eighteenth centuries.[10] Uprooted from their familiar work, religious, and familial environments, which normally served to police sexual behaviour, ever-greater numbers of young men were able to find others interested in same-sex contact. In London, these increased numbers supported the formation of a core group of 'mollies', men whose non-normative gender performance and semi-public displays of sexuality placed them most visibly at the centre of many loosely overlapping networks of queer men. These men would continue to seek each other for sex and sociability in the marginal spaces – parks, latrines, and certain taverns – of their growing cities.[11]

Although English secular law had prohibited sodomitical relations since the sixteenth century – a tradition transferred to the North American colonies – a series of laws passed over the course of the nineteenth century codified the 'unnatural' relations that were not to be named under offences like 'gross indecency' and 'indecent assault'.[12] In conversation with these legal attempts to demarcate the actions of men who sought sex with men as deviant, late-nineteenth-century sexologists – practitioners of a scientific study of sexuality that drew on such diverse specialties as public health, forensic medicine, and psychiatry – attempted to classify the abnormal sexual behaviours they witnessed, attributing them to an inherently deviant mental state. In doing so, they moved away from a history of analysing bodies for physical signs of infection. Previously, medico-legal authors such as Ambroise Tardieu in Paris and Johann Ludwig Casper in Berlin had considered whether infections like syphilis and growths around the anus constituted typical signs of those who practised anal intercourse.[13] Shifts away from physical signs towards psychological understandings of homosexuality may have contributed to a relative loss of the association between same-sex encounters and VD by the beginning of the twentieth century. Later, the work of some twentieth-century public health practitioners would represent a return of sorts to this positivist tradition, with attempts to define 'the homosexual' by appearance and comportment, by occupation, and by the presence of venereal lesions.

In the late nineteenth and early twentieth centuries, physicians, surgeons, and specialists in syphilis and other diseases drew on increasingly sophisticated techniques to differentiate between sexually transmissible infections, namely syphilis, gonorrhoea, and chancroid. In isolated reports and case studies, practitioners published examples of such infections being passed through 'unnatural' or 'perverted' practices. Gradually, experts attributed fewer cases of venereal infection to 'casual' contact, instead viewing syphilitic chancres in the mouth or around the anus as a product of oral-genital and penile-anal contact. Nonetheless, with VD a difficult topic for public conversation, many were very reluctant to acknowledge sexual practices other than vaginal intercourse. This silence contributed to a burgeoning belief on both sides of the Atlantic that oral and anal sex could not transmit infections and were therefore safer than vaginal intercourse.[14]

## A growing problem? Political and public health surveillance

The 1930s and 1940s brought increased political and public health scrutiny to non-normative forms of sex, through a retrenchment of gender conformity during the Great Depression, enhanced VD prevention drives, and a heightened attention to commercialized sexuality amid mobilization for war. *Sexual Behavior in the Human Male* (1948), by Alfred Kinsey and his associates, further concentrated public attention on the 'homosexual outlets' of American men – which appeared to exist in far greater numbers than previously imagined.[15] In synchrony with this increasing research and discussion of homosexuality, public health physicians increasingly reported outbreaks of VD spread through same-sex contact, or, as some saw it, 'from perversion'. The 'hitherto unsuspected source of the spread of venereal disease', as two Vancouver physicians referred to homosexuality in 1951, gradually drew more attention.[16] A subsequent shift occurred in the way observers discussed the same-sex transmission of VD in the 1950s. From earlier descriptions of isolated transmission incidents, where practitioners suggested the uniqueness and rarity of such events – particularly in the Anglo-American world – increasingly public health workers emphasized that such exposures appeared to be occurring more frequently and accounting for more cases than before.[17] In part this stemmed from the success of penicillin treatment in reducing cases of infectious syphilis from a post-war high in 1946 to sufficiently low levels that disease surveillance work could quickly detect new outbreaks; it was also due to a socio-political climate that was increasingly hostile towards homosexual behaviour.

The late 1940s and 1950s witnessed growing concern about the dangers posed by 'the homosexual', particularly for child protection and national security. In North America, legislators moved to address the perceived threat of dangerous sexual offenders in the 1930s and again in the immediate post-war years; these laws notionally targeted child molesters, yet in practice were often applied against gay men. Fears of homosexual men being Cold War security risks, either as members of socialist organizations or as blackmail targets, sparked high-profile media stories and purges of gay men from government departments, first in the USA, then later in England and Canada. In the 1960s, Canadian researchers – supported by the Royal Canadian Mounted Police, the Department of National Defence, and the National Research Council – attempted to create a 'fruit machine' that measured interviewees' pupillary responses to visual stimuli in order to determine their sexual orientation. The technology repeatedly failed to deliver conclusive results, and eventually the project was abandoned.[18]

Discussions of VD compounded these politicized concerns, and incorporated similar language suggesting a pervasive, secretive, and largely veiled threat. Drawing on recent research from one of its city clinics, the director of New York City's Bureau of Social Hygiene informed attendees of a 1953 conference that the homosexual contact was an 'important agent' in VD transmission who 'must be sought zealously'.[19] Homosexual men seemed to locate each other through some secret code, had sexual liaisons that transcended social strata, and were reluctant to name their partners. In articles for their peers, VD specialists sought to establish their expertise by explaining the difficulties of discerning a homosexual man and locating his hidden infections. One physician, who emphasized the 'highly organized' degree of the homosexual network on North America's West Coast and the widespread anal intercourse this facilitated, urged colleagues to perform a dark-field examination on every anal ulcer found in a male patient. Recasting an old epithet for syphilis with a new Cold War resonance, and linking the disease's renowned mastery of disguise with the homosexual's ability to evade detection, he wrote: 'We must be ever wary of the "great imitator" in dealing with lesions in the anorectal area.'[20]

Amidst the upswing of reported cases of venereal disease among homosexually active men in the late 1950s and early 1960s, some asked whether this was a real or apparent increase. One physician who had worked in a Baltimore syphilis clinic in the years leading up to the Second World War recalled how he and his colleagues had treated many early infectious syphilis cases, but had only occasionally noticed queer men in attendance. 'Is male homosexuality significantly more common to-day than 5 or 10 years ago?' he wondered.

> Is there more venereal disease in male homosexuals than there was in the past, or is it simply that venereal disease is more frequently diagnosed in such patients? Is it possible that male homosexuals attend [VD] clinics in greater numbers than in the past and are therefore subjected to a greater extent to contact interviews which reveal the existence of additional venereally-infected

homosexuals? Is the male homosexual to-day more promiscuous than he was in the past? And finally, since one is most likely to find what one is looking for, is the male homosexual with venereal disease more frequently identified to-day as a homosexual because of a higher index of suspicion?[21]

Other medical workers echoed and debated this physician's questions, as did members of the growing homophile movement. Some described their surprise in learning, following an infection, that syphilis and gonorrhoea could be spread between men.[22] The editor of *ONE*, a leading homophile journal, stated emphatically that this was a new development:

> We can remember the day when a venereal disease contact from an homosexual experience was highly unlikely – when, in fact, no one even remotely known to us, no matter how promiscuous, had ever picked up a disease through a homosexual source.

One writer suggested that gay men, for whom pregnancy was not a concern, were less likely to use condoms, particularly due to their reduction in erotic sensation, and thus experienced a commensurate reduction in their protection against disease. He also posited that gay men were likely to forego attending public clinics, expecting to find them lacking in understanding or confidence.[23] *ONE*'s editor advocated just such avoidance, advising readers that as long as laws against homosexual acts existed:

> Under no conditions, or for any reason, should a homosexual set one foot inside a public health office. If anyone of us needs a doctor, let him go straight to a private physician whose ethics should hold that which is between the doctor and his patient in confidence.[24]

Drawing on the work of historians tracing the post-war growth of lesbian and gay communities, we might hypothesize that the rise in reports *did* describe a real increase in syphilis, caused by an intersecting combination of factors. Urban gay communities experienced transformative boosts from the remarkable disruptions of mass-mobilization for the Second World War. Millions were uprooted from small towns, thrown together in all-male environments, and stationed far from home in their own countries and overseas. Many men who might previously have felt isolated by feelings of sexual difference were able to find others like themselves and participate in thriving war-time gay scenes in major cities like San Francisco and New York, Montreal and London. Many would choose to remain in these cities at war's end, or maintain connections with other gay men in their new networks. These men would have had a sense that they were part of a growing community of like-minded individuals, gained an increased awareness of that community's diversity – beyond enduring stereotypes that focused on 'the fairy', for example – and in many cases fostered a communal sense of normalcy in opposition to prevailing psychological theories of homosexuality as deviance. In many cities, more men were having sex with others in more tightly connected networks, with increased opportunities to meet more frequently, albeit in a less permissive social environment than during war-time.[25] The chances of an individual transmitting an asymptomatic infection to others increased, likely with a correlated rise in clinic admissions. Similarly, as a result of a delicate yet growing sense of communal confidence and solidarity, greater numbers of men were more likely to admit to sexual contact with other men when seeking treatment.

## The male homosexual and the female prostitute: rivals for sexual partners, public space, and public health attention

In an unpublished training manual written in 1951, an experienced American VD investigator explicitly compared the male homosexual and the female prostitute in an aside emphasizing the need to maintain confidentiality with teenaged VD patients:

> It will take only one breach of confidence with a youngster to ruin your repu-
> tation as a keeper of your word . . . Young people have a very efficient grape-
> vine ranking second to homosexuals and prostitutes tied in first place. You
> play ball with them; they'll play ball with you . . . It is not your job to cor-
> rect sex patterns already fixed or to save souls. It is your job to find cases of
> venereal disease.[26]

This brief comparison hints at a common association made by many individuals active in the fields of law enforcement, moral reform, and VD control: that the male homosexual and the female prostitute were roughly equals of one another in terms of their disregard for conventional morality and gender norms, habitual lack of consideration for laws regulating sexuality, and cause of public nuisance. In England, the grouping of the two in the Wolfenden Committee's deliberations in the mid-1950s most clearly demonstrates this connection – although the Committee's report and the evidence upon which it was based paid very little attention to VD transmission among men.[27] While a remarkable late-twentieth-century shift in public favour towards gays and lesbians has largely severed these conceptual links, their traces can be detected most easily in a shared argot, and become much more visible in the historical record prior to the 1970s. Frequently those responding to the threat of VD in earlier periods would equate the two in an uneasy binary.[28]

Historians have shown how queer men and female sex workers shared similar haunts and interacted with each other at the margins of respectable society, often in bars, cafes, and taverns with some criminal element, and that a common language emerged to describe sexual practices (for example, 'frenching' for oral-genital intercourse), sexual partners ('straight' for those preferring vaginal intercourse), and ways to meet them ('cruising' through public spaces looking to meet men). Furthermore, this sharing sometimes intensified into competition for the same men as sexual partners. For example, the sailor – a quintessential representative of working-class bachelor culture, often with extra money to spend while on shore leave – was a favourite of female prostitutes and also featured heavily in gay men's fantasies and real-life sexual culture.[29] Sailors also frequently featured among international efforts to regulate VD, although contemporaries tended to focus on the risks they faced from their contact with female prostitutes.[30] From the 1930s onward, expert and popular views of sexual orientation increasingly enforced a dichotomy between a majority of 'normal' heterosexuals and a minority of 'deviant' homosexuals. Before then, however, men – particularly working men – who slept with women could also have sex with men without necessarily losing social status nor their own sense of themselves as normal.

These sexual dynamics would evidently affect VD transmission patterns, and also, in times when infections were overwhelmingly associated with female prostitutes and vaginal sex, impressions of relative risk. Thomas Painter, a New York gay man studying male homosexuality and hustling in the late 1930s, indicated how this sexual economy might operate during wartime with reference to another mobile representative of the bachelor class – the soldier:

> Prostitutes are seldom very desirable . . . [and] are likely to be diseased, especially the kind that can be picked up on the streets, and finally they cost money, of which no soldier in the ranks ever has much. The homosexual then presents himself, providing free entertainment, drinks, and company, and offering a momentary fondness to the lonesome boy desperate to forget the weary boredom or the terrifying horror of the war or army life. And the homosexual offers the boy a form of sexual release without cost, and relatively free from the danger of disease. He often will pay the boy something in addition – which the boy then can, and often does, spend on the young woman or female prostitute of high quality, whom he really wants. No one will ever know – he is away from home – and anyhow, what the hell . . . he'll probably be killed day after tomorrow anyway. It is [for] these reasons that the soldier . . . succumbs to, and even seeks out, the advances of the homosexual – in 1865, in 1918, and in 1941 equally.[31]

Painter's class biases are expressed in his denigration of street-walking sex workers and assumptions about homosexual men's disposable income. Yet his description of shifting sexual dynamics draws attention to the 'normal' working man – not the female prostitute, nor the homosexual – as a lynchpin in VD transmission, and emphasizes the view, widespread before 1950, that homosexual men were 'relatively free from the danger of disease'.

During the 1950s, as word spread among health workers about the prevalence of VD infection among male homosexuals, some asked whether a double standard was at work in the legal system. Herman Goodman, a physician who worked for New York City's Bureau of Social Hygiene, had investigated a syphilis outbreak among a network of queer men in 1943. By the mid-1950s he was convinced that the male homosexual, population bore responsibility for a considerable amount of VD transmission. Based on his interpretation of statistical records from the 1957 city magistrate's report, Goodman concluded that roughly 1,000 female prostitutes and 6,500 male homosexuals were largely responsible for the city's current early syphilis and gonorrhoea infections. He pointed out, however, that while female sex offenders were charged under a penal code section that required them to receive a physical examination, blood test, and antibiotic treatment, none of the male sex offenders received this attention. 'No discrimination', he argued, 'for or against the sex offender either male or female', calling for an equitable application of screening and treatment measures to those in custody.[32] The jails of some cities, like Vancouver, had implemented routine testing of male inmates in the late 1940s, based on their older programmes'

445

successes in finding VD cases among female prisoners.[33] Others, like Los Angeles, demanded mandatory blood tests and treatment specifically for gay men sentenced in city courts as a VD reduction measure in the 1960s. This was shortly after one LA public health official announced to an international syphilis conference that 'the white male homosexual has replaced the female prostitute as a major focus of syphilis infection'.[34]

Where health measures ended and law enforcement began was often unclear to individuals on both sides of the equation. For example, the official responsible for VD control in Vancouver in 1942 protested the police department's efforts to obtain the names of those infected through illegal sexual acts. Just over a decade later, his successor promoted the value of extensive interdepartmental cooperation, to 'insure the exchange of all available information, to the advantage of both.'[35] As exemplified above by the 1962 *ONE* editorial, many gay men were, understandably, deeply concerned that information requested by public health clinic workers – the names, descriptions, addresses, and phone numbers of their sexual partners – might be transmitted to the police department, whereupon they and their partners risked entrapment, arrest, loss of livelihood, and registration as sex offenders. From their perspective, and that of health workers tackling VD, maintaining the confidentiality of this information was vital. All community-based VD education leaflets produced in the 1960s emphasized the commitment of public health agencies to meet this requirement, often in large, bold letters.[36]

### 'Perpetual spirals of power and pleasure': health workers and venereal disease

Michel Foucault's phrase linking pleasure to power neatly characterizes certain aspects of the relationships between health workers and queer men with VD in the mid-twentieth century – with the former working to seek out, examine, and apply diagnoses to these men, and the latter often attempting to evade and resist the effects of that gaze.[37] Although numerous health workers contributed to the emergent bureaucracy underpinning successful VD diagnosis and therapy – receptionists organizing their clinics' patients, nurses administering treatment, lab workers testing samples, and secretaries managing the expansive paper records detailing clinical and contact-tracing histories – this section concentrates on two professionals with whom these patients interacted: the contact tracer and the physician.

Contact tracing aimed to locate all sexual contacts of infected patients seeking treatment; the practice began in the United States in the mid-1930s, later spreading to Britain and Canada.[38] Initially a predominantly female workforce composed of nurses, almoners, and social workers, its ranks in some large North American cities eventually included male public health advisers.[39] At first, contact tracers focused almost entirely on locating male patients' female partners, reflecting widespread assumptions that VD was spread mostly by the professional female prostitute or by 'the amateur'. Following the widely discussed Kinsey Report, and the increasing acknowledgment of same-sex contacts by men attending VD clinics in England and North America, greater emphasis was placed on 'the homosexual'.[40] Refinements to VD investigators' interviewing techniques continued; by 1962, training courses instructed contact tracers how to identify homosexuals and encouraged them, through their questioning, to 'establish the fact that the patient is a sexually promiscuous person and that this promiscuity has developed into a continuous pattern from early in life'.[41] By providing examples of how homosexual men might respond to evade certain questions, and explaining that '[c]ertain known occupations may suggest deviant sexual activity', the instruction of contact tracers continued the positivist nineteenth-century sexological project of defining and identifying 'the sexual deviant', and succeeded in locating many cases of VD.[42] One Vancouver newspaper described a local epidemiologist, who had travelled to the US for his VD training, as being 'proud of his ability to flush out the "gay ones"'.[43]

Contact tracers lamented the fact that many private physicians failed to report the VD patients they treated; the detective trail often ran cold at the private clinic door. Sometimes this represented a potentially catastrophic spread of infection: reports suggested that certain practitioners' clientele consisted mainly of gay men, some of whom amassed dozens of partners while infectious. Often public health workers suspected that private physicians were shielding their patients, or were too busy for the time-consuming and unpaid task of interviewing them for information about their sexual partners. Health departments expended much effort educating private physicians about the value of their contact tracers and their absolute commitment to confidentiality.[44] It is also true that many physicians remained unaware of the possibility of same-sex VD transmission. Formal education about sexual matters was minimal – a persistent problem throughout the twentieth century – and if doctors learned about homosexuality at all it was often as a foreign perversion, or a practice that conferred a protective effect compared to sex with a female prostitute.[45] In some cases physicians may have held suspicions about certain patients' sexual orientation, yet were hesitant to ask prying questions or suggest rectal examinations for fear of losing clients.[46] It seems likely, though, that a Czech émigré physician accurately summarized the American situation in the mid-1950s when he wrote that 'the nearly total lack of reports of primary syphilis in the mouth and rectum due to homosexual practices can only be explained by the lack of awareness of this possibility by doctors'.[47]

From the mid-1940s onward, physicians were advised to suspect homosexual relations and overcome a reluctance to conduct oral and rectal examinations for hidden lesions. One doctor urged his colleagues to look further: 'Because homosexuals are notoriously imaginative in their sexual behavior, the varied lesions of venereal disease may be found anywhere on the body.'[48] Under the dominant sun of Freudian psychological theories, physicians learned that homosexuals' immature development made them self-loving, vain, cruel, amoral, risk-seeking, and untruthful, all of which fostered their characteristic promiscuity and propensity for VD infections.[49] These characterizations evidently shaped some practitioners' understanding of their patients, bringing mixtures of sympathy, disapproval, and disgust. One venereologist, who in 1965 took up a consultant position in a mid-sized English town, viewed most of his homosexually active patients as 'apprehensive and fearful persons desperately seeking sympathy and succour'. He recalled a middle-aged man from a small local town attending his clinic 'in tears':

> He was unable to sit down and almost unable to walk from pain. He had had little sleep for several days and had restricted his intake of food and drink for more than a week as his venereal condition caused intense pain on passing

urine and opening his bowels. The whole of his genital area was ulcerated as part of a widespread syphilitic rash and pus poured from his rectum which was infected with gonorrhoea. Trembling and humiliated he had almost lost the desire to live. A homosexual partner had driven him but only supported him as far as the threshold, lacking the courage or concern to stay with him.[50]

Increasingly vocal and visible lesbian and gay rights activism came to the fore in the late 1960s, drawing energy, inspiration, and individuals from the civil rights, women's health, and anti-war movements. Gay liberation's spirited activism fostered the development of more publicly apparent and politically active communities in many cities, some of which sponsored their own clinics and health outreach programmes. The Los Angeles Gay Community Service Center's VD clinic, Chicago's Howard Brown Memorial Clinic, New York City's Gay Men's Health Project, and Toronto's Hassle Free Clinic were among the better known of these new health initiatives. As part of this expanding medical infrastructure, gay and gay-friendly medical professionals, who were themselves occasionally VD patients, targeted poor physician training. They drew attention to the differential presentation of disease among homosexually active men, emphasizing the importance of oral and anal examinations. They also elucidated how different sexual practices, like oral-anal intercourse, had given rise to a new category of enterically spread VD, including amoebiasis, giardiasis, shighellosis, and hepatitis A. Sites of queer sex, like bathhouses, which had since the 1950s faced public health and civic scrutiny for facilitating VD transmission, saw gay doctors making new outreach efforts to test and treat those who attended, although some owners saw these efforts as bad for business. Perhaps the most significant outgrowth of the 1970s gay liberation health movement was the collaborative work between a number of gay VD clinics, the Centers for Disease Control, and the pharmaceutical company Merck to study the prevalence of hepatitis B and conduct clinical trials for a vaccine.[51]

## Patient experiences: from 'terribly embarrassing and terribly pitiful' to 'red badges of courage'?

In addition to enduring the widely felt stigma of medical conditions often associated with dirt and moral depravity, and often the need to pay for and juggle medical visits with work without raising employers' or family suspicions, men contracting infections through sexual liaisons with men encountered other significant challenges extending beyond their immediate physical health concerns. Few men with family doctors would previously have confided the specifics of their sexual attraction to their physicians; thus, they would have faced what for many would have been a humiliatingly frank conversation with their primary health-care provider, or the stress of concealing the source of their infection. Those who had gained access to networks of other gay men could inquire about local physicians who might be amenable to treating patients discreetly. In many cities, certain doctors – whether themselves closeted, non-discriminating, entrepreneurial or perhaps a combination of the three – became renowned for their expansive gay practices. For those who could not afford a private physician, the public clinic presented a mixed blessing. From the early twentieth century onward, these clinics increasingly provided affordable and sometimes free care with a minimum of moral condemnation, although even within the same clinic a patient's reception might differ drastically from one physician or nurse to another.[52] Nonetheless, significant numbers sought treatment from sources where disclosure could be minimized: as late as the 1960s, authorities would complain of patients seeking fraudulent cures from quacks and disreputable pharmacists.[53]

The location of a patient's physical complaint also coloured his experience, and to a considerable extent dictated the degree to which he might evade the health worker's scrutinizing gaze. While little about patients presenting with penile chancres or gonorrhoeal urethritis suggested homosexual contact, those with oral ulcers and certainly with anal disturbances would have aroused considerably more suspicion. That being said, the latter two locations would frequently go unexamined unless specifically prompted by the patient, so men with low-grade signs and symptoms might easily pass unnoticed – leading a Scottish committee reporting on sexually transmitted diseases to declare that 'passive homosexuals' were 'reservoirs of infection'.[54] One gay observer of the homosexual scene in 1930s New York remarked that, while certainly not widespread, among anally receptive homosexual men 'syphilis of the anus' was 'a really pitiful affliction, being terribly embarrassing and terribly painful at once'.[55] To this man, embarrassment was closely tied to the physical seat of the pain, a confirmation of one's demasculinization. Such humiliation caused some patients to delay seeking treatment, to their considerable distress, complicating the notion that venereal diseases were essentially minor irritations between penicillin's rise and the appearance of AIDS.

At public clinics, and to a lesser degree with private physicians, patients faced the ordeal of disclosing their infection's likely source. Many were deeply concerned about trusting government-employed contact tracers. In the 1950s, physicians realized that a good number of these patients reported female names for male sexual contacts. In some cases this was an ingenious balance between honesty and concealment whereby gay men provided the investigators with the widely employed feminine camp names used by their partners.[56]

In a letter to Donald Webster Cory, the pseudonym of one of the most widely read American authors on homosexuality in the 1950s and 1960s, one anonymous gay man in his early twenties wrote at length of his experience being diagnosed with syphilis at a New York City public clinic. Both he and Cory wished to highlight 'an old problem in a new form', one which had risen dramatically in importance since the Second World War, and Cory reprinted the letter in full. Several of the young man's points bear emphasizing here. First, the primary importance of personal networks over official channels: before attending the clinic, he consulted a gay friend who told him 'there was a lot of VD going around in New York, and that you can get it by having almost any kind of sex with an infected person'. Horrified that VD might have caused his anal discharge and spots on his torso, he visited a library to consult an encyclopaedia, but 'it didn't say anything about sores in the back'. Second, his lack of awareness: he admitted that he and most of his friends were 'pretty ignorant about it. How was I to know that that spot on that sailor's pipe was going to put me through such an ordeal?' Finally, he repeatedly expressed feelings of 'anguish and humiliation' in having to admit the source of his infection and submit to anal examinations by an intern and subsequently a team of medical students. It is likely that his sense of being scrutinized as 'a model specimen for their instructional purposes' was due in no small part to penicillin's success in reducing the prevalence of syphilis in the 1950s.[57] With clinical instruction on VD so limited, and infectious syphilis cases in such short supply, a number of forces compounded the scrutiny the young man received, and contributed to his sense of mortification.[58]

In many cases the availability of antibiotics reduced patients' concern about VD, and gay-run clinics succeeded in diminishing the stigma associated with infections during the 1960s and 1970s. With gay liberationist ideologies proposing that multiple sexual partners were the ties that bound the gay community together, some, like author Edmund White, were later quoted as saying that 'gay men should wear their sexually transmitted diseases like red badges of courage in a war against a sex-negative society'.[59] Still, the experiences of men whose sexual lives straddled these therapeutic innovations, and who fell ill with other diseases like hepatitis, invite more complicated interpretations. Samuel Steward, an American gay man whose remarkable career trajectory took him from university English professor to tattoo shop owner and artist, is one telling example. In his early twenties, Steward contracted syphilis from a casual male partner; the shock of the infection, his sense of pollution, and a lengthy and painful treatment experience scared him from sex for some time. Having returned to a vigorous pursuit of sexual encounters, the unsettling shock revisited him once more with a gonorrhoea diagnosis in 1950. Later he was scared by the prospect of police and health department investigation when one of his sexual partners was diagnosed with syphilis and named him as a contact, and he fell very ill with hepatitis shortly after a sexually active holiday spent in San Francisco in 1953. Though he never followed through with his occasional ideas to give up sex completely, Steward would go on to think of the trauma of venereal infection as being an important example of a 'dividing point', one of a series of ruptures organizing the lives of all homosexual men.[60]

## 'VD is no camp': education and prevention

Widespread reluctance to publicly discuss homosexuality fostered an environment where official efforts to address the transmission of VD generally ignored the possibility of same-sex transmission. This silence manifested in several ways. Many physicians, most of whom received very little education on sexual matters, remained unaware of this route of spread. Such ignorance of the possible risks of same-sex contacts also extended to queer men. Indeed, in times and places where the threat of VD was tied so closely to vaginal intercourse with female prostitutes, some interpreted the silences surrounding homosexual activity and male prostitution as suggestions that engaging in these realms conferred a *reduction* in risk, in the same way that they helped men avoid unwanted pregnancy. For instance, Richard von Krafft-Ebing, the Viennese psychiatrist who established himself as a late-nineteenth-century expert on homosexuality, interpreted cases and compiled editions of his *Psychopathia Sexualis* at a time when syphilis and fears of the disease were widespread. He hypothesized that among the reasons why some men might seek sexual contact with other men was a 'hypochondriacal fear of infection in sexual intercourse; or on account of an actual infection'.[61] Similarly, the expression 'Better a little shit than a chancre', which circulated in New York's Harlem district during the 1920s – an area with many unmet social and health needs during a period of more relaxed social mores and widespread prostitution – suggests how some men rationalized this safer-sex belief.[62] In the 1950s and 1960s, public health physicians continued to express concern that homosexual men were unaware that sexual contact could bring VD. Many seemed to view this particular consequence of sex, like pregnancy, as a concern solely affecting heterosexuals.[63]

As public health workers became more aware that sex between men could transmit disease, they were cautious in their attempts to promote this understanding. Information leaflets distributed by health authorities might cover all bases by indicating that syphilis or gonorrhoea could be spread from one infected 'person' to another, without specifying the sex of the persons involved – though of course this risked readers projecting their own assumptions onto the documents. As long as laws banned same-sex sexual contact, many health workers felt compelled to exercise caution, since there was a fear in some quarters that open discussion risked promoting the taboo – and illegal – practices. In 1963, at a time when the VD Program of the Communicable Disease Center (later the Centers for Disease Control) was otherwise encouraging its public health advisors to be assertively resourceful in their efforts to reduce VD transmission, its director chastised an advisor who spoke publicly, without prior clearance, on the issue of VD and homosexuality at a North Carolina medical society.[64] Although in 1957 the VD Program had relocated from Washington to Atlanta, the political reach of the capital remained strong over this stretch of distance and time. 'Washington is regarding VD education and behavioral studies as sensitive areas and screening for policy', wrote a representative of the CDC's Information Office in 1964.[65] There is no doubt that these political misgivings impeded the promotion of this knowledge, and required agencies seeking to make inroads to exercise strong discretion. That same year, as part of the nation's drive to eradicate syphilis by 1972, representatives of New York City's Department of Health teamed up with the Mattachine Society of New York, the nation's largest homophile organization, which sought to promote public understanding of 'sexual variants'. Their collaborative effort led to one of the earliest health leaflets created by and for gay men, entitled 'VD is No Camp'. Ten thousand copies were printed for distribution, though city workers were careful to insist that their assistance went uncredited.[66]

In the absence of official information and before homophile organizations began filling this void in the mid-1960s, queer men adopted numerous strategies to protect their health and safety. Many undoubtedly read official guidelines against the grain to find information that they could adapt to their own sexual circumstances.[67] Tabloid gossip columns, often the earliest published sources of community information, would occasionally warn readers of VD outbreaks.[68] Given their shared positions as sexual outlaws in overlapping social spaces until the mid-twentieth century, it is unsurprising to see queer men drawing on techniques employed by female prostitutes to reduce their risk. 'Frenching', a commonly employed phrase denoting oral-genital sex, was deemed by many sexually active individuals to be safer than vaginal or anal intercourse, not least because it allowed for close inspection of the partner's penis for chancres or discharge.[69] Experienced female prostitutes were known to reduce their infection risk by refusing partners who failed such visual examinations; similarly, some gay men carried small penlights with them to allow quick partner check-ups in a city's dark corners.[70] Among those familiar with the risks, men who enjoyed being the receptive partner in anal intercourse were deemed, like the female prostitute, to be at higher risk of acquiring venereal infections. Another technique some of these men borrowed from female prostitutes was the practice of post-intercourse douching with an antiseptic solution.[71]

451

Millions of enlisted men learned of the benefits of condom use and post-coital prophylactic disinfection through their Second World War training. It is not known how many would have thought to use condoms for protection in their same-sex encounters, though it seems plausible that some would have done so in the post-war years. Certainly the practice was rare by the 1970s; VD investigators remembered gay men laughing at the suggestion that they might consider using condoms, and one queer-identified man recalled thinking that men using condoms before the AIDS epidemic were fetishists.[72] By contrast, post-encounter genital washing formed part of standard healthy sex guidelines that gay men encountered in community-produced literature in the late 1960s and 1970s.[73] Finally, from the early 1960s some doctors regularly recommended their gay patients practise pre-exposure prophylaxis by taking penicillin pills if they foresaw the chance of an exposure with an infected partner. The fears of drug-resistance and sex without consequences raised in the ensuing conversations would presage many that would follow in the course of the HIV/AIDS epidemic decades later.[74]

## Conclusion

At the close of the Second World War, and throughout much of England and North America, many health workers and queer men were unaware that same-sex contact could transmit VD. This changed significantly between the late 1940s and the 1970s, as queer men gradually became the focus of heightened political and public health surveillance, and the incidence of VD among them appeared to rise. The increasing visibility of lesbian and gay communities and the rise of gay liberation had mixed results. By the 1970s gay men managed to shake off the shackles of medically defined deviance with a successful campaign against psychiatry's classification of homosexuality as a mental illness. On the other hand, much older links between same-sex activity and physical sickness were at the same time being reinscribed. These associations would become cemented for decades with the emergence of the HIV/AIDS epidemic in the 1980s.

Three brief concluding observations bear emphasizing here, to link with themes emerging elsewhere in this book. First, the shifting connections made between sexual activity among men and VD – ranging from tacit awareness to impressions of relative safety and then to increased risk – highlight the markedly contingent nature of beliefs about disease over time. Disease ecology, population movements, fears about prostitution amid rapid urbanization, changing axes of sexual orientation and identity, and shifts in physician education – these were but some of the factors whose changing configurations affected the comparative visibility or obscurity of same-sex VD transmission. Second, during a Cold War period that saw the widespread growth of highly developed technological systems in health and medicine, the bureaucratic technologies that rendered queer men most readily visible were remarkably simple. Developments in contact tracing and refined interviewing techniques were ultimately far more successful than failed hi-tech screening efforts like the Canadian fruit machine, and harkened back to administrative advances from earlier decades. This relates to the third observation: in the context of risk-factor epidemiology for chronic diseases, the dominant medical research paradigm of the Cold War period, there was something decidedly old-fashioned, even déclassé, about efforts to identify homosexual men and link them to the spread of VD. As the investigation of disease risk became focused on multifactorial webs of causation, health workers who concentrated on the social webs of well-established VD transmission moved to the periphery of professional practice; they remained there until the dramatic re-entry of infectious disease with the rise of the HIV/AIDS epidemic.[75]

Some issues from this earlier era would continue to feature strongly as HIV took hold. Gay activists' concerns about the confidentiality of their health information became one key battleground, and organised resistance to attempts to enact quarantine measures and mandatory testing for those infected with HIV ensured a legacy of special care given to the results of tests for the infection. Medical and popular understandings cast both queer men and female prostitutes as groups at risk for the new disease. However, the strength of the old conceptual and social affiliations linking the two groups together was rapidly dissolving. As the tide of public opinion against homosexuality peaked then began to fall in the twentieth century's last decade, those middle-class representatives of the lesbian and gay communities who survived the epidemic would see their social capital grow. Perhaps most importantly, in the absence of meaningful official assistance, and amid suggestions that they abstain from sex to avoid disease, gay men drew upon a historical legacy of pragmatic self-help. They appropriated the condom and developed a safer-sex ethos that built upon gay liberationist ideas while at the same time acknowledging the risks of acquiring infections through frequent partner exchange. By 'queering' the condom and pushing back against an often hostile world, gay communities ensured that same-sex encounters could continue, empowering pleasure and connection just as HIV presented a devastating new dividing point to the lives of many queer men.

## Notes

1 Authors addressing this subject have generally done so in passing: J. Cassel, 'Making Canada Safe for Sex: Government and the Problem of Sexually Transmitted Disease in the Twentieth Century', in C. Naylor (ed.), *Canadian Health Care and the State: A Century of Evolution*, Montreal: McGill-Queen's University Press, 1992, pp. 141–92; J. Oriel, *The Scars of Venus: A History of Venereology*, London: Springer-Verlag, 1994; G. Rotello, *Sexual Ecology: AIDS and the Destiny of Gay Men*, New York: Plume, 1998, pp. 38–64; A. Mooij, *Out of Otherness: Characters and Narrators in the Dutch Venereal Disease Debates 1850–1990*, trans. B. Jackson, Amsterdam and Atlanta: Rodopi, 1998, pp. 180–8; R. Davidson, '"The Price of the Permissive Society": The Epidemiology and Control of VD and STDs in Late-Twentieth-Century Scotland', in R. Davidson and L. Hall (eds), *Sex, Sin and Suffering: Venereal Disease and European Society since 1870*, London: Routledge, 2001, pp. 220–36; D. Evans, 'Sexually Transmitted Disease Policy in the English National Health Service, 1948–2000: Continuity and Social Change', in Davidson and Hall (eds), *Sex, Sin and Suffering*, pp. 237–52. Two exceptions are: M. Brown, '2008 Urban Geography Plenary Lecture: Public Health as Urban Politics, Urban Geography: Venereal Biopower in Seattle, 1943–1983', *Urban Geography*, 2009, vol. 30, 1–29; and C. Batza, 'Before AIDS: Gay and Lesbian Community Health Activism in the 1970s', D.Phil. dissertation, University of Illinois at Chicago, 2012.
2 'Before HIV: Homosex and Venereal Disease, c.1939–1984', Wellcome Trust grant no. 098705, Department of the History and Philosophy of Science, University of Cambridge, 2013–17.
3 L. Rupp, 'The Persistence of Transnational Organizing: The Case of the Homophile Movement', *American Historical Review*, 2011, vol. 116, 1014–39.
4 Florentine magistrate records from 1496 to 1497, for example, indicate that a 40-year-old man 'did not sodomize' a 17-year-old youth on account of the latter's anal venereal complaint, fellating him instead; see n. 26, in M. Rocke, *Forbidden Friendships: Homosexuality and Male*

*Culture in Renaissance Florence*, New York: Oxford University Press, 1996, p. 284. By the mid-sixteenth century, Lucca, Florence's neighbouring city-state, had firmly linked anal sex with the transmission of the 'French Disease' – a condition resembling modern-day syphilis; see M. Hewlett, 'The French Connection: Syphilis and Sodomy in Late-Renaissance Lucca', in K. Siena (ed.), *Sins of the Flesh: Responding to Sexual Disease in Early Modern Europe*, Toronto: Centre for Reformation and Renaissance Studies, 2005, pp. 239–60.

5 G. Rousseau, 'Policing the Anus: Stuprum and Sodomy According to Paolo Zacchia's Forensic Medicine', in K. Borris and G. Rousseau (eds), *The Sciences of Homosexuality in Early Modern Europe*, London: Routledge, 2008, pp. 75–91.

6 Trial of Mustapha Pochowachett (t16940524-20), May 1694, *Old Bailey Proceedings Online*, http://www.oldbaileyonline.org, accessed 18 June 2015.

7 J. Marten, *A True and Succinct Account of the Venereal Disease*, 4th edn, London, 1706, pp. 51–2.

8 J. Marten, *A Treatise of the Venereal Disease*, 7th edn, London, 1711, pp. 345–7.

9 K. Siena, 'The Strange Medical Silence on Same-Sex Transmission of the Pox, c.1660–c.1760', in Borris and Rousseau (eds), *Sciences of Homosexuality*, pp. 115–33; C. Berco, 'Syphilis and the Silencing of Sodomy in Juan Calvo's *Tratado Del Morbo Gálico*', in Borris and Rousseau (eds), *Sciences of Homosexuality*, 92–113.

10 C. Emsley, T. Hitchcock, and R. Shoemaker, 'London History: A Population History of London', *Old Bailey Proceedings Online*, http://www.oldbaileyonline.org, version 7.0, accessed 31 Aug. 2015.

11 M. McIntosh, 'The Homosexual Role', *Social Problems*, 1968, vol. 16, 182–92; G. Kinsman, *The Regulation of Desire: Sexuality in Canada*, Montreal: Black Rose Books, 1987, pp. 39–40; J. Weeks, 'The "Homosexual Role" after 30 Years: An Appreciation of the Work of Mary Mcintosh', *Sexualities*, 1998, vol. 1, 140.

12 H. Cocks, *Nameless Offences: Homosexual Desire in the 19th Century*, London: I.B. Tauris, 2003, pp. 15–49. For further discussion of deviance as disease, see Jana Funke's chapter in this volume.

13 J. Casper, *A Handbook of the Practice of Forensic Medicine, Based Upon Personal Experience*, 3rd edn, trans. G. Balfour, London: New Sydenham Society, 1864, vol. 3, pp. 328–41; A. Tardieu, *Étude médico-légale sur les attentats aux moeurs*, 5th edn, Paris: J. B. Baillière et Fils, 1867, pp. 171–263. Weeks suggests, however, that these European authorities were little known in nineteenth-century England: J. Weeks, *Sex, Politics, and Society: The Regulation of Sexuality since 1800*, 2nd edn, London: Longman, 1989, p. 101.

14 J. Comte, 'Syphilis and Sex: Transatlantic Medicine and Public Health in Argentina and the United States, 1880–1940', D.Phil. dissertation, University of Pittsburgh, 2013, pp. 27–120.

15 A. Kinsey, W. Pomeroy, and C. Martin, *Sexual Behavior in the Human Male*, Philadelphia: W. B. Saunders Company, 1948, pp. 610–66.

16 For example: G. Kulchar and E. Ninnis, 'Tracing the Source of Infection in Syphilis', *Journal of Social Hygiene*, 1936, vol. 22, 370–73; A. Jones and L. Janis, 'Primary Syphilis of the Rectum and Gonorrhea of the Anus in a Male Homosexual Playing the Role of a Female Prostitute', *American Journal of Syphilis, Gonorrhea, and Venereal Diseases*, 1944, vol. 28, 453–57; H. Goodman, 'An Epidemic of Genital Chancres from Perversion', ibid., 310–14; B. Kanee and C. Hunt, 'Homosexuality as a Source of Venereal Disease', *Canadian Medical Association Journal*, 1951, vol. 65, 138–40.

17 For an example of the earlier type, see C. Marshall, 'Anal Chancre Resulting from Pederasty', *American Journal of Syphilis*, 1919, vol. 3, 454–6.

18 D. Johnson, *The Lavender Scare: The Cold War Persecution of Gays and Lesbians in the Federal Government*, Chicago: University of Chicago Press, 2004; F. Mort, *Capital Affairs: London and the Making of the Permissive Society*, London: Yale University Press, 2010, pp. 139–96; G. Kinsman and P. Gentile, *The Canadian War on Queers: National Security as Sexual Regulation*, Vancouver: UBC Press, 2010, pp. 53–114, 168–90.

19 New York City Municipal Archives, Health Commissioner Records, Accession 90–116 (NYCMA MC), Box 289, folder: American Social Hygiene Association, 'Changing Aspects and New Challenges', session led by S. Frank, reported in the American Social Hygiene Association's 'Highlights of the National Conference on Social Hygiene', New York, 5–6 March 1953, p. 42.

20  C. Jackson, 'Syphilis: The Role of the Homosexual', *Medical Services Journal, Canada*, 1963, vol. 19, 638.

21  I. Schamberg, 'Syphilis and Sisyphus', *British Journal of Venereal Diseases*, 1963, vol. 39, 91–2.

22  Mr. D. G., 'Readers Write', *Mattachine Review*, 1957, vol. 3, no. 12, 4.

23  'Venereal Disease in the Homosexual', *Mattachine Review*, 1960, vol. 6, no. 5, 7.

24  D. Slater, 'Editorial', *ONE*, 1962, vol. 10, no. 10, 4–5.

25  J. D'Emilio, *Sexual Politics, Sexual Communities: The Making of a Homosexual Minority in the United States, 1940–1970*, 2nd edn, Chicago: University of Chicago Press, 1998; A. Bérubé, *Coming Out Under Fire: The History of Gay Men and Women in World War II*, 20th anniversary edn, Chapel Hill: University of North Carolina Press, 2010; Kinsman, *Regulation of Desire*, pp. 109–38; M. Houlbrook, *Queer London: Perils and Pleasures in the Sexual Metropolis, 1918–1957*, London: University of Chicago Press, 2005, pp. 195–6, 236–9.

26  R. Swank, 'Venereal Disease Contact Interviewing', training manual, Alto Medical Center, 5 February 1951, p. 39, courtesy of Fred A. Martich, retired CDC Public Health Advisor, Watsonian Society, Atlanta.

27  Mort, *Capital Affairs*, pp. 139–96.

28  Those aiming to repeal the Contagious Disease Acts in late-nineteenth-century Britain claimed that prostitutes were blamed for VD transmission that actually took place among men in the Royal Navy; see J. Walkowitz, *Prostitution and Victorian Society: Women, Class, and the State*, Cambridge: Cambridge University Press, 1980, p. 130.

29  G. Chauncey, *Gay New York: Gender, Urban Culture, and the Making of the Gay Male World, 1890–1940*, New York: Basic Books, 1994, p. 78; Houlbrook, *Queer London*, pp. 68–92.

30  L. Hall, 'What Shall We Do With the Poxy Sailor?', *Journal for Maritime Research*, 2004, vol. 6, 113–44.

31  Thomas Painter, 'Male Homosexuals and Their Prostitutes in Contemporary America', 1941, unpublished typescript, vol. 1, pp. 96–7, Thomas N. Painter Collection, courtesy of The Kinsey Institute for Research in Sex, Gender, and Reproduction.

32  H. Goodman, 'The Male Homosexual and Venereal Diseases', *Acta Dermato-Venereologica*, 1958, vol. 38, 274–82.

33  'Jail to Test All for VD', *Vancouver News-Herald*, 12 November 1948, 1.

34  R. Sachs, 'Effect of Urbanization on the Spread of Syphilis', in *World Forum on Syphilis and Other Treponematoses*, Washington, DC: US Government Printing Office, 1964, p. 155.

35  D. Cleveland, *Annual Report of the Division of Venereal Disease Control of the Provincial Board of Health for the year 1942*, Victoria, BC: 1943, p. 14; A. Nelson, 'Police and Health Cooperation in VD Control: The Vancouver Story', *Journal of Social Hygiene*, 1953, vol. 39, 393.

36  For example, New York Public Library, Mattachine Society, Inc. of New York Records, Box 11, Folder 4, 'VD is No Camp', leaflet, 1964.

37  M. Foucault, *The Will to Knowledge: The History of Sexuality, Volume 1*, 1978, trans. R. Hurley, 1998 edn, London: Penguin Books, p. 45.

38  A. Brandt, *No Magic Bullet: A Social History of Venereal Disease in the United States since 1880*, New York: Oxford University Press, 1987, p. 150; 'Editorial: The Venereal Disease Contact', *British Journal of Venereal Diseases*, 1945, vol. 21, 1.

39  B. Meyerson, F. Martich and G. Naehr, *Ready to Go: The History and Contributions of U.S. Public Health Advisors*, Research Triangle Park, NC: American Social Health Association, 2008, pp. 7–67.

40  F. Jefferiss, 'Venereal Disease and the Homosexual', *British Journal of Venereal Diseases*, 1956, vol. 32, 17–20; J. Tarr, 'The Male Homosexual and Venereal Disease', *GP* [*General Practitioner*], 1962, vol. 25, 91–7.

41  Venereal Disease Branch, *Field Manual*, Atlanta: Communicable Disease Center, Venereal Disease Branch, 1962, manual, pp. E15–21.

42  Ibid., E15; P. Davies et al., *Sex, Gay Men and AIDS*, London: Falmer Press, 1993, pp. 40–3.

43  A. Myers, 'VD Festers in the Haven of Skid Row', *Vancouver Sun*, 25 November 1964, 15.

44  Committee on Public Health of the New York Academy of Medicine, 'Resurgence of Venereal Disease', *Bulletin of the New York Academy of Medicine*, 1964, vol. 40, 802–23.

45  L. Potter, *Strange Loves: A Study in Sexual Abnormalities*, New York: National Library Press, 1933, pp. 4–9.

46 M. D., 'Readers Write', *Mattachine Review*, 1958, vol. 4, no. 3, 28.

47 H. Hecht, 'Venereal Diseases in Homosexuals', *Acta Dermato-Venereologica*, 1957, vol. 37, 183.

48 Tarr, 'Male Homosexual', 92.

49 J. Hinrichsen, 'The Importance of a Knowledge of Sexual Habits in the Diagnosis and Control of Venereal Disease, with Special Reference to Homosexual Behavior', *The Urologic and Cutaneous Review*, 1944, vol. 48, 469–86; E. Bergler, *One Thousand Homosexuals: Conspiracy of Silence, or Curing and Deglamorizing Homosexuals?*, Paterson, NJ: Pageant Books, 1959, pp. 243–9.

50 S. Laird, *Roses in December: Memories of the Early Antibiotic Age*, Braunton, Devon: Merlin Books, 1990, p. 320.

51 D. Ostrow, T. Sandholzer, and Y. Felman (eds), *Sexually Transmitted Diseases in Homosexual Men: Diagnosis, Treatment, and Research*, London: Plenum Medical Book Company, 1983; Batza, 'Before AIDS', pp. 233–65.

52 For example, E. Heaman, *St Mary's: The History of a London Teaching Hospital*, Montreal and Kingston: McGill-Queen's University Press, 2003, pp. 186–8.

53 Canadian Lesbian and Gay Archives, Toronto: B.C., 'Ask the Roving Reporter . . .', *ASK* [*Association for Social Knowledge*] *Newsletter*, Oct. 1964.

54 1973 Report of the Gilloran Committee on Sexually Transmitted Diseases, cited in Davidson, '"Permissive Society"', 224.

55 Painter, 'Male Homosexuals', vol. 1, p. 241.

56 A. Larsen, 'The Transmission of Venereal Disease through Homosexual Practices', *Canadian Medical Association Journal*, 1959, vol. 80, 24.

57 Anonymous author, quoted in D. Cory and J. LeRoy, *The Homosexual and His Society: A View from Within*, New York: The Citadel Press, 1963, pp. 176–88.

58 Committee on Public Health of the New York Academy of Medicine, 'Resurgence of VD', 818–19.

59 M. Callen, *Surviving AIDS*, New York: HarperCollins, 1990, p. 4.

60 J. Spring, *Secret Historian: The Life and Times of Samuel Steward, Professor, Tattoo Artist, and Sexual Renegade*, New York: Farrar, Straus and Giroux, 2010, quotation on p. 327.

61 R. von Krafft-Ebing, *Psychopathia Sexualis, with Especial Reference to the Contrary Sexual Instinct, a Medico-Legal Study*, 7th edn, translated by C. Chaddock, Philadelphia: F.A. Davis Co., 1894, pp. 187–9. This scenario was offered for men with 'acquired homo-sexual feeling'; see also Potter, *Strange Loves*, p. 9.

62 S. Watson, *The Harlem Renaissance: Hub of African-American Culture, 1920–1930*, New York: Pantheon Books, 1995, p. 134. See also the 1931 example described by Chauncey, *Gay New York*, pp. 85–6.

63 Tarr, 'Male Homosexual', 93.

64 Meyerson, Martich, and Naehr, *Ready to Go*, pp. 93–125; National Archives and Records Administration, Record Group 442 (NARA RG 442), Accession 70A470, Folder: 'Demchak, Maxim [/] Problem of the Homosexual in VD Control', William Brown to Maxim Dem'chak, 14 Nov. 1963.

65 NARA, RG 442, Accession 71A1708, Container 1, Folder: 'Forer, Raymond [/] Sociological Aspects of Sexual Behavior and VD', Helen O. Neff to Dick Deitrick, routing slip, 6 March 1964.

66 NYCMA HC, Box 440, Venereal, 'Progress Report, New York City Venereal Disease Education', 15 February 1964, p. 4.

67 P. Jackson offers the suggestive possibility that Canadian military medical authorities inserted coded text in VD guidelines during the Second World War to reach homosexually active men: Jackson, *One of the Boys: Homosexuality in the Military during World War II*, 2nd edn, Montreal & Kingston: McGill-Queen's University Press, 2010, pp. 60–1.

68 For example, T. Bain, 'The Big Beat', *Tab* [Toronto], 25 February 1961, 9, Tabloid Newspaper Collection, Canadian Lesbian and Gay Archives.

69 E. Clement, *Love for Sale: Courting, Treating, and Prostitution in New York City, 1900–1945*, Chapel Hill, NC: University of North Carolina Press, 2006, pp. 67, 141, 273–4.

70 Painter, 'Male Homosexuals', vol. 1, 241. An account of a female bordello operator employing a similar tactic with apparent effectiveness in the 1950s suggests that visual screening

was a well-used practice among those who were highly sexually active and aware of VD risks; see G. Moore and B. Moore, 'Fighting Venereal Disease in Fayetteville, North Carolina, 1951–1952', *Public Health Reports*, 2008, vol. 123, 236.

71  Painter, 'Male Homosexuals', vol. 1, p. 241.

72  William Darrow, interview with author, Miami, 28 March 2008; Spencer Macdonell [pseudonym], interview with author, Vancouver, 11 June 2008, recordings C1491/21 and C1491/27, British Library Sound Archive.

73  B. Lewis, 'The Ins and Outs of VD,' *The Advocate*, November 1968, 4, 30.

74  Anonymous author, quoted in Cory and LeRoy, *Homosexual*, pp. 187–8.

75  Andrew Moss, interview with author, 24 July 2007, recording C1491/09, British Library Sound Archive; M. Susser and Z. Stein, *Eras in Epidemiology: The Evolution of Ideas*, Oxford: Oxford University Press, 2009, pp. 166–7.

## Select bibliography

Anonymous author, quoted in Cory, D., and LeRoy, J., *The Homosexual and His Society: A View from Within*, New York: The Citadel Press, 1963, pp. 176–88.

Chauncey, G., *Gay New York: Gender, Urban Culture, and the Making of the Gay Male World, 1890–1940*, New York: Basic Books, 1994.

Clement, E., *Love for Sale: Courting, Treating, and Prostitution in New York City, 1900–1945*, Chapel Hill, NC: University of North Carolina Press, 2006.

Hinrichsen, J., 'The Importance of a Knowledge of Sexual Habits in the Diagnosis and Control of Venereal Disease, with Special Reference to Homosexual Behavior', *The Urologic and Cutaneous Review*, 1944, vol. 48, 469–86.

Jefferiss, F., 'Venereal Disease and the Homosexual', *British Journal of Venereal Diseases*, 1956, vol. 32, 17–20.

Kanee, B., and Hunt, C., 'Homosexuality as a Source of Venereal Disease', *Canadian Medical Association Journal*, 1951, vol. 65, 138–40.

Marten, J., *A Treatise of the Venereal Disease*, 7th edn, London, 1711.

Meyerson, B., Martich, F., and Naehr, G., *Ready to Go: The History and Contributions of U.S. Public Health Advisors*, Research Triangle Park, NC: American Social Health Association, 2008.

Ostrow, D., Sandholzer, T. and Felman, Y. (eds), *Sexually Transmitted Diseases in Homosexual Men: Diagnosis, Treatment, and Research*, London: Plenum Medical Book Company, 1983.

Schamberg, I., 'Syphilis and Sisyphus', *British Journal of Venereal Diseases*, 1963, vol. 39, 87–97.

Slater, D., 'Editorial', *ONE*, 1962, vol. 10, no. 10, 4–5.

Spring, J., *Secret Historian: The Life and Times of Samuel Steward, Professor, Tattoo Artist, and Sexual Renegade*, New York: Farrar, Straus and Giroux, 2010.

# Part IV

# NARRATIVES

# 25

# LEPROSY AND IDENTITY IN THE MIDDLE AGES

*Elma Brenner*

Although leprosy affected only a very small proportion of the population of medieval Europe, the disease had a significant cultural impact, manifested in literature, sermons, works of art, architecture and more. This cultural production reflects the strong responses that leprosy elicited, from fear and revulsion to compassion, charity and piety. Focusing on France, this chapter explores the identity of leprosy and lepers in Western Europe between 1100 and 1500, paying particular attention to the issues of narrative and language. While personal narratives of leprosy are almost completely absent for this period, especially in the earlier centuries, the perspectives of sufferers and onlookers can sometimes be deduced from the language of charters, testaments, episcopal visitation records and other documents.[1] These 'indirect' narratives testify to the diversity of responses to the disease, and also reveal that lepers and leprosy were not always referred to explicitly in this period. The chapter begins by considering how contemporaries identified the disease (or its absence) through examination and diagnostic procedures, the records of which also shed light on how lepers themselves were perceived. It goes on to survey the language associated with lepers and leprosy, and how such language changed over time. The final two sections consider the collective identities of groups of lepers, within and outside leper hospitals, and the individual identities of the leprous. The analysis considers 'lepers' to be those who were identified thus in the Middle Ages, since the perception that they had the disease – whether or not their illness was what is today termed Hansen's disease – influenced responses to them. Above all, it is clear that lepers had multiple social and religious identities, which were often shaped by their situation prior to contracting the disease. This suggests that leprosy did not fully transform an individual's identity, and that not only the disease itself, but also many other factors, affected the experiences of leprosy sufferers and responses to them.

Recent developments in the study of medieval leprosy have encouraged the consideration of questions of identity, language and narratives. Until the 1990s, the work of scholars such as Saul N. Brody and R. I. Moore emphasised the marginality of lepers in the Middle Ages, arguing that the disease was perceived as a punishment for sin, and that its sufferers were stigmatised and excluded from mainstream society.[2] François-Olivier Touati's publications presented a different perspective, pointing out that lepers were viewed as a special religious group, singled out by God to suffer on earth and thereby attain salvation.[3] Subsequently, Carole Rawcliffe and Luke Demaitre have further reassessed the status of lepers in medieval society, underlining the role of leprosy sufferers as a focus of Christian charity and as intercessors for the souls of their

461

benefactors, and examining the complex responses of medical authors to the disease and its sufferers.[4] The work of archaeologists and palaeopathologists has also shed light on living conditions and burial arrangements at leper hospitals, as well as confirming the presence of skeletons with bone changes indicative of Hansen's disease in cemeteries at leper hospitals and other locations.[5] This chapter aims to deepen our understanding of the complex identities of leprosy sufferers in the Middle Ages, and to encourage further cross-disciplinary research on this topic.

## Identifying leprosy: examination and diagnosis

The need to identify the presence of leprosy in an individual was linked to a number of factors, including concern about contagion and the belief that lepers should be accommodated in leper hospitals, monastic or quasi-monastic institutions where they could receive appropriate bodily and spiritual care. Formal examinations of leprosy suspects took place in Western Europe from at least the thirteenth century, and continued until the early eighteenth century in some locations. In many instances, especially in northern Europe, these examinations resulted in a legal judgement that determined the fate of the individual concerned. A certificate issued to the person either arranged for his or her entry to a leper hospital, or confirmed that he or she was free from leprosy.[6] Although members of the secular and regular clergy, especially parish priests, played a key role in earlier examinations, from the mid-thirteenth century medical practitioners and civic authorities increasingly dominated the process of diagnosis. This shift corresponds with the emergence of the professions of medicine and surgery, and the establishment of municipal governments in the towns of northern and western France and elsewhere. Nonetheless, examinations retained a religious flavour, particularly since many university-educated physicians, at least in the thirteenth century, also held clerical appointments. In May 1222, the election of a new archbishop of Rouen, Theobald of Amiens (1222–9), was challenged when a rumour circulated that Theobald was leprous. Pope Honorius III (1216–27) sent three judges delegate, the bishop of Sées, the dean of Amiens and the archdeacon of Rheims, to inquire carefully into the issue, 'having called faithful *medici* skilled in this matter'.[7] This case may mark a very early instance of a leprosy examination, which would have been overseen by high-status ecclesiastics but would also have involved physicians, who were expected already to have experience in diagnosing leprosy. In 1250 at Siena, Italy, a certain Pierzivallus was judged to be leprous by four physicians. A leprosy examination at Castiglione del Lago, Italy, in 1261 was conducted by two 'good men' who were advised by physicians. The latter examination was instructed jointly by the bishop's council and the civic government.[8]

By the fifteenth century, the leprous residents and administrators of leper hospitals also acted as judges at examinations, sometimes competing with the services offered by medical practitioners. The involvement of leprosy sufferers and those who oversaw their care suggests that it was recognised that skill in identifying the signs of the disease could derive from daily experience, in observing one's own symptoms and those of others. In fifteenth-century Cologne, Germany, many examinations took place at the city's chief leper house, Melaten, despite the fact that the medical faculty of the University of Cologne was also involved in leprosy diagnoses.[9] Since physicians and surgeons often charged high fees, an examination in a leper hospital may well have been a more affordable option, whether it was financed by the examinee, his or her parish, or the civic authorities.

Examinations took place not only in leper hospitals, but also in public squares, castle courtyards, and the houses of physicians and surgeons. Earlier on, churches also provided a setting; higher-status individuals were also sometimes examined in their own homes.[10] The degree to which the examination was public therefore varied, but in certain contexts suspected lepers were fully on display as their bodies were minutely examined. A woodcut depicting a leprosy examination printed in the 1551 edition of

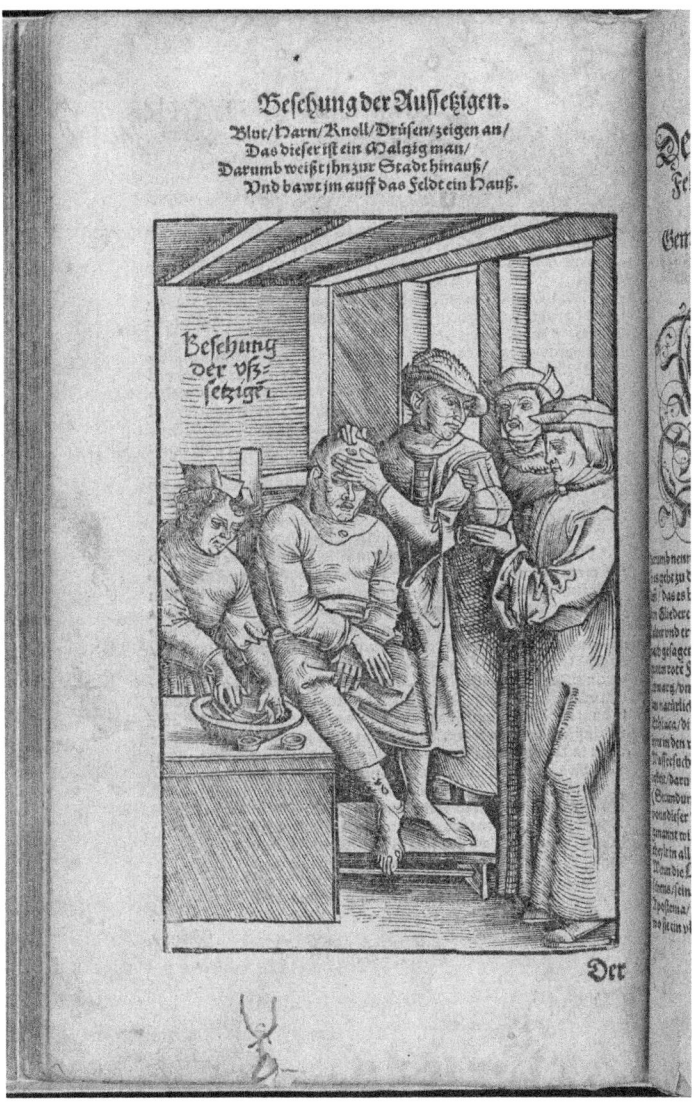

*Figure 25.1* Woodcut showing a leprosy examination in Hans von Gersdorff, *Feldtbuch der Wundtartzney*, Frankfurt-am-Main: Hermann Gülfferich, 1551, fo. LXXXv. Courtesy of the Wellcome Library, London.

a famous German surgical work (Figure 25.1) evokes a crowded setting in which the examinee was exposed, passive and somewhat vulnerable. The man being examined sits with his hands crossed over his lap as three medical practitioners to his left discuss his case. A physician is examining the contents of a urine flask, while a surgeon spreads his hand over the examinee's forehead, presumably feeling for the texture and temperature of the skin. The man on the examinee's right may be sieving a sample of blood; none of the figures in the scene makes eye contact with the examinee. Although no audience of the examination is shown, the view presented could reflect the perspective of one or more onlookers. Examinations in public squares, which were also the setting for executions of criminals and heretics, were especially open for all to see. A public examination could have a significant impact on the identity of the examinee, whether or not they were found to be leprous.

The examination involved the verbal questioning of the individual being examined and the scrutiny of the exterior of the body. Especial attention was paid to the face, but some examinations were conducted from head to toe. This physical examination involved touch, by surgeons and the non-professional examiners (leper hospital residents and staff, and civic officials), perhaps indicating that there was limited concern about contagion through direct physical contact. The blood and urine of the suspected leper were also examined, as evidence for the state of the interior of the body. Thirteenth- and fourteenth-century medical writers, such as Arnau de Vilanova (c. 1240–1311) and Guy de Chauliac (c. 1300–68), argued that the presence in the blood of grainy particles like sand confirmed a case of leprosy.[11] When undertaking leprosy examinations, physicians and surgeons searched for unequivocal signs of the disease, and referred to medical theory in identifying and interpreting these signs.[12] At an examination at Nîmes in 1327, for example, the physicians present, two masters in medicine from the University of Montpellier and one bachelor, formally stated that their diagnosis was based on the authority of medical authors. They examined six suspected lepers, confirming two cases, finding one to be free from leprosy, and the final three to be equivocal. They were assisted by three barber-surgeons, who palpated the suspects from head to toe and bled them, sieving the extracted blood and diluting it in water.[13] The fact that half of these cases could not be confirmed beyond doubt reveals that, for physicians and surgeons, diagnosing leprosy was a complex process which placed much at stake, in terms of both their own professional reputation and the fate of the examinee. Civic and parochial authorities may have preferred to commission non-professional examiners to undertake examinations not only because the latter's services were less expensive, but also because physicians and surgeons did not always provide a definitive verdict.[14]

In many instances suspected lepers were reported to the authorities by their neighbours or fellow parishioners, who had noticed symptoms or behaviour that caused concern. At Lausanne between 1398 and 1404, for example, Jean Giliet underwent an examination by a university-educated physician and a barber following the circulation of 'public rumour' that he was leprous; the diagnosis was confirmed.[15] In the late fifteenth or early sixteenth century, Jehan Jacquemin, a burgess residing in the suburbs of the town of Saint-Dié in the duchy of Lorraine, was 'maliciously' ('malveillans') accused of being infected with leprosy, and appealed to Duke René II of Lorraine (1451–1508) for assistance. Between 1487 and 1507 the duke issued a letter confirming that the ducal physicians had found that 'he is not infected with the said leprosy . . . he has been wrongly condemned'.[16] Provision was made for Jehan to be compensated financially, and it was established that he could freely move about as he had done prior to the accusation and come into contact with other people.[17] The fact that Jehan Jacquemin had actively solicited this certificate from the duke reveals that it was deeply important to an individual's social standing and situation to rectify matters when they were (as they claimed and others corroborated) misdiagnosed with leprosy. In certain instances the initial judgement could be appealed; here, ducal physicians intervened. The document evokes an atmosphere of accusation, rumour and the stigmatisation of lepers and leprosy. It refers to 'le menasseur' who had threatened Jehan, and concludes with the duke's expression of concern that his subjects should not be troubled and molested.[18] This suggests that an accusation of leprosy could be used as a means to attack an individual; indeed, in some instances local communities appear to have targeted undesirable persons by reporting them as suspected lepers. Penalties for false accusations, such as those outlined in the customary law code of 1562 of the Pays de Vaud region of Switzerland, indicate that this phenomenon was recognised.[19] Nonetheless, these accusations reflect the fact that, by the fifteenth and sixteenth centuries, concern about lepers and leprosy formed part of broader anxieties about contagious disease, the vagrant poor and public health.

## Language and narratives

The language associated with lepers and leprosy is significant, shedding light on how the disease was perceived by sufferers themselves and by those around them. In the twelfth and thirteenth centuries, the Latin label 'leprosus' ('leper') does not appear to have been problematic. A charter of 1183 recorded in the cartulary of the leper house of Le Grand-Beaulieu at Chartres, for example, refers to Renaud, son of Aubert *de Danumvilla*, as being 'voluntate Dei leprosus factus' ('made a leper by the will of God').[20] Another cartulary, that of the nunnery of Holy Trinity at Caen, records a gift by 'Willelmus de Caluz leprosus' ('William of Calix, a leper') in the late twelfth century, who granted a house in return for receiving the prayers of the nuns and for being received into a leper house which the nunnery probably oversaw.[21] This cartulary entry is written in the third person, reflecting the fact that cartularies were created by religious communities to copy the disparate charters they held into one volume. However, one would expect the text of the original charter to have been expressed in the first person, which suggests that William of Calix did refer to himself as a 'leprosus'. The size of William's donation indicates that he was a wealthy, high-status man; it is likely that he was synonymous with the William of Calix who was an affluent money-lender in Caen in the last decades of the twelfth century.[22]

More than 100 years later, a similar donation was made by another affluent man: in 1312 Peter of Saint-Gille granted 10 *livres* of Tours, two houses and a piece of land on the occasion of his entrance to the leper house of Mont-aux-Malades at Rouen. He did this 'by necessity of the disease with which I am occupied' and 'to be received in the community of the sick of the said place, to be with them all the course of my life, and to have the goods of the house like one of the sick brothers'.[23] Unlike William of Calix, Peter of Saint-Gille did not refer explicitly to his status as a 'leper', or to the fact that he had leprosy. Instead, he mentioned his 'ma[la]die' (disease or sickness) and described the community that he was joining permanently as the 'communauté des malades' ('community of the sick'). It is possible that he avoided making direct reference to leprosy because, by the fourteenth century, the language of this disease carried a stigma. Responses to leprosy did indeed become less positive from the fourteenth century onwards, even though they remained diverse and complex. At this time the disease became more closely associated with contagion, especially following the Black Death of 1348–50. But lepers and leprosy were also described using the more general language of 'sickness' much earlier on. Charters and papal bulls in favour of leper houses, from the twelfth century, are frequently addressed to 'the community of the healthy and the sick', and refer to the leprous residents as 'the sick'. A bull of Pope Alexander III (1159–81) issued on 10 May 1165 for the leper house at Hesdin in the Pas-de-Calais, for example, addresses the pope's 'beloved sons, the sick brothers of Hesdin'.[24]

Lepers may have been considered to be representative of the broader social category of 'the sick' in this period, which helps to explain this linguistic turn. Indeed, the theologian and preacher Jacques de Vitry (d. 1240) placed lepers and other sick people in the highest category of the laity in his hierarchical model of the structure of society. For de Vitry, the leprous and other sick persons had a similar status to other members of the laity who were suffering or undergoing a spiritual test, including pilgrims, those who had committed themselves to embarking on a crusade, and those in mourning. Among these lay categories, however, the leprous and the sick (as well as the poor) were closest to the religious above them in the hierarchy, especially those who served the sick in hospitals.[25] While lepers and the sick retained a lay status, therefore, their situation aligned them closely with people who held a religious status. Hospitals and leper houses, as settings for sickness and suffering, were spaces in which the line between lay and religious status became blurred. The language of Peter de Saint-Gille's donation of 1312, therefore, may reflect the association of leprosy with sickness more widely from at least the twelfth century. While the description of leprosy sufferers as 'sick' points towards the manner in which leprosy was emblematic of sickness in general, it also suggests that infirmity was a key marker of the identity of lepers themselves.

A number of documents associated with leper houses convey a sense of the passivity of leprosy sufferers. The donation charter of Peter of Saint-Gille, for example, states that he is 'occupied' (or 'inhabited') by the disease. The 1183 entry in the cartulary of the leper house at Chartres suggests that God's will had afflicted leprosy on Renaud, son of Aubert *de Danumvilla*. Another entry in the same cartulary, of 1188, expresses a similar perspective: in this instance, Nivelon, son of Geoffrey of Grand-Pont, was 'marked with leprosy by the will of God'.[26] The idea conveyed here of being marked by the disease underlines the fact that leprosy's disfiguring effects were highly visible, making God's intervention plain for all to see. In the late twelfth and thirteenth centuries, however, the notion of divine will did not necessarily imply punishment for sin: on the contrary, it could signify that an individual had been singled out by God to pursue a religious life within a leper house.[27] By suffering on earth and living devoutly, that person would attain salvation. Nonetheless, there is an implicit narrative in such documents that suggests the powerlessness and vulnerability of the afflicted person. The disease, or divine intervention, acted upon the person's body in a forceful way, and he or she was unable to counteract or overcome this action. Such a narrative recognises the gravity of leprosy, a disease which could not be cured and transformed a person's social identity. The embedded narrative also implies a compassionate response to sufferers.

A compassionate perspective is conveyed by a much later document, a report of 19 February 1524, by three civic officers of Rouen, Jean Le Roux, Guillaume Auber and Jean du Hamel, to royal officers on their inspection of the city's chief leper house, Mont-aux-Malades.[28] Although the incidence of leprosy had very much declined by the sixteenth century, leper houses continued to cater for the sick: there were three sick residents at Mont-aux-Malades at this time. This report brings us very close to the narrative testimonies of individual lepers, since it records the responses given by two of the three sick people, both men, when interviewed. These responses mainly provide factual information, in terms of the length of residence of the sick, and the provisions of food, money and clothing received by them. Nonetheless, the document reveals that the two men had been living with leprosy for a number of years (six or seven years), and that they had retained personal support, from a wife and a fiancée respectively.[29] It would appear that the third sick person, a woman, could not be interviewed, because she was bedridden and extremely ill. The report states: 'It should be noted that the sick woman's room does not have a fire'.[30] Given that the inspection took place in the cold winter month of February, this comment conveys a compassionate concern for the woman's plight on the part of the civic officers inspecting the leper house. Official or legal documents, therefore, still sometimes enable us to glimpse the narratives of both the onlookers who observed lepers and leprosy, and the leprosy sufferers themselves.

## Lepers: collective and individual identities

Although charters and cartulary entries sometimes shed light on the situations of individual lepers, the vast majority of medieval documentary references relate to groups of lepers, especially those residing in leper hospitals. The fact that lepers were viewed collectively suggests that the disease they shared bound them together, both internally (from their own individual perspectives) and externally (from the perspectives of others). According to this model, leprosy itself was the strongest component of the identities of sufferers. In certain instances, lepers explicitly surrendered at least a part of their individual identities upon entering a leper house, by donating their own person, as well as property, as their entrance gift. In 1188, for example, on entering the leper house at Chartres, Nivelon, son of Geoffrey of Grand-Pont, 'gave himself to the house', along with various revenues.[31] Such donations reflected a practice followed by certain healthy lay entrants to monastic houses, hospitals and leper houses, which marked a reciprocal arrangement whereby the entrant could expect to be supported by the house until he or she died.[32]

The prominence of groups of lepers in the records reflects practical circumstances as well as questions of identity. While the origins of many leper houses founded in the late eleventh and twelfth centuries remain obscure, a number of institutions appear to have emerged in order to organise a group of lepers who were already living collectively in a relatively informal manner. This communal living was necessitated by the need for a confirmed leper to leave his or her community. Canon 23 of the Third Lateran Council of 1179, which provided for every group of lepers to have its own church, cemetery and priest, refers to 'lepers, who cannot dwell with the healthy or come to church with others'.[33] Prior to being organised as a religious community residing in a leper house, a group of lepers could already have received charity, revealing that it was recognised as a social entity. For example, the grant of Henry I, king of England (1100–35), of 40 *sous* a month to 'the lepers of Rouen', made in the first part of the twelfth century, predates the earliest references to the city's main leper hospital, Mont-aux-Malades.[34]

The collective identification of the lepers of a particular town or city persisted long after leper houses became widespread. As late as 3 July 1402, for instance, a document relating to the leper house at Fécamp refers to its chapel as 'the chapel of the lepers of Fécamp'.[35] This suggests that geographical identity was important, both to groups of lepers and to their benefactors, who frequently chose to assist the leper house that served their own city, town or parish. At Pontoise, it was specifically those born within the town who had the right to enter the *leprosarium* of Saint-Lazare for life.[36] The statutes of the leper house at Les Andelys, drawn up before 1380, are explicit on this point: 'the said house is for the sick born from the burgess community of Les Andelys, in the parish [Notre-Dame] of Les Andelys, and not for others, even if they are burgesses'.[37] Here, the community of lepers was shaped by social status as well as geography: only lepers of burgess status were admitted. This restriction no doubt reflects the patronage of the leper house, which was presumably founded and supported by the local burgesses.

Within a leper hospital, the collective identity of lepers operated on a number of different levels. At the broadest level, lepers formed part of a larger religious community that also incorporated non-leprous constituencies, ranging from monks or nuns to priests, lay brothers and lay sisters. The language of charters confirms that all of the residents of leper houses were considered to constitute a single community. In June 1247, for example, Renaud, Bertin and Theobald du Châtel, brothers, made a grant 'to the prior and community' of the leper house of Mont-aux-Malades at Rouen.[38] The statutes of leper houses, like the charters for these institutions, convey the sense that all categories of residents were grouped together collectively. For example, the ordinance for the leper hospital at Amiens, confirmed on 21 July 1305, opens by stating that the rules it contains must be observed by 'the healthy and sick brothers and the healthy and sick sisters'.[39] The notion of a single community also spread beyond the walls of a leper hospital, since the hospital's benefactors, for whose souls the resident community prayed, were understood to form part of a fraternity of prayers that incorporated all those associated with the hospital, during life and after death. Extant memorial books recording the names of those to be commemorated by a hospital community, such as that begun c. 1296 by the community of the leper house at Gaywood, Norfolk, materially evoke the past existence of these complex networks of relationships and obligations.[40]

Yet, although the concept of one overarching community was key to the institutional model of the leper hospital, internally there were a number of distinctions that separated the leprous from other groups and, in various respects, from each other. The healthy and the sick occupied different accommodation, and often worshipped and ate meals separately. Furthermore, although most leper houses were mixed, there was firm segregation of the sexes, in order to ensure the chastity that one would expect in a religious community. In the early fourteenth century at the leper house at Sherburn, county Durham, according to the house's statutes, separate masses were performed on behalf of the leprous brothers and the leprous sisters, even though they came together for a great mass on Sundays and religious festivals. The leprous men and women lived separately and, following collective worship on Sundays and feasts, care was taken for the door to the sisters' accommodation to be closed.[41] The statutes of the leper house at Amiens, confirmed in 1305, forbade the leprous brothers from entering areas that were regularly used by the healthy brothers, including the kitchen, the cellar and 'the grange where wheat and oats are threshed'.[42] This provision, relating to areas where food and drink were stored and prepared, could imply concern about contagion. Nonetheless, it more likely reflects the view that members of each constituent group in the community should occupy their own distinctive space, according to identity and status.

Finely tuned distinctions were also made among the non-leprous residents of leper houses. Monks and nuns differed markedly in status from lay brothers and sisters. Broadly speaking, the former category, often following the Augustinian rule, devoted themselves to religious worship and the administration of the house and its landed possessions, while the latter category fulfilled practical tasks, including tending to the sick. Certain descriptions of leper house communities recorded in the visitation records of Eudes Rigaud, archbishop of Rouen (1248–75), compiled between 1248 and 1269, show a clear awareness of the different groups, both non-leprous and leprous. The account of the visit to the leper hospital at Bellencombre on 3 December 1255, for example, records the presence of four [Augustinian] canons, two lay sisters, four lay brothers, five male lepers and one female leper, as well as a gate porter.[43] Even in such a small community, it was considered important to make precise distinctions according to gender, state of health and religious status.

Not all people with leprosy entered a leper hospital: entry was frequently restricted according to social class, place of origin (for both restrictions, see the statutes of the leper house of Les Andelys), and the ability to offer an entrance gift. Wandering lepers who lived outside of any kind of institutional framework were a recognisable social category, and largely depended on begging. Although these lepers lacked the collective identity derived from membership of a leper house community, it is clear that they did possess a shared identity, and moved around both individually and in groups. In order to obtain a licence to beg, such lepers would willingly undergo an examination; in Nuremberg, an annual mass examination took place during Holy Week, known as the *Schau*. This event caused large numbers of itinerant lepers to gather together.[44] By the later Middle Ages, the boundary between itinerancy and residence in a leper house had become somewhat blurred, since a number of the major leper hospitals offered temporary shelter to wandering lepers. At Rouen's major leper house from at least the end of the fourteenth century, and at the leper houses at Saint-Omer, Bapaume and other towns, lodgings were assigned for lepers who were unable to join the community permanently.[45] Their occasional – or frequent – presence added a further layer of diversity to the leper house community.

While itinerant lepers represented the lowest social category of the sick within a leper hospital, it is clear that other lepers could hold a particularly elevated status. At Mont-aux-Malades, Rouen, leprous monks from Saint-Ouen, the city's largest abbey, had a special status, enjoying privileged accommodation as well as food provided by the abbey.[46] This preferential treatment reflected their status as Benedictine monks, which placed them in a higher religious category than their fellow lepers.

The arrangements also ensured that they retained a connection with the abbey community, of which they were technically still members. At a number of leper houses, certain lepers clearly came from a relatively comfortable social background. The agreement for the entry of Jehan Duquesnoy called 'le Bourguignon' to the leper hospital of Saint-Lazare at Aumône near Pontoise, issued on Tuesday 17 May 1412, provided for him to have a separate lodging, and to bring his possessions with him, which included a bed and linen sheets. Beds were important, often valuable objects in the later Middle Ages.[47] The document marking this agreement appears to acknowledge that Jehan was a man of a certain social status, since it refers to the utensils belonging to him 'pour son estat' ('for his estate').[48] It is clear that a leprous person's fate was to a great extent shaped by his or her social or religious identity prior to contracting the disease, and that he or she did not fully relinquish that identity upon being newly defined as a 'leper'.

## Conclusion

Discerning the identities of lepers in the central and later Middle Ages, as well as how leprosy itself was identified, is a difficult process. An examination of the language and implicit narratives of a range of documentary sources signifies change over time, but also certain striking continuities, such as the identification of lepers with the sick in general. Leprosy examinations, which probed the bodily condition of suspected lepers, reveal both the criteria for diagnosing leprosy, and the treatment of suspected and confirmed patients. While undergoing an examination, as well as entering a leper hospital, had a dramatic impact on a person's identity, the identities of lepers need to be understood within a very broad spectrum of different possibilities. Many sufferers had an itinerant existence, depending on the proceeds of begging and living beyond the confines of urban and rural society, as well as outside the institutional setting of the leper house. Membership of a leper house community offered the opportunity to assume a religious identity alongside healthy brethren, and also to assert a distinctive identity vis-à-vis one's fellow lepers. While discipline and obedience to a rule were expected within a leper house, some lepers had a more privileged position than others within these communities, testifying to the continuing importance of a person's previous identity after contracting the disease. Although the source material very much favours the exploration of lepers' collective identities, the investigation of individual identity sheds light on the complexity of responses to leprosy, by both the disease's sufferers and those who interacted with them.

The work of historians on leprosy in England, France and Germany, as well as that of archaeologists, demonstrates how much we can learn about the experiences of leprosy sufferers in medieval Europe, and their social and religious identities.[49] This chapter has demonstrated the many different attitudes and responses to leprosy that can be teased out from the documentary sources relating to France. While the French sources are very rich, especially in terms of leper house statutes and the charters issued and received by leper houses, the picture that these sources present is certainly not representative of Europe as a whole, a region within which there was some diversity in terms of the institutional provision for lepers and responses to this group. Our understanding of the identities of leprosy sufferers, and of perceptions of the disease itself, would undoubtedly be enriched further by comparative studies of different parts of Europe, as well as by more cross-disciplinary analyses that incorporate findings from documentary, literary, iconographic and archaeological sources. Broader studies along these lines would facilitate the exploration of hidden narratives, language, and individual responses to this disease. Such methodological developments in the study of medieval leprosy would in turn enhance the study of other diseases and experiences of sickness, disability and debility in the past.

## Notes

1 Personal narratives are similarly absent for other types of physical disability: I. Metzler, *A Social History of Disability in the Middle Ages: Cultural Considerations of Physical Impairment*, New York: Routledge, 2013, pp. 2–3.
2 S. N. Brody, *The Disease of the Soul: Leprosy in Medieval Literature*, Ithaca, NY: Cornell University Press, 1974; R. I. Moore, *The Formation of a Persecuting Society: Power and Deviance in Western Europe, 950–1250*, Oxford: B. Blackwell, 1987.
3 F.-O. Touati, 'Les Léproseries aux XIIème et XIIIème siècles, lieux de conversion?', in N. Bériou and F.-O. Touati, *Voluntate dei leprosus: les Lépreux entre conversion et exclusion aux XIIème et XIIIème siècles*, Testi, Studi, Strumenti, vol. 4, Spoleto: Centro Italiano di Studi sull'Alto Medioevo, 1991, pp. 1–32; F.-O. Touati, *Maladie et société au Moyen Âge: la lèpre, les lépreux et les léproseries dans la province ecclésiastique de Sens jusqu'au milieu du XIVe siècle*, Bibliothèque du Moyen Âge, vol. 11, Brussels: De Boeck Université, 1998.
4 C. Rawcliffe, *Leprosy in Medieval England*, Woodbridge: Boydell Press, 2006; C. Rawcliffe, 'Learning to love the leper: aspects of institutional charity in Anglo-Norman England', *Anglo-Norman Studies* 23 (2000), 231–50; L. Demaitre, *Leprosy in Premodern Medicine: A Malady of the Whole Body*, Baltimore: Johns Hopkins University Press, 2007.
5 C. A. Roberts, M. E. Lewis and K. Manchester (eds), *The Past and Present of Leprosy: Archaeological, Historical, Palaeopathological and Clinical Approaches*, Oxford: Archaeopress, 2002; J. Magilton, F. Lee and A. Boylston (eds), *'Lepers Outside the Gate': Excavations at the Cemetery of the Hospital of St James and St Mary Magdalene, Chichester, 1986–87 and 1993*, York: Council for British Archaeology, 2008.
6 Demaitre, *Leprosy*, chapter 2 and pp. 196–8.
7 'advocatis medicis fidelibus, et in hoc peritis'. *Honorii III Romani pontificis opera omnia*, ed. C. Horoy, 5 vols, Paris: Imprimerie de la Bibliothèque ecclésiastieue, 1879–82, iv, epistolae lib. vi, col. 151 (no. 177). In another letter, dated 18 May 1222, Pope Honorius instructed these three churchmen to examine Theobald's election and to establish that he was a suitable person to be made archbishop: idem, cols 150–1 (no. 176). See Vincent Tabbagh, *Fasti ecclesiae gallicanae: Répertoire prosopographique des évêques, dignitaires et chanoines de France de 1200 à 1500. Tome II: diocèse de Rouen*, Turnhout: Brepols, 1998, 80–1; Jörg Peltzer, *Canon Law, Careers and Conquest: Episcopal Elections in Normandy and Greater Anjou, c. 1140–c. 1230*, Cambridge: Cambridge University Press. 2008, 84–6. Theobald was found to be free from leprosy, and he was formally consecrated on 4 September 1222.
8 Demaitre, *Leprosy*, p. 37.
9 Ibid., pp. 41–3.
10 Ibid., pp. 36–7.
11 G. Dumas, *Santé et société à Montpellier à la fin du Moyen Âge*, The Medieval Mediterranean vol. 102, Leiden: Brill, 2015, pp. 291–3, 295–6.
12 Ibid., pp. 197–201.
13 Ibid., p. 295.
14 Demaitre, *Leprosy*, p. 42.
15 P. Borradori, *Mourir au Monde: les lépreux dans le Pays de Vaud (XIIIe–XVIIe siècle)*, Lausanne: Université de Lausanne, 1992, pp. 30, 31.
16 'qui par eulx a este trouve . . . non infecte de ladite lepre . . . quil nous |a fait exposciste a ester a tort condempne . . .'. London, Wellcome Library, MS 5133/3, fo. 1r.
17 Ibid.
18 Ibid., fos 1r., 1v.

19 Borradori, *Mourir au monde*, pp. 30–1.
20 *Cartulaire de la léproserie du Grand-Beaulieu et du prieuré de Notre-Dame de la Bourdinière*, ed. R. Merlet and M. Jusselin, Chartres: E. Garnier, 1909, p. 49 (no. 120); Touati, 'Les Léproseries', p. 14.
21 *Charters and Custumals of the Abbey of Holy Trinity Caen. Part 2: The French Estates*, ed. J. Walmsley, Oxford: Oxford University Press, 1994, p. 129 (cartulary document no. 17).
22 Ibid., p. 129 n. 5.
23 'pour la necessite de la ma[la]die dont jestois ocupe . . . pour moy recevoir en la communaute des malades du dit lieu pour estre avveques eulz tout le cours de ma vie et pour avoir des biens de lostel aussi comme un des freres malades'. Rouen, Archives départementales de Seine-Maritime, 25HP1; Bruno Tabuteau, 'Les Léproseries dans la Seine-Maritime du XIIe au XVe siècle', unpublished Mémoire de Maîtrise, University of Rouen-Haute-Normandie, 1982, p. 118.
24 'dilectis filiis infirmis fratribus de Hisdinio'. A. Bourgeois, *Lépreux et maladreries du Pas-de-Calais (Xe–XVIIIe siècles)*, Arras: [Commission départementale des monuments historiques du Pas-de-Calais], 1972, p. 267.
25 N. Bériou, 'Les Lépreux sous le regard des prédicateurs d'après les collections de sermons *ad status* du XIIIème siècle', in Bériou and Touati, *Voluntate dei leprosus*, pp. 39–41, 45.
26 'Dei voluntate lepra sigillatus'. *Cartulaire de la léproserie du Grand-Beaulieu*, ed. Merlet and Jusselin, p. 54 (no. 132); Touati, 'Les Léproseries aux XIIème et XIIIème siècles', pp. 14–15.
27 Touati, 'Les Léproseries aux XIIème et XIIIème siècles', pp. 11–19.
28 Paris, Archives Nationales (AN), S4889B, dossier 13, last document.
29 Ibid., fo. 1r.–1v.
30 'Et fait a noter que en la chambre de ladite mallade ny avoit point de feu', ibid., fo. lv.
31 'sese dedit domui'. *Cartulaire de la léproserie du Grand-Beaulieu*, ed. Merlet and Jusselin, p. 54 (no. 132).
32 On such lay entrants to religious houses, see: C. de Miramon, *Les 'Donnés' au Moyen Âge: une forme de vie religieuse laïque v. 1180–v. 1500*, Paris: Cerf, 1999; E. Brenner, *Leprosy and Charity in Medieval Rouen*, Woodbridge: Boydell Press, 2015, pp. 52–3.
33 'leprosis qui cum sanis habitare non possunt et ad ecclesiam cum aliis convenire'. *Decrees of the Ecumenical Councils*, ed. and trans. N. P. Tanner, 2 vols, London: Sheed & Ward, 1990, vol. 1, p. 222.
34 Paris, AN, *K23 no. 15/22 (charter of Geoffrey, count of Anjou, undated (1145–50), confirming the grant of King Henry I to the lepers of Rouen). See Brenner, *Leprosy and Charity*, Chapter 1.
35 Rouen, ADSM, G5235.
36 London, Wellcome Library, MS 5133/1 (agreement issued by Simon Paine, mayor of Pontoise, for the admission of Jehan Duquesnoy called 'le Bourgignon' to the *leprosarium* of Saint-Lazare at Aumône near Pontoise, 17 May 1412).
37 'ledit hostel est pour les malades nés de la bourgoisie d'Andeli, en la paroisse d'Andeli, et non pour autres, tout soient ilz de la bourgoisie.' 'Statuts de la léproserie des Andelys', in *Statuts d'Hôtels-Dieu et de léproseries: recueil de textes du XIIe au XIVe siècle*, ed. L. Le Grand, Paris: Alphonse Picard, 1901, p. 247.
38 'priori et conuentui'. Rouen, ADSM, 25HP1/2.
39 'li frère sain et malade et les sereurs saines et malades'. 'Statuts de la léproserie d'Amiens', in *Statuts*, ed. Le Grand, p. 224.
40 C. Rawcliffe, 'Communities of the living and of the dead: hospital confraternities in the later Middle Ages', in C. Bonfield, J. Reinarz and T. Huguet-Termes (eds), *Hospitals and Communities, 1100–1960*, Oxford: Peter Lang, 2013, pp. 125–9.
41 'Constitutiones Hospitalis domus leprosorum de Shirburne', translated in P. Richards, *The Medieval Leper and His Northern Heirs*, Cambridge: D. S. Brewer, 1977, repr. Woodbridge: Boydell Press, 2000, p. 125.
42 'le grange là où on bat le blé et l'avaine'. 'Statuts de la léproserie d'Amiens', p. 226.
43 *Regestrum visitationum archiepiscopi Rothomagensis: Journal des visites pastorales d'Eude Rigaud, archevêque de Rouen. MCCXLVIII–MCCLXIX*, ed. T. Bonnin, Rouen: A. Le Brument, 1852, p. 230; *The Register of Eudes of Rouen*, ed. J. F. O'Sullivan and trans. S. M. Brown, Records

of Civilization, Sources and Studies, vol. 72, New York: Columbia University Press, 1964, p. 253.

44 Demaitre, *Leprosy*, pp. 45–51.
45 Paris, AN, S4889B, dossier 13 (act of King Charles VI of France, 18 June 1393, mentioning the duty of the community of Mont-aux-Malades, Rouen, to receive the passing sick); Bourgeois, *Lépreux*, p. 66.
46 See Rouen, ADSM, 14H660(i).
47 On beds in aristocratic households in late medieval England, see C. M. Woolgar, *The Great Household in Late Medieval England*, New Haven: Yale University Press, 1999, pp. 63–5, 78–80.
48 Wellcome Library, London, MS 5133/1.
49 Rawcliffe, *Leprosy*; Touati, *Maladie*; Demaitre, *Leprosy*; Roberts, Lewis and Manchester (eds), *Past and Present of Leprosy*; Magilton, Lee and Boylston (eds), *'Lepers Outside the Gate'*.

## Select bibliography

Bériou, N., and Touati, F.-O., *Voluntate dei leprosus: les Lépreux entre conversion et exclusion aux XIIème et XIIIème siècles*, Testi, Studi, Strumenti, vol. 4, Spoleto: Centro Italiano di Studi sull'Alto Medioevo, 1991.

Brenner, E., 'Recent perspectives on leprosy in medieval Western Europe', *History Compass* 8, 2010, 388–406.

Demaitre, L., *Leprosy in Premodern Medicine: A Malady of the Whole Body*, Baltimore: Johns Hopkins University Press, 2007.

*Statuts d'Hôtels-Dieu et de léproseries: recueil de textes du XIIe au XIVe siècle*, ed. L. Le Grand, Paris: Alphonse Picard, 1901.

Metzler, I., *A Social History of Disability in the Middle Ages: Cultural Considerations of Physical Impairment*, New York: Routledge, 2013.

Peyroux, C., 'The leper's kiss', in Farmer, S., and Rosenwein, B. H. (eds), *Monks and Nuns, Saints and Outcasts: Religion in Medieval Society: Essays in Honor of Lester K. Little*, Ithaca, NY: Cornell University Press, 2000, pp. 172–88.

Rawcliffe, C., *Leprosy in Medieval England*, Woodbridge: Boydell Press, 2006.

Roberts, C. A., Lewis, M. E., and Manchester, K. (eds), *The Past and Present of Leprosy: Archaeological, Historical, Palaeopathological and Clinical Approaches*, Oxford: Archaeopress, 2002.

Tabuteau, B. (ed.), *Lépreux et sociabilité du Moyen Âge au Temps moderne*, [Rouen]: Université de Rouen, 2000.

Tabuteau, B., 'Historical research developments on leprosy in France and Western Europe', in Bowers, B. S. (ed.), *The Medieval Hospital and Medical Practice*, Aldershot: Ashgate, 2007, pp. 41–56.

Touati, F.-O., *Maladie et société au Moyen Âge: la lèpre, les lépreux et les léproseries dans la province ecclésiastique de Sens jusqu'au milieu du XIVe siècle*, Bibliothèque du Moyen Âge, vol. 11, Brussels: De Boeck Université, 1998.

# 26

# FRENCH MEDICAL CONSULTATIONS BY MAIL, 1600–1800

*Robert Weston*

> Medicine is a conjectural science, it is false because as an astronomer
> predicts when an eclipse will arrive, such and such year, such and such
> day, such and such minute, likewise a physician on sounding the pulse of
> a patient may predict such and such disease and decide that this day will
> be judged by such and such crisis, in this way it is certain and posited on
> some true principles compared to diagnoses.[1]

This contention by a professor at the Paris School of Medicine in the eighteenth century would seem at odds with today's perspective of medicine as a science-based discipline. In the early-modern period, however, medicine was regarded as an art, one that medically trained practitioners claimed was founded on a sound theoretical basis which justified their practices. Medicine as we understand it has three components, the disease, the patient and the healer. But what do we understand by disease? In 2004, bioethicist Jackie Scully suggested:

> At first sight, the answer to 'What is a disease?' is *straightforward*. Most of us
> feel we have an intuitive grasp of the idea, reaching mentally to images or
> memories of colds, cancer or tuberculosis. But a look through any medical
> dictionary soon shows us that articulating a satisfactory definition of disease
> is surprisingly difficult.[2]

This is just one of many comments on the difficulty of defining what is meant by disease. As Lester King observed over 50 years ago: 'Health and disease are value judgements'.[3] If making such a definition is problematic in the twenty-first century, how much harder is it to comprehend what was understood by the word in the early-modern period?[4] Charles Rosenberg has pointed out that:

> [disease] is at once a biological event, a generation-specific repertoire of ver-
> bal constructs reflecting medicine's intellectual and institutional history, an
> occasion for and potential legitimation of public policy, an aspect of social
> role and individual – intrapsychic – identity, a sanction for cultural values,
> and a structure element in doctor–patient interactions.[5]

This chapter explores the notion of disease and its representation in seventeenth- and eighteenth-century France, as it was expressed in written correspondence

between medical practitioners and their patients. Throughout this chapter I have
for the most part employed the nomenclature in the sources to avoid the complications inherent in engaging in retrodiagnosis.[6] Epistolary consultations were provided for the most part by university-trained physicians, who represented only a part
of the available resource that patients could turn to. Furthermore, those who could
afford recourse to these consultants were only a fraction of the population, an elite
section of the community who also sought medical advice from healers other than
the physician. The practice of requesting medical advice by letter dates back to at
least the twelfth century; Nancy Siraisi has identified an Italian consultation of this
type dated 1150.[7] The earliest French example I have encountered is that of Michel
de Nostredame – Nostradamus (1503–66) – who provided a written consultation
to Cardinal Laurent Strozzi (1513–71) in Beziers on 20 October 1559.[8] By the seventeenth century, this method of providing advice had become common practice
throughout Europe.[9]

This genre of correspondence has been employed by historians to focus on various
aspects of medicine. Philip Rieder has examined the role of the patient; Laurence
Brockliss the practices of the individual physician; Willemijn Ruberg has used Dutch
material to study historiographical methodology; Robert Weston has focused on the
practice in France; and Joël Coste on the role of rhetoric in such correspondence.[10]
None of these works, however, addresses the topic of disease per se; indeed, the history of disease where it has been mentioned has been quickly subsumed into the
'history of medicine' with the exception of studies of particular diseases or epidemiology. As Jonathan Andrews has remarked, 'We have profited from wide-ranging
analyses of epidemics, pandemics and fevers'.[11] Yet the range of diseases covered in
epistolary consultations, on which this chapter is largely based, was extremely diverse.
As Vincent Barras and Martin Dinges have described it, 'cutaneous eruptions, disturbed digestions, humoral excretions, uterine suffocations, nervous exhaustions,
troubled emotions, the most diverse disturbances of "the machine"'.[12]

The archival research employed here has focused on letters to and from physicians,
and to a lesser extent, surgeons who also engaged in the practice of consulting by letter.[13] It is unsurprising that the majority of surgeons' consultations involved physical
injuries, cuts, bruises, firearm wounds and the like, but this was not all. Traditionally,
internal diseases were the domain of physicians, while surgeons attended to external
ailments, but this demarcation was blurred. Treatment methods led to surgeons being
involved in cases of venereal diseases and when surgery was being contemplated. The
surgeon would be called upon, for example, if a patient suffered from hydropsy and
puncture of the abdomen was being contemplated to remove excess fluid.[14] Equally,
physicians dealt with external complaints, such as diseases of the skin.[15]

Epistolary medical correspondence typically comprised a letter, a *mémoire*, from a
patient, or more commonly his or her local medical practitioner, either a physician
or a surgeon usually termed *un ordinaire*; this outlined the symptoms, often what treatment had already been tried, and finally the response of the consulting practitioner.
In a French context the consultant was invariably an eminent practitioner, usually
associated with either the Paris faculty of medicine or the Montpellier University of
Medicine, the pre-eminent medical schools in France of the period. Some lesser lights
also engaged in consulting by letter, typically physicians located in smaller towns
but whom one can presume had a local reputation for competence. For example,

Vivant-August Ganiare (1698–1781) in Beaune and Marie-François-Bernadin Ramel (1752–1811) in Aubagne engaged in epistolary consulting among their local communities.[16] Many physicians and surgeons from around France wrote about unusual observations which were published in journals as individual cases.[17] Ramel, however, was exceptional amongst country physicians in that he published some of his consultations as a text. In composing their consultations, physicians took the opportunity to expound their medical knowledge, but, at the same time were cognizant of the likely (often not superficial) level of medical knowledge of their clients. As a consequence they did not propose challenging or innovative theories or treatments. On the contrary, if a patient proposed an unconventional treatment, the consultant would warn against it, such as when a patient had a preference for an Arab remedy over the consultants' recommendation.[18]

## Health and disease in epistolary consultations

How were the terms health and disease understood in the early-modern period? In the late seventeenth century, the French physician Anicet Caufapé wrote:

> I say firstly, on what constitutes health and illness (*maladie*) in a few words. I say that how those in a disposition or affection which interrupts the ordinary course of the natural functions; the other is a disposition or natural constitution which conserves the same actions, instead of their destruction.[19]

Contemporary medical dictionaries offer little guidance on physicians' understandings of disease. A 1741 medical dictionary defined *maladie* as 'when some organ was deranged or corrupt, or when the blood was not distributed as it must be, to those parts which need to be nourished'.[20] More general dictionaries were not a great deal more helpful. Diderot and d'Alembert's *Encyclopédie* was far more detailed in its exposition of what constituted *maladie*, although the entry commenced with the simple notion of that which was contrary to good health.[21] Ramel gave an explanation of health and disease which was based on the interaction between solids and fluids in the body, but which were idiosyncratic in nature:

> All men are born with a different temperament. The temperaments vary as much as the traits of appearance and it would be very difficult to find two temperaments exactly the same. This difference comes from the state, always different in all individuals, of the solids and of the fluids, and of their respective and reciprocal action. A certain degree of tension, of rigidity, and of energy in the solids relative to the fluids, a certain density in the fluids relative to the solids, their reciprocal action; their exact competition and balancing following the wish of nature, which constitutes that state which we call health; but if the fluids are altered, if they are too dense, or too diffuse, if the solids are out of balance by too much rigidity or by too much weakness or slackening, this condition which is not in the order of nature, the failure of equilibrium by which the animal body can no longer do with ease, and for a certain time, the functions which are its own, constitutes disease.[22]

Ramel was following contemporary theory of how the body functioned, or malfunctioned. At the same time he was emphasising the idiosyncratic nature of humoral balance, which necessitated each case being addressed on an individual basis.

Patients considered themselves ill if the ailment interfered with what they considered to be their normal daily activities to such a degree as to warrant seeking expert advice, or, if they were fearful that what ailed them could lead to a more serious, even fatal, condition. The most common condition which appears to have led to *fear* of death was melancholia, under its various names.[23] Time and again, physicians would advise their patients that this disorder was difficult to cure but never fatal. Montpellier professor Antoine Fizes (1690–1768) was writing directly to a patient, also suffering from melancholia, when he told him not to amuse himself reading medical books on the disease as it might only worsen his condition.[24]

For a disease to exist, it had to have a name, and herein lay one of the problems that potentially confronted the early-modern reader and the modern historian: inconsistency over nomenclature.[25] When analysing the correspondence of Parisian physician Etienne-François Geoffroy, Laurence Brockliss contended that there 'is little point in detailing the diseases from which Geoffroy's patients thought they were suffering, for the early eighteenth century was still a pre-nosological age.'[26] Despite this difficulty, it is necessary to consider what terms physicians and patients used for illnesses. There is little evidence that naming a disease constituted a source of confusion at the point of providing practical advice; the need for a common nomenclature was more of an issue at the academic level. François Boissier de La Croix de Sauvages (1706–67) set about constructing a classification of diseases. He was faced with a plethora of names used by authors for what he identified as a single disease, pointing out that an agreed nomenclature had been called for by Galen (AD 129–c. 200):

> I think that clarity demands that each thing has its correct name which serves one consistently, since the common names which do not mean one thing more than another, throws confusion and trouble in the mind of the reader, who cannot hear what one means, until one has removed the ambiguity of the term [Galen, le Dyspnæá, book 1].[27]

Perhaps an extreme case, but indicative of the problem, Sauvages referred to 24 names used for syphilis; *la syphilis* was his preferred term, although in French consultations the disease was most commonly referred to as *la vérole*.[28] Diseases had of course been listed before Sauvages' enterprise.[29] What made his work different from earlier attempts was his creation of a classification based on signs and symptoms, rather than a simple listing. Perhaps one of the more difficult arenas in defining diseases was fever. Today fever is for the most part regarded as a symptom (ignoring terms such as yellow fever or scarlet fever), whereas in earlier periods fevers were regarded as diseases. As a generalisation, fevers were distinguished by their periodicity, thus tertian fever returned every third day and quartan fever every fourth day.

Who proclaimed what disease a sick person was suffering from and did it matter to the patient what name was put to it? Patients only described symptoms, but *ordinaires* sometimes did state what they believed the patient to be suffering from. In August 1738 the father of a sick child asked of Parisian physician Louis-Jean Le Thieullier (d. 1751) the name of the disease from which the boy was suffering.[30] This request came, at least in part, because a number of local practitioners had already put different names to the child's illness. The parent was evidently disturbed by this, but more importantly, was concerned that the appropriate treatment for the illness should be provided. Such a request was extremely unusual amongst the many *mémoires* seeking expert advice. For the physician, designating the cause, and therefore implicitly naming it, was significant as this was how he justified to the patient his proposed therapy. In addition, this step stamped the consultant's authority on the process. In cases when physicians disagreed over the nature of a disease, differing names was the result. Marie-François-Bernadin Ramel was asked to provide a consultation to a man who had received advice from three other physicians. One termed his disorder *la dysenterie*, another *flux céliaque*, and the third *la maladie noire*, otherwise called *mélena*. Ramel, following Sauvages' nomenclature, declared the disorder *diarrhée vulgaire*.[31] These differing names and diagnoses must have caused confusion for the patient, let alone the consequential variation in the proposed curative regimens. However, it was not unusual at the time for patients to seek advice from more than one health provider. In one instance a woman had consulted and been treated by a charlatan, two surgeons and five eminent physicians.[32]

Who were the patients seeking help by letter? For the most part the clientele involved in epistolary consultations were drawn from the middle and upper classes of society, from the bourgeoisie to the aristocracy. They ranged in age from infants to the over-eighties, and both sexes were well represented. Probably an indication of the class of persons serviced by this type of consultation is the dearth of comment on disease arising from the client's occupation. An exceptional example is the obstruction of the bronchial tubes of a wig-maker, attributed to the powder he inhaled through his work.[33] Philip Rieder has pointed out that in early-modern France, unlike in the Anglo-Saxon world, the word 'patient' was not used, rather *le/la malade*.[34] This was because the French made a distinction between the notion of a 'patient' as a sick person, and the 'patient' in normal health. In this chapter I have used the word patient in the Anglo-Saxon sense of a sick person.

University-educated physicians were trained in and largely followed humoral theory, that is, that disease resulted from an imbalance of the innate humours in the body, black and yellow bile, blood and phlegm.[35] The notion that diseases were of a specific nature was not readily incorporated into this theory, although it would be false to suggest that there was no recognition of individual diseases; after all, some such as epilepsy were well known to the ancients. The *ordinaires*, who often interposed between patient and expert, were either surgeons, who did not benefit from a university education, or physicians who may or may not have studied at one of the leading medical schools.

The ailments encountered in epistolary consultations are varied, ranging from asthma to worms. First and foremost, epistolary consultations dealt mostly with chronic rather than acute diseases. Epidemics seldom appeared in epistolary consultations. The short-term nature of most epidemical disorders simply was unsuitable to be addressed by personal correspondence at an individual level.[36] The authors of *Consultations choisies*, a 10-volume collection of 739 such consultations, explained this on the basis that 'the course of acute diseases is usually so rapid that it is hardly possible to resort to advice other than those that one can undertake in the home of the patients.'[37] However, the distinction between acute and chronic diseases was ill defined. Jean-Claude-Adrien Helvétius (1685–1755) stated that acute diseases were those which 'ran into the third or fifth day but could extend to the fortieth'; chronic diseases he noted could continue 'for months even years'.[38] Similarly Théophile de Bordeu (1722–76) stated that all acute diseases had finished within 40 days.[39] According to Julian Martin, Sauvages' distinction between acute and chronic was:

> the principle upon which he made his first division of the species of diseases. He took Acute, conventionally enough, to refer to diseases which were likely to be very dangerous and perhaps fatal, and he took Chronical to encompass all those which appeared to be a constant but not fatal condition.[40]

This is very different from the time-based approach which had been the convention. On many occasions a patient's history indicated that the ailments he suffered from had persisted for months, even years. Acute attacks were different. Consider for a moment *apoplexie*, which was without question an acute event that could often be fatal. Catherine Storey has described it as 'the term used to describe a clinical presentation rather than a single disease entity'.[41] In humoral terms, *apoplexie* was attributed to an accumulation of phlegm in the brain. Despite its singularity of occurrence, it appears in letters on chronic diseases.[42] This apparent ambiguity is explained by the physician's (and the patient's) desire to deal with the consequences of such an event when not fatal, and to prevent a recurrence. Noël Chomel (1633–1712) described apoplexy as 'without contradiction the most cruel and most dangerous of all the diseases (*maladies*)'.[43] In contrast, many diseases were described as having afflicted patients for months, or years, before the help of an expert was sought.

A large proportion of the ailments complained about were more mundane than apoplexy – rheumatism and headaches for instance, the historical study of which, Jonathan Andrews contends, has been neglected.[44] The prosaic nature of such disorders does not appear to have attracted the attention of historians and only to a limited extent of physicians who have turned to writing history. No matter the nature of the ailment, if the patient considered it sufficiently worrisome to seek expert advice, one can reasonably conclude that he or she saw themselves as being ill.

The most significant diseases in terms of morbidity and mortality throughout the early-modern period were undoubtedly plague (*la peste*), smallpox (*la petite vérole*) and syphilis (*la vérole*).[45] These were also diseases that had a socially significant component. There can be little doubt that an outbreak of plague created fear amongst the population and resulted in state interference, such as the imposition of quarantine cordons around affected areas.[46] Apart from the high mortality associated with *la petite vérole*, those who survived were often disfigured and visually impaired. It appears in the correspondence in terms of its after-effects on patients, rather than incidents of the disease itself.[47] Unlike most chronic ailments, there were social implications with the endemic *la vérole*, as there was a moral dimension involved. The leading Dutch physician Herman Boerhaave (1668–1736) commented that *la vérole* 'throws the patient in the greatest dangers, and covers him with opprobrium and ignominy'.[48] Not only did it give rise to a social stigma amongst afflicted bourgeoisies, but also as infected females were often seen as the source, actions were at times taken to control prostitutes.[49]

Without doubt the disease most frequently encountered in the epistolary consultations examined was hypochondria, a term I use here to encompass hysteria, melancholia and the variety of other names used for what was much the same disorder.[50] Angus Gowland considered hypochondria as approaching epidemic proportions in the eighteenth century, although this was a figure of speech rather than a statistical construction.[51] Whether or not one accepts Gowland's viewpoint, hypochondria was neither infectious nor contagious.[52] Gilles Barroux has remarked, 'the word epidemic confers on disease an approach which is at the same time historical, social, cultural as well as physiological and pathological.'[53]

There were those diseases which without being life-threatening could have a significant impact on daily lives. Partial or total loss of sight was a common result of infection with *la petite verole*.[54] But other eye disorders were also referred to the consultants, such as *glaucome* and *cataracte*.[55] The patients simply recorded their loss of vision and it was left to the physicians and surgeons to diagnose the problem. In the sixteenth century the influential Parisian physician Jacques Fernel (1497–1598) provided a consultation for the wife of a M. de Rion whose sight was failing:

> The dimness of both the eyes and blindness thereby threatened, both by the remaining Opthalmia of the right Eye, and that which has since seized upon the left, have their first efficient Cause in the Brain. For therein store of extrementitious Flegm is wont to be bred and stored up, and that by reason of the cold and moist distemper of the sick GentleWoman and by her immoderate way of living. For as that is the chief cause of the Gout, when it goes into the hinder parts; so when it flows into so the former parts, and through the optick Nerves, it breeds in the Eyes such Symptoms as these hindering the sight.[56]

Hearing difficulties were another source of distress. In June 1737, a woman wrote to Le Thieullier:

> It is not the deafness which hampers me. I can put up with that, it is the noise of mills, of drums, above all at night, increase to the point that I cannot sleep, and the banging that rings from one ear to the other.[57]

Analysing the descriptions of symptoms in the *mémoires*, which were usually multiple, and the manner in which the majority of the consultants arrived at their diagnoses, one could think they were pursuing a syndrome. Yet the word was not used in the correspondence, perhaps because it was associated with empirics. Contemporarily the term syndrome (or sindrome) was understood to be a concourse of symptoms which collectively led to a diagnosis.[58] It had been argued that the problem with this approach was that if a physician waited until he had seen *all* the symptoms, it would be too late to apply remedies.[59] In reality the consultant took the information provided, seldom asked for more, and reduced any complexity of symptoms presented to a single cause. Finally on the subject of the diseases encountered in consulting letters, there remains those dealt with by surgeons, the most common of which were anal fistulae.[60]

## Attributions of cause

The physician had to attribute cause in order to provide a rational basis for therapy. According to Giles Barroux, the term aetiology 'existed in the eighteenth century, indicating the study of the causes of diseases, but it is like an instrument which lacked substance, or of which one does not fully understand the use to which it can be put'.[61] There were those who doubted that causes could be elucidated. For instance the English physician John Allen, in his widely and repeatedly published *Synopsis Medicinæ* stated:

> Concerning the Ætiology or Rationale of Distempers, The Authors Opinion is, that the origin or proximate Causes of Diseases are generally too abstruse for Human Nature to penetrate; and that except we dissemble, we must acknowledge, that they are so far hid from us that a Physician in this respect is often like one who sails in the deep, Ignorant of Longitude and Latitude.[62]

It would be incorrect to suggest that all medical practitioners sought to do was alleviate symptoms; these were only the case when a disease was deemed incurable. For instance, in 1725 François Chicoyneau (1672–1752) could only offer palliative treatment for a woman suffering from *une lepre*.[63] Indeed it was rare for the expert not to declare what was the cause of each particular illness. The treatments consultants proposed were derived from current medical theories. For the most part this was founded on the long-established Hippocratic–Galenic requirement of balancing the body's humours, as Malcolm Nicholson has explained:

> Morbid or potentially morbid, material was present in the body even in health. However, the healthy body was capable of rendering such material harmless and expelling it. This was the function of the normal processes of excretion. However if the normal channels of expulsion were overwhelmed or the internal organs weakened, morbid matter could accumulate in the body and cause ill health. If it was too long confined, this matter might change its character, perhaps becoming more acrid and injurious. In sickness, the body would labour to discharge morbid matter either by increasing normal discharge, hence diarrhoea or the profuse sweating of fever, or by making abnormal exits, such as the discharging pustules of smallpox. The aim of medical intervention was to facilitate this process of expulsion.[64]

In reality, humoralism was complicated by variations in theories of physiology and pathology which developed in the early-modern period. While iatrochemistry posited that the body functioned as a chemical plant,[65] iatromechanics or hydromechanics conceived the body as acting like a machine. Such was the case when François Chicoyneau and Nicolas Fournier (1703–1803) diagnosed a female patient's vertigo as being due to great a quantity of blood which rolled with difficulty in the tissues of the brain causing a compression of the nervous fibres.[66] Iatrochemical and iatromechanical theories appear occasionally in both patients' and practitioners' letters, but seldom to the extent that they completely overrode an excess or dearth of humours as the explanation for disease.

While humoral theory was dominant, humoral imbalance was not the sole primary cause of disease. Obstruction of the major organs was also considered to be a cause of disease, albeit intertwined with movement or blockage of humoral flow. This could be obstruction of the liver, the spleen or the kidneys, for instance, or the finer blood vessels. Antoine Deidier ascribed a man's ailments to blockages of the vessels of his liver and pancreas which prevented the bile and the pancreatic juice separating freely, leading to poor digestion that then caused fevers and vomiting.[67] Identifying the cause was not necessarily important to the patient or physician. In 1738, Le Thieullier provided a consultation to a 54-year-old suffering from kidney stones: 'it is useless to recall all the causes of the nephritis. We have found an active excess in his life up until now.'[68] The antecedent cause was overindulgence in wine and strong spirits from his youth onwards which had led to the formation of a tartarous deposit.

Diseases could be inherited, though the mechanism for this was not understood, and generally this was rolled out as a cause by the consultant when an *ordinaire* or the patient themselves reported that other family members had suffered the same or similar disorders. Diseases which were described as being heritable included asthma, deafness, epilepsy, mania, *phthsie*, sciatica and scrofula. Whether or not disease could be inherited and by what mechanism was a point of medical contention throughout the eighteenth century.[69] It was also a matter of concern to patients and their families.

One disease could give rise to another. An attack of *cholera morbus* was alleged by a Montpellier physician to be the cause of a woman's diabetes.[70] The diagnosis in this particular case was disputed by the Italian physician Francesco Torti (1658–1741), who had subsequently consulted the patient and did not believe this was plausible.[71] In another case two physicians and a surgeon authored a consultation in which they stated that because of poor treatment, a case of *gonorrhée virulente* had transformed into a *virus vérolique*.[72] Such a transformation invokes questions around the ætiology and diagnosis of both disorders.

Another source of disease was the wet nursing of infants. This could result directly from the wet nurses, as was the attribution in a 1743 case of *virus scrophleux*.[73] By far the most common complaint of infection through the wet nurse however was *la vérole*. It was even claimed that this process could be reversed, that the wet nurse could become infected by a diseased child.

The term *virus* occurs frequently as a causative agent and was contemporaneously understood as a 'Hidden fault of an unknown nature, which secretly infects the mass of our humours, and slowly changes the solid and fluid parts'.[74] A late-eighteenth-century definition was:

A Latin word . . . which signifies poison, venom. One means by virus, a maligne quality, pernicious, venomous enemy of nature. Such is the virus of syphilis, of scurvy, of scrofula, of scabies, of leprosy, of rabies. The venom of serpents, of the tarantula, etc.[75]

That the viper, the scorpion and the rabid animal could cause disease was commonly understood, and the presence of animals in the body in the form of worms was well known. The invention and improvements in microscopy had brought within the physician's purview small life forms that could not be seen with the naked eye, but although the idea of some living form being responsible for disease had been mooted, this was a disputed theory. Equally contentious was the idea of contagion. Nils Rosen Von Rosenstein (1706–63) remarked that 'I know of hooping-cough conveyed from a patient to two other children in a different house by an emissary. I have even myself carried it from one house to another undesignedly.'[76] Two means of disease transfer were recognised. There was person (or object) to person transfer, as expressed in the whooping cough case and believed to be the cause of the introduction of the plague in Marseille in 1720. The other was bad air, the notion that putrid air could somehow carry a disease which could infect people, as was the case with *cholera morbus*. Swampy exhalations in the form of miasma were regarded as another source of illness, but what was transferred remained a mystery prior to the nineteenth century. As Giles Barroux summarised the situation: 'One encounters [in epistolary consultations] pests, endemic fevers and other ills, which one often interrogates in vain for the mechanisms of contagion.'[77] Contagion was recognised even though the method of transfer was not understood. As early as 1546 Girolamo Fracastoro (1478–1553) had conjectured about the nature of contagion.[78] He suggested that there were seeds of contagion as opposed to the contemporary theory that contagion was caused by corrupted airs (miasmas), but he had no way of proving his ideas.[79]

Because the 'non-naturals' (food and drink, exercise and rest, voidages, air, sleeping and the passions) were considered an important part of maintaining health, it is hardly surprising that failure by the patient to address these issues could lead to diseased conditions. Derangements of the digestive system, gout, venereal diseases and melancholia commonly arose in the letters and were related to the lifestyles of the middle and upper classes. It was not unusual for excessive indulgence in alcohol to be blamed as the primary cause of illness. Thus, the jaundice and other liver-associated complaints of a 33-year-old wine vendor in the champagne region were attributed by both his *ordinaire* and consultant, Etienne-François Geoffroy, to excessive drinking stretching back to his youth.[80] In a not dissimilar light, a patient seeking relief from migraine wrote to Le Thieullier acknowledging that 'the table is my great enemy'.[81] Le Thieullier agreed that overindulgence was the prime cause of his ailments. On the topic of exercise, Montpellier physician Jacques Montagne (d. 1757) advised a 45-year-old man in 1745 that his various disorders were greatly contributed to by his immoderate hunting and his intemperance of all sorts of pleasures.[82] According to Erwin Ackerknecht the prescribing of diet in disease continued to become rarer and rarer in the eighteenth century.[83] However, the evidence of French consultants' letters suggests that diet was invariably and strictly prescribed for all ailments. With regard to air, in 1774 Paul-Joseph Barthez (1734–1806) attributed a case of epilepsy to breathing charcoal fumes.[84] A further causative agent of disease was climate. Exposure to inclement weather could be blamed for worse than simple *rheums* (colds). Le Thieullier warned that intemperate air could give rise to epidemic diseases which were for the most part fatal.[85] It was well recognised that certain epidemic diseases were seasonal.[86] Conversely, the Parisian Surgeon Antoine Louis wrote a dissertation on the most appropriate headwear for soldiers to wear in summer, as many fell ill and some died through over-exposure to the sun's rays.[87]

## Age and gender

Disease of children arose infrequently in the correspondence as many were acute infectious disorders, although childhood diseases were increasingly recognised as a separate field of medical endeavour through the early-modern period.[88] However, problems related to dentition, worms and rickets were raised as conditions which could be fatal. The origin of worms was a mystery. Jean Astruc (1684–1766) in *A*

*General and Compleat Treatise on all the Diseases incident to Children* (1746) realised they came from eggs: 'But whence spring these eggs in the human body? This is a difficulty I don't pretend to explain.'[89]

In 1732 Antoine Fizes declared the cause of rickets in an infant to be in part nutritional:

> a fault of nutrition generally widespread and in the bones, and in the soft parts that is to say muscles and membranes, the innards, the brain only is exempt for some time because of its uniform tissue, and that its consistency varies less by growing than the other parts . . . The general cause of this disease is a lymph of a badly mixed nature, mixed with water and phlegm, and loaded with concretions, which cannot distribute uniformily in the vessels nourishing the bones, and that relax the tissue of the soft parts, that give to them to their blockage, and to the obstructions which are formed later.[90]

Humoral for the most part, the conclusion that nutritional 'fault' could be a factor in disease causation was extremely unusual, notwithstanding the importance laid on diet in almost all consultations.

Women made up a large proportion for whom consulations were sought, and often for diseases particular to them. Most often these were associated with faulty menstruation, a plethora or diminution of flow, and with other vaginal discharges. In addition there many cases of hysteria, a disorder which had been in an earlier period seen as peculiar to women. By the eighteenth century this was regarded as analogous to hypochondria, and commonly an alternative term, vapours, was used for hysteria in women and hypochondria in men.[91] Diseases of the breast are infrequently encountered, most commonly descibed as *tumeurs*, a word that meant a lump, not necessarily cancerous, although cancers were occassionaly diagnosed.[92] The other disease of particular mention which recurred in the correspondence was *pâles couleurs* or chlorosis, an affliction associated with young women usually linked by physicians to hysteria.[93]

Beliefs that disease was caused by God, the Devil or astral influences were probably commonplace amongst the peasantry of France but were largely ignored by the elite of society, including university-trained physicians, as the period progressed. The Montargis physician Paul Dubé (1616–98) mixed theology with his medicine. In *Le Medecin et le chirugien de pauvres* he described a 1670 outbreak of *scorbut*.[94] He took it to be a new disease (despite the fact that *scorbut* was well known) as this particular outbreak had symptoms and signs he considered novel:

> We can say of the new disease of Scorbut in our climate that it has a cause Divine, to speak correctly, that is God, and incorrectly the constitution of the air, that God sends to men in order to chastise, and that the air takes on some qualities and some influences of the stars, also contributed to the production of this disease.[95]

However, no mention of godly causes arises in the correspondence examined, only that in severe cases a patient's future might remain in the hands of God. Astronomic influence did on rare occasions get a mention, mainly in regard to the timing of taking medication.

## Conclusion: treatment the last word

The final outcome of a consultation was how the disease was to be treated. It is beyond the scope of this chapter to debate this topic in depth. Suffice to say typically the expert would prescribe a regimen which comprised bleeding, and the use of vomicants, enemas and laxatives, all of which were intended to restore the humoral balance by supplementation, or more usually, by elimination. In addition he would prescribe *bouillons* (medicated thin broths), perhaps something to alleviate pain, and the taking of various mineral waters. Finally and invariably, lifestyle issues were addressed – exercise, diet, the type of company to keep and whether to be active or quiescent.

Whether or not the medicaments they prescribed were efficacious is difficult to ascertain. Philip Rieder has argued that the 'medical concept of efficacy is a foreign logic to the medical world of the *Ancien Régime*'.[96] This inevitably leads to the question of what is meant by 'cured'. In the period in question, it meant to regain health, which for the patient presumably meant an absence of symptoms. For example in 1747 a patient suffering from a cough, a heavy fever, loss of appetite, spots before the eyes and a bunion was treated by his *ordinaire* and commented that bleeding had reduced his fever, pains in his legs had not ceased, the spots before his eyes had lessened, his appetite was better and his bunion had improved.[97] Whether this was the sole consequence of following the physician's instructions is of course unknown. Such records of success or otherwise of treatment are rarely encountered. In *Consultations choisies* the authors acknowledged that not knowing the outcome of proposed treatments was a shortcoming, but claimed that this was because the patient had either been cured or had died. On other occasions in a printed text the consultant would add a note that the patient was cured. Ramel's text was rare amongst the sources examined, in that he included the outcomes of each of the consultations he had published. In the majority of cases the patient was said to have been returned to health, or had perished from some other ailment, but this could be the result of his selection of what to publish.

Epistolary correspondence provides a window to examine early-modern concepts of disease and their perceived causes. The distinction between being healthy and being ill was commonly understood but professionally ill-defined. The modern visualisation of disease entities being of a specific and definable nature simply does not accord with that of the seventeenth and eighteenth centuries. There were diseases readily identified with an agreed nomenclature, but even some of the most obvious were the subject of vernacular terms – for example *mal caduc* for *epilepsie*. Throughout the period imbalance of the humours was the mainstay of causation theory. But the fact that alternative explanations for ill-health were proposed is indicative of dissatisfaction with this theory to meet all the needs of a diagnosing physician. What is more, there were recognised causal factors outside this theory which the trained practitioners could not fully explain, such as heredity. Likewise, infection and contagion were recognised but could not be fully satisfactorily explained. For patients the letter was an established method of seeking relief from the varied diseases which troubled them, although we often do not know whether the advice proffered at a distance was followed or ignored. As the humoral theory underpinning physiology and pathology was widely understood by patients, physicians claimed that superior knowledge alone was often insufficient for him to establish his authority. This limitation was further accentuated by the relative social standing of patients and physicians. Where today's doctor encounters the occasional patient who is perceived to be recalcitrant, this was a recurring problem for his early-modern counterpart. Even the naming of a disease might be rejected if considered socially unacceptable. Equally, it was often difficult to get a patient to accept that lifestyle factors contributed to disease, and that patients needed to make changes to their everyday lives. The medical practitioner was thus forced to rely on rhetoric to convince his patients to accept his interpretation of symptoms, signs, disease causation and consequential prescriptions.

## Notes

1 Bibliothèque Interuniversitaire de Santé, (hereafter BIUS) MS 5155, Traité des maladies redigé d'apres les leçons de Monsieur Petit docteur regent de la faculté de medicine de paris, 1, pp. 5–6.

2 J. L. Scully, 'What is a Disease? Disease, Disability and Their Definitions', *European Molecular Biology Organisation Reports*, 5, 2004, pp. 650–3 at p. 650.

3 L. King, 'What is Disease', *Philosophy of Science*, 21 (1954), pp. 193–203 at p. 194.

4 The definition of health, illness and disease is a source of ongoing disputation amongst historians, philosophers and medical practitioners – see: R. P. Hudson, 'Concepts of Disease in the West', in *The Cambridge World History of Human Disease*, ed. Kenneth Kiple, Cambridge: Cambridge University Press, Part 2.1, pp. 45–52. Online. Available http://dx.doi.org/10.1017/CHOL9780521332866.007 (accessed 8.04.2014); and L. Nordenfelt, 'Health and Disease: Two Philosophical Perspectives', *Journal of Epidemiology and Community Health*, 41, 1986, 281–4.

5 C. E. Rosenberg, *Explaining Epidemics and Other Studies in the History of Medicine*, Cambridge: Cambridge University Press, 1992, p. 305; Scully, 'What Is a Disease? Disease, Disability and Their Definitions', pp. 650–3; M. Ereshefsky, 'Defining "Health" and "Disease"', *Studies in History and Philosophy of Biological and Biomedical Sciences*, 40, 2009, 221–7.

6 J. Riddle, 'Research Procedures in Evaluating Medieval Medicine', in B. Bowers (ed.), *The Medieval Hospital and Medical Practice*, London: Routledge, 2007, pp. 3–17; D. Harley, 'Rhetoric and the Social Construction of Sickness and Healing', *Social History of Medicine*, 12, 1999, pp. 407–35 at p. 419.

7 N. G. Siraisi, *Renaissance and Early Modern Medicine: An Introduction to Knowledge and Practice*, Chicago: University of Chicago Press, c.1990, pp. 115–16.

8 La Maison de Nostradamus, Salon de Provence, Le Fac-similé d'une consultation médicale de Nostradamus donnée à Béziers en le 20 Octobre 1559 à l'évêque Laurent Strozzi.

9 L. Brockliss, 'Consultation by Letter in Early Eighteenth-Century Paris: The Medical Practice of Etienne-François Geoffroy', in A. La Berge and M. Feingold (eds), *French Medical Culture in the Nineteenth Century*, Amsterdam: Rodopi, 1994, p. 79.

10 P. Rieder, *La figure du patient au XVIIe siècle*, Geneva: Libraire Droz S.A., 2010; L. Brockliss, 'Consultation by Letter in Early Eighteenth-Century Paris: The Medical Practice of Etienne-François Geoffroy', in A. La Berge and M. Feingold (eds), *French Medical Culture in the Nineteenth Century*, Amsterdam: Rodopi, 1994. pp. 79–117; W. Ruberg, 'The Letter as Medicine: Studying Health and Illness in Dutch Daily Correspondence, 1770–1850', *Social History of Medicine*, 23, 2010, 492–508; R. Weston, *Medical Consulting by Letter in France 1665–1789*, Farnham: Ashgate, 2013; and J. Coste, D. Jacquart and J. Pigeaud (eds), *La Rhétorique médicale à travers le siècles*, Geneva: Libraire Droz, 2012.

11 J. Andrews, 'History of Medicine: Health, Medicine and Disease in the Eighteenth Century', *Journal for Eighteenth-Century Studies*, 34, 2011, pp. 503–15 at p. 506.

12 V. Barras and M. Dinges, 'Maladies en lettres: An Introduction', in V. Barras and M. Dinges (eds), *Maladies en Lettres, 17e–21e siècles*, Lausanne: Éditions BHMS, 2013, p. 11; L.-Be.-B.

Jourdain, *Traité des maladies et des operations réellement chirurgicales de la bouche*, Paris: Valleyre, 1778.

13  H. Le Dran, *Consultations sur les plupart des maladies qui sont du ressort de la Chirurgie*, Paris: P. Fr. Diderot le jeune, 1765; Bibliothèque de Metz, MS 1321–1325, Manuscrits du docteur Antoine Louis, né à Metz, secrétaire perpétuel de l'Académie de Chirurgie à Paris, chirurgien en chef de la Salpêtrière; and M. M. Ruisinger, 'La Chirurgie en lettres. L'example de la consultation épistolaire de Lorenz Heister (1683–1758)', in V. Barras and M. Dinges (eds), *Maladies en Lettres*, pp. 113–22.

14  Le Dran, *Consultations sur les plupart des maladies qui sont du ressort de la Chirurgie*, pp. 184–91.

15  L.-J. Le Thieullier, *Consultations de médecine* 2 vols, Paris: Charles Osmont, 1739, 2, pp. 291–300.

16  Bibliothèque municipale de Dijon, MS 433. Dr. Ganiare de Beaune, *Consultations Médicales*; M.-F.-B. Ramel, *Consultations médicales et mémoire sur l'air de Gemenos*, La Haye: chez Les libraires associés, 1785.

17  *Recueil périodique d'observations de médecine, chirurgie, pharmacy*. Paris, 1754–1793, 157 vols.

18  *Consultations choisies de plusiers médecins célèbres de l'Université de Montpellier sur des maladies aigues et chroniques*, 10 vols (1750–1757), Paris, Durand et Pissot, 8 (1750), pp. 295–6.

19  A. Caufapé, *Methode pour prolonger la vie et conserver la santé*, Toulouse: J. Perch, 1686. BIUS Cote m.fiche 2137, p. 3.

20  N. Chomel and J. Marret, *Dictionnaire oeconomique, contenant divers moyens d'augmenter son bien, et de conserver sa santé*, Commercy: chez Henry Thomas & Cie 1741, p. 5.

21  *Encyclopédie, ou dictionnaire raisonné des sciences, des arts et des métiers, etc.*, ed. D. Diderot and J. le Rond d'Alembert, University of Chicago: ARTFL Encyclopédie Project (Spring 2013 Edition), R. Morrissey (ed.), 9, p. 930. Online. Available http://encyclopedie.uchicago.edu/ (accessed 01.04.2014).

22  Ramel, *Consultations médicales et mémoire sur l'air de Gemenos*, pp. 44–5.

23  Hypochondria was used interchangeably with melancholia, vapours and hysteria – Weston, *Medical Consulting by Letter in France 1665–1789*, pp. 152–9.

24  *Consultations choisies*, 8, p. 111.

25  If a 'new' disease appeared it needed a name, although it would appear that most so-called new diseases were in fact repeat episodes of disorders that had occurred previously and been forgotten about. For example, the French term coqueluche, whooping cough, was also used in outbreaks of other diseases – Robert Weston, 'Whooping Cough: A Brief History to the 19th Century', *Canadian Bulletin of Medical History*, 29, 2012, 329–49.

26  Brockliss, *Consultation by Letter in Early Eighteenth-Century Paris*, p. 90.

27  Cited in F. Boissier de la Croix de Sauvages, *Nosologie méthodique*, 2 vols, Lyon: Gouvion, 1772, 1, p. 149.

28  de la Croix de Sauvages, *Nosologie méthodique* 1, pp. 154 and 156.

29  J. Lommius, *Tableau des maladies ou l'on découvre leurs signes et leiurs evenemens de Lommius* (trans. l'Abése le Mascrier), Paris: chez Louis Silvestre, 1712. The original was published in Latin, Antwerp, 1562.

30  'Faire sçavoir le nom . . . de ce mal', Le Thieullier, *Consultations de médecine*, 2, p. 380.

31  Ramel, *Consultations médicales et mémoire sur l'air de Gemenos*, p. 228.

32  A. Deidier, *Consultations et observations médicinales de M. Antoine Deidier*, 3 vols, Paris: chez Jean-Thomas Hérissant, 1754, 1, pp. 201–36.

33  Ramel, *Consultations médicales et mémoire sur l'air de Gemenos*, pp. 35–40.

34  P. Rieder, 'L'histoire du «patient»: aléa, moyen, ou finalité de l'histoire médicale?', *Gesnerus*, 60 (2003), pp. 260–71, at p. 261.

35  Later in the period lymph was occasionally added as a humour; thus, in a 1733 consultation, *rheumatisme* was said to be caused by a 'muscular lymph swollen and salty' – *Consultations choisies*, 3 (1757), p. 263. For more on humoral theory, see the chapter in this volume by J. R. Hankinson.

36  *Consultations choisies* refers to two incidents of epidemics. The first described a visit in 1744 by two Montpellier physicians to Aigues Mortes to assist with a local epidemic. *Consultations choisies*, 2 (1757), p. 31; the second was in response to a 1749 request by an ordinaire addressed

to the Montpellier School of Medicine in 1749 seeking advice on what he considered to be a contagious viral venereal epidemic. *Consultations choisies*, 10 (1755), pp. 210–14.

37  *Consultations choisies*, 1 (1750), p. xvii.

38  J.-C.-A. Helvétius, *Idée generale de L'oeconomie animale, et observations sur la petite vérole*, Lyon: chez Freres Bruyset, 1727, pp. 1–3.

39  T. de Bordeu, *Œuvres complètes de Bordeu (1779–1840)*, ed. par Anthelme le Chevalier Richerand, Paris: chez Caille et Ravier, 1818, p. 210.

40  J. Martin, 'Sauvage's Nosology: Medical Enlightenment in Montpellier', in A. Cunningham and R. French (eds), *The Medical Enlightenment of the Eighteenth Century*, Cambridge: Cambridge University Press, 1990, pp. 111–37 at p. 123.

41  C. E. Storey, 'Apoplexy: Changing Concepts in the Eighteenth Century', in H. Whitacre, C. U. M. Smith and S. Finger (eds), *Brain, Mind and Medicine*, New York: Springer Science + Business Medias LLC, 2007, pp. 233–43 at p. 233.

42  Hélian, *Dictionnaire du diagnostic, ou l'art de connoître les maladies, et de les distinguer exactement les unes des autres*, Paris: chez Vincent, 1771, p. 19, in which apoplexy is described as nearly always fatal. Consultations choisies includes four cases of this disease.

43  N. Chomel, *Dictionnaire oeconomique, contenant divers moyen d'augmenter et conserver son bien, et même sa santé avec plusieurs remedes assurez et éprouvez, pour un trés-grand nombre de maladies, et de beaux secrets pour parvenir à une longue et heureuse vieillesse*, Paris: Le Conte et Montalant,1709, p. 57.

44  Andrews, 'History of Medicine', p. 508.

45  The last major European outbreak of plague, *la peste* to the French, occurred in the Marseille region in 1720–2, Messina in 1742, and Moscow in the 1770s. See the chapters in this volume by Michael Worboys and Samuel Cohn.

46  As were established in the Marseille outbreak for example.

47  An exception concerned a physician requesting advice about bleeding a woman suffering from smallpox: BIUS, MS 5241, Correspondance de Geoffroy, médecin de Paris, ff. 142–3; An adolescent male suffered from an ozène (a putrid ulcer of the nose) often consequential to smallpox – *Consultations choisies*, 2, p. 377.

48  H. Boerhaave, *Traité des maladies vénériennes, traduit du Latin*, Paris: chez Briasson, 1753, pp. 1–2.

49  C. Jones, *The Charitable Imperative: Hospitals and Nursing in Ancién Regime and Revolutionary France*, London: Routledge, 1989, Ch. 7.

50  Weston, *Medical Consulting by Letter in France 1665–1789*, p. 152.

51  A. Gowland, 'The Problem of Early Modern Melancholy', *Past and Present*, 19 (2006), pp. 77–120 at p. 83.

52  Gilles Barroux has remarked that 'the word epidemic confers on disease an approach which is at the same time historical, social, cultural as well as physiological and pathological' – G. Barroux, *Philosophie, maladie et médecine au XVIIIe siècle*, Paris: Honoré Champion Editeur, 2008, p. 21.

53  Barroux, *Philosophie, maladie et médecine au XVIIIe siècle*, p. 21.

54  *Consultations choisies*, 9 (1751), p. 95.

55  *Consultations choisies*, 4 (1757), p. 343.

56  J. Fernel [Johannes Fernelius], 'Select Medicinal Counsels of the Renowned Johannes Fernelius' in N. Culpeper, A. Cole and W. Rowland, *The Practice of Physick in Two Volumes, Very Much Enlarged, Printed by Peter Cole*, London: n.p., 1658, p. 331. In this case Fernel required more information to ascertain where the 'obstructing humour' was lodged so that he could better prescribe a cure.

57  Le Thieullier, *Consultations de médecine*, 1, p. 283.

58  The word first appeared in post-classical medicine in book four of G. de Chaulliac, *The questyonary of cyrurgyens*, London: n.p., 1542.

59  R. James, *A Medical Dictionary*, 3 vols, London, 1743–5, 3 (1745), n.p.

60  Le Dran, *Consultations sur les plupart des maladies qui sont du ressort de la Chirurgie*, p. i.

61  Barroux, *Philosophie, maladie et médecine au XVIIIe siècle*, p. 37.

62  J. Allen, *Dr. Allen's Synopsis Medicinæ: or a Brief and General Collection of the Whole Practice of Physick* 2 vols, London: n.p., 3rd edn, 1730, 1, p. 2.

63  *Consultations choisies*, 7 (1750), pp. 119–21.

64 M. Nicholson, 'The Metastatic Theory of Pathogenesis and the Professional Interests of the Eighteenth-Century Physician', *Medical History*, 32, 1988, pp. 227–300 at p. 280.

65 R. French, *Medicine before Science: The Business of Medicine from the Middle Ages to the Enlightenment*, Cambridge: Cambridge University Press, 2003, p. 206.

66 *Consultations choisies*, 2, p. 271.

67 Deidier, *Consultations et observations médicinales de M. Antoine Deidier*, 1, p. 472.

68 Le Theiullier, *Consultations de médecine*, 1, p. 355.

69 C. L. Beltran, 'Les Maladies héréditaires: Eighteenth Century Disputes in France', *Revue d'histoire de science*, 48, 1995, 307–50.

70 *Consultations choisies*, 4, pp. 71–2.

71 F. Torti, *The Clinical Consultations of Francesco Torti*, translated with an introduction by Saul Jarcho, Malabar, FL: Krieger Publishing Company, 2000, pp. 836–9.

72 *Consultations choisies*, 10, p. 210.

73 *Consultations choisies*, 10, pp. 179–90.

74 Mrs. le V ***, M*** et de la M*** / Le vacher de la Feutrie, Thomas / Moysant François / La Macellerie. *Dictionnaire de chirurgie contenant la description anatomique des parties humaine . . . le méchanisme des fonctions . . . le manuel des opérations chirurgicales, avec . . . les usages des différens instrumens et médicamens employés dans le traitement des maladies du ressort de la chirurgie*, 2 vols, Paris: Lacombe, 1767, 2, p. 685.

75 J.-F. Lavoisien, *Dictionnaire portatif de médecine, d'anatomie, de chirurgie, d'histoire naturelle, de botanique et de physique*, Paris: chez Théophile Barrois, 1793, p. 603.

76 N. Rosen von Rosenstein, *The Diseases of Children, and Their Remedies*, trans. A. Erikson Sparrman, London: T. Cadell, 1776, p. 191. [First published as Underrättelser om barn-sjuk domar och deras botemedel, Stockholm: Lars Salvius, 1764.]

77 Barroux, *Philosophie, maladie et médecine au XVIIIe siècle*, p. 20

78 G. Fracastoro, *De Contagione et Contagiosis Morbis*, Venice: Venetiis, 1546.

79 For a parallel Latin and French translation, see J. Fracastor and L. Meunier, *Les Trois livres de Jérôme Fracastor sur la contagion, les maladies contagieuses et leur traitement*, trad. L. Meunier, Paris: Société d'Éditions Scientifiques, 1893. See also V. Nutton, 'The Reception of Fracastoro's Theory of Contagion', *Osiris*, 6, 1990, 196–23.

80 BIUS, MS 5242, *Correspondance de Geoffroy, médecin de Paris*, fols 127–9.

81 Le Thieullier, *Consultations de médecine*, 1, pp. 270–7.

82 *Consultations choisies*, 6 (1750), p. 47.

83 E. Ackerknecht, *Therapeutics from the Primitives to the Twentieth Century*, New York: Hafner Press, 1973, p. 179.

84 P.-J. Barthez, *Consultations de médecine, ouvrage posthume, publié par J. Lordat*, 2 vols, Paris: chez Michaud Frères, 1810, 2, p. 112.

85 Le Thieullier, *Consultations de médecine*, pp. 219–28.

86 In 1777 Poitiers physician Jean-Gabriel Gallot reported that a cold spring had resulted in many cases of measles and scarlatina. *La Vie et les œuvres du Dr Jean–Gabriel Gallot (1744–1794)*, Poitiers: Societié des Antiquaires de l'Ouest, 1961. p. 200.

87 Bibliothèque de Metz, MS 1320, *Manuscrits du docteur Antoine Louis, né à Metz, secrétaire perpétuel de l'Académie de Chirurgie à Paris, chirurgien en chef de la Salpêtrière*, lettre R.

88 The earliest printed treatise on paediatrics was by Paduan physician Paolo Bagellardo (1425–ca.1494) – P. Bagellardo, *Opusculum de egritudinibus et remediis infantium*, Padua, [Paviæ]: Bar. Val. et Mar. de Septem Arboribus Prutenus, 1472.

89 J. Astruc, *A General and Compleat Treatise on all the Diseases incident to Children*, London: n.p., 1746, p. 185.

90 *Consultations choisies*, 2, p. 338.

91 P. Pomme, *Traité des affections vaporeuses des deux sexes*, 2nd edn, Benoît Duplain. Paris: n.p., 1765, p. 5.

92 Barthez, *Consultations de médecine, ouvrage posthume, publié par J. Lordat*, pp. 351–61. In 1773, Barthez considered surgery for a tumour of the breast.

93 BIUS, MS 2075, Chirac, *Recueil de consultations médicales*, pp. 93–4.

94 P. Dubé, *Le Medecin et le chirurgien des pauvres*, Paris: chez Edmé Couterot, 6th edn, 1683. pp. 311–25.

95  Dubé, *Le Medecin et le chirurgien des pauvres*, p. 320.
96  P. Rieder, *La Figure du patient au XVIIe siècle*, p. 34.
97  Le Thieullier, *Consultations de médecine*, 1, pp. 254–9.

## Select bibliography

Bibliotheque Interuniversitaire de Santé, *Correspondance de Geoffroy, médecin de Paris,* MS 5241–5.

*Consultations choisies de plusiers médecins célébres de l'Université de Montpellier sur des maladies aigues et chroniques,* 10 vols (1750–1757), Paris, Durand et Pissot.

Le Dran, H., *Consultations sur les plupart des maladies qui sont du ressort de la Chirurgie,* Paris: P. Fr. Diderot le jeune, 1765.

Le Thieullier, L.-J., *Consultations de médecine,* 2 vols, Paris: Charles Osmont, 1739.

Ramel, M.-F.-B., *Consultations médicales et mémoire sur l'air de Gemenos,* La Haye; Chez Les libraires associés, 1785.

Barras, V. and Dinges, M. (eds), *Maladies en Lettres, 17e–21e siècles,* Lausanne: Éditions BHMS, 2013.

Brockliss L. and Jones C., *The Medical World of Early Modern France,* Oxford: Clarendon Press, 1997.

Rieder P., 'L'histoire du «patient»: aléa, moyen, ou finalité de l'histoire médicale?', *Gesnerus,* 60, 2003, 260–71.

Rieder, P., *La Figure du patient au XVIIe siècle,* Geneva: Libraire Droz S.A., 2010.

Weston, R., *Medical Consulting by Letter in France 1665–1789,* Farnham: Ashgate, 2013.

# 27

# THE CLINICAL NARRATIVES OF JAMES PARKINSON'S *ESSAY ON THE SHAKING PALSY* (1817)

*Brian Hurwitz*

James Parkinson's contribution to understanding disorders of human movement in the late eighteenth and early nineteenth centuries is recognized in a well-known eponym.[1] This chapter considers the extent to which his *Essay on the Shaking Palsy* of 1817 successfully defined a new disease and incorporated a distinctively narrative view of the condition into the concept of it.[2] I will focus on Parkinson's approach to gathering and presenting his findings in the context of the observational culture of his time and on how the *Essay* exemplifies the literary and social aesthetic Tom Laqueur has called 'a new humanitarian narrative'. In depicting the clinical phenomenology of the condition, I will examine the dialectic between observation and abstraction, between individual case description and the general account of the condition which the *Essay* sets up,[3] as well as its author's debt to eighteenth-century sentimental writing and the literature of urban spectatorship.[4]

Laqueur designates a cluster of eighteenth-century scientific, medical and social inquiries 'humanitarian narratives', which speak of 'the aims and deaths of ordinary people' and seek to connect the sympathies and 'actions of . . . readers with their suffering subjects'.[5] Such accounts set out to persuade readers that the causes of pain and suffering should be ameliorated and if possible prevented, aims that were foremost in Parkinson's account of a condition he found began almost imperceptibly in the slightest of vibrations which worsen over time, and in the later stages disorganize many of the routine movements of daily living.

Roger Smith notes that the term story 'is the older word "history" with a lost syllable' and that 'the German "Gesichte", the Russian "istoria" and the French "histoire" all denote both history and story'.[6] In eighteenth-century medical cases, novels and biographical writings alike, historical narrative proved an effective and popular means in which to configure the events of a life and of an illness. A history could grasp the sequence and meaning of events and experiences unfolding over time, its narrative possibilities being well suited to depicting the effects of disorder, whether these arose from misfortune, physical or psychological trauma, altered identity or illness.[7] While accounts of disease drew on the language of patient experiences which came to frame specific clinical understandings, narrative gained legitimacy as a notion denoting both complicated medical cases and accounts of scientific experiments.[8]

Parkinson placed a 'History' of the condition at the head of his *Essay*, which sought to integrate sufferers' experiences into a distinct concept of a disease. In this he aligned himself with a shift that had taken place in German case descriptions: Carsten Zelle has documented how eighteenth-century German physicians had begun to incorporate into case descriptions information gleaned not only from 'exterior senses (*sensus exteriores: tactus, visus, auditus, olfactus, gustus*)' but also from their own 'internal emotional processes perceived by *sensus interiores*',[9] a double sensitivity apparent in the language of the *Essay*, which goes beyond the ocular to incorporate subjective perspectives on the effects of the malady on sufferers.

## The *Essay*

The *Essay* redefined a hitherto imprecise term, 'paralysis agitans', as the 'Shaking Palsy', designating it a slowly deteriorating disease of long duration by the conjunction of certain symptoms and disordered movements.[10] It was published when Parkinson was nearing retirement from practice as a surgeon-apothecary in the semi-rural hamlet of Hoxton, north east of the City of London (see Figure 27.1).[11] During his career Parkinson had pressed for public health reform and infection control in London workhouses, had been the medical attendant to a Hoxton madhouse, a radical political pamphleteer, a short-story-teller, geologist, fossilist and the author of several textbooks.[12] By the time he came to write the *Essay* he was an accomplished author, though by his own account he lacked a secure understanding of the cause of the Shaking Palsy which may have delayed the *Essay*'s appearance until long after his interest in the malady first took shape. In its 'Preface' he wrote: 'some conciliatory explanation should be offered for the present publication: in which, it is acknowledged, that mere conjecture takes the place of experiment; and, that analogy is the substitute for anatomical examination, the only sure foundation for pathological knowledge'.[13] Given this degree of caution the effect of the *Essay*'s appearance was almost epiphanic.

As soon as it appeared, the *Essay* was recognized to be an important description of a progressive condition. In the year of publication, sections from it were reprinted in *The London Medical and Physical Journal*, which announced the work worthy of 'universal perusal'.[14] John Cooke, physician to The London Hospital, in his *Treatise on Nervous Diseases* (1820), declared it 'highly deserving of our attention', and noted that neurologists hitherto had 'not classed *paralysis agitans* among the palsies'.[15] Parts of its 'History' were quoted verbatim in Thomas Graham's 1827 edition of *Modern Domestic Medicine*, one of many treatises of the time that set out symptoms to assist people unable to pay for doctors to diagnose and treat themselves.[16] The *Essay* was cited in the early pages of *The Lancet* by physicians at the height of their careers, including John Elliotson in 1830[17] and Marshall Hall in 1838,[18] and later in the century it proved influential on the work of international clinical authorities, such as the German physician Marshall Romberg in his *Manual of The Nervous Disease in Man* (1846),[19] the French physician Armand Trousseau, whose 1861 *Lectures in Clinical Medicine*[20] identified additional features of the malady, and Jean-Martin Charcot and Edmé Vulpian.[21] Charcot later named the condition '*La Maladie de Parkinson*'.[22] So impressed by the *Essay* was Charcot that he advised his students at the Salpêtrière to translate the work: 'it will provide you', he wrote, 'with the satisfaction and knowledge that one always gleans from a direct clinical description made by an honest and careful observer'.[23]

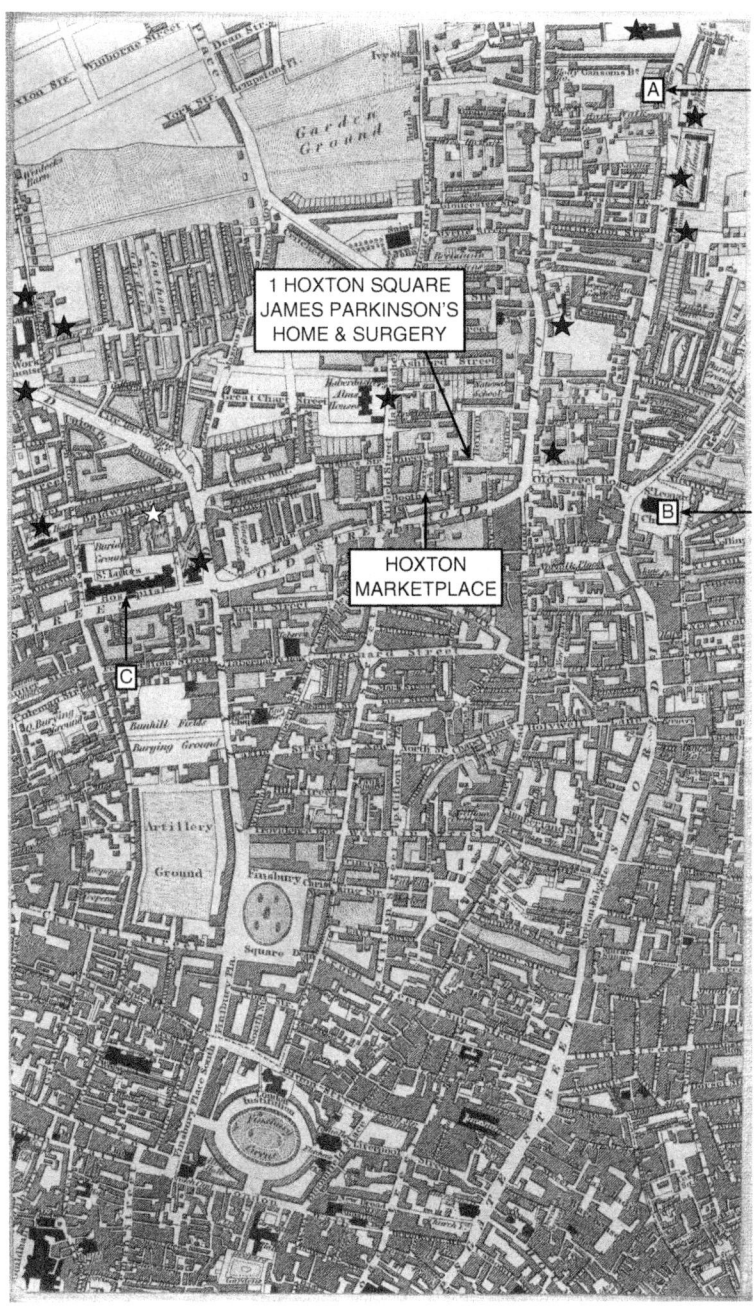

1 HOXTON SQUARE
JAMES PARKINSON'S
HOME & SURGERY

HOXTON
MARKETPLACE

*Figure 27.1* Greenwood Map of London from the City to London and Shoreditch 1827. Stars: almshouses, madhouses or workhouses. A: Parkinson's place of retirement in 1817; B: St. Leonard's Church, Shoreditch; C: St. Luke's Mental Asylum. Map hosted by Bath Spa University, UK, http://www.users.bathspa.ac.uk/greenwood/, accessed 25/08/2015. Web author: Mark Annand. Copyright permission has been granted.

493

The *Essay* was divided into five chapters, the first of which was in three parts, 'DEFINITION — HISTORY — ILLUSTRATIVE CASES' and defined the Shaking Palsy as:

> Involuntary tremulous motion, with lessened voluntary power, in parts not in action, and even when supported; with a propensity to bend the trunk forwards, and to pass from a walking to a running pace: the senses and intellects being uninjured.[24]

These various elements were not in themselves newly identified phenomena although this combination of them had not previously been advocated. At the time of his intervention these forms of movement had already been described and placed in different classes, species and genera of prevailing nosologies. Parkinson argued that they belonged together, the components necessary for diagnosis being *Tremor coactus* (a term referring to an involuntary trembling characterized by the Dutch physician and anatomist Franz de le Boë in the seventeenth century) and *Scelotyrbe festinans* (the conjunction of *scelotyrbe* – the ancient Greek term for hustle or totter – and *festinans* – the Latin signifying hurry).[25] Parkinson claimed these dysmobilities occurred together and constituted a single malady.

## The 'History'

The 'History' developed a generic account of the Shaking Palsy featuring details selected from six cases set out immediately after the first section of the *Essay*. Readers of the work therefore encountered an iterated clinical narrative in the form of a generalized story set out in the 'History', followed by a patchwork of case narratives whose unattributed features made up the generalized account of the 'History'. The trajectory of the malady presented in the 'History' was as follows:[26]

> So slight and nearly imperceptible are the first inroads of this malady, and so extremely slow its progress, that it rarely happens, that the patient can form any recollection of the precise period of its commencement. The first symptoms are a slight sense of weakness, with a proneness to trembling in some particular part; sometimes in the head, but most commonly in one of the hands and arms.
>    . . . [A]s the disease proceeds . . . the hand fails . . . to answer with exactness to the dictates of the will. . . . The legs are not raised to that height, or with that promptitude which the will directs . . . [W]riting can now be hardly at all accomplished . . . whilst at meals the fork not being duly directed frequently fails to raise the morsel from the plate. Commencing . . . in one arm, the wearisome agitation is borne until beyond sufferance, when by suddenly changing the posture it is for a time stopped . . . to commence, generally in less than a minute in one of the legs . . . Harassed by this tormenting round, the patient has recourse to walking, a mode of exercise to which sufferers from this malady are in general partial; owing to their attention being thereby somewhat diverted from their unpleasant feelings. . . .

... [A]s the malady proceeds ... the propensity to lean forward becomes invincible, and the patient is thereby forced to step on the toes and fore part of the feet, whilst the upper part of the body is thrown so far forward as to render it difficult to avoid falling on the face.... In some cases ... the patient ... is irresistibly impelled to take much quicker and shorter steps, and thereby to adopt unwillingly a running pace.

The power of conveying the food to the mouth is at length so much impeded that he is obliged to consent to be fed by others. The bowels, which had been all along torpid, now, in most cases, demand stimulating medicines ... he dares not venture on exercise, unless assisted by an attendant, who walking backwards before him, prevents his falling forwards, by the pressure of his hands against the fore part of his shoulders. He is not only no longer able to feed himself, but when the food is conveyed to the mouth, so much are the actions of the muscles of the tongue and pharynx impeded by impaired action and perpetual agitation, that the food is with difficulty retained in the mouth until masticated; and then as difficultly swallowed.... [T]he saliva ... hence is continually draining from the mouth, mixed with the particles of food ... As the debility increases ... the tremulous agitation ... becomes so violent as to shake the bed-hangings. The chin is now almost immovably bent down upon the sternum. The slops with which he is attempted to be fed, with the saliva, are continually trickling from the mouth. The power of articulation is lost ... and at the last, constant sleepiness, with slight delirium, and other marks of extreme exhaustion announce the wished-for release.[27]

Parkinson draws out the arc of the malady in an account of great pathos. Following an onset located in indistinctness the sufferer undergoes a prolonged deterioration whose course is unflinchingly charted in chronological and semi-biographical terms. The closing scene focuses down on the sufferer, cared for and tended to ('attempted to be fed'), tethered to his bed by the condition. A marked transformation is apparent: hitherto upright, independent and mobile (though becoming less so as the 'History' unfolds), the exhausted, shaking sufferer is forced into a recumbent position, chin-on-chest, surrounding furnishings also shaking, his movements ceasing only when 'constant sleepiness' and 'slight delirium' set in, and death finally supervenes.

This is not a conventionally good death: there is no hint of prayer other than the 'wished-for release'. The picture is of 'a train of harassing evils',[28] followed by an agitated transition to final repose. As we shall see, the sufferer of the 'History' is derived from the clinical cases that follow and is a composite constructed from observations of real people and their circumstances. The 'History' pictorially creates an amalgam of a sufferer whose daily activities, habits and propensities become progressively curtailed, Parkinson's use of impersonal constructions and the passive voice lending power to his depiction of frustrated volition. In taking parts of the man's body as the subject of his sentences ('the hand fails', 'the legs are not raised', 'the fork ... fails', 'the chin is ... bent down'), and in making the sufferer the subject of passive verbs ('the patient is irresistibly impelled' and 'obliged to consent to be fed'), the syntax mirrors the sufferer's condition of not being able to exert his will over his own body.

The fact that it is 'the hand' which the 'History' refers to, not 'his hand', is also telling as it portrays the alienation 'the sufferer' feels from his own body.[29] The end is a death that 'softens the heart',[30] even as it closes the section of the *Essay* which charts the trajectory of a disease *type*.[31]

All but the last of the six 'Illustrative Cases' that follow this 'History' are brief. The first of them reads:

> Almost every circumstance noted in the preceding description, was observed in a case which occurred several years back, and which, from the peculiar symptoms which manifested themselves in its progress; from the little knowledge of its nature, acknowledged to be possessed by the physician who attended; and from the mode of its termination; caused an eager wish to acquire some further knowledge of its nature and cause.[32] In this case, every circumstance occurred which has been mentioned in the preceding history.[33]

After the 'History', Case I is the first to be set out simply because it was the first Parkinson recognized as noteworthy, the first to stimulate his interest in characterizing the condition – the 'index' or exemplary instance of it. In comparison with the 'History', its prose is terse and pared down, and the clinical details are minimal. That this particular case could be coincident with the 'History', as Parkinson implies, seems doubtful because the account has been compounded from the details of several cases and includes a very distinctive scene of difficulty *initiating* walking taken from Case V, which could not also have belonged to Case I. However, taking account of Parkinson's reference to '*the mode of its termination*' (my emphasis), Case I would appear to have been the source of the deathbed scene.

It is surprising how little clinical content Case I contains, how *non*-'illustrative', in terms of features visualized, Case I turns out to be. The modern reader would expect the contents of Case I to be pivotal in convincing an audience that the general description of a condition is grounded in careful observation of clinical reality. But the generic description within the 'History' turns out to be *more* detailed than that of Case I, reversing modern expectations.

Parkinson introduced his second and third cases as 'the subject of the next case . . .' and 'the next case', indicating that they arose from consecutive observations. Both concerned people he stopped and questioned on the street. Each is short, and focused on the bowed posture, curious gait, and the use of a stick. The man in Case II, he reported, 'walked almost entirely on the fore part of his feet, and would have fallen every step if he had not been supported by his stick'.[34] The sufferer informed Parkinson that his condition had come on 'as the consequence of irregularities in mode of living and indulgence in spiritous liquors', an attribution faithfully reported but to which Parkinson gave little credence, because he maintained the trembling of the Shaking Palsy was quite distinct from that related to high intakes of alcohol.[35]

Case II concerned a man of 62 years of age 'whose life had been spent as an attendant at a magistrate's office. All the extremities were considerably agitated, the speech was very much interrupted, and the body much bowed and shaken'. The agitation of Case III's limbs, head and body was 'vehement' and, crucially, it was *coincident* with abnormal posture, gait and shaking limbs. However, the significance of linking these two distinct sets of clinical phenomena, which was an important new observation, is somewhat over-shadowed in the text by the detailed description of the man's changing posture and use of his stick:

> he was entirely unable to walk; the body being so bowed, and the head thrown so forward, as to oblige him to go on a continued run . . . and to employ his stick every five or six steps to force him more into an upright posture, by projecting the point of it with great force against the pavement.[36]

The street cases are closely observed and visually lively. Case III seems to have persuaded Parkinson that highly varied clinical features could belong together. Case IV, seen indoors, 'was that of a gentleman about fifty-five years', who consulted on account of a chest complaint, which turned out to be due to an abscess, 'a considerable degree of inflammation over the lower ribs on the left side, which terminated in the formation of matter beneath the fascia'. As Parkinson drained the abscess he noticed a 'trembling of the arms'.[37] These are the only clinical details provided and add nothing to the characterization of the Shaking Palsy already provided. However, the case conveys a strong sense of Parkinson at work and readers learn that trembling may not lead sufferers to complain of it, even when consulting on account of other conditions, and that the diagnosis of this malady may arise opportunistically.

Parkinson did not get the chance to speak to Case V because, as he put it, 'the lamented subject . . . was only seen at a distance'. But he offered the following sketch of how his mode of propulsion was initiated and sustained by an attendant who:

> standing before him with a hand placed on each shoulder, until, by gently swaying backward and forward, he had placed himself in equipoise; when, giving the word, he would start in a running pace, the attendant sliding from before him and running forward, being ready to receive him and prevent his falling.[38]

The visual impression of Case V is almost cinematic, as if this co-operative mode of locomotion had been seen from afar through binoculars.[39] The deftness with which the scene of this pair rocking to-and-fro in unison is encapsulated is striking. The sixth case Parkinson introduced as one 'which presented itself to observation since those above-mentioned', implying it was observed with the benefit of hindsight gained from the earlier case descriptions. Textually the longest case, it charted a sequence of symptoms in a man over a 12-year period. As before, the effects of the malady on the sufferer's life were contextualized with the patient's views about its vexations and causation: his bowels, we learn, were 'much retarded', a problem Parkinson thought common to many sufferers; and he noticed the man's trembling temporarily disappeared following a stroke, only to return again as the one-sided bodily paralysis resolved. Parkinson twice noted the sufferer's capacity to suspend the agitations by an act of will:

> At present he is almost constantly troubled with the agitation . . . when, by a sudden and somewhat violent change of posture, he is almost always able to stop it.[40]

> . . . [H]e, being then just come in from a walk, with every limb shaking, threw himself rather violently into a chair, and said, 'Now I am as well as ever I was in my life.' The shaking completely stopped; but returned within two minutes.[41]

The scene reveals not only the man's agency and ingenuity but also a note of triumph in being able to demonstrate a self-discovered method of auto-termination, albeit only transiently effective.[42] As Case VI reaches its close, another voice is heard, that of the patient's wife:

> . . . being asked if he walked under much apprehension of falling forwards? he said he suffered much from it; and replied in the affirmative to the question, whether he experienced any difficulty in restraining himself from getting into a running pace? It being asked, if whilst walking he felt much apprehension from the difficulty of raising his feet, if he saw a rising pebble in his path? he avowed, in a strong manner, his alarm on such occasions; and it was observed by his wife, that she believed, that in walking across the room, he would consider as difficulty the having to step over a pin.[43]

Parkinson's clinical interviewing was not only able to call forth informative responses from patient and carer but the affirmation felt by husband and wife at the grasp of the disease which he revealed in his questioning of them. In confirming the patient's fear of falling – apprehension at the prospect of stepping over small objects ('a rising pebble in his path') – and the tendency to forward running, Parkinson brings to the fore the man's psychological as well as physical vulnerability. The respectful relationship among those present is apparent, the report accommodating several viewpoints: the doctor's keen understanding and desire to learn more, the patient's subjective experience and vulnerability, and the wife's insightful observation of her husband's condition.

In the *Essay*'s subsequent chapters Parkinson sought to anchor the account of the malady in already known modules of human dysmobility. The pathognomonic features necessary to identify and diagnose the Shaking Palsy he located in *Tremor coactus* and *Scelotyrbe festinans*, each of which had been separately recognized in different classificatory schemata of the day. He distinguished the Shaking Palsy from other disorders of trembling and referred to a number of additional cases not included in his own series.[44] One of these in Chapter IV, another street case, concerned a man who had suffered agitation and convulsive movements of the legs following mercury treatment for a venereal affection many years before. 'Full ten years later', Parkinson wrote:

> the unhappy subject . . . was casually met in the street, shifting himself along, seated in a chair; the convulsive motions having ceased, . . . and the limbs having become totally inert . . . and insensible to any impulse of the will . . .[45]

Parkinson was at pains to distinguish this pattern of 'convulsive motions' from those of the Shaking Palsy. In Chapter V he called for trials of new treatments: if there was to be any chance of cure early diagnosis, he believed, was essential. Citing a successful report he found analogous (that of Count de Lordat), he suggested the condition be divided into an early phase – within two years or less of the first referable symptom – and a later phase. There was little hope of alleviating late-stage disease, but early on he believed it might be curable if blood was withdrawn from the upper part of the neck, or if blistering, scarifying and drainage of either side of the upper vertebrae were tried in order to reduce inflammation in the region. He referred to attempts by others to curtail the disease by use of mustard and caustic plasters, reflecting his provisional view that the source of the malady lay in congestion or inflammation of the spinal medulla.

## Medical observation in Parkinson's day

Parkinson believed his occupational position was key to his observational synthesis: 'The disease', he wrote, is of long duration: 'to connect, therefore, the symptoms which occur in its later stages with those which mark its commencement, requires a continuance of observation of the same case or at least a correct history of its symptoms, even for several years.'[46] But it is not only observational continuity that underpinned his achievement. Also important was his ability to select out and situate salient features of daily practice arising from conversation, questioning and prolonged follow-up, and a capacity to embody these narratively in a text that bore witness to evolving patterns of human movements developing slowly over years.

It is difficult to overstate the importance of personal witness in the clinical accounts of Parkinson's day. In his *Commentaries on the History and Cure of Diseases* of 1802, the accomplished physician William Heberden (1710–1801) wrote:

> The notes from which the following observations were collected, were taken in the chambers of the sick from themselves, or from their attendants . . . These notes were read over every month, and such facts, as tended to throw any light upon the history of a distemper, or the effects of a remedy, were entered under the title of the distemper in another book, from which were extracted all the particulars here given.[47]

Heberden's method was to chronicle daily work, mull over its details and abstract from these (in a separate volume) what he found salient about them, supporting his claim of a close match between the contents of the *Commentaries* and everyday clinical experience.[48] Heberden was a Fellow of St John's College Cambridge, the Royal Society and the Royal College of Physicians, the founder-editor of its *Medical Transactions*. Parkinson was a surgeon-apothecary born and brought up in the East End of London, a generation or so Heberden's junior, whose education had been one of apprenticeship to his father and six months' walking the wards of the newly established London Hospital in Whitechapel. Despite these marked differences in social and professional position, Heberden and Parkinson shared interests: each was an early member of the Humane Society (founded in 1774) from which Parkinson had been awarded a silver medal in 1777 for resuscitating a man who had tried to commit suicide; each had published case reports in the *Memoirs* of the Medical Society of London, and had an interest in the effects of lightning on people and on property.[49]

While no direct evidence connects the two men personally, Parkinson is likely to have been familiar with Heberden's *Commentaries*, which appeared in Latin – a language Parkinson read well – in 1802 and which by 1817 had attained wide currency, having appeared in three editions, including an English translation. Prior to the publication of the *Essay*, Heberden's textbook therefore offers a pertinent gauge of contemporary clinical understanding of disorders of shaking. 'A trembling of the hands or a shaking of the head', Heberden wrote:

> may be judged to have some alliance with paralytic, and apoplectic maladies; yet it has often continued for a great part of a person's life, without any appearance of further mischief; and therefore, if it have a tendency to palsies, it is a very remote one. . . . Hypochondriac persons are troubled with frequent fits of it; hard drinkers have it continually; and some degrees of it usually attend old age. This, like other affections of the nerves, is greatest in a morning, and is aggravated by any disturbance of mind. Coffee and tea make the hands of some persons shake; and yet I have known strong coffee drunk every day for forty years, by one who was remarkable for the steadiness of his hands even in extreme old age.[50]

This account is neither closely observed nor rendered in particular clinical detail and in pointing to associations of trembling with other conditions – old age, hypochondria, heavy intake of alcohol and coffee – Heberden did not allude to any new associations between shaking and lifestyle. No relationship between trembling and disorders of posture or gait are singled out and no reference made to classifications of tremor, such as those of Boissier De Sauvages or of William Cullen, both of whom Parkinson referred to in the *Essay*. De Sauvages's *Nosologia Methodica* of 1762 had defined tremor as a 'side-to-side movement without a feeling of being cold', and had listed 19 species of it in his fourth class of disease, 'Spasm and Convulsive Malady'.[51] Cullen's *Nosology* of 1800 defined tremor as 'an alternate and frequent motion of a joint to-and-fro', 15 species of which were outlined under his second class of disease, the 'Neuroses'. Both authors referred to this sort of shaking as *Tremor coactus*.[52]

The other component of Parkinson's definition, *Scelotyrbe festinans*, a tottering or hurrying gait, had been described in 1772 by Hieronymus Gaubius, Professor of Chemistry and Medicine at Leiden. Gaubius reported that 'it happens that the muscles by an involuntary agility and impetuosity accelerate their motions and hurry away against the determination . . . [he had] seen those who could run but not walk'.[53] De Sauvages had identified several forms of *scelotyrbe* but associated none of them with trembling: *Scelotyrbe of Galen*, he noted, was 'an impediment which prevents people walking in a straight line', which he likened to Thomas Sydenham's description of Saint Vitus's Dance. *Scelotyrbe festinans* De Sauvages defined as a disorder of people who can locomote only in haste, illustrating this with the case of a fast-walking painter who was unable to divert his path 'to the right or left, and on meeting an obstacle became almost fixed until, little by little, by shifting his position, he once again began to move forward in a straight line'.[54]

Parkinson's central claim was that a compound of these disorders of movement was a single disease: 'cases appear to belong to the same species: differing from each other, perhaps, only in the length of time which the disease had existed, and the stage at which it had arrived'.[55] His use of 'species' in this context is typical of a late-eighteenth-century physician-naturalist. Sydenham earlier had urged doctors to identify and classify diseases with the same exactitude which naturalists brought to bear on the characteristics of different creatures.[56] As one of the most experienced fossilists of his day and an adept in fossil types,[57] Parkinson would have recognized the Shaking Palsy to be a 'species of disease' in the same sense of the term's usage in relation to newly discovered plants and fossils of the period.[58]

## Urban observation

The *Essay* reported observations on people who consulted Parkinson as an apothecary and on those he saw and questioned on the street. Throughout the eighteenth century the busy-ness of London streets had increased as the metropolitan population grew and commerce expanded. A great deal of trade and transaction took place on the streets which accommodated many occupations.[59] The contrast between urban streets and those of the countryside became the subject of observation and comment.[60] In the mid-eighteenth century Dr Johnson found street postures, bearings and gaits signalled class, attitude and intention:

> Crowds that fill the streets . . . with motion . . . [are] difficult to behold without contempt and laughter . . . awakened by the sprightly trip, the stately stalk, the formal strut, the lofty mien; by gestures intended to catch the eye, and by looks elaborately formed as evidences of importance.[61]

The shape and attitude of human bodies in an urban landscape conveyed moral and morbid temperament,[62] and underpinned the physiognomist Casper Lavater's advice to travellers newly arrived in cities to take to the streets to gain the measure of their hosts:

> I would . . . plant myself in a public walk, or at the crossing of streets . . . would first consider the general form of the inhabitants, their height, proportion, strength, weakness, motion, complexion, attitude, gesture and gait. I would observe them individually; see, compare, close my eyes, trace in imagination all I had seen, open them again, correct my memory, close and open them alternately; would study for words, write, and draw, with a few determinate traits, the general form.[63]

Lavater's reference to combining sight, imagination, word retrieval, memory and the sketching of bodily appearances, in order fully to take in street appearances, is telling. In 1809 the short-lived journal *The London Medical and Surgical Spectator* announced that 'Medicine is, more than any other, the science of speculation',[64] speculation that took Parkinson beyond the confines of hospital, dispensary and the apothecary's shop and into the larger observatory of the urban landscape. Visitors to London were struck by the animation of its street life: in 1782, for example, the German author and essayist Karl Philipp Moritz included in his letters home vivid impressions of street scenes and descriptions of people's bearing and clothing.[65] Ten years after the *Essay*'s publication the poet, Heinrich Heine, visiting the capital, wrote:

> Send a philosopher to London . . . Send a philosopher thither and set him at the corner of Cheapside, and he will learn more than from all the books of the last Leipzig fair; and as the waves of human beings roar about him there will arise before him a sea of new thoughts . . . and . . . reveal to him the most hidden secrets of the social order.[66]

These reactions bring out how revelatory London street life was to visitors.[67] Cheapside was a fashionable, opulent thoroughfare, an 1823 view confirming its splendour and the degree of social mingling that took place on its pavements, which sported a mix of body types, shapes, postures and gaits. Figure 27.2 shows people stooping, lugging, pushing, pulling, carrying and stepping forth. London at this time had no public transport system and many of the streets – even major thoroughfares – were congested with foot and wheeled traffic and were generally much less orderly than this view of Cheapside suggests.[68]

The streets abutting Hoxton Square, where Parkinson lived and worked, were busy with activities of a mixed population. Hoxton had an unusually high number of almshouses and workhouses in its locality so many people on its streets would have been elderly (see Figure 27.1).[69] His home and surgery stood on a corner at Number 1 Hoxton Square and had a view to the front across the square and at its flank and rear over the eastern approach to Hoxton Market (see Figure 27.1). The apothecary's shop and surgery were on the ground floor with family accommodation above. To the north were market gardens, orchards and open fields and farms, this part of London being subject to daily tidal movements of people who entered the village by foot to purchase goods at Hoxton Market on their way to the city to work, retracing their footsteps at the end of the day.

The nearest major road junction to Hoxton Square, shown in Figure 27.3, was the four-way intersection of Shoreditch High Street. This engraving (c.1800) is dotted with portraits of pedestrians of distinctive posture: a man stooping, others crouching, carrying buckets and pushing a trolley. Another view of the same church (Figure 27.4) shows a figure with a dragging gait (with his back towards us) and a sailor strutting forth. In noticing postures and gaits in and around Hoxton, Parkinson was partaking in an observational urban culture shared by visual artists well practised in interpreting urban corporeality.

It was a culture shared by writers who reported on the capital's life in the tradition of early journalists such as Ned Ward, Richard Steele and Joseph Addison. In the novel *Humphry Clinker* (1771), the Glaswegian surgeon and novelist, Tobias Smollett evoked the chaotic impression urban life could make on country-folk when first arriving in the capital: people were 'rambling, riding, rolling, rushing, jostling, mixing, bouncing, cracking, and crashing in one vile ferment . . . one would imagine they were impelled by some disorder of the brain, that will not suffer them to be at rest'.[70] Urban appearances were also a pungent presence in Laurence Sterne's *A Sentimental Journey* (1768),[71] a novel structured around 'tableaux or "scenes" of those in distress'[72] that engendered 'feelings of a tender heart, the sweetness of compassion, and the duties of humanity'.[73] According to Henry Mackenzie, the author of *The Man of Feeling* (1771), sentimental

*Figure 27.2* Cheapside in 1823. Engraved by T.M. Baynes from a drawing by W. Duryer.
Author's Collection, author's own copyright.

texts fashioned emotional links between author and reader by depicting details of life and 'that pleasure which is always experienced by him who unlocks the springs of tenderness and simplicity'.[74] The 'man of feeling' was 'a man of seeing . . . moving through society', whose 'observation and introspection' were directed to objects of social inquiry.[75] More than just a 'humanitarian narrative', the *Essay*'s focus on special moments of poignancy can be read as a series of sentimental episodes whose truth to life was echoed time and again in the narratives of its six 'Illustrative Cases'.

*Figures 27.3–27.4* St Leonard's Church, Shoreditch, London, circa 1800. Author's
Collection, author's own copyright.

504

Jessica Riskin has argued that some modes of French science intermingled sensibility and empiricism in ways that acknowledged the felt responsiveness of scientists towards the objects of their inquiries, a responsiveness which generated observations 'not from sensory experience alone, but from a combination of sensation and sentiment'.[76] Parkinson's way of looking was not that of a spectator disconnected from the objects of his interest or subordinated to the taxonomic and voyeuristic drives of Walter Benjamin's later characterization of 'botanizing on asphalt',[77] but was rooted in the urban and medical geography of his working life and practice.[78] That the vast preponderance of people he identified with the Shaking Palsy were men was probably not the result of any lack of women out and about on the streets of London and the neighbourhoods of Hoxton.[79] These outdoor dimensions of Parkinson's practice have gone unnoticed in studies of the role of eighteenth- and nineteenth-century cities in supporting the conduct of scientific investigations, which have largely focused on their institutions.[80] To the museums, libraries and literary and philosophical societies of cities which provided the settings and intellectual contexts for experiments, expositions and public demonstrations of newly acquired knowledge should be added the metropolitan streetscape as a locus of clinical observation, reasoning and literary record in the development of nineteenth-century medicine.[81]

## Conclusion

Parkinson's *Essay* identified more than a loose collection of symptoms. The Shaking Palsy was 'a species of disease', a claim that has since been sustained over two centuries. The *Essay* reframed already recognized clinical phenomena and demarcated them from similar manifestations of different conditions. But without a proven aetiology or a causal account Parkinson could buttress his claim for the existence of 'a species' of disease only by binding together its diverse symptoms in an unfolding narrative shared by sufferers. Thus the generic account given in the 'History' entailed all the disease-related details of each case description, nothing of salience occurring in the cases that had not been pre-enacted – in a sense pre-destined – by the 'History'. The 'History' unveiled a pathological agency working itself out in sufferers, a 'flow of the nervous influence to the affected parts' through which 'the disease proceeds', a phrase Parkinson repeated on six occasions. But without an anatomico-pathological seat or originating process it was the condition's highly distinctive temporal trajectory that integrated its diverse symptomatology and manifestations, and conferred on it the identity of a distinct clinical ensemble.

The concept of disease set out in the *Essay* is of an abstract entity or 'species' operating through the grip it exerts on the invented composite of the sufferer in the 'History', who occupies a victimized position that stands for *all* sufferers of the condition. Whereas help-seeking is fully apparent in the *Essay*'s cases, the notional sufferer is abstracted from the world of medicine and treatment. Though he remains sensible and aware of his physical deterioration, and attempts to adapt to its worst ravages, he is different from the cases which are rooted in real people, signified by snippets of biographical information – age, occupation, aspects of their individual histories – and by reported conversation with Parkinson himself. The sufferer described in the 'History', by contrast, is not individuated and, apart from gender, lacks biographical referents. His role is to make fully apparent the effects of the malady by exemplifying a process Foucault referred to as 'subtraction', removal of individual and particular qualities to leave 'the negative element, the accident of the disease'. According to Foucault, medical understanding takes account of individual sufferers only to the extent that it places them 'in parentheses'.[82] Yet precisely because Parkinson's portrait of the malady lacks a convincing pathological mechanism, the *Essay* remains focused on people observed from many angles and appreciated as if from within, generating understanding in vernacular terms in a manner comparable to eighteenth- and nine-teenth-century lay writings on illness.

Mary Fissell has highlighted the extent to which autobiographical diaries and memoirs of the period showed great resourcefulness on the part of sufferers who attended closely to sources of information about their illnesses, and delved 'back in time to understand what may have befallen them', to glean any information that might prove useful in their search for treatments.[83] Parkinson's case descriptions do not 'overwrite'[84] such sufferers' perspectives or deny them legitimacy as implied by Foucault's formulation. Rather, the *Essay* succeeds as a *somatic roman* of profound human dysmobility, told through the bodies and accounts of particular people. In depicting something of the fabric of everyday life and the activities and adaptations of sufferers, Parkinson has his cases of the Shaking Palsy write their experience into his account,[85] which conferred on the concept of this particular disease a distinctive narrative pattern that has endured.

## Acknowledgements

This chapter is a development of my paper 'Urban Observation and Sentiment in James Parkinson's *Essay on the Shaking Palsy*', published in *Literature and Medicine*, 2014, vol. 32, 74–104. It owes much to my 2011 Fitzpatrick Lecture at the Royal College of Physicians, London, and invitations to present the work at the Universities of Oxford, Harvard, Bristol, the London School of Economics, and the Institute of Neurology at Queen Square, University College London. I thank Andrew Lees of the National Hospital, Queen Square, for much guidance over the years and for comments on an earlier draft. I am indebted to Ruth Richardson who has commented on earlier drafts and has been my guide, mentor and much more.

## Notes

1 Parkinson's name is invoked eponymously not only as 'Parkinson's disease' but also in the term 'Parkinsonian facies' or 'Parkinsonian sign', referring to the expressionless face characteristic of the condition and in 'Parkinsonism', meaning clinical signs of Parkinson's disease. Gerald Stern claims the eponym has made the *Essay*'s author the most famous of neurologists – G. Stern, 'The World's Best Known Neurologist?', *Practical Neurology*, 2011, vol. 11, 312–15.

2 J. Parkinson, *An Essay on the Shaking Palsy*, London: Sherwood, Neely and Jones, 1817.

3 David Turner charts a shift in eighteenth-century attitudes towards disability from being seen as marks of 'portent' to marks of 'pathology', a shift he characterizes as 'from religious to medical', in D. Turner, 'Introduction' in D. Turner, *Disability in Eighteenth Century England*, London: Routledge, 2012, p. 5–15 at 10.

4 For earlier accounts of paralysis agitans see http://perspectivesinmedicine.cshlp.org/content/1/1/a008862.full (accessed 25/08/2015) and M. Critchley, 'Introduction', in M. Critchley (ed.), *James Parkinson (1755–1824)*, London: MacMillan & Co., 1955, pp. 1–143,

and P. Kempster, B. Hurwitz, and A. Lees, 'A New Look at James Parkinson's Essay on the Shaking Palsy', *Neurology*, 2007, vol. 69, 482–5.

5  T. W. Laqueur, 'Bodies, Details, and the Humanitarian Narrative', in L. Hunt (ed.), *The New Cultural History*, Berkeley: University of California Press, 1989, p. 178. See also G. Sill, *The Cure of the Passions and the Origins of the English Novel*, Cambridge: Cambridge University Press, 2001, and P. M. Logan, *Nerves and Narratives: A Cultural History of Hysteria in Nineteenth-Century British Prose*, Berkeley: University of California Press, 1997.

6  R. Smith, *Being Human: Historical Knowledge and the Creation of Human Nature*, New York: Columbia University Press, 2013, p. 174.

7  N. Pethes, 'Telling Cases: Writing against Genre in Medicine and Literature', *Literature and Medicine*, 2014, vol. 32, 24–45. Historical narrative plays an important role in the novels of sensibility such as: S. Richardson, *Pamela*, London, 1741, and S. Richardson, *Clarissa*, London: S. Richardson, 1747–8.

8  B. Hurwitz, 'Narrative [in] Medicine', in P. Spinozzi, and B. Hurwitz (eds), *Discourses and Narrations in the Biosciences*, Göttingen: Vandenhoeck & Ruprecht Unipress, 2011, pp. 73–87. See also M. Holmes, 'A Narrative Concerning the Success of Pendulum-Watches at Sea for the Longitudes', *Philosophical Transactions of the Royal Society*, 1665, vol. 1, 13–15.

9  C. Zelle, 'Experiment, Observation, Self-Observation, Empiricism and the "Reasonable Physicians" of the Early Enlightenment', *Early Science and Medicine*, 2013, vol. 18, 453–70.

10  Parkinson preferred the term 'Shaking Palsy' to that of 'Paralysis Agitans', the Latin name for the condition. Parkinson, *Essay*, p. 1.

11  Although he occasionally used the designation 'physician' in some publications (see R. Richardson and B. Hurwitz, 'James Parkinson, Physician of Hoxton', http://spitalfieldslife.com/2013/11/20/james-parkinson-physician-geologist/ [accessed 25/08/2015]) by training and occupation Parkinson was a surgeon-apothecary. The frontispiece of Parkinson's *Essay* identified him as 'Member of the College of Surgeons'.

12  L. G. Rowntree, 'James Parkinson', *Bull Johns Hopkins Hosp.*, 1912, vol. 23, 33–45; W. H. McMenemey, 'James Parkinson (1755–1824): A Biographical Essay', in M. Critchley (ed.), *James Parkinson (1755–1824)*, London: MacMillan & Co, 1955, pp. 1–143; S. Roberts, *James Parkinson (1755–1824)*, London: The Royal Society of Medicine Press, 1977; C. Gardner-Thorpe, *James Parkinson 1755–1824, and a Reprint of 'The Shaking Palsy' by James Parkinson, Originally Published 1817*, Devon: Department of Neurology, Royal Devon and Exeter Hospital, 1987; A. D. Morris, 'James Parkinson', *Lancet*, 1955, 761–3; A. D. Morris, *James Parkinson: His Life and Times*, Boston: Birkhauser, 1989.

13  Parkinson, *Essay*, p. i.

14  Anonymous, 'An Essay on the Shaking Palsy; by James Parkinson, Member of the Royal College of Surgeons. 8 vols, pp. 66. Sherwood, Neely and Jones, 1817', *The London Medical and Physical Journal*, 1817, 38.

15  J. Cooke, *A Treatise on Nervous Diseases*, London: Longman, 1820–21, vol. 2, p. 207.

16  T. Graham, *Modern Domestic Medicine*, London: Simpkin and Marshall, 1827, pp. 479–80.

17  J. Elliotson, 'Clinical Lecture on Paralysis Agitans', *The Lancet*, 1830, 119–123.

18  M. Hall, 'Lectures on the Theory and Practice of Medicine', *The Lancet*, 1838, 41.

19  M. Romberg, *A Manual of the Nervous Diseases of Man*, trans. Edward H. Sieveking, London: New Sydenham Society, 1853, pp. 233–4.

20  A. Trousseau, *Lectures on Clinical Medicine*, trans. Victor Bazire, London: New Sydenham Society, 1868, vol. 1, 440–50.

21  J-M. Charcot, and E. Vulpian, 'La paralysie agitante', *Gazette Hebdomadaire Med Chir*, 1861, 765–8, 816–23, and 1862, 54–64.

22  J-M. Charcot, 'Du tremblement dans la maladie de Parkinson', in J-M. Charcot (ed.), *Oeuvres Complètes*, Paris: Bureaux du Progrès Médicale, 1880, vol. 1, pp. 414–42. Later physicians recognized some clinical features of the malady not recorded by Parkinson, contributions documented by E. D. Louis, 'Paralysis Agitans in the Nineteenth Century' in S. A Factor, S. A. and W. J. Weiner, (eds), *Parkinson's Disease: Diagnosis and Clinical Management*, New York: Demos Medical Publishing 2002, pp. 13–17 and by K. Hesselink, and M. Jan, 'Evolution of Concepts and Definitions of Parkinson's Disease since 1817', *Journal of the History of the Neurosciences*, 1996, vol. 5, 200–7; See also J. Elliotson, 'Clinical Lecture', *The Lancet*,

1831, 557; J. Elliotson, 'Clinical Lecture', *The Lancet*, 1831, 599; M. Hall, *On the Diseases and Derangements of the Nervous System*, London: H Baillière, 1841, pp. 320–1; and entries on paralysis agitans in *Lectures on Diseases of the Nervous System*, trans. G. Sierson, Philadelphia: HC Lea, 1879, pp. 105–27, especially J-M. Charcot, 'Leçon 21', June 12, 1888 and 'Leçons du Mardi', Policlinique à la Salpêtrière, Paris: Bureaux du Progrès Médicale, 1888.

23 C. Goetz, 'Historical Issues and Atypical Parkinsonian Disorders', in S. A. Factor, and W. J. Weiner (eds), *Parkinson's Disease: Diagnosis and Clinical Management*, New York: Demos Medical Publishing, 2002, pp. 19, 19–26. See also J-M. Charcot, 'Parkinson's Disease: A Case Without Tremor', in *Charcot the Clinician, The Tuesday Lessons*, trans. C. Goetz, New York: Raven Press, 1987, pp. 124–5 from the original by J-M. Charcot, *Leçons du Mardi à la Salpêtrière 1887–1888*, p. 155.

24 Parkinson, *Essay*, p. 1.

25 P. Koehler, and A. Keyser, 'Tremor in Latin Texts of Dutch Physicians: 16th–18th Centuries', *Movement Disorders*, 1997, vol. 12, 798–806.

26 This is a condensed version of the *Essay*'s 'History'.

27 Parkinson, *Essay*, pp. 3–9.

28 Ibid., p. 61.

29 I am indebted to John Holmes for suggesting this close reading of the *Essay* at this point in the text.

30 R. Knill, *The Happy Death-Bed*, London: Religious Tracts Society, 1836, p. 1.

31 The 'History' in Parkinson's *Essay* may be thought in the clinical realm to be a successor to the *Historia Naturalis* of the early-modern period, which Pomata and Siraisi characterize as aspiring 'to a truthful . . . narrative of the results of an inquiry' that straddles 'the distinction between human and natural subjects, embracing accounts of . . . the natural world as well as the record of human action . . . with . . . reference to the knowledge of particulars': G. Pomata and N. Siraisi, 'Introduction' in G. Pomata and N. Siraisi (eds), *Historia: Empiricism and Erudition in Early Modern Europe*', Cambridge, MA: MIT Press, 2013, pp. 1–38. See P. Jalland, *Death in the Victorian Family*, Oxford: Oxford University Press, 1996, p. 26; J. C. Lavater, *Essays on Physiognomy*, trans. C. Moore, London: Locke, 1793–4, p. 133; Knill, *Happy Death*, p. 1. R. Porter, *Flesh in the Age of Reason*, London: W. W. Norton, 2004, pp. 220–6.

32 Parkinson, *Essay*, p. 9.

33 Ibid., pp. 9–10.

34 Ibid., p. 11.

35 Ibid., p. 32.

36 Ibid., p. 12.

37 Ibid., p. 13.

38 Ibid., pp. 13–14.

39 Instruments such as telescopes were used by eighteenth-century artists to examine urban scenes, and when visiting London in the mid-eighteenth-century Canaletto used a *camera obscura* to trace outlines of human figures and buildings. See T. Hitchcock, *Down and Out in London*, London: Hambledon, 2004, p. 14. Such devices had a literary presence, signifying close-up views of goings-on from a comfortably safe and unnoticed distance – P. Egan, *LIFE IN LONDON; or, The Day and Night Scenes of Jerry Hawthorn, Esq., and his elegant friend Corinthian Tom, accompanied by Bob Logic, the Oxonian, in their Rambles and Sprees through the Metropolis*, London: Sherwood, Neely, & Jones, p. 46.

40 Parkinson, *Essay*, p. 16.

41 Ibid., p. 17.

42 Manoeuvres of this sort – devised by sufferers craving a moment's stillness – are familiar to patients and doctors in the present-day. See Kempster et al., 'A New Look', pp. 482–5.

43 Parkinson, *Essay*, pp. 17–18.

44 Note that 'series' is not a term Parkinson uses in the *Essay*. His cases are disposed in time and space in the sequential order of his clinical and observational encounters. For a consideration of the notion of scientific series from the seventeenth to the end of the long nineteenth century (though with little focus on medicine), see N. Hopwood, S. Schaffer and J. Secord, 'Seriality and Scientific Objects in the Nineteenth Century', *History of Science*, 2010, vol. xlviii, 251–85.

45  Parkinson, *Essay*, p. 37.

46  Ibid., p. ii.

47  W. Heberden, *Commentaries on the History and Cure of Diseases*, trans. William Heberden (Junior), London: T. Payne, 1802, pp. vii–viii.

48  Compare Heberden's approach to that of 'mastering on paper' in V. Hess and A. Mendelsohn, 'Case and Series: Medical Knowledge and Paper Technology, 1600–1900', *History of Science*, 2010, vol. xlviii, 288–314.

49  L. A. F. Davidson, 'Founders and Benefactors of the Royal Humane Society (1774–*c*.1808)', in *Oxford Dictionary of National Biography*, Oxford: Oxford University Press, 2013, http://www.oxforddnb.com/view/theme/101001?backToResults=%2Fsearch%2Frefine%2F%3Fdocstart=1%26themesTabShow=true (accessed 25/08/2015); and T. Lawrence, 'A Letter from Thomas Lawrence, to William Heberden, Concerning the Effects of Lightning, in Essex-Street, on the 18th of June, 1764', *Philosophical Transactions*, 1764, vol. 54, 235–8; J. Parkinson, 'Some Account of the Effects of Lightning, by Mr J. Parkinson, of Hoxton', *Memoirs of the Medical Society of London*, 1789, vol. 2, 490–500.

50  Heberden, *Commentaries*, 1802, p. 429.

51  B. De Sauvages, *Nosologie Methodique*, trans. M. Gouvion, Lyon: Jean-Marie Bruyset, 1772, vol. 4, pp. 36–50.

52  W. Cullen, *Nosology*, trans. William Cullen, London: Robinson, 1800, p. 105.

53  H. D. Gaubius, *The Institutions of Medicinal Pathology*, trans. Charles Erskine, London: Elliot & Cadell, 1778, p. 278.

54  De Sauvages, *Nosologie*, vol. 4, pp. 146–50.

55  Parkinson, *Essay*, p. 18.

56  T. Sydenham, *The Entire Works of Dr Thomas Sydenham*, trans. John Swan, London: Cave, 1753.

57  J. Parkinson, *Organic Remains of a Former World: Examination of the Mineralized Remains of the Vegetables and Animals of the Antediluvian World*, London: Sherwood, Neely and Jones, 1804–11, 3 vols; Martin Rudwick notes that this text is 'the first substantial illustrated book on fossils to be published in England'. See M. Rudwick, 'Encounter with Adam, or at least the Hyaenas: Nineteenth-Century Visual Representations of the Deep Past', in M. Rudwick (ed.), *New Science of Geology: Studies in Earth Sciences in the Age of Revolution*, Farnham: Ashgate, 2004, pp. 231–51 at 246.

58  B. Glass, 'Eighteenth-Century Concepts of the Origin of Species', *Proceedings of the American Philosophical Society*, 1960, vol. 104, 227–34.

59  D. George, *London Life in the Eighteenth Century*, London: Routledge, 1996, pp. 155–214.

60  Urban spectatorship refers to a multiplicity of looking practices. See: W. Mitchell, *Picture Theory*, Chicago: University of Chicago Press, 1994, pp. 11–34; P. Joyce, *The Rule of Freedom*, London: Verso, 2003, p. 199; D. Brand, *The Spectator and the City in Nineteenth-Century American Literature*, Cambridge: Cambridge University Press, 1991, p. 29; Anonymous, *Real Life in London or, the Rambles and Adventures of Bob Tallyho, Esq. and his Cousin, the Hon. Tom Dashall, through the Metropolis*, London: Jones, 1821, vol. 1, pp. 418–20, in A. O'Byrne, 'Art of Walking London Streets', *Romanticism*, 2008, vol. 14, 94–107.

61  *The Rambler* 1751, no. 179 (December 3rd) Reprinted in 1761, vol. 4, p. 94 accessed in F. Brady, and W. K. Wimsatt (eds), *Samuel Johnson: Selected Poetry and Prose*, Berkeley: University of California Press, 1977, p. 217; P. Tankard, 'Johnson and the Walkable City', *Eighteenth-Century Life*, 2002, vol. 32, 1–22; and J. Guldi, 'Walking and the Digital Turn: Stride and Lounge in London, 1808–1851', *Journal of Modern History*, 2012, vol. 84, 116–44.

62  Lavater, *Essays*, pp. 218–19.

63  Ibid.

64  *The London Medical and Surgical Spectator, or Monthly Register of Medicine in its Various Branches*, 1809, vol. 2, 9.

65  K. P. Moritz, *Travels, Chiefly on Foot, Through Several Parts of England, in 1782: A Literary Gentleman of Berlin*, trans. A Lady, London: Robinson, 1795, pp. 23–4.

66  G. Karpeles (ed.), trans. G. Cannan, *Heinrich Heine's Memoirs: From His Works, Letters and Conversations*, London: Heineman, 1910, pp. 192–3.

67  These observations evoke Foucault's notion of 'the bright, distant, open naivety of the gaze. Hence the two great mythical experiences on which the philosophy of the eighteenth-century

had wished to base its beginning: the foreign spectator in an unknown country, and the man born blind restored to light' – M. Foucault, *The Birth of the Clinic*, trans A. M. Sheridan, London: Tavistock Publications, 1976, p. 65. See also D. Hughson, *Walks Through London*, London: Sherwood, Neely & Jones, 1817, p. 77.

68 For vivid descriptions of street life in Georgian England, see S. Dickie, *Cruelty and Laughter*, Chicago: University of Chicago Press, 2011, p. 69.

69 Hoxton in Parkinson's day was an area known for its almshouses, workhouses and private mad-houses. See E. Murphy, 'Mad Farming in the Metropolis. Part 1: A Significant Service Industry in East London', *History of Psychiatry*, 2001, vol. 12, 245–82; and A. D. Morris, *The Hoxton Madhouses*, Cambridge: printed privately, 1958; and E. Murphy, 'Mad Farming in the Metropolis. Part 2: The Administration of the Old Poor Law of Insanity in the City and East London 1800–1834', *History of Psychiatry* 2001, vol. 12, 405–30; C. Miele, *Hoxton: Architecture and History over Five Centuries*, London: The Hackney Society, 1993.

70 T. Smollett, *Humphrey Clinker*, Oxford: Oxford University Press, 2009, p. 88.

71 L. Sterne, *A Sentimental Journey*, Oxford: Oxford University Press, 2008, 49.

72 B. Vickers, 'Introduction', in B. Vickers, (ed.), *Henry Mackenzie, The Man of Feeling*, Oxford: Oxford University Press, 2001, pp. vii–xxiv.

73 J. Hanway, *A Sentimental History of Chimney Sweepers*, London: Dodsley, 1785; G. Pinckard, *Notes on the West Indies, Including Observations Relative to the Creoles and Slaves of the Western Colonies*, London: Bladum, 1816, p. 52.

74 Horst Drescher (ed.), *Henry Mackenzie, Letters to Elizabeth Rose of Kilravock: On Literature, Events and People 1768–1815*, London: Oliver and Boyd, 1967, pp. 16–17.

75 C. Otter, *The Victorian Eye: A Political History of Light and Vision in Britain, 1800–1910*, Chicago: University of Chicago Press, 2008, p. 49.

76 J. Riskin, *Science in the Age of Sensibility: The Sentimental Empiricists of the French Enlightenment*, Chicago: University of Chicago Press, 2002, p. 5.

77 W. Benjamin, *Charles Baudelaire: A Lyric Poet in the Era of High Capitalism*, trans., Harry Zohn, London: New Left Books, 1979, p. 36.

78 A. Tomkins, 'Who Were His Peers? The Social and Professional Milieu of the Provincial Surgeon-Apothecary in the Late Eighteenth Century', *Journal of Social History*, 2011, vol. 44, pp. 915–35.

79 There seems little evidence that women were not a populous, day-light presence on London streets and Parkinson did not exclude the possibility of women suffering from the condition. In Chapter V he referred to a female case. See Guldi, 'Walking', pp. 116–44, and J. R. Walkowitz, *City of Dreadful Delight: Narratives of Sexual Danger in Late-Victorian London*, London: Virago, 1992.

80 S. Dierig, J. Lachmund, and J. Andrew Mendelsohn, *Science and the City*, Chicago: University of Chicago Press, 2003, pp. 1–19.

81 What, today, is referred to as 'field neurology' and 'bystander' or 'passer-by' diagnosis Parkinson employed in the context of the eighteenth- and nineteenth-century street. In the seventeenth century Sydenham had observed choreiform disorders in crowds in the open air, and had described the complex behavioral gestures of particular sufferers, but his outdoor observations were not made in the individualized, case-based manner in which Parkinson characterized dysmobilities on the streets of Hoxton. See T. Sydenham, *The Entire Works*, pp. 555–6. 'Field neurology' today continues to refer to neurological work undertaken in the open air, in military or other settings, away from recognizable institutes of medicine. See D. C. Gajdusek (ed.), *Discovery and Original Investigations on Kuru: The Smadel-Gajdusek Correspondence, 1955–1958*, Bethesda: NIH, 1975; and W. E. Mitchell, 'The Ethics of Passer-By Diagnosis', *The Lancet*, 2008, 85–7.

82 Foucault, *The Birth*, p. 8.

83 M. Fissell, *Patients, Power and the Poor in Eighteenth Century Bristol*, Cambridge: Cambridge University Press, 1991, p. 35.

84 Fissell, *Patients*, p. 150.

85 M. Class, 'Introduction: Medical Case Histories as Genre: New Approaches', *Literature and Medicine*, 2014, vol. 32, vii–xvi.

## Select bibliography

Charcot, J-M. and Vulpian, E., 'La paralysie agitante', *Gazette Hebdomadaire Med. Chir.*, 1861, 765–8, 816–23; 1862, 54–64.

Charcot, J-M., 'Du tremblement dans la maladie de Parkinson', in *Œuvres Complètes*, Paris: Bureaux du Progrès Médicale 1 (1880), pp. 414–20.

Charcot, J-M., *Leçons du Mardi: Policlinique à la Salpêtrière* {Leçon 21: June 12, 1888}, Paris: Bureaux du Progrès Médicale, 1888, 155–88.

Critchley, M., ed., *James Parkinson (1755–1824)*, London: Macmillan, 1955.

Goetz, C., 'Historical Issues and Atypical Parkinsonian Disorders', in *Parkinson's Disease: Diagnosis and Clinical Management*, ed. Factor, S. A. and Weiner, W. J., New York: Demos Medical Publishing, 2002, pp. 19–26.

Hesselink, J., 'Evolution of Concepts and Definitions of Parkinson's Disease since 1817', *Journal of the History of the Neurosciences*, 5, 1996, 200–207.

Hurwitz, B., 'Urban Observation and Sentiment in James Parkinson's *Essay on the Shaking Palsy*', *Literature and Medicine*, 2014, vol. 32, 74–104.

Koehler, J. P. and Keyser, A., 'Tremor in Latin Texts of Dutch Physicians: 16th–18th Centuries', *Movement Disorders*, 1997, vol. 12, 798–806.

Louis, E. D., 'Paralysis Agitans in the Nineteenth Century', in *Parkinson's Disease: Diagnosis and Clinical Management*, ed. Factor, S. A. and Weiner, W. J., New York: Demos Medical Publishing 2002, pp. 13–17.

McMenemey, W. H., 'James Parkinson: A biographical essay', *James Parkinson (1755–1824)*, ed. Critchley, M. London: Macmillan, 1955, pp. 1–143.

Morris, A. D., *James Parkinson: His Life and Times*, Boston: Birkhauser, 1989.

Mullan, J., *Sentiment and Sociability: The Language of Feeling in the Eighteenth Century*, Oxford: Oxford University Press, 1988.

Parkinson, J., *An Essay on the Shaking Palsy*, London: Sherwood, 1817.

Richardson, R. and Hurwitz, B., 'James Parkinson, Physician of Hoxton', http://spitalfieldslife.com/2013/11/20/james-parkinson-physician-geologist/ (accessed 25/08/2015).

Trousseau, A., 'On paralysis agitans', in *Lectures on Diseases of the Nervous System*, trans. Sierson, G., Philadelphia: HC Lea, 1879.

# 28

# DIGITAL NARRATIVES

## Four 'hits' in the history of migraine

*Katherine Foxhall*

I blog. I tweet. I email. I log in, search, sort, browse, download, type, copy, paste, hyperlink, delete, type again. Strange to say, then, that until recently, I had not considered myself to be a 'digital historian'. A social, cultural, and medical historian certainly, but not digital. As I began to contemplate how I might begin to shape my research project on the history of migraine into a book I realised that the sources that I had found, read and worked with online had changed the scope of the project, my understanding of the people in the past about whom I wrote, and the possibilities for the shape of the historical narrative that would result. Of course, all projects develop in ways that cannot be envisaged at the start, but the availability of digitised material has sent a project that I had initially conceived as modern in scope, creeping ever further back through the centuries. A stream of highly fragmented, but extremely rich sources, has recently become navigable through online catalogues, digitisation projects and Google searches. Alongside the archival research in medical collections that is at the heart of the project, images of glittering medieval and early-modern manuscripts – now freely available in high resolution – have repeatedly demanded my attention.

Since the millennium there has been an explosion of digitised information and material. Apart from scholarship in journals and books, these include catalogues and finding aids, digital reproductions of original documents and the 'born-digital' material that awaits historians of the future.[1] It is now possible to search thousands of items, and millions of pages, with a few keyboard-strokes and mouse-clicks. Keyword full-text searches instantly sift an otherwise unmanageable mass of information. Historians of medicine have benefited from some of the most generously funded and institutionally supported programmes of digitisation. The National Library of Medicine in the USA and the Wellcome Library in the UK provide single hubs for accessing a wealth of digital medical history resources. The entire back catalogues of *The Lancet* and *BMJ* are online. Amadeo.com promotes free access to a wide range of medical journals. High-profile medical libraries and archives contribute material to the online Medical Heritage Library. Hosted by the Internet Archive, by August 2015 the MHL collections numbered nearly 100,000 items. In July 2014, the Wellcome Library announced its intention to digitise 15 million pages in partnership with academic and medical college libraries, with funding from the Higher Education and Funding Council for England and JISC.[2] From huge international collaborations to more modest projects showcasing material in individual archives, digitisation projects have opened up an unprecedented diversity of sources that would have been beyond the scope of previous research projects. Domestic recipe books, case notes, private letters, image collections, film and sound archives, newspapers, provincial medical journals, pharmaceutical advertising, institutional records, even court proceedings can now be researched quickly and productively.

The creation of online resources – particularly when they relate to named individuals – raises important ethical issues around privacy and use.[3] Paradoxically, it seems that digitisation will hold the greatest challenges for historians of twenti-eth-century medicine. The UK-based Historic Hospitals Admission Records Project (HHARP) has indexed 120,000 admission records to four children's hospitals, including 8,000 pages of digitised case notes linked to 1,339 admissions. The regis-ters detail information about name, age, address, dates of admission and discharge, disease and outcome.[4] The creators of HHARP worked with archivists, NHS Trust Medical Officers, University Ethics committees and a steering committee of histori-ans in order to ensure the appropriate safeguarding of individual confidentiality; the database only releases records to the public once they are 100 years old.[5] In the USA, the complex requirements of the Privacy Rule within the Health Insurance Portability and Accountability Act (1996) have very real implications for historians wishing to access archives, whether in digital or hard-copy formats.

The move to an online world of research, and the increasingly electronic and virtual nature of our archives, reading material and outputs is one of the greatest methodological (and psychological) challenges the modern professional historical discipline has faced. Indeed, medical historians interested in public health have begun to consider the daunting practical issues of researching web material as pri-mary documentation.[6]

Historians have begun to take seriously the implications of an inevitable drive towards digitisation for their research, but in doing so have tended to focus on two problems: how to deal with an over-abundance of sources when we are used to archi-val scarcity being the norm; and the technical difficulties of preserving documents in a format that will not degrade or become obsolete as technologies change.[7] In much of this writing, there appears to be an implicit assumption that once we have come to terms with the changing availability and format of our sources, we will go back to the usual business of writing historical narratives. Yet, as Ludmilla Jordanova has argued, 'digital culture changes behaviour, expectations, patterns of work and mindsets'.[8] Doing digital history is also about changing the format of our outputs; good examples include the 'Casebooks' and 'Early Modern Practitioners' projects, run from Cambridge and Exeter Universities respectively.[9]

In a provocative article in *Cultural and Social History* (2013), Tim Hitchcock argued that 'by persevering with a series of outdated formats, and resolutely ignoring the proximate nature of the electronic representations we actually consult' historians have 'subtly downplayed' the impact of new technology in their work.[10] Toni Weller has implied that our reluctance to engage with the 'conceptual impact of the digital age' is affecting the very foundation of our scholarship: that of rigour.[11]

The issues are complex and extensive, but as medical historians it is imperative that we openly discuss digital challenges to our scholarship. In this chapter, I want to use digital history as a way to contribute to one particular debate that has long exercised medical historians: how to write about 'patients'. From the beginning of my migraine project I had envisaged writing a history of migraine 'from below' – a history that was about people with migraine, rather than just a chronology of changing scientific and medical conceptions of the disease and its physiological basis. In different times and places, I have been asking, who speaks about migraine, and how do different kinds of knowledge layer, accumulate, or disappear over time? Under what circumstances are illnesses such as migraine taken seriously?

In 1985, Roy Porter called for historians to take account of sufferers in their histories of medicine and healing, of how 'ordinary people in the past have actually regarded health and sickness, and managed their encounters with medical men'. Cast the net more widely than case histories and hospital records, he urged, to take note of sources such as folklore, superstitions, remedies, advice books and magazines, and interrogate the testimony of doctors themselves.[12] While many historians did turn to patients and their (often elite) bodies, recent reflections on this scholarship have suggested that much work remains to be done. Flurin Condrau argues that 'issues of how to write the patient's history, how to deal with subjectivity, experience and perhaps even choice is still very much uncharted territory'.[13] According to Jacyna and Casper, when historians of medicine have taken heed, they have 'too readily employed an essentialist and ahistorical conception of the "patient" in their historical analysis . . . historians failed in particular to consider the impact of illness upon the patient's economic, familial, legal and civic status'.[14] Digital sources challenge us to revisit these questions; their diversity often makes the categories of 'patient' and 'doctor' appear of little relevance. The boundaries of what can be considered 'illness narratives' (and where we find them) also expand.

In what follows, I use four online 'hits' from the history of migraine to reflect on how digitised methods affect our research, and to argue that digital sources have an important role to play in de-centring 'standard' medical historical narratives. My first 'hit' is from the late sixteenth century. An archival 'fragment' – a letter – raises the question: what are we searching for when we look for diseases in the past? And, how do decisions about terminology made by others affect what we are able to discover? The second 'hit', Jane Jackson's seventeenth-century recipe book, explores the possibilities opened up by digital abundance. Hit #3 suggests that the ease with which we can now search 'peripheral' sources can make us rethink standard nineteenth-century medical material such as *The Lancet*. Finally, I consider online comments attached to a YouTube video, and the significance of 'born-digital' sources for historians of the future.

Together, I hope that these four examples begin to demonstrate some of the challenges and opportunities that digital historical methodologies pose. Roy Rosenzweig, who has perhaps done most to explore the digital challenge, comments that 'the historical narratives that future historians write may not actually look much different from those that are crafted today, but the methodologies they use may need to change radically'.[15] Apart from the possibilities of presenting work in new forms, I think that doing history digitally changes both the way that we do research *and* the 'feel' of our narratives. Even for historians who choose to present their work in the traditional format of the journal article or printed monograph, the fragmentary and eclectic nature of digital sources challenges us to think more creatively about the historical narratives that we can piece together. One result may be that the advent of the digital may dissolve our historically contingent understanding of 'patients' as we know them.

## Hit #1: Monday morning. Francis Thomson's
## pigeon house, c.1590.

> I am much troubled so by the mygrame & sciatica in my hypp & I purposed
> this somer to goe to Buxton for it but now I know not what to doe for feare
> of Mr Toplyff.[16]

It is not clear exactly where and when Francis Thomson hid in his pigeon house, but
it was a Monday morning, in England, probably some time in the mid-1590s. As he
cowered, Thomson composed a letter to Sir Michael Hickes, the secretary to Lord
Burghley, Queen Elizabeth I's Lord Treasurer, to ask for his protection against 'Mr
Toplyff' who 'intendeth shortly to bringe me in truble'. He meant Richard Topcliffe,
the notorious persecutor of Catholics in Elizabethan England.[17] At the end of the
letter Thomson raised the problem of his health; he had planned to go to Buxton for
his sciatica and 'mygrame', but now he did not know what to do.

Thomson's letter, a single page written in the elaborate secretary hand common to
the formal writings of the Tudor period, is bound within a thick volume of petitions
and obsequious pleadings held by the British Library. There is little reason a medical historian researching a cultural history of illness would call this volume up from
the vaults but in fact Thomson's letter suggests much: he suffered from at least one
long-term, chronic condition; he anticipated that his infirmities would continue and
this had shaped his summer travel plans for the year ahead. Thomson's reference
to Buxton provides another lead: Buxton was the site of St Anne's well, its warm
spring waters reputed to cure a whole range of disorders. We know very little about
Thomson himself – where he lived, who he was, or what he understood by the word
'mygrame'. Yet here, in the midst of political, religious and social upheaval, we have
a late-sixteenth-century man writing to the highest authority in the land. His writing
indicates the level of discomfort that he was experiencing, and his expectation that
he would be taken seriously.

Once found, Thomson's case is illuminating, but the process of revealing this letter is by no means straightforward. Searching for 'mygrame' (the word Thomson
used in his letter) returns nothing at all in the British Library's online manuscript
catalogue.[18] 'Migraine' reveals a playscript for 'Pharmaceutical migraines' by Matthew
Westwood (1992), and a fifteenth-century copy (in Latin) of a three-page treatise by
Jean Spierinck, but again, no reference to Thomson. Instead, the letter only appears
after a search for 'megrim'; it is the only hit. So the BL online catalogue *does* make
Francis Thomson's existence visible, but only to someone who happens to be thinking about migraine in a particular way, that is as 'megrim'. Google reveals that we are
thinking like the nineteenth-century cataloguer of the BL's Landsdowne manuscripts
collection, for whom 'megrim' was the accepted contemporary term for migraine,
and from whose list (dated 1819) the exact text for the online catalogue is still taken
in the twenty-first century.[19]

It is also worth noting that whichever migraine-word we choose to search, the BL
online catalogue returns nothing about the contents of the Sir Hans Sloane collection
of over 4,000 manuscripts, at least half of which is strictly medical and includes thousands of letters from Sloane's patients.[20] The collection is considered one of the richest

sources for early-modern medical history by those who know about it, but it remains virtually invisible online to historians of disease because it is indexed by author, rather than the complaint for which they consulted Sloane. This situation is changing; when the online Sloane Letters Project is completed it will provide full-text access to transcriptions of 38 folio volumes of individual letters. The database will allow users to search by ailment (a list is included on the website), or by themes including 'aging', 'childhood' or 'illness narratives'. The database also enables a searcher to find all references to individual patients, which is, as the project's creator Lisa Smith suggests on the accompanying blog, where things 'start to get really interesting'.[21]

Finding Thomson in the BL catalogue raises an important methodological issue. Searching for key terms in online library and archive catalogues, or in the documents themselves, requires us to consider: when we research a disease or illness in the past, what exactly are we looking for? As historians we face a choice when we approach a disease's past. We can hunt for symptoms that correlate to current knowledge of a disorder or disease. In the case of migraine an oft-quoted example is a Sumerian epic poem dating from c.3000 BCE: 'The sick eyed says not I am sick eyed. The sick headed (says) not I am sick headed'.[22] This suggests a history of migraine with aura that is five millennia old, even though the diagnosis itself only dates from much more recently. Alternatively, we can accept that our own knowledge is historically contingent, and take historical evidence on its own terms, by tracing how concepts and models of disease change over time. Migraine is an apt example, because our modern words have developed directly from Galen's second-century term *hemicrania* to describe one-sided head pain. Looking for migraine words has become much easier since many digitised historical sources allow full-text word searches, but 'finding key terms is not in itself "history"', as Helen King has observed.[23] Historians have long debated how to conceptualise disease in the past; digitisation complicates the question by embedding these problems of terminology in the very metadata of digitised sources.[24]

As the case of Francis Thomson – whose letter is held hostage by the nineteenth-century concept of 'megrim' – shows, it is not just the words that we choose to search, which render visible (or not) historical documents, but the terms that cataloguers choose to define their sources. The same may apply in the future to our current cataloguing lexicon. Searching the nineteenth-century *British Medical Journal* (*BMJ*) for migraine words, for example, reveals a strange result – while the terms 'megrim', 'hemicrania', and 'sick-headache' return only the results containing those exact words in full-text, the majority of the 937 results for 'migraine' do not in fact contain that word anywhere in the full-text of the pieces, but have been retrospectively 'tagged' with the *BMJ* specialty topic 'Headache (including migraine)'. Using PubMed to search the same *BMJ* material for 'migraine' between the same dates returns 145 results – that is, only those items in which migraine appears in full-text, rather than as a retrospectively applied metadata tag. But PubMed has still searched for much more than 'migraine': historical research articles indexed by PubMed automatically receive a MeSH list created by PubMed's (human) indexers. The complete Boolean search string is: "migraine disorders"[MeSH Terms] OR ("migraine"[All Fields] AND "disorders"[All Fields]) OR "migraine disorders"[All Fields] OR "migraine"[All Fields]. The phrase 'migraine disorders' is a Medical Subject Heading (MeSH term) drawn from the *International Classification of Headache Disorders*, 2nd edition (2004).[25] A number of entry terms (that is, alternative terms likely to be used by authors) lead back to the main topic, but while this list for migraine includes hemicrania and sick head-ache, this time it does *not* include megrim.

The creators of the Carmichael Letters website 'Patient Voices in Early Nineteenth Century Virginia' have concluded that it is possible for historians to use MeSH terms creatively and in an interpretative rather than diagnostic capacity when constructing search data for online resources.[26] It is clear, however, that MeSH terms are not cre-ated to take account of medical history or historians. For instance, the terms 'slavery', 'slaves' and 'prisoners of war' only appeared in 2014.[27] Each year terms are added, updated, and deleted as they become 'old'. These lists of additions and exclusions are extremely difficult to find, because each year's 'new descriptors' simply override the last on the webpage, with archived lists being relegated to the archive of NLM Technical Bulletins. Understanding and navigating archival taxonomies is one of the historian's professional skills, but as Tim Hitchcock notes 'whereas the technologies of the hard-copy library and archive are intelligible to most humanist scholars . . . we are not similarly forearmed for working with digital sources'.[28] As a tool for think-ing about, organising, and crucially *rendering visible* medical data, we have yet to fully understand the historical significance of MeSH's control over the changing medical vocabulary historians have available, not just to think about the present, but about the past as well.

## Hit #2: Jane Jackson's recipe book, 1642

Take housleeke and garden wormes more of the housleek then of worms and stamp them together putt therto fine flower make it plasterwise spread it on a fine cloth and lay to the forehead temples and all.[29]

Searching for 'migraine' in the 'any text' box of the Wellcome Library's online Archives and Manuscripts Catalogue yields 85 results, dating from the seventeenth to mid-twentieth centuries. Among developmental rejects from Wellcome's market-ing archives, minutes of Advisory Board meetings, case histories and charity newslet-ters, the name of Jane Jackson stands out, with seven separate entries to her name.[30] Each 'hit' leads to one of six single remedy recipes in Jackson's 'receipt' collection, *A Very Shorte and Compendious Methode of Phisicke and Chirurgery* (1642). Anyone can access high-resolution full-colour images of Jane Jackson's collection instantly on the Wellcome Library website. The book is badly damaged by water, but still a beautiful object, its pages decorated with neatly ruled borders. A repeating calligraphy motif of a hand points to each new recipe, separated from its predecessor by a carefully cen-tred title. Each of the six recipes within Jackson's collection has a different spelling (migrin, migrine, migrime, mygrime, mygrim, meegreeme), but these become finda-ble with the addition of the standardised (MeSH) subject heading 'migraine' added to the catalogue record.

The Wellcome Library holds 270 volumes of manuscript recipe books dating from the sixteenth to the nineteenth century. Many of these volumes can be searched, viewed and downloaded from the Wellcome Library's online catalogue.[31] Social and cultural historians of medicine have become increasingly interested in these items: early studies focused on the economics of household medicine and gendered medi-cal practice, while more recent historians have interrogated particular disorders and states of health.[32] Digitisation has facilitated this shift in focus by making the remedy titles (and thus the names of conditions or diseases) an organising principle, rather than the authors or owners.[33]

The number of remedies for migraine in Jane Jackson's collection is unusual and it is tempting to wonder whether she used these remedies herself. Perhaps, but Jackson was certainly not alone in compiling migraine remedies. Searching published works in Early English Books Online (EEBO), as well as digitised and archival manuscript remedy collections dating from the late sixteenth to the mid-seventeenth century, reveals nearly 60 separate herbs and plants used as ingredients for migraine as well as over 20 different oil, resin and liquid bases including wine, ale, olive oil, frankincense and turpentine.[34] Recipes for plasters often contained egg white, honey and bread to help bind the elements together into a paste that would stick to a linen or leather patch to be placed on the head. Many treatments also utilised spices or strongly flavoured and aromatic ingredients such as nutmeg, cumin, cinnamon, ginger and mustard.

The ease with which we can find migraine remedies in early-modern everyday practice raises a number of questions. Was Jane Jackson's knowledge idiosyncratic, or common? Where did her recipes come from? Is it possible to correlate bodies of knowledge in manuscript collections with contemporary printed works, also now available in digital form via EEBO? With some persistence, searching EEBO for the key ingredients worms/wormes and housleek/houseleek certainly does allow us to locate Jackson's recipe in contemporary printed books. A recipe using garden worms and houseleek for migraine had been published in the anonymously authored *A Closet for Ladies and Gentlewomen* in 1608, 34 years before appearing in Jane Jackson's collection: 'Take Housleeke, and Garden wormes, the greater part being Housleeke, stampe them together and thereto fine flower, and make a playster in a fine cloth and lay it the forehead and temples'.[35] In 1655, 'Philiatros' reproduced the *Closet* recipe in *Natura Exenterata*, recycling the recipe for a new generation of medical readers.[36] The attraction of this recipe is understandable – it is short, simple, and the ingredients easily obtainable. Houseleek – like primrose, red rose, yarrow and blessed thistle – was considered a 'cold' herb, used to counteract a 'hot' disorder. Thomas Collins' recipe for a 'burning headach' (1658) used housleek mixed with woman's milk and a little rose-water.[37] Jackson included two versions of the houseleek and earthworm recipe in her own manuscript collection, as well as a third worm-based recipe that simply mixed worms with bread. Other manuscript collections contain similar instructions. Mrs Corlyon's collection (1606) used 'powder of the longe wormes of the earthe' in a recipe for Megreeme, and ointment of earthworms for strengthening or relieving an aching back. The College of Physicians of London's *Pharmacopeia*, published in Latin in 1617 and 1618, and later translated into English by Nicholas Culpeper, prescribed water of earthworms for complaints including consumption, jaundice, hectic fevers and diseases of the head and brain.[38]

It is notoriously difficult to assess how and to what extent ordinary people read books, let alone to determine how publications translated into common practice.[39] Nevertheless, the digitisation of such huge bodies of material makes it possible to reconstruct the traffic of a number of early-modern recipes in a way that simply would not have been feasible until recently. In one case, we can trace variations of a recipe using pellitory of Spain through manuscript and print across four centuries. In so doing we can construct a very different kind of migraine history than the sense of isolation that characterises Francis Thomson's pigeon-house letter. The compilers of seventeenth-century migraine remedies – mainly women – appear as participants in a rich body of shared vernacular knowledge that spanned decades, if not centuries. While Thomson contemplated a journey to Buxton fraught with danger, recipe books suggest that for many early-modern migraineurs, remedies, or at least the holders of remedy knowledge, were available almost on the doorstep.

### Hit #3: Patrick Murphy's lectures, 1854

> The sick headache, to which females are such martyrs when menstruation is ceasing, comes under this denomination of headache. It comes on at their usual period, but the menses either cease flowing, or escape very scantily; it is very distressing, and attended with an inclination to vomit, hence the expressive term sick headache.[40]

It takes most of a day to manually flick through the nineteenth-century indexes to the half-yearly volumes of *The Lancet* which are stacked, dusty from recent building work, on the Wellcome Library's open shelves.[41] I remember too late that researchers should not wear white when spending the day handling 150-year-old leather-bound volumes, and take note of entries for the words head, headache, hemicrania, megrim, migraine, and sick-headache. The resulting list is somewhat slim – out of 42 references, the majority (28) are instances of headache that hold potential relevance but on further inspection yield little. There are four mentions of hemicrania, one of megrim, nine of migraine, and one of 'migrainous headache'. The first mention of 'migraine' comes in 1833, with the publication of the French physician M. Andral's lectures on 'Perversions of Sensibility' in which he discussed hemicrania and migraine. In 1849 and 1850 there are a few correspondences about treating hemicrania with chloroform and caffeine. In 1854, a flurry of references lead to Patrick J. Murphy's five lectures on 'Headache and Its Varieties'. Does *The Lancet* online yield more? I can search the journal on Sciencedirect.com, but I cannot access the content; even as a lecturer in a research-intensive British university I come up against the publishing giant Elsevier's paywall, since 2012 the subject of calls for an academic boycott. A couple of text messages later and I log in with an online account from a different university. I search again. Between 1823 and 1890, a full-text search of *The Lancet* returns 72 results for megrim, 165 for migraine, 87 for hemicrania and 75 for 'sick-headache' – many more results than from the hard-copy indexes. Within these results the name of Patrick Murphy comes up again and again. Because his lectures contained multiple migraine-words (megrim, hemicrania, sick-headache), digital retrieval imbues his lectures with the appearance of a near-ubiquitous presence in different searches.[42]

Yet, Murphy is hardly a key figure. Until now, historians of migraine have commented briefly on the nosological technicalities of his classification, but little more.[43] Murphy's lectures *were* about classifying headache into five types, but he was concerned primarily with the headaches of 'young ladies'. He wrote that anaemic headache, commonly known as 'cephelea, vertigo, megrim or giddiness', often affected 'mothers in the lower classes of life' whose minds and bodies had been weakened by daily toil, disturbed sleep from the cries of their infants, and insufficient nourishment while their bodies were 'hourly drained by lactation'.[44] Megrim was a separate category of headache, the result of indigestion, but again a disease to which young women were 'such martyrs'. Murphy's third type (synonymous with the hemicrania of 'old authors') was 'neuralgic' headache. This was '*peculiar* to females', occurring during the period of puberty until the end of their periods, and 'undoubtedly hysterical' in its origin.[45]

Also revealed by digital searching is the extent to which Murphy's contemporaries often made casual asides and peripheral comments about sick-headache, megrim or hemicrania. In a piece on menstruation in the 1851 *Provincial Medical and Surgical Journal*, for instance, E. J. Tilt associated sick-headache with women's menstrual lives, identifying headaches, sick-headaches and hysteria as a result of the 'critical time' when menstruation ended.[46] In a report on etherization in medical practice (1847), Dr Ballard discussed the case of the charwoman Ellen L 'of highly nervous temperament . . . sometimes very badly off in her circumstances', who sought his expertise for hemicrania. F. G. Broxholme reported treating a case of 'quotidian hemicrania' with chloroform: the subject was 'Miss L—, aged thirteen years, of delicate constitution and appearance'. In a treatise on constipation, John Burne asserted that 'the popular term "sick-headach", is associated with a train of symptoms of every-day occurrence; and though not altogether confined to females, it is they who particularly suffer from this affection'.[47] Buried in peripheral material, these references show how assumptions about gender were embedded in English-language medical knowledge about migraine. In many cases these comments seem little advanced from William Buchan's eighteenth-century explanation of a disorder which was often seen among nurses 'who give suck too long, or who did not take a sufficient quantity of solid food'.[48]

One of the key challenges in writing the social and cultural history of migraine is to understand how, when, and why migraine came to be associated primarily with women.[49] Perhaps more important still is to understand how these modes of thinking affected the lives of ordinary people. In 1880, the laundress Ann Noakes was tried at the Old Bailey for the wilful murder of her youngest son, William. Amy Risbridger, who had worked with Noakes for four months, described how the defendant 'complain[ed] of her head very much at times – she had got a sick-headache – at those times she used to say that her trouble was too much for her to bear'. Another witness, 14-year-old Emma Dibstall, gave a similar testimony: Noakes 'said her head was so bad she could not bear her trouble, she used to seem very excited – she was a very hard-working woman, standing at the tub till late at night'. Called to give his professional opinion, John Walters M.D. diagnosed Noakes' illness as 'disease of the womb', an enlarged uterus that had failed to return to size following childbirth. He had impressed upon her the necessity of rest, he said, but Noakes had been 'very weak, in a low state of health, caused by the disease from which she suffered. She complained of great headache and restlessness at night; she was very pale and bloodless'. Walters concluded that Noakes' act in killing her child was one of homicidal mania, and that 'she would not know she was doing a guilty act'. Noakes was found not guilty, on the grounds of insanity.[50]

Sick-headache appears only a few times in the Old Bailey records. The value of these cases lies not in their numbers, but in the way that accounts of lived experience and vernacular understandings of illness reflect back on medical knowledge. In his study of disability, David Turner has made extensive use of Old Bailey records and observes that the stories told in court highlight 'the calamitous effects of disability, especially to those who became incapable of following their usual occupation'.[51] Ann

Noakes' case suggests profound differences in lay and medical perceptions of illness. The witnesses – friends, families and fellow workers – portrayed a hard-working young woman, dealing on a daily basis with the chronic problems of severe headaches, fainting fits and mental depression. Dr Walters did note the sick headaches, but only as a symptom of a deeper disorder of her reproductive system and mental health.

It would be easy, particularly given the abundance of nineteenth-century medical material online, to consider only professional narratives of disease and illness. Cases such as Ann Noakes challenge us to interpret afresh standard 'medical' sources such as case notes. If digital sources allow us to more easily place Murphy's lectures in the context of nineteenth-century thought about women and illness, Ann Noakes pushes us further still as a powerful example of how a highly gendered way of thinking had practical repercussions as medical men passed judgement on the lives of other people.

Illnesses appear in the historical record when, for some reason, they are taken seriously or gain legitimacy and meaning. Taking migraine seriously, even when it appears in the strangest of places, demands that we consider very carefully what legitimacy the disorder gained at that particular moment. When we juxtapose the evidence from a learned journal with the record of a criminal trial at the Old Bailey the boundaries of our history expand beyond the obvious realm of the 'patient'. Glimpsing the profound effect that migraines had on the lives of working people and the way that they experienced and talked about illness invites us to consider afresh what counts as knowledge and evidence and whose stories matter.

## Hit #4: jandr1ch's YouTube migraine aura video (or, migraine 2.0), 2011

Pretty good – I'm having one right now and the description is bang on . . . Nice to know I'm not about to keel over:)[52]

On 17 July 2011, the YouTube user jandr1ch uploaded a video depicting an animation of a migraine aura after a colleague rushed to the doctor in 'a (literally) blind panic' when experiencing a migraine aura for the first time. jandr1ch described the video as a 'first attempt to try to visualise the effect many people experience when having a visual migraine', and remembered his own 'total panic' the first time that he experienced a visual disturbance. Soon after the posting, one viewer complained that the video had triggered a migraine. jandr1ch acted quickly to warn potential viewers about the video's content, specifically the black and white flashing images: 'If you are susceptible to such effects triggering unpleasant consequences such as a migraine or epilepsy, please DO NOT WATCH IT'. jandr1ch was highly apologetic for the effect of his animation: 'The last thing I wanted to do was cause distress'. He had simply wanted to reassure those experiencing one of the visual disturbances that they were normal. jandr1ch had been 'amazed at the response to the video . . . I was going to try and make a better version at some point . . . I apologise if I caused you any pain'. By August 2015, 'Visual Migraine Animation' had gained over 214,000 views and 450 comments.[53]

More than a billion users, watching almost 6 billion hours of video, access the YouTube website each month.[54] While best known for music videos, advertisements and amateur footage, YouTube is also fast becoming a place where people living with chronic illness go for information, advice and support from the online community. We can understand YouTube as a major player in the rapidly emerging phenomenon known as 'Health 2.0'. This term first emerged in the first decade of the twenty-first century to describe a proliferation of online medical resources including blogs, podcasts, Wikis and discussion forums by patients, clinicians and scientists. One of the key features of 'Health 2.0' is the potential for web users to create online content.[55] For people with chronic conditions in particular, the web offers a way of sharing experiences, providing and receiving support, visualising disease, building relationships, accessing services and enabling collective advocacy. Twitter users with chronic conditions often use #spoonie to connect, sharing coping strategies, humour, frustration and insight.

While very little is known about how Internet use might affect mental and physical health, and there are real concerns about trust and accuracy of information, it is clear that the Internet has profoundly changed the way that we understand and even experience illness. Searching YouTube for 'migraine' returns videos explaining physiological mechanisms, discussions of symptoms and triggers, demonstrations of treatment techniques, guides for managing illness, and advertisements for remedies. Animations represent first-hand the experience of migraine aura, and the most popular of these have gathered over 100,000 views or around 60,000 'hits' per year.

jandr1ch's video has a fixed background image of a road as seen through a car windscreen. A point of light develops into a digital representation of an aura with pulses, flashes of light, and an expanding brightly coloured zigzag. Text instructs the viewer 'to get the effect . . . look at centre of road . . . not flickering image'. Even for someone who has never experienced a migraine aura the effect is uncomfortable to watch. Many viewers react to the video with gratitude because the animation provides a way to explain to friends and family what it is they see.[56] AntiMonopoly commented: 'That is what I'm dealing with at this very moment. Thank you for the effort you put in to make this, so I can explain what it's like to my friends.' RockinsaneInthebrain also expressed his thanks: 'I've show [*sic*] it to my friends and family so they can understand more what I am talking about when I experience these migraines!' The largest group of comments describe posters' own experiences. Many congratulate jandr1ch on the accuracy of his depiction. Christopher Reddick's comment encapsulates a number of common responses in the comments thread:

> OMFG!!!!!!! This just happened to me and it's insanely close to your video. It's happened before but no headache so I didn't realize it was migraine related. (scared me a bit) Nausea though, and yes I get gnarley headaches sometimes but not with the visual aura . . . weird. Thanks for helping give it a name. Peace out.

As in Reddick's comment, posters often express astonishment at finding their own deeply subjective experience depicted online; gratitude for providing knowledge; candid discussions of pain; and glimpses into lives lived with illness. One comment reveals the liberating diagnostic power of the Internet:

> omg – am i glad i FINALLY believe i found out what i've had on and off all of these years . . . the 1st time i got the same thing you are describing to a tee, i

was in my 20's and i KNEW it wasn't my eyes but the dr. told me to see an eye doc – i didn't do it and for MANY years i thought they were called hysterical blindness (which i read about in the library – no Internet then).

It is also striking how many of the comments have apparently been written *during* an attack. John Bertrand Russell writes: 'having an ocular migraine right now . . . first in almost a year. No pain, just cheap fireworks.' Kirsten 1977 explained what she was seeing: 'As I type, the migraine aura is like a back to front capital 'c' at the top right hand corner of my computer screen. Looking at the keyboard trying to type this comment has been interesting to say the least. Took me ages!' Wazmo100 writes 'This is great ! I've got one go'in right now. Annoying is'nt it.'

Emotions of fear and panic often feature. Noellenoelle90 explained 'I just suffered through one of these for the first time EVER not even 10 min ago. And at first I thought I was going blind so I Googled "jagged semi circle in eye sight" and this video popped up thank you so much for a second I really thought I was going blind!' 727lisa described driving:

> seeing multiple images such as repetitive road signs, lights from other cars coming at me while I was dizzy and not knowing if they were in my lane or not, and rays of colors shooting towards me all at the same time. Yes, I pulled over ASAP on a long, dark country road, scared to death and crying.

Posters also respond directly to each other. Some express concern or sympathy for others' pain, or make and respond to suggestions about possible triggers. Another theme through the comments is the reassurance that migraine auras, although frightening, are not serious; those who describe more serious symptoms such as vomiting blood are urged to seek medical advice. jandr1ch and other posters also police the boundaries of acceptable comment, responding to critical or derogatory posts. ValidEmailAddress responded angrily to one poster: 'Get a clue next time before you make an ass out of yourself. And it's NOT dumb shit. It's real, it happens, and its scary as fuck if you don't know what it is.'

At one point, a poster asks about gender. 'Question, I heard women suffer from migraines a lot more than men. Is it rare for a man to experience this?' jandr1ch responds 'I only know of other men that have this type of migraine personally. I can't tell by most people's tags what sex they are (I'm male by the way). Perhaps it would be helpful if comments could include this information so we can get an idea? I would guess it's 50/50 but I really have no idea.' Later, d.s. responded 'No, it isn't rare for a male to experience this. You know, this occurred to me after I had been spending a lot of time on the computer, and had been sleeping till real late in the am.'

Ambiguous usernames and photos make any accurate statistical analysis of gender almost impossible, but of the comments identifiable by gender, more than half have been made by men. By contrast, only around one in nine of the contributions to the Migraine Trust's Travelling Diary project come from men.[57] YouTube is not the only place on the Internet to discuss migraine, but it is providing a forum for men in particular to share their migraine experiences. This is significant because migraine is so strongly associated with women in modern culture (around three-quarters of people who have migraine are women). The comments of these men suggest that many of them have felt scared and isolated by their experiences and are deeply grateful and relieved to have found out that they are not, in fact, alone, or even unusual.

## Conclusion

Digitisation, and digitally enabled research, lifts the curtain on a whole new cast of characters and perspectives for our medical histories. Through the archival fragment of Thomson's letter we enter a world in which politics and religion were deeply intertwined with health. Jane Jackson's letter seems far less isolated, both in the sense that she clearly participated in a seventeenth-century culture of knowledge exchange, but also because of the way in which recipe collections such as hers can be accessed as a body of work, along with contemporary printed material. Patrick Murphy's nineteenth-century lectures in the *The Lancet* are more standard sources for medical history, but 'peripheral' documents, such as those in the Old Bailey, challenge us to re-frame his lectures and foreground assumptions about gender, rather than discussions of nosology. Finally, responses to jandr1ch's YouTube video unveil a new community of exchange, support and censure as the boundaries of illness talk shift online.

I have not attempted to construct a conventional chronological narrative here – Nancy Rose Hunt's conception of historical practice in the digital era as 'suturing' is a suggestive alternative way to think of the narratives that medical historians might write – but it is clear that in none of these four cases can we talk about 'patients' and 'doctors' in any standard sense of consultations, case notes or treatments.[58] A history of highly gendered assumptions about the kind of people who suffer from migraine shapes our modern understanding of this disorder. We need to take seriously men's experiences of illness by putting them back into the history but we also need to create a historical space in which to take seriously women's creation of medical knowledge. More broadly, if we are to write 'patients' into histories of illness and disease, we first need to understand the circumstances in which patienthood is possible, desirable or necessary. The long history of migraine, and particularly the way that digital sources have helped shape my project, shows how the very idea of the patient is historically contingent. It is, perhaps, far more interesting to ask 'who speaks', for whom, and how, about diseases in the past?

Digitisation offers up possibilities but also challenges. We must remember that the vast majority of material is *not* digitised. The ways in which sources become available to us shape the narratives that we write. Although the web is global, it is by no means universal: the story that I have written here is based on English-language sources from the UK and USA, reflecting how traditional biases in academic research are reinforced by digitisation policies. As Tim Hitchcock has observed, 'the very process of digitization is effectively reproducing a kind of Western cultural hegemony that would not be acceptable if it was a product of self-conscious policy'.[59] These circumstances determine whether words about illness gain authority, how different kinds of knowledge contradict, support, or efface others, and which kinds of knowledge can be rendered invisible.

For historians of the future, user-generated online content may provide an important source of evidence for understanding experiences and discourses of illness in the twenty-first century, but we have no idea how this material will be preserved, if at all, in the future. Online content is continually in flux, and this also goes for the taxonomies that organise historical sources, however stable or durable the source itself may be. At the same time as YouTube creates a new space for men to narrate their experiences of migraine, this fact should remind us that the Internet is not as democratic a space as it might appear. Scholars in elite universities might well be angered by the financial politics of publishing giants such as Elsevier who control access to *The Lancet*, but much more fundamental barriers exist. Socioeconomic status, gender, geographical location and disability all play a major part in accessing, consuming and creating online resources, preventing many people from accessing the Internet at all. As historians, we must continue to find ways of contextualising and historicising the dominant narratives that are emerging and will continue to emerge in the digital future. It is as important as ever that we continue to ask 'who speaks' in and for the history of illness? Who gets taken seriously? Whose experiences are silenced, and how can we access these?

## Notes

1 Edward Hampshire and Valerie Johnson, 'The Digital World and the Future of Historical Research', *Twentieth Century British History* 20, 2009, 396–414. Available: http://dx.doi.org/10.1093/tcbh/hwp036 (accessed 27 August 2015).

2 C. Henshaw, 'The UK Medical Heritage Library: Uniting Digitised Collections' (29 July 2014) Available: http://blog.wellcomelibrary.org/2014/07/the-uk-medical-heritage-library-uniting-digitised-collections/ (accessed 27 August 2015).

3 Suzannah Biernoff, 'Medical Archives and Digital Culture: From WW1 to *Bioshock*', *Medical History* 55, 2011, 325–30. Available: http://www.ncbi.nlm.nih.gov/pmc/articles/PMC3143874/ (accessed 27 August 2015).

4 Sue Hawkins and Andrea Tanner, 'The Historic Hospitals Admissions Project' (30 July 2009). Available: http://www.ariadne.ac.uk/issue60/hawkins-tanner (accessed 27 August 2015).

5 Susan E. Hawkins, 'RE: HHARP Query', email (25 July 2014).

6 Martin Gorsky, 'Sources and Resources into the Dark Domain: The UK Web Archive as a Source for the Contemporary History of Public Health', *Social History of Medicine* 28, 2015, 596–616. Available: http://dx.doi.org/10.1093/shm/hkv028 (accessed 4 August 2015).

7 Roy Rosenzweig, *Clio Wired: The Future of the Past in the Digital Age*, New York and Chichester: Columbia University Press, 2011, p. 7, Available: http://books.google.co.uk/books?id=4nivoPD7L40C (accessed 27 August 2015).

8 Ludmilla Jordanova, 'Historical Vision in a Digital Age', *Cultural and Social History* 11, 3, 2014, 343–8.

9 The Casebooks Project: A digital edition of Simon Forman's & Richard Napier's medical records 1596–1634. Available: http://www.magicandmedicine.hps.cam.ac.uk/; Early Modern Practitioners. Available: http://practitioners.exeter.ac.uk/ (both accessed 4 August 2015).

10 Tim Hitchcock, 'Confronting the Digital: Or How Academic History Writing Lost the Plot', *Cultural and Social History* 10, 2013, 9–23, p. 12.

11 Toni Weller (ed.), *History in the Digital Age*, Abingdon and New York: Routledge, 2013, pp. 1–2, 11–12. Available: http://books.google.co.uk/books?id=UZEfjUi1RyoC (accessed 27 August 2015).

12 Roy Porter, 'The Patient's View: Doing Medical History from Below', *Theory and Society* 14, 2, 1985, 175–98, pp. 182–3. Available: http://www.jstor.org/stable/657089 (accessed 27 August 2015).

13 Flurin Condrau, 'The Patient's View Meets the Clinical Gaze', *Social History of Medicine* 20, 2007, 525–40, p. 526. Available: http://dx.doi.org/10.1093/shm/hkm076 (accessed 27 August 2015).

14 L. S. Jacyna and Stephen T. Casper, *The Neurological Patient in History*, Rochester, NY: University of Rochester Press, 2012, p. vii.

15 Rosenzweig, *Clio Wired*, p. 23.

16 British Library, Lansdowne MS 108/56, Mr Francis Thomson, a persecuted Recusant, to Mr. Hicks, (here called Gabriel instead of Michael).

17 G. R. Smith, *Servant of the Cecils: The Life of Sir Michael Hickes, 1543–1612*, London: Cape, 1977, 54–67.

18 'Search our Catalogue: Archives and Manuscripts', Available: http://searcharchives.bl.uk (accessed 27 August 2015).

19 *A Catalogue of the Lansdowne Manuscripts in the British Museum*, London: 1819, p. 209. Available: http://books.google.co.uk/books?id=LRgQX7vQbMcC (accessed 27 August 2015).

20 The letters are mostly contained in the four volumes BL Sloane MS4075–4079.

21 Sir Hans Sloane's Correspondence Online. Available: https://drc.usask.ca/projects/ sloaneletters/doku.php (accessed 27 August 2015); Lisa Smith, 'Doctor Sloane and His Patients in Eighteenth-Century England' (12 June 2013). Available: http://www.sloanelet ters.com/ (accessed 27 August 2015).

22 F. Clifford Rose, 'The History of Migraine from Mesopotamian to Medieval Times', *Cephalalgia* 15 (Supplement 15), 1995, 1–3. Available: http://onlinelibrary.wiley.com/ doi/10.1111/j.1468-2982.1995.tb00040.x/abstract (accessed 27 August 2015).

23 Helen King, 'Second Opinion: Response to Shelton', *Social History of Medicine* 25, 2012, 232–8, at 233. Available: http://dx.doi.org/10.1093/shm/hkr168 (accessed 27 August 2015).

24 See Andrew Cunningham, 'Identifying Disease in the Past: Cutting the Gordian Knot', *Asclepio* 54, 2002, 13–34, at 17.

25 The concept of 'MeSH' derives from the first official list of 4400 Medical Subject Headings published by the US National Library of Medicine in 1960. This constantly evolving 'controlled vocabulary' has been widely adopted, and now contains more than 22,000 descriptors. Available: http://www.nlm.nih.gov/mesh/intro_hist.html (accessed 27 August 2015).

26 Alison White, 'MeSH in the Letters to Dr James Carmichael & Son', (2005). Available: http://carmichael.lib.virginia.edu/about/mesh.html (accessed 27 August 2015).

27 'New Descriptors – 2014' (24 July 2013). Available: http://www.nlm.nih.gov/pubs/tech bull/nd13/nd13_mesh.html (accessed 4 August 2015).

28 Hitchcock, 'Confronting the Digital', p. 14.

29 Wellcome Library, MS 373/16, Jane Jackson, *A very shorte and compendious Methode of Phisicke and Chirurgery Containeinge the Cures inwardly and also the Cureinge of all manner of woundes on the bodie together with remedies for the stone and the causes and signes thereof, etc. As also the making of Sirrups and all manner of oyles and waters whatsoever* (1642).

30 'Online Catalogue of Archives and Manuscripts'. Available: http://archives.wellcomeli brary.org.

31 'Recipe Books' (15 December 2014). Available: http://wellcomelibrary.org/collections/ digital-collections/recipe-books/ (accessed 27 August 2015).

32 Elaine Leong, 'Making Medicines in the Early Modern Household' *Bulletin of the History of Medicine* 82 (2008) 145–68. Available: https://muse.jhu.edu/journals/bulletin_of_the_ history_of_medicine/v082/82.1leong.html (accessed 27 August 2015).

33 Jennifer Evans, '"Gentle Purges corrected with hot Spices, whether they work or not, do vehemently provoke Venery'": Menstrual Provocation and Procreation in Early Modern England', *Social History of Medicine* 25, 2012, 2–19. Available: http://dx.doi.org/10.1093/ shm/hkr021 (accessed 27 August 2015); Seth Stein Lejacq, 'The Bounds of Domestic Healing: Medical Recipes, Storytelling and Surgery in Early Modern England', *Social History of Medicine* 26, 2013, 451–68. Available: http://dx.doi.org/10.1093/shm/hkt006 (accessed 27 August 2015). Conversations about this scholarship as it emerges are held on the highly successful Recipes Project blog: http://recipes.hypotheses.org/ (accessed 27 August 2015).

34 The recipes have been collected by systematically researching Wellcome library's collection of recipe books, both digitised and in archives, and searching EEBO for remedies with the keywords 'megrim', 'meagrim', 'meagrom', 'mygryme' and 'hemicrania'.

35 Anon. *A Closet for Ladies and Gentlewomen* (1608), p. 160. Early English Books Online. Available: http://gateway.proquest.com/openurl?ctx_ver=Z39.88-2003&res_id=xri:eebo&rft_id=xri:eebo:image:19518:86 (accessed 27 August 2015).

36 Philiatros, *Natura exenterata: or Nature unbowelled by the most exquisite anatomizers of her* (1655), p. 28. EEBO. Available: http://gateway.proquest.com/openurl?ctx_ver=Z39.88-2003&res_id=xri:eebo&rft_id=xri:eebo:image:115272:19 p. 28 (accessed 27 August 2015).

37 Thomas Collins, *Choice and Rare Experiments in physick and chirurgery, or, A discovery of most approved medicines for the curing of most diseases incident to the body of men, women, and of children* (1658), p. 4. EEBO. Available: http://gateway.proquest.com/openurl?ctx_ver=Z39.88-2003&res_id=xri:eebo&rft_id=xri:eebo:image:54371:10 (accessed 27 August 2015).

38 William Brockbank, 'Sovereign Remedies: A Critical Depreciation Of The 17th-Century London Pharmacopoeia', *Medical History* 8 (1964), 1–14, pp. 4, 6. Available: http://www.ncbi.nlm.nih.gov/pmc/articles/PMC1033332/ (accessed 27 August 2015).

39 Mary Fissell, 'The Marketplace of Print', in Mark Jenner and Patrick Wallis (eds), *Medicine and the Market in England and its Colonies, c. 1450–1850*, Basingstoke: Palgrave Macmillan, 2007, 108–32, p. 114.

40 Patrick J. Murphy, 'On Headache and Its Varieties', *Lancet* 63, 20 May 1854, p. 540.

41 It is no longer possible to browse the *BMJ* on the Wellcome open shelves, because it is so freely available online. *The Lancet*, by contrast, is also online, but the vast majority of its content is behind Elsevier's paywall, and requires institutional subscription: http://www.sciencedirect.com/science/journal/01406736.

42 Patrick J. Murphy, 'On Headache and Its Varieties', *The Lancet* 63, 1854, 182–3, 209–10, 300–1, 359–60, 540–1.

43 Mervyn Eadie, *Headache Through the Centuries*, Oxford: Oxford University Press, 2012, pp. 12–13.

44 P. J. Murphy, 'On Headache', *The Lancet*, 63, 2 February 1854, 209.

45 P. J. Murphy, 'On Headache and its Varieties', *The Lancet*, 20 May 1854, 540.

46 E. J. Tilt, 'On the Management of Women at, and after the Cessation of, Menstruation', *Provincial Medical and Surgical Journal* 15: 11 (28 May 1851), 281–7, p. 281. Available: http://www.jstor.org/stable/25492966 (accessed 27 August 2015).

47 John Burne, *A Treatise on the Causes and Consequences of Habitual Constipation*, Philadelphia: Haswell Barrington and Haswell, 1840, p. 42. Available: https://archive.org/details/treatiseoncauses00burn (accessed 27 August 2015).

48 William Buchan, *Domestic medicine: or, a treatise on the prevention and cure of diseases by regimen and simple medicines. With an appendix, containing a dispensatory for the use of private practitioners.* 11th edition, London (1790). Eighteenth Century Collections Online, pp. 352–4. Available: http://find.galegroup.com/ecco/infomark.do?&source=gale&prodId=ECCO&userGroupName=leicester&tabID=T001&docId=CW106969211&type=multipage&contentSet=ECCOArticles&version=1.0&docLevel=FASCIMILE (accessed University of Leicester, 27 August 2015).

49 See Joanna Kempner, *Not Tonight: Migraine and the Politics of Gender and Health*, Chicago: University of Chicago Press, 2014.

50 *Old Bailey Proceedings Online*, version 7.0. 26 April 1880, trial of Ann Noakes (36) (t18800426-428). Available: www.oldbaileyonline.org (accessed 27 August 2015).

51 David M. Turner, 'Disability and Crime in Eighteenth-Century England: Physical Impairment at the Old Bailey, *Cultural and Social History* 9, 2012, 47–64, p. 55. Available: http://www.tandfonline.com/doi/pdf/10.2752/147800412X13191165982953 (accessed 27 August 2015).

52 Boo Cook, YouTube user, comment on jandr1ch, 'Visual Migraine Animation' Online Video clip. YouTube. 17 July 2011 Available: https://www.youtube.com/watch?v=fo139jYAFzA (accessed 25 July 2014).

53 jandr1ch, 'Visual Migraine Animation' Online Video clip. YouTube. 17 July 2011. Available: https://www.youtube.com/watch?v=fo139jYAFzA (accessed 27 August 2015).

54 'Statistics'. Available: https://www.youtube.com/yt/press/en-GB/statistics.html (accessed 27 August 2015).

55 Wen-Ying Sylvia Chou et al., 'Cancer Survivorship in the Age of YouTube and Social Media: A Narrative Analysis', *Journal of Medical Internet Research* 13 (2011). Available: http://dx.doi.org/10.2196/jmir.1569 (accessed 27 August 2015).

56 All comments quoted below are attached to jandr1ch, 'Visual Migraine Animation' (as accessed on 21 July 2014).

57 Entries are shared by Facebook, Twitter and uploaded as images into the Trust's Flickr album. The Migraine Trust, 'Travelling Migraine Diary'. Available: https://www.flickr.com/photos/migrainetrust/sets/72157625121333665/ (accessed 27 August 2015).

58 Nancy Rose Hunt, *Suturing New Medical Histories of Africa*, Münster: LIT Verlag, 2007, pp. 13–14. Available: http://books.google.co.uk/books?id=4n6m91ERnF4C (accessed 27 August 2015).

59 Hitchcock, 'Confronting the Digital', 21, n. 3.

## Select bibliography

Biernoff, S., 'Medical Archives and Digital Culture: From WW1 to *Bioshock*', *Medical History* 55, 2011, 325–30.

Condrau, F., 'The Patient's View Meets the Clinical Gaze', *Social History of Medicine* 20, 2007, 525–40.

Eadie, M., *Headache Through the Centuries*, Oxford: Oxford University Press, 2012.

Hampshire, E. and V. Johnson, 'The Digital World and the Future of Historical Research', *Twentieth Century British History* 20, 2009, 396–414.

Hitchcock, T., 'Confronting the Digital: Or How Academic History Writing Lost the Plot', *Cultural and Social History* 10, 2013, 9–23.

Jacyna, L. S. and S. T. Casper, *The Neurological Patient in History*, Rochester, NY: University of Rochester Press, 2012.

Jordanova, L., 'Historical Vision in a Digital Age', *Cultural and Social History* 11, 3, 2014, 343–8.

Kempner, J., *Not Tonight: Migraine and the Politics of Gender and Health*, Chicago: University of Chicago Press, 2014.

King, H., 'History without Historians? Medical History and the Internet', *Social History of Medicine*, 25, 1, 2012, 212–21, http://shm.oxfordjournals.org/content/25/1/212.

Leong, E., 'Making Medicines in the Early Modern Household' *Bulletin of the History of Medicine* 82, 2008, 145–68.

Porter, R., 'The Patient's View: Doing Medical History from Below', *Theory and Society* 14, 2, 1985, 175–98.

Rosenzweig, R., *Clio Wired: The Future of the Past in the Digital Age*, New York and Chichester: Columbia University Press, 2011.

Turner, D. M., 'Disability and Crime in Eighteenth-Century England: Physical Impairment at the Old Bailey, *Cultural and Social History* 9, 2012, 47–64.

Weller, T. (ed.), *History in the Digital Age*, Abingdon and New York: Routledge, 2013.

Sir Hans Sloane's Correspondence Online. Available: https://drc.usask.ca/projects/sloaneletters/doku.php (accessed 30 July 2014).

# 29

# CASE NOTES AND MADNESS

*Alannah Tomkins*

The compilation and use of case notes as a component of clinical method date largely from the second half of the seventeenth century.[1] Before this period medicine prioritized the judgement of learned authority over observations of the present but Francis Bacon's (1561–1626) instructions for the creation of an empirical philosophy of science, which for practitioners entailed collecting histories of the diseases they treated, eventually had a practical effect.[2] Watching, recording and analysing numerous examples of disease was recognized as a valuable means to extract general principles and the practice made a significant contribution to the decline of humoral medicine.[3] An increasing reliance on both private and published case notes was characteristic of the early nineteenth century onwards, but from the later nineteenth century the focus on the laboratory as the site of medical advancement (and latterly concerns about patient confidentiality) have encouraged the excision of case material from outputs other than the individual patient record, and the generalization and anonymization of printed exemplars.

The historiography of the case note is much more recent, with debate circling around questions of power in the doctor–patient relationship and the historical value of therapeutic narratives.[4] Psychiatric case notes are possibly the most important variant of the genre, given the on-going role of patients' and carers' discursive accounts in current clinical practice for the diagnosis of mental illness and learning disability. This chapter will consider the significance of case notes for the course of Western medicine, with a particular focus on the processes of production and retention by British writers. It will explore the form and content of case notes generated by English asylums for the insane over the nineteenth century, and identify the case notes of patients who were also qualified medical practitioners as representing a distinctive group among asylum residents. Finally it will examine case notes generated on behalf of Charles Beard, a physician from Brighton, to illustrate both the potential yield from the genre and its limits.

## Case notes in Western medicine

The credit for embedding case notes in clinical medical practice has been conferred on Thomas Sydenham (c.1624–1689), who demonstrated the practical worth of case notes by the clarity of his writing and by reinforcing the theoretical validity of empirical evidence when conveyed as a narrative. Empiricism in medicine had often been regarded as mere quackery, but Sydenham's methodology showed the capacity for progress in medical science based on attentive observation and responsive therapy. Consequently Sydenham's example began to shift medicine's entrenched reliance on ancient texts. His subsequent impact across European medicine was assured when his practice was endorsed by Herman Boerhaave (1668–1738), whose students were instructed in practical, bedside medicine and routine note-taking.[5]

The case note, or in its modern published form the case report, is the prose account of an individual patient's disease or injury. It typically begins at the onset of symptoms, or at the time of the patient's presentation to a practitioner, and ends with either the dismissal of the case, the patient's recovery, or their death. Notes represent the conflation of the patient's own account of their case with the practitioner's reading of the same circumstances, which might involve editing details thought extraneous to the medical account and adding observations from physical examination. Typically arising from immediate personal contact, case notes could also accrue from correspondence.[6]

The content of early clinical notation could range from a brief account of symptoms and a running catalogue of the drugs prescribed or procedures conducted to a more heterogeneous mixture of patients' personal data, abbreviated jottings of treatment and lengthier, discursive prose about the progress of a case. The latter could include tangential commentary on the patient's character or behaviour and on the practitioner's own actions, feelings and responses unrelated to consultation or treatment. Contextual additions might also comprise some commentary on the type of disease or wound, including allusions to earlier similar cases, and might be supported by references to parallel medical literature. In this way appropriate content for notes was determined not by the patient's story but by the story surrounding the patient that the practitioner regarded as pertinent.[7] While the content of notes evolved, the process of distillation remained substantially unchanged as practitioners sifted and augmented the original patient narrative; a clinician-historian has recently characterized the account resulting from consultation as the 'deep, "true" history revealed by the skill of the physician'.[8] In a more equivocal but fundamentally convincing assessment, Hunter considers the case report as 'best understood as a practical response to medicine's radical uncertainty'.[9]

Case notes have been produced for individual, institutional and disciplinary purposes, but the initial generation of notes was motivated by practitioners' understanding of their medical role and responsibility. Individuals began to keep accounts of the ways that their patients presented for two reasons. First, they sought to learn from the repetitions and variations exhibited by the course of their patients' ailments, with a view to adjusting their practice and enhancing the efficacy of their treatments.[10] Second, they aimed to keep track of the debts owed by private patients. The early-modern practitioner was essentially an entrepreneur who, like any businessman of the period, found it expedient to extend credit to his customers but also struggled to keep unpaid debts to a minimum.[11] The professionalizing processes of the late eighteenth and nineteenth centuries encouraged the separation of clinical from financial notebooks, such that by 1900 the two genres had become largely discrete and practitioners had started to employ accountants among their support staff, but for most of that period practice notes might comprise a curious mix of medicine and accountancy.[12]

The proliferation and growth of medical institutions across the eighteenth century eventually gave rise to a new genre of case note. The infirmary movement aimed to offer charitable medical treatment to the labouring poor who suffered accidents or curable disease, yet individual patients' case notes are very thinly represented in the early survivals of infirmary archives. The notes that exist in relation to hospital admission cases rather than private practice are coincidentally present in institutional collections as records kept by individual men.[13] This does not necessarily mean that the limited number of paid medical employees of infirmaries did not take any case histories. The apothecary of the Leeds General Infirmary, the salaried and resident medical officer of the institution, kept notes of the cases which interested him in the early 1780s, but his records were not retained by the hospital.[14]

The public sharing of patient case notes for the purpose of advancing knowledge within the discipline of medicine emerged first as a medical subset of narratives contained in the *Philosophical Transactions of the Royal Society of London*. These elite scientific publications were followed by similar narratives for both lay and specialized readerships, with the growth of generalist periodical literature and the emergence of medical scientific publications from the mid-eighteenth century onwards.[15] Medical journalism was an established genre by 1800 and the ensuing century saw it expand dramatically, not least because it could function as a record of professional achievement or a *de facto* advertisement of skill.[16]

Case notes that were embedded in seventeenth- and eighteenth-century articles tended to focus on single occurrences of disease and gave prominence to the individual author by the use of 'I', and to the community of discourse by addressing and including readers as 'we'. In this way early published reports were frequently semi-autobiographical.[17] Following the lead of Robert Boyle (1627–91) and others, they also stressed the authenticity and authority of the writing by their apparently modest aims, density of description, or other conscious rhetorical strategies to incorporate the reader as an effectual witness.[18] Novelistic literary technique came to embellish a variety of scientific or medical notes to elicit another set of responses that were affective or empathetic, although subsequently notes in this vein which signposted the extraordinary or monstrous have conversely been found voyeuristic.[19] By the later part of the eighteenth century, cases were being presented in 'purpose-based, proto-sections' loosely characterized as case narratives, followed by *post mortem* reports, and concluding with a discussion.[20] This structure became more formalized with the imposition of devices such as subtitles, but from the mid-nineteenth century onwards articles including case information were more likely to condense and summarize, to offer multiple examples in composite 'case collections', or to present abstracted data from a large number of examples. Furthermore published cases gradually occluded the involvement of the author and depersonalized the subject of the case, so that the individuality of both doctors and patients was suppressed.[21] The trend for abstraction intensified in the twentieth century, aided by the rise of editorial boards and peer review to ensure standardization of content and quality. Case reviews increasingly offered surveys of tens or hundreds of cases, whereby patients were characterized in cohorts rather than as individuals. The result was a highly coded set of writings that might be explicitly presented to form 'part of a sequence aiming to build case lore' that could instruct practitioners in the use of probability in diagnosis.[22]

Ideas about the function of case notes in relation to medical power take one of two views. One identifies a historic change in the balance of medical power sometimes characterized as the decline of the patient narrative.[23] The rise of physical testing, and the associated insertion in case notes of technical language that could not easily be interpreted by their subject, meant that notes became dissociated from the patient as a person. Instead the patient became a bundle of symptoms or even inanimate tissue in the sub-genre of the autopsy report.[24] This distancing was undoubtedly a phenomenon that came to influence many practitioners' experience of professional life and their patients' encounters with medicine in the nineteenth and twentieth centuries. Alternatively, and more recently, there has been a modest restatement of case notes as comprising rich histories of individuals particularly when arising from patients in mental-health settings. Echoing Aleksandr Luria (1902–77), Oliver Sacks has credited nineteenth-century notes with 'an empathetic humanistic richness' that was only lost in the twentieth century.[25] In scrutinizing the case notes kept at Bethlem, Akihito Suzuki has concluded that 'doctors, in practice, did not possess the power to reconstruct the individual as a "case"' and that it is too simplistic to construe the balance of power in case notes as a tug-of-war.[26]

Furthermore there have always been expert patients of one sort or another.[27] Doctors' notes on their own illnesses put the observer and the observed in the same position, and so would seem to subvert the idea that case notes inevitably deprive the patient of narrative power. Early-modern practitioners were not necessarily coy about foregrounding thoughts about their own complaints and could be keen to extrapolate from their experience for the benefit of medical ideology, other patients, the body politic, their publishing careers, or a combination of the four. John Floyer's *A Treatise of the Asthma* drew heavily on his personal suffering of the complaint from early childhood onwards, and his treatment became a standard work.[28] Roy Porter has suggested that George Cheyne's some-time obesity was instrumental in shaping his thoughts on an inverse relationship between wealth, luxury, and immorality on one side and health and virtue on the other.[29] The most startling historic example of a medical man placing his own health in the service of the public must be that of John Hunter, who is alleged to have infected himself with syphilitic matter in an attempt to distinguish the disease from gonorrhea.[30] But Hunter's published notes on his 'patient' betray an early desire for researcher anonymity, reflected in some twentieth-century doctors' construal of their own case in the third person.[31] This tendency to distance the patient even when the patient is also the practitioner is an intriguing narrative prospect, but can still be read in two ways; either the need to objectify and assert power over patients is so strong that the prose technique is adopted even when the patient is the medically trained author, or the modesty necessary for authenticity which features throughout the case note's history demands the elision of the practitioner even when they are the subject. Therefore such cases cannot in themselves overcome the argument that notes have been used to subordinate patients' own accounts.

## Case notes of the mad in Britain

Case notes of patients with mental illness were initially found less frequently than those discussing physical complaints, but they too appear in individuals' practice records and in publications from the seventeenth century. Richard Napier (1559–1634) took notes on 2,483 patients deemed insane.[32] In the eighteenth century, as with infirmaries, case notes might have been generated for asylum patients but they tend to survive for individual practitioners attending or superintending asylums rather than among institutional collections.[33] Yet case notes of mental disorders proliferated in the early nineteenth century and by 1900 comprised one of the most voluminous sources of patient information that is now open to historians of medicine.[34] This profusion of material arises from the relatively sudden emergence of county and borough asylums for the pauper insane, which did not exist in 1800 but supplied 7,619 places for lunatics by 1851 and 72,396 places by 1900.[35] Case notes were compiled as a matter of routine in formatted ledgers, and following the 1845 County Asylums Act keeping such volumes became compulsory.[36] Ironically, though, the resulting archival riches were not initially generated so much for their therapeutic and educative purposes as they were for facilitating external scrutiny and eradicating asylum abuses.[37] The value of systematic note-taking for the study of insanity was recognized later.[38]

Notes deriving from nineteenth-century institutions were constructed by multiple authors and typically start with tabulated personal data including the patient's admission number and date, name, age, marital status, religious affiliation, occupation, address, and the tariff charged for the duration of their inpatient status, usually levied from the same person or organisation who sought the patient's admission. Ledgers (later *pro formas*) might also contain spaces to insert a patient's height, and voluntary additions could include a textual portrait of the patient comprising hair and eye colour, build, facial expression, level of nutrition and any physical tics. Personal information from patients themselves or families and friends was augmented by medical and diagnostic data including details from the admitting medical certificates. Asylums regularly observed whether the new patient was manic or melancholic, suicidal or epileptic, and whether there was any family history of insanity, alongside the alleged predisposing or exciting causes of the disorder, the patient's age at the time of their first spell of mental illness, the duration of this episode of illness, and whether or not they had already been under treatment at another institution. This preliminary material often features the further refinement of a regular or even slavish recording of the state of the patient's bowels.[39] Theories which identified the gut as the originary site of mental disorder were on the wane by the later nineteenth century, but having collected such details for decades, institutional procedure dictated that notes on easy or inhibited defecation should still be gathered. Few institutions compiled all of this information in all years, and entries for pre-set classes of information could become perfunctory or ignored, but most asylums gathered at least a subset of these items.[40]

The on-going, free-text narrative of patient cases typically comprised sequential, dated entries that modified the information that had been established during admission. For some patients this resulted in a tedious list of dates, all of which were written merely to confirm that there had been 'no change' in their condition. More typically, medical superintendents used the case notes to record the patient's bodily condition, evidence of their on-going mental disturbance including details of delusions, and the need for any special treatments, drugs or restraints. They also contain occasional comments on behaviour, verbal or written communications, leisure or work activities, visits from family and friends, permission to travel beyond the asylum, escape attempts, or violence. Any confession by patients that they recognized the error of their former delusions was regarded as a significant milestone towards recovery.[41] Notes for chronic patients tend to become more sparse and uninformative, while those for all long-stay patients might become dispersed across two or more ledgers, and can be tracked by contemporary notes to aid cross-referencing.[42] For patients who died in the institution, the final entry in the ledger of notes might be a narrative of their *post mortem*. This was typically a more self-consciously clinical sub-section of the case note, replete with abbreviated Latin, or weights and measurements of organs, alongside textual descriptions.

The content of these extended notes was probably guided if not wholly determined, at least at first, by the tenets of moral therapy. The early nineteenth century saw an increasing investment in moral therapy as a strategy for securing improvement or cure of the insane, devised as an alternative to vigorous or violent physical therapies.[43] It was predicated on the idea that social conversation and pursuits could be offered for 'good' behaviour and withheld for chaotic demeanour and irrational communication. The therapeutic optimism engendered by moral therapy went a long way to underpin statutory provision of asylum accommodation for the pauper insane, although the capacity of county asylums to offer anything akin to moral therapy was highly constrained in practice by staff-to-patient ratios.

Case notes are sporadically enhanced by other inclusions such as patient writings, newspaper cuttings, and photographs. Patient correspondence, poems, drawings, and other forms of inscription were occasionally selected for inclusion, and although the specific justification is rarely explained the material is usually illustrative of the writer's state of mind, perhaps providing instances of an entrenched or newly apparent delusion, and validating the judgement of clinicians.[44] Reports of the content of writings were more often summarized in the notes than original documents were pasted into the ledgers. Newspaper cuttings allude to episodes before admission or after discharge, and are most likely to be married with the notes where criminal action was given public notice. Consequently they are a rare but poignant addition to a small subset of cases. Photographs, in contrast, became quite common as a way to reinforce the case-note narrative. These might be used to track facial representations of the development or decline of cases or to assist in the tracing of patient escapees.[45]

The role of case notes in histories of mental illness or disease is problematized by changing aetiologies and the evident instability of symptomatic descriptions across centuries. John Monro's case book of 1766, for instance, distinguishes between true lunacy in contrast to mere fever or frenzy, but how are we to interpret this in hindsight? Similarly Monro noted that a higher proportion of women than men presented with depression and the same is the case in the early twenty-first century, but does this inevitably identify 'a transhistorical or illness-specific phenomenon'?[46] Case notes from the seventeenth to the twentieth centuries sometimes exhibit an emphatic confidence about supposed causation, or more often a confusing swirl of evidence that points to diagnostic uncertainty. The application of asylum case notes to a better appreciation of the nosology of mental illness was also problematic for contemporaries; 'symptoms and behavior noted in the case books are often left to stand for themselves instead of forming a diagnosis'.[47] At similar institutions even in later decades, case notes 'merely reaffirmed the capricious nature of disorders that were only dimly understood'.[48] Therefore, while there have been attempts to diagnose asylum patients retrospectively, the results must be viewed with scepticism.[49] Case notes remain valuable for histories of disease, though, where they permit sensitive charting of historically contingent clinical classifications. Gayle Davis has used Scottish asylum case notes to consider diagnoses of General Paralysis of the Insane. She notes the stability of this disease identity in the years before laboratory testing could be used to confirm its presence, and the relationship between laboratory and clinical assessments thereafter.[50]

Beyond the formal medical sphere, understandings have developed through the publication of patient narratives or fictional cases from well-informed authors, although lay narratives are much less likely to dwell on matters of causation. Instead these have served from the second half of the twentieth century to disseminate awareness of diverse conditions through popular cultural channels, alongside judgements of the treatments they have prompted and acceptance of behavioural difference among a broad constituency. Psychiatric institutions in North America have been subjected to repeated scrutiny.[51]

Aside from their complicated relationship to histories of disease, case notes have been used to secure two different views of asylum life: the administrative and medical superintendence of the institution; and the patient experience. Medical management of asylums over the nineteenth century generated changes that were advantageous for patients, such as the wholesale removal of physical restraint as an allegedly therapeutic intervention. Case notes had a powerful contemporaneous use for institutional promotion wherever they were converted to case histories and incorporated in annual reports. Histories were deployed to make points about medical expertise and education, curative success, rehabilitation of patients and other socially useful results of asylum activity.[52] Nonetheless it is important to move beyond case notes because institutions' own archives tend to support only positive readings of their activities, and critiques must be sought elsewhere.[53] In the early twentieth century, for example, the Leavesden asylum in Hertfordshire was regarded rather favourably in official reports, but a keeper employed there during the First World War thought it 'corruption from top to bottom'.[54]

The potential and pitfalls of case-note scrutiny for an analysis of patient experience have been surveyed for Gartnavel Royal Hospital in Glasgow by Jonathan Andrews. The first obvious problems arise from the incomplete condition of notes and their discrepancies with other institutional documentation. Information on the patient prior to admission can be thin or entirely absent, while the histories submitted in different forms, such as medical certificates or questionnaires completed by families, might be omitted or altered for the case notes.[55] Supplementary issues arise over the authorship of the notes, and the priorities of these writers that could govern what was annotated and what was rejected as vulgar or inadmissible; for example, in the Glasgow case notes as elsewhere masturbation and sexual fantasy were often treated in allusive language if at all, while patient articulacy or flamboyant delusion were given extensive coverage.[56] Davis concludes that 'patient testimony is only really found in those letters written by the patient and retained in their case notes' and even here institutional regulation of access to paper and pens filtered the production of viable patient writings.[57]

Therefore case notes provide only a partial view of what happened to patients, both before and after their admission to asylums. This could be the case regardless of the status of the individual patient. Pauper asylums admitted patients who were paid for by the poor law, but also private paying patients whose accommodation was determined by the level of their fees. This meant that county and borough asylums were treating middle-class, professional and wealthy individuals alongside the poor and destitute, but kept case notes to the same standard for all inmates. Consequently modern attempts to capture the nineteenth-century patient experience are no easier for private patients, even where the patient shared a medical background with his keepers.

## Doctors as patients

Nineteenth-century case notes used by historians to date, particularly published case reports, are quite narrow in their social reach; 'it is rare to find the case history of a professional patient or one from a "superior" walk of life . . . it would appear that medicine was practiced by one class upon another'.[58] This understanding has been underpinned by a reluctance to examine incidence of mental illness among middle- or upper-class men such as, for example, medical professionals.[59] Recent close work on institutional records has revealed that these men were not absent from asylums but that they have been overlooked.[60] Physicians, surgeons and other trained medical professionals were admitted to asylums including pauper establishments. On census day in 1881 there were 81 medical patients in asylums across 43 institutions in England and Wales. The institutions accounting for the largest numbers of medical patients were the York Lunatic Asylum (seven medical men) and St Andrew's Hospital in Northampton (six medical men), both of which were funded by charity rather than from county rates, but county and borough asylums collectively held 30 such patients.[61]

Closer investigation of the case notes compiled for medical patients reveals substantial parallels with the generalizations found elsewhere. Notes for chronic patients with protracted stays in hospital became sparse or petered out altogether, while notes for men with florid delusions or extremes of behaviour remained full. It is perhaps misleading, then, to draw anything firm from patterns of reportage on medical delusion, but one motif does crop up repeatedly across different institutional settings: doctors suffering false perceptions were described as particularly disturbed by invisible but tangible forces such as electricity and telegraphy, or associated aspects of fringe medicine such as mesmerism. This was never clearly a product of electrical therapy being applied to the same men, and some institutions went to great pains to account for any physical injuries among the men as arising from something explicitly other than electric shocks.

It is not at all surprising that medical men might be disturbed by the implications of unseen forces, or that in ill-health their thoughts turned to such forces in obsessive or fearful ways. It is similarly understandable that medical superintendents' observations of medical patients were particularly attentive to these alleged forces. Mesmerism presented a direct challenge to orthodox medical authority in the 1840s, one which was only overcome partially and piecemeal. Practical measures such as the development of chemical anaesthesia were insufficient to counter the powerful endorsements of Harriet Martineau and others.[62] Among mesmerism's many successors, electro-biology offered another set of apparent discoveries about the usurpation of human autonomy and consequential physiological, medical or mental impacts.[63] At the same time the incontestably concrete effects of telegraphy encouraged some men to develop morbidly complex theories of their own about electrical communication.

This contemporary swirl of scientific advance and widespread professional unease gave rise to two different medical responses among men with poor mental health that were regularly reported in case notes. The less common was to inspire a sense of dangerous power that must be contained. Angus Cameron, a physician admitted to the Warwickshire asylum in 1873, was alarmed about his own capacity to harm others through the inadvertent discharge of electricity. In characterizing his mental state, his case notes mention that on admission he complained 'of electricity flying about him & that he is obliged to point with his finger to discharge it', a phenomenon which a month later made Cameron fearful that he might set light to the building. He was also under the impression that he could telegraph all over the world in a minute.[64]

A more typical apprehension was one of helpless subjection to an external source of power. Frederick Wright had been a surgeon in Derby but his declining health in 1856 was marked by a belief that other practitioners in the town were practising mesmerism and other occult arts upon him. As a patient in the Derbyshire asylum, he perceived himself as in thrall to badly disposed individuals among the other patients: 'The parties against me now are an Inmate named Wells of nervous temperament, who acts the wide awake "clairvoyant", and Pegg the muscular, and hence between the two, and being prevented from taking the law in my own hands'. This helplessness was not strictly borne out by Wright's behaviour, since the notes go on to observe that 'he omits no opportunity to attack them either by violent language, or by actual blows'.[65] In the following decades fears of being controlled by others via electrical means were equally evident among the medical patients at Wonford House Hospital near Exeter. Silvanus Tucker fancied that people or parts of the hospital dining room discharged electricity at him, and that as a consequence he was driven to take measures to prevent 'communication thro' (such as sitting on pillows).[66] (See Figure 29.1.) Wonford's most tortured medical patient was Francis Spencer, since he apprehended persistent electrical persecution in the form of both chattering voices and the application of physical pain by unnamed assailants: 'the villains found out that they could talk to me & attack me wherever I was without necessarily seeing me and they have made exceedingly free use of their information'.[67] He was also the most active in publicizing his plight and seeking a practical remedy. In the 1880s he blackened his eyes with charcoal to encourage belief in his injury by electrification, and had leaflets printed on the subject which he distributed in the streets outside the hospital. In the 1890s he was reported to 'sleep with a pad in each ear, a tube in each nostril and a split cork between his teeth in order to stave off electric attacks'. His fear of subjection to external electrical manipulation remained undiminished in 1918.[68]

Systematic analysis of the diagnostic and symptomatic case notes kept for medical patients can therefore reveal aspects of concern to both the men who were sick and the men who treated them. The same notes can also serve to conceal biographical and diagnostic information of great importance, but these omissions can only be detected by painstaking cross-referencing of case notes with sources external to the institution or by serendipity. For example the case notes for Angus Cameron, mentioned above as prey to delusions about electricity, also describe his demeanour during recuperation as going on 'quietly but queerly'. He was a heavy smoker, who 'Amuses himself by playing Bagatelle by himself, or one day by scribbling all over the wall about Emily Burridge'.[69] This obscure reference to a female name inspires many possibilities but is not glossed further in the notes, and the name is too common to permit any narrowing of the field of candidates by reference to census records. Only

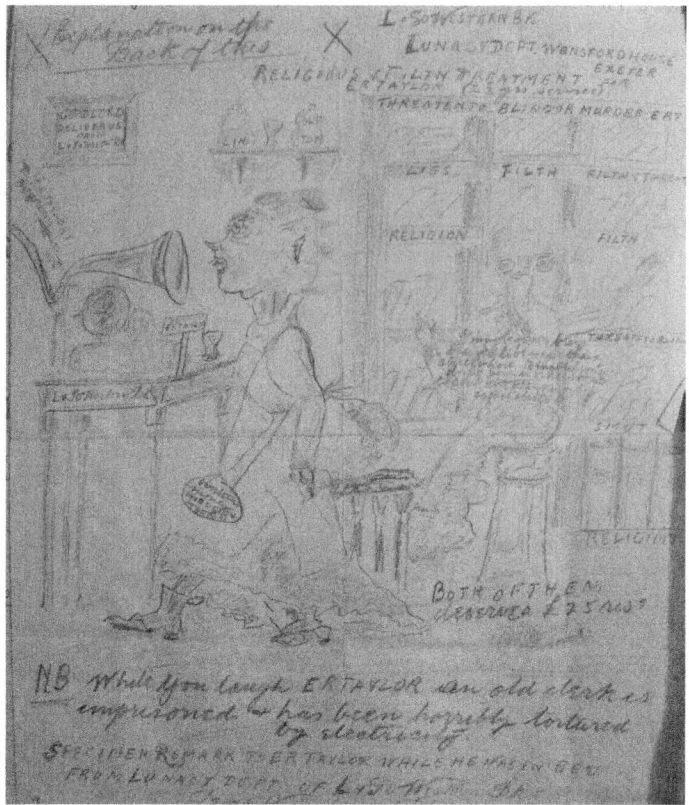

*Figure 29.1* Cartoon drawn by Ernest Taylor, patient in the Wonford House Asylum, 1913, giving a visual depiction of his being 'horribly tortured by electricity'; Devon Heritage Service, 3922F/H32/7 Wonford House case book 1885–1926, p. 415.

a chance order of a family death certificate revealed that Emily Burridge was a servant in the Cameron family household in Bristol who had been present at the death of one of Cameron's children.[70] She was therefore a viable candidate for the 'dear young girl who saved his life twice' who Cameron identified as Cleopatra when first admitted to the Warwick institution.[71] It is entirely unsurprising that the medical staff at Warwick should have remained mystified as to the role of Emily Burridge in Cameron's life, particularly given the fact that Cameron's wife was apparently never involved in supplying information for his admitting certificates or subsequent case notes. Asylum accounts of Charles Beard, on the other hand, demonstrate the potential for casenote silence even when pertaining to events of notoriety or infamy.

## The case notes of Charles Beard

Medical superintendency of patients who were also practitioners was perhaps more likely to incorporate some attempted discussion of professional life, given the shared experience of medical training and practice. Interactions of this type were reported in some case notes, employed either as a means to restore the patient's thoughts to a professional character or (in the case of recovering patients) to debate the practicalities of a return to work. For example John Hitchman (1816–1893), the superintendent of the Derbyshire asylum, lent books to Frederick Wright, mentioned above, in the hope of gaining his patient's confidence (a strategy which, sadly for both men, proved unsuccessful).[72] Yet these sorts of additional details are rare. The notes written for most medical patients, rather than being fuller or more textured, tended to betray no evidence of shared medical knowledge and were just as likely to suffer from absences, omissions, elisions and evasions as those for pauper patients. The case notes of one medical patient provide a particularly dramatic example of the way that institutional records could remain silent about relevant patient experiences prior to admission, even where these experiences had been the subject of national publicity.

Charles Beard was admitted to St Andrew's asylum in Northamptonshire on 6 August 1886. The 58-year-old physician had been born in Brighton and had practised there for most of his professional life. At the time of admission he was said to have been ill for the previous 15 years, but the cause of his poor health was allegedly unknown and he had not been treated elsewhere. The medical certificates that underpinned his admission drew a picture of Beard as labouring under a false impression of conspiracy against him, and the details emerge more fully as the case notes develop. Beard claimed that he was first unsettled by a large number of deaths arising from vaccinations some years before. He then claimed that accusations of bestiality had been made against him by another Brighton practitioner, and that the conspiracy had widened to include the police, the post office, and a number of titled men. His anxiety had been exacerbated by a subsidiary apprehension of having been accused of taking bribes, and by his wife's failure to perceive events in the same way.[73]

Beard made a good initial impression on staff at St Andrew's. He was described as 'courteous, affable, and highly intelligent. He converses freely and quite rationally upon all general topics & most of the incidents of his past life, describing his experiences as a Local Government Board Inspector, & General Practitioner'.[74] Thereafter and for over two years he presents in the case notes as a compliant if latterly reticent and unwilling patient. He had two interviews with Lunacy Commissioners (who on both occasions found him to be properly detained, despite his own conviction that he was not deluded), and occupied his days attending chapel, reading, smoking, and playing billiards. His determination in maintaining allegations of conspiracy did not inspire therapeutic optimism, however, and he developed a reputation for irritability, and unpopularity with other patients.[75]

Then in March 1889 Beard precipitated a crisis. He alleged to one of the hospital visitors that another patient, Theophilus Gist, had been maltreated by the asylum's staff. Beard claimed that two days before his death Gist had been thrown down, kicked, and otherwise assaulted by five attendants. This marked the start of a campaign by Beard to use all means possible to accuse the medical superintendent and staff of St Andrew's of gross mismanagement, assault and brutality.[76] The relationship between patient and institution broke down so thoroughly that Beard was transferred to another unnamed asylum on the advice of the Lunacy Commissioners.[77] Beard was allegedly living at home with his wife in the censuses of 1891 and 1901, but probably remained quite unwell; at some point he was admitted to Holloway Sanatorium and died there in 1916.[78]

Thus the case notes drawn from the St Andrew's archive reveal something of the mental state of a formerly active professional practitioner. They illustrate the value placed by the medical superintendent on the intelligence of Beard's general conversation when not distracted by 'delusion', and on the concurrent risk to the institution of potentially damaging allegations when coming from such a plausible and formerly authoritative source. What they do not do, however, is reveal the precipitating cause of Beard's ill-health.

Beard may indeed have been deluded in thinking himself the target of a conspiracy by fellow practitioners and officers of the state such as the policemen and postmen. He would have been entirely correct, however, in recalling that he had been the subject of a different campaign of covert violence against himself and his family, because 15 years prior to his admission to St Andrew's, Beard had been caught up in the trial of notorious poisoner Christiana Edmunds (1828–1907).[79] In 1871 Edmunds endangered the lives of her fellow Brighton residents by planting poisoned sweets in confectionary shops. When four-year-old Sidney Barker died of poisoning that summer she was questioned at the resulting inquest, but no further action was taken until Charles Beard came forward with new evidence, and a ghastly series of events was uncovered. Edmunds and Beard had been involved in some kind of relationship, although whether this ever moved beyond friendship on Beard's side is not clear. Beard was married, so Edmunds was apparently motivated to recommend herself more strongly to Beard, or even to remove the impediment to his second marriage, by poisoning his wife. Emily Beard was fed sweets by Edmunds in September 1870 and fell ill immediately afterwards. Mrs Beard recovered but Christiana Edmunds was, unsurprisingly, no longer welcome at the Beards' house. Therefore she seems to have adopted her subsequent practice of multiple, random poisoning to deflect blame for Emily Beard's illness. Edmunds was found guilty of the murder of Sidney Barker and sentenced to hang, but William Orange (1833–1916), the medical superintendent of Broadmoor, interviewed her and judged her insane; her sentence was commuted to imprisonment and she remained in Broadmoor until her death in 1907. Her own case notes reveal a woman fond of mystery, concealment, and convinced of her own attraction for men.[80]

Beard's experience of these events was probably acutely distressing, for three reasons. First, there is circumstantial evidence that he was at first a willing participant in his relationship with Edmunds. This may have been mere friendship on his side or it may have been more suggestive and romantic (even sexual, given the overheated nature of Edmunds' letters to him), but whatever the nature of their acquaintance Beard was relatively slow to name Edmunds as the possible source of Brighton's malicious poisonings. His reticence is understandable, whether he was merely embarrassed about an innocent friendship or whether he feared exposure as an adulterer, but there was a notable lapse of time before he came forward to the police. Second, his wife and the mother of his children had been emphatically placed in harm's way under his own roof. This was not just a challenge to the Victorian domestic ideal, but also to medical masculinity; a failure to treat or cure patients could be accommodated as part of normal professional life, but a failure to protect family members from physical harm was more difficult to accept.[81] Third, and in all likelihood quite devastating for Beard's sense of himself as a public persona, he was not allowed to give evidence against Edmunds and coincidentally defend himself at her criminal trial. Dr and Mrs Beard attended the Old Bailey trial of Christiana Edmunds for murder and Charles Beard took the stand, but was very quickly ordered to sit down on the basis that he could offer no direct evidence about the death of Sidney Barker; he could only speak to Emily Beard's illness for which Edmunds was not charged nor prosecuted.

This means that although Beard had received national exposure in the press during 1870 and 1871 as implicated in the background to Edmunds' trial and conviction, and although Emily Beard must have known that his illness of 15 years duration was at least coincident with, if not directly attributable to, Edmunds' impact on their lives, no word of this history reaches the case notes kept by St Andrew's asylum. It would presumably have been highly germane for the staff to appreciate the origins of Beard's delusions, rooted as they were in a literal conspiracy of sorts and a documented criminal prosecution. Is the absence of the case from the notes expressive of Emily Beard's failure to incorporate it in her husband's medical history, or of the medical superintendent's decision to excise it from the formal record? Emily Beard may have been exercising one of the few components of control left to her, since by her silence the Edmunds story was substantially restored to the status of family secret.[82] Was it elided from the notes on the grounds of discretion for the suffering and exposure of a former medical colleague? It is not easy to see why this should have been judged necessary, given the confidential nature of the notes, but it is not impossible. In either case, however, the limitations of asylum case notes are brought dramatically into view. Any history not revealed in the medical certificates, or by family testimony at the time of admission, is lost no matter how central it might have proved to the patient's onset of illness and regardless of pertinent information notionally held in the public domain.

## Conclusion

Practitioners' motives for keeping case notes from the seventeenth century were both theoretical and pragmatic, in that they aimed to assist clinical understanding and regularize patients' payments of their medical bills. The deployment of notes beyond the individual practice quickly developed a credibility and weight commensurate with a number of parallel developments such as the evolution of scientific method discussed briefly above and other phenomena such as rejection of supernatural causation, the emergent coalitions evident within the profession designed to raise the status of medicine, and the enhancement of clinical components in medical education. Yet subsequent variations in the retention or publication of notes were not enough on their own to promote the recognition of their worth by historians. Medical historians needed first to conceptualize notes as a rich source of factual and narrative data and, for this to take place, understandings of past events had to shift away from great men and technological change to refocus on the day-to-day mundanities of routine practice and of multiple individual patients.

The value of psychiatric case notes in particular lies in a combination of their volume and their relative avoidance of dogmatism or self-consciousness. Routine collection of notes by a vast asylum network gives rise to a data-rich set of sources, compiled by medical superintendents and others whose own diagnostic criteria were mutable. This encouraged note-writers to gather diverse forms of information and commentary on patients which can be scrutinized under a multitude of headings. This chapter has focused on the social history of an occupational group, but a more immediate use would address psychiatric history, through the presentation of patients with the same notional illness.[83]

Apprehending practitioners themselves as patients, and perhaps as the ultimate expert patients, is a very recent phenomenon. It is entirely plausible that the appetite for case notes could not readily overcome the shadow of pioneering practitioners' 'greatness' to entertain all at once the frailties of the same medical man. This reticence also remains a characteristic of some sections of the modern profession, despite the currency of questions around practitioners' fitness to practice. Unpicking the case notes of medical patients offers a variety of new perspectives, however, about annotation of the self, career stress, and intra-professional relations. Ideally it will also contribute to a more open and less guarded discussion about the private, less laudatory aspects of medicine in the past, which would surely benefit practitioners and patients alike.

## Notes

1   Earlier case notes were intended as didactic exemplars rather than as models for progressive practice; B. Hurwitz, 'Form and representation in clinical case reports', *Literature and Medicine* 25, 2006, 216–40.
2   B. J. Shapiro, *Probability and Certainty in Seventeenth-Century England*, Princeton: Princeton University Press, 1983, Ch. 2.
3   For an overview of the development of medical records see G. Brieger, 'The Historiography of Medicine', in W. F. Bynum and R. Porter (eds), *Companion Encyclopaedia of the History of Medicine*, London: Routledge, 1993, pp. 24–44, 32–3.
4   E. H. Ackerknecht, 'A Plea for a "Behaviourist" Approach in Writing the History of Medicine', *Journal of the History of Medicine and Allied Sciences* 22, 1967, 211–14, p. 212; G. B. Risse and J. H. Warner, 'Reconstructing clinical activities: Patient records in medical history', *Social History of Medicine* 5, 1992, 183–205, p. 184.
5   W. F. Bynum, 'Nosology', W. F. Bynum and R. Porter (eds), *Companion Encyclopaedia of the History of Medicine*, London: Routledge, 1993, p. 343; E. Ashworth Underwood, *Boerhaave's men at Leyden and after*, Edinburgh: Edinburgh University Press, 1977.
6   For this practice in the eighteenth century, see for example: D. Porter and R. Porter, *Patient's Progress. Doctors and Doctoring in Eighteenth-Century England*, Cambridge: Polity Press, 1989, pp. 76–8; W. Wild, *Medicine-by-Post: The Changing Voice of Illness in Eighteenth-Century British Consultation Letters and Literature*, Amsterdam: Rodopi, 2006. See also the chapter in this volume by Robert Weston.
7   The case note was one type among numerous 'realistic narratives of the lived body' which emerged or were consolidated in the eighteenth century: T. Laqueur, 'Bodies, details, and the humanitarian narrative', L. Hunt (ed.), *The New Cultural History*, Berkeley: University of California Press, 1989, 176–204, p. 201.
8   J. Gillis, 'The history of the patient history since 1850', *Bulletin of the History of Medicine* 8, 2006, 494.
9   K. M. Hunter, *Doctors' Stories: The Narrative Structure of Medical Knowledge*, Princeton: Princeton University Press, 1991, p. 106.
10   Not all practitioners changed their practice as a result of their note-taking; see L. M. Beier, 'Seventeeth-century English surgery: the case book of Joseph Binns', C. Lawrence (ed.), *Medical Theory Surgical Practice: Studies in the History of Surgery*, London: Routledge, 1992, p. 80.
11   See for example S. King, *A Fylde Country Practice: Medicine and Society in Lancashire, circa 1760–1840*, Centre for North West Regional Studies, 2001, pp. 82–95.

12 A. Digby, *The Evolution of British General Practice, 1850–1948*, Oxford: Oxford University Press, 1999, pp. 137–8, 314.

13 G. B. Risse, *Hospital Life in Enlightenment Scotland: Care and Teaching at the Royal Infirmary of Edinburgh*, Cambridge: Cambridge University Press, 1986, pp. 296–301 makes use of selective and discontinuous case notes compiled by students rather than infirmary staff.

14 S. T. Anning, 'A medical case book: Leeds, 1781–84', *Medical History* 28, 1984, 420–31.

15 D. Atkinson, 'The evolution of medical research writing from 1735 to 1985: The case of the *Edinburgh Medical Journal*', *Applied Linguistics* 13, 1992, 337–74, p. 342.

16 W. F. Bynum and J. C. Wilson, 'Periodical knowledge: Medical journals and their editors in nineteenth-century Britain', in Bynum, Lock and Porter (eds), *Medical Journals*, p. 30; C. Galley and R. Woods, *Mrs Stone & Dr Smellie: Eighteenth-Century Midwives and Their Patients*, Liverpool University Press: Liverpool, 2014, p. 47.

17 Hurwitz, 'Form and representation', 222.

18 S. Shapin and S. Shaffer, *Leviathan and the Air Pump. Hobbes, Boyle, and the Experimental Life*, Princeton: Princeton University Press, 1985, pp. 60–9; Laqueur, 'Bodies', pp. 182–3.

19 Kennedy, *Revising*, pp. 32, 35–7; R. Rylance, 'The theatre and the granary: Observations on nineteenth-century medical narratives', *Literature and Medicine* 25, 2007, 255–76, p. 263.

20 Atkinson, 'Evolution', 347–8.

21 Atkinson, 'Evolution', 346, 348, 350; Hurwitz, 'Form and representation', 225 and 228.

22 Rylance, 'Theatre', 258. See also the chapter in this volume by Brian Hurwitz.

23 M. Foucault, *The Birth of the Clinic: An Archaeology of Medical Perception*, London: Tavistock, 1973; M. Fissell, *Patients, Power and the Poor in Eighteenth-Century Bristol*, Cambridge: Cambridge University Press, 1991, Ch. 8.

24 Kennedy, *Revising*, pp. 57–61.

25 O. Sacks, *The Man Who Mistook His Wife for A Hat*, London: Picador, 1986, p. x; A. R. Luria, *The Man with a Shattered World*, Cambridge, MA: Harvard University Press, 1972. The quote comes from a consideration of Sacks' position in Rylance, 'Theatre', 255–6.

26 A. Suzuki, 'Framing psychiatric subjectivity: Doctor, patient and record-keeping at Bethlem in the nineteenth century', J. Melling and B. Forsythe (eds), *Insanity, Institutions and Society, 1800–1914: A Social History of Madness in Comparative Perspective*, London: Routledge, 1999, 115–36, pp. 119, 131.

27 There is extensive autobiographical literature from expert patients, but usually from an untrained perspective; see for example S. Hickey, *Finding Balance: Healing from a Decade of Vestibular Disorders*, New York: Demos Medical Publishing, 2011.

28 M. Jackson, *Asthma: The Biography*, Oxford: Oxford University Press, 2009, pp. 61–6.

29 G. Cheyne, *The English Malady*, [Dublin], 1733, pp. 222–51, 'The CASE of the author'; R. Porter, 'Civilisation and disease: Medical ideology in the Enlightenment', J. Black and J. Gregory (eds), *Culture, Politics, and Society in Britain, 1660–1800*, Manchester: Manchester University Press, 1991, 154–83, p. 165.

30 W. Moore, *The Knife Man: Blood, Body-Snatching and the Birth of Modern Surgery*, London: Bantam, 2005, Ch. 8.

31 Hurwitz, 'Form and representation', 231–3.

32 M. Macdonald, *Mystical Bedlam: Madness, Anxiety, and Healing in Seventeenth-Century England*, Cambridge: Cambridge University Press, 1981, Ch. 4.

33 J. Andrews, 'Case notes, case histories, and the patient's experience of insanity at Gartnavel Royal Asylum, Glasgow, in the nineteenth century', *Social History of Medicine* 11, 1998, 255–81, p. 256. n. 4.

34 G. Davis, *'The Cruel Madness of Love': Sex, Syphilis and Psychiatry in Scotland, 1880–1930*, Amsterdam: Rodopi, 2008, p. 23, reflects that the volume of clinical evidence itself may have deterred research.

35 P. P., *The sixth annual report of the Commissioners in Lunacy*, 1851, appendix A; P. P., *Copy of the fifty-fourth report of the Commissioners in Lunacy*, 1900, appendix A.

36 8 and 9 Vict. C.126.

37 Andrews, 'Case Notes', 256, 267. For wrongful confinement see N. Hervey, 'Advocacy or folly: The alleged lunatics' friend society, 1845–63', *Medical History* 30, 1986, 245–75.

38 Andrews, 'Case notes', 260.

39 For an early adopter of the 'alimentary Gospel' see Cheyne, *English Malady*, pp. 244–6.

40 Suzuki, 'Framing psychiatric subjectivity', p. 123, for flaws in the record-keeping at Bethlem.

41 Andrews, 'Case notes', 277.

42 Discrete case files for individual patients were not usually compiled until the twentieth century; Andrews and Scull, *Customers*, p. 18. There are exceptions to this rule, see for example the Broadmoor archive in the Berkshire Record Office.

43 A. Digby, *Madness, Morality and Medicine: A Study of the York Retreat, 1796–1914*, Cambridge: Cambridge University Press, 1985.

44 Andrews, 'Case notes', 277.

45 H. W. Diamond, 'On the application of photography to the physiognomic and mental phenomena of insanity', *Proceedings of the Royal Society of London* 8 (1857), reported p. 117; C. Gale and R. Howard, *Presumed Curable: An Illustrated Casebook of Victorian Psychiatric Patients in Bethlem Hospital*, Petersfield: Wrightson Biomedical Publishing, 2003. See also S. Sidlauskas, 'Inventing the medical portrait: Photography at the 'Benevolent Asylum' of Holloway, c. 1885–1889', *Medical Humanities* 39, 2013, 29–37. In contrast Davis, *Cruel Madness*, pp. 96–100, points to rather limited use of photography in Scottish asylums and the absence of a clear rationale for using the images.

46 J. Andrews and A. Scull, *Customers and Patrons of the Mad-Trade: The Management of Lunacy in Eighteenth-Century London*, Berkeley: University of California Press, 2003, pp. 62, 71.

47 T. Turner, 'Rich and mad in Victorian England', *Psychological Medicine* 19, 1989, 29–44, p. 43.

48 W. Jackson, *Madness and Marginality: The Lives of Kenya's White Insane*, Manchester: Manchester University Press, 2013, p. 86.

49 Turner, 'Rich and mad', notably published in a medical journal rather than an historical one.

50 Davis, *Cruel Madness*, p. 21, and chapters 3 and 4.

51 For example in K. Kesey, *One Flew Over the Cuckoo's Nest*, New York: Viking, 1962; S. Kaysen, *Girl Interrupted*, New York: Turtle Bay, 1993.

52 Andrews, 'Case notes', pp. 271–2.

53 A. Borsay and P. Shapely, 'Introduction', in A. Borsay and P. Shapely (eds), *Medicine, Charity and Mutual Aid: The Consumption of Health and Welfare in Britain, c. 1550–1950*, Aldershot: Ashgate, 2007, p. 3.

54 P. P., *Sixty Fourth Report of the Commissioners in Lunacy to the Lord Chancellor*, 1910, pp. 534–6; Brunel University Library, Burnett Archive of Working-Class Autobiographies, George Lloyd, 'The autobiography of Georgie Brawd', 3:108.

55 Andrews, 'Case notes', 262. Answers to questions about hereditary disposition were particularly likely to toy with the 'truth'; Turner, 'Rich and mad', 32.

56 Andrews, 'Case notes', 265–6.

57 Davis, *Cruel Madness*, p. 28.

58 Rylance, 'Theatre', 257.

59 M.F. a'Brook et al., 'Psychiatric illness in the medical profession', *British Journal of Psychiatry* 113, 1967, 1013–23, notes this reluctance, as does T. M. Hassan et al., 'A postal survey of doctors' attitudes to becoming mentally ill', *Clinical Medicine* 9, 2009, 327–32; L. Carpenter, 'Why doctors hide their own illnesses', *The Guardian* 17 May 2014. Online. Available: http://www.theguardian.com/society/2014/may/16/why-doctors-hide-their-own-illnesses (accessed 26 August 2015).

60 A. Tomkins, 'Mad doctors? The significance of medical practitioners admitted as patients to the first English county asylums up to 1890', *History of Psychiatry*, 23, 2012, 437–53; J. V. Kragh, 'Women, men and the morphine problem, 1870–1955' in T. Ortiz-Gómez and M. Santesmates (eds), *Gendered Drugs and Medicine: Historical and Socio-Cultural Perspectives*, London: Ashgate, 2014, pp. 177–98, for male physicians as morphine addicts in Denmark.

61 Tomkins, 'Mad doctors', 441–2.

62 A. Winter, *Mesmerized: Powers of Mind in Victorian Britain*, Chicago: University of Chicago Press, 1998, pp. 225–30.

63 Ibid., pp. 281–4.

64 Warwickshire Archives, CR 1664/623 Warwickshire asylum case book, 1872–75, patient number 2098, entries for 4 January, 24 January, and 13 February 1873.

65 Derbyshire Record Office, D1658/11/3, Derbyshire asylum case book 1856–57, patient number 41, entries for 11 March and 17 April 1856.

66 Devon Heritage Service, 3992F/H15/1-4 OR 3922F/H32/4-7, Wonford House case book, entries for 1 December 1865, 4 January, 30 June, and 25 October 1866.

67 Devon Heritage Service, letters interleaved with volume of case notes 3992F/H32/4.

68 Devon Heritage Service, 3992F/H32/4, Wonford House case book, entries for 24 September 1883, 21 October 1885, 5 March 1893; 3992F/H32/18 entry for January 1918.

69 Warwickshire Archives, CR 1664/617 Warwickshire asylum case book, 1872–75, patient number 2098, entry for 30 May 1873.

70 Death certificate for Alexander Cameron issued 14 June 1871.

71 Warwickshire Archives, CR 1664/617 Warwickshire asylum case book, 1872–75, patient number 2098, entry for 4 January 1873.

72 Derbyshire Record Office, D 1658/11/3, Derbyshire asylum case book 1856–57, entry of 16 February 1860; the case is discussed more fully in Tomkins, 'Mad doctors', 445–6.

73 St Andrew's archive, CL17 patient case notes 1885 to 1887.

74 Ibid. notes dated 6 August 1886.

75 Ibid. notes dated 15 November 1886, 10 December 1886 and 4 January 1889.

76 Ibid. notes of 22 March 1889 and 29 April 1889.

77 Ibid. notes dated 25 July 1889.

78 N.A. RG 12 census 1891 and RG 13 census 1901; death certificate issued 23 December 1916. A. C. Shepherd, 'Mental health care and charity for the middling sort: Holloway Sanatorium 1885–1900', in A. Borsay and P. Shapely (eds), *Medicine, Charity and Mutual Aid: The Consumption of Health and Welfare in Britain, c. 1550–1950*, Aldershot: Ashgate, 2007, 163–82, p. 177.

79 For a summary of Edmunds' life and criminal activity see her profile by the Berkshire Record Office. Available: http://www.berkshirerecordoffice.org.uk/albums/broadmoor/christiana-edmunds/ (accessed 3 June 2014).

80 Berkshire Record Office, D/H14/D2/1/2/1 case notes of Christiana Edmunds.

81 See Tomkins, 'Mad doctors', p. 449, for the depression and suicide of physician Angus Cameron following the deaths of his two children.

82 Suzuki, 'Framing', pp. 116–17, 129.

83 Davis, *Cruel Madness*; Kragh, 'Men, women and the morphine problem'; H. Marland, *Dangerous Motherhood: Insanity and Childbirth in Victorian Britain*, Basingstoke: Palgrave Macmillan, 2004.

## Select bibliography

J. Andrews, 'Case notes, case histories, and the patient's experience of insanity at Gartnavel Royal Asylum, Glasgow, in the nineteenth century', *Social History of Medicine* 11, 1998, 255–81.

D. Atkinson, 'The evolution of medical research writing from 1735 to 1985: The case of the *Edinburgh Medical Journal*', *Applied Linguistics* 13, 1992, 337–74.

G. Davis, *'The Cruel Madness of Love': Sex, Syphilis and Psychiatry in Scotland, 1880–1930*, Amsterdam: Rodopi, 2008.

C. Galley and R. Woods, *Mrs Stone & Dr Smellie: Eighteenth-Century Midwives and Their Patients*, Liverpool University Press: Liverpool, 2014.

J. Gillis, 'The history of the patient history since 1850', *Bulletin of the History of Medicine* 8, 2006, 490–512.

T. M. Hassan et al., 'A postal survey of doctors' attitudes to becoming mentally ill', *Clinical Medicine* 9, 2009, 327–32.

B. Hurwitz, 'Form and representation in clinical case reports', *Literature and Medicine* 25, 2006, 216–40.

G. B. Risse and J. H. Warner, 'Reconstructing clinical activities: Patient records in medical history', *Social History of Medicine* 5, 1992, 183–205.

A. Suzuki, 'Framing psychiatric subjectivity: Doctor, patient and record-keeping at Bethlem in the nineteenth century', J. Melling and B. Forsythe (eds), *Insanity, Institutions and Society, 1800–1914: A Social History of Madness in Comparative Perspective*, London: Routledge, 1999, 115–36.

A. Tomkins, 'Mad doctors? The significance of medical practitioners admitted as patients to the first English county asylums up to 1890', *History of Psychiatry* 23, 2012, 437–53.

# 30

# LITERATURE AND DISEASE

## A novel contagion

*Sam Goodman*

The relationship between literature and medicine and, indeed, between authors and sickness has often been a personal one. Much has been made within the field of medical humanities of Virginia Woolf's assertion that: '[c]onsidering how common illness is, how tremendous the spiritual change that it brings . . . it becomes strange indeed that illness has not taken its place with love, battle, and jealousy among the prime themes of literature'.[1] Suffering from illnesses both physical and mental throughout her life, Woolf recognised the struggle, drama, and inherent banality involved in being unwell, as well as the changing shape of medicine and health-care within early-twentieth-century society. As a range of critics including Thomas Szasz and Thomas C. Caramagno have recognised, Woolf sought to capture these elements in her fiction, alongside her own personal experience; for instance, in *Mrs Dalloway* (1925) Woolf famously addresses shell-shock and the inadequacies (as she saw them) of psychological treatment available under doctors of the Holmes and Bradshaw type. Woolf's novel damningly juxtaposes the alienating and patronising experiences of Septimus Smith, based on her own treatment for recurrent nervous breakdowns and depression at various points between 1897 and 1913, against the uninformed percep-tions of Peter Walsh who muses on how the efficient yet impersonal 'light high bell of the ambulance' was 'one of the triumphs of civilisation'.[2] The representation of sickness and health informs and influences the thematic character of *Mrs Dalloway*; illness fluctuates between the foreground and background of the narrative, much as it did in Woolf's own life.

However, whereas Woolf saw herself as perhaps singular in her focus on such themes, in the 90 years since her essay and novel, and particularly in the decades since the Second World War, a range of authors have taken up her mantle to the point where contemporary literary culture is suffused with accounts, testimonies, and rep-resentations of what it means to experience sickness and disease, to receive treatment, or simply to be 'ill'. Though Woolf would perhaps have been loath to acknowledge it, within this movement, popular and indeed, to use one of Woolf's most famously despised terms, 'middlebrow' culture has responded to her incredulous diagnosis by taking the situation to almost the opposite extreme. A cursory examination of the wider cultural milieu presents us with a proliferation of medical themes and medical settings across a range of media; from the variety of now familiar hospital-based TV dramas such as *ER*, *Holby City* or the long-running *Casualty*, through to the seemingly endless variations of 'true life' illness narrative that fill the shelves of booksellers and local supermarkets alike. Nearly anywhere we might look in contemporary culture we are presented with evidence of the frailty of the human body: from the myriad illnesses that might befall us and the difficult long-term consequences of dealing with them to the range of mental disorders we are all, if the literature is to be believed, likely to encounter at some point in our lives.

This body of medical literature and interest in representations of illness has received a great deal of critical attention in the last three decades, with the growth of medical humanities both as a sub-strand of various disciplines and as an interdisciplinary field of study in its own right. The development of medical humanities has been a process characterised by reciprocity and exchange, with authors, medical professionals, and scholars alike engaging with methodological approaches that themselves reflect the intersection of the medical and the literary. Landmark texts such as Susan Sontag's *Illness as Metaphor* (1978) and Rita Charon's *Narrative Medicine: Honoring the Stories of Illness* (2006) have sought to explore the potential of narrative formation and the depiction of illness through metaphor within clinical practice and the broader context of society, considering how representational strategies drawn from the literary sphere are responsible for a culture of victim shaming and obfuscation, or, more positively, are able to aid the development of empathy within health-care professionals.[3] Likewise, the medical exploration of literary texts has drawn focus on specific medical themes in high-profile, popular works of contemporary fiction, including, for instance, neuroscience in Ian McEwan's *Saturday* (2003) and medical ethics and organ donation in Kazuo Ishiguro's *Never Let Me Go* (2006), as well as prompting the critical reconsideration of medical themes in Western literature and beyond in a range of eras and literary movements dating back to the classical epics of ancient Greece.[4] In such instances, the scope and range of medical themes in each individual text have varied considerably, from the identification of medical topics and imagery, to allegorical and metaphorical uses of medicine, through to the more evaluative consideration of the complex integration and active criticism of medical knowledge within literary prose. Despite variations in method between these critical investigations, they all serve to suggest that Woolf's assertion that the sick body has long lain neglected by literature is now a much less credible one.

While this broad focus on illness and medicine has included the significant trope of disease within fiction of the last 150 years, the distinction between sickness, illness and disease has often gone unmarked, and many critical accounts either lack a necessary degree of specificity or choose to conflate these categories within their examinations of medical themes in literary texts. In his consideration of the vocabulary of medical humanities, Kenneth Boyd concludes: 'Disease then, is the pathological process, deviation from a biological norm. Illness is the patient's experience of ill health, sometimes when no disease can be found. Sickness is the role negotiated with society.'[5] Given the interrelation of these terms, some conflation of them in critical analysis is understandable, even inevitable. However, in this chapter I have sought to maintain Boyd's differences in my examination of the interplay between disease and literature as far as possible. Working within these parameters, the following sections will examine the various ways in which literary authors have chosen to represent disease and the effects of disease across a range of genres, but will also naturally involve some consideration of experiences of illness, and what it means to be 'sick' with disease. Also inevitably, and necessarily as a result of the limitations of this format, such an account will be a selective one, and I have chosen to dedicate the bulk of my examination of literature and disease to prose fiction of the twentieth century. Although I have included the nineteenth century, it is really in the twentieth century that the threat and symbolic potency of worldwide pandemic or infectious disease became foregrounded in the global literary imagination, alongside the concurrent growth of illness narratives and literary pathography. However, in reflection of the work on the phenomenology and experience of illness elsewhere in this volume, as well as that on AIDS narratives, this chapter will not focus on pathography in any extended depth, but rather acknowledge and allude to developments within the field where relevant.

The chapter will approach the representation of disease in three stages. First, it will present an overview of the relationship between literature and disease in the nineteenth century, historicising its development in relation to key shifts in historiography, historical context, and literary and critical paradigms. With particular emphasis on the representation of tuberculosis and cholera, this section will explore the changing signification of these diseases over time. Second, and against a theoretical backdrop of postmodernism and postcolonialism, the chapter will explore the representation of disease in terms of content. Focusing on a selection of novels published after the Second World War, this section will consider representations within fictional narratives of the historical past, particularly those set in the nineteenth century and concerned with the representation of cholera. This section will explore why, in an era of British national decline, authors turned to medicine and disease in order to examine the national past, and how such representations had shifted in comparison to those of nineteenth-century origin. Finally, the chapter will analyse the representation of disease as part of its role in developing and establishing literary form. With specific emphasis on science and speculative fiction published since 1945, this section will investigate the use of disease, real or invented, as a major narrative trope of post-apocalyptic fiction or those texts that seek to depict the near and imagined future. In particular, this section will present a close reading of disease within Margaret Atwood's *Oryx and Crake* (2003), arguing that the narrative function of disease within Atwood's novel is to act as a potent basis for Atwood's ecocritically informed message; disease, as much as genetics and over-population, becomes a key component of the novel, and permits Atwood to make a serious statement via a popular format. Atwood's novel mirrors many of the established discourses of science fiction and Cold War literature, but imbues their rhetoric with a vocabulary and imagery drawn from the language of disease. This chapter approaches these texts from a medical humanities perspective, subjecting them to both literary and historical analysis in recognition of the fact that while these are literary works they are nonetheless influenced by and respond to broader historical and contextual concerns within their respective periods.

## Against a canvas of mortality: literature and disease before the twentieth century

Contrary to Virginia Woolf's opinions, the connection between medicine and literature in the preceding 150 years, and particularly that of literature and disease, was not only well established but also widely recognised among authors and the reading public alike. Contemporary scholarship within the medical humanities has revealed that themes of illness, disease and sickness were present throughout canonical and non-canonical literature from the eighteenth century onwards, intensifying and multiplying in number throughout the Victorian age. For example, in *Passion and Pathology in Victorian Fiction,* Jane Woods states that the fact that Victorian writers were engaged in debates within the realms of medical science about health and disease is 'scarcely surprising' given the 'constant reality of sickness and the increasing incursion of medicine's authority into broader areas of intellectual and social life'.[6] Though Woods, like many other literary historians of nineteenth-century fiction focuses in detail on the great Victorian obsession with the diseased mind, both male and female, she nonetheless acknowledges the importance of recognising the figure of the sick body within prose writing of the period and recognises that the relationship between medicine and literature was one of 'elucidation and illustration'.[7] The representation of disease in fiction then was not only to fulfil the demands of verisimilitude or satisfy Victorian realism, but as much to explore the social meaning of sickness and to disseminate information.

The extent to which disease and ill-health played a part in Victorian fiction was a reflection of its visibility in Victorian society. Such regularity has enabled Miriam Bailin to observe that there is 'scarcely a Victorian fictional narrative without its ailing protagonist, its depiction of a sojourn in the sickroom'.[8] Bailin identifies a turbulent Victorian society riven by strife and political division, transposing personal experience against a far larger national canvas, but one united by the prospect of sickness and disease. Disease and ill-health become for Bailin a unifier, a leveller, in which characters from Victorian novels are brought together by their common experience of illness. Such an analysis sets up a contradiction in the representation of disease within Victorian fiction; namely that it appears both positive and negative development, as a curse and a blessing. Bailin identifies the sickroom on the one hand as a space of healing and emotional connection in which the soul of the protagonist is enriched; on the other, however, it is a site of anxiety and mortal danger. This contradiction is suggestive of how the meanings of disease and illness in Victorian fiction are malleable and changeable, reliant on social perception of specific conditions as well as shifting artistic and aesthetic sensibilities, tastes and fashions.[9]

The relationship between disease and meaning in early-nineteenth-century fiction possessed both a textual and an extra-textual dimension, especially with regards to those authors active in the first few decades of the 1800s. In *Disease and the Novel 1880–1960,* Jeffrey Meyers contextualises his analysis of Leo Tolstoy and André Gide by examining this period, and in particular the association between disease and the Romantic poets.[10] Meyers states that during the early part of the nineteenth century 'writers associated creativity with one specific disease: tuberculosis' and that this association and interest in the aesthetic aspects of consumption led to the 'extraordinary appearance of a whole series of tubercular artists' and writers.[11] Despite the wealth of critical literature on madness and the nineteenth century, especially after the influential work of Michel Foucault, Meyers' analysis is suggestive of the way in which disease and literature in this period is defined in terms of the physical embodiment of illness as much as any sense of mental instability. However, unlike later literary movements, at this point tuberculosis did not carry with it connotations of fear or panic, rather those of a tragic and sympathetic kind; in a quote redolent of literary fiction, Meyers mentions how John Keats, who had trained in medicine, became aware of his own infection: 'I know the colour of that blood . . . – it is arterial blood – . . . that drop is my death warrant. I must die.'[12] Again unlike a mental illness, tuberculosis in this instance does not alter or take over Keats' personality but rather gives his existing identity a sense of greater gravitas; Keats' terse response to his inevitable death aligns the poet with suffering, thus affording him the deeper spiritual insight necessary to create his art.[13] His disposition is not one of despair, but of nobility, affording his art a similar quality. The meaning of this disease to a contemporary readership was not necessarily understood in terms of the pathogenic effects of the disease on the individual, but, according to Meyers, the feeling attached to it, consequently producing sympathy for their condition more than any other response.

However, this perception of the romantic consumptive shifted over the course of the nineteenth century to become an altogether more active and much less positive entity within fictional narratives. Katherine Byrne argues that tuberculosis killed more people in the Victorian age than did cholera and smallpox combined, and was equal in contemporary imagination only to syphilis; consequently, the two diseases 'functioned as sites of social anxiety in Victorian times as cancer and HIV/AIDS do in ours'.[14] Byrne maintains that the consumptive was marked out in terms of difference, as Meyers does, but that the difference became one fraught with fears over femininity, sexual transgression, degeneration and more, playing on a familiar Victorian theme of self-control. Linking these concerns to a series of 'Condition of England' novels from the mid to late nineteenth century such as Charles Dickens' *Dombey and Son* (1848) and Elizabeth Gaskell's *North and South* (1855), Byrne's analysis demonstrates how the representation of disease in fiction began to develop its more modern form as a potent metaphor for the ills of the social body in this period. However, disease at the same time never lost its status, as Meyers puts it, as 'a great mystery: a visitation, a curse, a judgment'.[15] The judgements in the 'Condition of England' novels are those of providence, fate or some higher power on the British nation, as well as the individual, conferred through disease.

Aside from tuberculosis, the other disease most readily associated with the nineteenth century is that of cholera. Unlike tuberculosis, whose presence and death toll remained relatively constant throughout Victorian life, cholera appeared in four sudden epidemics across the course of the nineteenth century between 1832 and 1866. The effects of cholera on Victorian society, political action and social reform, and on practice within the medical profession, especially after 1850, are extensively documented in a range of historical and medical sources; the origin, cause, transmission, and remedies for the disease were debated in professional publications and private correspondence alike.[16] With so much activity in so many spheres of nineteenth-century life it would seem logical that cholera would be subject to the same process of representation within fiction as tuberculosis. However, as Pamela K. Gilbert argues, there is a 'surprising dearth of direct literary response to cholera' in nineteenth-century fiction, and the disease is almost never explicitly described in the Victorian novel.[17] Instead, continues Gilbert, it is euphemistically referred to as 'fever', which she believes would accomplish the same effect, one of tragedy or sorrow, without necessitating detailed explanation of the symptoms of cholera.[18] Gilbert's claim of euphemistic uses of disease is concurrent with Allan Conrad Christenson's contention that references to plague, cholera, typhoid, and smallpox throughout nineteenth-century fiction are hints of 'the ghostly presence of venereal infection', thus equating cholera with Byrne's understanding of tuberculosis as a harbinger of social anxieties.[19]

While it is compelling to consider all mentions of disease in Victorian fiction as oblique or explicit expressions of sexual deviance, such a position seems too narrow and misses the subtlety with which representations of disease are included in fictional narratives. One novel in which cholera does feature is that of George Eliot's *Middlemarch* (1872). Gilbert states that Eliot uses cholera as a form of 'historical background' and that she is far more interested in the relationship between the 'healthy body and the polis'.[20] However, when considered in the light of the social and political signification of disease within Victorian fiction, Eliot's interweaving of cholera with the main plot takes on a richer character than Gilbert's analysis gives it credit for. For instance, late in the novel Eliot writes:

> In the hundred to which Middlemarch belonged railways were as exciting a topic as the Reform Bill or the imminent horrors of Cholera, and those who held the most decided views on the subject were women and landholders.[21]

Though cholera does not appear paramount in importance here, receiving a seemingly throwaway mention despite its 'imminent horrors', the fact that it is juxtaposed with the two great social preoccupations of the period is nonetheless revealing of its place within both the context of the novel and the context of its writing 40 years after the events it describes. Eliot conglomerates within one sentence three elements responsible for dramatic and irreparable change to the physical, social, and political landscapes of the nation within the nineteenth century. The threat posed to the status quo by the Reform Act is here comparable to that wrought by disease; given that Gilbert observes how most medical texts on cholera from the period began by evoking imagery of the Great Plague, the intimation here is one of widespread upheaval to the social order, akin to that which occurred in the wake of the Black Death.[22] Similarly, there is a link between ill-health and the railways here, acknowledging the various anxieties over medical conditions that travelling by rail was expected to produce; though not explicitly associated with disease, the general thinking around railways was that they were in every way injurious to health, either by causing specific conditions or by spreading impure airs, evoking the miasma theory of disease transmission, itself associated with cholera until the end of the nineteenth century.[23] Further, all three elements in this extract perform a united function, namely the suggestion that they all bring into association social classes that would prefer to remain distinct from one another; a further reminder of the levelling power of disease. Ultimately, in this short extract and the sub-plot relating to cholera that runs throughout *Middlemarch*, Eliot calls attention to the manner in which everyday life in the nineteenth century was threatened by disease and the potential development of ill-health at every turn. Rather than just 'historical background', Eliot describes the feeling of living continually with the threat of disease; cholera is an often silent background in *Middlemarch*, as it was in Victorian life, but one with the potential to burst into the foreground.

## The sick man of Europe: British fiction and disease after Empire

It is no accident that Lytton Strachey chose to focus on Florence Nightingale as one of his 'Eminent Victorians', in his volume of the same name published in 1918. His decision to include Nightingale, both for her reforming zeal and for the familiar narrative of humanitarian kindness, innovation and generosity of spirit, is in itself indicative of the legacy of the nineteenth-century attitude to disease; that through reform and determination, near anything was possible. Strachey, however, took a notoriously dim view of such qualities and used his book to satirise the great and the good he selected as representatives of their various institutions, including Matthew Arnold and education, General Charles Gordon and the military, and Cardinal Manning and the Church. Strachey's was a subversive voice; he was a writer who discerned hypocrisy where others saw virtue, seeking to expose such pretence in his writing.[24] In the aftermath of the First World War, arguably the true end of the Victorian age, he was not alone. The legacy of the Victorians, in part maintained by the Empire they had secured, was resilient and continued to live on. Despite efforts by prominent literary critics on the political left, such as George Orwell and W. H. Auden for instance, it was not until after the Second World War, an era of global decline for Britain, that literature began the scrutiny of the nation's past and the legacy of its actions in earnest.[25]

A good deal of debate exists over when precisely this decline began. Some post-war historians have argued that the political and sovereign concessions made by Britain during the Second World War signalled the end of the British Empire; others have identified the point at which the Empire was renamed the Commonwealth in 1952.[26] However, as many again have suggested, Indian Independence in 1947 is equally as significant a milestone in the history of Empire as it precipitated a process that in just 20 years dismantled an empire three centuries in the making. With Britain's international status further degraded by the Suez Crisis of 1956–7, the rapid devolution and emancipation of British colonies around the globe began in earnest; Indian Independence through to the period after Suez has been assessed as the watershed moment of Britain's waning Empire, one in which the post-war era shifted irrevocably towards a post-colonial one.[27] Such a rapid shift in global position prompted critical re-examination of the legacy of Empire through literature, and a range of other media, which sought to explore the contemporary state of the nation through renewed focus on its past. Steven Connor has argued that the novel is an especially ductile format for exploring ideas of nationhood, a process that accelerates in the post-war period.[28] The genre of historical fiction in particular experienced a post-war resurgence, and, illustrating how the nineteenth-century approach to disease was a long-lasting one, many historical novels of this period used medicine or specific diseases such as cholera, as a lens through which to re-examine imperial history and explore anxiety over decolonisation.[29]

Medicine, and in particular disease, appeared fitting for scrutinising the seemingly rapid degeneration of the national position and the literary advantages to engaging such imagery are clear; as Jeffrey Meyers argues: 'the illness of the hero, who is both an individual and a representative of his epoch, is analogous to the sickness of the state'.[30] Although Meyers concludes his study in 1960, his analysis seems particularly appropriate given the embattled circumstances of Britain during decolonisation, and can be applied productively to novels that fall outside the chronological parameters of his study. One such novel is *The Siege of Krishnapur* by J. G. Farrell. Farrell's success during his own lifetime was fairly modest; while his novels, particularly *Troubles* (1970), *The Siege of Krishnapur* (1973) and *The Singapore Grip* (1978), which made up a loose collection informally titled 'The Empire Trilogy', received critical praise and recognition, including a Booker Prize in 1974, his novels have not begun to be widely known until the past decade, during which a range of critics have sought to restore his reputation.[31] In terms of the relationship between form and content, *The Siege of Krishnapur* is a deceptively simple novel. Set in the Indian Rebellion of 1857, its plot and structure follow that of a traditional 'Mutiny novel', with its characters representing a range of literary archetypes and its plot featuring the kinds of epic battle scenes and instances of British fortitude equated with such a storyline. However, as many critics including myself have argued, to read Farrell's work as nostalgic is to miss the parodic re-engagement with the clichés of Empire that his novel presents, and to overlook the importance of such a critique in an era of imperial crisis.[32]

*The Siege of Krishnapur* involves a British garrison, itself a thinly veiled fictional version of Lucknow, under siege from the Sepoy army; the plot is ostensibly no more complicated than documenting their fortunes while they hold out for relief. However, beneath this surface layer lies a range of satirical and critical positions, either on the British Empire and its legacy, or on those in Farrell's contemporary moment who still harboured such a rose-tinted view of Britain's imperial history.[33] Farrell's is a novel full of disease and ill-health, from malnutrition to fever, from sunstroke to cholera, and nearly all his characters are at some point ill. In part, his decision to include these illnesses is indicative of the disease he saw at the heart of Victorian society, linking his novel to those 'Condition of England' novels of the period he dramatises. However, Farrell's return to the nineteenth century does not involve a slavish mirroring of nineteenth-century narrative techniques, and disease, especially cholera, is explicitly foregrounded in the novel. Cholera and debates over its transmission and treatment drive a significant sub-plot in which the garrison's civil surgeon, Dr Dunstaple, and the army's regimental surgeon, Dr McNab, repeatedly clash. The proportion of the novel given over to cholera is indicative of its importance within the narrative and also Farrell's critique of Empire; rather than incidental mentions of cholera in order to reinforce period authenticity, Farrell uses cholera to illustrate the divisions inherent in the British garrison in a traditional narrative of unity. By doing so, he draws a parallel between historical and contemporary discord, as well as highlighting the fact that a view of the Empire as intrinsically more cohesive than the 1970s present is fallacious.

In terms of content, Farrell's approach to cholera differs greatly from that of preceding novelists. Rather than in the case of *Middlemarch* where the 'imminent horrors' of cholera are associative in meaning and effect, or in Gilbert's analysis whereby she argues that euphemisms are used in order to avoid mentioning the full details of the disease, Farrell sought to make his representation of cholera as accurate as possible. Along with his extensive research into the Sepoy Rebellion documented in his India Diary published after his death in 1979, Farrell's personal papers held by Trinity College, Dublin, reveal the extent to which his representation of mid-Victorian understanding of cholera, methods of treatment and physical symptoms of the disease itself were built on a basis of factual evidence.[34] In preparing the manuscript of *Krishnapur*, Farrell kept a series of index cards on which he transcribed themes, details, sources or quotes from texts of major importance. On the card marked 'CHOLERA', Farrell includes a number of quotations from sources ranging from published memoirs on the Siege of Lucknow through to the *Medical Times and Gazette* of 1854.[35] Some of these quotations are brief and intended as shorthand reminders of details Farrell clearly intended to include. Others are much longer and make their way into the novel in verbatim form. Some of these, such as when Dunstaple and McNab clash directly over the mode of transmission of cholera, offer a summary of leading medical thinking of the period, with this particular incident providing a précis of the history of cholera treatment from William O'Shaughnessy through to John Snow.[36] Other inclusions hold a more self-reflexive level of importance with regards to the representation of disease within literature. For instance, midway through the novel Dunstaple begins to distrust McNab's methods of treatment and copies notes from his medical diary that he views as proof of McNab's dangerous approach:

> She has almost no pulse,
> Body as cold as that of a corpse.
> Breath unbelievably cold, like that
> from the door of an ice-cavern.
> She has persistent cramps and vomits
> constantly a thin, gruel-like fluid
> without odour.
> Her face has taken on a terribly
> cadaverous aspect, sunken eyes, starting
> bones, worse than that of a corpse.
> Opening a vein, it is hard to get any
> blood . . . what there is, is of a dark
> treacly aspect.[37]

McNab's methods are representative of a shift in medical practice in which the disease becomes the focus of treatment rather than the human subject. This reprioritisation of the practitioner's role conflicts with Dunstaple's own paternalistic professional principles, and fuels his distrust in McNab; that the patient described in these notes is revealed to be McNab's wife only serves to compound Dunstaple's anger. The information that informs this section, including whole phrases on the 'dark, treacly aspect' of the blood, is included in Farrell's indexed notes as with so much of the material on cholera. Farrell's decision to include such extensive detail here is significant, and revealing of a range of concerns with regard to the literary depiction of disease. The level of detail included here far exceeds that of the nineteenth-century texts mentioned earlier in this study, and the level that is necessary for Dunstaple to make his point on the medical habits of McNab. Rather than allow the reader's imagination to provide this information, Farrell explains the horrors of cholera in explicit detail, either for a twentieth-century audience unfamiliar with them, or in order that his point be made clearly enough. When considered in relation to the structure of the extract, the extent of detail is significant within the analysis of the use of disease as content. Within the narrative, the extract is presented in a cribbed, truncated fashion as befitting notes in a medical diary; however, its appearance on the page gives it a stanzaic quality seemingly at odds with the intention and the level of detail included. To further this comparison, the account from McNab's diary is juxtaposed with snatches of song from a delirious soldier in the first stages of cholera infection. The soldier sings a ballad from the Crimea and acts to further underscore Farrell's association between disease and its representation in creative form; the Crimea, again evoking cholera and Nightingale's efforts at Scutari, is presented as a glorious defence of Queen and country, glossing over the human cost. Such a choice of juxtaposition suggests Farrell's intentions with regard to the representation of disease within his fiction; by providing so much information in this section and throughout the narrative, his novel removes any potential for tragedy and romance that the elision of a novel like *Middlemarch* or the poetry of Keats does. Disease and its treatment, like the myth of the 'heroic' defence of the British Empire in India, are stripped of their romance in Farrell's novel, and the reader is instead confronted with the stark reality of both disease and imperialism.

Ultimately, Farrell's novel re-engages with the legacy of nineteenth-century narratives both literary and historical. In an era of perceived social disintegration and global decolonisation, his return to the high point of the Empire and the Victorian century is deliberately chosen, as is his employment of cholera and a range of medical conditions as a lens through which to read the British past. His novel presents a British society and colonial project literally riddled with disease and ill-health; Farrell attempts to drive home the association between such a sickly society and the misguided assumption that a return to Empire would act as a cure for the perceived social ills of the contemporary moment. His final words in the novel are illustrative of this position; one of the central characters of the novel, the garrison's collector, muses on how the 'invisible cholera cloud had moved on' leading him to a consideration of how 'a nation does not create itself according to its own best ideas, but is shaped by other forces, of which it has little knowledge'.[38] Farrell's lasting message is that adherence to deeply-held beliefs, such as those over cholera that consume Dunstaple and McNab in his narrative or those that mourn the loss of Empire, confine a nation to an irrelevant past, restricting its ability to adapt to a changing world.

## Cold Wars and hot bioforms: representing global pandemic in popular literature

Concurrent with the decline of imperialism after the Second World War, and indeed a decisive factor in hastening decolonisation itself, was the development and prosecution of the Cold War. Existing scholarship on the Cold War has claimed variably that the conflict can be most productively read either in terms of science and technology, political ideology, intrigue and espionage, counter-cultural movements and social conflict, or as a clash of economic systems.[39] However, the medical dimensions of the Cold War and post-Second World War era more generally intersect with many of these existing branches of enquiry, especially those related to technological and scientific innovation and the growth of the pharmaceutical industry within the Military Industrial Complex.[40] This culture of innovation and the suggestion of cleaner and quicker treatment, social welfare developments such as the National Health Service in the United Kingdom in 1948 or new forms of clinical practice including arts therapies, alongside the growing state medical regulation of individuals' lives, means that the Cold War can be seen increasingly through a medical lens.[41]

In examining the social and cultural history of the era and how its politics were shaped and conceived in the popular imagination, the predominant language of the Cold War was one reliant on the nomenclature and imagery of disease, especially with regards to the spread or containment of 'aggressive' and 'virulent' Communism; phrases as commonly associated with the transmission of infectious disease as much as any ideology.[42] In further considering cultural constructivism and the reflection of contemporary worldview throughout the Cold War, a similar process of analysis can also be transferred to a consideration of Cold War literature. Two forms of literary genre most readily associated with the Cold War are those of espionage fiction and science fiction. The key difference between these two genres is that where spy fiction often uses the language of disease in its descriptions of the subterranean conflict between Cold War superpowers, science fiction more readily uses disease as a driving force of the narrative and is more able to represent the effects of disease within its pages; where espionage fiction must always prevent the final outbreak and restore narrative equilibrium (or homeostasis to the body of the state), science fiction has licence to depict what happens when such efforts fail, and do so in full imaginative detail. However, most often, the line between these genres is blurred, and David Seed identifies numerous Cold War sci-fi fictions from film, television and literature that employ the rhetoric, diction and imagery of disease, including *The Quatermass Experiment* (1953), Philip Wylie's *The Smuggled Atom Bomb* (1951), and many others.[43] Though of course the relationship between science fiction and disease stretches back much further, with such notable examples as the defeat of Martian invaders in H. G. Wells' *War of the Worlds* (1898) or the plague that destroys humanity in Mary Shelley's *The Last Man* (1826), it is in the Cold War that this association intensifies and has led to the identification of the genre as a vehicle for comment on a wide range of social and political issues. As a result of these interconnected factors, so many of the established nuclear anxieties associated with Cold War culture explored through literature mirror those of disease or sickness; for instance, fears of bodily disfiguration or obliteration, silent and invisible death via radiation aping the inexorable transmission or infection, and the threat of uncontrolled accident; often, the origin of the incident in question is the inadvertent responsibility of one of the superpowers, though culpability typically lies with the Soviet Union.

The disposition towards such outlandish scenarios meant that science fiction, like many other popular genres of writing, was for a number of years discounted as fantasy or entertainment; narratives that depict the currently impossible, such as interstellar space travel, or the implausible, such as extra-terrestrial invasion, were often critically marginalised as low cultural forms. However, as Robert Hewison, Seed and others have argued, science fiction was able to gain a sense of cultural legitimacy, either as a result of popularity, its ability to blend genres and thereby appeal across classes, or because of the relevance of its subject matter.[44] Similarly, the movement towards 'speculative fiction' that began in the 1960s gave science fiction a sense of respectability, enabling authors to make serious comment on social and political issues, but retain the imaginative freedom of fiction. Margaret Atwood's *Oryx and Crake* (2003) is one such novel, and a text that draws on a raft of techniques, themes and imagery from the genres of science fiction and Cold War literature. The narrative and plot of *Oryx and Crake* are driven by the fear of global pandemic and the responsibility of industry and the scientific/medical community with regard to the future state of the planet; much within Atwood's novel is currently impossible, but by no means implausible, and she has stated that *Oryx and Crake* is based on a body of well-documented research.[45]

Set in the first half of the twenty-first century (the protagonist of the novel is born in 1999 and the events take place when he is in his late twenties) and told partially through flashback, *Oryx and Crake* focuses on Jimmy, seemingly the last human survivor in a post-apocalyptic landscape he inhabits with 'The Crakers', a genetically enhanced group of humanoids that he is charged to protect. Over the course of the narrative, the reader learns that society and civilisation were destroyed by a combination of irresponsible attitudes to the natural world and a deliberately engineered and incurable pandemic concealed in a lifestyle drug called BlyssPluss created by Jimmy's childhood friend Crake. Equal parts science fiction, speculative fiction, and ecological parable, Atwood's novel is one suffused with the language and imagery of disease from the opening chapters. As well as the reason for the devastated environment in which he now lives, the spectre of disease forms one of Jimmy's earliest memories. Jimmy recalls aged five witnessing diseased livestock burnt in the secure OrganInc compound where his father, a biological engineer, worked:

> "They had to be burned to keep it from spreading." . . .
> "What from spreading?"
> "The disease"
> "What's a disease?"
> "A disease is like when you have a cough," said his mother.
> "If I have a cough will I be burned up?"[46]

The burning of livestock as the first memorable image for Jimmy and the first identifiable one for the reader is a deliberate narrative decision on Atwood's part; it not only sets a precedent for the theme of disease that becomes a preoccupation and continual thread running through the flashback sequences, but it also situates the events of the novel in a recognisably real-world context, being near contemporaneous, as Coral Ann Howells notes, with similar measures taken against Foot and Mouth Disease in the UK in 2001.[47] The attitude towards livestock also foreshadows the way in which the human race is regarded by Crake and the industry he represents: as cattle to be used for a particular purpose and then disposed of. Furthermore, the characterisation of disease in this quote links *Oryx and Crake* to the history of disease and fiction, with the cough a direct link to representations of tuberculosis in the nineteenth-century novel. Jimmy's fears of being 'burned up' as a result of a cough are reminiscent of the wasting effect of tuberculosis and the near-inevitable death that accompanies such an illness in traditional narratives; the irony in Atwood's novel of course, is that Jimmy will find himself the only one not burned up by Crake's engineered disease.

Atwood's connection of the theme of disease with clandestine research and laboratories demonstrates a clear link to the culture of the Cold War from which she draws her style and genre, as well as the post-Cold War world of late capitalism and globalisation. In terms of the Cold War, it is well-documented and widely known that vast amounts of funding were invested in combatting and advancing biological and bacteriological warfare, notably in facilities such as Porton Down in the United Kingdom and Fort Detrick, Maryland, in the United States.[48] The highly conspicuous yet highly secure nature of these facilities is reflected in the compounds of HelthWyzer, RejoovenEsense, and Paradice Labs in Atwood's novel; their purpose is common knowledge, but the details of what goes on there kept top secret. Atwood goes further to indict the university and college system as politically and practically complicit in such a process through her depiction of genetic and bacteriological research conducted at the fictional Watson-Crick Institute in the novel, which serves as a satirical mirror of Massachusetts Institute of Technology (MIT), a known recruiting ground, like many other universities, for military research personnel and the intelligence services throughout the Cold War; Crake is described as the subject of much bidding by various companies after his graduation, all of them eager to secure his talents.

In another link to context, an equally recurrent theme within *Oryx and Crake* is the threat of bioterrorism. Along with the overarching plot, a literary depiction of the aftermath of a biological weapon of mass destruction, the novel contains multiple instances in which disease is used as a weapon of terror on a smaller scale. Early on in the novel, the HelthWyzer compound is subject to an attack in which 'a fanatic . . . with a hostile bioform concealed in a hairspray bottle . . . nuked a guard who unwisely had his facemask off'.[49] The bioform is described as 'some vicious Ebola or Marbourg splice, one of the fortified hemorrhagics' that dissolves the guard into 'a puddle of goo'. Alongside the general rhetoric of the War on Terror and the exhortation towards vigilance against fanaticism, itself evocative of the Cold War fear of Communist fifth-columnists, Atwood's representation of disease as a weapon here is significant in conjunction with two contemporary events: firstly, the mention of Ebola recalls the outbreaks in the mid-1990s in Africa and the earlier American scare in Reston, Virginia, itself the result of imported macaque monkeys from the Philippines;[50] and secondly, the 'Amerithrax' incident, where in the week after 9/11 envelopes containing anthrax spores were sent to a number of United States news media offices and Democrat senators, resulting in the death of five people.[51] Atwood combines a range of contemporary fears and expresses them, tellingly, in the language of a previous age of anxiety; it is notable too that the guard at the gate is not simply killed, but 'nuked', once again reinforcing the link between a tense post-millennial world and that of the Cold War. The threat itself may change, but the fear remains the same.

Beyond such parallels, Atwood's novel expresses little overt interest in government, and never actually provides any information on how the future United States is governed. Instead, her targets are industry, and the irresponsibility she sees in late-capitalist society, one used to cheap fossil fuels and disposable products. Atwood asserts that such companies act with impunity and are allowed to do so by a complacent and apathetic public. In pursuing this theme, Atwood again mixes a range of genres, including detective story and conspiracy novel, as Crake finds that his late father had uncovered evidence that HelthWyzer had been developing new diseases in order to then market the cures, thus securing a constant demand for its products.[52] Given that this episode inspires Crake's final plan to destroy humanity, or take revenge through the same means, the novel is intended as a warning against experimentation without accountability, lest experimentation lead to circumstances beyond human control. Critics such as Shannon Hengen have argued that Atwood is concerned with the interconnectedness of life on earth and scornful of those who privilege the human animal above the natural world.[53] *Oryx and Crake* illustrates such an argument, and Atwood uses disease as a reminder that for all the advances made by the human race, we remain vulnerable animals within a fragile interconnected ecosystem; one that over-population, hubris and our own ignorance may serve to one day rebalance.[54] In the closing chapters of the book, Atwood describes through

Jimmy's eyes, the destruction of society as a result of the disease hidden in the BlyssPluss pills:

> At first Jimmy thought it was routine, another minor epidemic or splotch of bioterrorism, just another news item . . . then the next one hit, and the next, and the next, rapid-fire. Taiwan, Bangkok, Saudi Arabia, Bombay, Paris, Berlin. The pleeblands west of Chicago. The maps on the monitor screens lit up, speckled with red as if someone had flicked a loaded paintbrush at them. This was more than just a few isolated plague spots. This was major . . . It was a rogue hemorrhagic, said the commentators. The symptoms were high fever, bleeding from the eyes and skin, convulsions, then breakdown of the inner organs, followed by death. The time from visible onset to final moment was amazingly short. The bug appeared to be airborne, but there might be a water factor as well.[55]

The tense and hurried section in which the disease activates evokes Meyers' observation that disease 'has always been a great mystery; a visitation, a curse, a judgment', in this instance expanding beyond the nation, and beyond time and history, in order to encompass the whole planet.[56] Having been hinted at and suggested throughout, Crake's judgement on the human race, an extension of Atwood's own criticisms of unchecked scientific and medical engineering, is finally revealed; it is the mystery at the heart of the novel that the reader has been invited to solve as the narrative progresses, and one that Jimmy did not until it was too late. Atwood's description of the transmission of Crake's disease forms an ironic counterpoint to the spread of human civilisation and the achievements of scientific ingenuity; engineered to be near unstoppable and incurable, the infection works quickly through major cities across the globe. Atwood's point is clear; the long process of history and the accumulation of skill and knowledge in both medicine and science has led humanity to the point where its mastery of genetics, its most 'elegant concept' in the words of the novel, is the cause of its own destruction. Given the tawdry and immoral nature of the world that Crake's pandemic destroys, however, Atwood invites us to consider whether it is indeed a virus or a cure for a deadlier plague: humanity itself.

## Conclusion

In this chapter, I have hoped to illustrate that the relationship between literature and disease is constantly evolving and one whose ability to create meaning is always developing new modes of transmission. In the nineteenth century, this relationship shifted from the romantic popular perception of tragic tubercular artists, whose disease supposedly gave them special insight into the suffering of the human spirit, through to the use of disease and illness as a broad canvas on which to project drama or through which to develop character. In the twentieth century, narratives and representations of disease took on a political edge, either for specific critical movements such as postcolonialism and the reassessment of the age of nationalism and empire, or as part of the overt and covert political culture of the Cold War. Such projects lent a critical potency to the representation of disease that preceding fictions often lacked; while the novels of Dickens and Eliot sought to present social commentary in particular ways, until the twentieth century the use of infectious disease was an incidental part of this process, and it is only in the last half-century, alongside the growth of pathography and medical humanities, that literary representations of disease have become an overriding concern of fiction. Novelists like Farrell and Atwood recognised the advantages of disease both as metaphor and as vehicle for their respective messages, and not merely as accompaniment or plot device.

Unfortunately, there are multiple new fields of literary response to disease that I have not been able to include here, in particular the burgeoning fields of zombie fiction and disaster narratives that have been gaining ground at a prodigious rate since the end of the twentieth century.[57] Like Atwood, authors and filmmakers working in these genres are chiefly concerned with the perceived fragility of human civilisation, helping to make manifest fears that, for all its trumpeted advantages, the advent of globalisation serves not to strengthen the global community but makes it all the more vulnerable and susceptible to devastating catastrophe in the form of disease. Similarly, the work being done in the field of graphic pathography, fusing illness narratives and the graphic novel or comics format, is again taking the literary representation of disease in new directions and bringing interlinked visual and narrative forms of representation into greater currency and circulation.[58] With the resurgent interest in neo-Victorian fiction and post-human science fiction, it appears that the inquiries begun by authors such as Farrell and Atwood are helping to develop new strains of disease narrative all the time.

## Notes

1 Virginia Woolf, *On Being Ill: With Notes from Sickrooms from Victoria Stephen*, Paris: Paris Press, 2012, p. 1.
2 Virginia Woolf, *Mrs Dalloway*, London: Penguin, 1993, p. 116.
3 Rita Charon, *Narrative Medicine: Honoring the Stories of Illness*, New York: Columbia University Press, 2006; Susan Sontag, *Illness as Metaphor and AIDS and Its Metaphors*, London: Penguin, 2013.
4 Anne Whitehead, 'The Medical Humanities: A Literary Perspective' (pp. 107–28) in Victoria Bates, Alan Bleakley and Sam Goodman (eds), *Medicine, Health and the Arts: Approaches to the Medical Humanities*, London: Routledge, 2014, provides a thorough survey of the field.
5 Kenneth M. Boyd, 'Disease, Illness, Sickness, Health, Healing and Wholeness: Exploring Some Elusive Concepts', *Medical Humanities*, 26, 2000, 9–17, p. 10.
6 Jane Woods, *Passion and Pathology in Victorian Fiction*, Oxford: Oxford University Press, 2001, p. 1.
7 Ibid.
8 Miriam Bailin, *The Sickroom in Victorian Fiction*, Cambridge: Cambridge University Press, 1994, p. 5.
9 The notion of fashionable diseases is also pertinent to an analysis of nineteenth-century fiction and its representation of the sick body – see Heather R. Beatty, *Nervous Disease in Late Eighteenth-Century Britain: The Reality of a Fashionable Disorder*, London: Pickering and Chatto, 2011.
10 Jeffrey Meyers, *Disease and the Novel 1880–1960*, Hong Kong: Macmillan, 1985, pp. 4–5. The association of tuberculosis and ill-health in general with the second-wave Romantics is firmly fixed in the public consciousness; for instance, in 1987, a group of sickly poets, including Shelley and Byron, are featured in an episode of *Blackadder the Third* entitled 'Dish & Dishonesty'. The poets are found languishing in a pie shop unable to do any writing, lamenting their impending deaths, despite being obviously healthy.

11 Meyers, *Disease and the Novel*, p. 4.

12 Ibid., p. 6.

13 Ibid., p. 7.

14 Katherine Byrne, *Tuberculosis and the Victorian Literary Imagination*, Cambridge: Cambridge University Press, 2011, pp. 1–2.

15 Meyers, *Disease and the Novel*, pp. 1–2.

16 On the history of cholera, see: Mark Harrison, *Climates and Constitutions: Health, Race and Environment in British Imperialism in India*, Oxford: Oxford University Press, 2002; Christopher Hamlin, *Cholera: The Biography*, Oxford: Oxford University Press, 2009; and Pamela K. Gilbert, *Cholera and Nation: Doctoring the Social Body in Victorian England*, New York: State University of New York Press, 2008.

17 Gilbert, *Cholera and Nation*, p. 135.

18 Ibid., pp. 135–6.

19 Allan Conrad Christensen, *Nineteenth-Century Narratives of Contagion: 'Our Feverish Contact'*, London: Routledge, 2005, p. 22.

20 Gilbert, *Cholera and Nation*, p. 135.

21 George Eliot, *Middlemarch*, Ware: Wordsworth Classics, 1994, p. 513.

22 Gilbert, *Cholera and Nation*, p. 2. See also the chapters by Samuel Cohn, Mark Harrison, Robert Peckham and Akihito Suzuki in this volume.

23 Mark S. Micale and Paul Lerner, *Traumatic Pasts: History, Psychiatry, and Trauma in the Modern Age, 1870–1930*, Cambridge: Cambridge University Press, 2001, p. 35.

24 There are a number of excellent works on Modernism and medicine, including: Vike Martina Plock, *Joyce, Medicine and Modernity*, Florida: Florida University Press, 2012; and Mark S. Micale, *The Mind of Modernism: Medicine, Psychology, and the Cultural Arts in Europe and America, 1880–1940*, Stanford: Stanford University Press, 2004.

25 On political and literary subversion in the 1920s–30s see: James Smith, *British Writers and MI5 Surveillance 1930–1960*, Cambridge: Cambridge University Press, 2012.

26 Peter Clarke, *The Last Thousand Days of the British Empire*, London: Bloomsbury, 2009; Piers Brendon, *The Decline and Fall of the British Empire 1781–1997*, London: Vintage, 2008.

27 Keith Kyle, *Suez: Britain's End of Empire in the Middle East*, London: I.B. Taurus, 2011; W. M. Roger Louis, *Ends of British Imperialism: The Scramble for Empire, Suez and Decolonization*, London: I. B. Taurus, 2006.

28 Steven Connor, *The English Novel in History 1950–1995*, London: Routledge, 1996, p. 44.

29 Mariadele Boccardi, *The Contemporary British Historical Novel: Representation, Nation, Empire*, London: Palgrave Macmillan, 2009, p. ii.

30 Meyers, *Disease and the Novel 1880–1960*, p. 1.

31 Ralph Crane and Jennifer Livett, *Troubled Pleasures: The Fiction of J. G. Farrell*, Dublin: Four Courts Press, 1997; John McLeod, *J. G. Farrell*, Hornden: Northcote House Publishers, 2007; Hywel Dix, *Postmodern Fiction and the Break Up of Britain*, London: Continuum, 2010.

32 See Sam Goodman, '"A Great Beneficial Disease": Colonial Medicine and Imperial Authority in J. G. Farrell's *The Siege of Krishnapur*', *Springer Journal of Medical Humanities*, Vol. 36. Issue 2, June 2015, pp. 141–56. http://link.springer.com/article/10.1007%2Fs10912-014-9313-5.

33 Farrell outlined his intention in interviews at the time of *The Siege of Krishnapur*'s publication, stating: 'I hoped to say something . . . about how we, in our thriving modern world of the 1970s, hold our own ideas', Malcolm Dean, 'An Insight Job'. *The Guardian*, 1 September 1973.

34 My own inspection of Farrell's papers was made possible through generous support from a Wellcome Trust Small Grant (grant number: 100559/Z/12/Z) awarded in October 2012 and the kind permission of Trinity College Library, Dublin.

35 Papers of J. G. Farrell, #9142, Fols 1–98, Trinity College, Dublin.

36 Dunstaple's approach to treating cholera is largely drawn from an article entitled 'Statistics of the cases of the Cholera Epidemic 1853' by J. S. Pearse and Jeffrey A. Marston, but also includes information from 'Report on Epidemic Cholera' by Drs William Baly and William Gull, Royal College of Physicians (1854) included in the *Medical Gazette* from the same year.

37 J. G. Farrell, *The Siege of Krishnapur*, London: Phoenix Publishing, 2007, pp. 166–7.
38 Ibid., p. 313.
39 Melvyn P. Leffler and Odd Arne Westad (eds), *The Cambridge History of the Cold War*, Vol. I, Cambridge: Cambridge University Press, 2010.
40 Dwight Eisenhower with an introduction by Jesse Smith. *The Military Industrial Complex*, Basementia Publications, 2006, p. 5.
41 Leffler and Westad (eds), *The Cambridge History of the Cold War*, p. 13.
42 Tony Judt, *Postwar: A History of Europe Since 1945*, London: Vintage, 2010, p. 127.
43 David Seed, *American Science Fiction and the Cold War: Literature and Film*, Edinburgh: Edinburgh University Press, 1999, p. 19.
44 Robert Hewison, *In Anger: Culture in the Cold War 1945–60*, London: Weidenfield and Nicholson, 1981, pp. 114–15.
45 Coral Ann Howells, *Margaret Atwood*, 2nd edn, London: Palgrave Macmillan, 2005, p. 173. Atwood herself has addressed the classification debate in *In Other Worlds: SF and the Human Imagination*, London: Virago, 2011.
46 Margaret Atwood, *Oryx and Crake*, London: Virago, 2004, p. 22.
47 Howells, *Margaret Atwood*, p. 173.
48 See: http://www.theguardian.com/politics/2002/apr/21/uk.medicalscience (accessed 28/7/14).
49 Atwood, *Oryx and Crake*, p. 60.
50 Many film and television productions also dramatise such fears including *The X-Files* episodes 'Red Museum' (Series 2 Episode 10) and 'F. Emasculata' (Series 2 Episode 22) both from 1994, the film *Outbreak* (1995) starring Morgan Freeman and Dustin Hoffman and Terry Gilliam's *Twelve Monkeys* (1995). The 2014 Ebola outbreak in West Africa received extensive global news media coverage as a result of its potential communicability to Europe and the US. Flight bans were imposed between certain countries, and major UK airports ran screening programmes for passengers arriving from West African states affected by the spread of the disease. Despite the death of a Spanish aid worker, American and British cases recovered after repatriation and treatment, and coverage of the outbreak has largely vanished from mainstream Western media. However, the World Health Organisation continues to emphasise the potential risk on their website (http://www.who.int/en/), and the African death toll continues to rise.
51 The Federal Bureau of Investigation provides detailed information on the case on its website: http://www.fbi.gov/about-us/history/famous-cases/anthrax-amerithrax/amerith rax-investigation (accessed 30/7/14).
52 Atwood, *Oryx and Crake*, pp. 246–9.
53 Shannon Hengen, 'Margaret Atwood and Environmentalism', in Coral Ann Howells (ed.), *The Cambridge Companion to Margaret Atwood*, Cambridge: Cambridge University Press, 2006, p. 84.
54 Atwood was greatly influence by James Lovelock's *Gaia: A New Look at Life on Earth*, Oxford: Oxford University Press, 2000. The eradication of human life through disease has also undergone a resurgence in contemporary media: the board game *Pandemic* (2008) challenges players to control the global outbreak of four deadly diseases; the Channel 4 series *Utopia* (2013–14) depicts a shadowy organisation called The Network that intends to dramatically lower the world's population through administering a man-made flu-virus; and in a range of very popular tablet and smartphone based games such as *Plague Inc.* (2014), the objective is to develop a disease and infect the world.
55 Atwood, *Oryx and Crake*, pp. 379–80.
56 Meyers, *Disease and the Novel*, pp. 1–2.
57 Recent zombie narratives in which events have a pathologised cause include Max Brooks' novel *World War Z* (2006), later adapted for cinema in 2013, BBC3's *In the Flesh* (2013–14) written by Dominic Mitchell, and also the videogame *The Last of Us* (2013) by developer Naughty Dog.
58 On illness narratives, see Arthur Frank's chapter in this volume. Similarly, see the academic and 'comix' work of Ian Williams, including *The Bad Doctor* (2014), for examples of graphic medicine.

## Select bibliography

Atwood, Margaret. *Oryx and Crake*, London: Virago, 2004.

Bailin, Miriam. *The Sickroom in Victorian Fiction*, Cambridge: Cambridge University Press, 1994.

Boccardi, Mariadele. *The Contemporary British Historical Novel: Representation, Nation, Empire*, London: Palgrave Macmillan, 2009.

Byrne, Katherine. *Tuberculosis and the Victorian Literary Imagination*, Cambridge: Cambridge University Press, 2011.

Charon, Rita. *Narrative Medicine: Honoring the Stories of Illness*, New York: Columbia, 2006.

Christensen, Allan Conrad. *Nineteenth-Century Narratives of Contagion: 'Our Feverish Contact'*, London: Routledge, 2005.

Connor, Steven. *The English Novel in History 1950–1995*, London: Routledge, 1996.

Eliot, George. *Middlemarch*, Ware: Wordsworth Classics, 1994.

Farrell, J. G. *The Siege of Krishnapur*, London: Phoenix Publishing, 2007.

Gilbert, Pamela K. *Cholera and Nation: Doctoring the Social Body in Victorian England*, New York: State University of New York Press, 2008.

Hamlin, Christopher. *Cholera: The Biography*, Oxford: Oxford University Press, 2009.

Harrison, Mark. *Climates and Constitutions: Health, Race and Environment in British Imperialism in India*, Oxford: Oxford University Press, 2002.

Howells, Coral Ann. *Margaret Atwood*, 2nd edn, London: Palgrave Macmillan, 2005.

Meyers, Jeffrey. *Disease and the Novel 1880–1960*, Hong Kong: Macmillan, 1985.

Sontag, Susan. *Illness as Metaphor and AIDS and Its Metaphors*, London: Penguin, 2013.

# 31

# WHEN BODIES NEED STORIES IN PICTURES[1]

*Arthur W. Frank*

This chapter discusses how people represent their experiences of illness in graphic novels; the broader issue is how representations of suffering affect living with disease and finding forms of healing. Humanity's earliest known work of sustained storytelling, the Gilgamesh epic, is a kind of guide to the suffering of mortality.[2] It depicts the fear and anxiety of Gilgamesh's friend Enkidu as he realizes he is dying, and Gilgamesh's grief for Enkidu drives the plot of the epic's second half. Sophocles' *Philoctetes* is about a level of pain that can be expressed only in screams.[3] But Philoctetes suffers more than the pain of the festering wound on his foot. He suffers the isolation that humans have inflicted on him in response to his wound. Disgusted by the stench of his wound and distracted by his screams, Philoctetes' comrades in the war against Troy abandoned him on an isolated island, with only his bow to sustain him. Thus, the physical reality of disease, including dying, has always had a complement in the suffering caused by others' responses to that disease, injury, or disability. The representation of suffering – telling the story of how one came to suffer and how others' responses shaped this suffering – seems to be a fundamental human need. The media of telling often change, and today the graphic novel is increasingly popular. But the need to tell stories of suffering persists.

People tell stories not only to represent their sufferings and break out of the isolation that Philoctetes's abandonment epitomizes. Representations of suffering in stories teach people how to suffer; what to expect, how to interpret and categorize perceptions, and how to respond. This chapter is situated within a more general argument that humans live enmeshed in stories that circulate around and through us.[4] These stories have an independence from the corporeal human beings who tell them. We may speak of a story as mine or hers, but no story is ever really anyone's own. Even the most personal, autobiographical stories are made up from borrowings from other stories.[5] These other stories have taught us not only how to narrate our lives – how to frame a story that others will recognize and respond to – but also what is *worth* narrating about our lives. Our stories are our own, uniquely, but other people's stories set our parameters of narratability, and our story then sets parameters for others.

Although people have always told stories about their illnesses in one medium or another,[6] the contemporary illness narrative emerges in the 1970s, when people with considerable public profile began to write about their illness.[7] The term *pathography* is frequently used to describe this genre of first-person writing about illness.[8] I reject this term for several reasons.[9] Most important, the use of a faux-medical word implies that these writings exist to be available to a professional medical gaze, as patients make their bodies available in medical examinations. That is not at all my own experience of writing my own illness memoir,[10] nor is it how any other writer of an illness memoir has described his or her experience.

When he was dying of prostate cancer, Anatole Broyard gave a lecture with the title, 'The Patient Examines the Doctor'.[11] That reversal expresses the spirit in which most first-person writing about illness is undertaken. In some illness narratives, institutional medicine is marginalized entirely; if health-care workers enter the story at all, it is as peripheral actors, not given much specification as characters.[12] This narrative choice implies that the significant story is not medical; medical actors and institutions figure, if at all, in a story that is fundamentally personal and then cultural, in the sense that what happens reflects the community of which the ill person was part, at a particular time and place.

## Illness as demands and as types of narration

The problems of narrating illness derive from problems of being ill, or more exactly, problems of what serious illness does to the self. Here are three demands that illness makes from the self, expressed by three of my favorite prose memoirists, Audre Lorde, Reynolds Price, and Anatole Broyard.

Lorde wrote that she needed to tell about having breast cancer – she needed to narrate the experience – because 'I think perhaps I was afraid to continue being myself.'[13] In what for me is a memorable phrase, Lorde says that she writes 'In order to keep me available to myself'.[14] I think that applies to any memoirist of illness. Illness unmakes what Lorde calls her availability to herself; narration remakes the availability. The word *available* can be called poetic in its irreducible density. The word arrests the reader. Tools and books are what we call *available* to us. They are things we know where to find, and depend on being able to find, when we need them. Lorde puts the self in that category. Illness renders her no longer where she can find herself. Narration makes her once again findable and thus useable.

Reynolds Price, who was left paraplegic as a consequence of radiation treatments for a tumour in his spine, emphasizes the need to say goodbye to the person he was before illness, recognizing that 'you're not that person now'. The question the ill person must ask is: 'Who'll you be tomorrow?'[15] Price proceeds to describe the need to find what he calls 'the next viable you – a stripped down whole other clear-eyed person, realistic as a sawed-off shotgun and thankful for air'.[16] Telling one's story is a means of finding, perhaps discovering, that 'next viable you'. Price puts more emphasis on change than Lorde does. In her language, Price's old self is no longer available. Narration is the work of writing a new self into available reality. Words bring imaginations into a communicable, and thus shared, reality. A person can imagine whom she or he will be tomorrow by writing down that self. Action can follow invocation, which already is a form of action.

Anatole Broyard phrases the demand of illness in this way: 'for it seems to me that every ill person needs to develop a style for his illness. I think that only by insisting on a style can you keep from falling out of love with yourself as the illness attempts to diminish or disfigure you.'[17] On this account, the ill person narrates his or her experience because narration is a means of discovering a style through trial expressions of that style – it is an iterative process of revision. By finding one's own style of narrating illness, the ill person avoids 'falling out of love with yourself', which seems another variant expression of Audre Lorde's fear of ceasing to be available to herself.

The style that Lorde, Price, and Broyard seem to share is what I have previously called the *quest narrative*.[18] When I wrote *The Wounded Storyteller* in the mid-1990s, I had been listening to stories about illness – both the stories I told myself and stories other people told me – for close to a decade. I heard people telling stories that were uniquely their own as reflections of their embodied experiences, but I also heard people telling their stories based on narrative resources that, I realized, were widely shared. Central to these narrative resources were three skeleton plots or narrative structures into which people fit their own distinctive stories.

In the first of these, the *restitution narrative*, the narrator/protagonist becomes ill, is diagnosed and treated, and eventually is restored to a life that is at least a good enough analogue to life as it was before illness. The second narrative is the *chaos narrative*, which is really an anti-narrative, because there is no movement; the story, such as it is, is about the incapacity to move. In chaos narratives, things only get worse and other people's attempts to help only misunderstand how hopeless the narrator's life is. My third narrative type, referred to earlier, is the *quest narrative*. Here the ill narrator undergoes a crisis of self as well as a crisis of the body's health. That crisis is expressed by Lorde: 'I think perhaps I was afraid to continue being myself'. Responding to that fear requires asking Price's question: 'Who'll you be tomorrow?' And answering that question depends on finding the new style that Anatole Broyard refers to. The quest for that style is not for oneself alone. Instead, the narrator takes an 'Everyman' stance, seeking a generalizable template for living with illness, although narrators vary in how widely they claim the applicability of how they have learned to live.

Graphic novels may still be a niche genre – like science fiction – but in the past years they have achieved the objective respectability of being reviewed regularly in major newspapers alongside other new books. Within the graphic-novel genre, illness and the practice of medicine have become a notable sub-genre.[19] The initial question with which I began to read graphic novels was whether these three core narratives would be found there. What might these narratives look like in graphic form, and how might the graphic form generate different narratives and representations of illness?

## Graphic chaos

The clearest example of any of the narrative types that I described in *The Wounded Storyteller* is Allie Brosh's chaos story in a chapter entitled 'Depression Part II' in *Hyperbole and a Half*.[20] The page contains a block of text and several panels, in no sequential order, showing a small surreal creature with large eyes, pointed ears, stick arms and legs, a pink body, and a curious, nervous expression. What is remarkable is that such a simply drawn creature can convey a sense of expression. The creature exists in an undefined space that seems to tilt in impossible ways. The creature is depressed.

Brosh exemplifies chaos narration when she writes: 'It isn't always something you can fight back against with hope. It isn't even something – it's nothing. And you can't combat nothing. You can't cover it. It's just there, pulling the meaning out of everything.'[21] When I was writing *The Wounded Storyteller*, I never found a published memoir that narrated chaos that clearly, but my examples were drawn from oral and written narrations. The question is how the graphic format enables Brosh to express something that is far more difficult to convey in prose alone.

Brosh's book seems unusual among graphic novels in the extent of her use of printed text outside the frames of the drawings; in that difference, she seems closer to prose memoirists. But that raises the question of how her prose would work on its own. On my reading, Brosh is able to write prose that is so remarkably cool and straightforward because it is seen on the same page with drawings that give the prose an emotional jolt. Brosh's drawings make readers feel the claustrophobic lassitude of depression, the shrinking of one's world. If I imagine the prose by itself, it would seem dissociated and disembodied. Yet without the lengthy prose, Brosh's argument about depression would get lost, and Brosh seems concerned to make a specific point. She wants to bear witness to a reality that people who have never been literally *in* depression cannot imagine – until her pictures give us that imagination, and the prose articulates what the images provoke.

The combination of text and images allows Brosh to keep together, both on the page and in the reader's imagination, the fundamental contradiction that makes prose evocation of chaos near impossible. To be able to tell a story about being in chaos, one can no longer be in chaos. Brosh writes about chaos while putting us into that chaos: in her images, bodies sag and spatial geometry is askew. It is a graphic depiction of a world 'out of joint', in Hamlet's phrase.

Brosh does eventually find a way out of depression and again her drawing makes us believe it.[22] Like those who write quest narratives, Brosh refuses to turn what was chaos into a restitution narrative. 'Anyway,' she begins, 'I wanted to end this on a hopeful, positive note.' She then subverts this typical restitution ending. 'But seeing how my sense of hope is still shrouded in a thick layer of feeling like hope and positivity are bullshit' – which is how one feels while living in chaos – 'I'll just say this: Nobody guarantees that it's going to be okay, but . . .'.[23] She then refers back to the incident that was her turn-around point. Stories that depend upon a turning point are generally quest narratives. The hope that there can be a turn-around point is part of the solace that the story offers, and quest narratives balance solace with demand. Brosh refuses to turn depression into a quest; any solace is strictly limited. Those frames of pure chaos remain in the reader's memory, enlarging that reader's sense of life's possibilities.

## Restitution noir

If Brosh offers us chaos brightening, Harvey Pekar and Joyce Brabner's *Our Cancer Year* might be called *restitution noir*.[24] Illustrated by Frank Stack, who seems to me to deserve full authorship, the graphic format enables what is formally a restitution narrative – Harvey gets sick, is treated, and eventually returns to life as he has known it – to take on a darkness that such narratives usually minimize. *The Wounded Storyteller* presents the restitution narrative as the form of storytelling preferred by medical professionals. Doctors and nurses cue their patients to understand their experiences within a narrative that will culminate in cure, or 'walking out of here' as a common hospital phrase goes. The *telos* or expected end-state of cure thus justifies whatever the patient has to undergo during treatment, and complaints about treatment are minimized within a time frame that anticipates when cure will retrospectively erase bad memories. Patients are supposed to go along with that anticipation, telling their treatment stories framed by those narrative parameters.

*Our Cancer Year* presents the most disturbing images of the pain of chemotherapy that I have ever encountered.

The chemo protocol I underwent for testicular cancer was presented to me as one of the most aggressive, but my side-effects seem minimal compared to what Harvey goes through. Writing about how chemo feels is one of the most difficult problems of narration in writing a cancer memoir. As soon as I started chemo, I suddenly appreciated how effective drugs can be as a form of torture. Finding words to express that embodiment defies writers including me.

But the graphic novelist has an advantage, and here we get to what may be the singular formal difference between prose memoirs and graphic novels. In the graphic format, the protagonist can and must be represented as a drawn figure. Things happen

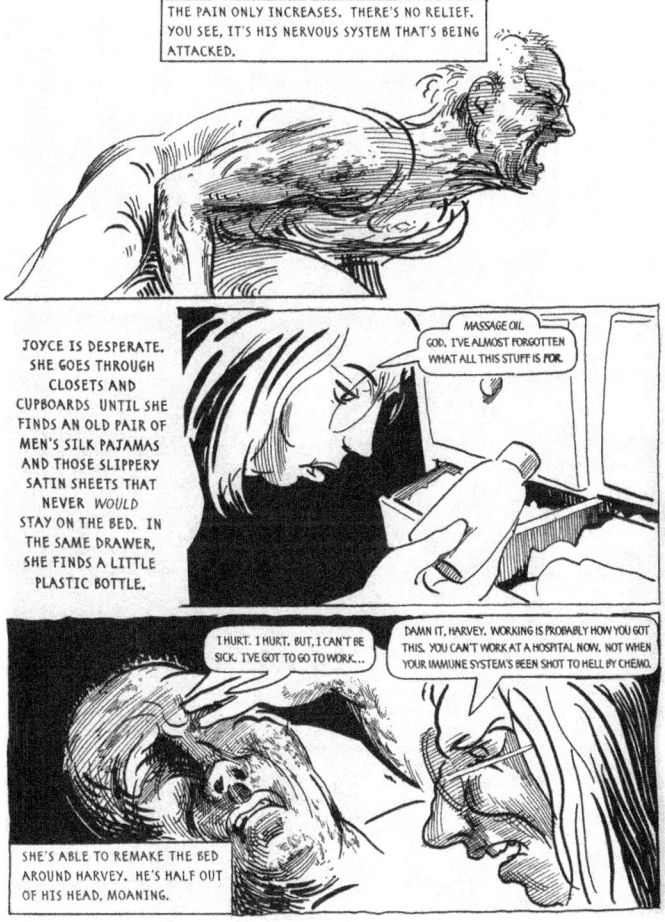

*Figure 31.1* 'The pain only increases . . .', from H. Pekar and J. Brabner, *Our Cancer Year*, art by Frank Stack. New York: Four Walls Eight Windows, 1994. Used by permission of Joyce Brabner.

569

to *that* figure, and those things can be shown, with the images liberated from the self-consciousness of a first-person voice. As with the chaos images in Brosh's depression story, the images linger in the reader's imagination. Of course words and phrases linger too and differences are easily over-emphasized. But images linger differently.

*Our Cancer Year* is a different kind of narrative insofar as the restitution plot is not framed within a telos of a happy recovery. At the end, Harvey's recuperation is in progress. He is living pretty much the same life he lived before cancer, but, after what has happened, to call his life-after-cancer a restitution betrays what readers have seen him go through. In *Our Cancer Year* nothing is forgiven as retrospectively understandable, as necessary for the ending that is now achieved. That's what I mean by it calling *restitution noir*; nothing is forgotten or forgiven.[25]

What is not forgiven is exemplified in the book's longest story segment, in which Harvey and Joyce tell a medical bad-treatment story; in this instance, a bad-nurse

*Figure 31.2* 'Time passes. He needs to take a leak', from H. Pekar and J. Brabner, *Our Cancer Year*, art by Frank Stack. New York: Four Walls Eight Windows, 1994. Used by permission of Joyce Brabner.

story. Harvey goes to the hospital for chemotherapy. The herpes blisters on his back burst, leaking through his shirt onto the chair; again, we have images of bodily pain.

The nurse, who seems to reincarnate Nurse Ratched from *One Flew Over the Cuckoo's Nest*,[26] is concerned only with cleaning off her chair; she abandons her patient lying on the floor.

The story is too dark to be what one expects within a restitution narrative, which is typically appreciative to health-care.

But neither is the chemotherapy-chair story presented as part of a quest narrative, which do typically include resentment stories.[27] What makes this story segment disjunctive with the quest narrative is that Harvey is not shown to be learning something – there's no quality of *bildungs* or the development of a self through trials and overcoming. To reverse Nietzsche's famous aphorism, what nearly kills him does not make him strong, or much of anything different; cancer is just a rotten thing to go

*Figure 31.3* 'She attacks the chair', from H. Pekar and J. Brabner, *Our Cancer Year*, art by Frank Stack. New York: Four Walls Eight Windows, 1994. Used by permission of Joyce Brabner.

*Figure 31.4* 'I'm not going to get pneumonia', from H. Pekar and J. Brabner, *Our Cancer Year*, art by Frank Stack. New York: Four Walls Eight Windows, 1994. Used by permission of Joyce Brabner.

through. In *Our Cancer Year* incidents *just happened*; for Harvey, it stays what it was. Yet for the reader, the mundane does take on greater significance.

## Qualifying the quest

I have already strayed into discussion of my third type of narrative, the quest. When I began to read graphic novels, it seemed to me that they do not do quest narratives, at least in the Lorde-Price-Broyard sense of quest as both personal realization and community coalescence. Graphic novels presented either *restitution noir* narratives exemplified by *Our Cancer Year*, or else they were what could be called *complicated restitution* stories like Ellen Forney's memoir of bipolar disorder, *Marbles*. A major complication in Forney's depression is that nobody, and her psychiatrist is candid about this, has much control over the trajectory of the illness.[28] Or, there are less complicated restitution stories like Marisa Acocella Marchetto's *Cancer Vixen*, in which the protagonist desperately wants to sustain a style, but unlike Broyard she seeks to maintain her former style rather than discovering a style distinctive to illness.[29] Then I realized that I was missing the point. A different kind of story was being told; again, a more ambivalent story than prose readily enables.

How I was missing the point is illustrated by a six-panel cartoon from Miriam Engelberg's *Cancer Made Me a Shallower Person*, which is definitely my favorite title among illness graphic memoirs.[30] Engelberg presents herself – and this may be self-satire – as thinking that various reflective exercises ought to produce what she calls an epiphany, by which I think she means some revelatory burst of meaning. That doesn't happen, so she goes back to watching television.

The story would seem to be an anti-quest satire. But I wonder which quest narratives Engleberg had been reading.

*Figure 31.5* Page from 'Stress' ('Wait a minute, that flashback reminded me . . .'), from *Cancer Made Me A Shallower Person: A Memoir in Comics* by Miriam Engelberg. Copyright © 2006 by Miriam Engelberg. Reprinted by permission of HarperCollins Publishers.

Neither Lorde nor Broyard claims having any epiphany; only Price does.[31] He describes a genuine mystical vision and an auditory vision, but those visions do not seem necessary to the advice he offers at the end of the book about the need to become the 'next viable you'. Price understands his visions as particular gifts of grace. He does not recommend that others seek, expect, or need such visions in order to live with debilitating illness. So epiphany or not, the quest in these three writers' memoirs is based on simple if difficult precepts, including: seek to be less afraid; remember that fear of illness readily turns into fear of yourself; the best revenge for being treated badly is to tell a good story about it; live fully with illness, pay attention to your life; do not dissociate, especially when things get ugly; let illness teach you what is essential, and let go of the rest without undue concern; and perhaps most important, paraphrasing Lorde, do not let your pain be wasted – others need what you can tell them about where they too are going.

What Engelberg led me to realize is that over the years I too had slid down the slippery slope into thinking of quest-as-epiphany; call it a cultural bias. I had forgotten what thin lines divide restitution from quest and quest from chaos. Every quest writer hopes his or her story will end in a form of restitution, *but* their good sense is not to expect that in the form of medical cure, which is always a contingency. The real restitution is that despite all the hassle, upset, real loss, and sometimes unbearable physical pain of illness – and those are the chaos segments – the ill person may be lucky enough to tell the tale. Through telling that tale, she or he realizes not only that during illness life has remained worth telling stories about, but also that life actually became *more narratable*. And maybe that is what really matters about any life. *Our Cancer Year* thus ends with wordless images of Harvey and Joyce, spending a normal day in a pleasant but unremarkable park.

### Arguing against impersonality: narrative as chronicle

Words to describe a narrative like *Our Cancer Year* came to me when reading a very long critical meditation on the extraordinary success of the Norwegian writer Karl Ove Knausgaard.[32] I call Knausgaard's success extraordinary because his books fail to fit any of the expectations for contemporary literary success. They have no particular plot, reveling instead in detailed descriptions of minutiae. The Canadian critic Ian Brown, who not coincidentally has written a parental narrative of extreme disability,[33] rehearses everything that ought to make Knausgaard unreadable and then explains his own absorption in Knausgaard's minutiae: 'an unavoidable tension ran through them, in which the specifics of his life were always arguing against the impersonality of the larger world.' That applies equally to Pekar and Brabner's writing. The specifics of their utterly mundane lives argue against the impersonality of the larger world. The graphic novel, as a medium, is not uniquely positioned to present that argument, but the argument is readily told in the graphic format.

Graphic-novel illness memoirs are written in three formats, which are not discrete types but form a continuum. One format collects cartoons of various length that stand alone. There is minimal over-arching plot and the order of cartoons could be rearranged without confusion to the reader or loss of meaning. *Cancer Made Me a Shallower Person* is that kind of book. At the other end of the continuum is David Small's *Stitches*, a memoir of childhood cancer in which every frame is in precisely the order it must be, for the story to have the effect it has.[34] Nothing seems superfluous and nothing could be rearranged. Then in the middle of this continuum are books like *Our Cancer Year* and *Marbles*, which tell multiple stories that could stand alone but accumulate their effect within the arc of a longer and looser chronicle of the author's life.

My usage of that word *chronicle* is specific to narratology. The historian Hayden White draws a narratological distinction between chronicles and stories, describing history as the 'transformation of chronicle into story'.[35] A *chronicle* is a recitation of events as they occur in chronological order, but a single event – what happens this year – has no necessary relation to previous or subsequent events. There is no plot to a chronicle. A *story*, however, depends on one event happening as a consequence of what happened before, although the precise terms of causality are unpredictable, and the characters are often less aware of how their present actions lead to subsequent effects.[36]

To call *Our Cancer Year* a chronicle that collects various stories is to take seriously the book's title. *Our Cancer Year* recounts a year during which Harvey and Joyce buy a house, Joyce is involved with some young people who live in countries affected by war, and, among these other events, Harvey has cancer. Those three clusters of events have effects on each other; for example, moving house involves a couple of crises due to Harvey's weakened condition. But the three clusters are also chronicle-like in that they just happened to happen in the same year – none depends on the others – and how each unfolds has its own logic that is separable from how the other events unfold. Moreover, that year is just one year within the longer chronicle of the *American Splendor* series, of which *Our Cancer Year* is one volume.

The narrative achievement of *Our Cancer Year* is its refusal to impose a story on events that are really a chronicle. This may sound simple, but few prose memoirists are able to do it. Pekar and Brabner are willing to tell a story when things happen that way, as in the story of Harvey's blisters bursting while he sat in the chemotherapy chair. But the whole book refuses to impose a plot logic of causality on events that do not fit together as a plot. That refusal is the authors' quest and the healing wisdom they offer their readers. Live as if nothing is a metaphor for anything else.

At issue here is avoiding what is possibly the fundamental danger of stories. We humans have this incredible capacity to tell stories, but that capacity becomes a liability when we turn what are best understood as chronicles into stories.[37] In the process of creating stories out of merely sequential events, we invent connections that simply are not there, and then endow these connections with causal force. Because we humans are creating our story as we live, connections that we believe in risk becoming self-fulfilling prophecies. Thus, a good deal is at stake in the refusal to impose plots on chronicles; it has a Zen quality of vision disciplined to see only what is. Even when Harvey-the-character falls into an imaginary narrative – when he is convinced he is paralyzed, but that paralysis remains unconfirmed either by Joyce or his physicians – Harvey-the-writer stands back and observes that excess of imagination as a lapse due to illness.

The only hint that there is a larger scheme of meaning in *Our Cancer Year* occurs in the authorial voice-over on the first page: 'This is a story about a year in which someone was sick. About a time when it seemed that the rest of the world was sick, too. It's a story about feeling powerless and trying to do too much. . . .' At the beginning of this chronicle, the events to follow are tied together by a very general theme, but that will be the only such authorial intrusion. After that first page's voice-over, we only see Harvey and Joyce living their lives.

## Learning to live with illness

The opening of *Our Cancer Year* takes me back to Brown's critical appraisal of Knausgaard, 'in which the specifics of his life were always arguing against the impersonality of the larger world'. We reach the core question of how graphic-novel memoirs do the fundamental work of any illness memoir: How do they help with healing both the writer and the readers? How, by narrating one's disease, does one bring about a form of healing for oneself and for others?

In graphic-novel memoirs, the specifics of the author's life argue against the impersonality of the larger world, and the quest is to keep that argument going, which is a struggle at any time, but especially during illness. In one panel of *Our Cancer Year*, Harvey reminds his readers that in the midst of all that was going on, he also got out the next volume of the *American Splendor* chronicle. We can also infer that he and Joyce were taking notes for what became *Our Cancer Year*. They were doing more than *coping*, whether with illness, or house buying, or supporting a group of young people who were living difficult lives. They were also creating the story of doing these things. In creating that story, they *humanized* what was happening, to borrow a word from the literary critic Frank Kermode.[38] And in Broyard's terms presented earlier, they *sustained a style*, which included being the kind of people who do not make many concessions to illness, even cancer.

The narrative distinctiveness of the graphic-novel form involves a good deal more than including pictures as illustrations for a story that could be told in words. The capacity of the graphic novel to mix story and chronicle give the genre a claim to be presenting a distinctive narrative form. The importance of this form is to keep open the question of whether and how one thing is connected to anything else – a crucial issue for seriously ill people. The reader of the graphic format must make constant decisions where to go next. Which panel follows which is not always self-evident, nor is it evident which panel is connected in what way to the one that only might be before. The reader's mind fills in connections between images. I often notice in graphic novels segments that would seem off-topic in prose, but flow right along in a book of images. The format can lull readers into the fictional/fantasy world of the story, but at best – as in the graphic novels cited in this chapter – the format requires readers to take responsibility for their active participation in the storytelling. That underlies these books' therapeutic contribution to the lives of people living with illness.

In mixing the distinctive narrative logics of chronicle and story, graphic novels can be more like life than prose narration. On any page, a lot is happening at the same time, and both characters in the book and readers of the book have to balance multiple focuses of attention. We constantly have to sort out which is which: what happens as part of a plotted story in which characters do things that eventually play out as having particular effects, versus what happens as part of a chronicle, in which *chronology* should not be confused with causality or consequentiality. I emphasize that this sorting is a high-stakes activity, fraught with real personal peril. If we confuse what is really a chronicle with being a story, then we invent linkages that are not there, and we risk demanding responsibility or attributing blame that is unwarranted. If we make the opposite mistake and think that what is really a story is only a chronicle, then we miss linkages and fail to attribute or assume responsibility.

What initially instigated my autobiographical writing about illness was my indignation at healthcare professionals positioning themselves as the protagonists in my story, relegating me to the bit-part of being a bystander. Restitution narratives are most pathetic when they involve one person consigning narrative responsibility for his or her life to another person – which is what medical patients are most often invited to do, especially in hospital. Yet I recognize that quest narratives risk becoming hyperbolic in their claims to self-realization, and they can distance illness experience from mundane demands of living. Graphic novels sustain the restitution plot but subvert its conventional narrative logic first by putting the ill person in the protagonist position. Then this subversion continues as the authors show things so awful that no restitution

*Figure 31.6* 'Two autobiographies became hugely important to me', from *Marbles: Mania, Depression, Michelangelo, and Me: A Graphic Memoir* by Ellen Forney, copyright © 2012 by Ellen Forney. Used by permission of Gotham Books, an imprint of Penguin Publishing Group, a division of Penguin Random House LLC.

is possible, like the chemo nurse throwing Harvey and Joyce out of her clinic for the Orwellian reason that he is too sick. In both of these narrative moves, the novels perform a significant quest.

Yet ultimately differences between the graphic format and prose narration dissolve in a common purpose. Forney makes explicit her relation to writers of memoirs of bi-polar disorder.

Forney writes about Kay Redfield Jameson's memoir: 'She was company.'[39] I wanted my own work, *At the Will of the Body*, to be company for people who had cancer; that is what I believe any teller of illness stories wants.

Disease, as any person lives it, only begins with pathology. A disease becomes a story that unfolds as it is lived, told, and responded to. How a person experiences disease depends on what other stories she or he knows and uses to make sense of experience. The history of disease is thus a history of people building on each others' stories. Human stories not only witness suffering, they accompany other sufferers, inviting others to become protagonists in their own stories.

## Notes

1 An earlier version of this chapter was presented as a keynote lecture at Comics & Medicine: From Private Lives to Public Health, Johns Hopkins University, June 26–28, 2014. My title echoes a chapter title in A. W. Frank, *The Wounded Storyteller*, 2nd edn, Chicago: University of Chicago Press, 2013 [1995], 'When bodies need voices.'

2 A. George (trans.), *The Epic of Gilgamesh*, London: Penguin, 1999.

3 Sophocles, *Philoctetes*, Phillips, C. (trans.), New York: Oxford University Press, 2003.

4 A. W. Frank, *Letting Stories Breathe: A Socio-Narratology*, Chicago: University of Chicago Press, 2010.

5 On the basis of a long career of translating and studying folk tales, Zipes writes: 'It is difficult to invent a tale; even a new creation will inevitably merge with the stream of tales heard before, and thus become a variant of what has already been around.' That narratological reality does not diminish the authenticity of the witness that a story can enact: J. Zipes, *The Irresistible Fairy Tale: The Cultural and Social History of a Genre*, Princeton, NJ: Princeton University Press, 2012, p. 38.

6 R. Porter and D. Porter, *In Sickness and in Health: The British Experience 1650–1850*, London: Fourth Estate, 1988; S.M. Rothman, *Living in the Shadow of Illness: Tuberculosis and the Social Experience of Illness in American History*, New York: Basic Books, 1994.

7 S. Alsop, *Stay of Execution: A Sort of Memoir*, Philadelphia, PA: Lippincott, 1973. Alsop was a weekly columnist for *Newsweek* and at the time of his illness, he was, with his brother Joseph, as well known as any American political journalist.

8 A. H. Hawkins, *Reconstructing Illness: Studies in Pathography*, West Lafayette, IN: Purdue University Press, 1993.

9 Frank, *Wounded Storyteller*, pp. 226–7, n. 34.

10 A. W. Frank, *At the Will of the Body*, New York: Mariner Books, 2002 [1991].

11 A. Broyard, *Intoxicated by My Illness*, New York: Clarkson Potter, 1992.

12 C. Wiman, *My Bright Abyss: Meditation of a Modern Believer*, New York: Farrar, Straus and Giroux, 2013. Wiman neither specifies exactly what the diagnostic label of his critical disease is, nor what his treatments were. Those medical details, given prominence in earlier illness narratives, would be a distraction in the story he wants readers to attend to. This gesture erases the *patho* in pathography.

13 A. Lorde, *The Cancer Journals*, San Francisco, CA: spinsters/aunt lute, 1980, p. 33.

14 Ibid., p. 65.

15 R. Price, *A Whole New Life: An Illness and a Healing*, New York: Atheneum, 1994, p. 182.

16 Ibid., p. 183.

17 Broyard, *Intoxicated*, p. 25.
18 Frank, *The Wounded Storyteller*.
19 I. C. M. Williams, 'Graphic medicine: The portrayal of illness in underground and autobi-ographical comics,' in V. Bates, A. Bleakley, and S. Goodman (eds), *Medicine, Health and the Arts: Approaches to Medical Humanities*, London: Routledge, 2014, pp. 64–84. While Williams provides one of the most comprehensive reviews, he writes from a medical perspective in which narratives of illness experience are 'pathographies'. Williams also writes about his own medical practice: I. Williams, *The Bad Doctor: The Life and Times of Dr. Iwan James*, University Park, PA: Penn State University Press, 2015. He moderates the website graphic-medicine.org. That site provides information about the annual Comics & Medicine con-ference, as well as other links. For another example of physician storytelling in graphic medicine, see M. Green, 'Missed it', *Annals of Internal Medicine*, 2013, vol. 158, pp. 357–61. See also J. McMullan, 'Cancer and the Comics', *Medical Anthropology Quarterly*, forthcoming.
20 A. Brosh, *Hyperbole and a Half*, New York: Simon and Schuster, 2013, p. 32. To view images, see, http://hyperboleandahalf.blogspot.ca/2013/05/depression-part-two.html, accessed September 22, 2015.
21 Ibid.
22 Ibid., p. 155.
23 Ibid.
24 H. Pekar and J. Brabner, *Our Cancer Year*, Art by Frank Stack. New York: Four Walls Eight Windows, 1994.
25 Among prose narratives, this same ending occurs in E. Handler, *Time on Fire: My Comedy of Terrors*, New York: Owl Books, p. 276. Now in stable remission from cancer, Handler refuses the idea of restitution: 'I've heard it said that whatever doesn't kill you makes you stronger. I've also heard it said that whatever doesn't kill you fucks you up for a really long time, and it's a miracle if you ever get it back together again.'
26 K. Kesey, *One Flew Over the Cuckoo's Nest*, New York: Signet, 1963.
27 A. W. Frank, 'How stories of illness practice moral life', I. Goodson, M. Andrews, A. Antikainen, and P. Sikes (eds), *The Routledge International Handbook on Narrative and Life History*, London: Routledge, forthcoming.
28 E. Forney, *Marbles*, New York: Gotham Books, 2012.
29 M. A. Marchetto, *Cancer Vixen*. New York: Alfred A. Knopf, 2006.
30 M. Engelberg, *Cancer Made Me A Shallower Person: A Memoir in Comics*, New York: Harper, 2006.
31 Price, *Whole New Life*, pp. 42–5, 80–1.
32 I. Brown, 'He put me on edge', *Globe & Mail*. Toronto, April 26, 2014, pp. F1, F6–7. K. O. Knausgaard, *My Struggle: Book 1*, New York: Farrar, Straus and Giroux, 2013.
33 I. Brown, *The Boy in the Moon: A Father's Journey to Understand His Extraordinary Son*, New York: St. Martins, 2011.
34 D. Small, *Stitches: A Memoir*, New York: W.W. Norton, 2009.
35 H. White, *Metahistory: The Historical Imagination in Nineteenth Century Europe*, Baltimore, MD: Johns Hopkins University Press, p. 5.
36 This idea is developed in Frank, *Letting Stories Breathe*.
37 A. W. Frank, 'The necessity and danger of illness narratives, especially at the end of life', in Y. Gunaratnam, and D. Oliviere (eds), *Narrative and Stories in Health Care: Illness, Dying, and Bereavement*, Oxford: Oxford University Press, 2009, pp. 261–75.
38 F. Kermode, *The Sense of an Ending: Studies in the Theory of Fiction*, New York: Oxford, 1973.
39 Forney, *Marbles*, p. 90.

## Select bibliography

Brosh, A., *Hyperbole and a Half*, New York: Simon and Schuster, 2013.
Broyard, A., *Intoxicated by My Illness*, New York: Clarkson Potter, 1992.
Engelberg, M., *Cancer Made Me A Shallower Person: A Memoir in Comics*, New York: Harper, 2006.
Forney, E., *Marbles*, New York: Gotham Books, 2012.

Frank, A. W., *At the Will of the Body*, New York: Mariner Books, 2002 [1991].

Frank, A. W., 'The necessity and danger of illness narratives, especially at the end of life' in Y. Gunaratnam and D. Oliviere (eds) *Narrative and Stories in Health Care: Illness, Dying, and Bereavement*, Oxford: Oxford University Press, 2009, pp. 261–75.

Frank, A. W., *Letting Stories Breathe: A Socio-Narratology*, Chicago: University of Chicago Press, 2010.

Frank, A. W., *The Wounded Storyteller: Body, Illness & Ethics*, 2nd edn. Chicago: University of Chicago Press, 2013 [1995].

Frank, A. W., 'How stories of illness practice moral life', in I. Goodson, M. Andrews, A. Antikainen, and P. Sikes, eds., *The Routledge International Handbook on Narrative and Life History*, London: Routledge, forthcoming.

Lorde, A., *The Cancer Journals*, San Francisco, CA: spinsters/aunt lute, 1980.

Pekar, H. and Brabner, J., *Our Cancer Year*, Art by Frank Stack. New York: Four Walls Eight Windows, 1994.

Porter, R. and Porter, D., *In Sickness and in Health: The British Experience 1650–1850*, London: Fourth Estate, 1988.

Price, R., *A Whole New Life: An Illness and a Healing*, New York: Atheneum, 1994.

Rothman, S.M., *Living in the Shadow of Illness: Tuberculosis and the Social Experience of Illness in American History*, New York: Basic Books, 1994.

Small, D., *Stitches: A Memoir*, New York: W.W. Norton, 2009.

# 32

# LIVING IN THE PRESENT

## Illness, phenomenology, and well-being

*Havi Carel*

> Of things, some are up to us, and others are not up to us. Up to us are
> opinion, impulse, desire, aversion and, in a word, all our actions. Not up
> to us are our body, possessions, reputation, offices and, in a word, all that
> are not our actions.
>
> (*Epictetus*, Enchiridion *1.1*)

The Stoics considered living in the present a way of securing one's happiness.[1] Indeed, they suggested a set of spiritual exercises (*askêsis*) to promote that existential stance and allow those who practise them to flourish. Some of the exercises focus on delineating a present moment, untarnished by regrets about the past and worries about the future. This sounds simple, but attaining such a moment of pure presence was, for the Stoics, a superb achievement. A moment of perfect happiness, achieved by securing relief from the agitations of the mind: *ataraxia*.[2] However, the Stoics have an unresolved problem plaguing this spiritual exercise: living in the present can only be a source of happiness if that present is pleasurable. But what if the present is full of pain, disability, or limitation? What if the present is a space of suffering? This is the first challenge illness confronts us with: I call it the *obstacle to happiness* challenge.

The obstacle to happiness challenge is frequently the first challenge ill people face upon symptom appearance or diagnosis. They may experience pain or feel anxious and alienated from their bodies. They may experience illness as bodily betrayal, or feel anger and fear. These experiences pollute the present moment and make happiness unattainable until they are addressed and tranquillity of mind can once again be pursued. I faced this challenge as a person diagnosed with a chronic progressive lung disease. As a patient and philosopher, I wanted to know if and how I would be able to flourish within the constraints of illness. Could we attain present happiness within the context of serious illness?

This chapter poses this question, and proposes a particular philosophical method, phenomenology, to enable us to (1) understand the experience of illness; and (2) reassess the possibility of happiness based on this understanding of illness. As a philosophical contribution to a volume on the history of disease, this chapter focuses on the present. Both the present time in which we live and experience illness, and the Stoic idea that the present moment is where happiness is to be sought. This is a chapter written by a philosopher, exploring the present and presence of illness. I use the term illness, rather than disease, to denote a useful phenomenological distinction between disease, a pathological process in the biological (objective) body, and illness, the lived experience of that disease.

## Illness and selfhood

I have the most pressing and personal interest in understanding the illness experience and in thinking how to make that experience less lonely, alienated and socially scripted. In the years since I was diagnosed I have navigated my rickety boat of selfhood between the Scylla of medicalisation and objectification of my illness and the Charybdis of social pressures to conform to the ideal of 'the good patient' or 'sick role'.[3] When subjected to philosophical analysis, this navigation, both philosophically salient and pragmatically important, can help us understand illness as part of selfhood; it is thus of prime philosophical value.[4]

Illness and selfhood have been prised apart for too long in the history of philosophy. The continuing journey towards understanding the connection between the two is unsettling; seasickness is inevitable. But if we manage to escape both the reductive effects of medicalisation and the restrictive effects of social stereotyping, we will gain considerable philosophical insight that also has practical import on health-care policy. Moreover, I suggest that not only our understanding of the illness experience improves through this philosophical analysis, but also our understanding of human values, human life, and philosophy stands to gain from this exchange.[5]

This divorce of illness from the philosophical conception of human life and value is baffling. The practice of philosophy as pure thought, according to Socrates and other thinkers of his period, is to teach the separation of the body from the soul, the point of which is to prepare us for our death, against which we have no defence: 'against other things it is possible to obtain security. But when it comes to death we human beings all live in an unwalled city'.[6] And yet we – and our modern-day Western society – remain unprepared for illness and death, denying our creatureliness, vulnerability and dependence on others. We still choose denial, not only of death, but also of frailty, ageing, and illness, as a long-term strategy and then stand bereft when confronted with such afflictions, despite their known inevitability. The first certainty I embraced when I became ill was the certainty of bodily decline and death. There is still nothing we can do about living in an unwalled city other than postponing death, not entirely successfully. We mostly still stand spiritually naked before the fact of our creation and annihilation.

As soon as one is born, one is old enough to die, writes Heidegger in his death analysis in *Being and Time*.[7] And yet we refuse to take that simple lesson seriously until forced to by a first- or second-person encounter with illness and death. We decline to think about the sloping path of ageing and disability, loss of autonomy, emerging dependency, and the concrete and unshakeable ways in which our body reveals to us our fallibility, limitedness and contingency. We decline to think about these unless forced to by illness, bereavement, or other trauma.[8] In this sense illness is a strict and demanding philosophical tutor: it *forces* us to confront such existential issues.[9] Noting this forced reflection points to the potential philosophical instruction and existential edification contained in illness.

Philosophers are no exception to this general tendency to deny our creatureliness and mortality; indeed, they have been seen to be vehement adherents of such denial. Alasdair MacIntyre has berated the discipline for this myopia:

> From Plato to Moore and since there are usually, with some rare exceptions, only passing references to human vulnerability and affliction and to the connections between them and our dependence on others. Some of the facts of human limitation and of our consequent need of cooperation with others are more generally acknowledged, but only then to be put on one side. And when the ill, the injured and the otherwise disabled *are* presented in the pages of moral philosophy books, it is almost always exclusively as possible objects of benevolence by moral agents who are themselves presented as though they were continuously rational, healthy and untroubled.[10]

In MacIntyre's view, philosophy has been doing a disservice to itself by covering over human vulnerability, dependence and affliction. But interestingly, it is not just a lack of reflection generating this denial. In fact what we see in the history of philosophy is a process of *forgetting*, forgetting the philosophical lessons that were the central *phronetic* goal of ancient schools such as the Stoics. These have now become obsolete, as part of a more general tendency to deny our bodily vulnerability, the fallibility of medicine, and the fact that 'we each owe nature a death', as Freud wrote.[11] This is an active process, requiring effort in order to cover up what was previously accepted as a universal truth about human existence: that life is unpredictable and contains unavoidable pain and suffering and that we have little control over the ways of the *cosmos*. However, our response to affliction is something we can control according to the Stoics, so our moral training should focus on our response to the causes of suffering and pain. That response, as you will see below, is the cultivation of virtue (or 'excellence of the soul') and the acceptance of our lack of control over the *cosmos*.[12]

One might object that medical technology and interventions have come a long way since the days of the Stoics. Doctors in the first century BC had little to offer, but in present times medicine can do what was once impossible: save lives, extinguish pain, and increase longevity. So, the objection might go, we need to revise the Stoic acceptance of suffering; we now have tools with which to combat suffering. I suggest that, on the contrary, we should constrain technological advancement with a normative philosophical analysis and consider reflectively how to utilise such advances. It is time for us to return to the Stoics and Epicureans, recover their attitude of reflective coping towards illness and death. Indeed, our illness experiences remain similar to those thousands of years ago in one important respect: illness still entails the loss of bodily certainty and control,[13] the loss of the familiar, homely world,[14] and the loss of wholeness and freedom.[15] These aspects of illness transcend any particular disease process and can be said to be generally descriptive of illness experiences, at least in modern Western civilisation.

The more medicine has developed and found temporary answers to some of the afflictions we face, the more we have turned away from the need to confront human vulnerability. Not only medicine has developed; also our faith in its omniscience and its ability to control pain, postpone death, and enable a dignified death has increased in tandem with medical technology and clinical improvement. This faith,

I suggest, has little grounding in the reality of health-care practice and embodied existence. In reality, medicine can control pain to an extent, but not always; it can postpone death, sometimes for decades, but not overcome it altogether; and it does not secure a dignified death, given the strong resistance to assisted dying and the tendency of modern medicine to over-treat. Most people die in hospitals rather than at home, and most people undergo invasive and ineffective treatments in the last days of their lives.[16]

The Stoics, intent on cultivating the right attitude to suffering, emphasise one's subjective demeanour towards illness. By contrast, in what might be considered a spiritual step back, we look to external interventions. Compare Epictetus, lecturing in the first century AD, with physician Atul Gawande's 2014 Reith Lecture. First Epictetus, telling us 'in what manner we ought to bear sickness':

> Now is the time for the fever. Let it be borne well. Now is the time for thirst, bear it well. Now is the time for hunger, bear it well. Is it not in your power? Who shall hinder you? The physician will hinder you from drinking; but he cannot prevent you from bearing thirst well: and he will hinder you from eating; but he cannot prevent you from bearing hunger well.[17]

And second, Gawande talking about his daughter's piano teacher, whose cancer treatments failed:

> I asked her, 'Well, what did the doctors say?' and she said, 'Well not much it doesn't seem. They're giving me blood transfusions to fight the blood cell counts dropping.' They were giving pain medications, they were giving steroids. But what next? They said there were no established chemotherapies that they were going to offer. And I think this is the moment that we continue to debate in many countries around the world: what is it that we think should happen?
>
> It's an expensive moment, it's a trying moment. And the way it feels reduced, even sitting there on the other end of that phone call, seemed to come down to: well, should we encourage her to try something, anything, is there an experimental therapy that she could try? . . . Or, she asked, should she just give up?[18]

The real challenge in illness is to grasp the false dichotomy inherent to the thinking expressed by the piano teacher in the quotation above. The question is not whether to continue to treat or give up, but to learn how to bear sickness well and ultimately how to die well. In this respect, our thinking about illness has not made much progress in the last 2,000 years.

Prioritisation of the individual response to illness once again became key in some quarters. In twentieth-century debates about coping with stress, for example, Hans Selye's emphasis on deviation and diversion in order to cope with stress, echoed Seneca's strategies for coping with pain and illness, namely inviting sufferers to divert their thoughts from current suffering to reminiscing about pleasurable times.[19] But individual responsibility is sometimes promoted in ways that are detrimental to the sick individual. This can be seen in the emphasis on 'positive attitude', culturally prominent in the USA, which Barbara Ehrenreich riled against in *Smile or Die*.[20] The sense of helplessness and the accompanying patiency that characterises many experiences of illness may be combated usefully, or non-usefully, by appeals to the individual's capacity to master the illness and other life events. However, this attitude differs in one important respect from the Stoic one: whereas 'positive thinking' suggests that disease can be halted or reversed through practices such as meditation, purges, spiritual purification, or prayer, the Stoics suggest no such thing. Responsibility lies with the individual only for what is 'up to us', namely, their state of mind and behaviour. They have no control over external events.

The Stoic understanding of illness as a moral challenge, calling on us to use our reflective skills to cope with it, is one that requires deep philosophical understanding of the experience of illness. In the next section I would like to demonstrate how useful philosophy is for exploring illness, how it changes one's way of being, what costs and opportunities it may bring with it, and what joyful and mysterious ways ill people have of obtaining that elusive moment of happiness: being undistractedly in the present. My methodological companion is phenomenology, the philosophical method for the study of lived experience. I begin by outlining what phenomenology is and how it illuminates illness. I show that when used to explore the experience of illness, this method returns us to the obstacle to happiness challenge. I then turn to a key question for all ill people, namely, can one be ill and happy? I end by returning to the Stoics in order to answer this question in the affirmative.

As this chapter is part of a volume on the history of disease, it is important to note that the examination of lived experience is also integral to historical research and to medicine. In mental health, for example, such an approach encourages clients and health professionals to foreground individual experience and meanings rather than clinical categories. Lived experience is also a key feature of some 'oral history' or 'life story' approaches to the past. What these approaches have in common is the prioritising of lived experience, which endows individual standpoints with the autonomy and weight they may otherwise lack, and viewing such standpoints as epistemically authoritative.

## Phenomenology of illness

Illness is a universal feature of human life, as well as having a dramatic impact on the ill person and those around her. Experiencing illness is a confrontation with our bodily vulnerability, our inability to control external events, and mortality.[21] Illness disrupts habits, expectations and abilities, often irreversibly. Because of this disruption, meaning structures are also destabilised and in extreme cases the overall coherence of one's life is destroyed. Illness radically disrupts the fundamental sense of embodied normalcy in which one's existence is rooted. Thus illness is a fundamental feature of human life, and its philosophical study reveals deeply philosophical aspects of human embodiment, the experience of space and time, and the interpersonal world.

How should we study illness philosophically? I propose that we study it phenomenologically, by illuminating the experience of illness from multiple perspectives (the ill person, her family, medical staff, etc.), allowing a rich understanding of it to emerge. Phenomenology is a philosophical tradition inaugurated in the early years of the twentieth century that focuses on *phenomena* (what we perceive) rather than on the reality of things (*pragmata*). It focuses on the experiences of thinking and perceiving: how phenomena appear to consciousness.[22] Phenomenology examines the encounter between consciousness and the world, and views the latter as inherently human-dependent; it is the science (*logos*) of relating consciousness to things as they appear (*phenomena*). The main aim of phenomenology is to study perception, cognition, and other aspects of our mental life, in a philosophical, rather than empirical, manner. Phenomenology is not a branch of empirical psychology and it does not ask, or attempt to answer, causal questions about specific mechanisms that give rise to experience (such as perception).

In order to study illness phenomenologically, we require an approach that emphasises the body's central role in human life and acknowledges the primacy of perception. Such an approach is found in the work of Edmund Husserl and Maurice Merleau-Ponty. Both thinkers developed a phenomenology of the body; Husserl in Division II of *Ideas II*[23] and Merleau-Ponty in his 1945 work *Phenomenology of Perception*.[24] Merleau-Ponty's understanding of phenomenology provides a robust account of human experience as founded on perception. Perception is itself an embodied activity. This is not just an empirical claim about perceptual activity but a transcendental view that posits the body as the condition of possibility for perception and action. For Merleau-Ponty the body is 'not merely one expressive space among all others . . . Our body, rather, is the origin of all the others, it is the very movement of expression, it projects significations on the outside by giving them a place and sees to it that they begin to exist as things, beneath our hands and before our eyes'.[25] As Gallagher and Zahavi write, 'the body is considered a constitutive or transcendental principle, precisely because it is involved in the very possibility of experience'.[26]

Both Husserl and Merleau-Ponty see the body and perception as the seat of personhood: 'Body and soul form a genuine experiential unity'.[27] To think of a human being is to think of a perceiving, feeling, and thinking animal rooted within a meaningful context and interacting with things and people within its surroundings. As Husserl writes in *Ideas II*, 'a human being's total consciousness is in a certain sense, by means of its hyletic substrate, bound to the body'.[28] The unity of mind and body is thus paramount to both thinkers. This view sees the body as the seat and *sine qua non* of human existence. To be is to have a body that constantly perceives the world. As such, the body is situated and intends towards objects in its environment. Human existence takes place within the horizons opened up by perception. Merleau-Ponty calls this unity the 'intentional arc':

> the life of consciousness . . . is underpinned by an 'intentional arc' that projects around us our past, our future, our human milieu, our physical situation, our ideological situation, and our moral situation, or rather, that ensures that we are situated within all of these relationships.[29]

The body is an enabler of action and experience, a condition for human experience, rather than an intentional object. But even when we engage in the most abstract of activities, the body is never completely effaced or even completely transparent, as some authors, such as Sartre and Zaner, suggest;[30] it is always felt and can move back into being an object of attention with a sensation as slight as an itch. More substantial and prolonged types of bodily demands call attention to the body in more distinctive ways. Such attention brings to light the body's practical and metaphysical significance: the body is inseparable from, and the condition for, any experience whatsoever. Its foundational status with respect to experience is key to an analysis of illness.

## Disease/illness, objective body/body-as-lived

It is useful to distinguish between disease and illness. Disease is a biological process affecting cells, tissues and organs. Illness is the experience of disease, the qualitative dimension of disease as it is experienced and made meaningful by the ill person. This includes the experience of receiving health-care, encountering social attitudes towards illness, pain, grappling with one's mortality and negotiating what may become a hostile world. To take an example: if someone has cancer that is at a very early stage, but has no symptoms and no knowledge of the cancer, then that person is diseased but not ill: they have no experience of illness. Conversely, if one suffers from migraine and cannot pursue everyday activities but has no corresponding brain lesion, then one is ill, but not diseased. There is no currently recognisable physiological causal process (although one may be discovered), but the experienced suffering and impairment are no less real than if a known pathology were present.

In other words, disease is to illness what our physical (objective) body is to our body as it is lived by us. This does not imply that illness that is not accompanied by disease is less real or less significant than an illness that is. Indeed, our ability to identify disease is contingent and changes over time. Hence I am not suggesting that an illness that is not accompanied by disease is less real but merely that we have not (yet) identified the disease. The epistemic restriction, namely our current inability to identify the disease, should not drive any ontological assumptions about the reality or severity of the illness.

Physiological dysfunction matters to us because it causes pain or discomfort, or prevents us from doing certain things. Our only point of contact with disease is via our *experience* of it. But rather than seeing illness as secondary to disease we ought to view it as a primary set of phenomena that invade one's life.[31] If we add these two claims to the notion that that illness is a profound and near-universal experience, we can further appreciate the importance of understanding how illness impacts upon one's life, how it changes experience, and how it shapes the life of the ill person, ultimately forming a 'complete form of existence'.[32]

This relationship relates to a broader distinction between the objective body and the body-as-lived. The objective body is the physical body, the object of medicine: it is what becomes diseased. Sartre calls this the 'body of Others': the body as viewed by others, not as experienced by me.[33] The body-as-lived is the first-person experience of this objective body, the body as experienced by the person. And it is on this level that illness, as opposed to disease, appears.[34] This distinction is fundamental to any attempt to understand the phenomenon at hand: the ill person is only and ever the one who experiences illness from within. Thus the experience of illness contains a measure of incommunicability that should be acknowledged.[35]

The relationship between illness and disease is not simple. Illness may precede one's knowledge of a disease: disease is commonly diagnosed following the appearance of symptoms. These symptoms are part of one's illness experience and are lived by the patient. Disease may exist without illness (as in the cancer example above).

Often we have both illness and disease, but the two do not perfectly cohere. For example, severe disease or disability may give rise to an illness experience that is experienced by the ill person as tolerable, due to adaptation.[36] So although the disease may be clinically classified as severe the illness experience is not as correspondingly negative as might be expected. This is of clinical importance because interventions ought to address patients' lived experienced, but are often designed to restore objectively measured function. In other words, interventions aim at disease, but the relationship between disease and illness is complex, non-linear and poorly understood.

The difference between the objective body and the body-as-lived emerges in illness because the body-as-lived is in large part habitual. Routine actions can be performed expertly and efficiently because they have become habit, and form a 'habitual body'.[37] While getting ready to go to work, one rarely notices the multitude of actions and the expertise required to have a shower and get dressed. It is only when we watch a novice that we can appreciate the complexity of the activity and the expertise it requires. The ease with which we perform habitual tasks often disappears in illness as a result of lost capacities, leading to a need to find new ways to perform routine tasks. While retaining the know-how, the ability to carry out a familiar action is lost.

The distinction between the objective and the lived body makes clear the fundamental difference between the two perspectives. The physician's perspective means that they can only ever perceive the disease through observation. The illness experience in its first-person form is not accessible to the physician,[38] by definition, other than via the patient's account. Taking the objective perspective may lead the physician to seek to treat the disease, sometimes with an inadequate understanding of the illness. The patient, on the other hand, can observe the objective indicators of disease but also has unique access to the first-person experience of the disease. In this sense the patient may have, at least in principle, an epistemic advantage because of access to her own illness experience *and* to the objective knowledge about the disease.

This epistemic advantage often goes unacknowledged and the patient experience may be subsumed under the medical view or discounted because the patient has no formal medical training.[39] The unique ability to oscillate between the two perspectives gives the patient a deeper understanding of the illness and the dual nature of the body, but this may also cause confusion and miscommunication. As the phenomenologist S. Kay Toombs notes, the physician's focus on disease may clash with the patient's primary interest in her illness, so although they may seem to speak of the same entity, they in fact refer to two different entities (disease, illness) and therefore have a communicative and interpretative gap.[40]

## The experience of illness

The experience of illness is diverse and constantly changing; it is bound with cultural and personal meaning; it can be idiosyncratic and difficult to describe, or even unshareable, as Toombs and others claim.[41] There are tens of thousands of diseases and, even within the same illness, each person is affected differently by symptoms, prognosis, pain, psychological impact and impact on daily life. It may therefore seem like a difficult task to try to distil shared features of illness.

Adopting a phenomenological approach, Toombs claims that the experience of illness exhibits a typical way of being.[42] Certain features of illness are manifest regardless of the particular disease state. These, claims Toombs, are typical characteristics of illness. These characteristics are integral to the illness experience and remain at its core regardless of varying empirical features.[43] The characteristics 'transcend the peculiarities and particularities of different disease states and consti- tute the meaning of illness-as-lived'.[44] Toombs does not say whether she thinks the characteristics of illness are stable across time. Indeed, their reliance on culturally determined features such as freedom and control would suggest that they are not transcendent or ahistorical. I suggest reading them as characteristics that capture the experience of illness as it is lived today, in Western cultures, but that may capture a more general tenor of such experiences, for example, the sense of loss of wholeness which relies on fractured bodily integrity.

Toombs lists five characteristics of illness: the perception of loss of wholeness, loss of certainty, loss of control, loss of freedom to act, and loss of the familiar world. These losses represent the lived experience of illness in its qualitative immediacy and are ones that any patient will experience. They cumulatively represent the impact of the illness on the patient's being-in-the-world. The loss of wholeness arises from the perception of bodily impairment, which leads to a profound sense of loss of bodily integrity. The body can no longer be taken for granted or be seen as transparent or absent, as it assumes an opposing will of its own, beyond the control of the self. The ill body thwarts plans, impedes choices, and renders actions impossible. In addition, illness disrupts the fundamental body-self unity, and the body is now experienced as other-than-me.[45] Thus illness is experienced as a threat to the self; the loss of integrity is not only of bodily integrity, but also of the integrity of the self.[46]

The second kind of loss, the loss of certainty, ensues from the loss of wholeness. The patient 'is forced to surrender his most cherished assumption, that of his personal indestructibility'.[47] The recognition of vulnerability and loss of certainty causes anxi- ety and this deep apprehension is difficult to communicate. Illness is experienced as a 'capricious interruption': an unexpected mishap in an otherwise carefully crafted life.

This experience of illness as an unexpected calamity leads to a sense of loss of control, which is the third kind of loss Toombs describes. The illness in its seemingly random unfolding is experienced like a stroke of fate. This makes the familiar world suddenly seem unpredictable and uncontrollable.[48] In addition, the ill person's abil- ity to make rational choices is eroded because of their lack of medical knowledge and limited ability to judge whether the health professional professing to heal can in fact do so.[49]

This leads to the fourth kind of loss, the loss of freedom to act. The ill person's ability to choose freely which medical treatment to pursue is restricted by her lack of knowledge of what the best course of action may be. In deciding whether to accept medical advice, the patient often assumes that the physician understands and shares her personal value system. However, the physician may often feel that it is inappro- priate, irrelevant or intrusive to enquire about the patient's values, and judges the clinical data alone to be sufficient for determining what is best for the patient. 'Thus patients not only lose the freedom to make a rational choice regarding their personal situation but additionally lose or abrogate the freedom to make the choice in light of a uniquely personal system of values.'[50]

Finally, the fifth kind of loss, the loss of the everyday world, arises from the dishar- mony of illness and its being a distinct mode of being in the world.[51] The ill person can no longer continue with normal activities or participate as before in the world of work and play. A large part of the familiarity of the world arises from its sharedness with other people, which is now lost.[52] The temporal dimension of one's world is also shaken because future plans have to be adjusted in the light of a medical prognosis and the healthy past is broken off from the ill present.[53] 'The future is suddenly disabled, rendered impotent and inaccessible'.[54] This loss of future further isolates the ill person from her hitherto familiar world.

Once experienced, the life domains incurring these losses can only be tenuously re-established, suggests Toombs. Even if the deficits are restored, any such re-establishment is always accompanied by a sense of its fragility and uncertainty.[55] The process Toombs describes is irreversible even if health is largely restored. This is not to say that a chronically ill person may not experience periods of wellness in between acute episodes or exacerbations. However, such periods are contextualised within the broader illness, and have been described as 'health within illness' or 'wellbeing within illness'.[56]

These five characteristics represent 'the "reality" of illness-as-lived. They reveal what illness means to the patient'.[57] For Toombs, this model of illness makes the primacy of the person explicit, and not secondary to an objective disease entity, as the biomedical model has it. In this sense the phenomenological model of illness can better serve not only patients but also physicians, whose ultimate goal is to improve individual patients' lives, not merely treat disease.

I suggest that the loss of freedom is a pervasive feature of illness, and can be understood more broadly. Toombs focuses on the loss of freedom to make rational decisions on the best course of action in response to the medical facts. However, this loss exists in a much broader loss of freedom brought about by illness. The loss of bodily freedom, freedom to make life plans, and freedom from anxiety about one's body is acute in both somatic and mental illness. For example, here is Arthur Frank's poignant description of his cancer diagnosis:

> What was it like to be told I had cancer? The future disappeared. Loved ones became faces I would never see again. I felt I was walking through a nightmare that was unreal but utterly real. . . . *My body has become a kind of quicksand*, and I was sinking into myself, into my disease.[58]

A similar closure of the future and of the freedom to choose one's course of action and future goals is a prominent theme in mental illness. John Stuart Mill, who suffered from depression, describes his illness in his autobiography: 'the whole foundation on which my life was constructed fell down . . . The end has ceased to charm, and how could there ever again be any interest in the means? I seemed to have nothing left to live for.'[59] In this situation of acute dejection the freedom to pursue goals is effaced by the loss of meaning of any goal. Although he is free, Mill cannot seize any particular goal because of his underlying feeling that the realisation of any goal would be pointless.

The loss of freedom is a pervasive loss, spanning the freedom to choose one's future, but also a loss of freedom in the present. Many routine activities easily performed are no longer possible and must be either given up or replaced by an alternative habit.[60] Toombs has herself broadened the conception of the loss of freedom in her later work, where she describes how in illness bodily intentionality is frustrated and the relation between the lived body and the environment is changed, and how possibilities become restricted.[61] The fuller account of loss of freedom in her later work stems from a more pronounced focus on embodiment as the source of meaning and locus of selfhood. In this view, illness disrupts the fundamental features of embodiment (being in the world, bodily intentionality, primary meaning, contextural organisation, body image and gestural display) and consequently, illness is experienced as a chaotic disturbance and disorder.[62]

An interesting question in the context of a volume on the history of disease is whether these typical features apply cross-culturally. It may be that in some cultures certain characteristics are not experienced because some values (such as freedom) do not exist in those cultures. Toombs' features of illness should be understood as offering a general characterisation of the experience of illness as lived by fully conscious adults in Western societies. Applying this framework to cases of children, mental disorder, diminished consciousness, and non-Western cultures, is one direction for future development of Toombs' seminal ideas.

## Health as transparency

In the smooth everyday experience of a healthy body, the body as object and the body as subject are aligned and experienced as harmonious. We do not experience the difference between the two orders most of the time; they cohere and make sense as a whole.[63] The fundamental bodily experience of health is one of harmony, control, and predictability. This has led some authors to describe the healthy body as *transparent*: we do not experience it explicitly as an object of our attention. When writing, we do not normally pay attention to a pen as long as it is functioning. We similarly do not pay attention to the hand that is writing. Our attention is focused on the task we are engaged at: writing a letter. Or, take a more explicitly embodied activity: if I prepare myself to catch a ball I do not focus on my body but on the ball, trying to anticipate its trajectory; my body simply 'follows me' to that point.

In such normal everyday experiences, the physical body is not prised apart from the lived body, and the experienced functioning of the body is natural, pre-reflective, and effort is experienced as normal or enjoyable. Sartre and Leder describe the healthy body as transparent or absent.[64] According to Sartre: 'consciousness of the body is lateral and retrospective; the body is the *neglected*, the *"passed by in silence"*'.[65] And Leder writes: 'while in one sense the body is the most abiding and inescapable presence in our lives, it is also essentially characterised by absence. That is, one's own body is rarely the thematic object of experience.'[66]

This transparency is the hallmark of health and normal function. We do not stop to consider any of its processes because as long as everything is going smoothly the body remains in the background: it is the vehicle through which we experience but not the thematic focus of experience: 'The body tries to stay out of the way so that we can get on with our task; it tends to efface itself on its way to its intentional goal.'[67] This does not mean that we have *no* experience of the body but, rather, that the sensations it constantly provides are neutral and tacit. An example is the sensation of clothes against our skin. This sensation is only noticed when we draw our attention to it or when we undress.[68]

Although we may have moments of explicit attention to the wellness of our body, for example, when a headache goes away or while exercising, it is when something goes wrong with the body that it moves from the background to the foreground of our attention. When functioning normally, our attention is deflected away from our body and towards our intentional goal or action. It is not that the body is absent but, rather, that our experience of it is in the background while the object of our focus is in the foreground. 'The body is in no way apprehended for itself; it is a point of view and a point of departure.'[69]

When we become ill our attention is drawn to the malfunctioning part. The harmony between the objective body and the body-as-lived is disrupted. Leder contrasts the healthy, absent body with illness and other situations when the body becomes an explicit object of negative attention and appears as a 'dys'(function) of sorts. 'In contrast to the "disappearances" that characterise ordinary functioning, I will term this the principle of *dys-appearance*. That is, the body *appears* as the thematic focus, but precisely as in a *dys* state'.[70] The body can dys-appear as ill, disabled, aesthetically flawed or socially awkward, objectified or sexualised, or as attracting negative attention from others.

The transparency of the healthy body is somewhat idealised in philosophical descriptions of health, since this transparency is often pierced by experiences in which the body comes to the fore. The first kind of such experiences is social experience of one's body as it is perceived by others. Sartre's analysis of the gaze (or look) as annihilating subjectivity and objectifying the gazed-upon person's body, which becomes an object in the other's (the subject's) field of vision, recognises the tension between the naïve, unthematised body and the social body.[71]

Transparency is lost in any encounter in which the other's gaze posits a subjectivity within which my own subjectivity is subsumed. This is 'transcendence transcended': my own being as transcendence is transcended by another consciousness.[72] 'My being for others is a fall through absolute emptiness towards objectivity . . . myself as object . . . is an uneasiness, a lived wrenching away from the ekstatic unity of the for-itself, a limit which I can not reach and which yet I am.'[73] And Leder writes, 'a radical split is introduced between the body I live out and my object-body, now defined and delimited by a foreign gaze'.[74] In this view, social existence of its very nature disrupts the transparent, effaced state of the body.

Even in normal everyday experiences, where objectification is not a primary mode, there are many ways in which the world resists us. Often the interaction between us and the world is smooth and regulated by familiar behavioural repertoires. But even in health the world may resist this smooth articulation and require conscious awareness. The small knocks and resistances that we encounter in little accidents, bodily failures, and bodily needs, disrupt bodily transparency in minor ways.

However, these experiences are contained within a normal everyday and are experienced on a spectrum of familiar, if frustrating, bodily failures. Illness, by contrast, creates areas of dramatic resistance in the exchange between body and environment. Even if the transparency of the healthy body is somewhat exaggerated, and that transparency is frequently disturbed by social interactions and bodily failure, it is still the case that our body serves as a medium through which we encounter the world while remaining in the background. The body 'plays a constitutive role in experience precisely by grounding, making possible, and yet remaining peripheral in the horizons of our conceptual awareness'.[75]

There are two ways of thinking about the relationship between the healthy transparent body and the conspicuous ill body. One might think of the two as lying on a continuum. It is simply a matter of degree of vulnerability and disruption. We can also think of the two as discontinuous. This view sees health and illness as distinctive states in which modes of being and experience are radically different. I suggest that the discontinuity view is more compelling. Although everyday experiences certainly include occasions when the body is explicitly thematised, and thematised negatively, these experiences are not the norm and do not fundamentally modify one's tacit sense of trust in one's body or disrupt the habitual body.

The ill body, which becomes conspicuous like Heidegger's broken tool,[76] takes over one's way of being by constricting the range of possible actions and hence restricting the number of projects available to choose from. It also constrains actions chosen. The activities of the healthy body enable projects, while the activities of the ill body disable or devalue projects. For example, if a healthy person goes sightseeing in London, they will experience hunger, fatigue and other bodily needs. But the sight-seeing will not be shaped by these needs. For an ill or disabled person the possibility of sightseeing must be conceived within the constraints of the illness or disability. Projects are not experienced perspicuously as pure projects, but as projects weighed down by concrete considerations and practical arrangements.

On the other hand, even a transient case of a headache can bring to light the tacit sense in which all projects ultimately rest on bodily abilities.[77] However, I suggest that such self-limiting ailments fall within, but do not modify, one's being in the world, whereas serious illness by definition modifies the ill person's way of being. It is defined as serious, in part, *because* of its phenomenology, that is, its life-changing and sometimes life-limiting impact. A headache will make my head conspicuous, and will be experienced as the frustration of an action, but it will not permanently and radically modify my bodily- and self-experiences and understanding in the way serious illness does.

Let us look at Sartre's example: I am reading a book and while doing so the body is given only implicitly. Then my eyes start hurting. The pain is not perceived separately to the project of reading. Rather:

> this pain can itself be *indicated* by objects of the world; i.e., by the book that I read. It is with more difficulty that the words are detached from the undifferentiated ground which they constitute . . . consciousness exists its pain . . . pain *is precisely the eyes* in so far as consciousness 'exists them' . . . pain in the eyes is *precisely my reading*.[78]

However, as Sartre himself points out, when the pain or illness recede, they disappear for good.[79] Minor illness and pain rise and then subside again, without fundamentally altering experience. But a fundamental change is brought about by serious illness: it changes one's embodiment, bodily habits, and ability to plan and pursue goals, and the freedom that normally accompanies choice.

Although the two modes of existence (the transparent body and the conspicuous body) are distinct and indeed contrasted, they still mutually imply one another. First, the appearance of the ill body (Leder's 'dys-appearance') is made possible because of the disappearance, or absence, of the healthy body. As Leder notes, 'it is precisely because the normal and healthy body largely disappears that direct experience of the body is skewed toward times of dysfunction. These phenomenological modes are mutually implicatory.'[80] Second, the contrast is not intended to deny that there are neutral and positive ways in which my body appears to me in health. The experience of dancing in front of an audience, for example, may be one in which pleasure is gleaned from the explicit thematisation of the performing body.

In such experiences, pre-reflective experience may be accompanied by an explicit appreciation of the positive feeling. Dys-appearance is qualitatively different from these experiences, as it has the character of a demand. The body does not appear simply to note its pleasurable state; it appears with a sense of urgency and a demand to do something about the pain, discomfort, or nausea through which the body comes to the fore. 'It would be a mistake to equate all modes of bodily thematisation with dys-appearance', Leder notes.[81] It would equally be a mistake to think that positive and negative ways of appearance have more in common than the explicit thematisation of the body.

## Conclusion: illness and happiness

Let us now return to the obstacle-to-happiness challenge. Could we attain such happiness within the context of serious illness? This question would be answered in the affirmative by the Stoics. Epictetus writes: 'What is it to bear a fever well? Not to blame God or man; not to be afflicted at that which happens, to expect death well and nobly, to do what must be done.'[82]

For the Stoics, given our lack of control over external things and the fact that the possession of virtue is necessary and sufficient for happiness, the absence of health does not preclude happiness.[83] Bearing illness well, and acting in accordance with nature, would enable even a severely ill person to be happy. Although we typically prefer health to illness, because it befits us, it is not a good in itself. Even in the absence of health, happiness is not only possible but graspable. There is no reason ill people cannot cultivate an 'excellent disposition of the soul'.[84] This can be done by cultivating a focus on the present moment, within which happiness can be achieved. This focus comes from an attitude of attention, vigilance and continuous tension concentrated on each and every moment. Marcus Aurelius trained himself to focus on the present moment by telling himself: 'this is enough for you'.[85] 'Enough' in two interrelated ways: enough to keep you busy and enough to make you happy. This spiritual exercise of 'delimiting the present' enables one to turn attention away from the past and the future, in order to concentrate upon the present moment.

The present suffices for our happiness because it is the only thing which belongs to us and depends upon us: recall the emphasis on virtue as self-sufficient for happiness.[86] The past does not belong to us, argues Gill, since it is definitively fixed; the future does not depend on us, because it does not yet exist.[87] The present, in contrast, does belong to us, because it depends entirely upon our will: as long as we harmonise our judgement, action, and desires with universal reason, we will be happy. And since it depends entirely on our will, and not on any external factor, success guarantees immediate and complete happiness.

Contemporary empirical work on happiness has shown that illness may be a challenge, and even an obstacle, to happiness, but that health is not a *sine qua non* for human flourishing. An example of this is a study of renal patients undergoing haemodialysis and healthy controls.[88] We would expect renal patients tethered to a dialysis machine and often incapacitated to be markedly less happy than healthy controls. But in fact, both groups reported a similar level of well-being.[89] Both dialysis patients and healthy controls overestimated the impact of haemodialysis on well-being and both focused too much on dialysis as affecting patients' well-being more strongly than it actually did.[90] The relationship between disease markers and illness experience is also less clearly correlated than one would expect. For example, there is a surprising lack of correlation between disease severity and the level of subjective well-being (happiness) that patients report.[91] This may be because life events in general impact less on our level of well-being than our genetic disposition, as 'set-point' theorists claim, or because of hedonic adaptation.[92]

Contemporary empirical work tallies with the Stoic view that happiness is possible even within the context of serious illness. Researchers who measure the impact of illness on the ill person's well-being find that in qualitative studies patients voice many positive, if unexpected, features of illness. They describe it as an opportunity for growth and self-discovery and as providing them with a new appreciation of life.[93] These unintended effects of illness surprise many, thus providing evidence for the importance of phenomenological accounts of illness and further elucidation of such accounts.

We can now point to an important future direction for a phenomenology of illness. It must broaden our conception of happiness by providing a rich account of flourishing within the context of illness. What remains for us to actively cultivate, and on this modern-day psychologists again concur with the Stoics, is our virtues, our response to life events, and our capacity for well-being. Compare contemporary practices aimed at increasing well-being, such as gratitude journals, with the Stoic *askêsis*. Both aim at modifying what is within our control: not life events, not disease, but how we respond to its challenges. Cultivating the capacity for reflective coping with illness is a pivotal skill for ill persons.

Epicurus concludes: 'he who says either that the time for philosophy has not yet come or that it has passed is like someone who says that the time for happiness has not yet come or that it has passed'.[94] Reflection is key to happiness and happiness is possible within the constraints of illness. What we need is a sensitive and comprehensive account of the richness and diversity of illness experiences, which phenomenology can provide. By marrying Stoic philosophy with the methods of phenomenology and applying both to illness, we can achieve a nuanced account of illness, allowing for the possibility of happiness within illness without belittling its harmful impact.

Returning to the historical focus of this volume, we can use this conclusion as a springboard for reflection on historical methods for the study of illness and disease throughout the ages. It suggests that attention to the ways in which illness, well-being and happiness are experienced, understood and narrated would help us in understanding the complex relationship between these concepts. A two-way dialogue can be opened up: the phenomenological analysis of illness would benefit from an increased awareness of how these conceptions change across time, cultures, and disciplines; and medical history would benefit from a robust methodology and conceptual framework through which to study illness. The relationship between the general claims of philosophy and the focus on the particular which characterises much historical research need not be one of tension. Indeed, it suggests a key methodological challenge to both disciplines: how can we combine awareness of difference and context with appreciation of sharedness and generality? I suggest that this challenge invites humanities disciplines to enter into dialogue that could promote the mutual illumination and questioning of core methodological assumptions held in each discipline.

## Notes

1 This chapter was written during a period of research leave funded by the Wellcome Trust, who generously funded a Senior Investigator Award held by the author (grant number 103340/Z/13/Z). See www.lifeofbreath.org.

2 J. Sellars, *Stoicism*, London: Routledge, 2006; P. Hadot, *Philosophy as a Way of Life*, London: Basil Blackwell, 1995.

3 T. Parsons, *The System of Modern Societies*, Upper Saddle River, NJ: Prentice-Hall, 1971.

4 H. Carel, 'The philosophical role of illness', *Metaphilosophy* 45(1), 2014, 20–40.

5 Ibid.

6 Epicurus, *The Epicurus Reader: Selected Writing and Testimonia*, trans. B. Inwood and L. P. Gerson, Indianapolis: Hackett, 1994, p. 37 (translation modified).

7 M. Heidegger, *Being and Time*, London: Basil Blackwell, 1962 [1927].

8 J. M. Bernstein, 'Trust: on the real but almost always unnoticed, ever-changing foundation of ethical life', *Metaphilosophy* 42 (4), 2011, 395–416.

9 H. Carel, 'Bodily doubt', *Journal of Consciousness Studies* 20(7–8), 2013, 178–97; Carel, 'The philosophical role of illness'.

10 A. MacIntyre, *Dependent Rational Animals*, London: Duckworth, 1999, pp. 1–2.

11 S. Freud, 'Thoughts for the times on war and death', in the *Penguin Freud Library* Vol. 12, *Civilization, Society and Religion*, London: Penguin Books, 1985 [1915].

12 Sellars, *Stoicism*, p. 123. This critique of the tendency of philosophers to deny the corporeal aspects of existence and their implications is mirrored by the 'cultural turn' and constructivist approaches within the history of medicine. Both approaches attempt to expose and conceptualise the desire to distance our subjective selves from the objective historical or philosophical analysis. See: L. Jordanova, 'The social construction of medical knowledge', *Social History of Medicine* 8 (3), 1995, 361–81; F. Huisman and J. Harley Warner, *Locating Medical History: Their Stories and their Meanings*, Baltimore: Johns Hopkins University Press, 2004

13 Carel, 'Bodily doubt'.

14 F. Svenaeus, *The Hermeneutics of Medicine and the Phenomenology of Health*, Linköping: Springer, 2001.

15 S. K. Toombs, 'Illness and the paradigm of lived body', *Theoretical Medicine* 9, 1988, 201–26.

16 According to surgeon Atul Gawande: 'The most common week of having surgery in your life is the last week of your life and the most likely day of having surgery during that week: the very last day . . . all we've given you are the complications and the pain and the suffering without adding any benefit', Reith Lecture 3, 2014, http://downloads.bbc.co.uk/radio4/open-book/2014_reith_lecture3_edinburgh.pdf (accessed 16 December 2014).

17 Epictetus, *A Selection from the Discourses of Epictetus with the Encheiridion*, trans. G. Long, 2004. http://www.gutenberg.org/files/10661/10661-h/10661-h.htm (accessed 15 December 2014).

18 A. Gawande, The 2014 Reith Lectures, lecture 3, http://downloads.bbc.co.uk/radio4/open-book/2014_reith_lecture3_edinburgh.pdf (accessed 14 January 2015).

19 M. Jackson, *The Age of Stress: Science and the Search for Stability*, Oxford: Oxford University Press, 2013.

20 B. Ehrenreich, *Smile or Die: How Positive Thinking Fooled America and the World*, London: Granta Books, 2010.

21 H. Carel, *Illness*, London: Routledge, 2013.

22 D. Moran, *Introduction to Phenomenology*, London: Routledge, 2000, p. 1.

23 E. Husserl, *Ideas Pertaining to a Pure Phenomenology and to a Phenomenological Philosophy. Second book*, Dordrecht: Kluwer, 1989 [1952].

24 M. Merleau-Ponty, *Phenomenology of Perception*, New York: Routledge, 2012 [1945].

25 Ibid., p. 147.

26 S. Gallagher and D. Zahavi, *The Phenomenological Mind*, New York: Routledge, 2008, p. 135.

27 Husserl, *Ideas*, p. 176.

28 Ibid., p. 160 (italics removed).

29 Merleau-Ponty, *Phenomenology*, p. 137 (translation modified).

30 J.-P. Sartre, *Being and Nothingness*, London and New York: Routledge, 2003 [1943]; R. M. Zaner, *The Context of Self*, Athens, OH: Ohio University Press, 1981.

31 Toombs, 'Illness and the paradigm of lived body'.

32 Merleau-Ponty, *Phenomenology*, p. 110.

33 Sartre, *Being and Nothingness*.

34 Some authors have suggested a third category, sickness, to denote the social dimension of human ailment.

35 Carel, *Illness*; H. Carel and I. J. Kidd, 'Epistemic injustice in healthcare: a philosophical analysis', *Medicine, Healthcare and Philosophy*, 2014. DOI 10.1007/s11019-014-9560-2.

36 H. Carel, '"I am well, apart from the fact that I have cancer": explaining wellbeing within illness' in L. Bortolotti (ed.), *Philosophy and Happiness*, Basingstoke: Palgrave, 2009, pp. 82–99; J. Haidt, *The Happiness Hypothesis*, London: William Heinemann, 2006.

37 Merleau-Ponty, *Phenomenology*.

38 Until she becomes ill herself.

39 Carel and Kidd, 'Epistemic injustice in healthcare'.

40 S. K. Toombs, 'The meaning of illness: a phenomenological approach to the patient–physician relationship', *Journal of Medicine and Philosophy* 12, 1987, 219–40. See also Carel and Kidd, 'Epistemic injustice in healthcare'.

41 S. K. Toombs, *The Meaning of Illness: A Phenomenological Account of the Different Perspectives of Physician and Patient*, Amsterdam: Kluwer, 1993, p. 23; Carel and Kidd, 'Epistemic injustice in healthcare'.

42 Toombs, *The Meaning of Illness*, pp. 90–8.

43 Toombs, 'The meaning of illness'.

44 Ibid., p. 229.

45 Ibid.

46 Ibid., p. 230.

47 Ibid., pp. 230–1.

48 Ibid., p. 231.

49 Ibid., p. 232.

50 Toombs, *The Meaning of Illness*, p. 96.

51 Ibid.

52 Ibid., p. 97.

53 M. Bury, 'Chronic illness as biographical disruption', *Sociology of Health and Illness* 4(2), 1982, 167–82.

54 Toombs, 'The meaning of illness', p. 234.

55 Carel, 'Bodily doubt'.

56 E. Lindsey, 'Health within illness: experiences of chronically ill/ disabled people', *Journal of Advanced Nursing* 24, 1996, 465–72; Carel, *Illness*.

57 Toombs, 'The meaning of illness', p. 234.

58 A. Frank, *At the Will of the Body*, New York: Mariner Books, 1991, p. 27 (my emphasis).

59 J. S. Mill, *Autobiography*, London: Penguin Classics, 1989 [1873], p. 112.

60 Carel, 'I am well'.

61 Toombs, *The Meaning of Illness*.

62 Ibid., p. 70.

63 We do experience the difference between the first and third order in some social situations, when we experience our body as experienced by others – Sartre discusses shame as one example.

64 Sartre, *Being and Nothingness*; Leder, *The Absent Body*.

65 Sartre, *Being and Nothingness*, p. 354.
66 Leder, *The Absent Body*, p. 1.
67 Gallagher and Zahavi, *The Phenomenological Mind*, p. 143.
68 M. Ratcliffe, *Feelings of Being: Phenomenology, Psychiatry and the Sense of Reality*. Oxford: Oxford University Press, 2008, p. 303.
69 Sartre, *Being and Nothingness*, p. 355.
70 Leder, *The Absent Body*, p. 84.
71 Sartre, *Being and Nothingness*, p. 276ff.
72 Ibid., p. 287. But see also Merleau-Ponty's criticism of Sartre's analysis in *Phenomenology*, p. 378.
73 Ibid., pp. 298–9.
74 Leder, *The Absent Body*, p. 96.
75 T. Carman, 'The body in Husserl and Merleau-Ponty', *Philosophical Topics* 27(2), 1999, 205–26, at 208.
76 Heidegger, *Being and Time*.
77 Sartre, *Being and Nothingness*.
78 Ibid., pp. 356–8.
79 Ibid., p. 360.
80 Leder, *The Absent Body*, p. 86.
81 Ibid., p. 91.
82 Epictetus, *A Selection from the Discourses*.
83 Sellars, *Stoicism*, pp. 112–13.
84 Ibid., p. 110.
85 Hadot, *Philosophy as a Way of Life*, p. 227.
86 C. Gill, *The Structured Self in Hellenistic and Roman Thought*, Oxford: Oxford University Press, 2006.
87 Ibid.
88 J. Riis, J. Baron, G. Loewenstein and C. Jepson, 'Ignorance of hedonic adaptation to haemodialysis: a study using ecological momentary assessment' *Journal of Experimental Psychology* 134 (1), 2005, 3–9.
89 Ibid., p.6.
90 For other examples, see: E. Angner, M. N. Ray, K. G. Saag and J. J. Allison, 'Health and happiness among older adults', *Journal of Health Psychology* 14 (4), 2009, 503–12; D. Gilbert, *Stumbling on Happiness*, London: Harper Press, 2006.
91 Angner et al., 'Health and happiness'; Carel, 'I am well'; Carel, 'The philosophical role of illness'; Gilbert, *Stumbling on Happiness*.
92 S. Lyubomirsky, *The How of Happiness*, London: Piatkus Books, 2007.
93 O. Lindqvist, A. Widmark a d B. Rasmussen, 'Reclaiming wellness: living with bodily problems as narrated by men with advanced prostate cancer', *Cancer Nursing* 29(4), 2006, 327–37.
94 Epicurus, *The Epicurus Reader*, p. 28.

## Select bibliography

H. Carel, '"I am well, apart from the fact that I have cancer": explaining wellbeing within illness', in L. Bortolotti (ed.), *Philosophy and Happiness*, Basingstoke: Palgrave, 2009, pp. 82–99.

H. Carel, 'Bodily doubt', *Journal of Consciousness Studies* 20 (7–8), 2013, 178–97.

H. Carel, *Illness*, London: Routledge, 2013.

Epictetus, *A Selection from the Discourses of Epictetus with the Encheiridion*, trans. G. Long, 2004, http://www.gutenberg.org/files/10661/10661-h/10661-h.htm (accessed on 15 December 2014).

C. Gill, *The Structured Self in Hellenistic and Roman Thought*, Oxford: Oxford University Press, 2006.

P. Hadot, *Philosophy as a Way of Life*, London: Basil Blackwell, 1995.

E. Lindsey, 'Health within illness: experiences of chronically ill/ disabled people', *Journal of Advanced Nursing* 24, 1996, 465–72.

M. Merleau-Ponty, *Phenomenology of Perception*, New York: Routledge, 2012 [1945].

D. Moran, *Introduction to Phenomenology*, London: Routledge, 2000.

J.-P. Sartre, *Being and Nothingness*, London and New York: Routledge, 2003 [1943].

J. Sellars, *Stoicism*, London: Routledge, 2006.

F. Svenaeus, *The Hermeneutics of Medicine and the Phenomenology of Health*, Linköping: Springer, 2001.

S. K. Toombs, 'The meaning of illness: a phenomenological approach to the patient–physician relationship', *Journal of Medicine and Philosophy* 12, 1987, 219–40.

S. K. Toombs, 'Illness and the paradigm of lived body', *Theoretical Medicine* 9, 1988, 201–26.

S. K. Toombs, *The Meaning of Illness: A Phenomenological Account of the Different Perspectives of Physician and Patient*, Amsterdam: Kluwer, 1993.

# INDEX

Printed in Dunstable, United Kingdom